P9-AEV-968

MAN-MADE FIBRES

MAN-MADE FIBRES

R. W. MONCRIEFF

Wiley Interscience Division
John Wiley & Sons, Inc. – New York

First published as Artificial Fibres 1950
Second Edition ,, ,, ,, 1954
Third Edition 1957
Reprinted with Revisions 1959
Fourth Edition 1963
Revised impression 1966
Fifth Edition 1970

Butterworth & Co: (Publishers) Ltd.

Published in the USA by Wiley Interscience Division
John Wiley & Sons Inc., 605 Third Avenue,
New York, N.Y. 10016

Library of Congress Catalog Card Number 70–127526

ISBN 0 592 06332 1

Printed in Great Britain by Richard Clay (The Chaucer Press), Ltd., Bungay, Suffolk

TO MY FATHER

WILLIAM MONCRIEFF

(1875–1946)

A PIONEER IN THE WEAVING

OF

ARTIFICIAL FIBRES

CONTENTS

PART 1 The Structure and Properties of Fibres

PART 2 Fibres Made from Natural Polymers

PART 3 Synthetic Fibres

PART 4 Processing

PREFACE

THE nylons, the polyesters and the acrylics continue to dominate the fibres scene. Their growth has been, and continues to be, amazingly rapid; so rapid in fact that the production of cotton has had to be cut back by legislation in the U.S.A., and there are large unsold stocks of wool throughout the world. These two effects are undesirable, and it seems reasonable to think and hope that they will be only very temporary. Cotton and wool are two of our finest fibres, and the proper function of man-made fibres is to supplement, not to supplant, them.

With the world-wide increase in the production of synthetic fibres there has come a sharp fall in their prices, particularly in those of the polyesters, which can now be made much more cheaply than polyamides. This fall in prices is to be welcomed. Just as cheap rayon brought down the price of cotton in the 'fifties, so now we are seeing the cheap synthetic fibres bring down the price of wool. Synthetic fibres made from alpha amino-acids, fibres that we have awaited so long, have not yet come out in strength, but there seem to be two or three on the threshold, and their coming may augur a new era for man-made fibres. They are, in fact, overdue, but nevertheless they will find the going rough to begin with.

Increased fibre production has led also to a spate of improvements in the dyehouse, and activity in this department has never been so great. The new developments in dyeing are to be welcomed by all the textiles industry. The synthetic-fibres industry grows apace. The world has never seen the like.

My thanks again to the manufacturers who have so generously supplied information and illustrations. Throughout the text of this book, various words are used which are Registered Trade Names. These are indicated by being printed with an initial capital.

R. W. MONCRIEFF

PART 1

THE STRUCTURE AND PROPERTIES OF FIBRES

CHAPTER 1

FUNDAMENTAL CONCEPTIONS

ALL ordinary yarns, either man-made or natural, consist of a number of fibres or filaments. In the case of man-made fibres this number is usually in the range 15–100—*i.e.*, most yarns will be composed of not fewer than fifteen and not more than 100 filaments. If a single thread of a man-made fibre yarn is broken, the individual filaments will usually be observed to splay out. They can always be pulled apart mechanically. The reason this multi-filament construction is adopted is that it confers pliability and flexibility on the yarn; a yarn composed of a number of fine filaments is much more flexible than a solid, thick filament of the same diameter as the yarn. The use of a large number of very fine copper wires in ordinary lighting-flex may be recalled as an analogy. The purpose is exactly the same: to obtain flexibility.

Monofils

For special purposes, monofilament yarns are made. One familiar application of this is when the monofilament is relatively thick and is used as bristle. Most readers will be familiar with the use of nylon bristle in tooth-brushes. Very strong synthetic fibres, such as nylon, are sometimes spun in very fine monofils to give the sheerest of stockings, and such fine yarns of 7, 12 and 15 deniers are usually monofils. They are made as monofils to give durability because coarse filaments are less easily damaged than fine, and as the monofil is itself so thin it is sufficiently pliable to be knitted without difficulty. It is easier, too, to make a coarse single 15 denier filament than a multi-filament 15 denier yarn, and the prices of monofil 15 denier nylon, for example, are lower than those of 3 filament 15 denier (each filament being of 5 denier) nylon. This is illustrated in the figures on p. 868. Sometimes, too, for special decorative or novelty effects, relatively thick monofilament yarns may be used, but the great majority of artificial-fibre yarns are spun in multi-filament form.

Continuous Filament

All the yarns that have just been discussed, multi-filament and monofil, are continuous filament yarns and are like real silk in the sense that the filaments are very long indeed, miles long in some cases. They are glossy, lustrous and silky, and were the first kind of artificial silk. At first, all artificial silk was continuous filament —long almost endless filaments. Nowadays, less than one-half of the fibres that man makes are continuous.

Staple

Most of the other half of man-made fibres are discontinuous; they have been chopped up into short lengths of a few inches. In this way they have become more like cotton, which has a fibre length of about 1–1½ in., or wool, which is usually within the range of 3–6 in. fibre length. All man-made fibres are spun continuously, but more than half of their weight is chopped up into short lengths, usually of from 1 to 6 in. Often this cutting is done continuously as the filaments are made; the chopping up is part of the process. The chopped up fibre is called " staple fibre " and is used for spinning (p. 644) on the cotton and worsted systems, being very largely used for blending. The staple man-made fibre and a staple natural fibre are spun together as a blend. That way the mixing of the two fibres is very intimate. Blending or mixing of fibres should be done at as early a stage in manufacture as possible. Of such blends, that of polyester (Terylene) and wool is as well known as any. Such a blend could not be made from continuous filament fibre; only by using the chopped up staple of the man-made fibre does perfect blending with wool (or cotton or linen) become possible. Sometimes, perhaps most often, the man-made staple is spun on its own without the admixture of a natural fibre; spun rayon fabrics for frocks, curtains and underwear have been ubiquitous for twenty years. Continuous filament fibres are long fibres, often miles long, and an end is reached only at the end of a bobbin or other package, or if a thread accidentally breaks. Staple fibres are very short, nearly always just a few inches. Sometimes for special reasons fibres of different lengths, say of 1½ in. and 3 in., are mixed in staple, but this is unusual. Nearly always all the fibres in a bale of staple fibre are the same in length. Staple fibres are mixed up in random arrangement and have to be sorted out, made parallel to each other and twisted in old traditional textile operations to make them into yarns.

Tow

Tow is fibre that is spun with tens or hundreds of thousands of filaments bundled together into a loose rope and wound up on to some sort of spool or package. It is used for direct-spinning operations (p. 648 *et seq.*) in which the filaments are cut or broken at intervals of a few inches and spun directly into yarn. Tow serves the same end as staple fibre, but the cutting into short fibres is done at a later stage. The disadvantage of staple fibre is that although the filaments are all spun parallel one to another, and although they end up parallel in the final yarn, in between they are mixed up, pointing in all directions. It seems wasteful to randomise parallel filaments and then sort them out and make them parallel once again. The use of tow avoids this; the filaments are spun parallel and are kept parallel right up to the finished yarn. Tow is often supplied in a rope of about 200,000 denier, say 40,000 filaments each of 5 denier.

Denier

The coarseness of a yarn or a filament is usually gauged as " denier ". The denier was the unit used in the real-silk industry long before man-made fibres came. It is defined as follows:

Definition. The denier of a yarn (or filament) is the weight in grams of a length of 9,000 metres of that yarn (or filament).

If, for example, 9,000 metres of a yarn weigh 100 grams, the yarn is said to be 100 denier; if 9,000 metres of another yarn weigh 45 grams, that yarn is 45 denier; if 9,000 metres of a single filament weigh 3 grams, that filament is of 3 denier. An instrument very suitable for rapid determination of denier is the Torsion Denier Balance (Fig. 1). A 9-metre length of the yarn is run off on a wrap-reel, hung on the hook of the balance, the pointer turned approximately to the expected denier on the dial, the balance released, a final adjustment of the pointer made, and the denier of the yarn read directly.

Conversion Factor for Cotton Counts

Many readers may be more familiar with the use of cotton counts —*i.e.*, the number of hanks of 840 yd. which will weigh 1 lb.—and it is useful to have a factor by which cotton counts may rapidly be converted into denier and vice versa. This may be obtained in the following way:

Consider a yarn of 1 s cotton counts. What would its denier be?

As the yarn is 1 s cotton counts 840 yd. weigh 1 lb.

or $$840 \text{ metres weigh } 453{\cdot}6 \times \frac{39{\cdot}37}{36} \text{ grams}$$

so $$9{,}000 \text{ metres weigh } 453{\cdot}6 \times \frac{39{\cdot}37}{36} \times \frac{9{,}000}{840} \text{ grams}$$

$$= 5{,}315 \text{ grams.}$$

Hence the denier is 5,315.

If the yarn is x s cotton counts, then 9,000 metres will weigh $\frac{5,315}{x}$ grams. Accordingly, to find the denier of a yarn from its cotton counts, divide 5,315 by the cotton counts. Conversely, to find the cotton counts of a yarn of known denier, divide 5,315 by the denier.

For example, 50 s cotton counts $\equiv \frac{5,315}{50}$ or 106 denier

2/60 s cotton counts $\equiv \frac{5,315}{30}$ or 177 denier

30 denier $\equiv \frac{5,315}{30}$ or 177 s cotton counts

150 denier $\equiv \frac{5,315}{150}$ or 35·4 s cotton counts

[*White Electrical Instrument Co. Ltd., Malvern.*]

FIG. 1.—Torsion Denier Balance.

Conversion Factor for Worsted Counts

Remembering that the basis of the worsted system of counts is the number of hanks of 560 yd. in 1 lb., the reader will find it easy to calculate that the conversion factor from worsted counts to denier is 7,972.

For example, 20 s worsted counts $\equiv \frac{7{,}972}{20}$ or 399 denier

200 denier $\equiv \frac{7{,}972}{200}$ or 40 s (nearly) worsted counts

It will be noted that approximations have been made in these calculations. The reason for this is that counts and denier are slightly variable; there is no point in describing the denier of a yarn as 398·6 when even the average of batches of it will vary by $\pm$ 2 denier.

Uniformity of Denier

Yarns are not absolutely uniform in denier. They are nearly so, and uniformity is a very desirable characteristic. It is not unusual for a tolerance of $\pm$ 3 per cent to be allowed—*i.e.*, in a batch of 100 denier yarn, extremes of 97 and 103 deniers may be found.

Filament Denier

Yarns of the same denier may be spun with different numbers of filaments. A yarn of 100 denier may be produced with twenty filaments and also with sixty filaments. The former would have a filament denier of $\frac{100}{20}$ or 5, and this is relatively coarse, and would be found in cheap " bread-and-butter " viscose. The latter would have a filament denier of $\frac{100}{60}$ or 1·67, and would be a speciality yarn suitable for the manufacture of high-class materials. The more filaments there are in a yarn for a given denier, the more soft and supple it is. Very fine filament yarns are suitable for fabrics where draping, anti-crease properties and softness of handle are required. Such yarns are, however, more liable to abrasion than those of coarse filament yarns, and for purposes where hard wear is a *sine qua non* coarse filaments are to be preferred. Men's linings are, for example, harder wearing if made from coarse filament yarns. There are uses for both coarse and fine filament yarns.

Tex and Millitex

Another unit which is sometimes used instead of denier is the tex, which is defined as the weight in grams of 1,000 metres (instead of

9,000 metres used in the denier definition). The millitex is one-thousandth of the tex, so that one and the same filament could be described as 9 denier or 1 tex or 1,000 millitex whilst others could be: 1 denier or 0·111 tex or 111 millitex; 4½ denier or 0·5 tex or 500 millitex; 100 denier or 11·1 tex or 11,100 millitex; 90 s cotton counts or 59 denier or 6·6 tex.

Another way to look at it is that the tex of a yarn or filament is its weight in milligrams per metre. Sometimes the term decitex or dtex is encountered; its relation to the others is:

$$0{\cdot}1 \text{ tex} = 1 \text{ d(eci)tex} = 100 \text{ millitex} = 0{\cdot}9 \text{ denier}$$

The tex is a universal unit, used for all fibres: man-made, wool, cotton, *etc.*

Twist

Nearly all yarns are twisted, as this protects the filaments from damage. Untwisted yarn is almost impossible to weave or knit without damage. The twist is expressed in turns per inch (t.p.i.).

[*Leonard Hill, Ltd.*]

FIG. 2.—The two directions of twist.

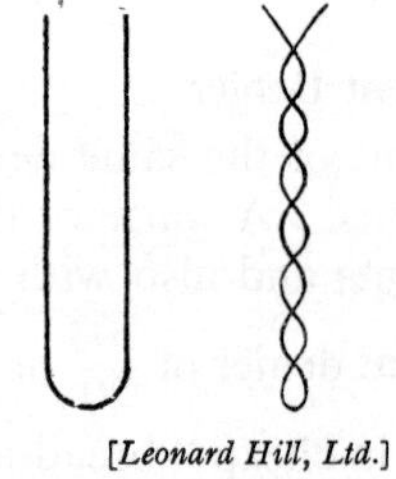

[*Leonard Hill, Ltd.*]

FIG. 3.—Balanced and unbalanced yarns.

Direction of twist is described as " S " or " Z ", the meaning of which will be clear from Fig. 2. This " S and Z " system is unambiguous, and fortunately has almost universally replaced many other mutually conflicting systems. Sometimes, too, yarns are doubled or folded, and then the doubling or folding twist should be such that the yarn is balanced. Fig. 3 illustrates balanced and unbalanced yarns; the latter tend to twist if held in U form.

Fig. 4 shows a simple type of twist-tester. One end of the yarn is held in a clamp, the other end is rotated until a needle can be drawn from one end to the other through the untwisted filaments. The number of turns that has been required to untwist the yarn is read on the engraved dial. Stroking the filaments with the needle will often electrify them, cause them to balloon, and so simplify their separation.

The addition of twist always increases the denier of any fibre. For fairly low twists (up to say 30 t.p.i. for 70 denier) the increase in denier is not great, but at one point secondary twist develops (the

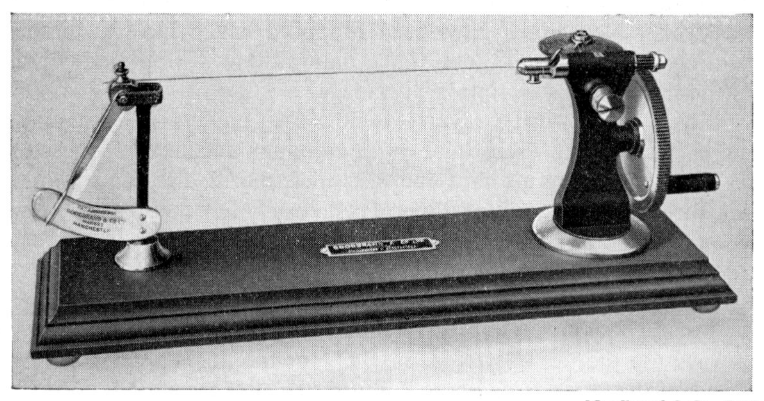

[*Goodbrand & Co., Ltd.*]

FIG. 4.—A twist-tester.

already twisted fibre starts to " corkscrew "), and then the denier rises quickly. This effect may be seen in the following table of figures for Dacron polyester yarn.

Initial denier of Dacron.	Denier with twist (t.p.i.) of							
	1.	8.	14.	24.	36.	48.	70.	100.
40 (34 fils) . .	40	40	40	40·5	41	42	44	49
70 (34 fils) . .	70	70	71	72	74	77	83	—
210 (34 fils) . .	210	213	216	226	246	271	—	—

This behaviour is not peculiar to Dacron; all fibres behave somewhat similarly. As twist rises strength falls and elongation at break rises, as shown by the following figures for 210 denier 34 fils Dacron.

Twist (t.p.i.).	Tenacity (gm/denier) (see p. 11).	Elongation at break (per cent) (see p. 13).
1	7·1	11
23	6·6	15
36	5·3	16
47	3·9	17

Nomenclature

" Artificial silk " was the early name for all man-made fibres. When such fibres were first spun the avowed aim of the manufacturer was to make something that would replace real silk, and the name was then quite apposite. The description " artificial silk " is less appropriate to-day, as artificial fibres have long passed the stage of imitating real silk. New fibres have been produced which have properties entirely different from those of any natural fibre. Man-made fibres is the most usual description nowadays.

Rayon. The word " rayon " is now used universally to describe fibres made by the viscose and cuprammonium and acetate processes.

Whilst there are no hard-and-fast rules, the reader is advised to use the word " rayon " for fibres of cellulosic origin (viscose, cuprammonium, and cellulose acetate), and the term " synthetic fibres " for the true synthetics, nylon, Vinyon, *etc.* All of them can be referred to collectively as " man-made fibres ".

Viscose Rayon. One other point to note is that manufacturers always use the term " viscose rayon " for the fibres. They do not speak of the fibres as " viscose ", but restrict this term to describe the solution from which the fibres are spun. Converters and merchants (in fact most people who do not manufacture it) often describe the fibre as " viscose ", but, to be quite correct, the reader should refer to the fibre or yarn or fabric as "viscose rayon " and not as "viscose".

Azlon. The name " Azlon " has also been tentatively approved by the Federal Trade Commission as a generic term for reconstituted protein fibres, irrespective of the type of protein used as the base of the fibre. This term appeared to include soya-bean, casein, groundnut and zein fibres which comprise a group with diverse compositions and properties, but " Azlon " seems never to have become popular. Most of the Azlons have failed and been discontinued. Ardil (pea-nut protein) and Vicara (maize protein) were the best, but they are no longer made.

Acrylics and Modacrylics

In America, the Federal Trade Commission has defined rules for the enforcement of the federal Textile Fiber Products Identification Act (March, 1960). According to these, acrylic fibres are those that comprise at least 85 per cent of acrylonitrile. Modacrylics are those that contain from 35 to 85 per cent acrylonitrile. There are many acrylic fibres: Orlon, Acrilan, Courtelle, Creslan are typical. There are only four modacrylic fibres: Verel, Dynel, Kanekalon and Teklan.

" **Spinning** ". It is unfortunate that the verb " to spin " and its derivatives are used in the following two quite different senses.

1. To spin viscose rayon, by extruding a viscose solution through a fine hole into a coagulating bath, or to extrude a solution of cellulose acetate in acetone through a fine hole into an air-drying medium, or to extrude a melt of nylon through a fine hole into a cooling atmosphere. All man-made fibres are " spun " in this way, just as a spider spins its web or a silkworm spins its cocoon. This always results in a continuous filament which may, however, immediately or later be cut into staple or short lengths.

2. To spin cotton fibres into a yarn by arranging them in parallel formation in a bundle and gradually pulling out and thinning the bundle and at the same time twisting it. A similar process is carried out on wool, mohair, flax, schappe, *etc.* Man-made fibres which have already been " spun " in the first sense (extruded) are sometimes then cut into staple fibre, and spun in the second sense, *i.e.*, parallelised, thinned and twisted into a yarn.

Both meanings are " to make a yarn ", but in the first case it is from a solution or a melt, in the second it is from fibres. A staple fibre yarn has been " spun " in both senses of the word.

No one who has worked amongst textiles will find confusion in the two meanings of the word " spinning ", but it may well be confusing to the student.

Tenacity

The tenacity or strength of rayons is usually expressed as grams per denier. If a load of 250 grams will just break a 100-denier yarn, the tenacity is said to be 2·5 grams per denier. Some fibres such as the regenerated proteins have tenacities of the order of 1 gram per denier; others, like nylon and Fortisan, of 6–7 grams per denier.

Sometimes the tenacity or tensile strength is expressed in pounds per square inch instead of in grams per denier, and we can derive the relation between the two methods of expression as follows:

Suppose we have 1-denier filament of specific gravity 1, then its cross-sectional area will be:

$$\frac{1}{900{,}000} \text{ cm.}^2$$

because 9,000 m. or 900,000 cm. of it weigh 1 gram.

If the specific gravity of the filament is not 1 but 2, then the cross-sectional area of the 1-denier filament will be

$$\frac{1}{900{,}000 \times 2} \text{ cm.}^2$$

and if the specific gravity is d then the cross-sectional area of a 1 denier filament will be

$$\frac{1}{900{,}000d} \text{ cm.}^2$$

or

$$\frac{1}{900{,}000d \times 2{\cdot}54^2} \text{ in.}^2$$

This by definition has a breaking load of 1 gram, so that a " filament " 1 in.² in cross-sectional area would have a breaking load of:

$$900{,}000d \times 2{\cdot}54^2 \text{ grams, or}$$

$$\frac{900{,}000d \times 2{\cdot}54^2}{453{\cdot}6} \text{ lb.}$$

or

$$12{,}800d \text{ lb.}$$

Accordingly to convert tenacity expressed as grams per denier into tensile strength expressed as lb./in.² we can use the expression

$$\text{Tensile strength} = (12{,}800d \times \text{gm./denier}) \text{ lb./in.}^2$$

Thus Courlene with a specific gravity of 0·92 and a tenacity of 2 grams per denier has a tensile strength of:

$$12{,}800 \times 0{\cdot}92 \times 2 \text{ lb./in.}^2 \text{ or}$$
$$23{,}500 \text{ lb./in.}^2$$

Glass with a specific gravity of 2·54 and a tenacity of 6·3 grams per denier has a tensile strength of:

$$12{,}800 \times 2{\cdot}54 \times 6{\cdot}3 \text{ lb./in.}^2 \text{ or}$$
$$205{,}000 \text{ lb./in.}^2$$

It would need a load of about 90 tons to break a glass " filament " of 1 in.² section. Fortisan with a tenacity of 7 grams per denier and a specific gravity of 1·51 has a tensile strength of 136,000 lb./in.²; nylon with a tenacity of 5·5 grams per denier and a specific gravity of 1·14 has a tensile strength of 80,000 lb./in.²; cuprammonium rayon with a tenacity of 2 grams per denier and a specific gravity of 1·53 has a tensile strength of 39,000 lb./in.²

Although in engineering practice the terms " tenacity " and " tensile strength " are used synonymously, it is convenient in dealing with fibres to confine the expression " tenacity " to that property (cohesiveness) which is measured in grams per denier and " tensile strength " to that which is measured as lb./in.². There is the density factor that divides one from the other and two fibres with similar " tenacity " values may have very different " tensile strength " values; *e.g.*, glass and nylon both have tenacity values of about 6 grams per denier but the tensile strength of glass (because it is more dense) is more than double that of nylon.

Elongation at Break

The elongation at break is an important characteristic of a yarn. If a length of 100 cm. of a yarn can be stretched to 112 cm. before it breaks, it is said to have an " elongation at break " or an " extension " of 12 per cent. Except for special purposes, such as tyre-cord, a minimum extension of 10 per cent is desirable to facilitate textile working. Ordinarily, there is no advantage in an extension being higher than 20 per cent. Often, in the manufacture of yarns, it is possible to make a yarn with either a high tenacity and a low elongation at break, or a moderate tenacity and a reasonably high elongation at break. As a rule, the manufacturer compromises, and produces a yarn that will have the highest strength compatible with an extension high enough not to give trouble in winding, weaving and knitting. Often two kinds of filament will be spun; a very high tenacity yarn with a fairly low extension intended for industrial use, and a yarn with a somewhat lower tenacity and higher extension and generally a softer handle for apparel use. As an example, Tergal, the French equivalent of Terylene, is made by Société Rhodiaceta in high tenacity of 6·5 grams per denier with 10 per cent extension and also in normal filament with 5 grams per denier tenacity and 20 per cent extension. Rhodiaceta even make a third variety their staple fibre, which has a tenacity of only 3·5–4 grams per denier but an extension at break of 40–25 per cent. Such modifications are usually dependent on the degree of stretch that is applied to the fibre either during the actual spinning operation, or in the drawing (stretching) operation that follows it, or in both. The greater the stretch, the higher the tenacity and the lower the extensibility. Stretches are usually quite high: 4 or 5 times the spun length is quite usual in some of the synthetic fibres and 10 or 12 times is not unknown. The natural fibres cannot be stretched much and it is the ability of man-made fibres to take high stretches in the plastic state that enables fibre manufacturers to make yarns with surprisingly good physical properties.

The conditions under which tenacity and extension are measured must be kept standard. An instrument very suitable for making determinations is Goodbrand's Single Thread Tester shown in Fig. 5. Fibres should be conditioned (p. 17) before test.

Elasticity

The elasticity of a fibre is its ability to recover from strain. If a fibre is stretched 10 per cent—*i.e.*, 100 cm. becomes 110 cm., and then, on release of the tension, it reverts to its original length of 100 cm.—it is perfectly (or 100 per cent) elastic. If, on the other hand, it

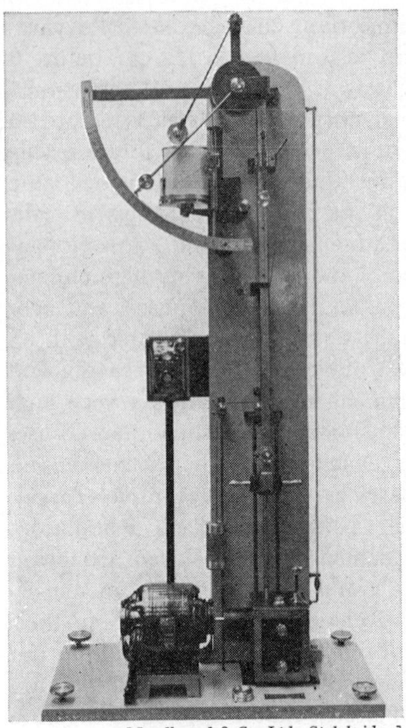

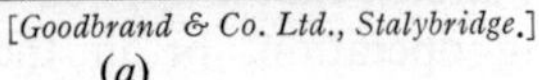

[*Goodbrand & Co. Ltd., Stalybridge.*]

(*a*)

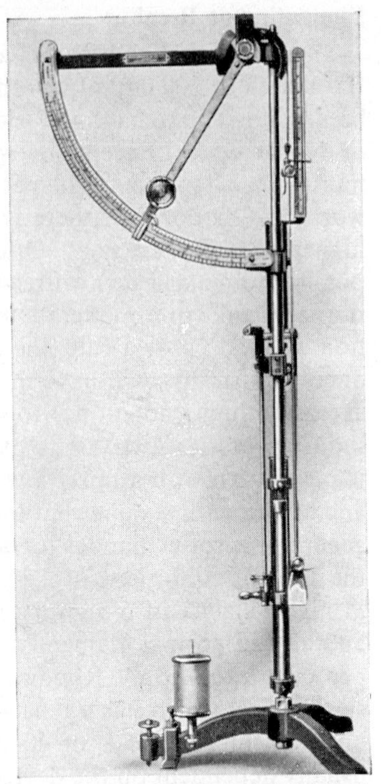

[*Goodbrand & Co. Ltd., Stalybridge.*]

(*b*)

FIG. 5.—Single Thread Tester (constant rate of extension). This machine simultaneously measures breaking load and elongation at break: (*a*) This up-to-date model is driven by an electric motor. (*b*) An earlier model, which depended not on a motor but on the gravitational slow fall of a plunger into an oil bath, was used for decades and played a significant part in the development of new fibres.

contracts to 102 cm., then it is 80 per cent elastic; if it contracts to 104 cm., it is 60 per cent elastic. Usually fibres have a high elasticity for low stretches—*i.e.*, if stretched 5 per cent they regain very nearly their original length, but they may have only a relatively low elasticity for high stretches. A fibre with a low elongation at break may have a high elasticity, and, conversely, a very extensible fibre may have only a low elasticity. It is important to distinguish clearly between " elongation at break " and " elasticity ".

Breaking Length

Occasionally the strength of a fibre will be expressed as " breaking length ". This is the length of a fibre which will just break under its own weight, and it is usually very long. If we consider a viscose filament with a tenacity of 2 grams per denier, we imply that a length of 18,000 metres will just break under its own weight—*i.e.*, the breaking length is 18 km. Similarly a yarn with a tenacity of 4 grams per denier will have a breaking length of 36 km. The strongest yarns hitherto made on a manufacturing scale have breaking lengths of about 80 km. It will be appreciated that the breaking length of a 100-denier yarn will be the same as that of a 2-denier single filament, because, although the former is stronger, it is proportionally heavier.

Stress–Strain Diagrams

The stress–strain diagrams of fibres and yarns are often very informative. In them the strain—*i.e.*, the elongation undergone by a fibre or yarn per unit length—is plotted against the stress, or load applied to it.

The stress–strain diagram—or " characteristic curve ", as it is also called—is obtained by using a strength-testing machine of autographic type. The testing machine shown in Fig. 5(*a*) is supplied as an autographic model; the axes of the diagram it records are at an angle of 45°. Other machines, including the Inclined Plane Tester, shown in Fig. 6, give diagrams in which the axes are at 90°. The latter machine operates at a constant rate of loading, the former at a constant rate of extension. For everyday use in the laboratory and for plant control the writer's preference is for the machine shown in Fig. 5, but for prototype testing of fibre and yarn characteristics the Inclined Plane model is preferred.

If the stress–strain curve is a straight line, then the fibre is obeying Hooke's Law, and so long as the " curve " is straight, the fibre will be elastic. Usually, after a very few per cent extension the elongation of the fibre becomes greater than would be required by the straight-line curve. This generally means that plastic flow is setting in, and that the fibre is no longer truly elastic. The *yield point* corresponds to the break in the curve where it departs from the straight line. For most fibres this break takes place very early.

A fibre which will be most satisfactory for textile purposes is one that does not stretch much for small loads, so that during winding it does not stretch appreciably; the advantage of this is that even if the winding tension varies slightly from spindle to spindle, the yarn will not suffer seriously different degrees of stretch from spindle to spindle, as it would if it were susceptible to high stretches for very small loads. Consequently a stress–strain diagram which for small loads up to, say, 0·2 or 0·3 gram per denier is straight, is desirable.

A fibre which gave the stress–strain diagram shown in Fig. 7 would, other things being equal, be preferred to another fibre which gave that shown in Fig. 8 in so far as freedom from liability to winding strain was concerned. But it is impossible to have every desirable characteristic in any one kind of fibre, and there are other good reasons for preferring fibres with the Fig. 8 type of characteristic curve to those with the Fig. 7 type. It is usually found that the Fig. 8 type have a better and softer handle and that they drape better. Most fibre curves have inflexions as in Fig. 8 but in some it is much more marked than in others; cellulose acetate (Fig. 93)

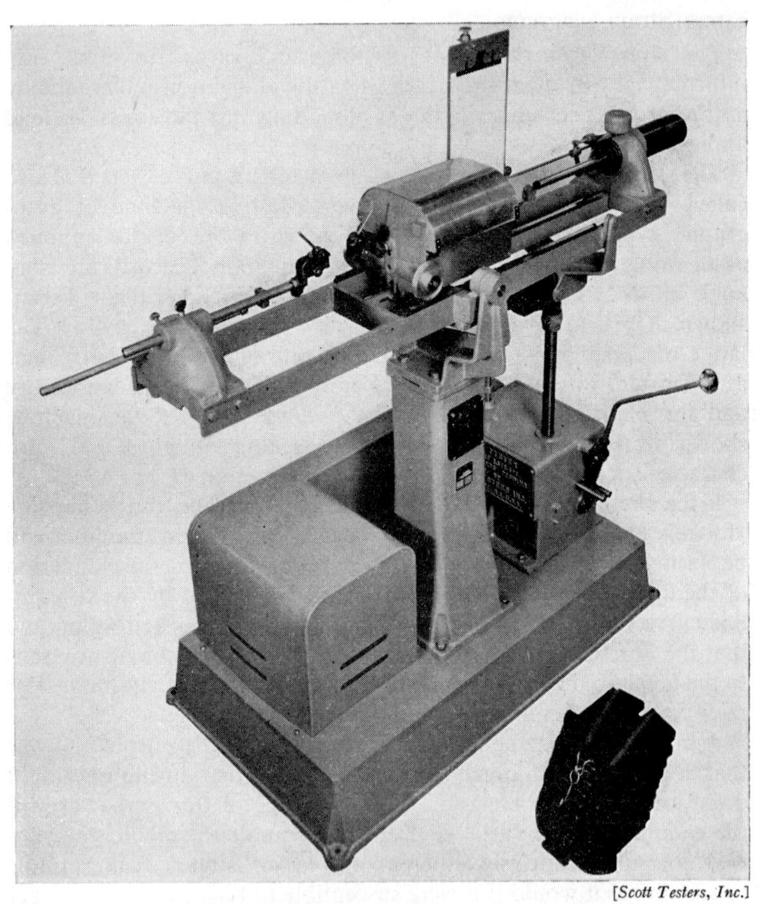

[*Scott Testers, Inc.*]

FIG. 6.—Inclined Plane Tester (constant rate of loading). This machine records the stress–strain diagram for a yarn.

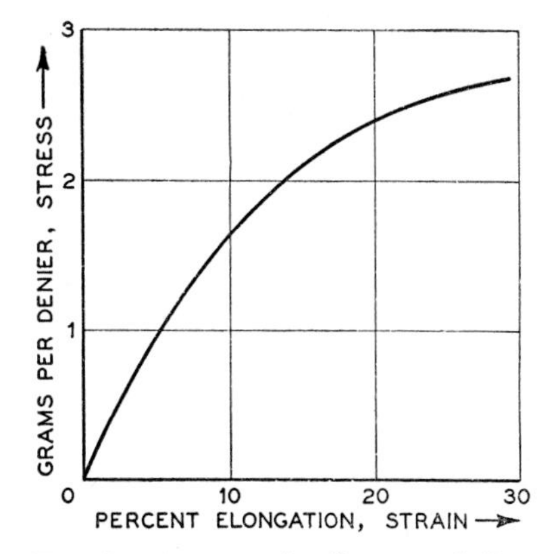

FIG. 7.—Stress–strain diagram of fibre with desirable winding characteristics.

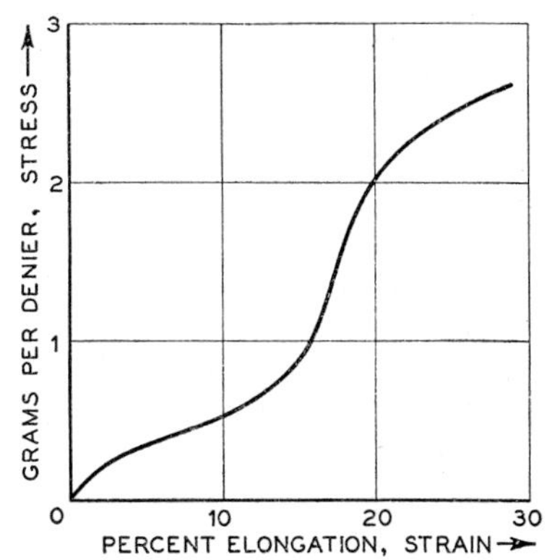

FIG. 8.—Stress–strain diagram of fibre which might easily be strained during winding.

has a more pronounced inflexion than viscose rayon (Fig. 72) and is softer to the touch. Wool and the regenerated protein fibres have the best handle of all and they all have a low initial modulus (cf. p. 309) and a high compliance ratio (cf. p. 310) which appears on the characteristic curve (Fig. 182) as a flattening for moderate extensions. It is probably true to say that the fibres with the softest handle are those that are most easy to damage by excessive winding tensions. The fibre that is the best for soft and comfortable underwear is unlikely to be the best for industrial uses such as ropes and belts where high strength and good recovery are essential.

The following fibres have their stress–strain diagrams recorded in this book; viscose rayon (Figs. 72, 99), cellulose acetate rayon (Fig. 93), Zantrel (Fig. 108), nylon 6 (Fig. 141), nylon and Terylene (Fig. 146), Kodel (Fig. 160), Dynel (Fig. 165), Orlon (Figs. 175, 176, 182), Acrilan (Fig. 187), and Darvan (Fig. 190). They should be compared.

Regain

All fibres exposed to the atmosphere pick up some moisture. The quantity of moisture picked up varies with the relative humidity and with the temperature of the atmosphere. Measurements are therefore made at standard conditions, which are arbitrarily fixed at 65 per cent relative humidity (R.H.) and 70° F. It takes some time—often a matter of hours—for a bundle of fibres to become in equilibrium with an atmosphere; if fibres are to be tested they should first be " conditioned " for several hours in a conditioning room maintained at 65 per cent R.H. and 70° F. If the fibres are in the form of a thick, highly twisted yarn or a heavy fabric they may require two days to reach equilibrium; for small bundles of loose fibres two hours is usually sufficient. Usually the moisture content of a fibre or yarn is described as regain.

Definition. The regain of a fibre (or yarn) is the percentage weight of moisture present, calculated on its oven-dry weight. (The " oven-dry weight " is the constant weight obtained by drying the fibre at a temperature of 105–110° C.)

Thus, if 100 parts by weight of a textile material contain 10 parts by weight of moisture, the regain is $100 \times 10/90 = 11{\cdot}1$ per cent, and the equivalent moisture content is 10 per cent. Note carefully that the regain is always calculated on the oven-dry or bone-dry weight of the fibre or yarn. It follows that the percentage regain is always greater than the percentage moisture content.

Some fibres absorb moisture much more readily than others; those that pick up more moisture than others are said to be more hygroscopic, and under similar conditions have higher regains. The molecular structure of a fibre largely determines its hygroscopicity. Cellulose fibres such as viscose rayon, which are characterised by the possession of a large number of hydroxyl groups, have high regains; polyvinyl fibres, which have no strongly polar groups, have very little affinity for water and have low regains.

Regain varies greatly with atmospheric conditions, and indeed the determination of the weight of wool under standard atmospheric conditions has always been a matter of commercial importance and of need for care; a consignment of wool charged for as 100 lb. may appear to weigh only 96 lb. when received, and care has to be taken to ascertain the weight of a sample under standard conditions (often simplest to weigh bone dry and calculate what the weight with the agreed regain would be). Nylon has a low regain compared with the natural fibres but a high one amongst the group of synthetics. Variation of nylon's moisture regain with relative humidity of the atmosphere at 25° C. is as shown in the following table; the measurements were made on undrawn nylon.

R.H. of atmosphere (%).	Moisture regain of undrawn nylon.
25	1·7
40	2·9
55	3·7
70	5·2
85	7·4
90	8·3

It has sometimes been observed, for example with acrylic fibres (p. 527), that the moisture regain is higher when equilibrium is approached from the wet end, *i.e.*, by drying a wet fibre, than when approached from the dry end, *i.e.*, by wetting a dry fibre. This affords an example of hysteresis.

Until recently the usual method for determining the moisture content of a fibre has been to dry it and observe the decrease in weight on drying; nowadays electrical and other devices are available which enable very rapid determinations to be made of moisture content.

Soviet Numbering System

The numbering system used by the Russian industrialists is a metric one; it indicates the length in metres of a yarn or filament that weighs one gram. The equivalent deniers (grams per 9,000 metres) and tex (grams per 1,000 metres) and millitex (milligrams per 1,000 metres) values for some metric numbers are shown in the following table:

Soviet metric number	Denier	Tex	Millitex
1	9,000	1,000	1,000,000
10	900	100	100,000
50	180	20	20,000
100	90	10	10,000
500	18	2	2,000
1,000	9	1	1,000
1,500	6	0·67	667
3,000	3	0·33	333
4,500	2	0·22	222
9,000	1	0·11	111

A convenient method of conversion is:

Russian metric number $= \frac{9{,}000}{\text{denier}}$. The higher the Russian metric number, the finer is the yarn or filament and in this respect it is in line with our own cotton and worsted numbering systems.

FURTHER READING

N. Eyre, " Testing of Yarns and Fabrics ", pp. 27–48. Textile Manufacturer Monograph No. 4 (1947) (denier, tenacity, twist, etc.).

M. V. Forward and S. T. Smith, " Moisture regain of 66 nylon continuous filament yarn ", *J. Text. Inst.*, **46**, T158–T160 (1955).

CHAPTER 2

THE STRUCTURE OF FIBRES

WHAT is the one feature common to all fibres? Let the whole gamut of them pass through the mind: cotton, flax, hemp, jute, ramie, silk, tussah, wool, rabbit, alpaca, mohair, llama, camel, viscose, cellulose acetate, nylon, Vinyon, Velon, Pe Ce, Ardil, Vicara, alginate, glass, asbestos, stainless steel, aluminium, and all the others. Most are organic, a few, like glass and asbestos, are inorganic; some are of animal origin, some of vegetable; some have continuous filaments, others have short filaments; some are transparent, others are opaque; some will burn, others will not; some are weak, others are strong. With almost any property but one the diversity of fibres will readily provide examples in which the property is well-marked—or in which it is to all intents and purposes absent.

THE SHAPE OF FIBRES

One property, however, is common to all fibres—that is, that all are very long in relation to their breadth. A little reflection will show that a high length : breadth ratio is the essential characteristic of a fibre. The material of which the fibre is constituted does not determine that it will be a fibre; there are many substances—of which mention can be made of nylon and of cellulose acetate—that are used in the form of moulded plastics as well as in fibre form. It is purely and simply the *shape* that is characteristic of a fibre. One would not think of a nylon moulding or of a cellulose acetate sheet as a fibre. The primary characteristic—the one indispensable prerequisite for a fibre—is a high length : diameter ratio.

Length : Breadth Ratios of Natural Fibres

The universal possession by fibres of the high length: diameter ratio may be observed by examination of the following typical figures for some of the natural fibres:

Fibre.	Typical length.	Typical diameter.	Length : diameter.
Cotton . .	1 in.	0·0007 in.	1,400
Wool . . .	3 in.	0·001 in.	3,000
Flax (ultimate) .	1 in.	0·0008 in.	1,200

The three fibres just listed are usually described, so far as their staple length and diameter are concerned, in inches. To familiarise ourselves with fibre dimensions in the metric system, too, we will use this system for the following fibres:

Fibre.	Typical length (mm.).	Typical diameter (μ) (= 10^{-3} mm.).	Length : diameter.
Hemp . . .	20	22	900
Jute . . .	2·5	15	170
Ramie . .	150	50	3,000
Manila . .	6	24	250
Sisal . . .	3	24	125
Pineapple . .	6	6	1,000

In the case of real silk, which is a *continuous* filament (actually the silkworm spins them paired), the length may be about 500 metres and the diameter about 15 μ, so that in such a case we have the enormous length to breadth ratio of $\frac{500}{15 \times 10^{-6}}$ or $33 \times 10^6 : 1$; the silk filament is 33 million times as long as it is broad.

These figures have been given in some detail because it is desired to emphasise that a high length: breadth ratio is the essential characteristic of natural fibres. The most useful of our natural fibres—cotton and wool—both have length: breadth ratios of more than 1,000: 1. Even the very coarse natural textile materials such as jute, manila and sisal, which are used mainly for cordage and for packing materials, have length: breadth ratios of over 100: 1.

Length : Breadth Ratios of Man-Made Fibres. We shall find that a high length: breadth ratio is equally characteristic of man-made fibres. The diameters to which such fibres are spun are of the same magnitude as those of the natural fibres. Thus viscose and cellulose acetate filaments are usually spun with average diameters varying between 10 and 30 μ, which are similar to those found in the natural fibres. As, too, man-made fibres are for the most part used in continuous filament form, and even when they are cut to staple, the staple is seldom less than 1 in. in length, it will be clear that a high length: diameter ratio is equally as characteristic of man-made fibres as it is of natural fibres.

Relation between Denier and Filament Diameter. It will provide an interesting exercise to calculate the diameter of a nylon filament of 3 denier, from a knowledge of the specific gravity, which is 1·14, and the shape of the cross-section, which is circular.

Example. Inasmuch as the denier is three, 9,000 metres of filament weigh 3 grams or 3,000 metres weigh 1 gram. But 1 gram also, since the specific gravity is 1·14, occupies a volume of $\frac{1}{1\cdot14}$ c.c. Therefore 3,000 metres of yarn occupy a volume of $\frac{1}{1\cdot14}$ c.c.

If d is the filament diameter, the volume of a given length l of a filament is given by the expression $\frac{\pi}{4} d^2 \times l$. In this case (1 gram of filament) the volume is $\frac{1}{1\cdot14}$ c.c. and l is 3,000 metres, so that

$$\frac{\pi}{4} d^2 \times 3{,}000 \text{ m.} = \frac{1}{1\cdot14} \text{ c.c.}$$
$$= 0\cdot877 \times 10^{-6} \text{ m.}^3$$
$$\text{so } d^2 = \frac{0\cdot877 \times 10^{-6}}{3{,}000 \times 0\cdot7854} \text{ m.}^2$$
$$\text{or } d = 19\cdot3 \times 10^{-6} \text{ m.}$$
$$= 19\cdot3\ \mu$$

It will be found convenient to remember that a 3-denier nylon filament has a diameter of about 20 μ. Correspondingly, a 5-denier filament will have a diameter of $\sqrt{\frac{5}{3}} \times 20\ \mu$, or about 26 μ.

Viscose rayon has a higher specific gravity (1·51) than nylon, and in its case the mean diameter of a 5-denier filament, which is a very usual size, will be $\left(26 \times \sqrt{\frac{1\cdot14}{1\cdot51}}\right)$ or 22·5 μ.

If this filament was magnified 500 times its diameter would be $22\cdot5 \times 10^{-3} \times 500$ mm.—*i.e.*, 11·2 mm.

Now, Fig. 9 is a photomicrograph of a 5-filament denier viscose

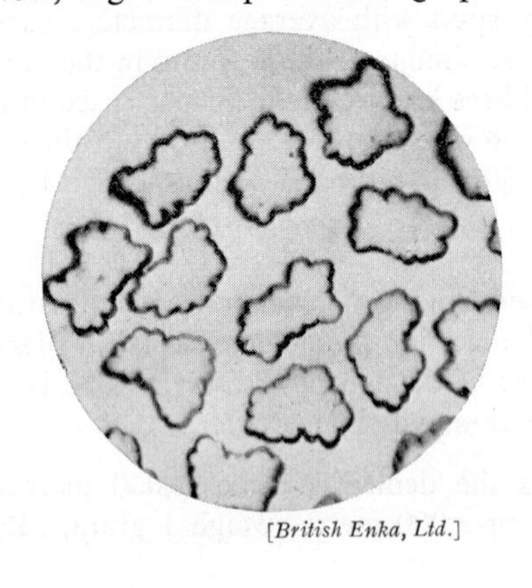

[*British Enka, Ltd.*]

FIG. 9.—Photomicrograph of cross-sections of Lustre viscose rayon filaments (×500).

rayon which has been magnified 500 times. It is not easy to measure the mean diameter exactly of filaments which have such irregular cross-section, but the reader will have no difficulty in verifying by measurement that the mean diameter of the filaments shown in Fig. 9 is about 11 mm.

Clearly, the diameter of a filament will vary as the square root of the denier; if the filament diameter is doubled, the denier is quadrupled.

One more example will serve to illustrate the maintenance of the high length: diameter ratio of man-made fibres, even in the case of cut staple.

Example. One of the most widely used of the staple fibre rayons is Fibro, of 3-filament denier and $1\frac{7}{16}$ in. staple. Knowing that the specific gravity of the fibre is about 1·51, and assuming that the cross-section is approximately circular, we can calculate the length: diameter ratio as follows:

$$\text{Length} = 1\tfrac{7}{16} \text{ in.} = 36{\cdot}5 \text{ mm.}$$

Because

$$\frac{\pi}{4} d^2 \times 3{,}000 \text{ m.} = \frac{1}{1{\cdot}51} \text{ c.c.} = 0{\cdot}662 \text{ c.c.}$$

$$d^2 = \frac{4 \times 0{\cdot}662 \times 10^{-6}}{3{\cdot}1416 \times 3{,}000} \text{ m.}^2$$

So

$$\text{Diameter } (d) = 16{\cdot}8 \times 10^{-6} \text{ m.}$$
$$= 16{\cdot}8 \times 10^{-3} \text{ mm.}$$

Therefore

$$\frac{\text{Length}}{\text{Diameter}} = \frac{36{\cdot}5 \text{ mm.}}{16{\cdot}8 \times 10^{-3} \text{ mm.}}$$
$$= 2{,}173$$

It will be seen that when man makes fibres he keeps to approximately the same dimensions and the same length: breadth ratio as Nature uses. Amongst the best staple fibres for textile purposes this ratio usually lies between 1,000 and 4,000.

Filament Cross-Sectional Shapes

The filaments of some man-made fibres are very different in cross-section from those of others. Some, such as nylon and Dacron, are practically circular; in fact, on first seeing a cross-sectional view of a nylon yarn under a magnification of 300 or 400 diameters, one is amazed at the apparently perfect symmetry of the circular filaments. Remembering that they are formed by extruding molten nylon through circular holes, it is not surprising that the filaments should be almost circular in section. Other fibres, such as those of Vinyon and Orlon, have dumb-bell or dog-bone sections. Cellulosic fibres vary considerably: cuprammonium filaments are almost circular; viscose rayon are usually serrated (although not nearly so highly serrated as they were twenty years ago); cellulose acetate are lobed.

Are differences in cross-section of much consequence? Those fibres that have been spun from a melt have round sections; those that have been spun from a solution have lobed or dog-bone sections formed as the filaments, circular in section as they emerge from the

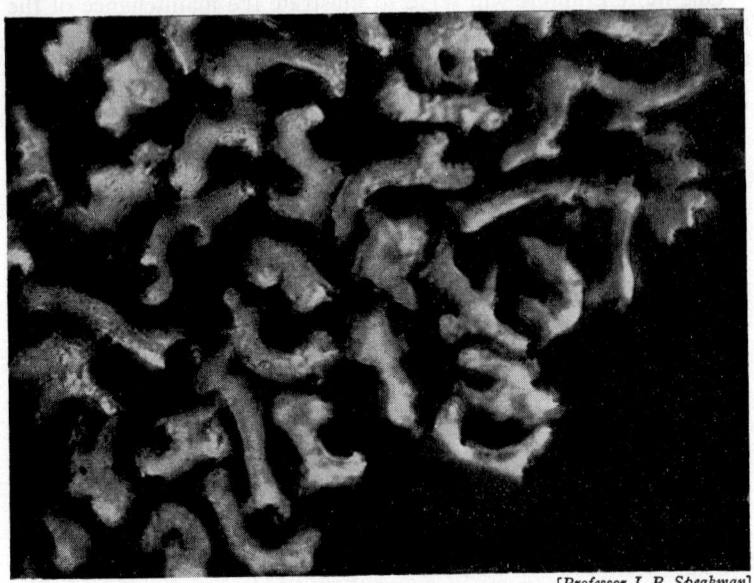

[*Professor J. B. Speakman*]

FIG. 10.—Photomicrograph of cross-sections of calcium alginate filaments spun from 5 per cent solution of sodium alginate (×450). Flat and harsh to handle.

jet, collapse as solvent is removed either by evaporation in dry spinning (*e.g.*, cellulose acetate and Orlon) or by solution in wet spinning (*e.g.*, viscose or alginate). But if wet-spun fibres are greatly stretched during spinning whilst still plastic, their cross-sections will approximate to round (*e.g.*, cuprammonium rayon and Acrilan). Shape is important and the first requirement of a man-made fibre is that all its filaments shall have similar cross-sections. In the early days of rayon it was no uncommon experience to find pieces of fabric quite spoilt in appearance by the presence of bright picks or "shiners"; apparently an end of yarn would vary along parts of its length from its normal to a very high lustre. When such yarn was examined it was found that the very bright parts of the yarn were characterised by the filaments having much flatter cross-sections than usual; the flatter section reflected much more light. Very careful control of the spinning machinery is necessary to prevent the sudden appearance (and disappearance) of " flats ".

Aside from the rather obvious necessity of a fibre having a *uniform* cross-section, is there any advantage to be derived from particular sectional shapes?

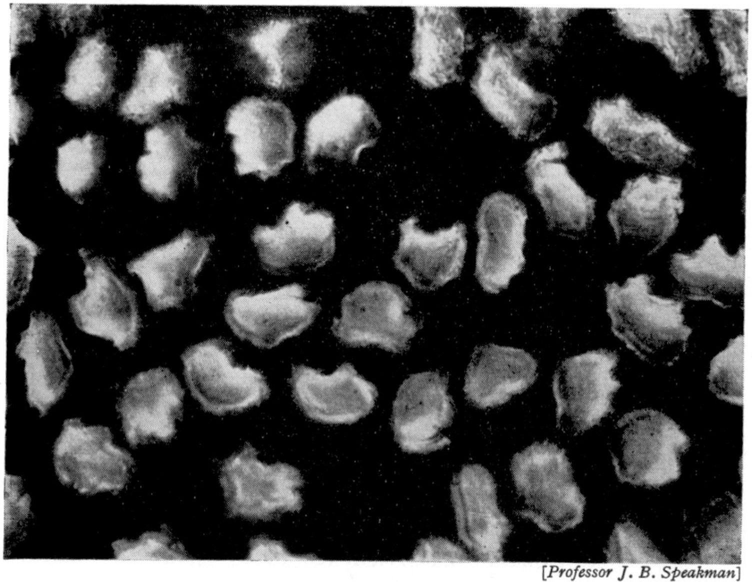

[*Professor J. B. Speakman*]

FIG. 11.—Photomicrograph of cross-sections of calcium alginate filaments spun from 8 per cent solution of sodium alginate (×450).
Round and kind to handle.

In general, flat sections have a high lustre and excellent covering power, but have a harsh, unpleasant handle. An illustration of this defect of harsh handle being due to flatness of filament section is to be found in calcium alginate yarns. Fig. 10 shows the flat section of calcium alginate spun from a 5 per cent spinning solution of sodium alginate; this yarn had a harsh handle. Fig. 11 shows the nearly round section of calcium alginate spun from an 8 per cent spinning solution of sodium alginate; this yarn had a soft handle.

Circular sections give a yarn with a kindly handle and feel, but with a poor covering power. Remembering that the circular section offers the minimum surface for a given volume, this is not surprising.

It is interesting to remember that amongst the natural fibres wool and the goat fibres have nearly circular sections and have very good handle, whereas cotton, which is harsher, has a flat section.

The greater the surface area, the better the absorption of dyestuff, and it was probably for this reason that the serrated section of viscose rayon was developed; this type of section is produced by the presence of zinc salts in the coagulating bath.

The attractive handle of cellulose acetate is due in part to its lobed structure, which in this respect is not very different from a circular section, but affords a better cover.

Trials have been made to determine the effect of a non-circular section on nylon, by spinning not through circular jets but through rectangular slots or triangular holes. A *sine qua non* for experiments of this kind is that the melt which is extruded must be of very high viscosity, because if a relatively thin melt is extruded through holes of any shape whatever, the filaments will take up a circular section. But when a very viscous melt has been extruded through rectangular slots (instead of the usual round holes) fibres of elliptical shape have been made, and in some trials the departure from circularity has been so great as for the major axis of the ellipse to be 1·95 times the minor axis. There was no marked difference in physical properties, nor indeed in cover. The covering power of the yarns was determined as follows: yarn was wound on to a bobbin, the barrel of which was covered with graph paper, until the lines could just no longer be seen. The yarn was then removed and weighed.

Comparison of the Covering Powers of Nylon Yarns of Filaments with Circular and Triangular Cross-Section

Denier (filaments).	Circular.			Triangular.		
	150 (50).	60 (20).	30 (10).	150 (50).	60 (20).	30 (10).
Covering weight .	3·23	2·96	2·91	2·81	2·52	2·33
Covering power .	47·3	20·2	10·4	54·6	23·8	13·1
Percentage by which covering power of triangular is greater than that of circular product	—	—	—	15	18	26

A low "covering weight" represented a good covering yarn. "Covering power" was defined as 1/(metres required to mask). Whilst primarily a measure of the reciprocal of the covering weight, this expression clearly takes account of denier also; heavy denier yarns have higher covering powers than light deniers. Elliptical and circular section nylons did not show much difference, but some yarn that was spun with triangular section filaments did show significant differences; these are tabulated on p. 26.

There is no doubt that the triangular section increased the covering power. It also made a definite difference to the handle of fabric made from it; some observers thought it was more like that of real silk than is ordinary nylon and when it is remembered that real silk is composed of filaments which approximate to triangles in cross-sectional shape, this is not altogether surprising. One other effect noticed was that the fabrics from filaments of triangular section were lustrous, and the lustre was associated with pin points of light or flecks. Following this investigational work, nylon has been spun commercially with a trilobal, approximating to triangular, cross-section (p. 346); so has Dacron (p. 426). A similar principle has already been applied to polypropylene (p. 565), one of the newest synthetic fibres. In all of these cases there is a resemblance to the approximately triangular cross-section of "boiled-off" real silk. Although circular cross-section filaments are the easiest to melt-spin, they are not necessarily the best.

Yarns of mixed filaments, some elliptical in section, others circular, have also been envisaged. It is interesting that wool fibres, although nearly circular in section, do vary a little and most of them are slightly elliptical; the average value of their major axis/minor axis ratio is 1·22.

Amongst heavy monofils, flat sections are already available commercially; often they give an appearance of a fine Panama straw; Clorene (p. 469) and polystyrene (p. 567) provide some examples.

Fineness of filament is associated with softness of handle and with the necessity for an increased application of dyestuff, due to the tendency for very fine filament yarns to dye paler shades than normal yarns—they offer a greater surface area to be covered by the dyestuff.

The Polymer—Natural or Man-Made

Eighty years ago all fibres were natural fibres. Then in about 1883 the first successful man-made fibres were introduced. These fibres were regenerated—*i.e.*, they were obtained by dissolving a natural fibrous material, such as cotton or a derivative of it, and then extruding the solution into an atmosphere or a solution which reprecipitated or regenerated the fibrous material. The original cotton had consisted of short, hairy, nearly opaque fibres, and after regeneration it consisted of very long, smooth, transparent filaments, but it was still the same essential material. This natural fibrous material was essential for the production of a regenerated fibre. Man could make a fibre only if he started with a fibre (or with some equivalent fibrous material such as wood pulp). This state of affairs lasted (with the partial exception of glass fibres) until 1938, when nylon was introduced. Nylon was not a regenerated fibre, but a true synthetic fibre; it was made not from some original natural fibrous material, but from simple chemicals which were built up to give a synthetic fibrous material.

A clear distinction should be made between regenerated fibres such as viscose rayon, cuprammonium, cellulose acetate, alginate, Tricel, Tufcel, *etc.*, and synthetic fibres such as nylon, Vinyon, Dacron, Orlon and glass. The former are all made from some natural fibrous material; the latter are not—they are synthesised from simple chemical substances.

It seems strange that the production of a true synthetic fibre should have been so long delayed. The reason for this delay was that before a substance can be synthesised it must be analysed; it was, for example, useless to attempt to synthesise penicillin until analysis had revealed its constitution. In the case of fibres the same delay attendant on their analysis characterised the efforts to synthesise them. Forty years ago fibres were thought to be complex and irregular bodies with no simple design; they were not amenable to chemical treatment, and a student could pass through an honours school of chemistry without hearing fibres mentioned.

Fibrous Molecules. The first real insight into the structure of solids came from the X-ray diffraction experiments carried out by Laue and by W. H. and W. L. Bragg; these experiments were made on simple crystalline bodies, and X-ray diagrams revealed the dimensions of the molecule, and, by inference, the arrangement of the atoms in the molecules. This method was extended by O. L. Sponsler and later by W. T. Astbury and others to fibres, and the outstanding discovery emerged that the molecules that make up wool, cotton and other fibrous materials were themselves very long and narrow; in other words, they were fibrous. Just as it needs reasonably long fibres to spin into a good yarn (cotton with a fibre length of less than $\frac{1}{4}$ in. is almost impossible to spin), so a prerequisite for a good fibre is that the constituent molecules must themselves be very long. The long, narrow shape of the natural fibres is a reflection of the long, narrow shape of their molecules. The cellulosic fibres, which, including as they do cotton and jute, are the most abundant of any that we use, were all characterised by being composed of long chains of glucose residues, linked up in this way:

```
                     |
                     O - - - - - - - ↑
                     |               |
                    HC               |
                   /  \              |
      CH2OH—HC     HOCH              |
             |        |              |
             O      HCOH             |
              \     /                |
                CH                   |
                 |                   |
                 O                   |  10·28 Å
                 |                   |
                CH                   |
               /  \                  |
           HCOH    CH—CH2OH          |
             |      |                |
          HOCH      O                |
              \    /                 |
               HC                    |
                |                    |
                O - - - - - - - - - -↓
                |
               HC
              /  \
  CH2OH—HC     HOCH
         |        |
         O      HCOH
          \     /
            CH
            |
```

The glucose molecule is:

```
                 HO
                  |
                 HC
                /  \
     CH2OH—HC     HOCH
            |        |
            O      HCOH
             \     /
               CH
               |
               OH
```

and after the elements of water have been split off from it, the residue (what is left) is termed a " glucose residue ", and is:

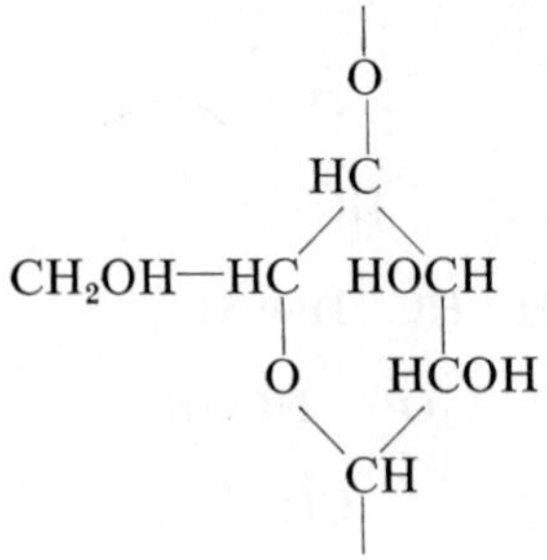

```
                    |
                    O
                    |
                   HC
                  /  \
   CH2OH—HC   HOCH
             |         |
             O     HCOH
              \      /
                CH
                 |
```

Note that although in the cellulose chain all the glucose residues are alike, they alternate in right- and left-hand arrangement, so that the " repeat " occurs on two glucose residues (or on one cellobiose residue) as shown above, and that the length of the repeat is 10·28 Å.

$$\text{Å} = \text{Ångström units} = 10^{-10} \text{ m.}$$

As

$$\mu = 10^{-6} \text{ m.}$$

$$\mu = 10^{4} \text{ Å or } 10{,}000 \text{ Å} = \mu$$

Accordingly 10·28 Å is just a little more than 10^{-9} metres, or than a millionth part of a millimetre. Picture, then, the cellulose molecular chain built up of about 1 million cellobiose residues per millimetre. Clearly, therefore, there will be about 2 million glucose residues in 1 mm. of the molecular chain of cellulose.

Why Molecular Chains cannot be Seen. Recent determinations of the molecular weight of native cellulose suggest that the number of glucose residues in one molecular chain may be about 10,000 in the case of cotton, so that the molecules of cellulose in cotton are about $\frac{1}{200}$ part of a millimetre long. This brings us into the realm of visible objects, and the reader may reasonably inquire why such long molecules cannot be distinguished under the microscope. The answer is, of course, that their other dimensions are very much too small; one can imagine a hair line made progressively finer and finer; it might still be quite long, but as it became finer and finer it would become more and more difficult to see.

Degree of Polymerisation. Until recently it was believed that there were only about 2,000 glucose residues in the cellulose molecule, but it is now known that a value of 10,000 is closely approached in the case of cotton, at least. This figure is known as the *degree of polymerisation.* We say that the degree of polymerisation of glucose in cotton cellulose is 10,000, whereas it was formerly thought to be about 2,000.

The width of the cellulose molecule is known to be 7·5 Å, measured as shown below

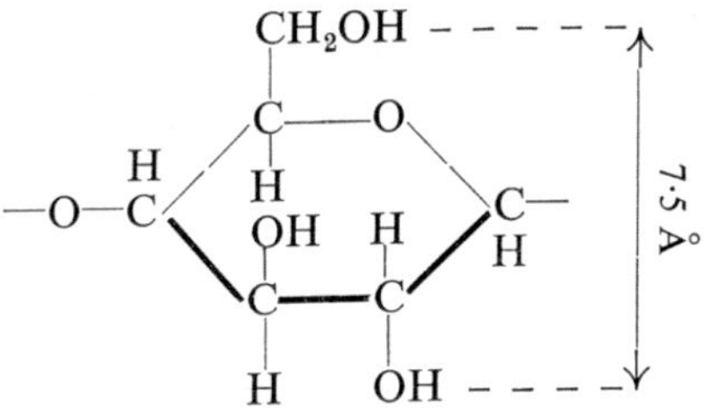

```
                 CH₂OH - - - - -↑
                   |             |
                   C——O          |
           H      /|   \         |
      —O—C/    H      \C—       | 7·5 Å
           \   OH  H   /H        |
            \  |   |  /          |
             C———C               |
             |   |               |
             H   OH - - - - -↓
```

We are now in a position to arrive at a tentative figure for the length: breadth ratio of the cellulose molecule. This value is

$$\frac{\frac{10{,}000}{2} \times 10\cdot28\ \text{Å}}{7\cdot5\ \text{Å}} \quad \text{or } 7{,}000 \text{ (approx.)}$$

Clearly, therefore, the essential characteristic of a high length: breadth ratio, which is so typical of fibres, is also typical of their molecules. In fact, in the molecular region this ratio may be even higher than in the macroscopic region. It is of interest that the ratio is of the same *order* in both regions. That is all that can be said, for there is still too much uncertainty about the degree of polymerisation in natural polymers to assign any precise figures with confidence. Usually when different methods are used for the determination of degree of polymerisation—*e.g.*, viscosity, osmosis, chemical end groups, and rate of sedimentation in the ultra-centrifuge—very different results are obtained. Uncertainty about the precise degree of polymerisation still exists, but there is no possible doubt that this runs into thousands, and that the length: breadth ratio of the molecules of fibrous materials such as cotton also runs into thousands. The molecules of fibrous materials are, in effect, themselves fibrous. Often these fibrous molecules are likened to strings of beads, in which illustration a glucose residue is the counterpart of a bead. Whilst the reader should perhaps be familiar with the analogy, it is not one that has much to commend it, because it completely fails to illustrate the hold that each residue has for the next. The daisy-chain provides a sounder analogy. Let each glucose residue be represented by a daisy, and then the cellulose molecule can be likened to a daisy-chain. For the purposes of illustration, note that if the width of the cellulose molecule is taken as equivalent to the height of one of the capital letters in which THIS is printed, the cellulose molecule would be as long as a cricket pitch. Truly, the molecules of fibres are themselves fibrous.

Cellulose. Some work that was carried out at the Shirley Institute in 1955 was directed to determine the degree of polymerisation and its distribution in cellulose rayons. Four rayons were used, standard viscose, Tenasco, which is a medium high tenacity stretched viscose, cuprammonium, and old-type Durafil, which is a very high tenacity highly stretched viscose rayon. All four were nitrated with a mixture of

75 parts nitric acid
20 parts acetic acid
5 parts acetic anhydride

The cellulose nitrate so formed was dissolved in acetone and precipitated in fractions—a bit at a time—by the progressive addition of small lots of hexane to the acetone solution. The hexane is not a solvent for the cellulose nitrate. The first fraction precipitated contains the longest molecules, those with the highest degree of polymerisation; the second fraction contains the next longest and the last fraction contains the shortest. The fractionation scheme was complicated and the number of fractions was considerable. Each precipitated fraction was purified and then dissolved in butyl acetate and the intrinsic viscosity of the solution was measured. From the intrinsic viscosity it was possible to calculate the average degree of polymerisation. The results can be arranged as in the following table, to show how the content of low, medium, and high degrees of polymerisation vary amongst the fibres:

Fibre.	Percentage by weight with D.P.		
	Below 500.	500–1,000.	Over 1,000.
Viscose rayon	72·6	24·5	2·9
Tenasco	55·2	38·7	6·1
Cuprammonium rayon	49·3	35·0	15·7
Durafil (old type, p. 260)	34·0	34·6	31·4

Cuprammonium is made by a process fundamentally different from the other three and it is one in which depolymerisation is obviously less severe than in the viscose process. Of the other three the Durafil, which is very strong, has a much bigger percentage of very long molecules than the other two, and the viscose rayon which is the weakest has a very high percentage of short molecules. This points to a sound generalisation—to make good fibres avoid depolymerisation as far as possible; it is often fairly easy to obtain a high tenacity by stretching but real fibre strength, transverse as well as longitudinal, with good durability comes from an intrinsically fibrous structure in the molecular dimensions. Nature provides long linear polymeric molecules and when these are converted into man-made fibrous form, it behoves man to depolymerise them as little as possible. When truly synthetic fibres are built up it is essential to carry the polymerisation far enough to get a high degree of polymerisation running into hundreds if good fibres are to be made.

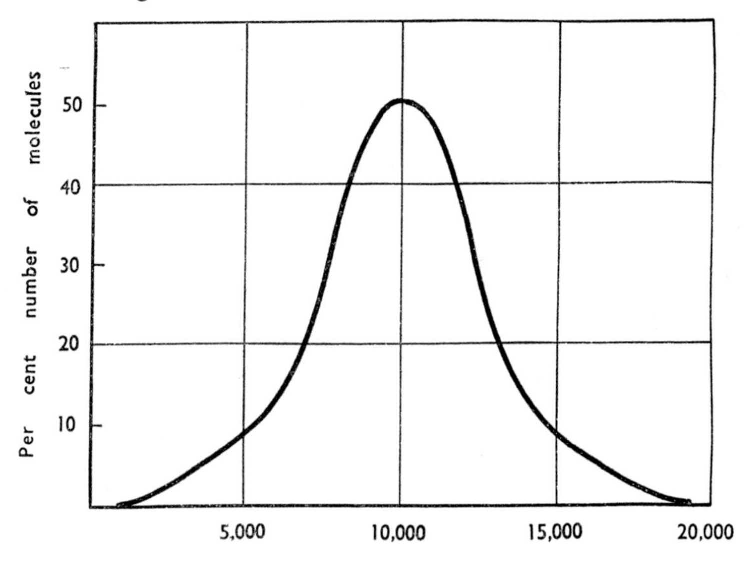

FIG. 12.—Relative abundance of long and short molecules in a fibrous polymer.

MOLECULAR WEIGHTS OF FIBRES

The molecules of which a monomeric material such as benzene or butyl alcohol is composed are all of exactly the same weight and size (except in respect of very small weight differences due to the presence of isotopes—a matter which we can forget, as it has no bearing on our present considerations). But the molecules of a polymer vary very considerably. Some will be very much longer than others. A synthetic fibre that has perhaps half its molecules within the range of 15,000–18,000 molecular weight may well have a few as low as 1,000, and a few as high as 50,000. The abundance of molecules of different molecular weights will vary as shown in Fig. 12; most will cluster round the mean. In cases where a polymer contains a significant number of very short molecules (*e.g.*, of the order of 1,000 molecular weight) the average molecular weight will be considerably reduced by these, and the most frequent molecular weight may be more significant—*i.e.*, the molecular weight that occurs most commonly in the group of molecules being considered.

Regenerated Cellulosic Fibres. Until recently it was considered that native cellulose had a mean molecule of 2,000 glucose residues, which corresponds to a molecular weight of about 320,000, but recent work has suggested that native cellulose may have a mean molecule of about 9,000 glucose residues, with a molecular weight of about 1½ million. Undoubtedly, in manufacture these long molecules are seriously reduced in length. Cuprammonium rayon has about 500 glucose residues in its cellulose molecule, viscose rayon about 250, and cellulose acetate rayon about 300. The manufacturer knows that desirable fibrous properties depend on the preservation of the long molecular-chain backbone, and that as the length of this is reduced so do strength and durability fall away, and he makes every effort to avoid degradation, but some is inevitable. The molecules in rayon may be only 3 or 4 per cent as long as those in the starting material. In the case of viscose rayon from wood pulp, it is likely that the greater part of the degradation takes place during the conversion of the timber into the pressed board—*i.e.*, the dry sheet pulp.

Synthetic Fibres. The molecular weights of synthetic fibres are high. So much is certain, although little reliability can be placed on " exact " figures that are quoted, so long as different methods give different results. The molecular weight of nylon, for example, is about 12,000–20,000. Vinyon is made from a co-polymer of vinyl chloride and vinyl acetate, and it has been stated that polymers with satisfactory tensile properties are limited to a molecular weight range of 9,500–28,000.

In the synthesis of fibres it is found that as polymerisation starts there is at first no evidence of fibre-forming properties; as it proceeds, fibres which are weak and brittle can be formed; then, as polymerisation proceeds still further, high strength is obtained. By way of illustration the fibre-forming properties of nylon may be related to its molecular weight somewhat as follows:

Molecular weight.	Fibre-forming properties.
Below 4,000	None.
4,000–6,000	Weak, brittle, fibres formed.
6,000–8,000	Longer, less brittle but still weak fibres formed.
8,000–10,000	Increasing ease of fibre formation and increasing strength.
Over 10,000	Excellent fibres formed.

In order to obtain fibres with satisfactory tensile properties, a minimum degree of polymerisation, a minimum mean molecular weight, must be reached. One cannot, however, go on indefinitely, as beyond perhaps 30,000 molecular weight, solubility and fusibility fall off; and how is one to spin a fibre from a material that can be neither dissolved nor melted? This has been done with Teflon (p. 569) but it is difficult. The highest molecular weight fibres of all are the stereoregular polyolefines (pp. 62, 556); in this class molecular weight (but not average molecular weight) can run into millions.

Kinds of Molecular Weights of Polymers

Polymers are heterogeneous in that they consist of mixtures of long chain molecules of different chain lengths. Accordingly the molecular weight of any polymer is an average molecular weight. There are two kinds of average molecular weight according as one is primarily concerned with the numbers of molecules or with their weights; they are known respectively as " number average molecular weight " and as " weight average molecular weight ".

Number Average Molecular Weight. This is the weight of the polymer divided by the number of molecules. It is the molecular weight average that is arrived at by measurements such as those of osmotic pressure which are determined by the *number* of molecules (or of chemical titration methods for end groups). It can be defined as:

$$\overline{M}_n = M_o \times \frac{\Sigma l n_l}{\Sigma n_l}$$

Where $\overline{M}_n$ is the symbol used for the number average molecular weight

M_o is the molecular weight of the monomer or repeating unit

n_l is the number of molecules of chain length l.

For purposes of illustration, let us suppose that we have a speck of polymer which is built up of units with a molecular weight, M_o, of 44. This is actually the molecular weight of the repeating unit

```
 H  OH
 |  |
—C—C—
 |  |
 H  H
```

in polyvinyl alcohol. Suppose that our speck of polymer consists of:

2 molecules of molecular weight 44 or chain length 1 unit
3 molecules of molecular weight 88 or chain length 2 units
4 molecules of molecular weight 132 or chain length 3 units
3 molecules of molecular weight 176 or chain length 4 units
3 molecules of molecular weight 220 or chain length 5 units
1 molecule of molecular weight 264 or chain length 6 units

Its composition is represented by Fig. 13.

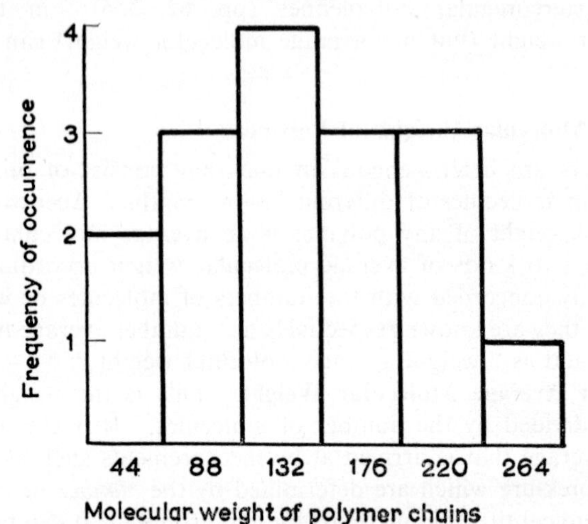

FIG. 13.—Distribution of molecules of various lengths in the polymer discussed on pp. 35–37.

In order to find the number average molecular weight we first sum the products of l and n_l thus:

$$\begin{aligned}\Sigma l n_l = & \quad 2 \times 1 = \quad 2\\ & +3 \times 2 = +\ 6\\ & +4 \times 3 = +12\\ & +3 \times 4 = +12\\ & +3 \times 5 = +15\\ & +1 \times 6 = +\ 6\\ & \overline{\qquad\qquad 53}\end{aligned}$$

Next we sum the values of n_l thus:

$$\Sigma n_l = 2 + 3 + 4 + 3 + 3 + 1 = 16$$

Therefore $\overline{M}_n = 44 \times \frac{53}{16}$ or 146 (about)

Clearly, if we had postulated that our speck of polymer had contained not 2 molecules of molecular weight 44 and 3 molecules of molecular weight 88 and so on, but 2 million and 3 million respectively and so on, we should have arrived at the same number average molecular weight.

Weight Average Molecular Weight. This is the average molecular weight which is obtained from measurements of light scattering by polymer solutions, because the scattering is determined not only by the number of polymer molecules but also by their size. The weight average molecular weight can be defined as:

$$\overline{M}_w = \frac{\Sigma M_l w_l}{\Sigma w_l}$$

where $\overline{M}_w$ is the symbol used for the weight average molecular weight.

M_l is the molecular weight of molecules of chain length l, and where

w_l is the weight of molecules of chain length l (so that $w_l = n_l \times M_l$ where n_l is the number of molecules of chain length l).

For illustration let us consider the speck of polymer already discussed and illustrated in Fig. 13.

In order to find the weight average molecular weight we first sum the products M_l and w_l thus:

$$\begin{aligned}
\Sigma M_l w_l = \quad & 44 \times (2 \times 44) = 3872 \\
& + 88 \times (3 \times 88) = 23232 \\
& + 132 \times (4 \times 132) = 69696 \\
& + 176 \times (3 \times 176) = 92928 \\
& + 220 \times (3 \times 220) = 145200 \\
& + 264 \times (1 \times 264) = 69696 \\
& \overline{\qquad\qquad 404624}
\end{aligned}$$

Next we sum the values of w_l thus:

$$\begin{aligned}
\Sigma w_l = \quad & 2 \times 44 = 88 \\
& + 3 \times 88 = 264 \\
& + 4 \times 132 = 528 \\
& + 3 \times 176 = 528 \\
& + 3 \times 220 = 660 \\
& + 1 \times 264 = 264 \\
& \overline{\qquad\qquad 2332}
\end{aligned}$$

Therefore $\overline{M}_w = \dfrac{404624}{2332}$ or 173 (about)

The weight average molecular weight is different from the number average. Generally, the number of small molecules present mainly determines the number average molecular weight because they increase " number of molecules " more than " weight of molecules ". They have little effect on the weight average molecular weight, which can, however, be raised quickly by an increase in the number of large molecules.

Methods of Determining Molecular Weights

Methods of estimating molecular weights of fibres are highly specialised, and very great disparities are to be found in the results obtained from them. One of the main difficulties is that a solution has generally to be prepared for estimating the molecular weight, and there is often grave doubt as to whether the mere act of solution has not degraded the molecules, so that a low value is obtained. Nylon, for example, is usually examined in *m*-cresol solutions, and, in order to avoid depolymerisation during solution, a low (room) temperature should be used. This makes the act of solution a tedious business, and it is most conveniently carried out in a mechanical shaker for perhaps twenty-four hours. Cellulose is soluble only in highly concentrated electrolytes, and in such cases the question of depolymerisation is bound to introduce a little uncertainty. The Haworth method (p. 43) is, however, free from this defect.

The principles of the methods used for molecular-weight determination of fibres are as follows:

1. Measurement of viscosity of the solutions. As the length of molecule increases, so does the viscosity of its solution; the long molecules of solute become entangled and inhibit the flow of solvent. Viscosity measurements of solutions of nylon in *m*-cresol can be used as a control of the degree of polymerisation reached during manufacture. (See Fig. 14.)

2. Sedimentation using the ultra-centrifuge. If solutions of a polymer are made and allowed to stand, then after perhaps twenty years there might be some significant separation of the heavier molecules to the bottom. In the ultra-centrifuge an acceleration equal to half a million g (g = acceleration due to gravity) has been developed. Observation of the rate of sedimentation—which under these conditions is fast enough to be measured conveniently—can be made, and a chart prepared showing the number of molecules of one length, the number of another, and so on, similar in principle to the staple-fibre analysis frequently made by a fibre-sorter on wool or cotton.

3. Determination of osmotic pressure. This method is limited in scope, but is useful for molecular weights below 100,000.

4. The scattering of light by solutions of high polymers. This method is comparatively new, but seems to be gaining widespread adoption.

5. Chemical determination of end groups. If a fibre is, for example, a polyamide, the terminal groups on each chain molecule will be free amine or free carboxylic groups; if both

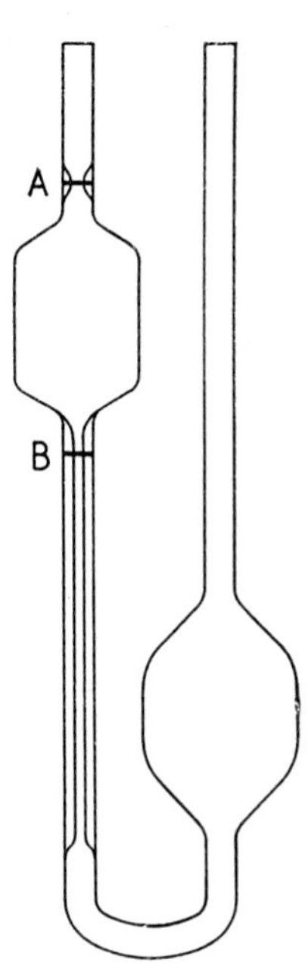

FIG. 14.—Ostwald Viscometer.

The time taken for the level of a solution of polymer to pass from A to B is observed. The higher the molecular weight of the polymer, the higher the viscosity of its solution and the longer the time of flow.

of these can be determined, the " equivalent weight " and thence the molecular weight of the fibre can be calculated. If the polyamide has been stabilised (Chapter 19) with acid, more of the end groups will be carboxylic than amine groups, but there will still be some of each and if both can be determined (as they can) then their combined equivalent weight will still be a measure (one half) of the molecular weight of the polyamide.

Viscosity

Absolute viscosity. This is defined in c.g.s. units and is expressed in poises. It is not much used in fibre work, but it may be noted that for liquids of similar densities, their times of flow through a capillary, *e.g.*, the viscometer shown in Fig. 14, are proportional to their absolute viscosities. The following derived expressions are more commonly used; all are based on this relation between absolute viscosity and time of flow.

Relative viscosity. The relative viscosity of a solution of a polymer is the ratio of the absolute viscosity of the solution to that of the solvent. If it takes t_1 seconds for the solution to flow through a viscometer and t_2 seconds for the solvent alone, then the relative viscosity, η_{rel}, is t_1/t_2 (the Greek letter eta is commonly used to indicate viscosity). The term " relative viscosity " as applied to nylon polymers was defined by E. W. Spanagel of du Pont in U.S. Patent 2,385,890 (1945) as: the ratio of the absolute viscosity of an 8·4 per cent by weight solution of the polymer in 90 per cent formic acid (10 per cent water), to the absolute viscosity of the 90 per cent formic acid. Moisture content of the polymer is disregarded; measurements are made at 25° C. How many times does 8·4 per cent polymer increase the time of flow of the 90 per cent formic acid in which it is dissolved—that is the relative viscosity of the polymer solution. In 1943 it was usually about 33 or 34 for nylon; now it may be 50 or more, especially if non-round filaments are to be spun.

Specific viscosity. This is the increase in viscosity of a solution over that of the pure solvent. We can write:

$$\eta_{sp} = \frac{\eta_{solution} - \eta_{solvent}}{\eta_{solvent}} \text{ or } \frac{\eta_{solution}}{\eta_{solvent}} - 1$$

which is clearly $\eta_{rel} - 1$. In fact, the specific viscosity is obtained by subtracting unity from the relative viscosity. Thirty years ago Staudinger developed an equation relating specific viscosity to molecular weight, one which has been of great use in work on fibrous polymers. It is

$$\frac{\eta_{sp}}{c} = kM$$

where c is the concentration of the solution used, k is a constant and M is the molecular weight of the polymer, this being the quantity usually required to be found. This relation was used to determine the molecular weight of cotton (in cuprammonium solution); it was about 320,000 corresponding to about 2,000 glucose residues; that of viscose rayon was found to be about one-fifth of this value, an indication of the considerable reduction in D.P. that takes place in the processes of regenerating the cellulose. More recent determinations suggest that typical values of D.P. are 4,000 for cotton and 250 for viscose rayon.

Intrinsic viscosity. The intrinsic viscosity is the specific viscosity divided by the concentration when the concentration is very small. When concentration is small, specific viscosity is then also small. The best way to obtain the intrinsic viscosity is to determine the specific viscosity at several low concentrations, to plot the results in the form η_{sp}/c against c, where c is the concentration, and to extrapolate the " curve " until it cuts the axis corresponding to $c = 0$; the value of η_{sp}/c at this point is the intrinsic viscosity. An example may make this clear.

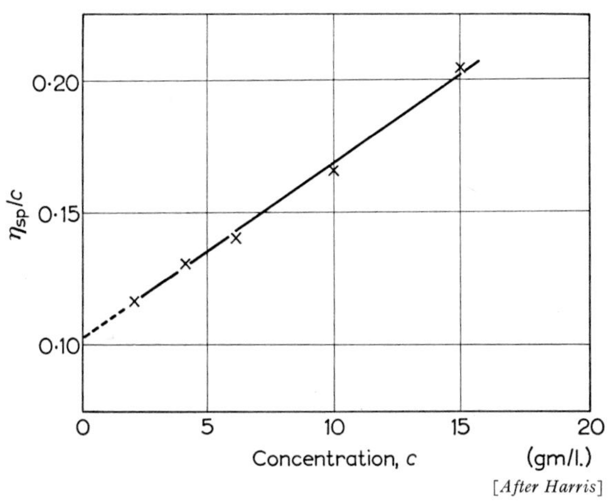

[*After Harris*]

FIG. 15.—Illustrating the way in which intrinsic viscosity is determined. It is indicated by the point where the dotted extension of the curve intersects the vertical axis.

I. Harris of I.C.I. determined the intrinsic viscosity of one polyethylene (amongst others) known as Alkathene 2. The polymer was dissolved in xylene at 75° C. (it is not soluble cold) and the relative viscosity determined at the concentrations of 2, 4, 6, 10 and 15 gm/l. The specific viscosity was obtained by subtracting unity from the relative viscosity and the quantity η_{sp}/c was plotted against c (concentration) as shown in Fig. 15. A straight line results and if this is extended (extension is " broken " in Fig. 15) to the axis where $c = 0$, we can read that the intrinsic viscosity, η, at this point of intersection is 0·101. Harris was able to derive an equation:

$$\eta = 1{\cdot}35 \times 10^{-4}\, \overline{M}_n^{\,0{\cdot}63}$$

which related the intrinsic viscosity η of polyethylenes to $\overline{M}_n$, their number average molecular weight. Because, for Alkathene 2,

$\eta = 0{\cdot}101$, $\overline{M}_n$ must be about 36,000. Other methods of molecular weight determination (boiling point and osmotic pressure) showed that Alkathene 2 had a number average molecular weight of 32,000, so that the agreement is good. A comparison of Harris's equation with Staudinger's much earlier one is interesting; they are different of course, but not very different. When polypropylene came along, R. Chiang of Hercules Powder Co. made a somewhat similar investigation for it and found the relation:

$$\eta = 10^{-4}\ \overline{M}_n^{\,0{\cdot}80}$$

This is even closer to Staudinger's early equation. Another worker, Ciampa, has suggested the following relation between intrinsic viscosity and number average molecular weight for polypropylene:

$$\eta = 2{\cdot}5 \times 10^{-5}\ \overline{M}_n$$

which brings us right back to Staudinger. In the experimental work which brought this relationship to light, the number average molecular weight of the polypropylene was determined by osmotic pressure measurements and the intrinsic viscosity was determined in tetralin solution at 135° C., the high temperature being necessary to dissolve the polypropylene. Once the relationship has been established, all that is necessary to obtain the molecular weight of any sample of polypropylene is to measure its intrinsic viscosity. It is of interest that I.C.I. in the control of their polypropylene manufacture use Ciampa's relation, but the solvent they use is decalin at 135° C. Usually they find that the intrinsic viscosity of polypropylene so determined lies between the values of 1 and 2 which correspond, respectively, to number average molecular weights of 40,000 and 80,000. These values are lower than would be given by Chiang's expression.

Melt extrusion method. The use of a high temperature (135° C.) for determining the intrinsic viscosity of polypropylene is not very desirable because of the danger of degradation of the polymer occurring in the rather lengthy preparation of the solution; it has, for example, to be protected against oxidation. Accordingly, another method known as melt extrusion is also used by I.C.I. and although this too requires a high temperature, even higher at 190° C., the time for which the polymer is heated is very much shorter indeed than in the lengthy preparation of a solution. The rate of extrusion of the molten polypropylene, at 190° C. under a standard pressure exerted by a piston loading of 2 kg. applied over 0·71 $cm.^2$ of polymer, through an extrusion die of 0·082 in. diameter and $\frac{5}{16}$ in. long is measured in grams over a 10 min. period. From this " melt index " the intrinsic viscosity and thence the molecular weight are calculated.

Intrinsic viscosity of nylon. The methods described above could equally be applied to nylon. Carothers in his basic nylon patent (U.S. Patent 2,130,948 of 1938) defined the intrinsic viscosity of his nylon polymers by the relation:

$$\eta = \frac{\log_e \eta_{rel}}{c}$$

where c was 0·5 per cent. Instead of determining the relative viscosity of a series of dilute solutions of nylon in *m*-cresol and extrapolating to zero concentration, he approximated by using only one fairly low (0·5 per cent) concentration. All that is needed experimentally is to observe the times of flow at 25° C. (Ostwald viscometers are always nearly completely immersed in a thermostatically controlled bath) of (*a*) *m*-cresol and (*b*) a 0·5 per cent solution of the nylon polymer in *m*-cresol.

If the polymer has an intrinsic viscosity below 0·4, it will not form satisfactory fibres; in the early days the intrinsic viscosity of commercial nylon was about 0·6; nowadays it is usually within the range 1·0–2·0. When non-round filaments are being spun, a high viscosity is essential; otherwise, if the melt is too fluid, surface tension will pull the filaments to a circular section before solidification has taken place.

End Group Determinations in Cellulose. One method that has proved to be of great value in the determination of chain-length is the " end group " method which was devised by W. N. Haworth and his associates. The end groups of the cellulose molecule are different from the intermediate glucose residues in the chain, in that they have an extra hydroxyl group. The method used is to methylate all the hydroxyl groups in the cellulose chain:

The methylated glucose is next hydrolysed to give:

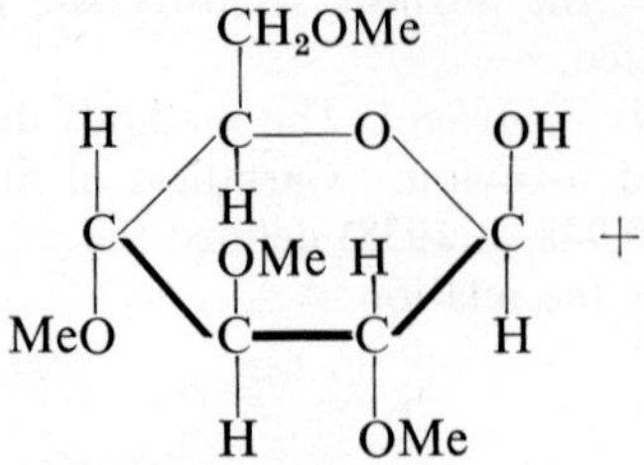

CH_2OMe

2 : 3 : 4 : 6
Tetramethyl glucose *

2n [2 : 3 : 6 Trimethyl glucose] + [1 : 2 : 3 : 6 Tetramethyl glucose]

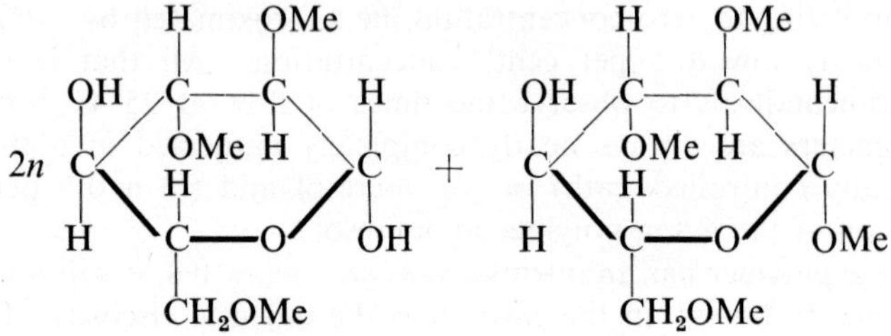

2 : 3 : 6
Trimethyl glucose

1 : 2 : 3 : 6
Tetramethyl glucose

But the 1 : 2 : 3 : 6-tetramethyl glucose is unstable, and during hydrolysis it loses the methyl group on position 1 and is converted into 2 : 3 : 6-trimethyl glucose.

Accordingly a cellulose chain which contains $(n + 2)$ glucose residues will on methylation and hydrolysis give:

1 molecule 2 : 3 : 4 : 6-tetramethyl glucose
$(n + 1)$ molecules 2 : 3 : 6-trimethyl glucose.

If, therefore, the quantity of tetramethyl glucose which results from the methylation and hydrolysis of a known quantity of cellulose is estimated it is possible to calculate what the chain length was. If, for example, 100 parts of cellulose yield 1 part of tetramethyl glucose, it is clear that the average chain contains about 100 glucose residues. Suitable corrections must be made for changes in weight

* Key to numbering system of carbon atoms

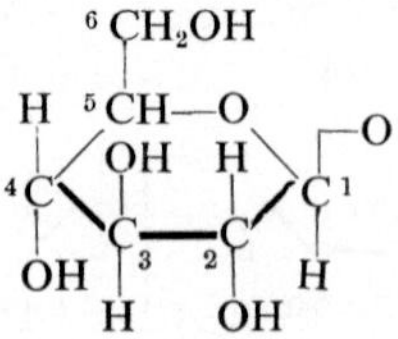

on methylation and hydrolysis, so that in fact the ratio becomes about 69. For example, the molecular weight of a cellulose chain which contained 100 glucose residues would be 16,236 because the molecular weight of $C_6H_{10}O_5$ is 162 and the two end groups contain additionally two molecules of water. The weight of the molecule of tetramethyl glucose is 236 from $C_6H_8O_2(OMe)_4$. Hence 16,236 parts by weight of cellulose (containing 100 glucose residues per molecule) yield 236 parts tetramethyl glucose. The conversion factor, which will be reasonably constant for long molecules, is clearly $\frac{16,236}{236}$ or about 69.

Nature of End Groups in Nylon. The end groups on nylon molecules are either amino or carboxylic, and such groups are much the most reactive of any in the nylon molecule; they are, of course, relatively few, and the longer the molecules the relatively fewer they are. Current standard 66 nylon yarn is stabilised with acetic acid which reduces (in practice not to zero) the number of amino end groups, and increases the number of carboxylic acid end groups. Nylon is usually dyed with disperse dyestuffs and as the dyeing mechanism is mainly one of solution the nature of the end groups does not affect the dyeing. If acid dyes are used, the terminal amino groups on the nylon become of great importance in the dyeing operation. Acid dyestuffs are inclined to show up physical non-uniformities (due to the cold drawing) and give barriness in nylon and in the early days of nylon they could not be used to advantage. But nylon, nowadays, is much more uniform and acid dyestuffs are commonly used to dye it in dark shades such as marron, navy and maroon. The more amino and the fewer carboxylic acid end groups there are in the nylon, the greater will be the affinity for acid dyestuffs; the molecules with terminal amino groups will have a much higher affinity for the acid dyestuff molecules than molecules with acid end groups. Experimental work has been carried out to investigate the effect of using a basic instead of an acid stabiliser.

Two bases that have been proved suitable are N-amino propyl morpholine and N-amino hexyl morpholine; these do not interfere unduly with spinning extrusion performance (although they increase the number of stoppages due to drips at the spinneret), nor are they too readily volatile to be lost. Yarns made with these substituted morpholines instead of acetic acid as stabiliser have had very similar properties to ordinary nylon, but their acid dye uptake, using Naphthalene Scarlet 4 RS as a test dyestuff, has been twice as high. Amine-stabilised nylon may well eventually supersede acid-stabilised;

the advantage of being able to use acid dyes which are commonly used to dye wool and which are fairly fast would be quite valuable.

Determination of Equivalents of End Groups. A method for determining amino and carboxyl end groups in nylon has been described by Waltz and Taylor; it is in principle as follows:

For amino end groups. A 2·00 gram sample of dry nylon is dissolved in 50 ml. of purified phenol; then 25 ml. 95 per cent ethanol and 25 ml. water are added; the polymer is retained in a single-phase

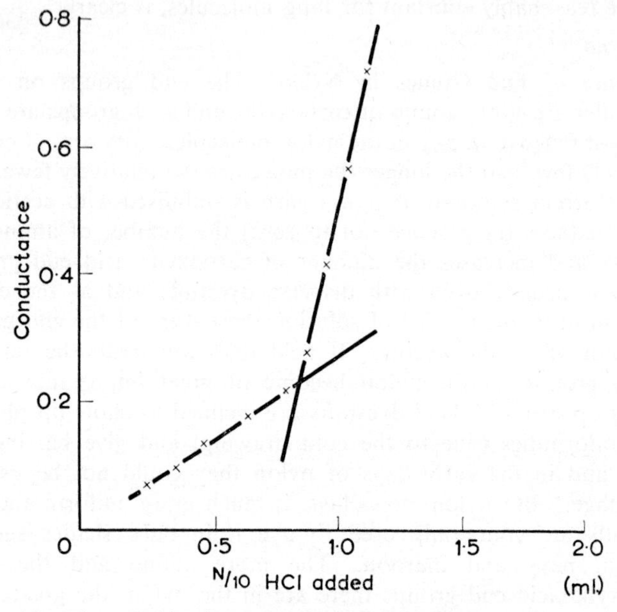

Fig. 16.—Plots resulting from chemical titration of end groups of nylon.

solution and does not separate. With constant slow stirring the solution is titrated with decinormal hydrochloric acid fed in from a micro-burette. The electrical resistance of the solution is measured at frequent intervals and the conductance (the reciprocal of the resistance) is plotted against the quantity of hydrochloric acid. The plots fall on two straight lines (Fig. 16) and their point of intersection is the equivalence point of the reaction, *i.e.*, the hydrochloric acid that has been added at that point is equivalent to the amino groups that were present on nylon. The theory of the method is that the polymer molecules have so low a mobility that they need not be considered; as acid is added, the chlorine ions of moderate mobility are liberated and the conductance rises slowly, then *after the equi-*

valence point is passed, highly mobile hydrogen ions are liberated and the conductance then rises steeply; that point where the increase in conductance changes from slow to fast marks the introduction of the hydrogen ions and therefore marks the equivalence point.

For carboxylic end groups. Decinormal caustic potash dissolved in 10 per cent methanol/90 per cent benzyl alcohol is used as the titrant. Phenol cannot be used as the solvent for the nylon, because it is sufficiently acidic itself to react with the caustic potash. Actually benzyl alcohol at 175° C. is used to dissolve the nylon, and then the solution of caustic potash is added using phenolphthalein as an indicator—this is colourless at first but reddens in excess alkali. The conductance method can be used but it is not so good as for the determination of basic end groups, and the colorometric titration is very simple.

Determination of Number Average Molecular Weight. These methods of determining the chemical equivalence of the end groups on nylon consequently offer a means of determining the " number average molecular weight ". Let us suppose that a 2 gram sample of nylon gives a titration corresponding to 0·06 milliequivalents of amine ends (that is, it needs 0·6 ml. of decinormal hydrochloric acid to bring it to equivalence) and 0·10 milliequivalents of carboxyl ends (it needs 1·0 ml. of decinormal caustic potash to bring it to equivalence). Accordingly 1 gram of the nylon would contain 0·03 milliequivalents of the amine end groups and 0·05 milliequivalents of carboxyl end groups, a total of 0·08 milliequivalents. Hence 12,500 grams of nylon would contain 1 gram-equivalent of end groups and 25,000 grams of nylon would contain 2 gram-equivalents of end groups. We know that each nylon molecule must have two end groups (there are two ends to the linear molecule), so that the " number average " molecular weight of the nylon must be 25,000.

Commercial nylon usually has a saturation value of 0·04 milliequivalents per gram (sometimes expressed as 0·04 equivalents per kilogram) free amine groups, and a slightly higher figure for carboxyl groups, say about 0·06. If these figures are used as a basis the number average molecular weight of commercial nylon will be seen to be about 20,000. For nylon 6 end groups, see p. 368.

Variations of Chain Length and Fibre Length. One final point that may be noted is the analogy between the uneven fibre length of a natural fibre such as of wool or of cotton, and the irregularity of the molecular length that exists in all fibres. Just as a tuft of fibres taken at random will contain fibres of very different lengths, so will a single fibre be composed of fibrous molecules of very different lengths.

Possibly the reason that different methods of molecular weight determination give such widely different results is that they do not measure the same property. There may be weak links that hold one molecule to another, and one method of molecular-weight determination may break these links, another may not.

FURTHER READING

W. T. Astbury, " Fundamentals of Fibre Structure " (1933).

L. Simmens and A. R. Urquhart, " The Structure of Textile Fibres ", *J. Text. Inst.*, **40**, P 3 (1949).

F. Howlett, " Fibres and Plastics ", *ibid.*, **40**, P 241 (1948).

J. M. Preston, " Fibre Science ". Manchester (1949).

W. N. Haworth and H. Machemer, *J. Chem. Soc.*, 2270 and 2370 (1932); 1888 (1939).

W. G. Harland, " Fractionation of Nitrated Rayons ", *J. Text. Inst.*, **46**, T483–T499 (1955).

I. Harris, *J. Polym. Sci.*, **8**, 353 (1952).

R. Chiang, *ibid.*, **28**, 235 (1958).

CHAPTER 3

THE SYNTHESIS OF FIBRES

WHEN it was appreciated that molecules must be very long for fibrous properties to be manifest, chemists envisaged the possibility of building up long molecules, and so making new fibres—synthetic fibres. They remembered how Emil Fischer, attempting to build up proteins, had combined nineteen molecules of glycine (amino-acetic acid) to give a polypeptide. The term *polypeptide* to-day indicates any polymer which contains an abundance of amide—CONH— linkages; real silk, wool and nylon are all polypeptides; the word has no deeper significance, no lingering union with the process of digestion. It constitutes a misnomer in the elegant and precise phraseology of organic chemistry; one can always replace it by polyamide, and, as it is a word that is likely to be encountered frequently, it is well to remember its precise meaning—*viz.*, "polyamide".

Synthesis of Long Molecules

Could this work of Fischer's be extended to give very long molecules? Fischer had found it by no means easy to build up his relatively short polypeptides, and thirty or forty years later, despite the tremendous advances in knowledge of chemical structure and of synthetic methods, the problem of building up these long molecules was still very difficult. One reason for this was that the chemistry of very big molecules had been neglected—if an experiment had produced an insoluble, infusible, non-crystalline, and unreactive material, it was considered to have failed, and the product from it was usually consigned to the waste-bin.

With true genius, W. H. Carothers recognised that these rejected products usually resembled much more closely than did the average laboratory reagent those materials, such as wood, rubber and fibres, with which we came in contact in our daily life. Carothers was employed by E. I. du Pont de Nemours & Co., and was given a free hand in the selection of a subject for study. Carothers, with great perspicacity and courage, set out to work on polymers, to make and study the infusible, intractable materials which hitherto, when they had been unwittingly made, had been rejected.

Formation of Linear Polymers by Condensation

Carothers reasoned that if he took a substance which had a reactive or *functional* group at either end of its molecules, then two of its molecules could combine together to give a molecule twice as long, which also would have a functional group at either end of its molecule. Then two of these molecules would combine with each other, and, by such a repeated process of doubling, very long molecules would soon result.

If, for example, the compound ε-hydroxycaproic acid was taken, the hydroxyl group on one molecule would react with the carboxylic group on another molecule in the following way:

$$\underset{\text{Monomer.}}{HO(CH_2)_5COOH} + \underset{\text{Monomer.}}{HO(CH_2)_5COOH} \longrightarrow \underset{\text{Dimer.}}{HO(CH_2)_5CO{\cdot}O(CH_2)_5COOH} + H_2O$$

Then two of the molecules of the dimer would react to give a larger molecule:

$$\underset{\text{Dimer.}}{HO(CH_2)_5CO{\cdot}O(CH_2)_5COOH} + \underset{\text{Dimer.}}{HO(CH_2)_5CO{\cdot}O(CH_2)_5COOH}$$

$$\longrightarrow \underset{\text{Tetramer.}}{HO(CH_2)_5CO{\cdot}O(CH_2)_5CO{\cdot}O(CH_2)_5CO{\cdot}O(CH_2)_5COOH} + H_2O$$

It should be noted that the initial compound, the hydroxycaproic acid, is referred to as the *monomer*, the product resulting from the condensation of two molecules of this body is known as the *dimer*, that from three molecules as the *trimer*, from four molecules as the *tetramer*, and from many molecules as the *polymer*.

When the reaction with hydroxycaproic acid is allowed to go much farther, repeated doubling of the molecule results in the formation of a polymer of formula:

$$H[-O(CH_2)_5CO-]_nOH$$

formed by the reaction

$$nHO(CH_2)_5COOH \longrightarrow H[-O(CH_2)_5CO-]_nOH + (n-1)H_2O$$

Such a process of polymerisation is known as *condensation* polymerisation; the characteristic of the condensation type of polymerisation is the splitting off of the elements of water or of some other simple compound, such as ammonia or hydrogen chloride. If, for example, the acid chloride of ε-hydroxycaproic acid was polymerised, the polymer would be essentially the same as was obtained from the acid itself; but, instead of water being split off, hydrogen chloride would be evolved, in this way:

$$nHO(CH_2)_5COCl \longrightarrow H[-O(CH_2)_5CO-]_nCl + (n-1)HCl$$

All polymerisation reactions in which simple substances, such as water, hydrogen chloride or ammonia, are evolved are known as

" condensation " polymerisations, and should be clearly distinguished from " addition " polymerisation, in which no substance is split off.

Linear Polymers. The polymeric substances that result from the condensation polymerisation of substances such as hydroxycaproic acid are known as *linear* polymers, because they are long, straight-chain compounds—they have, in fact, only one dimension—or, more truly, one dimension is very much greater than the other two. Here we stop for a moment at this significant statement.

One dimension is much greater than the other two—that is exactly the outstanding characteristic of the fibrous molecules of fibre-forming substances. We have therefore an approach here to a method of making fibrous molecules which in bulk would constitute a fibre-forming material.

Carothers put the matter to the test, and tried various methods of heating hydroxy-acids to give a fibre-forming material. He found that when such compounds were polymerised they gave fibre-forming substances—*i.e.*, if touched when molten with a glass rod, which was rapidly drawn away, the rod carried away a filament. At first these filaments were weak, but Carothers found that if he pushed the polymerisation still farther the filaments that were formed were strong, and he referred to those polymers which gave strong fibres as *superpolymers*.

The logic behind the work was vindicated. Once the ground had been broken, once it had been shown that it was possible to synthesise a fibre-forming compound, developments quickly followed.

Varieties of Linear Condensation Polymers

It was soon shown that just as a hydroxy-acid would give a polyester, so an amino-acid when heated would give a polyamide.

$$n\mathrm{HO(CH_2)_5COOH} \longrightarrow \mathrm{H[-O(CH_2)_5CO-]}_n\mathrm{OH} + (n-1)\mathrm{H_2O}$$

Hydroxy-acid. Polyester.

$$n\mathrm{NH_2(CH_2)_5COOH} \longrightarrow \mathrm{H[-NH(CH_2)_5CO-]}_n\mathrm{OH} + (n-1)\mathrm{H_2O}$$

Amino-acid. Polyamide.

Then, because amino-acids were relatively costly to synthesise, whereas diamines and diacids were less costly, the device was used of making a polyamide by condensing together a diamine and a diacid.

$$n\mathrm{NH_2(CH_2)_6NH_2} + n\mathrm{COOH(CH_2)_4COOH} \longrightarrow$$

Diamine. Diacid.

$$\mathrm{H[-NH(CH_2)_6NHCO(CH_2)_4CO-]}_n\mathrm{OH} + (2n-1)\mathrm{H_2O}$$

Polyamide.

Similarly it was shown that a dihydric alcohol—*i.e.*, a glycol—and a diacid would give a polyester:

$$nHO(CH_2)_6OH + nCOOH(CH_2)_4COOH \longrightarrow$$

Dihydric alcohol. Diacid.

$$H[-O(CH_2)_6O\cdot CO(CH_2)_4CO-]_nOH + (2n-1)H_2O$$

Polyester.

Doubtless most readers of this book will readily understand the formulæ, but in case there may be some to whom they are not clear, the following extension of one of them may be helpful. It is absolutely essential at this stage to grasp clearly the significance of the general chemical method of building up linear polymers from difunctional reagents. The polyamide which is written

$$H[-NH(CH_2)_6NHCO(CH_2)_4CO-]_nOH$$

would, if $n = 3$, signify

$$NH_2(CH_2)_6NHCO(CH_2)_4CONH(CH_2)_6NHCO(CH_2)_4 \\ -CONH(CH_2)_6NHCO(CH_2)_4COOH$$

If $n = 50$, which is about the lowest value at which useful fibres can be formed, the molecule would clearly be nearly seventeen times as long as that just written at length. To extend it in part, it could be represented:

$$NH_2(CH_2)_6NHCO(CH_2)_4CO-[-NH(CH_2)_6NHCO(CH_2)_4CO-]_{48} \\ -NH(CH_2)_6NHCO(CH_2)_4COOH$$

It is essential to appreciate that the molecules of fibre-forming materials are themselves very long in comparison with their other dimensions.

Ring Formation—A Danger in Fibre Syntheses

The polymerisation reaction already noted of ε-hydroxycaproic acid may be taken as typical of condensation polymerisation. The first stage is the condensation of two molecules, again, as already noted,

$$HO(CH_2)_5COOH + HO(CH_2)_5COOH \longrightarrow \\ HO(CH_2)_5CO\cdot O(CH_2)_5COOH + H_2O$$

The reaction does in fact proceed in this way, but one could reasonably inquire why a similar reaction of ester formation does not take place between the head and tail of *one* molecule in this way

$$HO(CH_2)_5COOH \longrightarrow \begin{matrix} & CH_2-CH_2 & \\ CH_2 & & CH_2 \\ | & & | \\ CH_2 & & O \\ & CO & \end{matrix} + H_2O$$

The reason is that such a 7-atom ring is not very stable or easily produced. But if the monomeric starting material is one that can produce a 6-atom ring (or a 5-atom or a 4-atom ring), then ring closure will take place. Thus δ-hydroxyvaleric acid will give the lactone by ring closure.

$$HO(CH_2)_4COOH \longrightarrow \begin{array}{ccc} & CH_2 & \\ CH_2 & & O \\ | & & | \\ CH_2 & & CO \\ & CH_2 & \end{array}$$

It is essential, therefore, in choosing monomeric materials from which to synthesise fibre-forming materials, to select those which cannot give rise to a stable ring system. Provided that any ring system that could be formed would contain more than six atoms, the reaction will go mainly to form a linear polymer, although there will be a small amount of ring formation. If the ring would contain eight or more atoms the reaction will go practically exclusively to the linear polymer.

Thus aminocaproic acid will give a linear polymer, but aminovaleric acid will form a six-membered ring compound instead; adipic acid and hexamethylene diamine give a linear polymer (nylon), the first stage of the reaction being:

$$\underset{\text{Hexamethylene diamine.}}{NH_2(CH_2)_6NH_2} + \underset{\text{Adipic acid.}}{COOH(CH_2)_4COOH} \longrightarrow$$

$$\underset{\text{Linear dimer.}}{NH_2(CH_2)_6NHCO(CH_2)_4COOH} + H_2O$$

but oxalic acid and ethylene diamine will give not a linear polymer, but a ring compound:

$$\underset{\text{Ethylene diamine.}}{NH_2(CH_2)_2NH_2} + \underset{\text{Oxalic acid.}}{COOH{\cdot}COOH} \longrightarrow \underset{\text{A ring compound.}}{\begin{array}{ccc} & CH_2 & \\ CH_2 & & NH \\ | & & | \\ NH & & CO \\ & CO & \end{array}}$$

Therefore, when choosing reactants for the synthesis of fibre-forming materials care should be exercised to choose sufficiently long molecules to avoid the danger of ring formation. Within recent years, though, several examples of short molecules reacting to give linear polymers have come to light. Glycollic acid, for example,

$HOCH_2COOH$ will condense to a linear polyester which when melted yields good fibres.

$$H[-OCH_2CO-]_nOH$$

Polymeric substances which will yield fibres can be made just as well from intermediates which are themselves composed of relatively long molecules, although naturally immensely shorter than the molecules of a polymer. In 1968 some Russian workers, Fedotova *et al.*, described the preparation of polyamides from the C_{14} and C_{18} dicarboxylic acids. The C_{14} acid is dodecane dicarboxylic acid, $COOH(CH_2)_{12}COOH$, and the other acid has four more methylene groups in the chain. Both acids were condensed with ethylene diamine, $NH_2(CH_2)_2NH_2$, hexamethylene diamine, $NH_2(CH_2)_6NH_2$, and decamethylene diamine, $NH_2(CH_2)_{10}NH_2$. The polyamide from the C_{14} acid and ethylene diamine would be called nylon 2,14, that from the C_{18} acid and decamethylene diamine would be called nylon 10,18 and so on. The method of polymerisation used was to make the salt of the diacid and diamine, presumably by mixing methanol solutions of the two, the salt being precipitated. The salt was then heated under an inert atmosphere for 5 hr. at 260° C. followed by 2 hr. at the same temperature under a vacuum (a residual pressure of 5–7 mm. Hg).

These conditions were those found suitable for and used for the C_{14} acid. It was an odd finding, and a very interesting one indeed, that when the C_{18} acid was one of the monomers, the polymerising temperature needed to be only 180° C. The melting points of the polyamides that were obtained were as follows:

Acid.	Amine.	Melting point (° C.).	Name of polymer.
C_{14} (dodecane dicarboxylic)	Ethylene diamine	244	Nylon 2,14
	Hexamethylene diamine	195	Nylon 6,14
	Decamethylene diamine	154	Nylon 10,14
C_{18} (hexadecane dicarboxylic)	Ethylene diamine	206	Nylon 2,18
	Hexamethylene diamine	173	Nylon 6,18
	Decamethylene diamine	149	Nylon 10,18

It is instructive to note how lengthening of the molecules of the monomers reduces the melting point of the polyamide.

Commercial Fibres Made by Condensation Polymerisation

The following fibres made by condensation polymerisation processes will be studied in this book: nylon and Perlon, which are polyamides, and Terylene, Dacron, Kodel, Fortrel and Vycron which are polyesters. The essential characteristic in all cases is a linear polymeric constitution. This gives the characteristic fibre-forming property, irrespective of whether the compound is an amide or an ester. Commercially, polymerisation is nearly always carried out by melting the two components together; some diluent such as water in the synthesis of nylon, or excess glycol in the synthesis of polyesters, may be present in the early stages to improve heat conduction and prevent unevenness of heating of the mix, but it is soon boiled away and the high polymer is made in the melt form.

Interfacial Reaction. Much has been heard lately about interfacial reactions. For example, nylon can be made by reaction of an aqueous solution of hexamethylene diamine and an organic solution of adipyl chloride both of them cold, the two reacting at their common surface:

$$NH_2(CH_2)_6NH_2 + ClCO(CH_2)_4COCl \longrightarrow \\ -HN(CH_2)_6NHCO(CH_2)_4CO-(+ 2\,HCl)$$

Chemically, there is nothing new in this; it is simply another version of the old Schotten–Baumann reaction in which acid chlorides are reacted with alcohols and amines to give esters and amides. So far as fibres are concerned, interfacial polymerisation is not used for the bulk of manufacture; that is still done in the melt. Nevertheless, the process has points of interest which might possibly give it some industrial importance in the future.

It is extremely rapid; whereas condensation polymerisation usually involves heating for several hours at a temperature in excess of 250° C., interfacial polymerisation takes place apparently instantaneously at room temperature. This is because the acid chloride which is used is so very reactive.

The Nylon Rope Trick. This exceptional reactivity has led to the introduction of the nylon rope trick, a demonstration experiment of making a polyamide, nylon 6,10, continuously and winding it up as fast as it is formed. The process as described by Morgan and Kwolek runs along the following lines: 2 ml. sebacoyl chloride (the 10 carbon atom diacid chloride) is dissolved in 100 ml. carbon tetrachloride as an inert solvent and is put in a 200-ml. beaker. Then 4·4 gm. hexamethylene diamine (the 6 carbon atom diamine) is dissolved in 50 ml. water, and this is poured on top of the first solution. A film of nylon 6,10 forms at once at the interface due to the reaction:

$$NH_2(CH_2)_6NH_2 + ClCO(CH_2)_8COCl \longrightarrow \\ -HN(CH_2)_6NHCO(CH_2)_8CO- + 2HCl$$

The film is grasped at the centre with tweezers and is pulled out of the beaker in rope form; the rope can be wound up continuously as fast as 1 ft./sec. Thus, from a beaker containing only two clear solutions, a rope is pulled continuously (Fig. 17) until one of the reactants is exhausted. If the draw-off is stopped no new interface is exposed, and the reaction stops. It can be restarted at will by drawing off the rope again. Although the polymer forms in a matter of seconds, or more likely in a fraction of a second, the rope

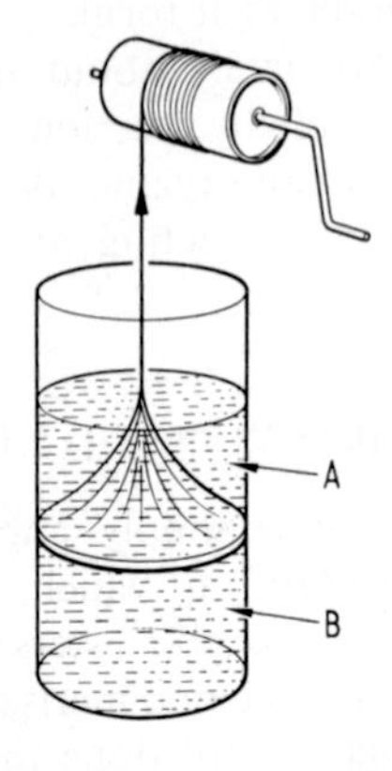

FIG. 17.—The Nylon Rope Trick. The nylon is wound up manually on a suitably supported (support not shown) roller, polyamide being removed from the interface at which it is formed.

A = solution of diamine in water.
B = solution of diacid chloride in organic solvent.

[*after* J. Chem. Educ., **36**, 182–184, 1959.]

that is drawn off is quite strong and may have a molecular weight of 20,000, as great as that of some commercial nylons. It is much the simplest demonstration, *e.g.*, for a lecture, of the formation of nylon that is possible. If one tries to do it by melting the polymer there is trouble with oxidation and charring; if the melting is done under a nitrogen atmosphere in a pressure vessel, then as a rule the audience cannot see what is happening.

Variables. There are several variables, and these are as follows:

1. The two solvents, which must be immiscible so that there is an interface (where the two solvents meet) at which the polymerisation reaction can take place. Usually, one of the solvents will be water; the other can be methylene chloride or carbon tetrachloride—something that will be a solvent for a low polymer so that it does not throw it out of solution until a really high polymer has been formed.

2. The concentration of the diamine and the diacid chloride in their solvents: if the solutions are too dilute the acid chloride will tend to be hydrolysed by the water instead of reacting solely with the diamine; if they are too concentrated, the solid polymer, as fast as it is formed, may absorb so much of the liquid

that proper mixing of the residual part of the reactants cannot take place, and consequently only a low polymer can be formed. Probably the concentrations indicated by the inventors of the nylon rope trick, *viz.*, 2 per cent in the organic solvent and 9 per cent in the water, are as good as any.

3. The presence or absence of a surfactant. The anionic detergent, sodium lauryl sulphate, has been recommended. It must be one that will not react with either of the monomers.

4. To stir or not to stir? For demonstrating the formation of a nylon rope, it is better not to stir. If a high yield of a powdered or granular nylon (not yet in fibre form) is aimed at, then stirring is desirable. It may lead to a fierce reaction between diacid chloride and diamine so that the whole heats up; such heating should be avoided, as it tends to hydrolyse the diacid chloride thus:

$$ClCO(CH_2)_nCOCl + H_2O \longrightarrow ClCO(CH_2)_nCOOH + HCl$$

This, of course, will stop the interfacial polymerisation.

Wide Chemical Range. The rope trick is more easy to carry out with nylon 6,10 than with 6,6, probably because of the lower compatibility for water of the former. But if conventional methods of precipitation (*i.e.*, not in rope form) are used, then almost any nylon polymer can be so made. Beaman and his collagues made 13 different kinds of polyamide by interfacial polycondensation. One of these was made from an aqueous solution of *iso*propylene diamine and from a methylene chloride solution of terephthalic acid chloride. The polymer had a molecular weight of 30,000 and a repeating structural unit of

$$—NHCH(CH_3)CH_2NHCO—C_6H_4—CO—$$

The process was as simple as it well could be: stir, filter off the polymer, wash it in boiling water and dry. Polyterephthalamides such as this have such high melting points that they cannot be melt spun; they have, however, been spun into fibres from solution in trifluoracetic acid. Other fibres made in the same way from ethylene diamine, which has a very short chain, and terephthalic acid had remarkably high moisture regains of 15 per cent or more and this property might be very valuable indeed. Not only polyamides but also polyesters, polyurethanes and polysulphonamides have been made by interfacial polymerisation.

Advantages. (1) Interfacial reaction may make possible fibres with properties that the synthesist has hitherto regarded as highly desirable, but impossible to attain. (2) Polymers which cannot be prepared by the ordinary melt condensation process, because the necessary temperatures are so high that the monomers decompose. For example, ordinary (*ortho*) phthalic acid, which is very much cheaper than terephthalic acid and is a most unpromising candidate for normal polymerisation, can be used as its acid chloride, phthaloyl chloride, in organic solution to react with piperazine in aqueous solution to yield a polymer with the recurring unit

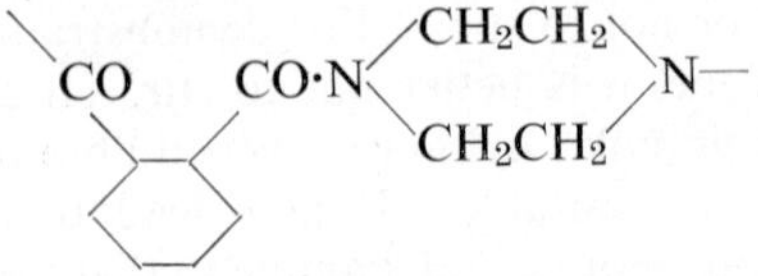

$$\diagdown\mathrm{CO}-\mathrm{C_6H_4}-\mathrm{CO{\cdot}N}\begin{matrix}\diagup\mathrm{CH_2CH_2}\diagdown\\ \diagdown\mathrm{CH_2CH_2}\diagup\end{matrix}\mathrm{N}-$$

Such condensations may ultimately be of great industrial consequence. (3) The reaction takes place at room temperature and in a matter of seconds. These conditions reduce the likelihood of the formation of by-products which might sully the purity of the polymer. (4) The equipment required is only simple. (5) There is a possibility that the method will help to produce ordered co-polymers in which the arrangement of the different chemical groups can be regularised. If, for example, a co-polymer contains two recurring units, A and B, then melt polymerisation usually yields a product in which the two units are randomly distributed, *e.g.*, —ABABBAABABBBAA—, but interfacial polycondensation, with its low temperature of reaction and ease of control, can be made to yield a regularly alternating structure, *e.g.*, —ABABABABAB—. Thus an alternating co-polyurea was prepared by Lyman and Jung by the interfacial polycondensation of 1 : 4-piperazine dicarbonyl chloride with hexamethylene diamine, the polymer having the molecular repeat

$$\left[-\mathrm{HN}-(\mathrm{CH_2})_6-\mathrm{NH{\cdot}CO{\cdot}N}\begin{matrix}\diagup\mathrm{CH_2CH_2}\diagdown\\ \diagdown\mathrm{CH_2CH_2}\diagup\end{matrix}\mathrm{NCO}-\right]_n$$

This product melted at 265° C., whereas the similar random co-polyurea prepared by melt polymerisation had a polymer melt temperature of only 194° C.; the random co-polymer showed greater solubility than the alternating co-polymer. Because the temperature of interfacial condensation is low, redistribution of groups does not occur. The ideal or perfect alternation of groups is retained, although it is not usually completely retained in melt high-temperature polymerisations.

When a manufacturer produces a new condensation polymer and withholds information about its constitution and manufacture the student might give a thought to the possibility that it has been made by interfacial polymerisation.

Acid Chloride Reaction in Solution

It will be appreciated that the common method of polymerising, *e.g.*, nylon 6 or nylon 66 at a high temperature, depends for its success on the mixture being molten; if the mixture were solid, uniform heating throughout the mass would not be practicable. Both these common nylons melt well below the temperature of 280° C. that is used for polymerisation; but if one is dealing with a polymer that does not melt at 280° C. or 300° C. the possibility is lost of high-temperature polymerisation. There are two other possibilities, either interfacial polymerisation or solution polymerisation. The three basic methods of polymerisation are: (1) melt; (2) interfacial; (3) solution.

Interfacial polymerisation as described above is a good method of polymerisation, but hardly, despite the rope trick, one of fibre formation. If interfacial polymerisation is used, as it can be, to make polymers which cannot be made at the melt for the simple reason that they do not melt but just decompose, the problem of spinning the polymer still remains. Melt spinning is impossible, and all that remains is solution spinning. It is sometimes difficult to dissolve high-melting polymers with their compact molecular structures, once they have been isolated. It may be more practicable to perform the polymerisation in the solution state, without the polymer coming out of solution, and then to spin from the solution directly. This reference to the polymer coming out of the solution during polymerisation may need explanation; it is simply that as the polymer chain grows and grows, it becomes less and less soluble, and at some stage which may be short of the desired degree of polymerisation the polymer may be precipitated. A suitable choice of solvent can do much to avoid this difficulty. Dimethylacetamide is one solvent that is used. If there is a little of its hydrochloride, $CH_3CONMe_2{\cdot}HCl$, present it is more effective. Often, too, the presence of a little lithium chloride will improve its solvent properties. The hydrogen chloride given off in the reaction of an acid chloride with a diamine may provide the hydrochloric group necessary, so that the addition of lithium hydroxide is all that is required.

With really high-melting polymers, *e.g.*, all aromatic polyamides, it has to be accepted that melt polymerisation is out of the question, and only the low-temperature condensation of acid chlorides with diamines is a satisfactory method of obtaining polymers high enough to give good fibres.

Whereas polymers can be made in the melt, interfacially or in solution, the fibres can be spun from the melt (nylon 6, nylon 66) and from solution. In the latter event the solution may be made from: (1) polymer prepared in the melt (this is not of practical interest—if a melt can be made at all, it is better and cheaper to spin from it); (2) polymer prepared interfacially; or (3) polymer prepared in the solution, *i.e.*, the solution in which the polymerisation has been made is used for spinning. When a solution is used for spinning it may be either wet spun or dry spun, *e.g.*, nylon 6-T has been wet spun from cold concentrated sulphuric acid solution and dry spun from trifluoracetic acid solution. Heroic measures are necessary for the high-melting polymers, the aromatic polyamides.

The possibilities are considerable and in number perhaps sufficiently so to be confusing. In practice, the two important methods are:

1. Polymerise in the melt and spin from the melt (perhaps with the intermediate isolation and purification of the polymer in chip form). If this method is possible it will be used.
2. Polymerise in solution and spin straight from this solution This method is used only when (1) is impossible because the polymer will not melt.

Addition Polymerisation

There is another general method of polymerisation, as well as condensation polymerisation. It is known as *addition* polymerisation and consists simply of the union of several molecules with a redistribution of the valency bonds, but with no splitting off of water or hydrogen chloride. This feature is that which distinguishes " condensation " from " addition " polymerisation; in the former a simple compound is split off, in the latter nothing is split off. As its name implies, addition polymerisation consists of the addition of one molecule to another. Many unsaturated compounds, particularly ethylene and its derivatives, will polymerise under suitable conditions. The simplest case, in theory, is the polymerisation of ethylene to give polythene, which may be represented:

$$CH_2{:}CH_2 + CH_2{:}CH_2 + CH_2{:}CH_2 + CH_2{:}CH_2 \; . \; . \; . + CH_2{:}CH_2 \longrightarrow$$

Ethylene.

$$—CH_2CH_2CH_2CH_2CH_2CH_2CH_2CH_2 \; . \; . \; . \; CH_2CH_2—$$

Polythene.

The reaction—known as addition polymerisation—is purely and simply one of addition; it is brought about by the application of high pressures and high temperatures, usually in the presence of a catalyst.

Mechanism of Addition Polymerisation

There are three stages in the polymerisation of vinyl compounds; initiation, propagation (growth) and termination.

Initiation

Some vinyl compounds, notably vinyl acetate and styrene, will polymerise slowly at room temperature, but most of them require the application of heat or light or a catalyst. In practice vinyl polymerisations for fibres are usually initiated by heat and the presence of a catalyst together. Two kinds of catalyst are used: free radical and ionic.

Free radical initiation. A free radical is a broken molecule usually very unstable and ready to combine to a more complete state. Some unstable compounds such as peroxides and diazonium compounds and persulphates are known to produce free radical intermediates when they decompose and these intermediates are powerful catalysts for vinyl polymerisation. Benzoyl peroxide is often used; at about 80° C. it decomposes in solution into free phenyl and benzoate radicals:

$$C_6H_5COO{-}OOCC_6H_5 \longrightarrow C_6H_5COO\cdot + C_6H_5COO\cdot$$
$$C_6H_5COO\cdot \longrightarrow C_6H_5\cdot + CO_2$$

(The dot at the side is used to indicate the free radical nature of the compound, sometimes it is placed on top. Its significance is that the electron structure of the dotted atom is incomplete.) When the free radical comes in contact with a vinyl compound, it disturbs the electronic structure of the carbon atoms which are at either end of the double bond and we represent this by saying that the double bond opens. If R· is the free radical:

$$R\cdot + CH_2{=}CHX \longrightarrow RCH_2\dot{C}HX$$

The two have combined together to form another free radical, which itself is capable of further growth; the reaction has been initiated. It is noteworthy that the catalyst (more properly called the initiator) enters into the reaction and its parts, for example benzoate radicals, can often be detected in the final polymer.

Ionic initiation. The catalysts used for this kind of initiation are Friedel–Crafts type: aluminium chloride, boron trifluoride, titanium tetrachloride and stannic chloride. Sometimes they will initiate a polymerisation that a peroxide will not start, for example the polymerisation of *iso*butene and vice versa. The ionic catalysts are all strong electron acceptors. First the catalyst will dissociate into negative and positive ions in an organic environment: for example, boron trifluoride hydrate dissociates thus:

$$BF_3{\cdot}OH_2 \rightleftharpoons H^+ + [BF_3OH]^-$$

The hydrogen ion attaches itself to the vinyl compound at the double bond and leaves the carbon atom at the other side of the bond in a highly active state and ready to combine

$$H^+ + \underset{\substack{|\\H}}{\overset{\substack{H\\|}}{C}}{=}CHX \longrightarrow H{-}\underset{\substack{|\\H}}{\overset{\substack{H\\|}}{C}}{-}\underset{\substack{|\\X}}{\overset{\substack{H\\|}}{C^+}}$$

Propagation

In a condensation polymerisation the small polymers that are formed early in the reaction, such as the dimers and trimers, react with each other, so that after a very short time there is none of the original monomer left, even although the average degree of polymerisation may still be very low. In vinyl polymerisation matters are very different and a partially polymerised mixture will consist of high molecular weight polymer and of unchanged monomer with virtually no constituents at intermediate stages of growth; some individual polymer molecules grow to maturity whilst most of the monomers still remain intact. A given molecule of very high D.P. is formed by consecutive steps of a single chain process initiated by a free radical or an ion. This is very important; if it were otherwise it would be almost impossible to prepare the isotactic and other stereoregular polymers that we shall discuss later—polymers characterised by a high degree of geometric regularity. The possibility that such polymers can be made derives from the fact that vinyl polymers build themselves up on the basis of: one and one and one and one and one. . . . Even although the number of molecules joined together in the chain may ultimately run into the tens of thousands, they will all have been joined in the same way: one and one makes two, two and one makes three, three and one makes four, four and one makes five . . ., and so on up to the thousands. One at a time—no short cuts. We can illustrate the process: using a free radical R· as the initiator (R· may for example be $C_6H_5COO{\cdot}$ from benzoyl peroxide)

$$R\cdot + CH_2{=}CHX \longrightarrow R{-}CH_2{-}CHX\cdot$$

$$R{-}CH_2{-}CHX\cdot + CH_2{=}CHX \longrightarrow R{-}CH_2{-}CHX{-}CH_2{-}CHX\cdot$$

$$R{-}CH_2{-}CHX{-}CH_2{-}CHX\cdot + CH_2{=}CHX \longrightarrow R{-}CH_2{-}CHX{-}CH_2{-}CHX{-}CH_2{-}CHX\cdot$$

and so on until the chain contains several thousands of the original monomer units.

If the initiator has been of the ionic type just the same sort of reaction proceeds

$$H^+ + CH_2{=}CHX \longrightarrow CH_3{-}CHX^+$$

$$CH_3{-}CHX^+ + CH_2{=}CHX \longrightarrow CH_3{-}CHX{-}CH_2{-}CHX^+$$

$$CH_3{-}CHX{-}CH_2{-}CHX^+ + CH_2{=}CHX \longrightarrow CH_3{-}CHX{-}CH_2{-}CHX{-}CH_2{-}CHX^+$$

and so on. The chemical mechanism is the same whether free radical initiators (from peroxides for example) or ionic initiators (Friedel–Crafts type) are used—the electron rearrangements involved are different but the end products may be the same (*cf.* p. 112).

Termination

The monotonous " one and one and one " process goes a long way but is not endless. There comes a time when the free radical at the end of one chain meets that at the end of another, the two combine and that long molecule is finished: it cannot grow any more:

$$R{-}(CH_2{-}CHX)_y{-}CH_2{-}CHX\cdot + R{-}(CH_2{-}CHX)_z{-}CH_2{-}CHX\cdot \longrightarrow R{-}(CH_2{-}CHX)_{y+z+2}R$$

Alternatively, instead of joining together, the two active chain ends may disproportionate, one giving up a hydrogen atom to the other, so that both molecules are finished and can grow no more:

$$R{-}(CH_2{-}CHX)_y{-}CH_2{-}CHX\cdot + R{-}(CH_2{-}CHX)_z{-}CH_2{-}CHX\cdot \longrightarrow R{-}(CH_2{-}CHX)_y{-}CH{=}CHX + R{-}(CH_2{-}CHX)_z{-}CH_2{-}CH_2X$$

Either way, whether by combination or by disproportionation, the growth of those particular molecules is stopped.

Differences between Condensation and Addition Polymerisations

Vinyl or addition polymerisations differ from condensation polymerisations in the following ways:

1. No small molecular substance is split off.

2. They are faster. The reaction, especially when ionic as opposed to free radical initiators are used, may proceed very rapidly, and usually at a much lower temperature than is necessary for the condensation type.

3. The molecular weight attained can be much greater; it is unusual in a linear condensation polymerisation for molecular weights greater than 25,000 to be achieved, but in vinyl polymerisations the molecular weight can run up to a few million.

4. They are more likely to include minor side reactions leading to branching or to cross-linking.

5. If polymerisation is terminated at an early stage, the vinyl mixture will consist of some very long completely polymerised molecules and a lot of unchanged monomers, with nothing intermediate; the condensation mixture will contain practically no monomers and probably no very long polymeric molecules but will consist of all sorts of low polymers: dimers, trimers, tetramers, *etc.*

Commercial Fibres Made by Addition Polymerisation

There are many: polyethylene and polypropylene which are polyolefines; polyvinyl chloride; polyvinylidene chloride such as Saran and others; all the polyacrylonitrile fibres Orlon, Acrilan, Courtelle, and the modacrylics Dynel and Verel; polyvinyl alcohol Vinylon fibres. These indicate the range; there are many variants of those specifically mentioned. The most important single group is the acrylics.

The Consumer Synthesist

When the first edition of this book was published there was a touch of magic about the manufacture of synthetic fibres. It was inevitable that the established spinners, the men who had been spinning rayon in enormous quantities, would find themselves rather at a loss when confronted with the task of synthesising diamines and dicarboxylic acids and building them up into polymers which they would be able to spin. The chemical manufacturers looked at the possibilities that synthetic fibres offered to them and were favourably impressed, but they had very little idea of how to spin fibres. The two groups of men pooled their resources, and in England I.C.I. made polyamide, Courtaulds contributed the spinning technique and British Nylon Spinners was born of the union. In other parts of the world, too, somewhat similar arrangements were made. The outcome of such arrangements has been generally that the chemical manufacturers have picked up the spinning techniques very quickly;

there was not so much magic in it as might have been thought. At any rate it is the chemical manufacturers, such as I.C.I., du Pont, and Badische, who have obtained command of a large part of the industry. Their customers have had to buy what was available and to pay what was asked for it, and for the best part of twenty years the chemical synthesists were batting on a very good wicket.

The position is rather different today, because consumers of fibres are now able to install equipment on which they can make at least their own polyamide and polyester fibres. Vickers-Zimmer, mainly of London and Frankfurt/Main, are designing, making, and installing fibre-producing plants for consumers of fibres so that these latter can make themselves more or less independent of the big spinners. It is estimated that if a textile manufacturer, *e.g.*, a weaver or a knitter, is using two tons of polyamide or polyester fibre a day it will be worth his while to install a plant and make the fibre himself. The old magic, the esoteric know-how, has become a do-it-yourself job. The attractions are cost savings, no waiting for deliveries and additionally a flexibility of production, *e.g.*, denier and filament denier changes to meet immediate requirements. The experimental work which got Vickers-Zimmer started on this now very large business was carried out in a garage. There are still opportunities in the world of fibres for those who enjoy working and are not too easily put off by reports of other people's difficulties. A final point of interest is that Vickers-Zimmer in their Research Centre at Frankfurt/Main maintain pilot plants for demonstration or for the investigation of customers' specific requirements.

FURTHER READING

" Collected Papers of Wallace H. Carothers on Polymerisation ". Interscience, New York (1940).
This book contains Carothers' early work on linear polymers, which led to the discovery of nylon.

W. H. Carothers and J. W. Hill, *J. Amer. Chem. Soc.*, **54**, 1559–87 (1932).
The original papers on fibre-forming linear polymers.

C. E. H. Bawn, " The Chemistry of High Polymers ". London (1948).
This book deals excellently with polymers and their physical chemistry. It does not discuss fibres at great length.

R. W. Moncrieff, " The Genesis of Synthetic Fibre ", *Textile Manufacturer*, **75**, 285 (1949).

O. Ya. Fedotova, M. I. Shtilman and G. S. Kolesnikov, " New Fibre-forming Polyamides ", *Khim. Volokna*, **3**, 5–7 (1968).

" Interfacial Polymerisation ". See a series of papers in *J. Polymer Science*, **40**, 289–407 (1959).

R. W. Moncrieff, " Interfacial Polymerisation as a Means of Producing Textile Fibres ", *Text. Recorder*, 70–73 (Sept. 1962).

R. W. Moncrieff, " Polymer Formation at the Interface ", *Text. Weekly*, 1262, 1272 (14 Dec. 1962).

CHAPTER 4

ORIENTATION AND CRYSTALLINITY

WE were able to draw a close analogy between the shape of fibres and the shape of the molecules of which they were composed, and to show how an appreciation of this resemblance of fibre-molecules to fibres had enabled the synthesis of fibre-forming substances to be achieved. The analogy can, without undue strain, be taken a significant step farther.

The Arrangement of the Fibres in a Yarn

The materials synthesised according to the methods outlined in the previous chapter consist of agglomerations of long, narrow molecules, and they correspond (on the molecular scale) to the bale of loose wool or loose cotton (on the fibre scale) from which a yarn has to be made. The material is there, but some work has still to be done on it before a textile material that is of any use has been produced. In the case of loose wool or cotton, the mass of fibres is spun into yarn by traditional processes, which, although carried out in many stages, consist essentially of two processes:

(1) parallelisation of the fibres;
(2) insertion of twist into the fibres.

Attempts have been made in spinning cotton yarns from loose fibre to substitute a " resin-bonding " or " sticking of the fibres together with a resin " for the insertion of twist, but the process is still in its infancy.

Clearly, any yarn that is inspected, whether taken from silk dress, cotton frock, woollen pullover, jute sacking, or sisal rope, shows all the fibres to be arranged roughly parallel to each other, and parallel, apart from the angle of spirality, to the longitudinal axis of the yarn. This parallelisation, or orientation of the fibres parallel to the length of the yarn, is a feature that is universal in textile yarns, threads, twines and cords.

The main function of the twist is to prevent the fibres sliding over each other when pulled. If fibres slide or slip over each other when pulled there is very little strength in the yarn; a worsted sliver affords an illustration of this. But if the fibres are prevented from slipping, either by being twisted together or by being bonded together by a resin, then they will not slide over each other, but, when subjected to tension, they remain together until the tension is increased to that which breaks the individual fibres.

The Arrangement of the Molecules in a Fibre

Evidence from several sources has shown that in natural fibres the long molecules are arranged more or less parallel to the longitudinal axis. There is a close and real analogy between the arrangement of the fibres in a yarn and the arrangement of the fibrous molecules in a fibre.

The main sources of evidence of this state of affairs are: (1) anisotropic swelling, (2) birefringence, (3) X-ray analytical data, and it will be advisable to consider each of these a little more carefully. It should at first be noted that the parallelisation of the molecules in the natural fibres is not perfect, and that it varies considerably from one kind of fibre to another. This variation plays a notable part in determining the properties even of the natural fibres.

FIG. 18.—Orientation of molecules in cotton and flax.

Tenacity of Cotton and Flax. Consider, for example, flax and cotton. Both are cellulose, both are practically free from lignin; chemically they are almost indistinguishable, but physically they are very different indeed. The cotton fibre has a tenacity of about 2·5–3·0 grams per denier and an elongation at break of about 6–7 per cent, whereas flax has a tenacity of 5·5–6·0 grams per denier and an elongation at break of about 1·5–2·0 per cent. The molecules of cellulose which constitute both cotton and flax are similar—possibly they may have a slightly greater average length in flax than they have in cotton, but this is not a very serious difference; the reason for the difference in the physical properties of cotton and flax lies in the *arrangement* of the molecules. In flax they are highly oriented, they are very well parallelised, and they lie side by side along the length of the fibre; in cotton some of the fibres lie parallel to the fibre axis, but quite a large proportion of them lie at an appreciable angle to the fibre axis. This difference is illustrated in Fig. 18. When tension is applied to the fibre nearly all the molecules in flax take their fair share of the load, and a high breaking load is the result; but in the case of cotton the strain has to be taken by those molecules that are facing in the right direction—*i.e.*, roughly parallel to the fibre axis—those molecules lying approximately at right angles to the fibre axis take little or none of the load. Consequently the tenacity of cotton is lower than that of flax. Probably in the case of cotton those molecules which initially lie along the fibre axis are actually ruptured before the other molecules have been aligned with the axis.

Formerly the idea of breaking or rupturing individual molecules was inclined to be scouted, but not so to-day. If the load required to break a molecule is less than the sum of the loads required to separate it from those molecules to which it is contiguous, then the molecule will break (the branch of science which deals with phenomena such as this is known as Tribo-chemistry). If the load required for separation of the molecules is the lesser, then the molecules will slide.

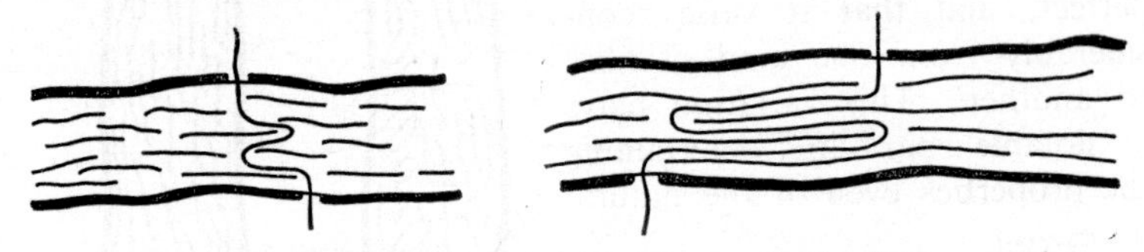

FIG. 19.—To separate long molecules, greater cohesive forces must be overcome than with short molecules.
Long molecules are essential for fibre strength.

Fig. 19 illustrates how the area of attraction that has to be broken is greater in the case of long than of short molecules. We are thus able to see why fibres which consist of very long molecules will be stronger than those consisting of shorter molecules.

Chain Length and Molecular Slip. It is, however, important to note that once the chain length has reached that stage at which the energy required to make a molecule slide over its fellows is greater than that required to break a molecule, then further increase in molecular chain length will be without effect on the fibre strength. There is an idea frequently expressed that the strength of a fibre increases with the length of its constituent molecules. Until the stage is reached when molecular slip is stopped the idea is correct, but beyond that it is incorrect.

Weaknesses in the Fibre Structure. It should be noted, too, that the strength of fibres is always very much lower than the calculated strength of the fibre based on the known strengths of the bonds. Apparently there are " faults " (used in the same sense as in geology) in the fibre structures, and these constitute weak places. It has, for example, been calculated that a perfectly oriented bundle of cellulose molecules should have a tenacity of the order of 50 grams per denier, whereas the strongest form of cellulose that has been marketed has a tenacity of only 8 grams per denier.

Perhaps it should be added that in both flax and cotton the molecules are arranged spirally along the fibre axis; in the case of flax the angle of spirality is about 5°, and in the case of cotton about 30°. This has no marked bearing on what has just been discussed, except that it accounts for the higher elongation of cotton at break. Before the fibres break they will doubtless be pulled more into line with the axis of the fibre. This spiral structure is not uncommon in some of the natural fibres, and is doubtless a characteristic of growth, but it does not occur in the man-made fibres. Some man-made fibre *molecules* are themselves spiral in nature but they are not arranged spirally in the fibre.

FIG. 20.—Random arrangement of molecules in fibre.

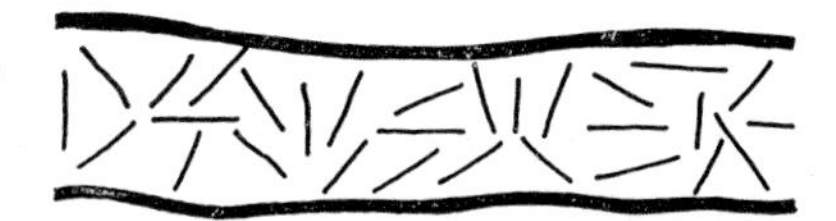

ORIENTING THE MOLECULES

Definition. The molecular orientation of fibres is the alignment of long-chain molecules relative to the fibre axis.

The arrangement of the molecules in a natural fibre—*i.e.*, the proportion of them that lie along the fibre axis—and the average inclination to it of the remainder is a characteristic of each natural fibre and is for all practical purposes fixed and unalterable. Ramie and flax both have extremely high degrees of orientation, with high tenacities and low elongations at break, whereas cotton has a much lower degree of orientation, with a lower tenacity and higher elongation at break.

In the case of man-made fibres it is often possible to control the degree of orientation. The means by which the orientation is increased is invariably that of stretching. Consider a filament which has been extruded without any stretch being applied to it; the molecules will approximate to a random arrangement, as shown in Fig. 20, although there will probably be a slight degree of orientation of the molecules parallel to the fibre axis, which has been caused by the direction of flow of the spinning solution through the spinning-jet. Remember, as an analogy, that sticks floating down a stream usually turn head on; they do not for long float broadside to the stream. But the degree of orientation due to this effect will not be great, and may perhaps be equivalent to that shown in Fig. 21.

If, however, a material which consists of randomly arranged

molecules is stretched in one direction, the molecules tend to orient themselves in that direction.

Nylon, for example, when spun has a near-random arrangement of molecules, the only deviation from the random arrangement being that caused by entrainment in the flow of the melt (molten nylon) through the spinning-jets. But nylon can be stretched or "cold-drawn" to five times its original length, and when this is done the molecules slip over each other and arrange themselves increasingly in line with the fibre axis (Fig. 22).

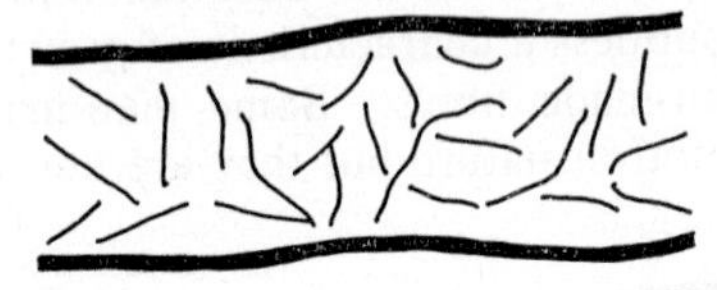

FIG. 21.—Nylon. Random arrangement of molecules before cold-drawing.

FIG. 22.—Nylon. Orientation of molecules after cold-drawing.

Effects of Stretch. The changes that take place during the stretches of several times (*e.g.*, 2×, 4×, or 10×) to which many artificial fibres are subjected are as follows:

1. The molecules slide over each other; it would be quite impossible to stretch any fibres in which the molecules were cross-linked (like rungs in a ladder) by more than about 100 per cent.
2. The molecules turn into the direction of stretch, so that after being stretched they are oriented parallel to the fibre axis.
3. The crystallinity of the fibre is increased.
4. The properties of the fibre are modified, as will be described, by the increase in the degree of orientation.

Anisotropic Swelling

Anisotropy is a word that is frequently encountered in fibre studies. Inexplicably missing from the average dictionary, it implies a state of having unequal properties in different directions. We ordinarily test, for example, the tenacity of yarns or fibres in the direction of their long axis, but tenacity tests at right angles to this length are not ordinarily made. They would be very difficult to make—*i.e.*, to test the strength of a fibre across its width—and for many everyday purposes the measurements would have little value. Nevertheless, the strength of fibres across the fibre is usually different from that along the fibre, and the more the fibres have been oriented the greater is the difference.

There are, however, other properties which it is comparatively easy to measure both along and across the fibre. One of these is the ability of the fibre to swell when wetted either with water or other liquids. If a substance is taken, such as a disc of gelatin, which is not anisotropic, but which swells in water, it will be found on wetting to swell equally in all dimensions—a circular piece will still be circular, although with a larger diameter after swelling; a thin strip cut in fibre shape will extend proportionately in each dimension: if the breadth increases by 50 per cent, so will the length.

If the same experiment is made with a fibre—*i.e.*, if it is carefully measured in respect of length and diameter—and is then wetted in water and re-measured, it will be found that the increase in diameter is very much greater than the increase in length when both increases are compared with their respective original dimensions. When a cotton fibre is immersed in water the diameter increases by about 14 per cent, but the length by only about 1·2 per cent. Other fibres behave similarly—wool, viscose rayon, flax, cellulose acetate and the regenerated proteins all increase very considerably in diameter on wetting, but only very slightly in length. There are, however, very considerable differences between fibres, as the following table illustrates.

Fibre.	Increase in diameter (per cent).	Increase in length (per cent).
Viscose rayon	26	3–5
Silk	19	1·7
Wool	16	1·2
Cotton	14	1·2
Nylon	5	1·2

The comparatively high increase in length in the case of viscose rayon is attributable to the relatively poor orientation—many of the molecules lie at a considerable angle to the fibre axis, and when these are pushed apart the fibre lengthens.

The relatively low increase in diameter on wetting nylon is an indication not that it is poorly oriented, but that it is extremely resistant to the penetration of water. All fibres swell much more in section than longitudinally.

Clearly this behaviour, which it will be found instructive to confirm experimentally, is a proof that the structure of the fibre along its axis is different from that across its axis. The arrangement of the linear molecules of which it is composed must be different in one direction from that in the other.

Explanation of Swelling Phenomena. The conception that we already have of an oriented fibre with the long molecules lying roughly parallel to the fibre axis affords an acceptable explanation of swelling phenomena. It is believed that when a fibre is immersed in water, the water molecules infiltrate into the fibre and find their way in between the long-chain molecules, so pushing them apart (Fig. 23). As the fibre-chain molecules are oriented in line with the axis of the fibre and as they are pushed apart there will be a considerable increase in the diameter of the fibre, but very little increase in length. Unless we imagine that the ultimate constituent parts of a fibre are oriented lengthwise along the fibre, it is difficult to conceive a mechanism which satisfactorily explains why the fibres swell so very greatly in diameter and only so very slightly in length when wetted.

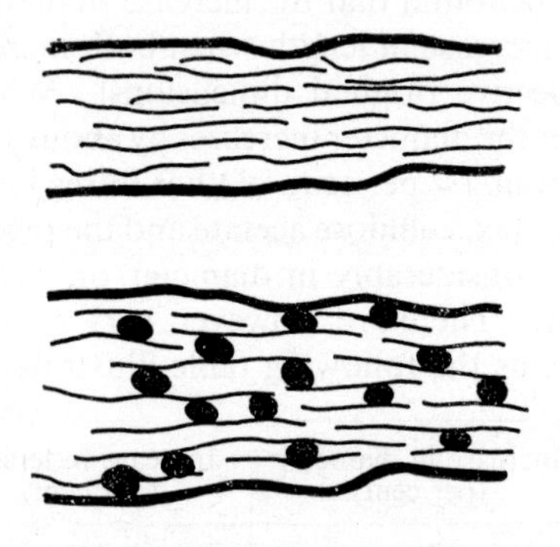

FIG. 23.—Swelling of oriented fibres on wetting is due to lateral separation of the chains by infiltration of water molecules (shown disproportionately large).

A substance such as gelatin, which has similar properties in all directions, is said to be *isotropic*; fibres which have different properties in different directions are *anisotropic*. Consider how very different is the swelling of a fibre from the expansion on heating of a metal. In the latter case it is the increase in length which is most noticeable, because the length—*e.g.*, of a railway line—is much the greatest dimension; length and breadth both increase according to the coefficient of expansion of the metal, which is equal in all directions.

Birefringence or Double Refraction

Every substance has a refractive index. This is a measure of the deflection that a ray of light, not normal to the surface, suffers on passing into it. It is also a measure of the velocity of light in the medium. Amorphous materials such as glass and liquids such as water have a constant refractive index; that of water is the same irrespective of the direction through which light traverses it; so is that of glass (although different from the refractive index of water). Cubic crystals have the same properties in all directions and are isotropic, and they, too, have a constant refractive index. But other forms of crystals—for example, those that are needle-shaped—have two refractive indices and when a ray of light enters such a crystal it is, in general, split up into two separate components which travel with different velocities. The resistance to the passage of light is greater in one direction than in the other and this is a clear indication that the structure of the crystal is different in one direction from another. Similarly, in the case of fibres the refractive index along the fibre is different from that across the fibre. The property of having two refractive indices is known as " birefringence " or as " double refraction ".

When Nägeli observed in 1858 that fibres were birefringent, he concluded that they must be crystalline in structure, and he invented the name of *micelles* for the structural crystalline units. His views that fibres were crystalline, although correct, were not accepted until they were confirmed in the 1920–30 period by X-ray analysis.

The difference between the refractive indices of a fibre will clearly be greater if the orientation is greater, and a measure of the specific index of birefringence is often taken as a measure of the orientation.

Definition. The specific index of birefringence is the difference between the index of axial refraction and the index of transverse refraction.

Some values which illustrate how this difference is greatest for the most highly oriented fibres are:

Fibre.	Index of axial refraction (refractive index along the fibre). E.	Index of transverse refraction (refractive index across the fibre). W.	Difference. E − W.
Ramie .	1·596	1·528	+0·068
Silk . .	1·591	1·538	+0·053
Nylon . .	1·580	1·520	+0·060
Polythene .	1·556	1·512	+0·044
Wool . .	1·554	1·544	+0·010

All these fibres are strongly birefringent; all, too, are positively birefringent. Note that the highly oriented fibres, like ramie and nylon, have much higher specific indices of birefringence than the poorly oriented wool. It appears that positive birefringence is a general characteristic of long-chain molecules, the refractive index for light vibrating along the molecules being greater than that for light vibrating across the molecules.

The birefringence is mainly due to orientation; *e.g.*, oriented nylon has a specific index of birefringence of 0·06 but undrawn nylon has the low value of 0·002. Undrawn Urylon (a discontinued polyurea fibre) had a specific index of 0·003–0·012 depending on the spinning conditions but the drawn fibre had the much higher value of 0·09. Melt-spun fibres have higher values than fibres spun from solution.

Melt-spun.		Solution-spun.	
Fibre.	Specific index of birefringence.	Fibre.	Specific index of birefringence.
Terylene	0·162	Orlon	0·010
Perlon	0·095	Rayon	0·007
Urylon	0·090	Acrilan	−0·006

In addition to Acrilan, two other fibres—cellulose triacetate and polyvinylidene chloride—are weakly negatively birefringent. It may perhaps be significant that in each of these fibres most of the mass of the fibre is in the side-chains and not in the long-chain backbone molecules.

Sakajiri of the Teikoku Rayon Co. has reported that undrawn polyester (Terylene type) has a low birefringence value of 0·005 which compares with about 0·162 for drawn Terylene. Furthermore if the undrawn Terylene is heated, the greater freedom of the molecules enables internal stresses to be reduced, and the birefringence index falls accordingly; when the glass transition temperature (p. 88) of 80° C. is reached the birefringence index has fallen to zero, and if heating is thereafter continued to 110° C. the sign of the index changes and the undrawn polyester fibre has a birefringence value of −0·002.

More and more attention is being paid to the property of birefringence as a key to knowledge of fibre structure. I.C.I. have disclosed in B.P. 762,190 a method to control the uniformity of such melt-spun continuous filaments as Terylene by a continuous determination of the birefringence index; if anything should go wrong with the structure of the filaments the fault would show up as a change in the birefringence.

Dichroism

Dichroism is an optical effect which it has been possible to link with fibre orientation. There are certain dyestuffs which absorb polarised light anisotropically; if polarised light passes through their crystals along one axis it is absorbed more than if it is passed along another axis—the absorption of polarised light is greater in one direction than in the other and dyestuffs that behave in this way are said to be " dichroic ". Congo Red is one of these dyestuffs, and if cellulose fibres such as viscose rayon are dyed with Congo Red the fibres themselves exhibit dichroism; such fibres exhibit maximum absorption of light when the light vibrates in a direction parallel to the axis of the fibre, and least when the light vibrates perpendicular to the fibre axis.

The explanation of this behaviour is that the dyestuff elongate molecules are adsorbed on to the fibre not in a random way, but in some pattern, *e.g.*, the dyestuff molecules may lie with their length parallel to that of the fibre. They will lie in the spaces between the microfibrils of the fibre. If the molecules in the fibre are randomly arranged, so will be the dyestuffs which are adsorbed on them and the fibre will not show dichroism, but if the molecules in the fibre are oriented, then so will be the dyestuff particles that are adsorbed on them, and consequently the fibre will exhibit dichroism. Clearly, too, the greater the degree of orientation, the more marked will be the dichroism, and it is possible to observe the degree of dichroism in a fibre which has been suitably dyed and to calculate from it the degree of orientation. In a comparison of the methods of dichroism and of X-ray reflections to determine the orientation of cellulose acetate fibres, it was found that the method of dichroism yielded lower values of orientation than the X-ray method, but there is sufficient correlation between the results given by the two methods to make it certain that the dichroism of the fibre is a real measure of the orientation of the molecules in the fibre.

X-ray Analysis

Visible light can be used to show that there is a rough orientation of the giant molecules in fibres, and that the degree of orientation varies from one fibre to another, but it cannot tell us much more about the internal structure of fibres—any more than the use of a surveyor's chains and measures could tell us much about the dimensions of an apple. We must seek a smaller measure. X-rays are suitable because their wave-length is about one-thousandth part of that of visible light and is of the same order as interatomic distances.

About forty years ago it was discovered that when a narrow beam of X-rays was directed on to the face of a crystal, the beam was split into two; one beam passed straight through the crystal, whilst the other was diffracted from it. The beams could be detected by means of a photographic film. The arrangement used is shown diagrammatically in Fig. 24. The crystal of (say) sugar is shown at *C*, and an X-ray beam of wave-length λ (= 0·71 Å), after passing through a slot S, strikes the crystal. Part of the beam is undeflected and passes straight through to point P, where it is detected on the curved photographic film PR, but another part of the incident beam is reflected from the crystal to point R, where it is also detected photographically. Then if the angle between the two beams is 2θ, the following expression holds:

$$\lambda = 2d \sin \theta$$

where λ = wave-length of the X-rays
$\theta = \frac{1}{2}$ angle subtended by the arc PR
and d = distance between the layers of atoms which lie *parallel* to the face of the crystal from which the beam was reflected.

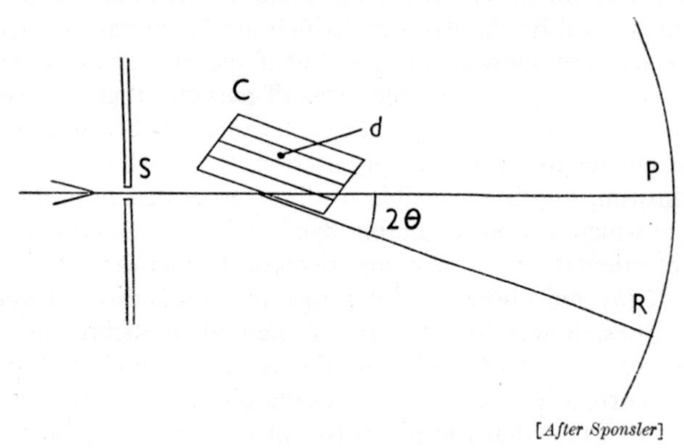

[*After Sponsler*]

FIG. 24.—Diagrammatic representation of X-ray analysis of crystal structure.

As λ and θ can be measured, it is easy to calculate d. Next the crystal is rotated so that its other faces are presented at the proper glancing angle to produce beams reflected from the other atomic layers, and their angular deflections can be measured and the atomic spacings calculated. From these measurements it may be possible to build up a three-dimensional model of the crystal.

Much of the early work on X-ray examination of crystal structure was carried out by Laue and by W. H. and W. L. Bragg. Then O. L. Sponsler attempted to use the method for the examination of natural products such as starch and protoplasm. After a little while he found it advantageous to use natural fibres because their elongated cells could be so placed that only those atomic layers which lay along the length of the fibre would be suitably placed to reflect the X-ray beam. Sponsler found that when he used ramie (a very highly oriented natural cellulosic fibre) and put the fibres parallel to the X-ray beam, there were three reflected beams as indicated in Fig. 25. Furthermore when the fibres were turned so that only those layers which extended across the fibre were in a position to deflect the beam, *i.e.*, when the fibres were at right angles to the X-ray beam, then five sets of reflecting planes were indicated, as shown in Fig. 26. From the first picture (Fig. 25) the spacings of the three sets of atomic layers running lengthwise of the fibre could be calculated, and from the second picture (Fig. 26) the spacings of the atomic layers which extended across the fibre could be calculated. The problem, rather like sorting out a three-dimensional jig-saw, of fitting the spacings into a three-dimensional lattice, then had to be solved. When it was solved it was found that the unit in the lattice was a group of atoms which occupied a volume of 169 cubic Ångström units. This is exactly the volume

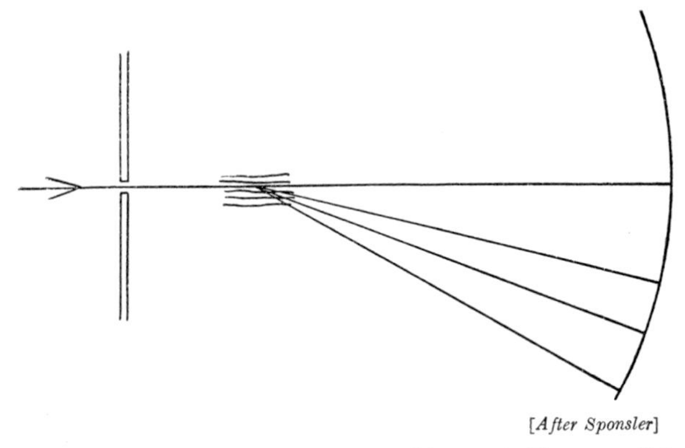

[*After Sponsler*]

FIG. 25.—Three X-ray reflections caused by ramie fibres parallel to the incident radiation.

of a glucose unit, and this, together with the known fact that the hydrolysis of cellulose yields glucose, showed that the glucose molecule was the unit or building brick of cellulose. Additionally, the model that was made from the knowledge of the spacings of the atomic layers showed that the glucose molecules were fastened together through an oxygen atom (ether linkage) which acted as a bridge between each two molecules and resulted in the formation of a long chain. Each glucose molecule was a link in the chain, and the chains ran along the length of the fibre to form the parallel layers from which the X-ray reflections were obtained.

Researches carried out by Mark and Meyer on other forms of cellulose and related substances showed that all were composed of long chain molecules of which a sugar, usually glucose, constituted the link in the chain. The same method was applied by Astbury to determine the structure of the wool molecule.

The type of picture that is obtained in practice is shown in Figs. 27–30. The centre spot is due to the undeflected beam, but it can be seen that the deflected beams give a quite well defined pattern.

The fact that a sharp pattern is produced is evidence of three-dimensional order, because one- or two-dimensional order would result in discrete scattered beams in some directions only; in others diffuse scattering would occur. The patterns that are given by such highly oriented fibres as flax and ramie (Fig. 27) are comparable in sharpness with those given by simple crystalline substances. In

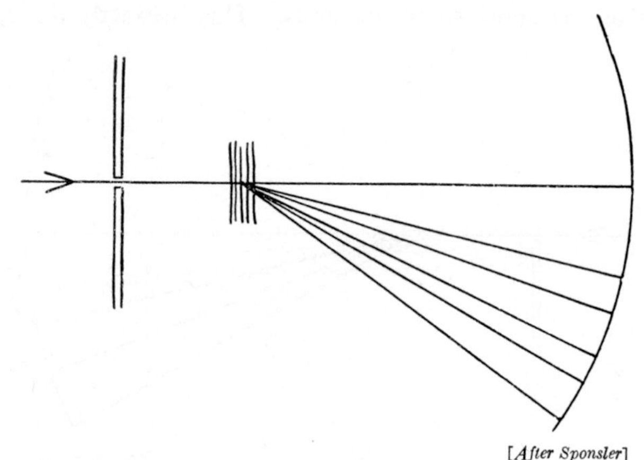

[*After Sponsler*]

FIG. 26.—Five X-ray reflections caused by ramie fibres at right angles to the incident radiation.

general, however, the sharpness of the diffracted beams or " reflec tions " from fibres is less than that from monomeric crystal probably because in fibres the crystals are small, because they are not perfectly aligned with the axis, and because there are non-crystalline or amorphous regions in fibres which cause diffuse scattering (Fig. 28).

Nylon (Fig. 29) and some other synthetic fibres give sharp diagrams, and the degree of sharpness is a reliable indication of the degree of order, or regularity of internal arrangement, of the linear molecules in a fibre; but before being cold-drawn nylon is amorphous and the X-ray diagram diffuse (Fig. 30). Regenerated protein fibres in which the degree of orientation is known to be low give extremely diffuse patterns, like those of amorphous substances. Ardil provides an example of this behaviour.

Information Supplied by X-ray Diffraction Diagrams. To the layman the X-ray diffraction diagrams are not very informative, but

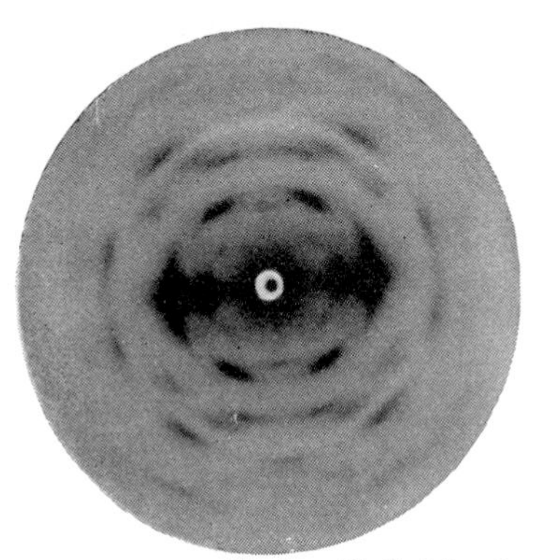

[*Professor W. T. Astbury*]

FIG. 27.—X-ray diagram of ramie.

Sharpness indicates good crystallinity and orientation.

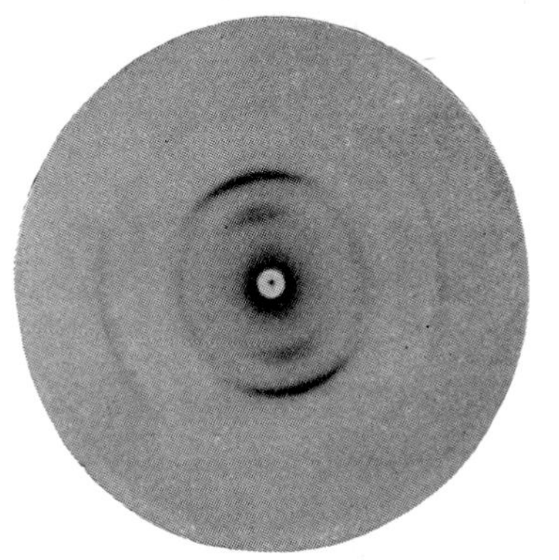

[*Professor W. T. Astbury*]

FIG. 28.—X-ray diagram of cotton, more diffuse than Fig. 27.

Cotton is not so highly crystalline as ramie, and the molecules not so well oriented.

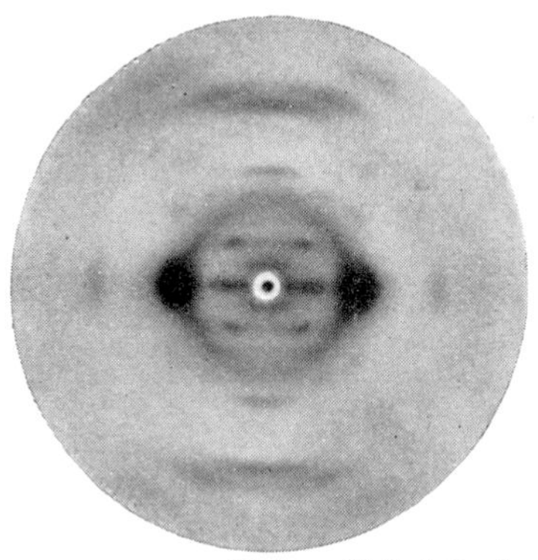

[*Professor W. T. Astbury*]

FIG. 29.—X-ray diagram of nylon after having been cold-drawn, indicative of crystallinity and high degree of orientation.

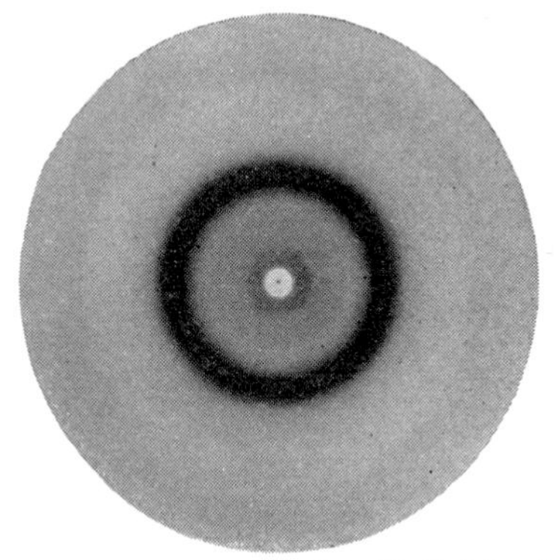

[*Professor W. T. Astbury*]

FIG. 30.—X-ray diagram of nylon before having been cold-drawn, indicative of amorphous nature and random arrangement of molecules.

to the physicist they afford a good deal of information that the fibre scientist finds very useful when it is passed on to him. This information runs along the following lines:

1. Even the layman can appreciate that the appearance of a pattern on the X-ray diagram is indicative of order or regularity in the fibre. When he is also told that similar diagrams are given by crystalline simple substances, but not by amorphous (non-crystalline substances) he infers that fibres are crystalline. But because the X-ray diagrams from fibres are not so perfect, not so sharp, as those from monomeric crystalline materials, he can infer that the crystallinity of fibres is by no means perfect. Further, because some fibres—usually those that are very strong and inextensible, such as ramie—give much sharper diagrams than others, such as wool and regenerated proteins, that are weak and highly extensible, the fibre scientist can infer that some fibres are much more crystalline than others, and that highly crystalline fibres are characterised by high strength and low extensibility. Order, regularity of arrangement, orientation and crystallinity in respect of fibres mean very much the same thing. This will be referred to again, but it will be clear even now that those fibres that give X-ray diagrams nearly as sharp as do monomeric crystals must have a very orderly arrangement or orientation of their molecules.

2. A characteristic feature of X-ray diagrams taken on cylindrical film is that the " reflections " are arranged in rows parallel to the " equator ". From the spacing of these rows the fibre repeat distance —*i.e.*, the length of the repeating unit that makes up the fibre—can be calculated. In the case of cellulose, native or regenerated, it is 10·28 Å, so that we can write:

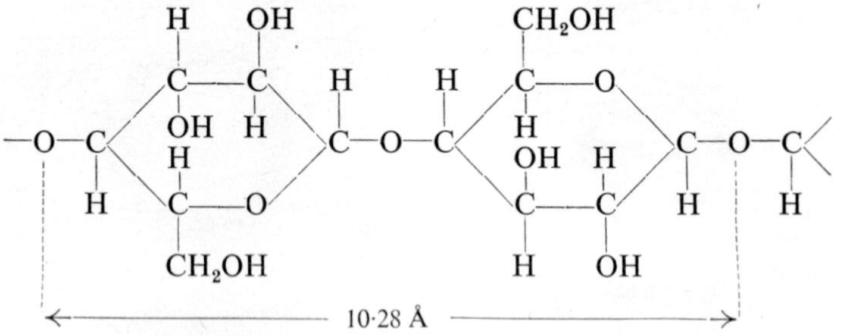

```
             H    OH                      CH2OH
             |    |                        |
             C————C     H     H            C————O
           /  |   |  \   |     |        /  |      \
 —O—C    OH   H    C—O—C       H           C—O—C
     |     H              \            OH   H       |     |
     H     C————O                     C————C      H     H
           |                          |    |
          CH2OH                       H    OH
   <——————————————— 10·28 Å ———————————————>
```

Similarly the fibre repeat distance for nylon is 17·2 Å:

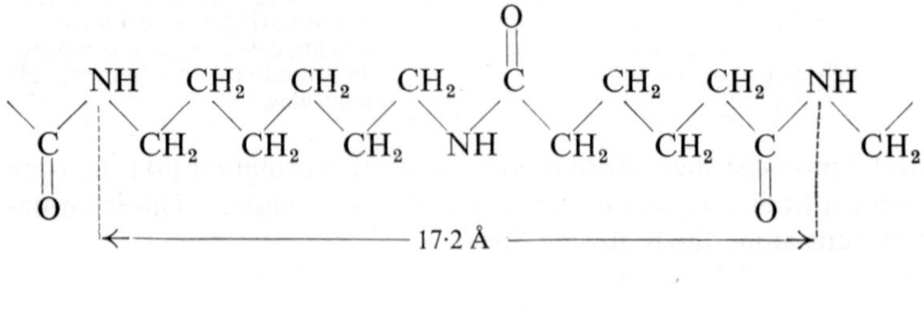

```
                                     O
                                     ||
      NH   CH2   CH2   CH2    C      CH2   CH2   NH
  \  /  \  /  \  /  \  /  \  /  \  /  \  /  \  /  \  /
   C     CH2   CH2   CH2   NH    CH2   CH2   C      CH2
   ||                                        ||
   O                                         O
      <—————————————— 17·2 Å ——————————————>
```

On the same basis, the repeat distance for Urylon (p. 396) would be expected to be about 15·0 Å but it is actually 13·6 Å:

```
      H     H                                         H
      |     |                                         |
      N     N     CH2   CH2   CH2   CH2   N
 \   / \   / \   / \   / \   / \   / \   / \   /
  CH2    C     CH2   CH2   CH2   CH2   CH2    C
  :      ||                             :     ||
  :      O                              :     O
  <-------------- 13·6 Å --------------->
```

This may indicate that in this fibre the molecular chains are not quite straight. The chain repeat distance for Polythene is only 2·53 Å:

```
      CH2       CH2    CH2    CH2
  \  /  :  \   /  :  \  /  \  /  \
   CH2  :   CH2   :   CH2    CH2
        <-2·53 Å->
```

But the chain repeat distance for polyvinyl chloride fibre (p. 475) is 5·0 Å, showing that the chlorine atoms lie alternately on each side of the zig-zag plane.

```
  Cl              Cl
  |    CH2    CH2 |   CH2
 \|  /    \  /   \| /    \  /
  CH      CH      CH      CH
  :       |       :       |
  :       Cl      :       Cl
  <----- 5·0 Å ---->
```

Polyvinylidene chloride (basis of Saran) has a similar value of 4·7 Å. As it is slightly shorter, it is clear that in its case the carbon chains are not fully extended.

3. Some knowledge of the internal arrangement of the atoms in the repeat unit can be deduced from the intensities of the reflections; but in the main they are determined by empirical methods. Knowing, from chemical evidence, the chemical structure of a fibre, and knowing the length of its repeat, it is possible, by using well-established interatomic bond lengths and angles, to construct molecular models. It will be found that some of them are not compatible with the " repeat ", and these can be rejected; if there is only one model that has the correct repeat it can be accepted with confidence (there is, for example, only one molecular configuration for nylon which is in agreement with the experimentally determined repeat distance of 17·2 Å). If more than one model can be built with the correct " repeat ", some uncertainty will remain as to the

actual molecular configuration. Very similar considerations apply to the determination of the other dimensions of the repeat; models have to be made, and whatever other evidence can be adduced brought in to decide between them.

The other dimensions of the repeat unit may be important. Thus the dimensions of native cellulose—*i.e.*, cellulose that occurs naturally in cotton, wood, plants, *etc.*—which is called Cellulose I, are very different from those in regenerated cellulose—*i.e.*, viscose and cuprammonium rayons—which is called Cellulose II. They are:

Cellulose I (native)	10·28 × 8·35 × 7·9 Å
Cellulose II (regenerated)	10·28 × 8·14 × 9·14 Å

X-ray analysis therefore shows clearly that there is a considerable difference in the arrangement of the cellulose molecules in, say, cotton and viscose rayon, and offers, to some extent, satisfactory explanations of the different properties of the two. Native cellulose and viscose rayon are both pure cellulose, chemically the same (except that in the manufacturing processes the molecules have been shortened by " degradation "), but they are very differently arranged and oriented with respect to each other. Both have a simple repeating pattern corresponding in length with that of a cellobiose unit. Not only cellulose, but all other fibres, including wool and the synthetics, have been shown to have simple repeating units.

4. The probable size of the crystalline regions has been estimated from the degree of diffuseness of the discrete spots on the X-ray diagrams.

Micelles

Nearly 100 years ago, when Nägeli discovered the optical anisotropy of fibres, he endeavoured to account for it. He conceived the fibre as being built up of an immense number of " micelles "; these were agglomerations of molecules, which could be oriented, and so confer anisotropy like that of a crystal on the fibre. Perhaps Nägeli thought of them as tiny crystals of the fibres.

That there was a need for some such concept is shown by the long life that the term *micelle* enjoyed, despite the fact that its meaning was very vague, and that different workers attached to the term whatever properties they found convenient.

When it became known, about 1930, that fibres were actually crystalline—although of course the crystals are much too small to be seen with the human eye or with the microscope—attention was immediately again riveted on the micelles. Once again the term came in for a good deal of use and misuse, but to-day it is simply identified with the crystals or crystalline regions in a fibre. The crystalline micelle of cellulose has been calculated by Mark from the dimensions of the X-ray diagram spots to be about 500 Å long by 50 Å wide—*i.e.*, to be about fifty times as long as the cellobiose repeat and about seven times as wide. Given an increase in depth equal to that of width, the micelle contains some 2,500 cellobiose units; in any case it would be about 100 glucose units long.

Crystalline Regions

Evidence of birefringence, dichroism and X-ray analysis has been adduced to support the view that fibres are crystalline, but it is supposed that they are not entirely crystalline. One modern conception of a fibre, in fact, is of an amorphous, non-crystalline, continuous medium or matrix in which are embedded micelles or crystallites. To-day the terms *micelles* and *crystallites* are used synonymously. Both refer to crystalline regions.

The evidence for the existence of amorphous regions in fibres is based largely on the imperfections of the X-ray diagrams. If the fibres were wholly crystalline one would expect much sharper diagrams with far less diffuse scattering of the " reflections " than are actually obtained. An explanation of these is provided by the postulate that amorphous regions exist.

As in many other matters, it is possible to obtain a clearer picture by going back to first principles.

A fibre, as we have seen, is made up of a large number of very long fibrous molecules. We can imagine that if care were taken to impose no stretching or flow during spinning, the molecules would be in random order as shown in Fig. 31, but that, as soon as the fibre is stretched, the molecules turn into line with the fibre axis. Whereas at first they lack any order, the time will come, as they are turned into line, when they will arrange themselves one with another in an orderly way. Arrangement in an orderly and regular way is crystallisation; crystals are simply ordered arrangements of molecules. If the molecules are long and flexible (as fibre molecules are known to be), and have no cumbersome side-chains, then they will fall readily into place with one another, as shown in Fig. 32, and the crystals will soon form and grow. The amorphous or non-crystalline state represented in Fig. 31 is really typical of the internal condition of a plastic material; as the molecules are oriented and as crystallisation develops it becomes typical of a fibrous material. But even if the shape of the molecules favours orderly arrangement, it is too much to hope that all the molecules will be able to click into position all the way along the fibre; there will be places where they are entangled as shown in Fig. 33, and where they cannot arrange themselves in such an orderly way. There will therefore be crystalline regions and non-crystalline regions; these latter will vary in nature from almost perfectly crystalline to completely amorphous regions, although it is

FIG. 31.—Random arrangement of molecules in an amorphous fibre.

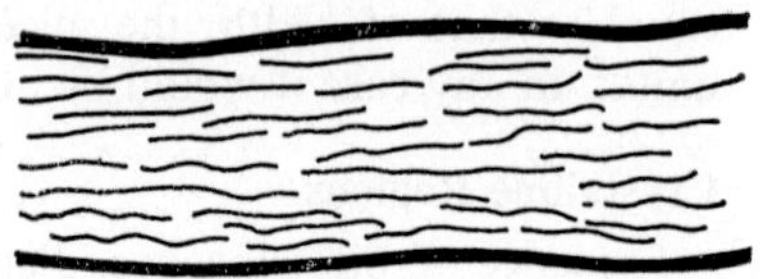

FIG. 32.—Orderly arrangement of molecules typical of crystallinity in a fibre.

unlikely that in any fibre that has been oriented there will be any regions where a completely random arrangement of molecules still prevails. Most probably, the most irregular regions may be no more irregular than as shown in Fig. 33 (lower).

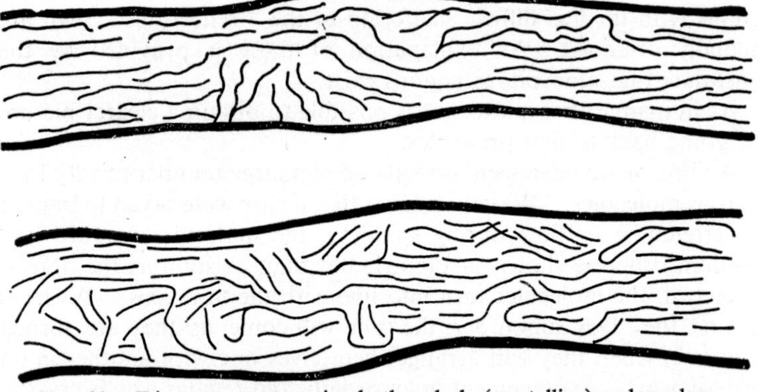

FIG. 33.—Diagrams representing both orderly (crystalline) and random (amorphous) regions in the same fibre.

Lower diagram represents lower degree of crystallinity than upper. Note that in either case the longest molecules pass from one crystalline region right through an amorphous region to another crystalline region.

The Fibre as a Sliver of Fibrous Molecules. There was a tendency current not long ago to consider a fibre as an amorphous matrix in which are embedded crystals, as if there were a sharp line of distinction between crystalline and amorphous materials. This view has been abandoned and a truer picture is now obtained by visualising the oriented fibre as a mass of nearly parallel, fibrous molecules which here and there are fortuitously in perfect alignment, in neighbouring patches in nearly perfect alignment, and in gradually descending order of regularity to some degree of randomness. The fibre should be visualised not as some hypothetical matrix in which are studded, like gems, crystals of perfect regularity, but as a sliver of fibrous molecules, somewhat like a worsted top. If a worsted top is examined it will be found that parallelisation of fibres is much more perfect in some places than in others, but that there are no plainly separated, nicely arranged and random regions; on the contrary, the top embodies a continuous gradation from perfect regularity to considerable confusion.

It has already been found that the length of molecules is much greater than that of micelles. We can accept the reality of the molecules without any hesitation; various methods have been worked out and used for estimating the molecular weight of fibres. In the case of native cellulose it is found that the molecule may comprise some 9,000 glucose units, which corresponds to a length of 45,000 Å, whereas the estimated length of a micelle is about 500 Å. This presented a difficulty which has been overcome by abandoning the view that the micelle is a simple discrete particle, and admitting that the single long molecules pass from one crystalline region (micelle) through an amorphous region, right through another micelle, and so on; in fact it has been suggested that the strength of the fibre largely depends on this feature of the long molecules joining the strong crystalline regions to each other, for it would be expected that the amorphous regions would have only a low strength. The conception of the micelle as a discrete particle was an over-simplification that has had to be abandoned or modified. How much simpler to conceive of the fibre as a diminutive sliver, incorporating every stage of order and disorder.

Effect of Side-chains

If each fibrous molecule is long, straight and flexible there will be every opportunity for an orderly arrangement to be arrived at, but if the molecules have bulky side-chains, as shown in Fig. 34, these may considerably impede the orderly arrangement of the molecules. Some sort of order may be arrived at when the fibres are stretched, but there will be no stability, because the side-chains will be under considerable strain, so that when the tension is released they will pull back the long-chain molecules into their original position. In other words, the fibre will retract, and will behave like rubber. It has been shown that if side-chains are introduced into nylon before it is cold-drawn, it acquires a rubber-like elasticity, and when the tension is released after stretching it retracts to its original length. The intro-

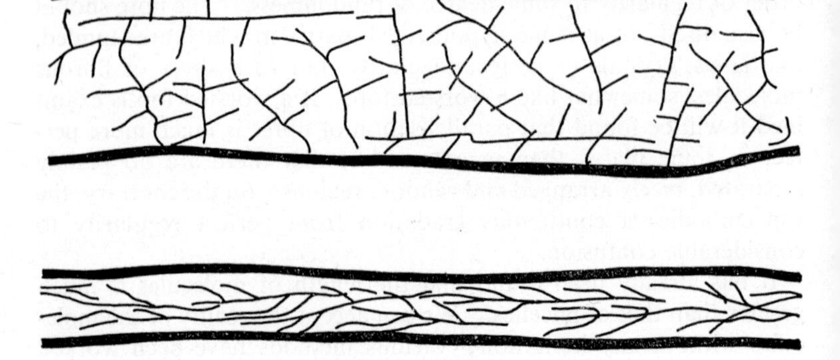

FIG. 34.—Diagrams representing molecular arrangement of polymer with side-chains.

Upper (unstretched) shows random arrangement, lower (stretched) shows that orientation can only be brought about by distortion of side-chains. On release of tension these return to their original position so that the fibre retracts, *i.e.*, it is rubbery.

duction of side-chains into the linear polymer of nylon is the basis of one method whereby an early elastic nylon was made.

Cross-Linking

The effect of cross-linking the long molecules of a fibre is at first good. It prevents the molecules sliding over each other, and gives good recovery from strain, so that cross-linked fibres do not readily crease. Wool and the natural hair fibres are the best example of cross-linked fibres, and cross-linking is one of the normal resin finishing processes for viscose rayon crease-resistant fabrics. If, however, cross-linkages are allowed to form in large numbers, giant three-dimensional molecules are formed, and the material which they constitute becomes infusible and insoluble, and therefore unsuitable for the preparation of fibres. For this reason, care has to be taken to use only difunctional molecules in the synthesis of fibre-forming materials; if, for example, an attempt were made to make a nylon out of hexamethylene diamine (a normal difunctional constituent) and citric acid (or any other tribasic acid), three-dimensional instead of linear molecules would be formed, and the polymer would soon become infusible and insoluble, and it would be an impossible task to attempt to convert it into a fibre. The use of trifunctional intermediates may be regarded as a case of cross-linking carried to absurd and useless limits. But it should be remembered that a modicum of cross-linkages may well confer very desirable properties on a fibre, notably ability to resist deformation and power to recover from deformation, as well, usually, as enhanced chemical stability and better resistance to biological attack.

Phenolic and urea-formaldehyde plastics are typical giant three-dimensional structures in the molecular domain.

Fibres, Rubbers, Plastics

Fibres are composed of long, linear molecules, oriented to varying degrees parallel to the fibre axis.

Rubbers are composed of molecules which are of a shape too awkward to form stable orderly (crystalline) arrangements, but which

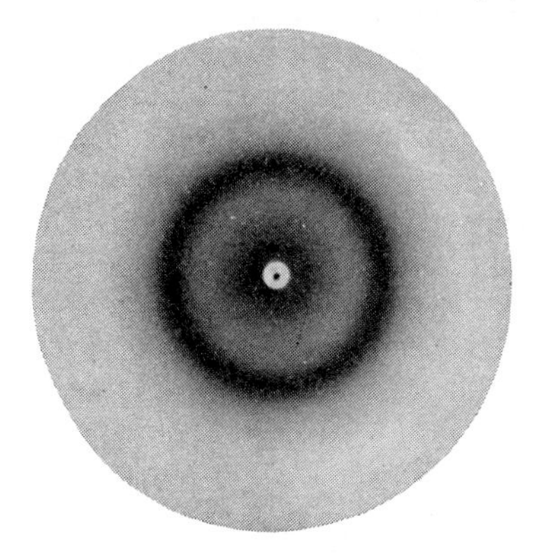

[*Professor W. T. Astbury*]

FIG. 35.—X-ray diagram of unstretched rubber.

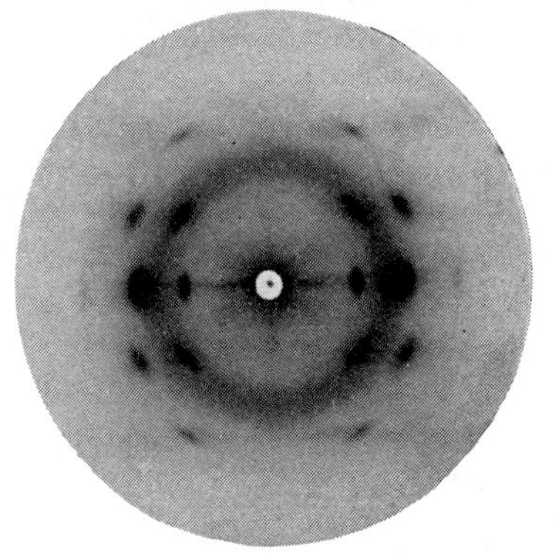

[*Professor W. T. Astbury*]

FIG. 36.—X-ray diagram of stretched rubber.

will, under tension, temporarily adopt such arrangements. When the tension is released, the molecules return to their random arrangement. Unstretched rubber gives a diffuse scattering of X-rays indicative of an amorphous, structureless, molecular configuration; stretched rubber gives a sharply patterned X-ray diagram, showing that the molecules have accepted an orderly arrangement (Figs. 35, 36). The main difference, from the physical standpoint, between nylon and rubber is that in the former case the molecules can slide over each other and readily fit alongside each other, so that when nylon is cold-drawn the new molecular arrangement arrived at is stable, whilst with rubber the orientation is difficult to achieve and the molecules are so ungainly that they cannot line up in an orderly manner—they are pulled into a state of strain, and when the stretching tension is released they return (like extended springs) to their former position.

Plastics are usually divided into two classes: (*a*) thermoplastic, (*b*) heat setting. The former are similar in molecular structure to fibres, but are not oriented: the molecules are randomly arranged. Most of them—*e.g.*, cellulose acetate, cellulose acetate-butyrate and nylon—can be spun into fibres.

The second class of plastics—the heat-setting resins—are made up of giant three-dimensional molecules, which it would be quite impossible to orient: one might as well try to spin a yarn out of grains of wheat.

Second Order Transition Temperatures

Melting or freezing is a main transition—a first order transition. At temperatures lower than the melting point another change can take place; it is less definite, less sudden and is known as a second order transition. It is that, over a range of temperature (not at a sharp temperature like freezing), polymers as they are cooled lose some rubbery characteristics and become more like a glass. The change (rubbery nature to glassy nature, or the reverse) is the second order transition. Always, of course, it is at a temperature lower than the melting point. Molecularly it is probably the transition temperature at which the chains lose their freedom of rotation and become fixed; correspondingly the " substance " loses its elasticity.

Second order transition points for some high polymers are:

P.V.C.	−75° C.
Saran	−17° C.
Nylon 66	47° C.
Terylene	80° C.*
Polypropylene	−35° C.
Polyethylene	−100° C.

The glass transition temperature of linear polyethylene was at one time thought to be −21° C. but this may have been due to chain branching; the true second order transition point is probably below −100° C. The second order transition temperatures of atactic and isotactic polymers of the same chemical composition, *e.g.*, atactic and isotactic polypropylene, are not usually very different. Generally, those fibre-forming polymers that have high melting points have relatively higher second order transition temperatures than the others. In any one kind of fibre the glass transition temperature rises with increase of crystallinity. According to Thompson and Woods, the glass transition temperature of Terylene can rise from 80 to 125° C. with increasing crystallinity.

* See p. 74 for change in birefringence of undrawn fibre at the glass transition temperature.

Degree of Crystallinity

The various methods used to determine the proportions of a fibre which are crystalline and amorphous are in outline:

1. In crystalline regions the molecules are packed more tightly, therefore the higher the density the greater the degree of crystallinity provided that the fibres being compared are chemically similar, such as those of viscose rayon and cotton or of old and Ziegler polyethylenes.
2. X-ray analysis offers a knowledge of the orientation of a fibre, and from this the degree of crystallinity can be calculated. The amorphous region of a fibre gives a diffuse scattering of the reflections, and the intensity of the " amorphous ring " in comparison with the crystal reflections gives an indication of the proportions of amorphous and crystalline matter in the fibre.
3. To boil cellulosic fibres with dilute hydrochloric acid in the presence of ferric chloride, and to estimate the carbon dioxide evolved. The argument in this case is that only the amorphous regions will be attacked.
4. Measuring the water absorption, in the belief that water is absorbed in the amorphous regions only. It may penetrate a little into the crystalline regions, but this penetration is thought to be insignificant.

Some figures which have been quoted in Preston's " Fibre Science " for regenerated cellulose are as follows:

Method.	Percentage of crystalline material.
Density	25
X-ray	40–50 (upper limit)
CO_2 evolved	79 (stretch spun viscose rayon)
	76 (saponified cellulose acetate)
Water sorption	25 approx.

Whilst the methods are of interest, the widely scattered results seem to lend support to the view already expressed, that there are no clear divisions between crystalline and amorphous areas, and that, in fact, all degrees of order and disorder (down to a fixed limit, few fibres will escape some orientation even in their most random patches) exist.

If one method is used consistently, and different fibres are examined by it, the results may be taken as representative of the degree of orientation, but to attempt to use the results as a measure of the proportion of a fibre that is crystalline is, in the author's view, quite unjustifiable.

Degree of Orientation

Of the methods just described, that of X-ray analysis is most commonly used to assess degree of orientation. When the orientation is poor, birefringence can be taken as a measure of orientation; the higher the degree of orientation, the greater is the difference between the two indexes of refraction. As, however, there is very little change of birefringence for angles of tilt (from the fibre longitudinal axis) up to 10°, the method is not very sensitive when the orientation is nearly perfect. In measuring degree of orientation the average angle of inclination of the chain molecules to the fibre axis is estimated. In a perfectly oriented fibre this would be zero; in a perfectly random fibre—*i.e.*, one not oriented at all—it would be 45°, because there would be as many molecules at right angles to the fibre axis as parallel to it, and intermediately accordingly.

Dielectric Anisotropy and Fibre Orientation

Another method, one different from the others in that it does not depend on some optical property, is that of measuring the dielectric (instead of the optical) anisotropy. This was described in 1965 by some Russian workers (Eifer *et al.*).

The capacity of an electrical condenser varies with the nature of the substance between the plates. This substance, known as the dielectric, is often air or mica. If when some given dielectric is used the capacity of the condenser is twice that of what it was when air was used, then the dielectric constant of the dielectric or insulating material is 2; if with some other substance the capacity is y times what it is with air as the dielectric, then the dielectric constant of that substance is y.

What the Russian workers found was that the capacity of a condenser, with ribbon-like plates and a fibre dielectric, varies according to the way in which the fibre is arranged; the dielectric constant (E_r), measured radially, is different from that (E_a), measured axially; the fibre has two different dielectric constants, it is anisotropic in respect of its dielectric constant.

Measurements are first made on unstretched fibre; it is true that even unstretched fibres usually possess some measure of orientation which has been imposed in the spinning process, but it is not normally very great. Next, the measurements are made again on stretched fibre; in the stretching process the orientation of the fibre will have been increased very greatly, and correspondingly the fibre anisotropy as revealed by measurements of the dielectric constants will be found to have increased considerably.

When measurements were made on Lavsan fibre, which is very similar to our Terylene and is the main Russian polyester fibre, the results were as shown below:

State of fibre.	Dielectric constant		Ratio E_a/E_r or n.
	Axially (E_a).	Radially (E_r).	
Unstretched	3·43	2·97	1·16
Stretched ×4	3·68	2·62	1·40

If the ratio E_a/E_r is called n, as it is in the last column, we arrive at two values of n, *viz.* n_u for unstretched yarn, which is 1·16 in the above table, and n_s for stretched yarn, which is 1·40 in the table. If we now let $N = n_s - n_u$ then N is a measure of the difference of the anisotropy of stretched fibre from that of unstretched fibre, so that N is also a measure of the increase in degree of orientation that has been brought about by the stretching process.

Tests were made on four kinds of fibre and they gave the results shown in the following table. It will be observed that also given are the tenacity and the breaking extension of each fibre.

Effect of Stretching Orientation on Fibre Properties

Fibre.	Type.	Stretched?	E_a.	E_r.	n	N	Tenacity (gm./den.).	Elongation (%).
Lavsan	Polyester	Unstretched	3·43	2·97	1·16	—	—	—
Lavsan	Polyester	Stretched ×4	3·68	2·62	1·40	0·24	5·2	11
Polypropylene	Polypropylene	Unstretched	2·05	2·05	1·00	—	—	—
Polypropylene	Polypropylene	Stretched ×8	2·39	2·03	1·18	0·18	4·8	25
Nitron	Polyacrylonitrile	Unstretched	4·09	3·45	1·19	—	—	—
Nitron	Polyacrylonitrile	Stretched	4·56	3·30	1·39	0·20	3·8	22
Teflon	Polytetrafluoroethylene	Unstretched	2·12	1·84	1·15	—	—	—
Teflon	Polytetrafluoroethylene	Stretched	2·00	1·73	1·16	0·01	0·6	42

There is a relation evident between N, the increase in anisotropy due to stretching, and the fibre strength; Lavsan and Teflon provide the extreme examples. Why is the value of n, for unstretched polypropylene, unity? Because the fibre is isotropic and the molecules are non-polar and flexible. It is only in anisotropic fibres that the dielectric constant measured axially (E_a) is different from that measured radially (E_r). However, when it has been subjected to an

8 times stretch polypropylene has become anisotropic. It is a nice piece of work and something that deserves development.

Unfolding of the Molecule

In at least two fibres it is known that unfolding of the molecule takes place when the fibres are stretched and oriented. This knowledge is derived from X-ray photographs and their mathematical analysis. When wool is stretched, the α-keratin, of which it is composed, is converted into β-keratin, and this is accompanied by an unfolding of the molecules as shown below.

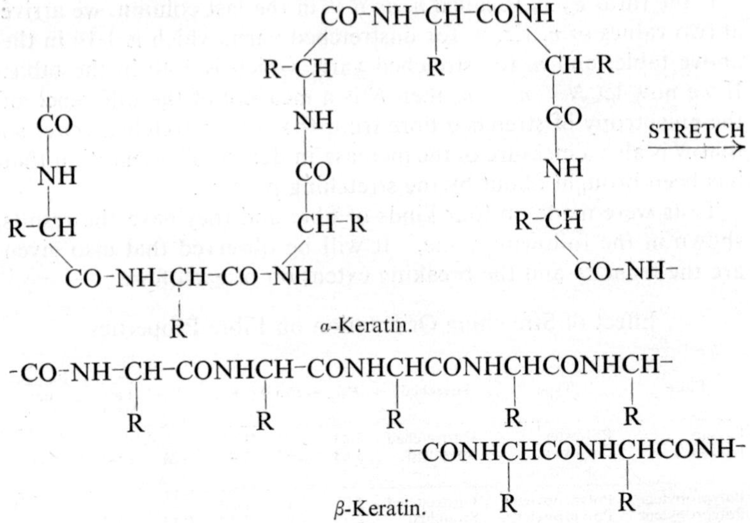

```
                              CO-NH-CH-CONH
                             /      |      \
                        R-CH        R       CH-R
                          |                  |
  CO                      NH                 CO          STRETCH
  |                       |                  |          -------->
  NH                      CO                 NH
  |                       |                  |
R-CH                      CH-R             R-CH
    \                    /                     \
     CO-NH-CH-CO-NH                             CO-NH-
           |
           R              α-Keratin.

-CO-NH-CH-CONHCH-CONHCHCONHCHCONHCH-
       |       |      |      |     |
       R       R      R      R     R
                       -CONHCHCONHCHCONH-
                              |     |
                 β-Keratin.   R     R
```

The intramolecular fold in the polypeptide (polyamide) chain which is present in the α-keratin opens out when the wool is converted, by stretching, into β-keratin. Because the elongation at break of (wet) wool and allied keratinous fibrous materials, such as goat hairs, approximates to 100 per cent, it is thought that the folds in the α-keratin must be approximately square; the length of square has been determined at 5·1 Å. On release of the tension, the molecule folds itself up again and the β-keratin changes back to α-keratin. Doubtless the cystine cross-linkages which join together the polypeptide chains in wool, just like the rungs in a ladder, act as springs; they are themselves deformed during the stretching operation, and as soon as the tension is released, like springs, they recover from their deformation; in so doing they fold up the polypeptide chains.

The other fibre in which molecular folding has been shown, principally by Wrinch, to take place is nylon. The stretching may perhaps be explained in the following way, although this represents an over-simplification. The amide linkages in nylon are $-CONH-$ groups, and these, in contradistinction to the remaining parts of the molecule, which consist of polymethylene groups $(CH_2)_n$, are strongly polar. Electrical fields of force surround the polar amide linkages and attract to themselves other polar amide linkages, so that the molecules fold to accommodate these polar amide linkages in close proximity in this way:

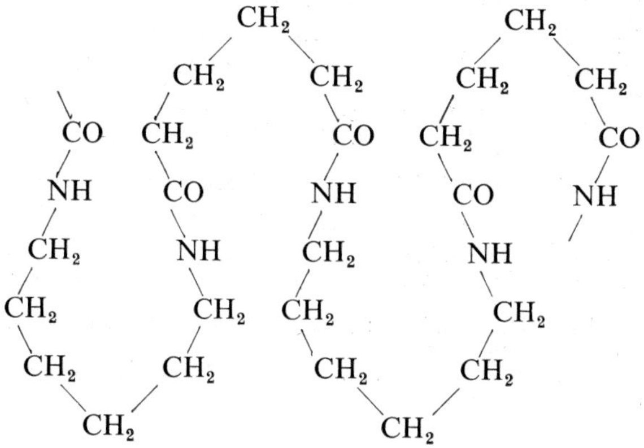

```
                          CH2                       CH2
                         /   \                     /   \
                    CH2       CH2              CH2      CH2
       \           /            \             /           \
        CO      CH2              CO        CH2             CO
       /          \             /             \           /
     NH            CO         NH               CO       NH
    /                \        /                  \      /
  CH2                 NH    CH2                   NH
  /                     \   /                       \
CH2                     CH2 CH2                      CH2
  \                     /     \                      /
  CH2                CH2       CH2                CH2
      \             /              \             /
         CH2                           CH2
```

Then, when the nylon is cold-drawn—*i.e.*, stretched—the chains are unfolded and long, straight molecules result.

$$-CONH(CH_2)_6NHCO(CH_2)_4CONH(CH_2)_6NHCO(CH_2)_4CONH-$$

Even when they have been cold-drawn, the molecules are still crimped like this

and this " crimp " corresponds to the 22 per cent elongation at break which the finished nylon possesses. When it has been stretched 22 per cent the molecules are straight; any greater stretch results in breakage of the filaments.

FURTHER READING

O. L. Sponsler and W. H. Dore, " Structure of Ramie Cellulose ", Colloid Symposium Monograph IV, 174–202 (1926).

H. K. Meyer and H. Mark, " Der Aufbau der hochpolymeren organischen Naturstoffe ", Leipzig (1930) and New York (1940).

W. T. Astbury and A. Street, " X-ray Studies of the Structure of Hair, Wool and Related Fibres ", *Phil. Trans. Roy. Soc.*, **230** *A*, 75–101 (1931).
P. H. Hermans, " Physics and Chemistry of Cellulose Fibres ". London (1949).
J. M. Preston, " Fibre Science ". Manchester (1949).
" Regain and Crystallinity ", *Fibres*, **10**, 218 (1949).
R. W. Moncrieff, " Crease Resistance in Textiles " (due to Cross-linking), *Textile Mercury and Argus*, **120**, 1011 (1949).
E. L. Lovell and O. Goldschmid, *Ind. Eng. Chem.*, **28**, 811 (1946) (determination of crystallinity by means of hydrolysis).
P. H. Hermans, J. J. Hermans and D. Vermas, *J. Polymer Science*, **1**, 149 (1946) (relation of crystallinity and density).
J. M. Preston, " Modern Textile Microscopy " (1933).
P. H. Hermans, " Contribution to the Physics of Cellulose Fibres ". London (1946) (pp. 158 *et seq.* give a good explanation of the use of X-rays in fibre analysis).
D. V. Morley and E. V. Martin, " Structural Orientation in Cellulose Acetate Filaments and its Relation to the Dichroism of Adsorbed Dyes ", *Text Res. J.*, **21**, 607 (1951).
G. Natta, *J. Polymer Sci.*, **16**, 143–154 (1955) (isotactic polymers).
J. F. Rudd and R. D. Andrews, *J. Appl. Phys.*, **29**, 1421–1428 (1958) (birefringence).
I.C.I. Ltd., *British Patent* 762,190 (1956) (birefringence as a control of melt spinning).
S. Sakajiri, *J. Polymer Sci.*, **28**, 452–453 (1958).
A. B. Thompson and D. W. Woods, *Trans. Faraday Soc.*, **52**, 1383 (1956).
R. W. Moncrieff, " Dielectric Anisotropy ", *Man-Made Textiles*, 38, October (1965).
R. W. Moncrieff, " The Illuminant: Its Effect on Colour Perception ", *Canadian Text. J.*, 41–2, 30 August (1963).
I. Z. Eifer, E. Z. Fainberg and N. V. Mikhailov, " Dielectric Anisotropy ", *Khim. Volokna*, **2**, 48–50 (1965).

CHAPTER 5

THE INFLUENCE OF ORIENTATION ON FIBRE PROPERTIES

WHEN a fibre is oriented by stretching, five processes simultaneously take place. These are:—

1. The molecules align themselves parallel, or nearly so, to the fibre axis.
2. The molecules, because of their improved alignment, are able to pack themselves in a more orderly manner, so that the fibre becomes more " crystalline ".
3. In some cases the molecule unfolds, particularly in the case of wool and of nylon.
4. The molecules slide over each other; molecular slippage cannot, however, take place to any considerable extent if the fibres are cross-linked.
5. As a result of the improved packing there is more opportunity for interatomic attractive forces to be exerted; in particular, hydrogen bonding—*i.e.*, bonds of mutual attraction between hydrogen and other atoms on adjacent molecules—come into play. These act like weak cross-links, but as they are very numerous, in spite of the weakness of the individual bond, together they confer a big increase on the strength of the fibre.

The last four of these processes are the accompaniment of the alignment of the molecules. The alignment, or orientation of the molecules, and the resultant increased crystallinity of the fibre, were discussed in the previous chapter, and it was shown that what are usually spoken of as separate phenomena are in fact identical.

The unfolding of the molecule was also discussed earlier, but inasmuch as a better appreciation of the changes in fibre properties that accompany orientation will be made possible by a better understanding of the molecular processes that accompany stretching; those that have not already been discussed will now be briefly considered.

Molecular Slip

Fibre slip is well known to take place when a soft-spun (lightly twisted) yarn is stretched. In order to build a strong yarn, fibre slippage is prevented either by the insertion of sufficient twist or by the bonding of the fibres together by means of a resin; Positex—a suspension of rubber latex in an aqueous medium to which has been added a cationic surface agent—is an example of such a bonding agent; owing to the presence of the cationic agent the latex particles are attracted on to wool (or other protein) fibres that are immersed in the suspension. The rubber particles adhere to the fibre at countless points, but not as a continuous film; no elasticity is imparted to the yarn, but the fibres are bound together so that fibre slippage is prevented and so that the strength of a soft-spun yarn is brought up to that of a well-twisted yarn. The purpose is to prevent the fibres slipping, so that the yarn will break only when its constituent fibres break.

Much the same considerations apply in the molecular region. When the fibres are stretched, the molecules slide over each other because they are not bonded together; but as the molecules become oriented they do become bound together, due to the attractive forces between them. These attractive forces, in the unoriented fibre, have to operate over distances so great that they are weak and ineffective, but as the molecules are brought into line and packed closely together, the distances are greatly reduced and strong attractive forces are set up. These forces offer very considerable resistance to fibre slippage, and in the oriented molecule it may well be that less force will be required to rupture the molecules than to separate them from their fellows. Once the molecular arrangement has been oriented so that the fibre can be broken only by rupture of molecules, greater orientation (or, as already noted, increase in length of molecular chain) may not increase the fibre strength. Molecular slippage will take place during the orientation of the fibre, but is more and more unlikely to obtain as the molecular orientation improves.

Hydrogen Bonding

In addition to the covalent forces in the main chains, there are secondary (van der Waals) forces acting between the chains and joining them together. In addition, in well-oriented fibres there is powerful hydrogen bonding. Hydrogen bonding plays a large part in determining the properties of fibres. This bonding depends on the ability of hydrogen atoms to behave as if they had a valency greater than one. Classical organic chemistry assigns a valency of one, and only of one to the hydrogen atom; it can, for example, form such compounds as

H–O–H
Water.

C_2H_5–O–H
Alcohol.

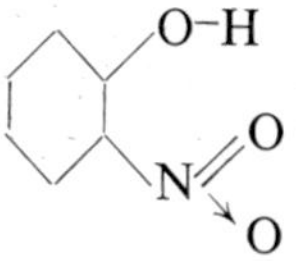

[o-Nitrophenol structure: benzene ring with O–H and N(=O)→O substituents in ortho positions]
o-Nitrophenol.

In each case the hydrogen atom is the terminal group of a chain of atoms, and it was thought that a hydrogen atom could not attach itself to more than one atom. But since 1920 a great deal of evidence has been collected to show that in certain circumstances the hydrogen atom can link two other atoms together; it can, for example, give rise to the following hydrogen bond types:

–O–H····O=, –O–H····N≡, =N–H····N≡, –O–H····S=, =N–H····S=

In all of these bonds the hydrogen atom behaves as if it is divalent and the bonds are known as hydrogen bonds (shown dotted). Their existence has been established by both chemical and physical means, and they afford an accepted explanation of many phenomena that previously seemed very odd. Two examples are as follows:

1. Ice contracts when it melts. This is because the structure of ice is a network of oxygen atoms each of which is hydrogen bonded in the following way:

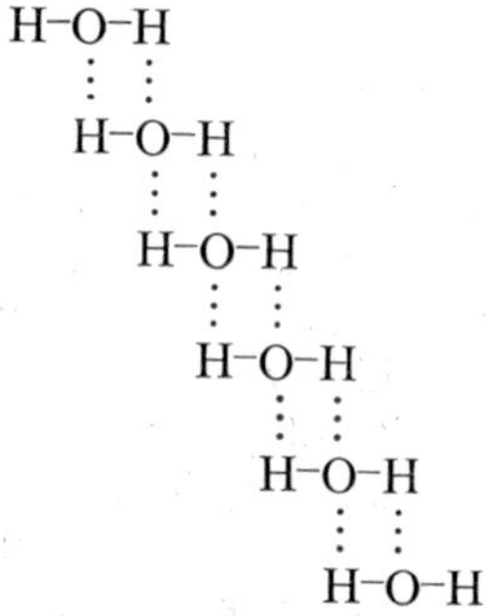

```
H–O–H
:   :
  H–O–H
  :   :
    H–O–H
    :   :
      H–O–H
      :   :
        H–O–H
        :   :
          H–O–H
```

In effect an ice crystal is a huge molecule in which the identity of any particular water molecule is lost. When ice melts, some of the hydrogen bonds break and the atoms pack themselves more closely. Until the temperature rises to 4° C. this closer packing overcomes the thermal expansion. If it were not for hydrogen bonding the water masses of the earth would freeze from the bottom upwards.

2. *ortho*-Nitrophenol differs from its *meta* and *para* isomers in having lower melting and boiling points, lower solubility in

water, higher volatility in steam, higher solubility in non-polar solvents and in being non-associated. The phenolic characteristics are depressed by hydrogen bonding

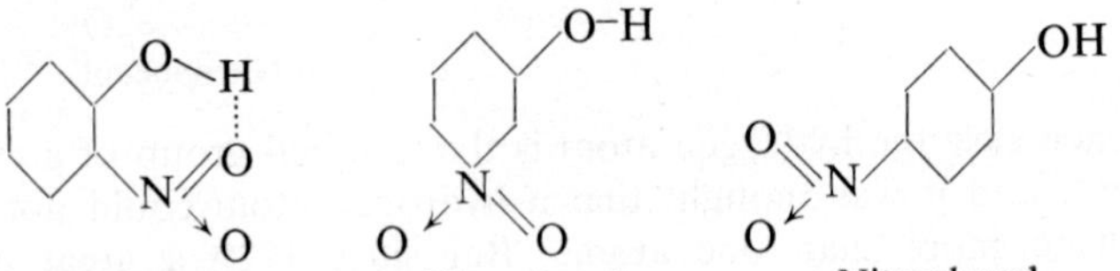

o-Nitrophenol. *m*-Nitrophenol. *p*-Nitrophenol.

Hydrogen bonding can take place in *ortho*-nitrophenol but not in the *meta* and *para* isomers, because in these the oxygen atom of the nitro group is too remote from the hydrogen atom of the phenol group. Hydrogen bonding can take place only when the two atoms to be linked through hydrogen can approach each other within distances less than about 3·4 Å; it is for this reason that it occurs in *ortho*-nitrophenol but not in the *meta* and *para* isomers.

Hydrogen bonding exerts considerable influence on fibres in the following ways:

1. When fibres are oriented adjacent chain molecules are lined up and pack closer together; they pack so closely that hydrogen bonds can be formed. As they are formed they prevent molecular slip, so that the extensibility of the fibre is reduced; highly oriented fibres always have a low extensibility. The hydrogen bonds do, however, impart such mechanical strength to the molecular structure that the elasticity of the fibres is increased; most highly oriented fibres have high elasticity.

2. The multiplicity of polar groups in many fibres confers on them a high water sorption. Thus cotton, viscose rayon, cuprammonium rayon, and other cellulosic fibres can absorb water by virtue of their hydroxyl groups and chain oxygen atoms thus:

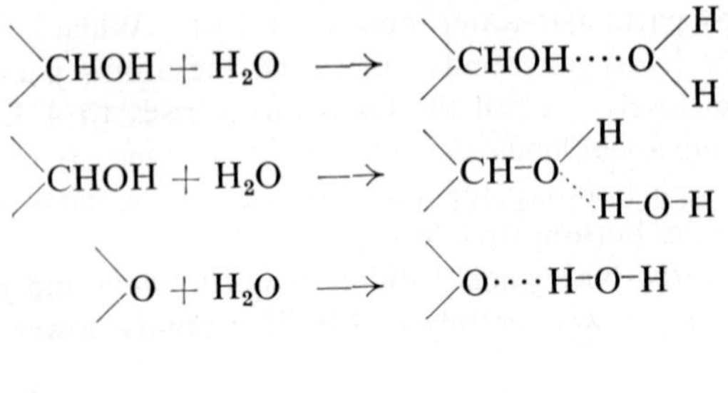

$$>CHOH + H_2O \rightarrow >CHOH \cdots O\begin{matrix} H \\ H \end{matrix}$$

$$>CHOH + H_2O \rightarrow >CH{-}O\begin{matrix} H \\ \cdots H{-}O{-}H \end{matrix}$$

$$>O + H_2O \rightarrow >O \cdots H{-}O{-}H$$

Protein fibres such as wool and casein and Ardil can combine with water by virtue of their polypeptide groups as follows:

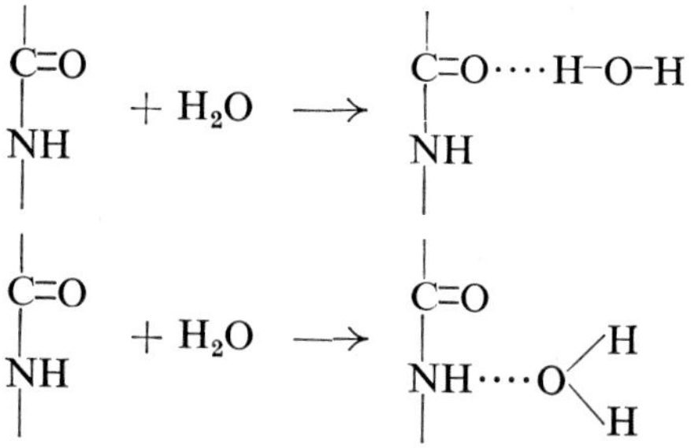

$$\begin{array}{c} | \\ C{=}O \\ | \\ NH \\ | \end{array} + H_2O \longrightarrow \begin{array}{l} | \\ C{=}O \cdots H{-}O{-}H \\ | \\ NH \\ | \end{array}$$

$$\begin{array}{c} | \\ C{=}O \\ | \\ NH \\ | \end{array} + H_2O \longrightarrow \begin{array}{l} | \\ C{=}O \\ | \\ NH \cdots O\begin{smallmatrix} \diagup H \\ \diagdown H \end{smallmatrix} \\ | \end{array}$$

It is this capacity for absorbing water which makes protein fibres particularly suitable for underwear; they can absorb the water chemically without engendering a feeling of wetness.

A corollary of this is that if protein fibres are stretched and highly oriented they absorb water less readily, because hydrogen bonds can form between the atoms of adjacent chains so that fewer atoms are available for hydrogen bonding with molecules of water.

3. Hydrogen bonding as it increases during the orientation of a fibre will reduce the accessibility of the fibre to other molecules. Highly oriented fibres are often difficult to dye, and this difficulty is partly due to hydrogen bonding, which not only reduces accessibility, but also reduces the number of active groups which are available as sites for dye molecules to anchor to. Not only does proper orientation of fibre molecules reduce their accessibility to water, but under suitable conditions it may make the difference between solubility and insolubility, the poorly oriented fibre being soluble and the highly oriented fibre insoluble. Polyvinyl alcohol is ordinarily soluble in water, and in the Japanese fibre Vinylon it has been treated with formalin to insolubilise it by the introduction of cross-linkages; it now appears (U.S. Patents 2,610,359 and 2,610,360) that if very good orientation is obtained in manufacture, then without chemical after-treatment, a polyvinyl alcohol fibre resistant to boiling water results.

The preparation of elastic nylon by reduction of hydrogen bonding is described in Chapter 24 (p. 441). Note also how hydrogen bonding reduces the solubility of polyacrylonitrile (p. 498).

Properties of Highly Oriented Fibres

When fibres become highly oriented through being stretched under suitable conditions, they usually acquire certain properties. In the main these are:

High Tenacity. The better the orientation, the better, in general, is the tenacity. This is the natural outcome of the stretching process in which the denier is very greatly reduced, whereas the breaking load is substantially unaffected. The tenacity of nylon is about 6 grams per denier, that of Fortisan about 7 grams per denier and that of Vinyon about 4 grams per denier. All these values are much higher than those for poorly oriented fibres.

The following table well illustrates the effect of orientation on the tenacity of fibres. It is taken from Fierz-David's "Abriss der chemischen Technologie der Textilfasern" (p. 57, Verlag Birkhäuser, Basle, 1948). Nowadays it should be studied in conjunction with the properties of the polynosic rayons (p. 273).

Fibre.	Orienta-tion.	Tenacity (gm/denier).	Elongation at break dry (per cent).	Ratio wet tenacity : dry tenacity (per cent).
Fortisan H.	Extreme	8–9	8	83
Fortisan	,,	6	6–7	70
Durafil (old)	,,	5·5	6–7	70
Tenasco	High	3·6	18	61
Viscose rayon	Moderate	2·0	18	50
Viscose monofil.	Slight	1·0	40–50	30

Low Elongation. Highly oriented fibres are usually relatively poor in extension at break. Inasmuch as they have undergone a stretching process to bring about their orientation, this is not altogether surprising. Nylon, with an extension of about 22 per cent, is relatively good, but Fortisan, with 7 per cent, is poor, and this low extension militates considerably against its usefulness. But even the best of these figures is low compared with the extension of wool, which is poorly oriented.

Brittleness. Many oriented fibres are brittle. They have acquired anisotropic properties through stretching—*i.e.*, their physical properties are different along the fibre from across the fibre. Sometimes great strength is achieved at the expense of flexibility and pliability.

Increased Lustre. When nylon is spun, it is a dull-looking, semi-opaque fibre which acquires its brilliant lustre only on cold-drawing —*i.e.*, on orientation. Many fibres are, of course, highly lustrous, although they are not highly oriented; viscose rayon is an example.

But, in general, high degrees of orientation are accompanied by increased lustre.

Low Moisture Absorption. As the molecules of the fibre pack closely together during orientation, it becomes increasingly difficult for molecules of water to penetrate between them. Highly oriented fibres therefore have only low moisture absorption. Vinyon, Dacron and Saran absorb less than 0·5 per cent, and even nylon takes up only 4·2 per cent moisture at standard conditions compared with a moisture content of 12 per cent possessed by viscose. Whilst it is true that the hydroxylic nature of cellulose contributes towards this, it may be noted that native cellulose, which is more highly oriented than viscose, also has a lower regain.

High Chemical Stability. Probably largely for the same reason—*i.e.*, difficulty experienced by external molecules in penetrating the intermolecular spaces in the highly oriented fibres—these fibres have extremely good resistance to chemical attack. Nylon is almost unaffected by 10 per cent caustic soda at 85° C., and Dynel and Saran are both used as chemical filtering media and for making protective clothing for chemical workers.

Low Dyeing Affinity. Because of the difficulty that water finds in penetrating the oriented fibre molecules, it is no easy matter to dye such fibres. In the case of Dacron, dyeing assistants whose function is the temporary softening of the filament surface to permit dye absorption have been used. In addition, the stretching of fibres results in the yarns having a very low filament denier—*i.e.*, the filaments, after having been stretched, are finer than those that are ordinarily spun or are found in nature. The fineness of filaments means that the total surface of the filaments in a yarn will be much greater than usual, just as an ounce of fine sand has a very much greater surface area than an ounce of coarse sand. The great surface area of the stretched filaments causes them to reflect more light than a similar weight of coarse filaments, and this extra light which is reflected has a " diluting " effect on the colour. Two 100-denier yarns of cellulose acetate, one of coarse filament and one of extra fine filament, both dyed with 1 per cent of the same dyestuff, will look very different; the coarse filament yarn will appear to be much darker than the other. In an exactly similar way, a fine wool such as 70 s botany requires a considerably greater application of dyestuff than does a coarse 46 s crossbred to give the same hue. In a sense this diluting effect is due to the fineness of filament, and not to the molecular orientation; but it seems legitimate to include it in this discussion, because fineness of filament nearly always results from the stretching process which is used to orient the fibre molecules.

Unattractive Handle. Oriented fibres sometimes have unattractive " handle ". They are often a little harsh. This quality is not an invariable accompaniment of orientation, but is characteristic of it.

Properties of Poorly Oriented Fibres

It is instructive to consider, and to compare with the above, the properties of some fibres which are poorly oriented. Of these, we may take the regenerated proteins as typical. They are made from natural proteins which occur as " globulins "—*i.e.*, in spherical rather than linear form—and which are made into fibres by dissolving the protein and extruding its solution into a coagulating bath, followed by various hardening treatments. All these chemical processes have the effect of uncoiling the globulins, building cross-links between their molecules and generally introducing some sort of fibrous orientation into materials that were never intended by Nature to compose fibres. But even after all these attempts to improve matters, the regenerated protein fibres such as Lanital, Merinova and Fibrolane (all casein) are only very poorly oriented. In this respect they resemble wool, which itself is poorly oriented. Let us consider the chief characteristic properties of such fibres.

Low Tenacity. They are very weak—mostly about 0·8 gram per denier—rising exceptionally to 1·7 grams per denier. On an average they are five times weaker than highly oriented fibres.

High Elongation. The elongation at break of these fibres is often as high as 50 per cent, and in this respect they are greatly superior to the highly oriented fibres.

Pliability. They are extremely pliable and flexible.

Lustre. In general, the regenerated proteins, in common with wool, have only a subdued, not a high, lustre.

High Moisture Absorption. All the regenerated protein fibres have a high moisture content, usually about 12–14 per cent under standard conditions.

Low Chemical Stability. Because of their low order of orientation and the haphazard arrangement of the molecules, it is easy for external molecules of water and solvents to penetrate the intermolecular spaces in the fibre, and the fibres are peculiarly susceptible to chemical attack. The low wet strength of the regenerated protein fibres is their greatest defect, and considerable care has to be exercised to see that they do not come to grief during scouring owing to the presence of alkali, even although it is only very dilute, in the scouring-bowls. Another factor contributing to their poor chemical stability is their high content of polar groups, which are chemically reactive.

High Dyeing Affinity. Doubtless because of the ease of penetration by the dyestuff solutions, the regenerated protein fibres have an extraordinarily high avidity for dyestuffs, and recall the high affinity of chlorinated wool, which they generally resemble in dyeing properties. Polar groups contribute to this affinity.

Warm and Soft Handle. Those properties of regenerated protein fibres which are of overriding importance and which confer on them most of their commercial importance are their warmth to the touch and their soft handle. In these respects they approach most closely of all the artificial fibres to wool. Warmth and softness of handle seem to be associated with reasonably high moisture absorption, with pliability, with high extension and low tenacity; in general, such properties are associated with a low degree of orientation. It seems clear, therefore, that a certain set of properties is characteristic of highly oriented fibres, and that the exact opposites, the antitheses of these properties, are to be found in those fibres which are most poorly oriented.

The Skin Effect

It has been demonstrated that the surface of a fibre is different from the core or inside. This difference has been variously described

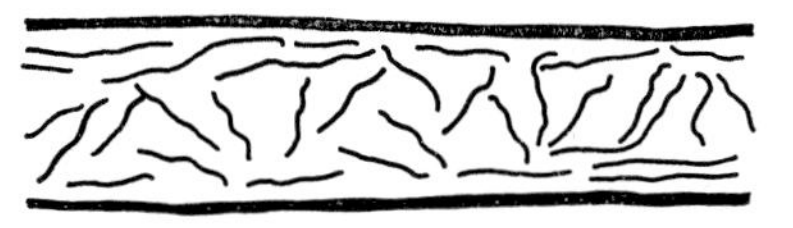

Fig. 37.—The skin effect.
The molecules in the skin of the fibre are more highly oriented than those in the middle.

as a greater degree of orientation or a greater degree of crystallinity obtaining in the skin of the fibre. It is easy to see why this should be so, for as the spinning solution passes through the spinnerets, that part which is in contact with the sides of the orifice will be subject to more frictional resistance than the solution in the centre of the orifice. Consider, as an analogy, how sticks flow faster in the middle of a stream than near the banks. The extra resistance encountered by the spinning solution in contact with the edge of the orifice lines up the long molecules and increases orientation or crystallinity (Fig. 37).

The effect of this skin orientation is rather important. Although it protects the fibre from too easy wetting, it also reduces the speed of wetting during dyeing and makes the first stages of dyeing slow; once the dye liquor has penetrated the skin, dyeing will proceed more rapidly. The effect is most serious in the printing process, where the initial application of colour to the fibre is mainly to the surface.

The dyestuff applied to the more highly oriented skin is not held so fast by the fibre as that which has penetrated to the core, and as a result the colour sometimes marks off easily.

FURTHER READING

J. M. Preston, " Fibre Science ". Manchester (1949).
R. W. Moncrieff, *Silk and Rayon*, **23**, 676 (1949).
P. H. Hermans, " Physics and Chemistry of Cellulose Fibres ". London (1949).
" Contribution to the Physics of Cellulose Fibres ", 130–87. London (1946).

CHAPTER 6

STEREOREGULAR FIBROUS POLYMERS

TWO-DIMENSIONAL REGULARITY IN POLYMERS

Regularity of Alternation of Methylene Groups. Most fibrous polymers have two-dimensional regularity. When vinyl compounds are polymerised, they usually produce a polymer that is regular in two dimensions. Thus:

$$nCH_2{\cdot}CHR \longrightarrow$$
$$—CH_2{\cdot}CHR{\cdot}CH_2{\cdot}CHR{\cdot}CH_2{\cdot}CHR{\cdot}CH_2{\cdot}CHR{\cdot}CH_2{\cdot}CHR—$$

and not

$$\longrightarrow —CH_2{\cdot}CHR{\cdot}CHR{\cdot}CH_2{\cdot}CH_2{\cdot}CHR{\cdot}CH_2{\cdot}CHR{\cdot}CHR{\cdot}CH_2—$$

The regular alternation of the $—CH_2—$ and the $—CHR—$ groups occurs naturally; polyvinyl chloride has the structure:

$$—CH_2{\cdot}CHCl{\cdot}CH_2{\cdot}CHCl{\cdot}CH_2{\cdot}CHCl{\cdot}CH_2{\cdot}CHCl{\cdot}CH_2{\cdot}CHCl—$$

and is not an irregular assembly of $—CH_2—$ and $—CHCl—$ groups, such as:

$$—CH_2{\cdot}CH_2{\cdot}CH_2{\cdot}CHCl{\cdot}CHCl{\cdot}CH_2{\cdot}CHCl{\cdot}CHCl{\cdot}CHCl{\cdot}CH_2—$$

and polyvinyl cyanide (polyacrylonitrile) has the regular structure:

$$—CH_2{\cdot}CHCN{\cdot}CH_2{\cdot}CHCN{\cdot}CH_2{\cdot}CHCN{\cdot}CH_2{\cdot}CHCN{\cdot}CH_2{\cdot}CHCN—$$

and does not have cyanide (nitrile) groups located on adjacent chain carbon atoms. Regularity such as this—*i.e.*, regular alternation of the methylene groups, is common to the addition polymers and is not that which is known as stereoregularity.

Irregularity of Group Arrangement in Co-polymers. If there are two or more monomers that are co-polymerised, there is not as a rule any reason to expect regular alternation of the two side groups. For example, vinyl cyanide and vinyl chloride, when co-polymerised, will give a polymer in which every other carbon atom in the chain is a part of a methylene ($—CH_2—$) group, in accordance with the rule explained above, but there will be no regularity of alternation of the cyanide and chloride groups. What we shall get will be:

$$—CHCl{\cdot}CH_2{\cdot}CHCN{\cdot}CH_2CHCN{\cdot}CH_2{\cdot}CHCl{\cdot}CH_2{\cdot}CHCl{\cdot}CH_2{\cdot}CHCN—$$

and not

—$CHCl{\cdot}CH_2{\cdot}CHCN{\cdot}CH_2{\cdot}CHCl{\cdot}CH_2{\cdot}CHCN{\cdot}CH_2{\cdot}CHCl{\cdot}CH_2{\cdot}CHCN$—

There will ordinarily be an irregular distribution of the —Cl and the —CN groups although, by chance, regularity may obtain in some parts of the chain molecules.

Exceptional Regularity in Some Co-Polymers. The fibre Darvan affords an exception to this ordinary incidence of irregularity; it is a polymer made from the monomers vinylidene dicyanide and vinyl acetate. We should expect that every alternate carbon atom in the polymer chain would be part of a methylene group and so, of course, it is; we should not, however, expect that the dicyanide and acetate groups would alternate regularly, but so apparently they do and give:

—$CH_2{\cdot}C(CN)_2{\cdot}CH_2{\cdot}CH(OCOCH_3){\cdot}CH_2{\cdot}C(CN)_2{\cdot}CH_2{\cdot}CH(OCOCH_3)$—

and not

—$CH_2{\cdot}C(CN)_2{\cdot}CH_2{\cdot}C(CN)_2{\cdot}CH_2{\cdot}CH(OCOCH_3){\cdot}CH_2{\cdot}CH(OCOCH_3)$—

or any other irregular arrangement of the monomer groups. It is, in fact, this regularity of alternation that is responsible for the good qualities (p. 541) of the fibre. But, although Darvan and some related polymers have a high degree of two-dimensional regularity in their molecular structure, they are not what is known as stereoregular. There is still room in their molecules for some steric or three-dimensional irregularity. Thus, the acetyl groups in Darvan may be either above or below the plane of its main carbon chain. It is the position of the substituent groups (chloride, cyanide, or acetyl, for example), relative to the plane of the main carbon chain, that determines whether a polymer is stereoregular. If they are all above, or all below, or if they alternate regularly, the polymer is stereoregular; if, on the other hand, they are arranged some above, some below irregularly, the polymer is stereo-irregular. It is important to understand what is meant by stereoregularity and in order to gain this understanding it is desirable to go back to first principles.

Three-Dimensional Regularity in Polymers

The Zig-Zag Carbon Chain. Carbon is tetravalent, each atom has four valency bonds. In saturated aliphatic compounds the four bonds are equally disposed in space. If the carbon atom is pictured to be at the centre of a regular tetrahedron (a four-sided solid in which each side is an equilateral triangle) then the four bonds from the carbon atom will point to the corners of the tetrahedron. Fig.

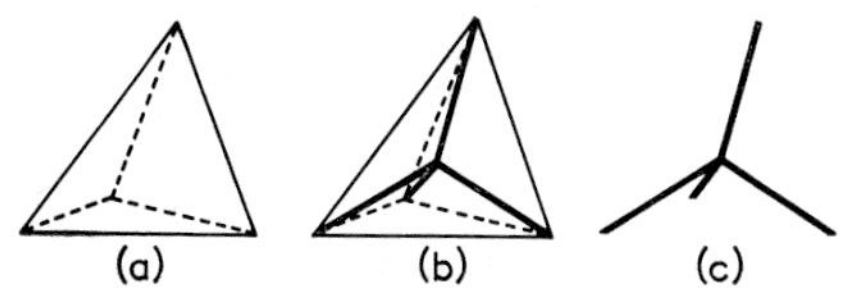

FIG. 38.—(*a*) The carbon atom is situated at the centre of a tetrahedron; (*b*) with its bonds directed towards the corners of the tetrahedron; (*c*) Removing the imaginary tetrahedron, there remain the four bonds radiating from the central carbon atom.

38 (*a*) shows the tetrahedron (the dotted lines show the corner that is hidden by the front face). Fig. 38 (*b*) shows a similar tetrahedron with the bonds indicated boldly; they radiate from the centre of the

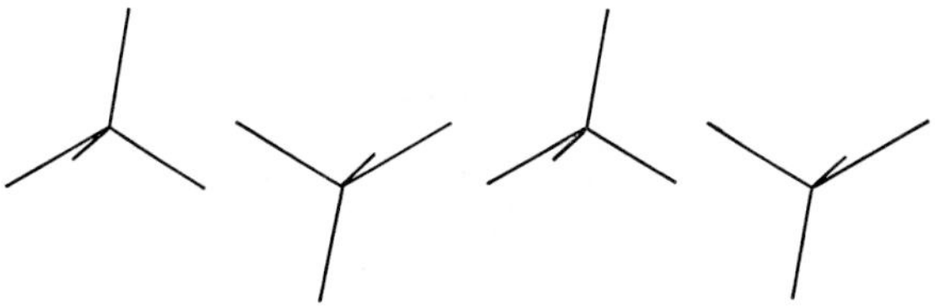

FIG. 39.—Representation of four separate carbon atoms with their radiating bonds.

tetrahedron which is where the carbon atom is supposed to be. Now let us take away the tetrahedron which was only a device to locate the position of the bonds and we are left only with the bonds

FIG. 40.—The four carbon atoms shown in Fig. 39 are shown here moved together into the combined positions. This results in the formation of the zig-zag carbon chain which constitutes the molecular backbone of linear polymers.

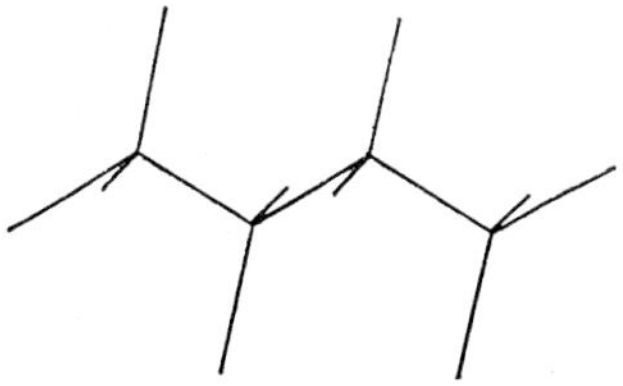

as shown in Fig. 38 (*c*). Now imagine four of the carbon atoms with their radiating bonds arranged as shown in Fig. 39. Move these up together so that the carbon atoms (at the centre of each group of 4 bonds) are joined by the bonds and we have Fig. 40. This can be extended to Fig. 41 which is the usual paraffinic carbon

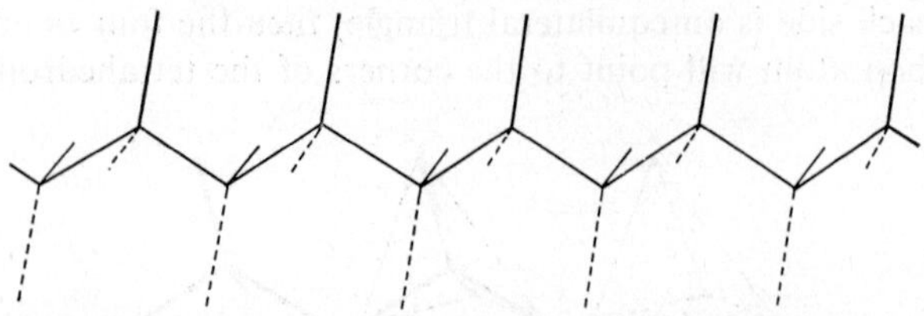

FIG. 41.—An extension of Fig. 40, but now those bonds that lie above the zig-zag plane are drawn solid, those below it dotted. This structure, which has been derived from the tetrahedron in Fig. 38(*a*), is the one commonly used to indicate molecular stereoregularity or otherwise.

chain zig-zag. It is this which constitutes the molecular backbone of linear polymers.

Stereoregular Polymers

The method of its derivation makes it clear that:

1. The zig-zag can lie in one plane (it is easy to satisfy oneself about this by making a few model atoms; use a little ball of plasticine to represent a carbon atom and stick four matches in it to represent the bonds, pointing them towards the corners of the imaginary tetrahedron; then line up a few atoms by superimposition of the bonds).

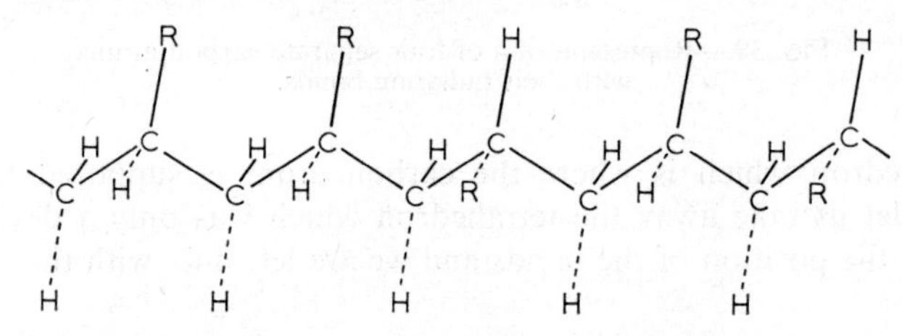

FIG. 42.—An atactic molecular structure. The substituent R groups are arranged irregularly above and below the plane of the zig-zag carbon chain. If R = CH_3 then the polymer represented here would be atactic polypropylene, which is only a grease at room temperature, and which does not form fibres. It is stereo-irregular.

2. Some of the free bonds project above the zig-zag plane, some below it. Those that project above are shown in Figs. 41–44 in solid line and those below in dotted line.

These considerations enable us to derive Figs. 42–44. Fig. 42 depicts an atactic structure, the groups R being arranged irregularly above and below the plane in which the carbon zig-zag lies (if R = CH_3 then the polymer is polypropylene).

The isotactic structure has all the groups R on the same side of the plane of the main chain (all above or all below) as shown in Fig. 43. In the syndyotactic structure the groups R alternate above and below the main chain as in Fig. 44. (In addition there can be block polymers in which blocks of isotactic and atactic polymer arrangements alternate.)

If the group R is not too small, if it is bigger than H, F or OH,

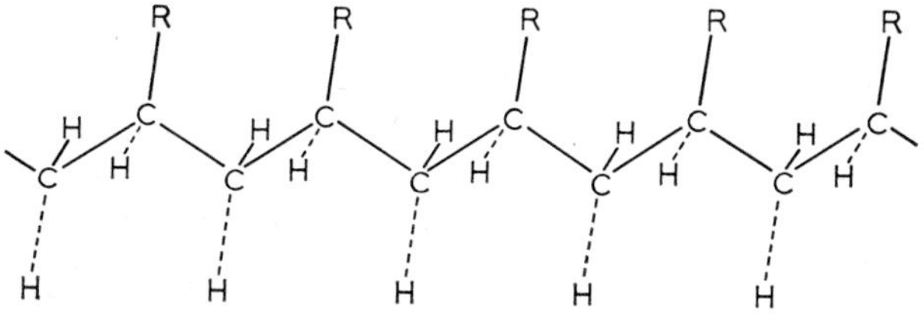

FIG. 43.—An isotactic molecular structure. All the substituent R groups are arranged regularly above (or below if the diagram is inverted, but either all above or all below) the plane of the zig-zag carbon chain. If R = CH_3 then the polymer represented here would be isotactic polypropylene, the basis of the fibres Ulstron and Meraklon.

there is a considerable difference between the types of polymer: isotactic, syndyotactic and atactic. The atactic polymers are amorphous and do not crystallise and they have low softening temperatures. Completely isotactic or syndyotactic polymers generally show high melting points, high crystallinity and good mechanical properties.

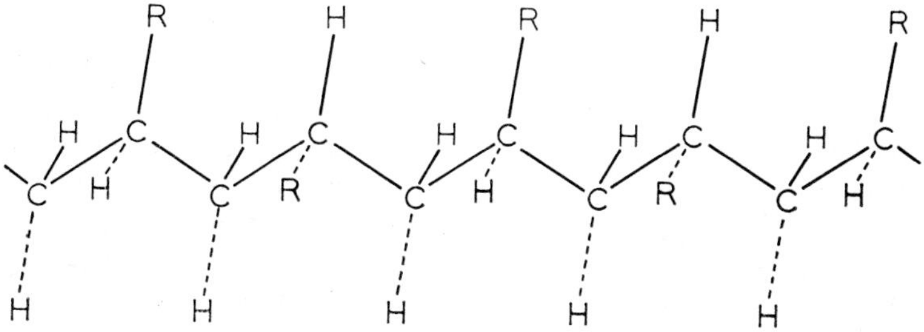

FIG. 44.—A syndyotactic molecular structure. The substituent R groups alternate regularly above and below the plane of the zig-zag carbon chain. If R = CH_3 then the polymer represented here would be syndyotactic polypropylene.

Properties of stereoregular polymers. As we have seen, stereoregular polymers are those in which the side-chains are arranged above or below the backbone of the molecule in a regular pattern. Because such molecules are regular in shape and size they can pack together nicely, so that the density of the polymer is high; furthermore, the softening and melting points are higher than usual because the better packing gives an interlocking and more stable molecular structure which will withstand some extra energy of molecular oscillation (due to a higher temperature) without disintegrating. Other properties that derive from the stereoregularity are better strength and better chemical resistance. The most important of these properties is undoubtedly the rise in melting point and the classic example concerns the polymer made from propylene (methyl ethylene). High molecular weight polymers from propylene were reported in 1952 by Fontana who made them with an aluminium bromide–hydrobromic acid catalyst, but these polymers were viscous oils or soft greases; they were in fact atactic or disordered although this was not then known. Yet when in 1954 a Ziegler catalyst (aluminium trimethyl and titanium chloride) was used by Natta, the polymers from propylene were hard colourless solids melting at temperatures higher than 150° C. Starting material (propylene) was the same, molecular weights of the final polymers were about the same, yet one was semi-liquid at room temperature, the other was a good strong solid. What was the difference? Stereoregularity; the solid polymer consisted of isotactic molecules that were regular and that would fit together nicely, but the semi-liquid greasy polymer consisted of atactic molecules that were irregular. What brought about the difference? What caused the stereoregularity? The choice of catalyst.

Stereospecific Catalysts

The catalyst is not, in the polymerisation of olefines, just an onlooker that by its mere presence encourages the participant molecules to react; on the other hand, it engages in the reaction and every single molecule of propylene is ordered, one might say handled, by the catalyst molecules to engage in the reaction in an orderly way. Stereoregular polymer propagation (growth) depends entirely on the use of stereospecific catalysts. The main types of stereospecific catalysts are: Ziegler, Phillips and Standard Oil.

Ziegler catalysts. Ziegler was the man who discovered the first catalysts which had this unique property of directing olefines to polymerise in an orderly way; he found that his new catalysts speeded up the reaction of olefine molecules enormously and that they sometimes unexpectedly produced solid polymers. Ziegler worked at the Max Planck Institute for Coal Research at Mülheim, Ruhr. At the time he made his discovery he was working on the use of aluminium hydride to catalyse the polymerisation of ethylene;

in 1949 he discovered that the first thing to happen, at 60–80° C., was the formation of aluminium triethyl:

$$3CH_2{=}CH_2 + AlH_3 \longrightarrow Al\begin{cases} CH_2CH_3 \\ CH_2CH_3 \\ CH_2CH_3 \end{cases}$$

Thereafter, he used aluminium triethyl instead of the hydride as the catalyst. He found that if the reaction temperature was raised to 100–120° C., more ethylene molecules interposed themselves between the aluminium atom and the ethyl groups, building up chains, thus:

$$Al\begin{cases} C_2H_5 \\ C_2H_5 \\ C_2H_5 \end{cases} + \begin{matrix} a\,CH_2{=}CH_2 \\ b\,CH_2{=}CH_2 \\ c\,CH_2{=}CH_2 \end{matrix} \longrightarrow Al\begin{cases} (C_2H_4)_aC_2H_5 \\ (C_2H_4)_bC_2H_5 \\ (C_2H_4)_cC_2H_5 \end{cases}$$

The values of a, b and c were considerable: they could run from 200 to 1,000. When water or acid was added to the aluminium polyalkyl it decomposed and gave polyethylenes of different chain-lengths according to the values of a, b or c:

$$Al\begin{cases} (C_2H_4)_aC_2H_5 \\ (C_2H_4)_bC_2H_5 \\ (C_2H_4)_cC_2H_5 \end{cases} + 3H_2O \longrightarrow \begin{matrix} H(C_2H_4)_aC_2H_5 \\ H(C_2H_4)_bC_2H_5 \\ H(C_2H_4)_cC_2H_5 \end{matrix} + Al(OH)_3$$

Ziegler was well embarked on the use of these aluminium alkyls which were remarkable catalysts for polymerising the olefines. Then he had a stroke of luck. A batch of ethylene was polymerised with aluminium triethyl as the catalyst in an autoclave that had just previously been used for a hydrogenation using nickel as a catalyst; a trace of the nickel still remained in the autoclave, which had not been properly cleaned out. In this contaminated vessel a new reaction proceeded rapidly (aluminium triethyl present) and gave a nearly quantitative yield of buten-1 from the ethylene:

$$2CH_2{=}CH_2 \longrightarrow CH_3CH_2CH{=}CH_2$$

Although the reaction had stopped at the dimer (butene is the dimer of ethylene) and this was not altogether what Ziegler wanted, he realised that when a trace of nickel had been present the reaction had proceeded much faster and at much lower temperatures, even below 100° C. He then started a systematic search for other elements which might act like nickel as a second catalyst with aluminium triethyl. After some disappointments he tried zirconium acetylacetonate as the second catalyst with aluminium triethyl; this was very successful and the ethylene in the autoclave yielded a great white mass of polyethylene. Subsequently many metals were found to be effective, and titanium was particularly so. It has become customary to refer to a Ziegler combined catalyst as catalyst and co-catalyst:

Catalyst: Compound of one of the transition elements from Groups IV–VIII in the Periodic Classification, typically titanium trichloride.

Co-catalyst: Organo-metallic compound from Groups I–III of the Periodic Classification, typically aluminium triethyl.

The specification given above of catalyst and co-catalyst will be seen to cover most of the Periodic Classification, and if a study is made of the patent literature of the last few years it will be found that there is now an embarrassment of riches so far as Ziegler catalysts are concerned. The subject advances so rapidly that even the specialists find it impossible to keep abreast of all the new developments.

However, the basic principle of Ziegler catalysts is simple enough: the use of a catalyst of the titanium chloride type with a co-catalyst of the aluminium triethyl type. The use of Ziegler-type catalysts forms the basis for the synthesis of most stereoregular polymers. Ziegler catalysts are anionic; free radical and cationic catalysts are not as a rule capable of producing stereoregular polymers.

Aside from the diverse materials that constitute Ziegler catalysts, there are three other kinds of catalyst that have stereospecific properties. They were discovered through their ability to make really linear (high density) polyethylene. One was discovered by the Phillips Petroleum Co., the other two by Standard Oil of Indiana and they are usually referred to by the names of their discoverers.

Phillips catalysts. These consist of calcined chromium oxide supported on a base of silica and alumina. No other metallic oxide will work like that of chromium; other earths can sometimes be used as the catalyst support, but their selection is important because the solid phase plays a very important part in aligning the monomer. The best support of all contains 90 per cent silica and 10 per cent alumina; in practice it is impregnated with about 2 per cent of chromium in the form of oxide. Such a Phillips catalyst will polymerise either ethylene or propylene, but if zirconia and thoria are used as the support instead of silica–alumina, then the catalyst will polymerise only ethylene satisfactorily, not propylene. The solid surface has a very specific action. The Phillips chromium oxide catalyst on its support has to be " activated " before it can be used and this is done by treating it for 10 hr. with air containing

5 per cent steam at 650° C. In use, the catalyst is in pellet form. The olefine (ethylene or propylene) is used dissolved as about a 4 per cent solution in a hydrocarbon such as *iso*butane. For ethylene the temperature of polymerisation is about 130–150° C.; for propylene the best temperature is 105° C. In either case a pressure of about 30 atm. is necessary.

Phillips catalysts yield remarkably linear polymers; they are known as Marlex; polyethylene is Marlex 50. It is free from side-chains, has a molecular weight of about 20,000 and each long molecule is believed to have a methyl group at one end and a vinyl group at the other:

$$CH_3CH_2CH_2CH_2CH_2 \cdots CH_2CH{=}CH_2$$

The polymerisation of propylene by Phillips catalysts gives a 50 per cent yield of isotactic polymer. Two of the early (1954) Phillips' patents that relate to the work are Belgian 530,617 and 535,082. British Patents 790,195/6 of 1958 and U.S. Patent 2,825,721 are equivalents.

Standard Oil nickel catalysts. Standard Oil have found that nickel or cobalt which is supported on activated carbon will catalyse the polymerisation of ethylene or propylene or other α-olefines. The nature of the support is very specific; nickel or cobalt supported on diatomaceous earth or on silica or alumina will not convert ethylene to a solid polymer; yet mounted on activated carbon either will do so admirably. The activity of these mounted nickel and cobalt catalysts can be promoted by the presence of sodium or sodium hydride so that they work more easily and quickly.

Polymerisation of the olefine is carried out in benzene or xylene or a similar liquid medium. Temperature of polymerisation of ethylene may be about 100° C. at a pressure of 80 atm. and the yield is polyethylene of m.p. 120° C., of density 0·95, and 80 per cent crystallinity. Some of the earlier patents that relate to these catalysts are U.S. 2,658,059 (1953), 2,692,261 (1954) and British 721,046 (1954).

Standard Oil molybdenum oxide catalysts. These afford one more instance of a specific support being essential; molybdena alone and activated alumina alone will not polymerise ethylene. Yet molybdena supported on alumina will catalyse its polymerisation to give high molecular weight solids. The action of molybdena–alumina catalysts can be modified advantageously if they are used in conjunction with a promoter.

Polymerisation of ethylene with a molybdena–alumina catalyst is carried out at about 200° C.; a pressure of about 80 atm. is used and its main function is to maintain the ethylene in solution in the *iso*butane or whatever other hydrocarbon is used as the liquid phase. The polymer so made may be 90 per cent crystalline and may be extremely strong.

Polymerisation of propylene in *iso*-octane solution has been carried out (B.P. 753,350) with 8 per cent molybdena mounted on alumina, using lithium aluminium hydride, $LiAlH_4$, as the promoter and has given a tough flexible isotactic polypropylene. Some of the earlier patents that relate to these catalysts are U.S. 2,692,257/8 (1954) and British 734,501 (1955).

Appreciation of Stereoregularity. Apparently neither Ziegler nor Phillips Petroleum, nor Standard Oil at first, appreciated that the polyolefines that they had made with their catalyst systems were stereoregular. Ziegler's work on his catalysts was communicated to Montecatini and to their consultant Natta at the Milan Polytechnic. Natta took up work with Ziegler catalysts and prepared many polymers; in particular he polymerised propylene using aluminium triethyl and titanium tetrachloride in tetralin; the organo-metallic compound reduces some of the titanium tetrachloride to the trichloride. Natta fractionated, by differences in solubility, the polymer of propylene that he made and found that there were some high fractions of unusual insolubility, and that these fractions possessed a distinct tendency to crystallise; they also had a higher density than usual, 0·91 compared with 0·85 for the lower fractions, and a higher melting point, sometimes as high as 176° C. When their X-ray diffraction diagrams were made, they were rather sharper than those from the lower soluble fractions; furthermore, if the high fractions were melted and drawn into fibrous form and the fibre was *stretched* it would then give an X-ray diagram that was entirely different and was just as sharp as the X-ray diagrams from such fibres as silk and nylon which are known to be largely crystalline. This study of X-ray diffraction diagrams established that the high-melting and relatively insoluble polypropylene was partly crystalline. By studying the X-ray diagrams it was also possible to calculate the dimensions of the unit cell. Just as the cellulose unit cell (p. 77) had much earlier been found to have a volume of 169 cubic Å, so the very insoluble high polymers of propylene were found to have a unit cell with the dimensions $6{\cdot}7 \times 21{\cdot}0 \times 6{\cdot}5$ Å or about 906 cubic Å. There are twelve monomers per unit cell. The period of 6·5 Å is that along the chain or fibre axis and it indicates a repeating length of three (see p. 81, the ethylene unit is only 2·5 Å long) monomeric units arranged in the form of a helix, the helixes lying parallel to the fibre axis. It was also possible to calculate the exact position of all the carbon atoms in the unit cell. The structure that was deduced from this evidence was that which was called isotactic (Fig. 43).

Tactic Polymers (Isotactic and Syndyotactic)

Although natural tactic polymers (notably rubber and wool) have always been with us, it was only when Ziegler catalysts came along that we were enabled to make our own polymers tactic. Examination of the X-ray diffraction diagrams from the very insoluble high fractions of polypropylene showed that these polymers were isotactic; all of the methyl side-chains were on the same side of the plane of the main carbon chain. We may represent the formula for isotactic polypropylene:

$$\begin{array}{cccccccccc}
 & CH_3 & & CH_3 & & CH_3 & & CH_3 & & CH_3 \\
 & | & & | & & | & & | & & | \\
-CH_2 & CH & CH_2 & CH & CH_2 & CH & CH_2 & CH & CH_2 & CH-
\end{array}$$

or by Fig. 43; all of the methyl groups are above the plane which includes the carbon backbone, none is below.

Another arrangement of atoms within the molecule that has been found to occur in man-made polymers is known as syndyotactic and is characterised by a regular alternation of the side-chains on the polyolefine, *e.g.*, on a syndyotactic polypropylene, above and below that plane which includes the zig-zag paraffinic carbon backbone, thus:

$$\begin{array}{cccccccccccc}
 & CH_3 & & & & CH_2 & & & & CH_3 & & \\
 & | & & & & | & & & & | & & \\
-CH_2 & CH & CH_2 & CH & CH_2 & CH & CH_2 & CH & CH_2 & CH & CH_2 & CH- \\
 & & & | & & & & | & & & & | \\
 & & & CH_3 & & & & CH_3 & & & & CH_3
\end{array}$$

or as shown more clearly in Fig. 44. A syndyotactic polymer is capable of crystallising, and of giving well-defined X-ray diagrams and relatively high melting points in the same way as an isotactic polymer. A mainly syndyotactic structure has so far been found to exist in both poly-1,2-butadiene and polymethyl methacrylate neither of which is fibrous. To a lesser degree syndyotactic structures occur in the following fibre-forming polymers (in addition to polypropylene already shown):

polyvinyl alcohol

$$\begin{array}{cccccccc}
 & OH & & & & OH & & \\
 & | & & & & | & & \\
-CH_2 & CH & CH_2 & CH & CH_2 & CH & CH_2 & CH- \\
 & & & | & & & & | \\
 & & & OH & & & & OH
\end{array}$$

polyvinyl chloride

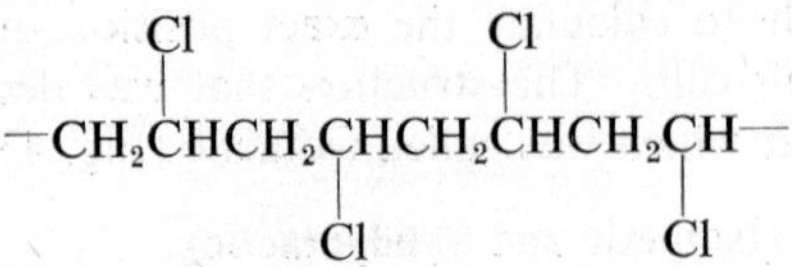

$$-CH_2\underset{}{\overset{Cl}{C}}HCH_2\underset{Cl}{C}HCH_2\overset{Cl}{C}HCH_2\underset{Cl}{C}H-$$

(see p. 81)

and polyacrylonitrile

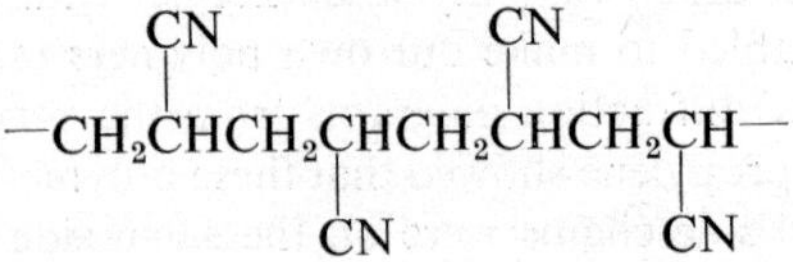

$$-CH_2\overset{CN}{C}HCH_2\underset{CN}{C}HCH_2\overset{CN}{C}HCH_2\underset{CN}{C}H-$$

It is the polymerisation conditions and particularly the nature of the reaction medium and of the catalyst which determine whether an isotactic or a syndyotactic polymer will result. If methyl methacrylate is polymerised with an organo-lithium compound at $-60°$ C. in a medium consisting of the dimethyl ether of ethylene glycol then an isotactic polymer results; if the reaction medium is toluene, then with all other conditions the same a syndyotactic polymer results. If the reaction medium has an intermediate solvating power, for example if a little dioxan is added to the toluene, then an intermediate type of polymer results; it may consist of alternating sequences or blocks of isotactic and syndyotactic structures. Isotactic polymethyl methacrylate has a melting point of 160° C. and a specific gravity of 1·22; the syndyotactic polymer melts at 200° C. and has a specific gravity of 1·19; the supposed isotactic–syndyotactic co-polymer melts at 170° C. and has a specific gravity of 1·21. This co-polymer appears to be a real one because it cannot be fractionated into separate isotactic and syndyotactic components. It is noteworthy that in this example of polymethyl methacrylate (prepared by Rohm and Haas) it is the syndyotactic polymer that has the highest melting point. This seems to be not unexpected: the syndyotactic has a regularity of structure and a crystallisability as great as the isotactic polymer; furthermore, because its side-chains project in both directions, up above and down below the plane that includes the carbon backbone, the molecules should interlock better and give a more cohesive structure which should be more stable and resistant to the growth of molecular irregularity induced by the application of heat. Syndyotactic fibrous polymers may increase in importance and eventually outstrip the isotactic fibrous polymers.

Incompleteness of Control of Stereoregularity

In general, with our present technology, one must not be surprised to find incompleteness of control of propagation so that the resulting polymer is only partly stereoregular and may consist of parts of:

(1) quite random atactic polymer,
(2) quite regular isotactic (or syndyotactic) polymer,
(3) stereoblock polymers which have isotactic (or perhaps syndyotactic) blocks of definite length, in between atactic blocks.

It may be necessary to use the concept of " degree of tacticity " to characterise such incompletely ordered linear molecules.

Fractionation of polymers of different degrees of tacticity. Differences in solubility between low amorphous atactic polymers and crystalline isotactic polymers afford a basis for separating the low molecular weight amorphous material from the highly crystalline. For example, an isotactic polypropylene reaction mixture will contain some atactic material; the mixture can be treated first with refluxing acetone which will remove low molecular weight amorphous, often oily, polymer; the residue can then be treated with refluxing ether which dissolves out solid amorphous material; then it can be treated with boiling *n*-heptane which dissolves out partially crystalline polypropylene, probably about 5–6 per cent of the polymer. What is left should be highly crystalline isotactic polypropylene.

This process of removing by one solvent or another low molecular weight material from a polymer before it can be spun is common to many polymers and is not confined to the preparation of isotactic polymers; for example there is usually some low molecular weight material that must be removed from nylon polymer before it can be spun. In the isotactic preparations, there are, however, more opportunities for unwanted materials to be present and their removal constitutes more of a major operation than usual. It will be appreciated that the isotactic and syndyotactic arrangements are limiting; they represent the two extremes with all atactic arrangements intermediate. But this way of looking at them should not be allowed to hide the fact that both isotactic and syndyotactic polymers represent states of maximum order; all atactic arrangements are inferior to them both in respect of order.

Side-chains in Polyolefines

It is easy enough to see why isotactic polypropylene prepared with a Ziegler or other stereospecific catalyst has a higher density, higher melting point and better strength than atactic polypropylene; in this last the side-chains are arranged irregularly and the polymer cannot crystallise because its constituent molecules will never fit together nicely. The factor that mainly determines the crystallinity or otherwise of polypropylene is the regularity or the irregularity of arrangement of the methyl side-chains, and every propylene monomer molecule carries one such side-chain. But ethylene has no methyl or other side chains at all. Why, then, should polyethylene prepared with a Ziegler catalyst have higher density, higher melting point, and better strength than " old " polyethylene prepared by the application of very high temperatures and pressures? The answer is that although there are no side-chains in the monomer, there are some in the polymer made by the old method, and there are in fact some but only very few indeed, in the polymer made by Ziegler catalysts. The physical differences (density, melting point and strength) between " old " polyethylene and Ziegler polyethylene are entirely due to the presence of a considerable proportion of side-chains in the " old " polymer. They were unexpected and they are unwanted. High density polyethylene differs from the low density material in the absence of side-chains; this absence enables the polymer molecules to pack together more closely into a more stable arrangement. Molecular closeness gives high density, molecular stability gives high melting point.

Chain branching in polyethylene. When infra-red absorption spectra of " old " polyethylene films are compared with those of branched chain paraffins there are points of resemblance. Analysis of these shows that polyethylene which has been made by high temperature and high pressure polymerisation contains about one methyl group per fifty methylene groups; most of the methyl groups appear to be present in ethyl side-chains. How do they come about? It is postulated, to explain their occurrence, that during the polymerisation of ethylene a free radical can transfer its activity not only to a monomer and so initiate the chain-building propagation process, but also alternatively to a long molecule that has already grown and stopped growing; furthermore, although it does this but rarely, when it does do it, it can do it not only at one of the chain ends, but even at any position along the molecule. The formation of a branch results when the so-activated part of the chain reacts with a monomer. This may be represented:

$$-CH_2CH_2CH_2CH_2CH_2CH_2- + R\cdot \longrightarrow$$
$$-CH_2CH_2CH_2\dot{C}HCH_2CH_2- + H + R$$

$$—CH_2CH_2CH_2\dot{C}HCH_2CH_2— + CH_2{=}CH_2 + H \longrightarrow —CH_2CH_2CH_2\underset{\displaystyle CH_2CH_3}{\underset{|}{C}H}CH_2CH_2—$$

Ziegler type polyethylene (linear polyethylene) is very much freer from these side-chains than is " old " polyethylene. The mechanism just described for their formation is not yet proved beyond doubt, but there is no doubt that the side-chains are there. Additionally, slight traces of $>C{=}O$ groups have been detected and these may be concerned in the formation of side-chains, for a trace of oxygen is the catalyst used for old high pressure polyethylene.

Polymethylene: A Very Linear Polymer

It might be thought that one of the best ways of making a really linear polyethylene would be to make polymethylene by removal of the nitrogen from diazomethane:

$$nCH_2{=}N{\equiv}N \longrightarrow —(CH_2)_n— + nN_2$$

Diazomethane can in fact be polymerised to polymethylene by copper or silver catalysts. Or, if boron trifluoride etherate is added to an ethereal solution of diazomethane the reaction flashes to give polymethylene of very high molecular weight (S. W. Kantor *et al.*, *J. Amer. Chem. Soc.*, **75**, 931 (1953)). Polymethylene turns out to be a linear polymer with fewer than 1·5 methyl groups per 1,000 carbon atoms; it has, too, fewer than 0·1 double bonds per 1,000 carbon atoms and it melts at 132° C. (for comparison Ziegler polyethylene melts at 124° C. and Marlex 50 at 134° C.); its specific gravity is 0·97 compared with 0·95–0·96 for Ziegler polyethylene and Marlex 50.

Assessment of Irregularities in Polyolefines

A comparison of some properties of related polymers is shown in the following table:

Polymer.	Per 1,000 carbon atoms.		Specific gravity.	Melting point (° C.).
	Methyl groups.	Double bonds.		
Polymethylene	<1·5	<0·1	0·97	132
" Old " polyethylene	23	0·7	0·92	107
Ziegler polyethylene	4	0·5	0·95	124
Marlex 50	<1·5	1·5	0·96	134
Isotactic polypropylene	330	0·4	0·92	176
Atactic polypropylene	Not known		0·85	Semi-liquid

The variations in specific gravity and melting point have already been dealt with sufficiently to enable the figures in the last two columns to be understood. So far as concerns the number of methyl groups, which can only be present as terminal groups or as side-chains, polypropylene necessarily has a lot because each monomer molecule has one methyl group and three carbon atoms; so we should expect about 333 methyl groups per 1,000 carbon atoms. It is noteworthy that isotactic polypropylene does not contain a greater number than this. Old polyethylene contains quite a lot of methyl groups; three-quarters of these are found to exist as methyl and ethyl side-chains, and the other quarter as methyl end groups on the polymer. Ziegler polyethylene has many fewer methyl groups and three-quarters of what it has are terminal groups, with less than 1 ethyl side-chain per 1,000 carbon atoms. Polymethylene and Marlex 50 are almost entirely free from methyl groups, these polymers are really straight-chain.

Just as the quantity and location (terminal or side-chain) of the methyl groups in a polyolefine can give information about the structure of the polymer, so the quantity and location of points of unsaturation (double bonds) can yield information about the mechanism of the polymerisation, and particularly about its termination step. Marlex 50 has a relatively high proportion of double bonds with 1·5 per 1,000 carbon atoms and 95 per cent of these are terminal vinyl groups $CH_2{=}CH\cdots$. " Old " polyethylene has about half as many and only 10 per cent of them are terminal vinyl groups, the other 90 per cent are vinylidene

$$\begin{array}{c} CH_2 \\ \| \\ -C- \end{array}$$

side-chains; so old polyethylene carries not only ethyl and methyl side-chains, but also quite a few vinylidene groups. Ziegler polyethylene contains rather fewer double bonds and they are mainly terminal vinyl groups and vinylidene side-chains.

These considerations show that the purest, most nearly perfect, polyethylene is that made from diazomethane and because of its method of preparation called "polymethylene". Next comes Marlex 50 which is substantially free from side-chains. Next, not very far behind, comes Ziegler polyethylene with an odd ethyl side-chain but a considerable number of terminal methyl groups. Old polyethylene is a long way from being a pure linear polymer; it contains far too many side-chains and these give to it bulkiness (low specific gravity), low melting point and only moderately good mechanical properties.

FURTHER READING

P. J. Flory, " Principles of Polymer Chemistry ". Cornell University Press, Ithaca (1953).

N. G. Gaylord and H. F. Mark, " Linear and Stereoregular Addition Polymers ". Interscience, New York (1959).

C. E. H. Bawn, " The Chemistry of High Polymers ". Butterworth, London (1948).

CHAPTER 7

CHEMICAL CONSTITUTION AND FIBRE PROPERTIES

ALL fibres are long and narrow, all have a modicum of strength, all are constituted of fibrous molecules, which again are long and narrow.

Characteristics which are common to all fibres, in particular the possession of a structure composed of fibrous molecules, have rightly attracted great attention. The discovery of the *fibrous nature of the molecules of fibres* was perhaps the most important that has been made in connection with textiles since some primitive genius hit on the idea of shedding a warp. As the discovery of fibrous molecules has come in our own lifetime, it has been widely hailed and every emphasis has been laid on it. The importance of the discovery cannot be questioned—it opened up the way for the synthesis of fibres; if the physicists had not demonstrated the long linear nature of fibre molecules it is extremely unlikely that we should at the present time have had nylon, Vinyon, Orlon, Terylene or any other of the synthetic fibres.

From the theoretical standpoint, too, the gain that has accrued from the discovery has been immense, for without an understanding of the molecular structure, how could we possibly have rationalised our knowledge of fibre behaviour?

Although the writer would be the first to subscribe to the overriding value of the discovery of the basis of fibre structure, it must be pointed out that the emphasis which is laid on the unity of fibres and on their common structural features has led of late years to a neglect of appreciation of the diversity that exists amongst fibres.

Variation. Fibres vary very greatly from one kind to another, and in some cases the variation is dependent on chemical structure. But by no means always, for, as we have already seen, there are very considerable differences between Fortisan, Durafil, Tenasco, viscose rayon and viscose monofil, all of which consist of regenerated cellulose, and there are considerable differences amongst the natural fibres between flax and cotton, both of which in their final form are almost pure cellulose. Still greater differences are evident between the native cellulose fibres such as cotton on the one hand and the regenerated cellulose fibres such as cuprammonium and Durafil on the other.

Then again, turning to the proteins, although the differences between wool, mohair and alpaca are not very marked, rabbit wool is quite distinctly different, and natural silk is very different.

Even amongst the natural fibres there are great and outstanding differences between the cellulosic and the protein fibres; cotton and flax are very different from wool and silk. Amongst the artificial fibres the differences are even more manifest. Consider on the one hand a mineral-glass fibre, on the other a seaweed alginate; consider the difference, even in one type of fibre, between a viscose rayon and a cellulose acetate. Consider the applications to which fibres are put: frocks, stockings, overcoats, blankets, sheets, shoe-laces, carpets, toothbrushes, parachutes, mosquito nets, ropes, filter-cloths, string, twine, felts, velvets, tarpaulins, awnings, curtains, upholstery. It is not easy to think of times when we make no use of textile fibres, except perhaps when we are in the bath, and even from the bath we step directly on to a bath-mat, dry ourselves with towels and are once again in the world of fibres. Some unusual applications of fibres are shown in Figs. 45 and 124.

In modern life, fibres are everywhere, and fibres of every kind abound. The diversity of fibres depends in some measure on chemical constitution. How, we shall now examine.

Small Ring Structures have Little Effect. Students of chemistry will remember that the study of organic chemistry is commonly divided into that of straight-chain compounds on the one hand and that of ring compounds on the other. The differences between the two groups are very considerable, and are well known. Surprisingly, when a small ring system occurs repeatedly in fibre molecules it seems to make no very great difference to the properties of the fibres. The glucose ring, which is the molecular " brick " from which all cellulosic fibres are built, hardly reveals its presence; its only function seems to be that of lending greater chemical stability than would obtain if the carbon and oxygen atoms were lined up in a straight chain. Terylene contains a benzene nucleus in its repeat, thus:

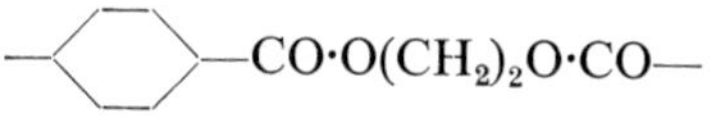

$$—C_6H_4—CO{\cdot}O(CH_2)_2O{\cdot}CO—$$

yet its properties are very similar in most respects to those of nylon, which is the most perfect example of a straight-chain compound that it would be possible to adduce. Fibre-forming polymers which appear to incorporate the ring system

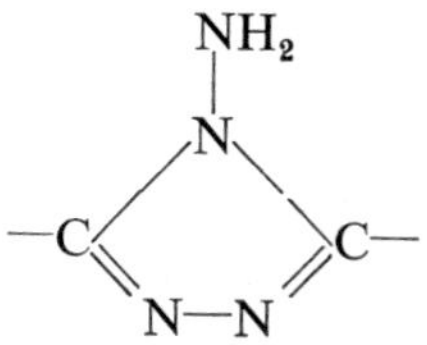

$$\begin{array}{c} NH_2 \\ | \\ N \\ —C \quad\quad C— \\ N—N \end{array}$$

[The Firestone Tire & Rubber Co.]

FIG. 45.—Forest green Velon cloth providing shade for acres of tobacco growing at Windsor, Connecticut. Tobacco is best grown in the shade and usually fabric has to be renewed each year. Velon should last several years.

in their chain have been described by the author in British Patent 612,609 and U.S. Patent 2,512,627. It does therefore seem that the presence of a small ring system in the molecular chain has no detrimental effect on fibre properties.

The Hydroxyl Group

The cellulosic fibres are characterised by the possession of an immense number of hydroxyl groups. Apart from the alginates, which chemically are related not very distantly to cellulose, no other fibres share this constitutional feature.

The glucose residue is the basis of the cellulose molecule, and each glucose residue contains three free hydroxyl groups, so that if there are about 300 glucose residues in a molecule of viscose rayon, each of the latter contains 900 hydroxyl groups.

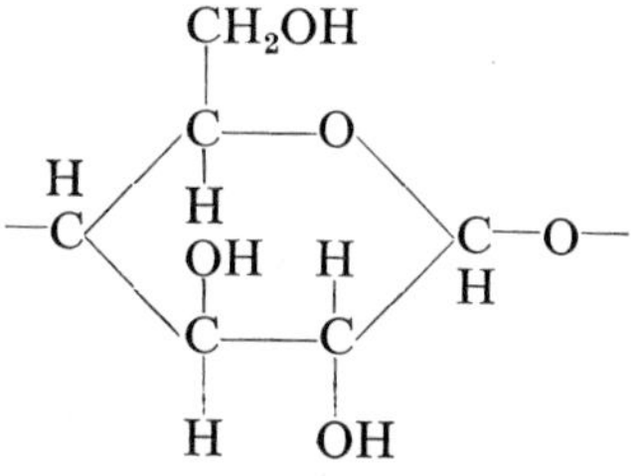

Glucose residue (cellulose building brick). Note possession of 3 hydroxyl groups.

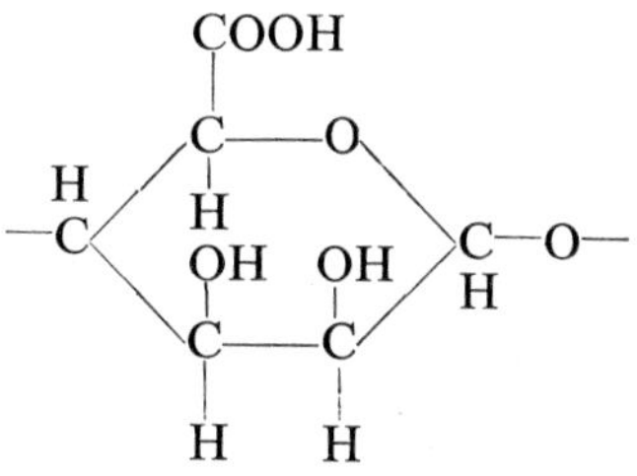

Mannuronic acid residue (alginate building brick). Note possession of 2 hydroxyl groups and 1 carboxyl group.

Solubility in Water. The outstanding feature of substances rich in hydroxyl groups is usually solubility in water. Alcohols, glycols and sugars dissolve readily in water; even the benzene nucleus, under the kindly influence of a hydroxyl group (phenol), becomes soluble in water. Cellulose fibres are, however, fortunately insoluble in water, and this must be ascribed to two causes.

1. The very high degree of polymerisation. High polymers are always increasingly insoluble.

2. The mutual attraction between the hydroxyl groups on one cellulose molecule and those on other cellulose molecules aligned alongside it. These attractive forces between the hydroxyl groups constitute an example of hydrogen bonding (p. 96), and because of the very high frequency of occurrence of hydroxyl groups in cellulose, the sum of all the numerous attractive forces is considerable. These forces confer stability on the structure and make it difficult for the

water molecules to find their way into it and disintegrate it by means of solution. The main reason for the insolubility of cellulose in water is the size of the molecule; it is unusual for substances made up of such giant molecules to be soluble.

The Attractive Forces and the Stability of the Fibre. The influence of the attractive forces between the hydroxyl groups must, however, be quite considerable, because when the hydroxyl groups are etherified (as in methyl and ethyl celluloses) or esterified (as in cellulose acetate) products result which are soluble. The relatively large ethyl or acetyl groups hinder the very close approach of adjacent molecules, which is a necessary preliminary to hydrogen bond formation. When an ethyl cellulose is prepared in which only a few of the hydroxyl groups are etherified in this way

$$-\overset{|}{\underset{|}{C}}-OH \longrightarrow -\overset{|}{\underset{|}{C}}-OC_2H_5$$

the product is soluble in alkali; then, as more of the ethyl groups are introduced, the product becomes soluble in water; finally, when a considerable proportion of the hydroxyl groups are etherified the product is soluble in organic solvents, although no longer soluble in water. Cellulose esters are well known to be soluble in organic solvents, and that the structure of the fibre is greatly weakened by acetylation is shown by the ease with which cellulose esters can be softened and stretched in hot water and steam. Viscose rayon cannot be stretched to a large extent in hot water or steam, but cellulose acetate can be so stretched. This behaviour, taken together with the known fact that viscose rayon contains many more hydroxyl groups than does cellulose acetate, so that it should be more susceptible to the action of water, is a clear indication that the attractive forces between the hydroxyl groups in regenerated cellulose, contribute quite considerably to the stability of the fibre.

It is probably for this reason that the natural cellulosic fibres are so difficult to dissolve, when compared with the fairly ready solubility of the majority of the synthetic fibres in organic solvents. The presence, too, of the strongly polar hydroxyl groups would considerably lessen the chance of cellulosic fibres being soluble in non-polar organic solvents, such as carbon tetrachloride, chloroform, methylene dichloride, benzine and benzene, and it is, of course, only when a considerable proportion of the hydroxyl groups in cellulose have been converted into other groups—*e.g.*, by esterification or etherification—that solubility in such non-polar solvents commences to manifest itself.

Just as one hydroxyl group will attract another on a second molecule of cellulose to itself, so will it also attract to itself the hydroxyl groups of water. For this reason, the cellulosic fibres easily absorb water, and have high regains. They are, too, readily dyed from aqueous solutions of dyestuffs, because the aqueous dye-liquors can easily penetrate the fibres. That this is due to the presence of the hydroxyl groups in the fibres cannot be doubted, for cellulose acetate, which is similar to cellulose except for the fact that the great majority of the hydroxyl groups have been esterified, has a much lower water content than viscose rayon, and is much more difficult to dye. Because of its poverty of hydrophilic (water-liking) groups, special methods had to be devised to dye cellulose acetate.

The moisture regains of viscose rayon and of cellulose acetate rayon under standard atmospheric conditions are about 12 and 6 per cent respectively.

Accessibility

It is usually maintained that moisture can penetrate into the cellulose fibre in the amorphous regions. It is believed that it can never penetrate into the crystalline regions of native cellulose (Cellulose I), the reason for this belief being that the dimensions of the repeat of Cellulose I as measured by X-ray analysis do not change when the cellulose is wetted. But those of Cellulose II (regenerated cellulose) do change, so that evidently water can penetrate into the crystalline regions of this modification.

If a bone-dry Cellulose II fibre is wetted, the X-ray diagram shows that the lattice is widened from 7·32 to 7·73 Å.; the following figures have been given by P. H. Hermans:

Relative humidity of air.	Cellulose II water content.	Distance apart of 101 planes (Å.).
0	0	7·32
35	6·6	7·52
85	17·0	7·73

It is because water molecules can penetrate even into the most orderly regions of Cellulose II and force the chain molecules apart that viscose rayon and other regenerated forms of cellulose are weaker wet than dry.

Attempts have been made to calculate the amount of amorphous material in a fibre on the basis of the water absorbed. Figures that have been arrived at on this basis are:

Fibre.	Percentage of amorphous material.	Percentage of crystalline material.
Cotton	32	68
Wool	56	44
Silk	20	80
Nylon	15	85

It is doubtful if much significance can be attached to such figures, except perhaps that the percentage of crystalline material may be taken as a measure of orientation and regularity of arrangement.

These regions which are amorphous are considered to be accessible to water molecules, whereas crystalline regions are not. It has, however, been found necessary to modify this view by pointing out that the hydroxyl groups on the edges of the crystalline regions will be accessible, and if, therefore, the crystals are very small and very numerous, the proportion of fibre accessible to water molecules will be greater, because the ratio of area to volume is greater for small than for large particles. Thus, two chemically similar fibres, both 50 per cent crystalline, might be capable of picking up different amounts of water, simply because the crystals in one were smaller than in the other; the former would pick up more water and would have greater accessibility.

As already explained, the conception of sharply delineated crystalline areas may be quite erroneous. In that case the values obtained for percentages of crystallinity may be usefully adopted as a measure of the regularity of arrangement.

The hydroxyl group is the hall-mark of the cellulose fibres; it determines their moisture absorbency, their ease of dyeing and of washing.

The Carboxyl Group in Alginates

Alginates are the most similar of other fibres to cellulose—the molecular configuration bears some points of similarity, although the dimensions, including the linear repeat, are different. The properties of alginates appear to differ from those of cellulose mainly because of the carboxyl groups possessed by the alginates. These acidic groups readily combine with even weakly alkaline materials, and hitherto the only really satisfactory way of preventing alginate fibres from forming water-soluble sodium salts has been to convert the carboxylic acid groups into metallic carboxylates. Such metallic salts have the virtue of flameproofness, but they are not sufficiently stable to weak alkali (which converts them to the soluble alkali metal, *e.g.*, sodium, salt) to withstand repeated washing. The carboxylic groups give to alginates both their virtue and their weakness, and afford a good illustration of the important influence that chemical groups, quite apart from considerations of arrangement and orientation, may have on the properties of fibres.

THE PROTEIN FIBRES

The protein fibres, wool, the goat-hairs, and the regenerated proteins, are all characterised by the presence of free amino- and free carboxylic acid groups, which combine internally to form salt linkages. Such linkages are electrovalent, not covalent, and are very easily broken. If, as we have seen, the presence of free carboxyl groups in alginate fibres causes such instability and ready solubility of the fibre, why does it not do so in the case of wool? The answer to this question probably lies in (1) the much greater frequency with which carboxylic groups occur in alginic acid than in wool and in (2) the possession by wool of the cystine cross-linkages which join the long molecular chains together. Harris has shown that wool in which the disulphide (cystine) linkages have been ruptured by reduction is much more liable to chemical damage and to biological attack, and that if the cross-linkages are rebuilt, the former good properties of the wool return. Further than this, if the cystine linkages are reduced and are rebuilt as bisthioether linkages, the chemical stability and resistance to biological attack are improved—in fact, the modified wool is mothproof. All the regenerated proteins lack these cross-linkages, and all are sensitive to dilute alkali. This sensitivity is a sure indication that the source of attack is at the carboxylic acid groups. The difficulty is, in a sense, fundamental, because although cross-linkages probably occur in the natural protein material from which the fibres are originally made, it is quite impossible to dissolve these natural proteins until their cross-linkages are severed. Wool, for example, can be dissolved only in agents such as caustic alkali or sodium sulphide, which break the cross-linkages.

Remedies for Sensitivity to Alkali. When, therefore, the protein has been dissolved and reprecipitated as a regenerated protein fibre, this protein fibre is extremely sensitive to alkali; it dissolves or is badly weakened even on contact with warm soapy water. Steps have to be taken to remedy this state of affairs, and the most common and successful step is to introduce new cross-linkages, usually by treatment with formaldehyde, which forms cross-links in this way:

$$\overset{|}{\underset{|}{\mathrm{NH}}} + \underset{\underset{\mathrm{CH_2}}{\|}}{\mathrm{O}} + \overset{|}{\underset{|}{\mathrm{HN}}} \longrightarrow \overset{|}{\underset{|}{\mathrm{N}}}-\mathrm{CH_2}-\overset{|}{\underset{|}{\mathrm{N}}} + \mathrm{H_2O}$$

Probably all regenerated proteins are treated with formaldehyde to cross-link and harden them.

One other way in which the regenerated proteins can be made more resistant to the effects of water and soap solutions is to acetylate them. This makes them more organic and less hydrophilic; it has been reported that the substitution of 2 per cent acetyl group in zein effects a considerable improvement in its properties.

Polyamides and Polyesters

There are very few (only those at chain ends) free hydroxyl, amino or carboxylic groups in nylon or Terylene or Perlon; consequently these fibres show the following differences from the cellulosic rayons and the regenerated proteins.

1. They have lower moisture contents, because they contain so few strongly hydrophilic groupings. The following figures illustrate this difference.

Fibre.	Regain at standard conditions.
Terylene	0·4 per cent
Nylon	4 ,,
Cellulose acetate	6 ,,
Viscose rayon	11 ,,
Silk	11 ,,
Wool (loose fibre)	16 ,,

It is clear that those fibres that have most of the chemically reactive groups in their molecules have the highest water absorption.

2. The polyamides and polyesters are more difficult to dye. Aqueous dye-liquors do not swell them greatly nor are there many reactive groups to attract the dye molecules. As a result nylon and similar fibres are usually dyed by applying dispersed cellulose acetate-type dyestuffs, which, although they do not combine with the fibre, dissolve in it and colour it in that way.

3. Nylon, Terylene and Perlon will all melt. Cellulose, whether native or regenerated, will not melt. It is believed that the reluctance, or indeed inability, of cellulose to melt is, like its insolubility, due to the great size of the molecule and to the very strong attractive forces that exist between the hydroxyl groups on adjacent molecules. Protein fibres, too, will not melt, those that are cross-linked being prevented from doing so by this feature, and the others decomposing on heating, probably because the free amino and carboxyl groups are unstable at high temperatures—if they have nothing else with which to react, they will react with each other. Nylon, on the other hand, has no free reactive groups such as exist in the proteins, and no such intensely powerful cross-attractions as are furnished by the

hydroxyl groups in the cellulose fibres. When therefore it is heated, the molecules vibrate more and more rapidly, tear themselves loose from their anchorages, and the nylon melts. Nylon melts under nitrogen at 263° C., Perlon U (polyurethane) at 175–180° C., and Terylene (polyester) at 249° C.; in air both nylon and Terylene melt at about 248–250° C.

The Polymer Stick Temperature

The polymer stick temperature is defined as being that at which a portion of the polymer, *e.g.*, a thread, when drawn over a heated copper pin leaves a trail, *i.e.*, the lowest temperature of the copper pin which causes a noticeable trail to be left. In the example on p. 388 the stick temperature of the piperazine phthalamide discussed is 350° C. It is easy with a little practice to measure accurately the melting point of most pure chemicals; they usually melt sharply, and the exact temperature at which they do so is taken as evidence of identification. But polymers are much more difficult; they (the fibrous kind) all decompose if heated in air, and even if heated under nitrogen they often decompose and discolour due to the high temperatures required. The melting point given for even such a familiar polymer as nylon 66 varies considerably from one observer to another, according to the method he has used to determine it. Various values for the melting point of nylon 66 will, in fact, be found in the literature, in manufacturers' leaflets and in ironing instructions.

Melting Point of Nylon. It is noteworthy that if the chain length between the polar groups of nylon is increased, the melting point falls. Nylon 610, made from hexamethylene diamine and sebacic acid, melts at 214° C., and this is what might be expected. The –CONH– groups in nylon are polar, although not nearly so strongly polar as the hydroxyl groups in cellulose, or as the free $-NH_2$ and –COOH groups in the proteins. Nevertheless, some polarity exists, and some attractive forces obtain between the amide linkages on adjacent molecules. This intermolecular bonding is not nearly so strong in nylon as in cellulose and the proteins, but it does exist, and has to be reckoned with. Clearly, the shorter the chains between the amide-groupings, the more frequently will these cross-attractions occur, and the more difficult it will be to separate the linear molecules. Melting consists of separating the molecules, and it can naturally be accomplished more easily if there are fewer intermolecular attractive forces. There are fewer of these in nylon 610 than there are in standard nylon 66, so that the former has a lower melting point. One other feature that calls for attention here is that the nylons made from intermediates containing an even number of chain atoms, such as hexamethylene diamine and adipic acid, melt at higher temperatures than do those made from intermediates, one of which has an odd number of chain atoms. In the latter event hydrogen bonding can occur only at every other repeat of the amide-linkage along the molecule, the intermolecular forces are reduced and the melting point is lowered. The melting point of nylon 56 made from pentamethylene diamine and adipic acid is 223° C. The following table illustrates how melting point is reduced by (*a*) extension of chain unit between amide linkages; (*b*) the occurrence of a chain with an odd number of carbon atoms.

Nylon.	Intermediates.	m.p. of polymer (° C.).
46	Tetramethylene diamine and adipic acid	276
56	Pentamethylene diamine ,, ,,	223
66	Hexamethylene diamine ,, ,,	263
76	Heptamethylene diamine ,, ,,	235
86	Octamethylene diamine ,, ,,	235
96	Nonamethylene diamine ,, ,,	198
106	Decamethylene diamine ,, ,,	230

The effect of introducing an aromatic group into nylon is discussed on p. 137, and in Chapter 21; it can greatly increase the melting point.

Melting Point of Terylene. Terylene melts under nitrogen at about 249° C. compared with nylon's 263° C., although melting in air with decomposition there is not much to choose between them; the advantage that nylon has is more concerned with conditions in a non-oxidising atmosphere or *in vacuo* where decomposition does not take place simultaneously with melting. The polyesters as a group have lower melting points than the polyamides (see p. 137) and the relatively high melting point of Terylene is due to its possession of an aromatic nucleus in its linear structure. Polyesters made with aliphatic constituents analogous to those in nylon have very low melting points and moreover are easily hydrolysed; if, however, the chain-length of the repeat is reduced, the melting point rises. Thus hydroxypivalic acid, $HOCH_2C(CH_3)_2COOH$, will condense to give a fibre-forming polymer melting at 130° C.; the formula of this polyester is:

$$H-\left[-OCH_2\overset{\overset{\displaystyle CH_3}{|}}{\underset{\underset{\displaystyle CH_3}{|}}{C}}\cdot CO-\right]_n OH$$

where n has a value of several hundreds. Similarly glycollic acid,

$$\begin{array}{l} CH_2OH \\ | \\ COOH \end{array}$$

will condense internally to give a fibre-forming polyester, $H[-OCH_2CO-]_nOH$, melting at 220° C.; this is the simplest possible polyester and will probably have economic advantages that may promote its development. The same polyester can be made from the internal anhydride of glycollic acid which is known as glycollide, formed thus:

$$\begin{array}{ccc} CH_2O\fbox{H \quad HO}\,CO & & CH_2O{\cdot}CO \\ | \qquad\qquad\quad | & \longrightarrow & | \qquad | \qquad + 2H_2O \\ CO\fbox{OH \quad H}O\;CH_2 & & CO{-}O{\cdot}CH_2 \end{array}$$

Glycollic acid. Glycollide.

Whereas glycollic acid is water-soluble with m.p. 79° C. and is a stronger acid than acetic, the fibre-forming polymer is insoluble in water, melts at 220° C. and is neutral.

4. The chemical stability of the polymer is increased. Clearly the reactivity of any fibre will depend on its possession of reactive groups, typical of which are hydroxyl, amino and carboxylic acid groups. Wool and cellulose fibres are reactive; they dye easily and are easily attacked by chemical agents. Nylon, on the contrary, has no intensely reactive groups, for the reactivity of the $-CONH-$ amide groupings is of a very inferior nature; it is difficult to dye, and picks up moisture rather poorly, but correspondingly is extremely stable chemically, and, as will be seen later in the text, can be used for many purposes where chemical stability and resistance to attack are a *sine qua non*. The polyester and polyacrylic fibres also exhibit outstanding chemical resistance.

It will be clear that considerable differences in the properties of fibres can sometimes be brought about by only slight changes in the chemical composition of the intermediates from which a synthetic fibre is made. When the chemical differences are great—as, for instance, between cellulose and nylon—then the differences in the properties of the fibres are correspondingly great.

Dynel, Saran and Orlon

The poverty of polar groups, noticeable in the case of nylon, Terylene and Perlon, and already markedly influencing the properties of these fibres, becomes considerably more acute in the vinyl and vinylidene fibres, such as Dynel, Saran (Velon), Pe Ce and Orlon.

It is perhaps rather strange at first sight that the very reactive vinyl compounds should, on polymerisation, give such very unreactive compounds, but when it is remembered that the reactivity of the vinyl compounds is due to their unsaturation, which is entirely lost in polymerisation, this can be understood.

The polymeric vinyl and vinylidene fibres are very unreactive chemically. They have very low moisture contents—often below 1 per cent at standard conditions; they are not easily dyed from aqueous liquors and the most satisfactory way of colouring them, a way usually adopted, is to incorporate a pigment in the spinning solution. They are, however, very stable chemically, and have found special uses in the manufacture of protective clothing for chemical workers, and as filter cloths. They are equally resistant to biological attack, and have been used successfully as insect-netting in tropical countries; as, too, their strength is practically unaffected by immersion in water, they have, as expected, found considerable application for marine cordage and fishing-nets. They are unlikely to be universally used for apparel, because of their low hygroscopicity. This defect might eventually be overcome, but so long as there are so many fibres that are already suitable for use in apparel, it seems pointless to try to modify unsuitable fibres. One imagines that the best way to make them similar to orthodox apparel materials would be to incorporate strongly polar groups in their constitution—groups such as hydroxyl or amino. This has already been done (Acrilan, Pan, Zefran) by co-polymerisation, which will shortly be discussed.

The very good non-flam properties of vinylidene chloride fibres, due to their high chlorine content, should be noted, likewise those of polyvinyl chloride (Rhovyl).

POLYOLEFINES

Polyethylene—the polymer from ethylene—is a hydrocarbon, and typifies the least polar and least reactive type of fibre that has so far appeared. Its composition results in the very low specific gravity of 0·92—the lowest for any fibre. It finds most use as an electrical insulator, and is unlikely to be used on a large scale for traditional textile purposes. Despite the fact that it is oriented and consists of long, fibrous molecules, it is unsuitable, because of its chemical constitution, for many fibre uses. The complete absence of polar groups gives no opportunity for attractive forces to operate between the molecules; as a result, it has a very low melting point of 110–120° C. It provides a useful illustration of the importance of chemical constitution in fibre structure. Linear polyethylene, with

Marlex 50 as a superb example, made with catalysts which include a solid phase and build the molecules' stereoregularity, has a rather higher melting point. Polypropylene, with an isotactic configuration, is even better. But it still remains that polyolefines, even if stereoregular, feel the lack of polar groups and are unkindly, water repellent, and difficult to dye.

The polytetrafluoroethylene fibre, Teflon, is characterised by an unreactivity even greater than that of polyethylene, and, moreover, it has a much higher melting point, which is attributed to very close packing of its molecules and resultant inter-chain bonding. Cellulose, proteins, nylon, vinyl fibres, polypropylene; truly these illustrate the diversity of fibres.

Co-polymers

The synthesis of fibres has been an achievement that has brought with it the possibility of making fibres with specified properties. Naturally, it is not always possible to meet a specification; it is one thing to draw up a specification of perfection, and usually quite another thing to match it. Nevertheless, it is possible to some extent to modify the physical properties of synthetic fibres by suitably adjusting their chemical composition. It may be found that one substance, such as vinyl chloride, when polymerised gives fibres unsatisfactory in respect of suppleness, and another substance, vinyl acetate, when polymerised has a very low softening temperature, and is mechanically weak. Yet when the two substances are polymerised together, or *co-polymerised*, the properties of the fibre are satisfactory in all these respects. In the case of vinyl chloride (88 per cent) and vinyl acetate (12 per cent), the co-polymer Vinyon results. Other improved products—Verel and Dynel—are co-polymers of vinyl compounds and acrylonitrile. So are most of the acrylic fibres.

In the case of the polyamides and other compounds similar to them, it is possible to modify their properties by co-polymerising different materials. When this is done the usual perfection of molecular arrangement cannot be obtained, the strength of the co-polymers is usually lower, and the melting point lower, because the polar groups on different chains are farther apart and the attractive forces between them are weaker. Co-polymers have lower melting points and increased solubilities, and there is more tendency to shrink on heating; on the other hand, the softness and flexibility are improved, as would be expected by a reduction in the degree of crystallinity, and often too the dyeability.

When, for example, co-polymers are made from ω-aminocaproic

acid (basis of nylon 6), and from hexamethylene diammonium adipate (salt of nylon 66), the melting points of the co-polymers vary according to their composition in this way:

Nylon 6 base, per cent.	Nylon 66 base, per cent.	m.p. (° C.).
100	0	203
80	20	166
60	40	158
40	60	168
20	80	214
0	100	263

Although the co-polymerisation of another component with the nylon 66 components ordinarily brings about a reduction in melting point, there is an interesting exception to this generalisation. If the added component is of the same molecular length as the component it is partially replacing, then it can effect the substitution without disturbing the molecular packing and without diminishing the forces between one molecule and another, and accordingly it does not reduce the melting point. A specific example is of the partial substitution of the adipic acid (length of molecular unit 5·5 Å) by *tere*phthalic acid (length of molecular unit 5·8 Å); the lengths of the two molecules are sufficiently close for there to be little or no interruption in the arrangement of the molecular chains. In fact, the substitution of *tere*phthalic for a portion of the adipic acid in nylon does not reduce the melting point. It is safe to say, though, that the substitution of *tere*phthalic acid for a small portion of the sebacic acid in 610 nylon would reduce its melting point, because the *tere*phthalic acid molecule would be much shorter than the sebacic acid molecule.

It is expected that by suitable variation in the constitution of co-polymers it will be possible to modify the properties of nylon in a large measure as may be required; it should, for example, be possible to increase the hygroscopicity—which from many points of view would be an improvement—by incorporating polar groups or by reducing the crystallinity.

It has already been found possible to modify the elastic properties so that, by suitably selecting the intermediates, elastic nylon—nylon with a retraction after being stretched, similar to that of rubber—has been made. One form of elastic nylon, described in Chapter 24, is a co-polymer of hexamethylene diammonium sebacate (salt of nylon 610) and of N-isobutyl hexamethylene diammonium sebacate. These elastic nylons were of academic interest but commercially they have been superseded by the spandex fibres (p. 442).

Influence of Aromatic Groups

The influence of an aromatic group in a linear polymer is to make the molecule more rigid, to increase the cohesive forces, and thereby to raise the melting point. Now, three of the commercial fibres, the polyesters and the snap-backs and Nomex (Chapter 21), contain an aromatic nucleus in the chain as a major constituent. The first polyesters were fibre-forming, but the fibres they gave had low melting points and were too easily hydrolysed to withstand normal washing and wearing conditions. When a polyester was made from the aromatic *tere*phthalic acid and ethylene glycol, a better fibre was obtained, one with a more rigid and cohesive molecular structure and with a higher melting point and good chemical stability. Of the three isomeric phthalic acids:

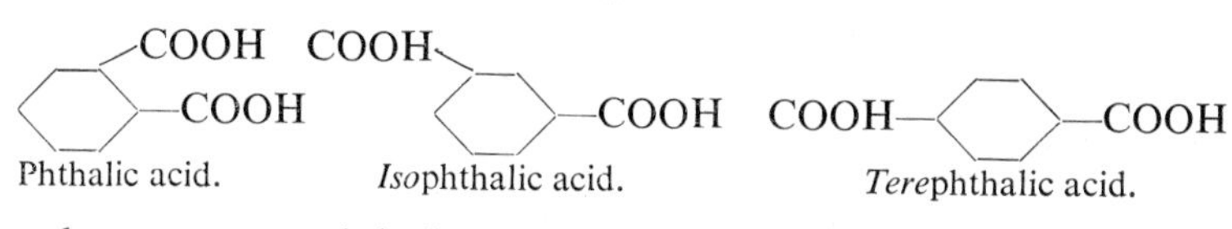

Phthalic acid. *Iso*phthalic acid. *Tere*phthalic acid.

only *para* or *tere*phthalic acid is suitable for the formation of linear polymers, the *ortho* (ordinary phthalic acid) and the *meta* (*iso*phthalic acid) acids are prone to ring formation and if the attempt is made to polymerise them with a glycol, cyclic compounds will be formed as well as linear polyesters. In the same way linear polyesters can be formed from *pp*′-diphenylene dicarboxylic acid, $COOHCH_2$—⟨ ⟩—⟨ ⟩—CH_2COOH, but not from its isomers; so they can be formed from 2 : 6-naphthalene dicarboxylic acid, COOH—(naphthalene)—COOH in which the carboxyl groups are symmetrically disposed.

In Terylene it is the aromatic nucleus which makes the molecule more rigid and thereby increases its melting point. If, for example, a polyamide is made from *tere*phthalic acid and hexamethylene diamine, that is by substituting the aromatic *tere*phthalic acid for the aliphatic adipic acid in nylon, the resulting polymer, now known as "nylon 6-T", has a melting point of 400° C. compared with nylon's 263° C. This would be very useful if the polymer did not, as it unfortunately does, decompose before melting so excluding the possibility of melt-spinning. There is, however, no obvious reason why such a polymer should not be spun from solution as it would almost certainly be soluble in phenol or xylenol. If *tere*phthalic acid is one of the components for a polyamide then in order to bring down the melting point of the polyamide to 275° C. there must be used a long chain diamine—octadecamethylene diamine, $NH_2(CH_2)_{18}NH_2$, instead of hexamethylene diamine, $NH_2(CH_2)_6NH_2$.

High thermal resistance has always been a desired attribute of polyamide fibres. Car and truck tyres run very hot, and as they are often built on a nylon carcase, resistance to fairly high temperatures —say up to 140° C.—for long periods without weakening or deformation has been very desirable. So it has for brake parachutes and their straps and shrouds, which are subjected to the hot gases from jet planes when they are used to retard the plane on landing. Above all, the sputniks and the space rockets have made demands for fabrics that would withstand high temperatures, and these demands have been so insistent that very large sums of money have been available to sponsor the necessary research.

It was clear from the earlier work just outlined that the presence of an aromatic instead of an aliphatic group would greatly increase the thermal resistance of linear polymers. If, for example, a polyester is made from ethylene glycol and suberic acid, this being the straight-chain acid which contains eight carbon atoms, the same number as *tere*phthalic acid, it melts at 45° C. compared with 249° C. for Terylene. The corresponding adipate was described by Carothers as melting at 50° C. The respective structural repeats of these three polymers are:

$$-CH_2CH_2O\overset{\displaystyle O}{\overset{\|}{C}}CH_2CH_2CH_2CH_2CH_2CH_2\overset{\displaystyle O}{\overset{\|}{C}}O-$$

Polyethylene suberate, m.p. 45° C.

$$\begin{array}{l} \qquad\qquad\qquad\quad\; /\!\!/CH-CH\backslash\!\!\backslash \\ -CH_2CH_2O\overset{\displaystyle O}{\overset{\|}{C}}-CH \qquad\qquad\quad CH-\overset{\displaystyle O}{\overset{\|}{C}}O- \\ \qquad\qquad\qquad\quad\; \backslash CH{=}CH/ \end{array}$$

Polyethylene *tere*phthalate, m.p. 249° C.

$$-CH_2CH_2O\overset{\displaystyle O}{\overset{\|}{C}}-CH_2CH_2CH_2CH_2\overset{\displaystyle O}{\overset{\|}{C}}O-$$

Polyethylene adipate, m.p. 50° C.

The adipate has the same number (4) of carbon atoms between the two carboxylic groups as does the *tere*phthalic acid; the suberate possesses the same total number (8) of carbon atoms as does the *tere*phthalic acid, yet their polymers (polyethylene esters) are very low melting. The influence of an aromatic nucleus in such polyesters is to increase the melting point enormously. It was the appreciation of this simple fact which led Whinfield and Dickson to the discovery of Terylene; earlier polyesters such as those which Carothers had synthesised had been made mostly from aliphatic monomers. It was the use of an aromatic monomer which made Terylene right. Ten or fifteen years later du Pont were still bitterly reproaching themselves for having missed this point and the consequent discovery of Terylene.

When it was required to produce very high melting fibres for the space programme the natural line of attack (in view of the then existing knowledge) was to make aromatic polyamides or nylons. It was, as noted just above, already known that hexamethylene diamine, when condensed with *tere*phthalic acid, would give a nylon which melted with decomposition in the region of 400° C. compared with 263° C. for nylon 66. The high melting point of the polymer from hexamethylene diamine and *tere*phthalic acid (nylon 6-T) is a result of high molecular stability, particularly to heat. An evident way of increasing the thermal stability of the fibres was to make them not just partly aromatic, but wholly so. Thus:

nylon 66 made from hexamethylene diamine and adipic acid is wholly aliphatic $[\text{—NH(CH}_2)_6\text{NHCO(CH}_2)_4\text{CO—}]_n$

nylon 6-T made from hexamethylene diamine and *tere*phthalic acid is aliphatic-aromatic

$$[\text{—NH(CH}_2)_6\text{NHCO—C}_6\text{H}_4\text{—CO—}]_n$$

para-phenylene *tere*phthalamide made from *p*-phenylene diamine and *tere*phthalic acid is wholly aromatic and does not melt below 500° C., where it decomposes severely

$$[\text{—NH—C}_6\text{H}_4\text{—NHCO—C}_6\text{H}_4\text{—CO—}]_n.$$

It is all very logical, and now that some of our best fibre chemists and engineers have carried out the experimental work it all seems very easy and it looks as though it must have seemed pretty obvious all through. Those who have carried out experimental work on making condensation polymers, as the author did in the 'thirties, know that the work is beset by pitfalls, not the least being the tendency of the polymers to degrade at high temperatures. How Carothers ever found the skill and patience to blaze the trail defies conjecture. The logical interpretation of experimental results can be a nightmare unless one can work very cleanly and think very clearly. The best advice one can receive or give is: "Keep it simple."

Block Co-polymers

When co-polymers are made they usually have lower melting points than normal polymers (p. 136). If the second component, which makes the polymer into a co-polymer, has a high molecular weight, a reasonable proportion of it can be introduced without much detriment to the melting point. The weight percentage added may be considerable but the molar percentage may be small. If, for example, a polyethylene glycol of the type

$$HOCH_2CH_2OCH_2CH_2OCH_2CH_2OCH_2CH_2OH$$

with a longer chain so that it can be written $HO(CH_2CH_2O)_{50}H$ is introduced into a polymer of *tere*phthalic acid and ethylene glycol, so that it is equivalent to 5 per cent of the ethylene glycol, the weights of substances that will be used will be:

1. For normal Terylene: 166 parts *tere*phthalic acid
   62 parts ethylene glycol

   (because the molecular weight of *tere*phthalic acid $COOH—C_6H_4—COOH$ is 166 and that of ethylene glycol $CH_2OH{\cdot}CH_2OH$ is 62).

2. For the modified Terylene: 166 parts *tere*phthalic acid
   59 parts ethylene glycol

   $$3 \times \frac{2218}{62} \text{ or 107 parts polyethylene glycol}$$

   (because 2218 is the molecular weight of the polyethylene glycol and because 3 is 5 per cent of 62 (nearly)).

Although the polyethylene glycol has a molar proportion of only 5 per cent of one-half (*i.e.*, of all the glycol molecules, the other half being the *tere*phthalic acid molecules) or 2·5 per cent molar, yet it has a weight proportion of 107/332 or 32·2 per cent. The reduction in melting point of the co-polymer is determined by the molar proportion of the second constituent (the polyethylene glycol) and as this is small, the drop in melting point is small, but there are other properties that are determined by the weight proportion. One of these is the moisture regain. The more ether groups there are in the final co-polymer the higher will be the moisture regain, just as it is high in cellulose which is built up by ether groups. Accordingly when a "Terylene" is made in which polyethylene glycol replaces a part of the ethylene glycol, the fibre has a higher moisture regain and dyes more easily from aqueous solutions. The formula of such a polymer could be:

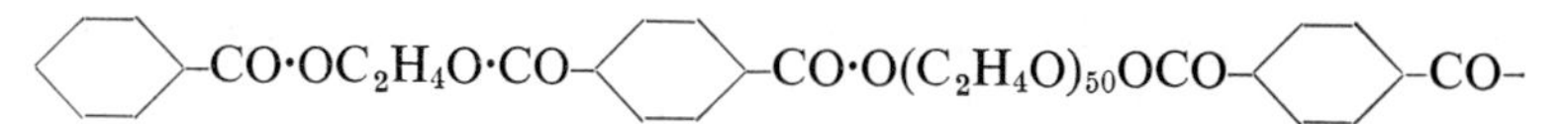

At intervals along the long polymeric molecules there would be blocks of a succession of $-CH_2CH_2O-$ groups, fifty, one after the other, and this causes the polymer to be known as a block polymer. Although the original intention of increasing the moisture regain and the dye receptivity of the Terylene was accomplished by making the block polymer, these advantages were offset by a reduced stability to ultra-violet light, leading to fading of dyeings and degradation of the polymer—an unfortunate and unforeseen new property. Block polymers are common in Nature; thus, natural silk consists of polypeptide chains of which the constituent amino acid residues are mainly alanine (A) and glycine (G) with small amounts of bulkier residues like tyrosine (T). It appears that the polypeptide chains in real silk are not made up with a random arrangement of these amino acid residues such as –A G A G G A A T G A G A G A G T G A A G A G A G A G A G A A T G A G A A G A A G G G– but that there are long parts which alanine and glycine have to themselves and that then there is a part, less orderly and less crystalline, where the bulkier amino acids congregate, then another length of the simpler acids, thus: –A A A A A G G G A A G G G G G G A A A A G A G G A G A G A G G G T T T A A A A A G G –. Silk is a block co-polymer of alanine, glycine, tyrosine and other amino acids. The block co-polymer differs from the random or normal polymer of the same ultimate composition in that the block polymer consists of longish parts where the molecules can bed together well with occasional humps where conditions are confused and the molecules probably lose orientation. In the nomenclature of today this can be expressed as having long crystalline regions with occasional amorphous places; by comparison, the random polymer would be less crystalline throughout but more or less uniform throughout with no humps of disorientation. If we try to magnify these states mentally we see that the block polymer has a coarser, more spongy and more accessible structure than the random polymer, and that it may well be rather weaker.

Graft Co-polymers

It may at first be a little difficult to distinguish the concept of a graft polymer from that of a block polymer. In the block polymer there need be only two kinds of monomer as there are in the modified Terylene, but one of these monomers is first polymerised alone and then this polymer is built into the block polymer. If, for example, *tere*phthalic acid residues —CO—⟨ ⟩—CO·O are represented by T and ethylene glycol residues $-CH_2{\cdot}CH_2O-$ by E, we have:

Terylene	–T E T E T E T E T E–	Normal polymer
Modified Terylene with higher moisture regain	–T E T E E E E T E T E T E–	Block polymer.

In a graft co-polymer a third or different polymer forms the block. For example, a *tere*phthalic acid-ethylene glycol polymer might be formed in the presence of a polyvinyl alcohol (an already made polymer) using, with respect to the acid, sufficient deficiency of ethylene glycol as was chemically equivalent to the added polyvinyl alcohol. Then if V represents the vinyl alcohol residue $-CH_2CHOH-$ we have:

Terylene modified with polyvinyl alcohol	–T E T V V V V T E T E T E–	Graft co-polymer

The block polymer has the extended formula

```
  O        O   H H    O          O   H H    H H    H H    H H
  ‖        ‖   | |    ‖          ‖   | |    | |    | |    | |
 -C-⟨  ⟩-C-O-C-C-O-C-⟨  ⟩-C-O-C-C-O-C-C-O-C-C-O-C-C-O-
               | |                   | |    | |    | |    | |
               H H                   H H    H H    H H    H H
           ←——————————→
      Unit same as units in block.               Block.
```

whilst the graft polymer has the extended formula

```
  O        O    H  H     O          O    H  H  H  H  H  H  H  H
  ‖        ‖    |  |     ‖          ‖    |  |  |  |  |  |  |  |
 —C—⟨  ⟩—C—O—C—C—O—C—⟨  ⟩—C—O—C—C—C—C—C—C—C—C—
                |  |                     |  |  |  |  |  |  |  |
                H  H                     H OH  H OH  H OH  H OH
          ←————————————————→
      Unit different from units in graft.              Graft.
```

Clearly the method of grafting opens up possibilities of making large numbers of different polymers from the same basic materials, *e.g.*, A, B and C can be co-polymerised all at once, or C can be polymerised alone first and the polymer of C graft-polymerised with A and B or even A and B can be polymerised together, C separately and then the two polymers can be co-polymerised to give a graft polymer. Some of the acrylic fibres are graft co-polymers (p. 523).

Another kind of graft polymerisation is one in which side-chains are grafted on to the backbone of an already made polymer. So far as published material is concerned, this method has been more used for rubbers and plastics than for fibres. Some examples that apply to fibres are:

1. Teflon which is chemically inert can be made to adhere to other surfaces by grafting side-chains of polyvinyl alcohol on to the Teflon.
2. Ethylene oxide side-chains, especially polyethylene oxide, have been grafted on to polyamides to improve water absorption.
3. Polar side-chains have been grafted on to polyolefines by irradiation with cobalt-60 (p. 560) to improve their dye affinity.

Ideally, the molecular framework of a polymer to which side-chains had been grafted on would look something like Fig. 46. The preparation of such side-chain graft polymers is not yet easy.

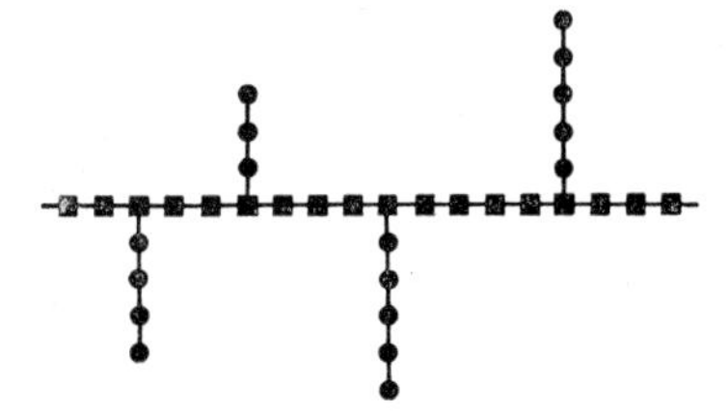

FIG. 46.—Representation of ideal molecular framework of a polymer to which side-chains have been grafted on—a special kind of graft polymer.

It will be clear that there are very many ways in which fibre properties can be modified. Probably the next few years will see many new fibres; thousands will be made in the laboratory; dozens on pilot-plant scale; a few with properties that lend themselves with peculiar suitability to specific uses will survive to the production and manufacturing stages.

FURTHER READING

A. J. Turner, " Natural and Man-made Fibres ", *J. Text. Inst.*, **38**, P 411 (1947) (a dissertation on the diversity of fibres).
J. M. Preston, " Fibre Science ". Manchester (1949).
R. W. Moncrieff, " Wool Shrinkage and its Prevention ". National Trade Press, London (1953).
C. G. Overberger and A. Katchman, " Graft and Block Polymerization ", *Chem. Eng. News*, **36**, No. 45, 80–85 (1958).

PART 2

FIBRES MADE FROM NATURAL POLYMERS

CHAPTER 8

THE FIRST RAYONS

CHARDONNET SILK

MUCH publicity has been given to Robert Hooke's suggestion for making artificial silk, contained in his book " Micrographia ", published in 1664. Hooke speculated on the possibility of making " an artificial glutinous composition much resembling the substance out of which the silkworm wire-draws his clew ". But Hooke did nothing about it, so far as is known; he did not make any artificial silk. Certainly, too, the idea was not novel, for the Chinese had speculated along similar lines two thousand years earlier. It was, in fact, not until the nineteenth century that real efforts were made to put these speculations into more solid form.

After attempts had been made along different lines, success was finally achieved by Count Hilaire de Chardonnet, who combined in his personality the fire of an enthusiastic inventor with the perseverance and drive of an industrialist. His silk consisted of de-nitrated nitrocellulose. Nitrocellulose had been obtained by Braconnet in 1832 by treating all kinds of cellulosic materials (cotton, wood, paper, *etc.*) with nitric acid. Braconnet called his product " xyloidine ", and noted that it was very inflammable. He is not known to have tried to make filaments from it. Note that " nitrocellulose " and " gun-cotton " are other names commonly given to the substance which chemically should be referred to as " cellulose nitrate ".

Audemars' Silk

In 1846 Schonbein made gun-cotton, and in 1855 Audemars of Lausanne took matters a little further. He obtained bast fibres from mulberry twigs, purified and bleached them and then treated them with nitric acid. He dissolved the cellulose nitrate in a mixture of ether and alcohol together with caoutchouc, and from the mass of " collodium " that resulted he drew, with a steel needle, threads that solidified in air, and finally wound them on a spool. This process is described in British Patent 283 of 1855, but it was too early then for the process to be developed practically.

Hughes' Silk

In 1857 (British Patent 67) E. J. Hughes spun a mass obtained from starch, glue, resins, tannins, *etc.*, into something resembling silk.

Ozanam's Jets

Ozanam in 1862 (*C.R.*, **55**, 833) described how a solution of real silk could be spun from orifices. He invented spinning jets; he did not make any artificial silk, but his contribution was considerable in opening up the way for the discovery of rayon. A modern spinning jet is shown in Fig. 47.

[*Courtaulds, Ltd.*]

FIG. 47.—A modern viscose spinning jet (×5).

Swan's Silk

In 1883 (*B.P.* 5978) J. W. Swan produced threads by squirting a solution of cellulose nitrate in glacial acetic acid through holes, and he exhibited some of these fibres, which he called " artificial silk ", to the Society of Chemical Industry in London in 1884. One year later fabrics made in the form of crocheted nets from these threads were shown at the Exhibition of Inventions in London. These fabrics were made from denitrated filaments, and this shows that Swan had already discovered a practicable method of converting the dangerous and very highly inflammable cellulose nitrate into cellulose hydrate, which is harmless. Swan's filaments were used in early electric lamps. Swan's processes really anticipated those of Chardonnet, but whereas Swan never developed his process, Chardonnet persevered with his until he had put it on a commercial basis.

Gerard's Silk

About the same time as Swan and Chardonnet were working, other inventors were engaged on similar problems. Gerard spun a mixture of gelatin and cellulose nitrate dissolved in acetic acid.

Soie de France

du Vivier invented a process in which three solutions—(1) guncotton in glacial acetic acid, (2) fish glue in glacial acetic acid and (3) guttapercha in carbon disulphide—were mixed and spun. Despite its complexity the process was utilised by the *Société Général du Dynamite* of Rouen, which for a time marketed the product as " Soie de France ".

Chardonnet

With true insight, Chardonnet avoided the use of complex mixtures. His silk was cellulose nitrate, which later he denitrated. To-day Chardonnet silk is of historical interest only, but it was the first artificial silk to be produced in any considerable quantity, and was at one time used extensively. To Count Hilaire de Chardonnet is universally accorded the honour of having invented the first practicable rayon process. His process was patented in 1885, but a year earlier the *Société Anonyme pour la Fabrication de la Soie de Chardonnet* was founded. Not unnaturally, it passed through bad times, but by 1895 it was paying dividends. Samples of Chardonnet silk were shown at the Paris Exhibition in 1889, and aroused considerable interest.

Manufacture. A great number of variations of the general methods were patented, and it is certain that the methods of manufacture were modified as experience was gained, but the general outline of the methods was as follows:

Nitration. Although in some of his earliest experiments Chardonnet had used mulberry fibres, and although he envisaged the possibility of using other cellulosic material, such as wood-pulp, on a large scale, in practice he used only cotton as his raw material. This was nitrated with a mixture of nitric acid 40 per cent, sulphuric acid 40 per cent, water 20 per cent, at a temperature of 45° C. for from one to two hours. The cellulose nitrate so formed was dissolved in a mixture of 40 per cent ether and 60 per cent alcohol.

Wet Spinning. The alcoholic solution of the cellulose nitrate was spun through a mouthpiece which dipped into cold water. As the stream of collodion (solution of cellulose nitrate) passed into the water, the filaments solidified, first on the outside. As they passed through the fluid they were stretched, and finally became solid all through. At a later stage the wet spinning process was superseded by one of dry spinning.

Dry Spinning. The solution of cellulose nitrate in ether and alcohol was extruded by high pressure through small nozzles of about 0·07 mm. diameter. The filaments hardened very quickly. They were wound, twisted, and reeled into hank form.

Denitration. This process was necessary, inasmuch as cellulose nitrate is so extremely inflammable that its use is dangerous. The hanks of cellulose nitrate were treated with ammonium hydrosulphide for about twenty hours at 65° C., or with a solution of a sulphocarbonate for twelve hours at 35° C. The alkali saponified the cellulose nitrate to regenerated cellulose, which was no more inflammable than cotton, and was quite safe to use.

Purification. The denitrated rayon was washed with water, and was then a dirty greenish-grey colour, due to the presence in it of iron sulphide. It was washed with hot very dilute hydrochloric acid to remove the iron, washed again and bleached. Bleaching was carried out either with hypochlorites, which caused some yellow discoloration, or with hydrogen peroxide, which gave better results but was at that time very expensive. After a final rinse the yarn was centrifuged and dried.

Properties

Cellulose nitrate has a specific gravity of 1·66. It dyes well with basic colours, even without a mordant. Its dangerous inflammability makes its widespread use as a textile material impossible.

After denitration, the product consists of regenerated cellulose, but it has undoubtedly suffered degradation—*i.e.*, the cellulose molecules have been broken and are shorter than in the raw cellulosic material—probably, too, oxycellulose has been formed during the nitration process. The denitrated product has some affinity for basic dyestuffs unmordanted; good dyeings, deep in shade and of reasonable fastness, can be obtained if a tannin-antimony mordant is used. (See Chapter 9.)

The reader should appreciate that Chardonnet silk was of great importance, in that it was the first commercial rayon, and because, in a sense, it started the tremendous rayon industry that exists to-day. The process is, however, obsolete, and has been replaced by the viscose and cuprammonium processes, both of which, like Chardonnet's process, finally give a regenerated cellulose.

FURTHER READING

M. H. Avram, "The Rayon Industry", 544–58 (nitrocellulose). London (1930).

R. W. Moncrieff, "The Genesis of Synthetic Fibre", *Textile Manufacturer*, **75**, 285 (1949).

J. T. Marsh and F. C. Wood, "An Introduction to the Chemistry of Cellulose", 307. London (1945).

M. D. C. Crawford, "5,000 Years of Fibres and Fabrics", The Brooklyn Museum (1946).

This little book gives an outline of the developments that finally led to man-made fabrics.

CHAPTER 9

VISCOSE RAYON

THE process of making viscose yarn was discovered by C. F. Cross and E. J. Bevan—two chemists who brought about a much better understanding of the chemistry of cellulose. The process was discovered in 1891, and patented by Cross, Bevan and Beadle in 1892, but took some considerable time to establish itself.

Development of Viscose

The greatest single factor in the development of the viscose process has undoubtedly been the support given to it by Courtaulds, Ltd., although there have naturally appeared other viscose producers. The pioneer work was undoubtedly carried out by Courtaulds, Ltd., who not only founded and developed an important new industry, but also introduced it to America under the name " The American Viscose Co." During the 1939–45 War this American Company was sold to American interests in order to provide dollars for Britain. Other names prominent in viscose manufacture are du Pont de Nemours & Co. (but see p. 153), the Industrial Rayon Corporation and the American Enka Co. in America; Snia Viscosa in Italy; the Glanzstoff Corporation in Germany and Toyo in Japan.

In this country Coventry was the home of the first viscose spinning-plant, and immense quantities of yarn still come from there. The present happy position of the viscose industry not only in this country, but throughout the world, is undoubtedly largely due to the industrial genius of the late Mr. Samuel Courtauld.

In 1912 it was unusual to see artificial silk in a Lancashire mill. I remember, as a boy at about that period, seeing a few bobbins of it, and being considerably impressed; it was still so unusual as to be a novelty. Nor was it very good, for when washed the fabrics into which it was made nearly fell to pieces on account of its very low wet strength. However, even at that time it found employment as effect threads in stripes for dress materials and shirtings. In later years the wet tenacity was greatly improved.

Contrast that state of affairs with a world production of nearly 3 million tons of viscose rayon in 1964. The growth in production may be realised from the following figures:

Year.	World production of viscose rayon.
1900	1,000 tons
1910	8,000 ,,
1920	15,000 ,,
1930	200,000 ,,
1940	1,100,000 ,,
1950	1,300,000 ,,
1960	2,300,000 ,,
1963	2,700,000 ,,
1967	2,700,000 ,,

Such a consumption will eventually make inroads on the timber supplies. However, cellulose is abundant in the vegetable kingdom and there are many other possible sources of it than timber. Wheat straw, esparto grass and bamboo all contain about 50 per cent cellulose. Bamboo is being developed in India as a cellulose source. Timber does however have the big advantage as a raw material for fibres that it is abundant in the arctic and sub-arctic forest regions, so that its growth does not interfere with the world's food supply.

One factor that has had a considerable bearing on the growth of the consumption of viscose rayon is the success which has attended the introduction of viscose staple fibre. This material which, at first, had a rough and unattractive appearance and handle and was easily crushed and creased, has been vastly improved, and is now an attractive and deservedly popular material.

Viscose rayon has always been the cheapest of the artificial silks, and probably for that reason, more than any other, its production is much greater than that of all other artificial fibres taken together.

It is surprising to find nowadays that some of the older spinners of viscose rayon, presumably dazzled by the benefits of the synthetic fibres, are giving up viscose spinning. Du Pont, one of the biggest, have discontinued viscose rayon. Teijin, in Japan, has closed its rayon staple plant at Iwakuni, which had an output of 58 million lb. of yarn per annum. This plant had been built before the war, and it latterly required so much maintenance that it had become uneconomic to operate. We shall possibly see in the next decade much of the cellulosic fibre manufacture drift to India and China. Japan, too, will probably reduce her production, and spend more and more of her skill and industry on the synthetic fibres.

The increase in the production of rayon, its distribution between continuous filament and staple, and the relation it bears to the production of natural fibres are discussed in Chapter 49.

Chemical Nature

Viscose rayon is regenerated cellulose. The cellulose comes from wood, of which it is the major constituent. It is purified, treated with caustic soda, which converts it into alkali cellulose, then treated with carbon disulphide, which converts it into sodium cellulose xanthate, and then dissolved in a dilute solution of caustic soda. This solution is then " ripened ", the solution becoming at first less viscous and then increasing nearly to its original viscosity; it is then spun into an acid coagulating bath, which precipitates the cellulose in the form of a viscose filament. The cellulose which constitutes the final filament differs chemically from the original cellulose of the wood in only one respect—that it has suffered some degradation during the manufacturing processes: the very long cellulose molecules have been partly hydrolysed and have been broken down into shorter, although still very long, molecules. When artificial fibres are made by chemical processes from naturally occurring polymers, as in the case of viscose from wood cellulose, or Bemberg or cellulose acetate from cotton cellulose, or alginate from seaweed, it is desirable so far as possible to keep the inevitable degradation of the materials to a minimum. If the natural long-chain polymers are broken down considerably there is a significant loss in strength and fibrous properties. In the present state of our knowledge some breakdown is inevitable, but ways have been found of reducing this depolymerisation very considerably, compared with that which obtained in the earlier days of the industry. It should never be forgotten that good fibres can be made only from long polymeric molecules and that, as a result, both in the manufacture of the fibre and in processes to which the yarns and fabric made from it are subjected, conditions which encourage hydrolysis should not normally be allowed to obtain. In viscose, the cellulose molecules are about one-quarter as long as those in the wood cellulose from which it was made.

Chemical Reactions for Manufacture. The chemical reactions that are involved in the manufacture of viscose are as follows:

1. The cellulose is treated with a 17·5 per cent solution of caustic soda which converts it into soda cellulose. The reaction is usually expressed:

$$\underset{\text{Cellulose.}}{(C_6H_{10}O_5)_n} + \underset{\text{Caustic soda.}}{n\text{NaOH}} \longrightarrow \underset{\text{Soda cellulose.}}{(C_6H_9O_4ONa)_n} + nH_2O$$

and, according to this equation, 162 parts by weight of cellulose should require 40 parts of caustic soda. In practice it is found that twice as much caustic soda as this is required.

The reaction between the soda cellulose and the carbon disulphide may be represented:

$$(C_6H_9O_4ONa)_n + nCS_2 \longrightarrow \left[SC\begin{matrix} \diagup SNa \\ \diagdown OC_6H_9O_4 \end{matrix} \right]_n$$

Soda cellulose. Carbon disulphide. Sodium cellulose xanthate.

Each glucose residue in the cellulose polymer chain must be thought of as reacting in the following way.

```
        CHOH·CHOH                          CHOH·CHONa
       /3       2\                         /        \
 —O—CH 4        1 HC—    →       —O—CH              HC—
       \5        /                         \        /
        CH——O                              CH——O
        |                                  |
        CH2OH                              CH2OH

     CHOH·CHONa                          CHOH·CHO·CS·SNa
     /        \                          /         \
—O—CH          HC— + CS2       —O—CH               HC—
     \        /             →            \         /
     CH——O                               CH——O
     |                                   |
     CH2OH                               CH2OH
```

The reaction takes place on the hydroxyl group in position 2 in the glucose residues in the cellulose polymer. According to the above equation 162 parts by weight of cellulose should require 76 parts of carbon disulphide. In practice it is found that only about 70 per cent of the theoretical quantity, *i.e.*, 52 parts of carbon disulphide, is required. The measurement of the requisite quantity of carbon disulphide is shown in Fig. 48.

When the sodium cellulose xanthate is dissolved in weak caustic soda to form the viscous solution known as "viscose" the xanthate radicles probably combine loosely with the molecules of caustic soda.

When the viscose solution is first made, it is very thick. On standing, it becomes thinner and then, later, more viscous again. This process is known as "ripening", and whilst it is going on, some of the sodium cellulose xanthate decomposes, regenerating cellulose which is maintained in emulsion form by that part of the sodium cellulose xanthate which is still undecomposed, this latter part acting as a protective colloid. If the ripening process was allowed to proceed far enough, the cellulose would eventually be precipitated when there was no longer a sufficient amount of the undecomposed sodium cellulose xanthate to hold it in emulsion; but in practice, just before this precipitation is due to begin, the solution is spun. It

[*Courtaulds, Ltd.*]

FIG. 48.—The requisite quantity of carbon disulphide being measured for addition to a batch of " crumbs ".

is spun into a coagulating bath of sulphuric acid which completes the conversion of the sodium cellulose xanthate to cellulose:

$$\begin{array}{c} \text{CHOH·CHO·CS·SNa} \\ \text{—O—CH} \qquad\qquad \text{HC—} \\ \text{CH——O} \\ | \\ \text{CH}_2\text{OH} \end{array} + \tfrac{1}{2}\text{H}_2\text{SO}_4$$

Sodium cellulose xanthate.

$$\downarrow$$

$$\begin{array}{c} \text{CHOHCHOH} \\ \text{—O—CH} \qquad\qquad \text{HC—} \\ \text{CH——O} \\ | \\ \text{CH}_2\text{OH} \end{array} + \text{CS}_2 + \tfrac{1}{2}\text{Na}_2\text{SO}_4$$

Viscose.

This completes the series of chemical reactions by which viscose is made. Essentially they consist of the following stages:

1. Wood cellulose and concentrated caustic soda react to form soda cellulose.
2. The soda cellulose reacts with carbon disulphide to form sodium cellulose xanthate.
3. The sodium cellulose xanthate is dissolved in dilute caustic soda to give a viscose solution.
4. The solution is ripened.
5. It is extruded into sulphuric acid which regenerates the cellulose, now in the form of long filaments (viscose rayon).

It will be appreciated that natural cellulose is built up of glucose residues in this way:

```
    CHOH–CHOH                 ┌  CHOH–CHOH        ┐
   /         \                │ /         \       │
nCHOH        CHOH   →   HO –CH             CH–O– H + (n − 1)H2O
   \         /                │ \         /       │
    CH——O                     │  CH——O            │
    |                         │  |                │
    CH2OH                     └  CH2OH            ┘n
   Glucose.                         Cellulose.
```

The first step in this natural polymerisation is the formation of one molecule of cellobiose from two molecules of glucose

```
            CHOH–CHOH
           /         \
      2 CHOH         CHOH
           \         /
            CH——O
            |
            CH2OH
             Glucose.

          ↓  CH2OH
             |
  CHOH–CHOH  CH——O
 /         \ /     \
CHOH      HC–O–CH   CHOH + H2O
 \         /    \   /
  CH——O         CHOH–CHOH
  |
  CH2OH
             Cellobiose.
```

When the polymerisation proceeds further the molecules of cellobiose condense and form a long string of glucose residues. In wood cellulose the molecule contains about 1,000 of these glucose residues. In viscose rayon the cellulose contains about 270. The cellulose molecules have been broken down in the chemical processes by which the viscose is made, but apart from the fact that the viscose cellulose molecules are shorter than the wood cellulose molecules, they are similar. With that one exception, viscose yarn is the same chemically as wood cellulose. There is also a difference in the *packing* of the molecules between native and regenerated celluloses.

Manufacture

Preparation of the Wood Pulp. The starting material is timber, usually spruce. After it has been felled it is floated on rivers to the mills. Here the bark is removed and the wood is chipped into pieces about $\frac{7}{8} \times \frac{1}{2} \times \frac{1}{4}$ in. The chips are treated with calcium bisulphite, and the treated chips then cooked with steam under pressure for fourteen hours. This treatment does not greatly affect the cellulose, but it decomposes and solubilises the encrusting substances, so purifying the cellulose. After cooking, the mass is diluted with water on which the pulp floats; the pulp is then sucked through slots about 0·2 mm. wide. It is concentrated to 30 per cent cellulose content, bleached with hypochlorite and converted into paper board. The bleaching that the cellulose receives in these preparatory stages greatly reduces the need for bleaching of the rayon, and in fact some rayon is not bleached—*i.e.*, the only bleaching it has undergone is the pulp bleach.

Steeping and Pressing (Formation of Soda Cellulose). The flat, white sheets of board contain about 90–94 per cent of pure cellulose. They are first conditioned by storing in a room at a definite humidity and temperature. This conditioning is necessary so that a known weight of cellulose can be weighed out for each batch (Fig. 49). The sheets are then stacked vertically in the press (Fig. 50), and then soaked in a 17·5 per cent solution of caustic soda for from one to four hours. This process is known as steeping or mercerising. The boards become greatly swollen, and the hemicelluloses are dissolved in the caustic soda, and turn the liquor brown. About 8·2 per cent of the original wood pulp dissolves. The cellulose itself is swollen, but not dissolved.

Then the excess alkali is pressed out by a hydraulic ram in the press—the unit in which the soaking has been carried out. The pressing leaves a moist mass of soda-cellulose, which passes straight into a shredding machine.

Recovery of Caustic Alkali. The alkaline liquors that are ex-

[*Courtaulds, Ltd.*]

FIG. 49.—The sheets of wood pulp are carefully weighed for each batch.

[*Courtaulds, Ltd.*]

FIG. 50.—The wood pulp is placed in the press prior to running in the caustic soda.

pressed are allowed to diffuse through parchment-paper membranes. The caustic soda passes through the membrane, and, after being made up to strength by the addition of fresh alkali, is re-used. Ultimately, when it is so foul that it can no longer be used, it is disposed of to soap-works. The recovery of alkali is an important factor in the economic operation of the viscose process.

Shredding. The shredding machine consists of a drum inside which revolve a pair of blades with serrated edges. The shredders are water-cooled, and will take 200 lb. of pressed soda-cellulose at a time. In two to three hours they break it up into fine " crumbs ".

Ageing. After shredding, the " crumbs " are aged. They are contained in a galvanised vessel with a lid, and through oxidation with atmospheric oxygen, degradative changes set in. Some depolymerisation occurs, and the degree of polymerisation (number of glucose residues in a cellulose molecule) falls from about 800 to 350 in this process. Of recent years it has been found possible, by using a rather higher temperature, which is automatically controlled, to complete the ageing in from one to one and a half days. (If it were allowed to continue for a matter of months the crumbs would become water-soluble when the degree of polymerisation fell below 100.) Until recent years it was customary to age for about three and a half days at 22° C.—*i.e.*, in a comfortably warm room. The higher the temperature, the more rapidly does ageing—which is mainly a process of depolymerisation—proceed.

The American firm, Oscar Kohorn & Co., Ltd., who are primarily machinery designers and manufacturers, have developed a " Rapid Ageing Process ", in which the ageing of the soda-cellulose is carried out during the shredding process, by careful control of temperature and other physical conditions. This saves a process in practice, together with labour costs, and cuts down the time necessary to make viscose. It is claimed that in the Kohorn process the quality and filterability of the final viscose are better than usual because of the reduced oxidation and degradation that takes place during ageing.

Ageing consists essentially of storage under controlled atmospheric conditions. Oxidation and depolymerisation occur, and these changes considerably influence the properties of the viscose yarn that will eventually be made.

Churning (Xanthation or Sulphidising). After ageing, the soda-cellulose crumbs are introduced into rotating air-tight hexagonal churns (Fig. 51). About 10 per cent of their own (crumbs) weight of carbon disulphide is added (Fig. 48), and the crumbs and disulphide are churned up together; a deep orange, gelatinous mass of sodium cellulose xanthate is formed. Note that theoretically 162 parts of cellulose require 78 parts carbon disulphide. The crumbs contain about 30 per cent cellulose, so that 162 parts crumbs should need 23 parts carbon disulphide—*i.e.*, 100 parts should need 14 parts carbon disulphide, so that the quantity actually used, 10 parts, is about 70 per cent of the theoretical quantity.

[*Oscar Kohorn & Co., **Ltd.***]

FIG. 51.—Xanthating barattes in a plant installed by Oscar Kohorn & Co., Ltd., New York.

The " crumbs " are churned with a definite quantity of carbon disulphide.

Churning is continued for about three hours, the churns rotating at about 2 r.p.m. Each churn (which is water-jacketed) holds a charge of about 200 lb. At the completion of churning the vessels are exhausted by vacuum to remove ill-smelling vapours and the charge is dumped in mixers. (Sometimes churning and mixing may be carried out in the same, specially designed, vessel.)

Mixing (Solution). In the mixers the sodium cellulose xanthate is stirred with dilute caustic soda solution for four to five hours, the vessel being cooled. The xanthate dissolves to a clear-brown, viscous liquid, similar in appearance to honey. This liquid is known as " viscose ", but is still too impure, too aerated and too young to spin. It contains about 6·5 per cent alkali and 7·5 per cent cellulose. It is transferred to a secondary mixer or blender, which takes the charges from eight primary mixers. Note that it is always desirable to blend the materials from which rayon will eventually be made, as inequalities are thus balanced out and there is a better chance that the rayon will be uniform in quality and characteristics.

In the secondary mixer or blender the viscose is stirred and pumped round. As it still contains some undissolved fibres from the original wood pulp, fibres which have resisted all the chemical treatments, it is filtered. The first filtration is carried out through cotton-wool, and the viscose is then twice filtered through cotton filter-cloth. If a delustred yarn is eventually required, a pigment, perhaps 2 per cent on the weight of the cellulose of titanium dioxide, will be added to this solution and dispersed by stirring. Air-bubbles are then removed by exposing the viscose in air-tight tanks to a vacuum.

Ripening. The viscose solution is stored for 4 to 5 days at 10–18° C., and during storage it ripens. During the ripening the viscosity at first falls, and then rises, so that by the time the solution is ready to spin, the viscosity has risen almost to its original value. Ripening is an essential part of the viscose process; " young " viscose cannot be spun satisfactorily; neither can viscose that is even a few hours too old. Two tests which are available to determine when viscose is ripe enough to be spun are as follows:

1. *Acetic Acid Test.* Before it is ripe the viscose solution will dissolve in 40 per cent acetic acid, but when it is ripe the viscose will precipitate in this acid.
2. *The Hottenroth (Ammonium Chloride) Test.* Twenty grams of the viscose to be tested are mixed with 30 c.c. distilled water, and to this solution is then added from a burette with stirring, a solution of 10 per cent (wt/wt) ammonium chloride. The ripeness figure is the number of c.c.s of ammonium chloride solution necessary to bring about coagulation of the viscose to a gelatinous condition. The lower the number, the riper the viscose.

Of these two tests the acetic acid test is no longer used; the Hottenroth, called after its originator, is most universally employed.

When the viscose solution is ripe, it is drawn off to settling tanks, where all air-bubbles are removed from it by exposure to a vacuum for twenty-four hours. This is essential, for any bubbles in the spinning solution would arrest the spinning and there would be discontinuities in the filaments.

The normal time interval that elapses between the start of the steeping operation and the viscose solution being ready for spinning is about a week, most of the time being taken up in ageing of the alkali-cellulose and ripening of the viscose solution. Probably, in most modern plants, improvements have been effected to reduce this time to about five days. The Kohorn Process, in which the ageing is carried out simultaneously with the shredding, and in which the normal ripening time has apparently been reduced, cuts down the total time considerably, so that viscose ready for spinning is produced in approximately forty hours from the start of the steeping operation.

Spinning. When the viscose solution is ripe, it is forced by compressed air (2·6–5 atmospheres) to the spinning frames, and there distributed. It is metered by a small viscose pump for each spinning-

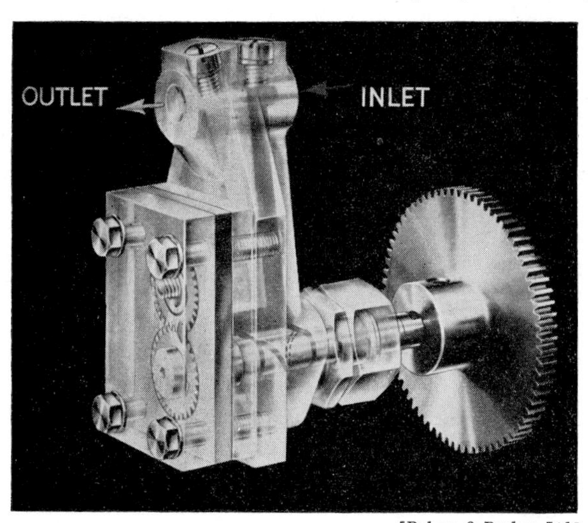

[*Dobson & Barlow, Ltd.*]

FIG. 52.—Spinning pump of gear wheel type.
This pump meters as well as pumps viscose solution to the filter candle and thence to the jet. Its metering action ensures uniformity of denier.

head. The viscose pump is usually of the gear wheel type, and the wheels and plates are constructed of stainless steel to resist the etching effect of the viscose with which they come in contact. Pumps used in the early days of rayon spinning did not approach the performance of to-day's gear pump. The pump used on Courtaulds' original spinning machine at Coventry in 1903 was a single plunger type. Later came double plunger types, and ultimately the gear pump which has the outstanding advantage that it gives a constant output (so that denier will not vary) irrespective of the input pressure. The viscose, which the pump (Fig. 52) meters continuously, lubricates the pump-wheel shaft. The metering pump's action ensures constant and regular delivery of viscose to the filter and spinning jet, at the exact rate required for the denier being spun. Alteration in the speed of the pump drive enables the denier to be changed. The pump illustrated delivers 0·6 c.c. per revolution.

From the pump the viscose passes to the filter, which may be either of the candle or disc type, either of which filters the viscose through fabric.

After passage through the filter the viscose passes into the glass "rounder-end", which carries a spinneret (see Fig. 47). The spinnerets are submerged in the acid bath, and are usually arranged so that the filaments emerge almost at right angles to the surface of the bath.

The jet is precision-bored with orifices of diameter usually between 0·05 and 0·1 mm. diameter. The number of holes in the jet, together with the rate of delivery from the spinning pump, and the godet speed (*i.e.*, take-up speed) determine the filament denier. The spinning jet is made from tantalum, or one of the noble metals—*e.g.*, gold, platinum or rhodium—on which the bath liquids have very little action. The drilling of these jets is, as may be imagined, a highly specialised art.

The spinning bath is made of sheet lead. Fresh acid is fed into it by a pipe which runs along its bottom, and a weir overflow controls the level of the bath. A channel carries away overflow from the bath, as well as acid centrifuged from the spinning pot and drips from the godet rollers. The bath contains a solution which consists essentially of 10 per cent sulphuric acid plus sodium sulphate. As the viscose solution passes through the jets into the acid bath, it solidifies into filaments, owing to regeneration of the cellulose. The bath, at a temperature of about 40–55° C., may be made up as follows, although the exact composition will vary from one manufacturer to another:

Sulphuric acid . . .	10 parts by weight
Sodium sulphate . . .	18 ,, ,,
Glucose	2 ,, ,,
Zinc sulphate . . .	1 ,, ,,
Water	69 ,, ,,

Sometimes, too, magnesium sulphate is a constituent of the bath.

At one time it was usual to use 5 per cent glucose, but for reasons of economy more than 2 per cent is not as a rule used now, and in fact some first-class viscose rayon is spun without any glucose in the bath. The composition of the spinning bath has been arrived at empirically, and it is not safe to describe too boldly the functions of the ingredients, but, with this reservation, they may be indicated as follows: the sodium sulphate precipitates the sodium cellulose xanthate from the viscose solution into the form of filaments, and the sulphuric acid converts it into cellulose; the glucose gives pliability

[*Courtaulds, Ltd.*]

FIG. 53.—Godet rollers.

These can be seen on the spinning machines in Figs. 56 and 57. Stretch is imparted to the filaments by running the top godet at a higher speed than the bottom godet.

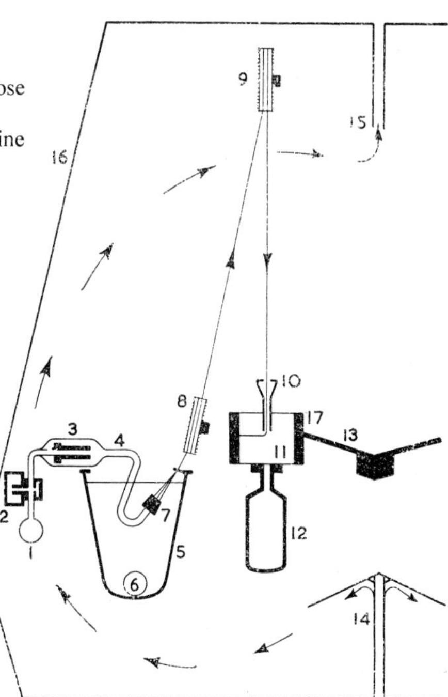

FIG. 54.—Diagram showing viscose rayon spinning essentials.

One side of a two-sided machine is shown.

1. Section of main viscose feed.
2. Pump.
3. Candle filter.
4. Glass rounder end.
5. Coagulating bath.
6. Section of acid replenishing pipe.
7. Spinneret.
8. Bottom godet.
9. Top godet.
10. Funnel.
11. Topham Box.
12. Drive for Topham Box.
13. Traverse.
14. Fresh air inlet.
15. Fumes sucked out.
16. Sliding windows.
17. Cake.

and softness to the yarn, probably because it increases the viscosity of the coagulating bath, and thus retards a little the chemical change from sodium cellulose xanthate to cellulose; the zinc sulphate is responsible for added strength, an important feature, and also for the serrated cross-section, for if it is omitted the filaments are round or oval in section and do not dye so well as when they are serrated or indented.

Factors which affect the quality of the viscose rayon are:

1. temperature and composition of the bath;
2. speed of coagulation;
3. length of immersion;
4. speed of spinning;
5. stretch imparted between the godets.

As the filaments emerge from the jet (they do not stick together) they are led to an eye at the surface of the bath, and thence guided round the bottom godet rollers. The godet rollers are illustrated in Fig. 53. The filaments pass from the bottom godet round another guide, made of glass or other acid-resisting material, round the top godet, then over a second glass guide and again round the top godet rollers. The filaments (bunched together as a yarn by the first guide they meet at the bath surface) pass round the godets at the speed of the godets—*i.e.*, they do not slip—and as the top godet is driven faster than the bottom, the filaments are stretched, usually about 100 per cent, between the bottom and top godets. The yarn passes from the top godet downwards through a glass funnel to the centrifugal spinpot, known as the Topham Box or Topham Pot. The diagram in Fig. 54 illustrates the essentials of the viscose spinning process. Fig. 55 shows a complete Viscose Rayon Spinning Machine. Figs. 56 and 57 show Viscose Rayon Spinning Machines in operation.

Spinning Speeds. There has been a gradual increase in the spinning speed as the viscose process has developed. Probably most viscose rayon is spun within the range 60–80 metres per minute. Attempts to modify the process to run at higher speeds have been made; the von Kohorn high speed spinning system for viscose filament yarns incorporates some modifications to the precipitation baths and enables speeds of 140–150 metres per minute to be obtained and it can be used for any denier.

The Topham Box. This is the take-up device for cake-spun viscose rayon. It consists essentially of a bucket on a spindle driven by an electric motor. The complete assembly is shown in Fig. 58. Fig. 59 shows a cut-away view of the bucket of a Topham Box. Fig. 60 shows a cut-away view of another type of Topham Box; in this the cake of viscose rayon can be seen inside the bucket.

[*Dobson & Barlow, Ltd.*]

FIG. 55.—A centrifugal (Topham Box) Viscose Rayon Spinning Machine.

[*Courtaulds, Ltd.*]

FIG. 56.—Close-up view of the viscose spinning operation.

Note the yarn passing upwards from bottom to top godet rollers, thence downwards to the funnel which guides it into the spinning pot.

[*Oscar Kohorn & Co., Ltd.*]

FIG. 57.—View of another Viscose Rayon Spinning Machine in operation.

This particular machine will spin at speeds up to 110 metres per minute, and will make deniers ranging from 50 to 600.

[*Courtaulds, Ltd.*]

FIG. 58.—Topham Box with spindle and driving motor.

Note bucket in which the yarn is collected in cake form at the top of the illustration.

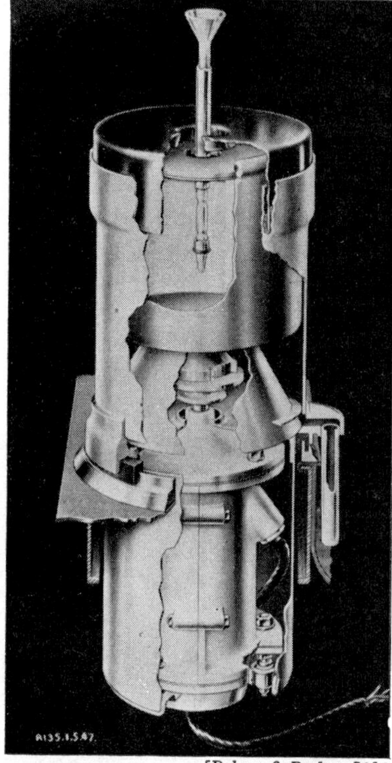

[*Dobson & Barlow, Ltd.*]

FIG. 60.—Cut-away view of another type of Topham Box and driving unit.

Note the funnel which leads the yarn into the pot (or bucket).

[*Oscar Kohorn & Co., Ltd.*]

FIG. 59.—Cut-away view of Westinghouse spinning bucket of a Topham Box moulded in Textolite.

The Topham Box is a plastic or metal perforated vessel; it dates back to the early days of rayon spinning. The type used by Courtaulds in 1908 consisted of an ebonite-lined aluminium box perforated with relatively few fairly big holes. By 1918 the design had changed to an ebonite box with a large number of small holes and by 1952 a reinforced Bakelite spinning box with four times the capacity of the earlier ones had taken over. It is nearly cylindrical, but slightly tapered in shape, about 7 in. in diameter, and rotates at about 7,000 (in some cases up to 12,000) r.p.m. The yarn is led from the top godet roller, through the glass funnel which may be seen in Fig. 60, into the pot, and is thrown against the side of, and taken up by, the rapidly rotating pot. A traversing arrangement is provided, and the yarn builds itself in the box into the form of a " cake ". Twist is put into the yarn automatically as winding takes place; the amount of twist is dependent on the speed of yarn delivery and on the spindle speed. If, for example, the yarn delivery speed is 100 metres per minute and the spindle speed of the Topham Box is 7,000 r.p.m. a twist of $\frac{7{,}000}{100 \times 39{\cdot}37} = 1{\cdot}8$ turns per inch (t.p.i.) will be inserted. According to the direction of rotation, either " S " or " Z " twist will

[*Courtaulds, Ltd.*]

FIG. 61.—A doffer removing a cake of yarn from a Topham Box. Note the covering lid and ring in his right hand.

[*Courtaulds, Ltd.*]

FIG. 62.—The Cakewash Machine.

Cakes of viscose rayon are being loaded on to carriers on which they will be washed, desulphurised, bleached and washed again.

be inserted. Topham Boxes are made so that they can be rotated in either direction.

Topham Boxes of different sizes are made, but in a typical instance after running for eight hours a cake will be built up which weighs 3 lb., of which about 1 lb. will be viscose rayon and the remaining 2 lb. coagulating solution.

The main advantage of the Topham Box is that as it takes up the yarn it inserts twist, and so enables the yarn, still wet and in a vulnerable condition, to be built up into a package without damage. Once twist has been inserted, yarn is very much less likely to be damaged.

Note that on its way from the surface of the coagulating bath to the Topham Box the yarn has been stretched. This orients it to some extent and increases the tenacity. It is, of course, essential for the degree of stretch to be kept constant and to be the same from spindle to spindle. Differences in tension greatly affect the lustre of the yarn.

Fig. 61 shows a doffer removing a cake from a Topham Box; several other cakes already removed may be seen standing on the table.

Purification. The yarn collected in the Topham Box is impure and relatively weak; it has to be purified. Purification consists of four operations, *viz.*, (1) washing; (2) desulphurising, a process which is sometimes called sulphiding; (3) bleaching; (4) washing.

At one time the yarn was wound into skeins, and these were washed, sulphided, bleached and washed, and then re-wound on to bobbins or cones. These operations were carried out with the hanks suspended from ebonite-covered rods, which as they were carried down a long machine were sprayed with the purification liquors. It is, however, a costly operation to wind yarn into skeins and then re-wind it, and an operation, moreover, that requires highly skilled and practised operatives.

Nowadays it is more usual to carry out the purification with the yarn still in cake form, and this effects a considerable economy, besides avoiding the possibility of damage (hairiness due to broken filaments) that is likely to occur in skein-winding and backwinding. The cakes are mounted on perforated spindles, as shown in Fig. 62, and the purification liquors are fed to the spindles and forced through the cakes.

The first wash is with water; the second is with sodium sulphide solution to remove sulphur, and possibly compounds of sulphur, residual from the xanthate. The yarn is still not quite white, so it is next treated with a slightly alkaline solution of sodium hypochlorite to bleach it; this is followed by a wash with dilute hydrochloric acid and a final wash with water. The cakes are then dried.

Owing to the improved quality of the wood-pulp that is available, it has been found possible in some cases to cut out the hypochlorite bleach, and a considerable proportion (although probably not the bulk) of viscose rayon is not bleached at all.

Again, some manufacturers purify their cakes without taking them out of the pots. The full pot is removed from the spinning machine, and is rotated and simultaneously sprayed with the purification liquors, which leave the pot through the perforations in its sides.

Textile Processing. Some consumers will take delivery of viscose rayon in cake form, and twist directly from the cakes. For others it is first wound on to cheese, cone, or bobbin before despatch.

Bobbin Spinning. Not all viscose rayon is collected in cake form in the Topham Box. Some is wound, without twist, on to spools or bobbins with perforated barrels. These are then put in a pressure-dyeing type of machine, and the purification and bleaching carried out with the yarn on the perforated bobbins. Finally, the yarn is dried, and then oiled, twisted and wound on to ordinary bobbins for despatch.

Continuous Spinning

Three ways of purifying viscose have been described: (1) in hank form, the hanks having been wound from the cakes in which the viscose was spun with twist; (2) in cake form with twist in the yarn; (3) on perforated bobbins without twist. Clearly methods (2) and (3) represent a great saving of labour compared with method (1), for in this last method the hank-winding and subsequent backwinding from hank to cone or bobbin are expensive in labour and therefore in cost. At one time all viscose was purified in hank form, but to-day more and more viscose is being purified in cake form. All these three processes are, however, discontinuous or batch processes, and usually manufacturers prefer for any process a continuous rather than a batch method. Usually a continuous process is well suited to a large production at an economical cost. If viscose rayon has to be collected in cake form, wound into hanks, given several wet processes of purification and then re-wound on to cones or other packages for textile use, such processing will involve heavy labour costs. If the process is shortened by treatment of the yarn in cake form, much is saved, but still each cake has to be handled, put in a pressure purification plant, treated, taken out and as a rule re-wound. Considerable labour is still involved, and the manufacture of viscose rayon even by this method is a process much more complicated than the dry spinning of cellulose acetate. Accordingly, various methods have been devised for continuous processes in which viscose could be extruded, purified, dried and wound in the first instance on a bobbin on which it could be supplied to textile manufacturers. Those processes which have reached production dimensions are:

1. The Nelson Process which Lustrafil Ltd. announced in 1947 that they had been operating for some years. The sole licence for the manufacture and supply of the machinery for this process to all countries except N. and S. America and the U.S.S.R. has been acquired by Dobson and Barlow Ltd., to whom I am indebted for much of what follows regarding this process.
2. The Industrial Rayon Corporation of America Process.
3. The Filamatic, a process announced by the American Viscose Corporation.

The Nelson Process

The cardinal difficulty in the continuous spinning of viscose is that the wet treatments which it has been usual to give to viscose spun into cakes have occupied considerable time. If, for example, viscose were spun at a speed of 70 metres per minute and it required 30 minutes (washing, bleaching, etc.) to purify it, then a run of 2,100 metres—more than a mile—would be required in which to give the treatment. Floor space has to be kept within reasonable confines. In the Nelson Process three factors contribute to keep the floor space within very reasonable confines. They are:

1. The method of presentation of the yarn to the treating liquors. It takes very much less time to wash a single end of yarn than it does to wash a large hank or cake. This single factor may cut down the time necessary for purification to one-tenth of what would otherwise be required.
2. The desulphurising and bleaching processes are omitted, so that the only treatment the yarn has to undergo is (*a*) completion of the coagulation process which started as soon as the yarn was extruded, (*b*) removal of the acid liquors from the yarn by washing, (*c*) drying.
3. The use of thread storing and thread advancing rollers.

Elimination of Desulphurising. The elimination of the desulphurising process (treatment with sodium sulphide) was the crucial step in the development of the process. Long ago it was known that if sulphur (and possibly some sulphur compounds) was left in viscose rayon and the usual sodium sulphide desulphurising wash omitted, yarn of dull or subdued lustre was obtained; this method was, in fact, one of the earliest used for making delustred yarns. It was at first suspect on the grounds that sulphur or sulphur compounds left in the yarn would oxidise to sulphur dioxide or sulphur trioxide and that the sulphurous and sulphuric acids so formed would tender the yarn or fabric. In the Nelson Process it is found that the sulphur residual in the viscose rayon is usually from 0·2 to 0·3 per cent, although it can be kept down to 0·1 per cent. There is no tendering, not even if much larger quantities of sulphur are left in the viscose rayon. Evidently the early fears that residual sulphur would cause tendering were groundless. Nearly all fabrics made from rayon receive some wet treatment for de-sizing, scouring or dyeing purposes and in any such process the residual sulphur in Nelson Process viscose rayon is removed, so that there is nothing to fear in this respect. Very occasionally, fabrics are woven for use, *e.g.*, in upholstery, in the loom state, *i.e.*, they receive no wet processing and in these relatively rare cases, Nelson Process viscose might not be suitable unless the yarn were given a scour before weaving.

Elimination of Bleaching. Once the decision to omit the desulphurising process had been taken, little difficulty was foreseen in omitting the bleaching. Actually, too, considerable quantities of cake-spun viscose rayon are not bleached. This has been made possible by the improved quality of the wood-pulp available, and by the fact that if it is necessary to obtain really perfect whites, then the viscose rayon can be bleached in fabric form. The decision to omit the desulphurising and bleaching processes clearly cut down the time necessary for the purification of viscose rayon to a fraction of what would otherwise have been necessary. Even so, if we assume that, as a result of the presentation of the yarn in single-end form and of the omission of these just-discussed processes, the time of treatment is reduced from, say, thirty minutes to two minutes, a considerable problem still remains. In two minutes a machine spinning at 70 metres per minute will produce 140 metres of yarn. How is this to be disposed during the purification processes? The answer in the case of the Nelson Process is by the provision of thread storage and thread advancing rollers.

Thread Storing and Advancing Rollers. The thread storing and thread advancing rollers are the essential feature, the heart of the Nelson Process. They consist of two rollers skew to each other, as shown in Fig. 63. The yarn passes round these rollers more than 100 times, and, as the diameter of the rollers is $6\frac{5}{16}$ in.—which corresponds to a circumference of nearly 20 in.—the time taken to make this journey, even if the rollers were close together, would be about one minute at a speed of 70 metres per minute; because, as is shown in Fig. 64, the rollers are some distance apart, the time taken by the yarn to pass from one end of the rollers to the other is between two and three minutes. Here, then, is a solution to the problem of purifying viscose rayon as it moves continuously from coagulating bath to take-up bobbin in a reasonable space: by " storing " the continuously moving yarn on

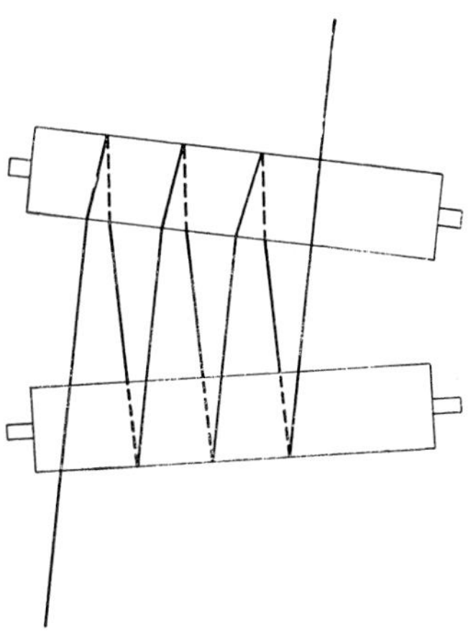

Fig. 63.—Showing how two rollers skew to each other cause a thread to advance along them.

two rollers, a length of time running into minutes is provided for the purification process. The rollers are each 3 ft. 8 in. wide, and are made of ebonite, except for the drying end, which is of corrosion resistant metal. Note that the spiral passage of the yarn along the roller is achieved by having two rollers a little out of parallel, as shown (exaggerated) in Fig. 63. In the Nelson Process the two rollers are a little out of parallel in two planes.

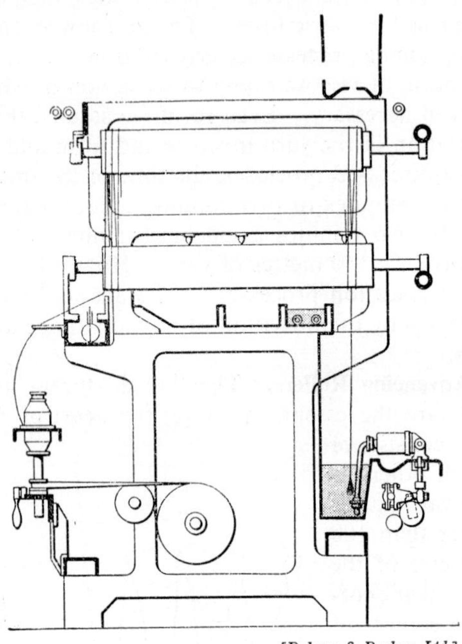

FIG. 64.—Diagrammatic sketch of the Nelson Continuous Rayon Spinning Machine.

Note filaments emerging from coagulating bath at right-hand side, passing round the thread-advancing rollers, and being taken up as dry yarn on the bobbin at the left.

[*Dobson & Barlow, Ltd.*]

Manufacture. The viscose is metered by a pump through a filter and through the spinnerets which are submerged in a lead-lined tank containing the usual coagulating liquor. The thread emerges at right angles to the surface of the bath, and is drawn up almost vertically to the top roller; it descends to the bottom roller and travelling round the lower half of its circumference, returns to the top roller. During the process it advances itself along the axes of the rollers (due to the slight inclination of the axes of the two rollers towards each other), and in this manner travels around the rollers and across the machine from the wet side to the dry side. The standard machine has forty-eight spindles and is 45 ft. long and 6 ft. wide. The rayon is spun on one side of the machine, called the " wet side ", and by its journey along the tilted rollers reaches the " dry side "; it passes on the rollers across the width of the machine. During the first fifty turns round the rollers (about 75 metres of yarn) coagulation is completed, partly by acid carried by the yarn from the bath, partly by acid sprayed on to it on the roller; in the next thirty turns the yarn is washed with two jets of water directed on to the top surface of the lower roller. The yarn now passes to the metal section of the rollers, which are electrically heated, so that in another thirty turns the yarn is dried. In 110 turns the yarn travels about 165 metres, so that if the spinning speed is about 70 metres per minute it takes the yarn nearly two and a half minutes to pass from coagulating bath to take-up bobbin. On leaving the drying section of the rollers the thread is guided over an oiling device, then through a lappet, and down to a cap-spinning or ring-spinning bobbin. The cap bobbin holds 12 oz., the ring bobbin 3–3½ lb. yarn. This process is an elegant solution of the problem of continuously spinning viscose, and is being extensively adopted by some of the leading spinners. It has already been used on a large scale, and gives a good product which can be made into the finest fabrics.

The Yarn Spun. In practically every respect continuous spun viscose rayon is similar to that spun by the older methods. Because the viscose emerging from the spinneret is converted into a thread which is not again touched by hand until it is dry, it is claimed that the quality is greatly improved and that this results in fewer warp breaks in weaving and in comparative freedom from bright picks, bars and stripes in the finished fabric.

A stretch can be applied by having a reduced diameter on the end of that roller to which the thread passes directly from the spinning-bath: as the thread next passes round the normal diameter of the roller, it is stretched. Yarn of 2-filament denier with a tenacity of 2·5 grams per denier and an extension at break of 20 per cent can be spun.

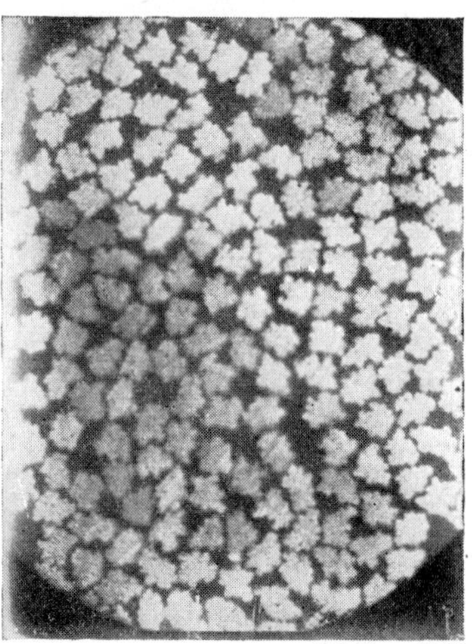

[*Harbens, Ltd.*]

FIG. 65.—75 denier 19 fils viscose rayon spun by the Nelson Continuous Process.

The most important physical difference from batch-treated yarn is that Nelson Process yarn, having been dried unrelaxed, shrinks on re-wetting; an allowance of 4 per cent must be made for this in respect of expected cloth widths, warp lengths, hosiery lengths, etc. Fig. 65 shows a cross-section of viscose rayon spun by the Nelson Process. The filaments are uniform and normal for viscose rayon (compare Figs. 75, 76).

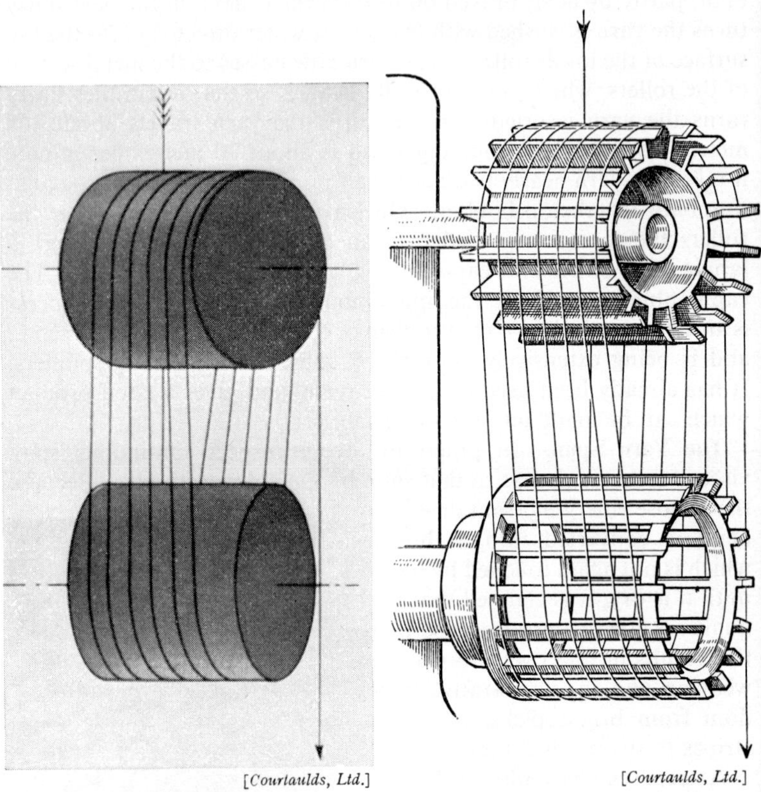

[*Courtaulds, Ltd.*]

FIG. 66.—A diagrammatic reproduction showing the principle of the advancing thread by two reels set eccentrically.

[*Courtaulds, Ltd.*]

FIG. 67.—Two cages (a sprocket above and a spider below) set eccentrically will also advance the yarn.

The Industrial Rayon Corporation Process

The Industrial Rayon Corporation of America have developed and successfully operated a continuous-spinning process for viscose rayon. The same problem—*viz.*, providing sufficient time for the purification of the rayon—had to be solved. Their method was to store the continuously moving yarn on helical reels. These are small and made of moulded plastic materials which resist the attack of the chemicals used in the spinning process. Each treatment can be isolated on one or more small advancing reels. The advancing helical reel was first suggested in some German patents as early as 1910, was then forgotten for a long time, was taken up again in America in connection with the spinning of cuprammonium rayon and finally adopted and used successfully for viscose rayon by the Industrial Rayon Corporation.

The Thread Advancing Reel. The thread advancing reel, which is the " heart " of the Industrial Rayon Continuous Spinning System, consists of two skewed reels, one inside the other. The principle on which it works is as follows: If a belt passes round two pulleys that are skew to each other, the belt worms its way off as they rotate; similarly, yarn passed round the pulleys travels or advances along them (Fig. 66). If reels or cages are used instead of pulleys, as shown in Fig. 67 the yarn still travels along them. If now one cage is placed inside the other, skewed, the thread advancing reel is made (Fig. 68).

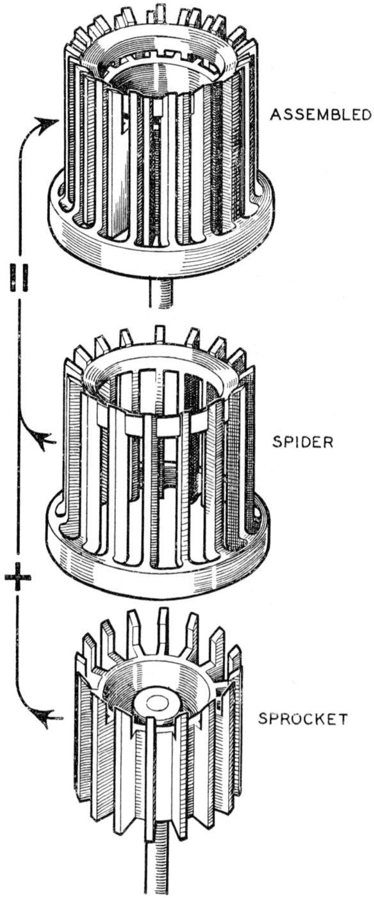

[*Courtaulds, Ltd.*]

FIG. 68.—Diagram showing how the sprocket fits into the spider on an eccentric axis to form the complete thread advancing reel.

The Industrial Rayon Corporation started their plant at Painesville, Ohio, in 1938; the introduction of this process marked a revolutionary change in the methods of spinning viscose in America, although even at that date Lustrafil Ltd. were already using the Nelson Process in this country. World rights (excluding U.S.A. and South America) of the Industrial Rayon Process have been purchased by Courtaulds, Ltd., who have built a factory in which Industrial machines have been installed.

The preparatory processes used are those common to all viscose plants, but as the yarn emerges from the coagulating bath it is drawn up to the first of the thread advancing reels, on which the yarn after many revolutions is advanced from the back to the outer end, and then falls to the next section, in which eight more advancing reels (made of plastic) carry the yarn through a like number of processing treatments for washing, desulphurising, bleaching and conditioning. The arrangement is shown diagrammatically in Fig. 69. Note that whereas the Nelson Process omits the desulphurising and bleaching processes, the Industrial Rayon Corporation Process includes them. The treating liquors are applied through glass tubes placed immedi-

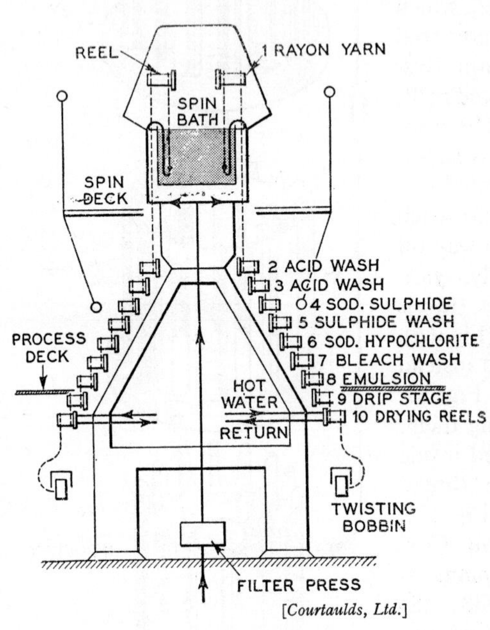

[*Courtaulds, Ltd.*]

FIG. 69.—Diagrammatic arrangement of Industrial Rayon Corporation Continuous Spinning Process.

Note that the filaments are formed in the coagulating bath at the top of the picture, and work their way downwards round ten thread advancing reels to be taken up (dry) on the twisting bobbin at the bottom of the picture.

ately above the individual reels, so that the liquor flows on to the yarn as it passes over the surface of the reels in a single layer. As a result, every inch of every filament receives the same uniform treatment. Finally the yarn passes to an aluminium drying reel which is internally heated with hot water or steam; then the dry yarn is cap-twisted on to plastic bobbins which hold about 3 lb. yarn. It is then ready for shipping. During the process the yarn has passed round ten reels; its time of processing is five and a half minutes, and more than 400 metres of yarn is travelling between jet and bobbin at the spinning speed, which is probably of the order of 80 metres per minute. Each machine has 200 spindles; it has three operating levels, whereas the Nelson Process is operated entirely from floor level. The Industrial Rayon machine is double-sided, the Nelson is single-sided.

The quality of the yarn produced by the Industrial Rayon Corporation Process is very good, and especially it contains relatively few knots and broken filaments. Before the 1939–45 War this process was producing 9,000 tons per year of viscose rayon; during the War the production was largely diverted to 1,100 denier high-tenacity yarn for tyre cord. The stretch necessary to give the high tenacity was applied by merely changing the gear ratio of two of the ten thread advancing reels. The process is in operation under licence throughout the world.

Conclusion

It seems clear that two processes of spinning viscose continuously have shown themselves satisfactory. A huge production of viscose yarn has been made by them, yarn that is claimed to be superior in quality to that made by batch processing. In addition, the Filamatic Process, although not yet publicly described, doubtless makes a third satisfactory process.

It has been reported that the Avisco Filamatic Continuous Spinning Process gives in a matter of seconds a dried and finished yarn of uniform physical properties and dyeing affinity, yarn which is said to be spun at a high speed.

Note that the main differences between the Nelson and the Industrial Rayon Corporation Processes are that the former uses one pair of thread advancing rollers; the latter uses ten single thread advancing reels; the former omits desulphurising and bleaching; the latter includes them.

Chemical Crimp

There is no need for continuous filaments to be crimped, but staple fibre is made a little more like wool by the insertion of crimp. The usual method is to pass the fibre through fluted rollers or some such device which will crimp it mechanically just before it is cut into staple. The finer qualities of wool such as 70 s merino have more crimp than coarser qualities, and a good crimp seems to enhance the quality of any yarn; even coarse *wool*, straight and free from crimp, has been crimped to make better carpets.

About 1939 it was observed in Japan that when viscose rayon was spun into a coagulating bath which was not, by ordinary standards, strongly enough acid, the viscose filaments were finely crimped. The process has been developed well and to-day most of Japan's staple viscose is sold chemically crimped. If the spinning coagulation bath contains 120 grams per litre of acid, the filaments are normal and do not crimp, but if the acid content is only 90 grams per litre then the filaments show a fine crimp; if the acid is reduced to 70 grams per litre they show a lot of crimp. In order, therefore, to make the crimped yarn, the viscose is spun into a bath with a low acid content and a high salt content. The coagulated filaments are immediately stretched 40 to 50 per cent in a second bath at 90° C., they are stretched a little more in air as they leave the bath, and are then squeezed and cut into staple. The fibres are still straight, but when wetted further by immersion in water they crimp. Then they are dried and baled. If tow is being made it is straight whilst it is dried under tension but crimps on wetting.

When the fibres that crimp are examined it can be seen that they are asymmetrical; all the serrations are on one half of the filament and the *skin* on this half is thinner than on the nearly smooth half (Fig. 70). It is the asymmetry that is responsible for the crimping but the mechanism of the formation of the asymmetry is less clear; it has been suggested that the reduced acid concentration retards the decomposition of the xanthate, but that the dehydration and coagulation of the viscose which are brought about mainly by the salts, which are high in concentration, in the bath proceed at their usual speeds. The coagulation process causes a skin to form round the filament whilst decomposition is incomplete; decomposition goes on and the products of it burst the skin at one side; coagulation and serration start again but the viscose is already too far decomposed to do more than form a very thin skin. This results in a fibre with one half nearly smooth in section and covered with a thick original skin (top half of Fig. 70) and with the other half serrated but with the thinnest of (secondary) skins (lower half of Fig. 70).

FIG. 70.—Cross-section showing asymmetry of "chemical crimp" fibre.

When a fibre is wetted it swells, and when a fibre with thick skin on one side and thin skin on the other is wetted, it will, if free, coil in such a way that the thick skin is on the inside of the coil—on the inside because it swells less than does the thin skin. If held more or less straight in yarn or fabric form it is not free to coil, but makes the first movement to do so, and this results in crimp. The two factors which contribute to the crimping power are (1) the asymmetry of the serrations, (2) the irregularity of the skin thickness—the filaments are called " broken-skin filaments "; either will give crimp if produced alone, ordinarily the two causes of asymmetry are produced simultaneously.

There is another way of producing skin asymmetry which also gives a crimping fibre. This is by making " conjugate filaments ", spun from a special jet (Fig. 71) wherein two different viscose solutions, one young and one old, are deliberately combined to give a filament with one side made from young and the other from old viscose. The young viscose gives a thick skin and the old viscose a thin skin. In Fig. 71 young viscose is fed into space C on one side of the dividing septum B, and old viscose is fed into D; both are squeezed side by side through the spinneret hole A; there is no turbulence, but only viscous flow so that the two solutions emerge side by side from the same spinneret.

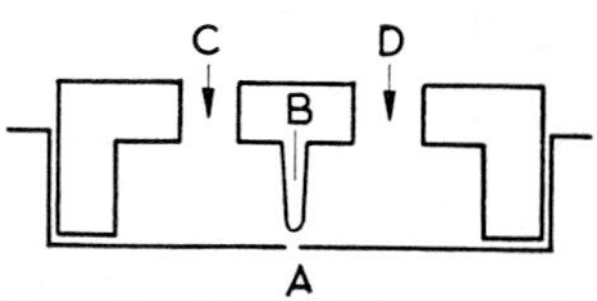

FIG. 71.—Type of jet used for making conjugate (crimping) filament.

This principle is the basis of the method used for making Orlon Sayelle, an acrylic fibre which develops a spiral crimp when finished (p. 516).

SARILLE

Sarille is a crimped viscose staple made by Courtaulds; its aim is to produce wool-like fabrics, but with its cellulosic structure it cannot proceed very far along this path. It has been used for dress fabrics with a full soft handle. In the knitwear trade it finds its way into interlock and double jersey. The success of the candlewick bedspreads that are made out of rayon staple has been attributed to Sarille. The crimp is due to asymmetry of structure of the fibre and is permanent in the sense that it will reassert itself in the absence of tension. If maximum relaxation of fabrics made from Sarille is allowed, the crimp exerts its full action, shrinking the fabric and making it bulky. Shrinkage should be about 10–12 per cent, and accordingly fabric must be woven sufficiently wide to allow for this. Sarille is usually spun on the cotton system; the coarsest filament denier possible for the required yarn counts should be used in order to yield crispness and voluminosity. For example, a 3-denier 2-inch staple can be spun into a 22 s (cotton counts, *i.e.* 240-denier) yarn. Very often the manufacturer of man-made fibres has his eye on the uses to which wool is put—how fine to make a " woollen " pullover or cardigan or dress out of cellulose or polyacrylonitrile. The purchasing public is perhaps not so discriminating as it once was.

Properties of Viscose Rayon

Tenacity and Elongation. Ordinary viscose rayon is reasonably strong; usually, in these days, its tenacity is about 2·6 grams per denier. This compares very favourably with cellulose acetate, which has a dry tenacity of only 1·3–1·7 grams per denier. The wet strength of viscose rayon is about 1·4 gram per denier. A great deal of work has been carried out to increase the wet strength of viscose rayon, and when it is remembered that as recently as 1934 the average wet strength was only 0·7 gram per denier, it will be realised that a considerable measure of success has attended the work. The elongation at break (dry) is about 15 per cent and (wet) about 25 per cent. The characteristic curves for viscose rayon dry (*i.e.*, with standard regain) and wet are shown in Fig. 72.

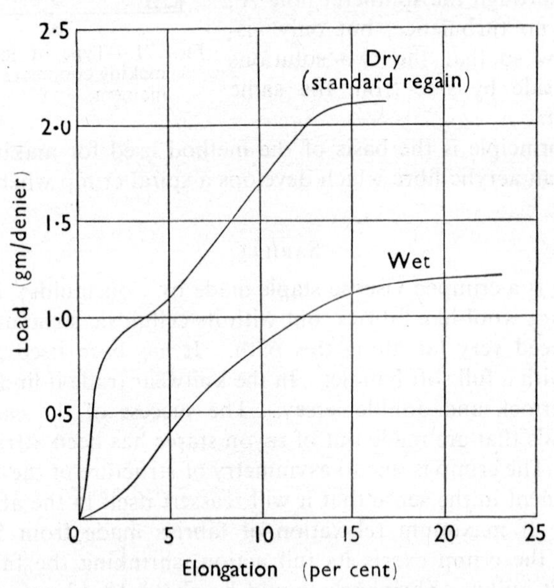

Fig. 72.—Stress–strain diagram of viscose rayon. The stress is measured by the applied load in grams per denier, the strain by the elongation as a percentage of the original length. Tenacity (1962) is usually rather higher (up to 2·5 grams per denier) than shown.

Moisture Content. Under standard conditions (65 per cent R.H. and 70° F.) the moisture content of viscose rayon is 12–13 per cent. The lower the humidity of the atmosphere the lower the moisture content of the yarn and *vice versa*. For example, at 20 per cent R.H. the moisture content of viscose rayon is about 5 per cent and at 90 per cent R.H. it is about 22 per cent.

Creep. The elasticity of viscose rayon is not high. If stretched and then released from strain it does not return quite to its original length, although for some time afterwards it continues to shrink towards, but not completely to, its original length. This phenomenon has been variously described as " delayed elasticity ", " creep " and " plasticity ". The effect of this behaviour is that if ends of yarn are during weaving exposed to sudden strains they may be permanently stretched, and will result in " shiners " and streaky dyeing.

Density. The specific gravity of viscose rayon is 1·52, which is high compared with wool and cellulose acetate, both of which have values about 1·32.

Electrical Properties. Owing to its high moisture absorption, viscose rayon does not lend itself particularly well to insulation purposes. When quite dry it is a good insulator, but the moisture that it inevitably picks up considerably reduces its value for electrical uses. Viscose is not so liable to develop static charges in textile working as is cellulose acetate.

Resistance to Light. On exposure to light, photocellulose is formed and weakening takes place. The loss in strength is less than that which occurs with real silk, but more than in the case of cellulose acetate.

Resistance to Heat. The ironing properties of viscose rayon are satisfactory, although on lengthy exposure to high temperatures it turns yellow.

Chemical Properties. Acids attack viscose rayon more quickly than they do cotton. The most important single factor is temperature. Cold solutions of acids and short times of treatment do not usually cause undue tendering. For example, 2 per cent acetic or formic acid is safe at room temperature, and 2 per cent oxalic acid may be used at temperatures up to 60° C. to remove (with luck) iron stains. At high temperatures acids carbonise viscose.

Biological Resistance. Moulds and mildew discolour and weaken viscose rayon; the initial attack is often made on the farinaceous size applied to viscose warps. Once the goods have been scoured and dyed, and the size thereby removed, the danger of attack is less. Moth larvae will eat viscose rayon mixed with wool, but they cannot digest it, and will excrete it unchanged; they do not attack all-viscose fabrics.

Susceptibility to Bleach. Sodium hypochlorite, neutral, is the most satisfactory bleach for viscose rayon. Potassium permanganate followed by a clearing solution of sodium bisulphite may also be used. Hydrogen peroxide can be used at temperatures not exceeding 55° C. Bleaching of piece-goods is not always, or usually, necessary, especially when the viscose yarn has been already bleached, as much of it has, during manufacture.

Action of Solvents. Dry-cleaning solvents do not attack viscose.

Nuclear Radiation. The effect of nuclear radiation on fibres is a subject that has acquired interest in the last few years. Four kinds of radiation have to be considered.

α-Particles—These consist of the nuclei of helium atoms and are characteristic of the emanation from *naturally* radioactive materials such as radium. α-Particles are heavy and have only poor penetrating power, only a few centimetres in air. They are not commonly encountered, and because of their poor penetrating power would not cause serious damage; because they are positively charged they are repelled by the nuclei of atoms in cellulose and only very occasionally will a direct hit be made. They would damage cellulose in sufficiently high exposure but the incidence of damage would be very slow.

β-Particles—These are high-speed electrons negatively charged; their mass is low and they have only very limited powers of penetration. Substances which emit β-particles as many of the radioisotopes do, *e.g.*, Carbon 14 and Phosphorus 32, are very useful for tracer work for mechanism and transfer studies. They have little effect on fibres; in practice their effect is negligible.

γ-Rays—These are similar in nature to X-rays and cosmic rays; they have very good powers of penetration and will damage animal tissue, causing cancerous growths and will also damage fibres. Experimentally a bright viscose rayon was exposed to γ-radiation from Cobalt 60, a radioisotope which is a powerful γ-emitter, for 42 hours and in that time its dry strength was reduced by more than half. At low loads the irradiated yarn stretched no more than the control, yet at high loads, it stretched more, as indicated by a flattening of the characteristic curve (Fig. 73). The changes in the wet strength were similar as shown in Fig. 74. It is of interest that when cotton is similarly irradiated its dry strength is reduced by 40 per cent and its wet strength by about 60 per cent so that although normal cotton (unlike viscose rayon) has a higher wet than dry strength, yet γ-irradiated cotton (like viscose rayon) has a lower wet than dry strength.

Neutrons—Neutrons are one of the three elementary particles of nature from which atoms are built up, the other two being the proton and the electron (various other particles such as the mesons usually have very short lives; their nature and formation are not well understood but they play only a minor part in the constitution of matter). Neutrons are like protons in mass but they carry no charge; they are heavy and have good penetrating powers, and because they are uncharged they are not repelled by the nuclei of atoms so that they frequently collide with and shatter them. Some elements, *e.g.*, beryllium, if suitably irradiated will emit neutrons, and the neutrons being heavy and uncharged are ideal missiles to aim at atoms which are required to be disintegrated and they are used for this purpose in atomic piles. When viscose rayon was bombarded in an atomic pile with neutrons for 26 hours in which time $2 \cdot 3 \times 10^{17}$ neutrons hit each square centimetre of the viscose rayon, the radiation

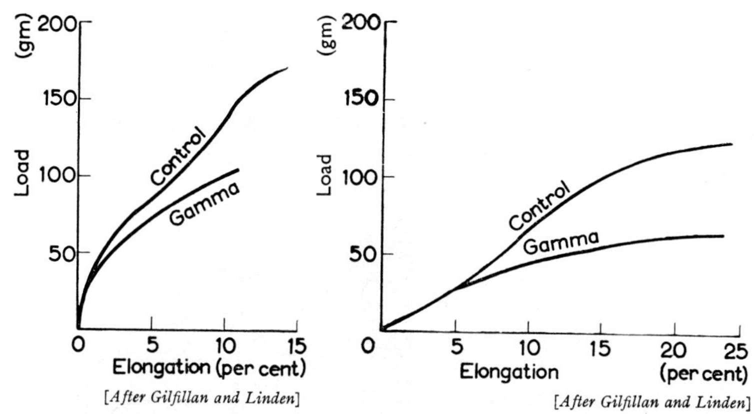

[*After Gilfillan and Linden*]

FIG. 73.—Effect of irradiation of viscose rayon with γ-rays on its stress–strain curve (normal regain).

[*After Gilfillan and Linden*]

FIG. 74.—Effect of irradiation of viscose rayon with γ-rays on its stress–strain curve (wet).

completely destroyed the strength of the yarn. As will be seen later (pp. 334 and 504) nylon and especially Orlon Type 81 stand up to neutron bombardment much better than viscose rayon. Cotton behaved nearly as badly as viscose rayon and retained only 2 per cent of its strength after a similar bombardment.

Morphology. Under the microscope the longitudinal view of viscose rayon is that of a striated cylinder, whilst the cross-section is highly, and characteristically, serrated (Figs. 75, 76).

AUTOMATION IN SLASHER SIZING

Rayon threads like cotton threads are wound on to beams which are put into looms, and so provide the warp threads for the shuttle carrying the weft to interlace together. Warps are often about 1,000 yards long; their preparation is a skilled business, and given a good warp the weaver's task of making good cloth is made much easier; a bad warp with loose ends, missing ends and perhaps

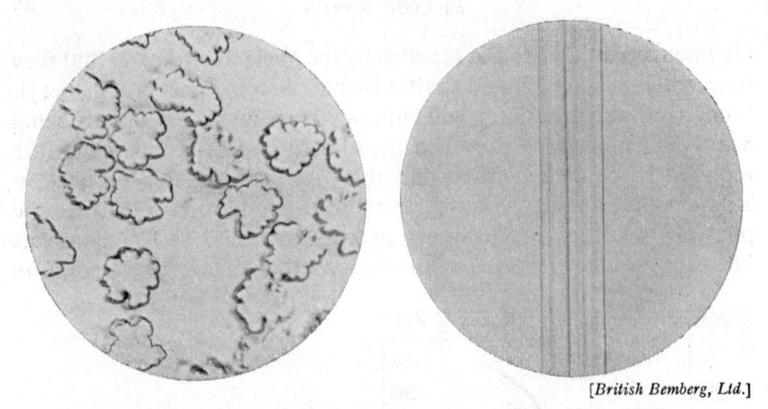

[*British Bemberg, Ltd.*]

FIG. 75.—Photomicrograph of viscose rayon filaments (× 300).

Note the highly serrated outline of the cross-section, and the striated appearance of the longitudinal view. Filament denier 5·5.

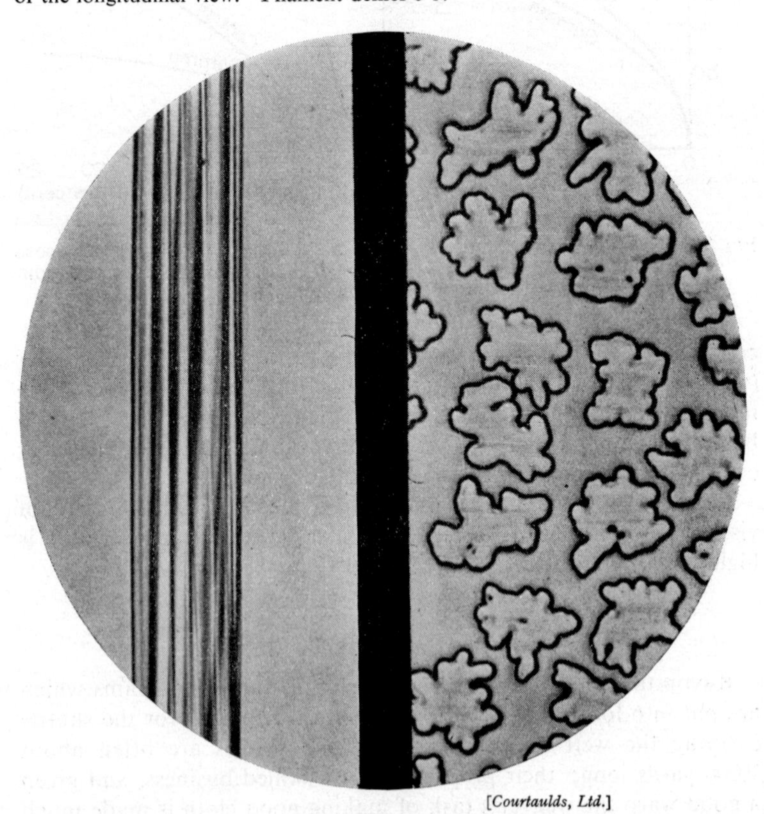

[*Courtaulds, Ltd.*]

FIG. 76.—Another photomicrograph of viscose rayon (×500).

threads stuck together, will defeat the best of weavers. At one time, in the early thirties, the art of making good beams had to be learnt, and during the learning process it was all too common for weaving factors who had bought the beams to find they were bad and send them back. Things have been better in this respect for a long time.

The last operation before a beam is sent to a weaver is known as slasher sizing. The threads, perhaps 5,000, wound parallel on the beam, are run through a solution of some colloid and then dried to

[*Foxboro-Yoxall, Ltd.*]

FIG. 77.—A Foxboro controlling installation on a Slasher Sizer operating at the Sheraton Mills Corp. N.C.

coat them and make them more able to withstand the ups and downs (quite literally in the healds or heddles) of life in a loom. At one time the colloid was usually starch with some tallow to act as a lubricant; at a later period gelatine was used especially for acetate warps, then modified starches, gums, glues and more recently polyvinyl alcohol. The size must necessarily be something that will wash out easily when the fabric which is made from the warp comes to be scoured in the dyehouse. Three exactly similar slasher sizers (dip bath, mangle and steam-heated drying drums) could in the old days be used to size warps under carefully controlled conditions. Each of the three slashers was looked after by an experienced man, but one slasher operative always got better warps than the other two, although all three were careful and industrious. The origin of the difference was never easy to pin-point. Nowadays such processes are automatically controlled, and the introduction of automation has reduced the likely differences between the work of different operatives. Figure 77 shows a set of Foxboro controllers on a 9-cylinder slasher. The following are the things that these instruments control on the slasher sizer: (1) the level and temperature of the colloid solution in the size box, so ensuring steady unvarying pick-up of size by the warp; (2) pneumatic loading (instead of screw pressure) of squeeze rollers, compensating for any out-of-roundness of the roller itself—this improves the uniformity of penetration of the fibre by the size solution; (3) automatic control of the temperature of each drying drum, including an optimum temperature gradient through the dryer, *i.e.*, from first to last drum; (4) automatic control of the pressure of the steam used for heating the size box and the drying drums.

In some mills the size-cooking, *i.e.*, the preparation of the size solution that will be used on the slashers, is also controlled automatically.

Automation nowadays plays a big part in textile processes, more especially (p. 746) in the wet processes of dyeing and finishing.

Dyeing

The high affinity of viscose rayon for water would lead one to expect it to dye easily, and in fact it can be dyed with direct cotton colours, for most of which it has an affinity higher than cotton has. Accordingly, dyeing of viscose rayon is usually carried out at temperatures lower than those used for cotton dyeing; less salt is needed to bring about exhaustion, and sometimes retarding agents are used.

The one great difficulty which has been met in dyeing viscose rayon is that slight variations in physical properties of viscose yarns affect the dyeing. Yarn or fabric which appears to be homogeneous before dyeing will often come up streaky and variable in shade. Even although the viscose yarn is of good quality and has been well spun, if, when wetted, it is submitted to irregular strain it is almost sure to show differential dyeing properties. This is probably because on wetting it swells, softens and loses its elasticity, so that irreversible changes take place in the structure of the viscose. Naturally, every care must be taken during fabrication to avoid tight ends and picks. The most satisfactory way of overcoming this defect of irregular dyeing has proved to be by selection of dyestuffs;

some dyestuffs emphasise slight irregularities in viscose yarn, whilst others seem to level up. Naturally the latter are preferred, and various criteria which have enabled these dyestuffs to be selected will be described. The dyeing differences are least noticeable when basic dyestuffs are used, but, as will be seen later, these suffer from other defects. Some direct dyestuffs are very good; others are very poor. The reason for this appears to be due to differences in rate of dyeing, some dyestuffs dyeing much faster than others. Conditions which minimise irregularity of dyeing on viscose rayon are :

1. choice of suitable dyestuffs—this is the most important;
2. high temperature;
3. long dyeing time;
4. concentrated dyebaths, *i.e.*, short liquor ratios.

Selection of Level-dyeing Dyestuffs

Criteria which enable level-dyeing direct dyestuffs to be segregated from unlevel dyeing colours (all on viscose rayon) are:

1. Capillary test. Viscose threads are suspended in the dyestuff solution. Even-dyeing colours rise only a small distance up the thread, but uneven dyeing colours rise considerably. Those dyestuffs which rise most are said to have a low " suction number " (this being a measure of the length of the *undyed* portion of the thread), *e.g.*, Sky Blue FF, which dyes unevenly, has a suction number of only 9, whereas Chrysophenine G, which dyes evenly, has one of 19. This capillary test was used in the period 1927–31; thereafter it was superseded.

2. Dyestuffs which redistribute themselves readily between dyed and undyed viscose rayon are level-dyeing. If, for example, a piece of dyed viscose rayon is boiled with some undyed material in a soap-bath, some of the colour will be transferred from the dyed to the undyed material. The more the two pieces of fabric or yarn " level up "—*i.e.*, the nearer together are the shades after boiling together for a standard time—the more level-dyeing is the dyestuff.

3. The "temperature range test". It was found that those direct dyestuffs which had their maximum affinity for viscose rayon, *i.e.*, gave the greatest tinctorial value, at temperatures of about 20° C. dyed evenly on viscose rayon, whereas other dyestuffs which had their maximum affinity at 90° C. dyed unevenly on viscose rayon. Those dyestuffs in fact which have their maximum affinity in the cold are the best. This cold-dyeing was used only to *select* suitable dyestuffs; they were actually

dyed on production at 85–90° C. in order to obtain good *fastness*, which cannot be obtained by low-temperature dyeing.

4. "Time of half-dyeing test". This is based on the knowledge that even dyeing of viscose rayon results when equilibrium is reached in the distribution of the dyestuff between fibre and dyebath. First of all, experiments are made to determine dyeing conditions under which at equilibrium the fibre absorbs 50 per cent of the dyestuff originally in the dyebath. Then, using these conditions but varying the time, the time required for a 25 per cent absorption of dyestuff is determined, and this is known as the "time of half-dyeing". Those dyestuffs which dye viscose rayon unevenly have high times of half-dyeing, whereas those which dye it evenly have low times of half-dyeing, *e.g.*, Chlorazol Fast Orange AG, which dyes unevenly, has a time of half-dyeing of over two and a half hours, whereas Chlorazol Fast Orange G, which dyes evenly, has a time of half-dyeing of less than five seconds.

The "temperature range test" gave good results on most dyestuffs and failed only when a range of faster than usual to light dyestuffs was introduced. This led to the introduction of the time of half-dyeing test, which, being based on the physical equilibrium that exists between fibre and dyebath for the dyestuff, gives sound results.

Classification of Direct Dyestuffs. Whittaker has classified direct dyestuffs into three classes:

Class A. Those which exhaust well on to viscose rayon without salt and which dye level. Most of these have times of half-dyeing of less than five minutes, and often of only seconds. These dyestuffs can be applied to viscose rayon throughout at 90° C. The addition of salt increases the degree of exhaustion of the dyebath.

Class B. Those which exhaust only poorly without salt and which level badly. Most of these dyestuffs have times of half-dyeing of more than eight minutes and often of some hours. Dyeing can be conducted throughout at 90° C., but the salt used must be added very gradually, otherwise unlevel dyeing will result.

Class C. Those which exhaust well on to viscose rayon without salt, and which dye unlevel. Most of these have times of half-dyeing between six and thirty minutes. One method of applying these dyestuffs is to start dyeing at a low temperature, to raise to 90° C., and then gradually add salt to increase exhaustion. Alternatively they can be dyed throughout at 90° C. without the addition of any salt.

This classification of dyestuffs is useful to the practical dyer; when he is making compound shades, he tries to select all his dyestuffs from one class, *i.e.*, all from Class A or Class B or Class C, not one from Class A and another from Class B, *etc.* (see also p. 212).

Application of Direct Dyestuffs. Direct colours are applied from a bath containing dyestuff, water and either common salt or Glauber's salts. A dyeing temperature of 85–95° C. gives the most uniform results. If sample dyeings are uneven, even after selection of the best dyestuffs available, the salt or Glauber's salts should be reduced in quantity, or even omitted altogether for the bulk lots, as this will probably improve the levelness.

If viscose yarn is dyed in hank form, soap or a sulphated fatty alcohol should be added to the bath, to act as a lubricant and reduce the friction and chances of hairiness in the subsequent winding of the dyed yarn.

The following dyestuffs will give reasonably level dyeings on viscose rayon which is itself not entirely uniform.

Yellow.	Chrysophenine GS *
	Chlorazol Yellow 6GS
Orange.	Chlorazol Orange POS *
	Icyl Orange GS
Red.	Durazol Red 2BS
	Chlorazol Fast Scarlet 4BS *
Violet.	Durazol Helio BS *
	Icyl Violet RS
Blue.	Diazo Sky Blue 3GL
	Durazol Blue 2GNS
Brown.	Chlorazol Brown CVD
	Chlorazol Brown BS
Green.	Benzo Viscose Green B
Black and Grey.	Durazol Grey NS *
	Icyl Blue Black 6BS

Those dyewares that are marked with an asterisk in this list are suitable for dyeing not only continuous filament viscose rayon, but also for staple fibre viscose to which a crease-resistant finish has been applied. This finish considerably reduces the light fastness of some dyestuffs, but not of those indicated above.

The method of direct dyeing viscose rayon consists of these essentials:

1. Use selected level-dyeing colours and common salt.
2. Dye at 85–95° C.

3. Use a fairly short liquor : goods ratio.

4. Enter at about 20–30° C., raise the temperature slowly to about 90° C. and maintain there for about forty-five minutes.

Various after-treatments may be given to direct dyed viscose rayon. These are:

1. Diazotisation and development of the dyestuff on the fibre, coupling with diazotised *p*-nitraniline.

2. After-treatment with chromium compounds to improve the fastness to washing, especially of those dyestuffs with free hydroxyl groups. Treatment is for thirty minutes at 95° C. with 3–4 per cent chromium fluoride and 1–3 per cent acetic acid (30 per cent) on the weight of the goods.

3. After-treatment with copper compounds, which improves the fastness to light, particularly of blues. Treatment is for thirty minutes at 60° C. with 0·5–3 per cent copper sulphate and 2–3 per cent acetic acid.

4. Combined copper and chromium treatments may be given.

5. After-treatment with formaldehyde. Treatment is for thirty minutes at 90° C. with 3 per cent formaldehyde (40 per cent) and 2 per cent acetic acid. In cases where the dyestuff molecule contains two amino-groups or two hydroxyl groups *meta* to each other reaction may take place in this way:

$$\text{R—, R}'\text{—C}_6\text{H}_2(\text{—OH})(\text{—OH}) + \text{OCH}_2 \longrightarrow \text{R—, R}'\text{—C}_6\text{H}_2(\text{—O—CH}_2\text{—O—}) + \text{H}_2\text{O}$$

The groups R and R′ are meant to represent the remainder of the dyestuff molecule.

Basic Dyestuffs without Mordant. Brilliant shades, that are quite impossible to attain in any other way, can be obtained with basic dyestuffs. The dyestuff is dissolved in water, and from 2 to 5 per cent acetic acid (30 per cent) on the weight of the goods is used. The goods are entered cold, and the temperature is raised to 68° C. and dyeing continued at that temperature. Defects of this method, which largely offset the advantage of the brilliance of shades obtainable, are:

1. The shades are of only poor fastness to washing and light.

2. Only light shades can be obtained, without the use of a mordant.

Basic Dyestuffs with a Mordant. This method has to be used if medium or heavy shades are required from basic dyestuffs on viscose rayon. Two kinds of mordant are used:

1. tannic acid and antimony;
2. synthetic mordants such as Taninol BM.

When tannic acid and antimony are used, the antimony is conveniently applied in the form of tartar emetic, which is potassium antimonyl tartrate

$$\begin{array}{ll} COO(OSb) & \\ | & \\ CHOH & \tfrac{1}{2}H_2O \\ | & \\ CHOH & \\ | & \\ COOK & \end{array}$$

The goods are entered in a bath containing water and 1–5 per cent tannic acid, according to the depth of shade required; for light shades the smaller amount of tannic acid is sufficient, but for heavy shades up to 5 per cent will be used. The bath is raised to 60–70° C., and the goods are worked for one to two hours and then hydro-extracted without a preliminary rinse. This process simply impregnates the viscose goods with tannic acid, and this acid is next fixed on the fibre by working for thirty minutes in a cold bath containing 0·5–2·5 per cent tartar emetic. The mordant being now fixed, the goods may be safely rinsed.

When the mordanting is completed, the goods may be dyed. The basic dyestuff is dissolved and added to the bath with 2–5 per cent acetic acid (30 per cent). The presence of the acid retards the dyeing, and in this case the lighter shades require more acid than the darker shades. The goods are entered cold, and opportunity is given for the dyestuff to dye on to the fibre in the cold; the temperature is then raised to 50–60° C.

When Taninol BM, which is typical of the synthetic mordants, is used the procedure is as follows: the Taninol BM is pasted with cold water and 10 per cent of its weight of soda ash in water is added. Boiling water is added to dissolve the paste. From 2 to 6 per cent Taninol BM is used for one hour at 90° C. with a short liquor ratio, together with 20–40 per cent sodium chloride. The goods are rinsed in water and then in 1 per cent acetic acid. Dyestuff is applied exactly as in the case of the goods mordanted with tannic acid and tartar emetic.

Some basic dyestuffs, used with a mordant, dye viscose rayon much more evenly than others and, of these, some of the best are:

Yellow.	Auramine OS
Orange.	Acridine Orange RS
Red.	Safranine TS
Violet.	Methyl Violet 2BS
Blue.	Methylene Blue 2BS
	Victoria Blue BS
Green.	Malachite Green AS cryst.
Brown.	Bismarck Brown

Before leaving this subject of mordants, it should be added that a direct dyestuff will itself sometimes function as a mordant—*i.e.*, fabric is dyed nearly to shade with a direct colour, and then topped up with a basic dyestuff. In this event neither the tannic acid-tartar emetic mordant nor the synthetic mordant is required.

The poor fastness of basic dyestuffs has always been a serious disadvantage and has greatly restricted their use. Within the last few years improved methods of applying vat dyes have been provided and nowadays basic dyestuffs are not often used on viscose rayon.

Sulphur Colours. The fourth method of dyeing viscose rayon is to use sulphur colours. These colours are used mainly when very good fastness to washing and wet processes is required. They are fast to cross-dyeing. They do not level up variations in viscose yarn, but tend to dye unlevel, and are consequently more suitable for application to viscose staple fibre in slubbing form than for application to yarn or fabric. They are useful for blacks; dinitrophenol black is an important member of this class. When fabric is dyed it is usually done on the jig. Less salt is required than when sulphur colours are used on cotton.

Azoics. The azoic colours of which the Brenthols are typical have a much greater affinity for viscose rayon than for cotton. The colours are reasonably fast, but do not approach the fastness of vats. Brenthols are, however, used for furnishings and casements. Azoics can be dyed in an open beck on yarn, and on winch or jig for piece-goods. Any Brenthol can be combined with any base, and different shades result from different combinations. Brenthol FR (*o*-anisidide of β-oxynaphthoic acid), when coupled with Fast Red RC Base (4-chloro-2-aminotoluene), gives a red which is fast to bleaching. If the same base is coupled with Brenthol BT (*p*-chloranilide of 3-carboxylic-2-hydroxycarbazole) a brown results. The method of application is as follows: The Brenthol, which is insoluble, is first pasted with caustic soda and dispersed in water with Turkey-red oil; boiling water is then added and the dyestuff goes into solution.

The goods are impregnated for twenty to thirty minutes with the solution and hydro-extracted. The Fast Red RC base is diazotised with excess sodium nitrite and hydrochloric acid, and the goods are immersed in this solution. Finally, a soaping off is given at 65° C. to remove surface colour, and so leave the goods fast. The colour that is dyed on the goods is fast; it is necessary only to remove the loose surface colour by soaping to obtain a fast-dyed product. It is interesting to compare this azoic dyeing of viscose with the azoic dyeing of a modacrylic fibre (p. 763).

Vats. For the highest fastness to light and washing, vat dyestuffs are suitable; they really are superbly fast. Viscose for window hangings, for furnishings and for effect threads in expensive fabrics should be vat-dyed. Compared with cotton, viscose rayon has a much greater affinity for vats, but dyes more unlevel, so that precautions have to be taken.

There are four main ways in which vat dyestuffs may be applied to viscose rayon yarns and fabrics; these are:

1. dyeing (vat, jig or winch);
2. pigment padding;
3. the Standfast molten metal method;
4. use of solubilised vat dyestuffs.

Of all the artificial fibres, viscose rayon is the one best suited to the application of vat dyes, and it is on this fibre that they are mainly (apart from their use on natural fibres) used. Their use will therefore be discussed in this chapter.

Dyeing (*Vat*, *Jig or Winch*). The vat dyestuff is purchased usually in paste, but sometimes in powder form. It is quite insoluble in water and is dispersed (not dissolved) by the dyer in water with a little Monopol Soap; then for each 1 lb. of dyestuff paste about 1 pint of 70° Tw. caustic soda and 4 oz. Monopol Soap are added, the mixture is thinned by the addition of more warm water and then about 4 oz. sodium hydrosulphite is stirred in. This gives the stock solution of the alkaline *leuco* compound which is later used for making up the dyebath. This alkaline *leuco* compound is soluble in water and the solution so prepared should be clear; the colour of this solution bears no relation to that of the original vat dyestuff, most often the alkaline *leuco* compound solution being yellow or colourless. Cloudy vats can be cleared by adding a little more alkali or hydrosulphite, or both.

The dyebath is made by filling the vat with warm soft water and adding a little hydrosulphite and caustic soda. Then the correct quantity (the quantity of dyestuff varies according to the depth of shade required) of the stock solution of the *leuco* compound is added through a sieve and thoroughly stirred—this completes the preparation of the dyebath.

The viscose rayon is introduced into the dyebath and moved about in it (hanks of yarn are manipulated on poles) for about one hour at the specified temperature. Then the rayon is squeezed, thoroughly oxidised by exposure to air, and soured in dilute (0·2 per cent) sulphuric acid. This oxidation re-converts the soluble *leuco* compound to the insoluble vat dyestuff; during oxidation the rayon attains the colour of the pigment. When oxidation is complete, the rayon is soured, rinsed in cold water, and then treated in a soap solution at 75° C. to remove surface colour; it is rinsed in cold water and dried.

The chemical reactions that take place are typically the conversion of the $>C{=}O$ group which is characteristic of vat dyestuffs to the *leuco* compound in which this group has become $\geqslant C{-}OH$, having been reduced by the hydrosulphite; it is then converted by the caustic soda to the group $\geqslant C{\cdot}ONa$, which is characteristic of the soluble *leuco* compound. The soluble sodium *leuco* compound dyes uniformly on to the rayon, for which it has a marked affinity; after dyeing the $\geqslant CO{-}Na$ group is re-converted to the original $>C{=}O$ group as the original vat dyestuff is reformed by the oxidation and souring processes.

The use of vat dyestuffs gives excellent fastness to light and to washing (because dyestuffs of this class are so insoluble in water) and to crocking or rubbing (providing that loose surface colour has been effectively removed by adequate soaping and rinsing after the oxidation stage). The whole rather complicated vat-dyeing process is simply a means of converting the vat dye into a soluble form with which the fabric can be uniformly " dyed " or impregnated, and then re-converting it (now safely in the fabric) to the insoluble form.

One point to which attention should be directed is that the whole range of vat dyestuffs includes many members of very different composition. Some need strongly alkaline baths; some vat dyestuffs need much more hydrosulphite than others, and finally there are considerable variations in the temperatures and times of application. All these variations make the selection of suitable mixtures for compound shades a matter for great care.

Pigment Padding. This is a modification of the process described above. The yarn or fabric is first impregnated with a very fine dispersion of particles of the unreduced vat dyestuff; the dyestuff is not " dyed " on at this stage but is simply held mechanically. The impregnated yarn or fabric is then passed into a caustic soda–sodium hydrosulphite bath in which the dyestuff is solubilised and the soluble *leuco* compound " dyes " on to the fibre, which is soured, oxidised and soaped as usual. The *leuco* compound has an affinity for viscose rayon and can truly be said to "dye" on to it, even though the final colour does not develop until the oxidation stage is reached.

Pigment padding was originally introduced to deal with very tightly woven materials such as ducks and drills which were difficult to penetrate in the vat. It has, however, been found to give such excellent results that the process has been adopted for finer fabrics such as spun rayons and viscose rayon/cotton unions. The dyestuffs used for this process must be in an extremely finely divided form and suitable grades are manufactured especially for this process; a wetting agent such as Calsolene Oil HS is often used to improve the dispersion of the padding bath. A padding mangle is normally employed for the impregnation or padding operation and a jigger for the reduction with caustic soda and hydrosulphite of impregnated fabric.

The American Pad-Steam Continuous Dyeing Process is a modification of the Pigment Padding process just described. In it, fabric is first impregnated with a dispersion of the unreduced vat dyestuff and is air-dried and cooled; it then passes through a caustic soda–hydrosulphite bath, and from this to a steam cabinet where the high temperature rapidly (ten to sixty seconds) reduces the vat dyestuff to the *leuco* compound which rapidly penetrates and dyes on to the fibres. The whole series of operations is conducted continuously. Finally the fabric is oxidised and soaped as usual. The process gives good results in respect of brilliance, penetration and fastness.

The Standfast Molten Metal Method. This process represents a radical departure from traditional textile processes and as such is to be welcomed. The day has come when the organic chemist has prepared such a multiplicity of dyestuffs that they will meet any reasonable requirement and the attention of research workers is accordingly being directed more and more to the development of improved methods of *applying* the dyestuffs; the Standfast process is an excellent example of this trend. So far it has been used for viscose rayon fabrics, for viscose rayon/cotton mixtures, and cottons.

It consists of the passage of the fabric, which is pre-heated by passing it over hot rollers, through a bath of the reduced vat dyestuffs, *i.e.*, vat dyestuff, caustic soda and sodium hydrosulphite; in its passage through this bath the fabric picks up some of the liquor. Then the fabric passes immediately through a bath of molten metal at about 95° C. The molten metal has three functions:

1. it provides a rapid and even supply of heat *in the absence of air* to fabric which has been impregnated with dyestuff;

2. it exerts a pressure on the fabric which keeps in it the amount of moisture necessary for the reduced dyestuff to be fixed on the fabric;

3. it facilitates the use of an extremely small dyebath.

The molten metal consists of a fusible alloy with the approximate composition:

Tin	13·3
Lead	26·7
Bismuth	50·0
Cadmium	10·0
	100·0

This alloy melts at about 70° C., and is maintained in the dyeing process at about 95° C. The fabric passes through the molten metal in about ten to fifteen seconds and is thereby dyed; subsequently it requires only the usual oxidation and soaping. The novel feature of the process is the use of the molten metal bath which, owing to the intimacy of contact of the heat source (metal) with the fabric, enormously accelerates the rate of dyeing. The dye liquor at 70–75° C. is supported directly by the molten metal.

The apparatus used is shown diagrammatically in Fig. 78. The undyed fabric passes round the steam-heated rollers, *A*, to pre-heat

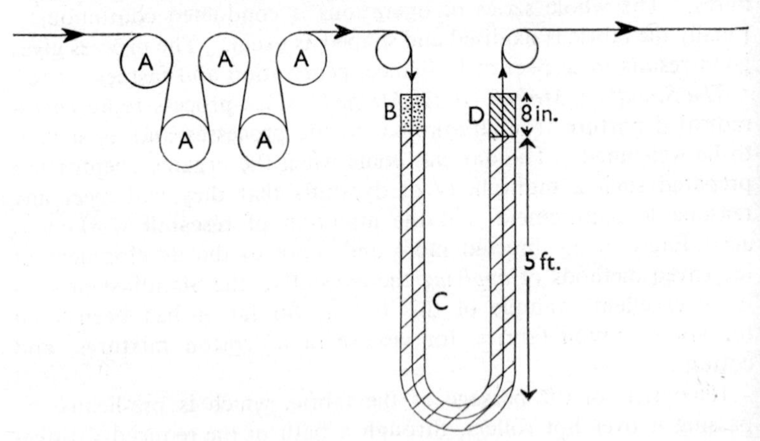

FIG. 78.—Diagrammatic representation of Standfast molten metal continuous dyeing process. Undyed fabric enters at the left, whilst dyed fabric, requiring only oxidation and soaping, leaves at the right.

A. Pre-heating rollers (steam heated). B. Reduced vat dyebath floating on C. C. Molten metal bath. D. Strong salt solution floating on C.

it, so that it will not cool the dyebath, *B*, into which it next runs; this dyebath, *B*, contains the vat colour in the reduced form with caustic soda and hydrosulphite. The fabric then runs round the U-tube of molten metal, *C*, at about 95° C. and then through the strong salt solution, *D*, which floats on top of the molten metal. Oxidation and soaping are accomplished in the conventional open width soaper through which the cloth runs continuously. Fabric can be run through this apparatus very fast, *e.g.*, pale and medium shades can be dyed at about 100 yards per minute and heavy shades at 50 yards per minute.

When using traditional methods of dyeing, it is found that different vat dyestuffs require different concentrations of alkali and hydrosulphite and different conditions of temperature and time, but with the molten metal process, practically all vat dyestuffs can be used with similar conditions, so enormously simplifying the matching of compound shades.

Two modifications of the molten metal process which may be noted are:

1. the fabric may, if desired, be pigment padded in a dispersion of the vat dyestuff and then passed through an alkaline reducing bath floating on the top of the molten metal, *i.e.*, the dyestuff is applied to the fabric from a padding mangle instead of being mixed with the alkaline reducing liquor;
2. the fabric can be *printed* with the vat dyestuffs (unreduced) and then passed through the alkaline reducing agent floating on the top of the molten metal.

As yet, the molten metal process has been used only for viscose rayon (and cotton, *etc.*); a limited amount of acetate and cotton brocades, in which the cotton is fully dyed but the acetate only tinted, giving a two-tone effect on the fabric, is molten metal dyed. But the method has not been generally applied to the dyeing of acetate, nylon and Terylene owing to the lack of affinity of these fibres for vat dyestuffs (a high affinity is necessary because of the shortness of the time of dyeing).

The complete train of equipment is shown in Fig. 79, wherein System 1 utilises a dyebath containing the vat colours in the reduced form with caustic soda and hydrosulphite, and System 2 uses a padding mangle containing the pigment dyestuff prior to the dyeing machine. System 1 is most suitable for light fabrics, System 2 is best to give adequate penetration on densely woven cloths or cloths containing highly twisted yarn such as heavy poplins.

Solubilised Dyestuffs. In all the three methods of vat dyeing

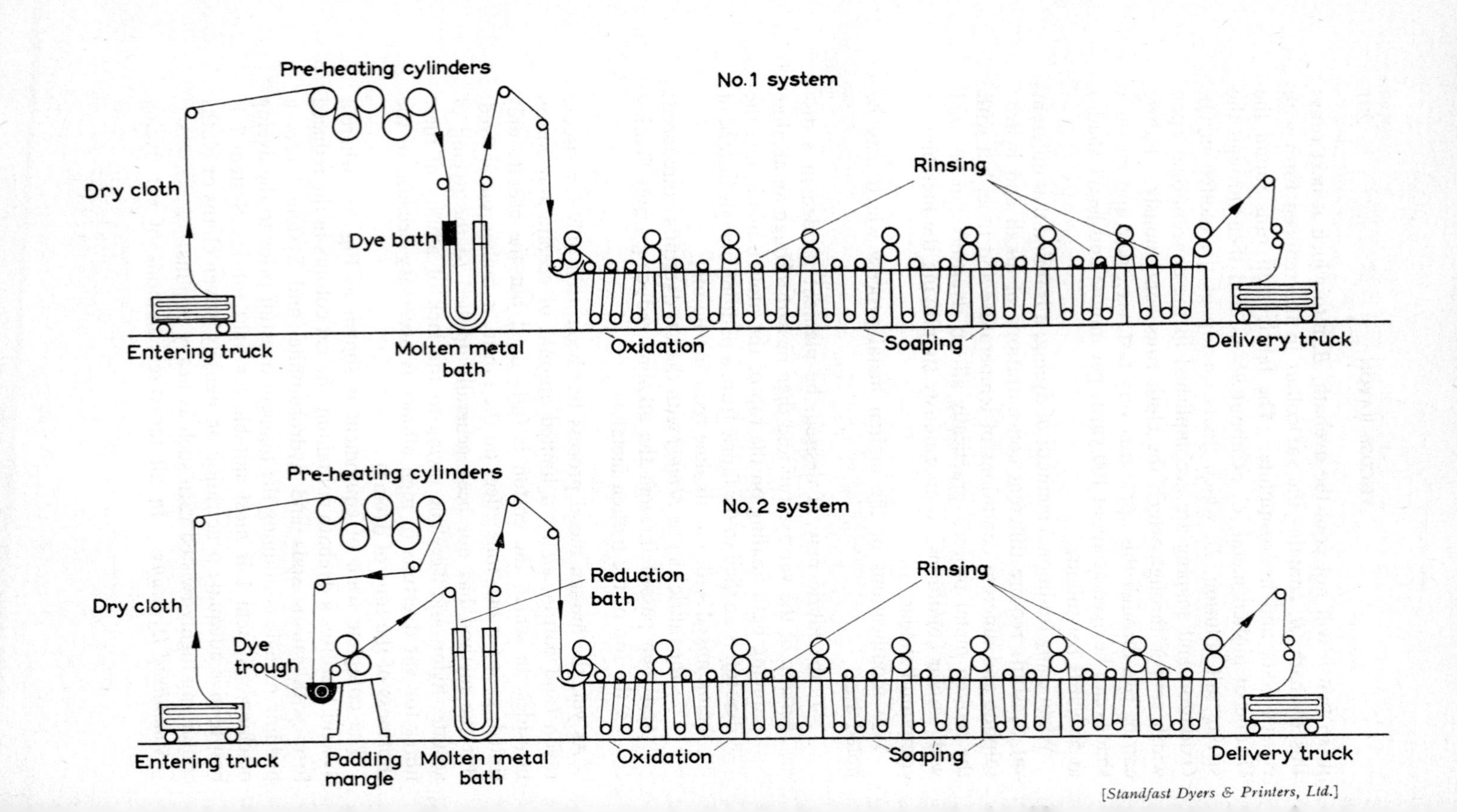

[*Standfast Dyers & Printers, Ltd.*]

which have just been described, the vat dyestuff is purchased in the insoluble form and has to be converted to the soluble sodium *leuco* compound by treatment with caustic soda and sodium hydrosulphite just before the dyeing operation. There is, however, another method of rendering a vat dyestuff soluble in water and this method, which consists of converting it to a sulphuric ester, has the advantage that the dyestuff in the soluble form, *i.e.*, as the ester, is stable and can be marketed as such by the dyestuff manufacturer. The Indigosols and Soledons are dyestuffs of this type; they are characterised by the possession of $-OSO_3Na$ groups which confer water solubility.

They are used extensively on viscose rayon fabrics and on viscose rayon/cotton unions, particularly in pale shades. Solid effects on unions are not easy to obtain in heavy shades because the viscose rayon has a much higher affinity than the cotton for the dyestuff.

In application, the dyestuff is dissolved in water (no caustic soda or sodium hydrosulphite is necessary) and some sodium nitrite and Glauber's salts are added. This solution can be made up in the jigger and fabric dyed in it at about 50° C. for about forty-five minutes (conditions vary from one dyestuff to another). Then the fabric is squeezed and developed on a second jigger in dilute sulphuric acid; this development changes the soluble ester with which the fabric has been impregnated in the first bath to the original insoluble vat dyestuff, the ester group splitting off. After development the fabric is rinsed, neutralised with a dilute sodium carbonate (soda ash) solution, rinsed and soaped at about 70° C. The winch may be preferred to the jigger for delicate fabrics.

These soluble dyestuffs (Indigosols and Soledons) give, as a rule, better penetration than can be achieved with the normal vat dyestuffs. Probably because they are so easy to apply they have become very popular particularly for dyeing lingerie shades on viscose rayon.

Procion Dyestuffs. In 1956 I.C.I. introduced the Procion dyestuffs; their chemistry and general characteristics are discussed on p. 771. From the practical standpoint the following considerations are relevant to their application to viscose rayon:

1. No special equipment is required.
2. Processes of many kinds can be used: from a simple cold batch to continuous high speed, *e.g.*, the Standfast Molten Method Process.
3. Fixation of dyestuff to the fibre is by direct covalent linkage.
4. Shades are brilliant and a wide range is available.

5. Fastness to washing and to rubbing are excellent (one must rub away the fibre). Fastness to light and perspiration are good.

Procions can be applied batchwise to viscose rayon: dyestuff and salt (to urge the dye out of solution on to the fibre) in neutral solution are padded on to the fabric say at 25° C. for 30 min., then sodium carbonate is added to the dyebath, still cold, to fix the dyestuff, to make it combine with the fibre. Dyeing is continued for another 30 min., followed by good washing and drying. Yarn in hank form can be satisfactorily dyed with Procions on the Hussong machine.

That the process can be carried out cold is useful for tufted carpet dyeing; the crimp is preserved and the Procions give uniformity of shade.

The Procions can be used for printing, but when so used the thickening agent must be sodium alginate for the following reason: most of the thickening agents used for textile printing are carbohydrates chemically similar to cellulose. The Procion dyes react with such thickening agents and in most cases the reaction renders the thickening agent insoluble and difficult to remove from the fabric. Sodium alginate thickenings, however, retain sufficient solubility after printing to enable them to be removed in the usual way. The low chemical reactivity of the hydroxyl groups in the alginate radical is well known. Rather unexpectedly, calcium alginate (which is insoluble) is dyed by the Procions which combine with it. Apparently, when alginates are in solution (as they are in printing paste) the cations associated with the carbonyl group defend the hydroxyl groups and make them inaccessible to the reactive chlorine atoms on the Procions.

Uses

The uses of viscose rayon are manifold. Practically all the textile field is open to it. Hose, underwear and dress-goods made from viscose are ubiquitous. The old defect of poor washability has practically gone. Linings are particularly good in viscose rayon; they are resistant to abrasion, bright, shiny and slippery; the garment lined with viscose rayon slips readily into position. Rayon staple fibre, made crease-resistant, has provided enormous new outlets for viscose; these will be considered separately, as will also high-tenacity viscoses and the polynosic rayons.

Much of the ready acceptance of viscose rayon is due to the system of quality control which some of the leading fabric manufacturers have adopted. They have specified minimum requirements for non-slip, seam resistance, colour fastness and strength. This policy served to overcome a dangerous tendency which was becoming widespread about 1930 of making and marketing fabrics thinner and thinner, weaker and weaker, poorer and poorer, cheaper and cheaper. Each manufacturer tried to sell fabric cheaper than anyone else, and impoverished fabrics flooded the market. The system of quality control killed that evil.

Viscose rayon is suitable for all normal textile needs, including those of apparel, but it is not suitable for such purposes as sea-ropes, fishing-lines and nets, insect netting, or for materials subject to chemical contact. Typical of the uses to which viscose rayon is put are the following: curtains, chair coverings, transport furnishings, table-cloths, cushions, bedspreads, quilt covers, lace, fine fabrics for bridal and evening gowns, day and afternoon dresses, beach and sports wear and underwear.

One of the fundamental advantages that man-made fibres have over natural fibres is that the fibre length is controllable, *i.e.*, it can be whatever is desired. But much of cotton is too short in staple length to be spun properly, and the same can be said even more cogently of asbestos fibre. This last is relatively scarce and expensive, and it is imperative to make the best use of what can be found. Accordingly, a substantial poundage of rayon staple is used as a carrier fibre to assist in the spinning of asbestos and short-staple cotton. The rayon staple, cut to a desirable length, is easy to spin, and the natural short fibre is run on with it.

Evlan

Evlan is a viscose rayon staple fibre used by the carpet trade. It has a smoother cross-section than standard viscose rayon, and this reduces soiling that may take place on the carpet. It has a higher crimp than usual, and this gives good cover; it has a higher strength than usual, and this gives durability. Evlan has been widely used since its introduction in 1962. Evlan M is a similar fibre, but has better resistance to abrasion.

Hollow Filament Viscose Rayon

One of the most desirable characteristics that a fibre can have is that of good covering power; those fibres that have low specific gravities will, after they have been converted into fabrics, give better cover than heavier fibres. Of two fabrics each weighing 4 oz. per yard, one made from cellulose acetate (specific gravity 1·30) would have a much better cover than another made from viscose rayon (specific gravity 1·53). Another factor which determines covering power is the cross-sectional shape of the filaments; such fibres as nylon and Terylene are extruded from the melt and their filaments have circular cross-sections and these have only a low covering power; other fibres such as those of viscose rayon and cellulose acetate have irregular or lobed cross-sections, and these fibres have greater surface per unit volume than fibres with circular sections. Flat fibres have the greatest surface per unit volume and reflect most light and have the greatest covering power. Although flat fibres are not usually liked, mainly because they have a harsh feel, round filaments are by no means ideal, and some of the important synthetics are now melt spun with trilobal section.

One other method of increasing the covering power of a fibre has received attention, and that is to make the fibre hollow. A viscose rayon of this type was first manufactured in France and marketed in 1922 under the name of Celta. The filaments are not continuous tubes, but contain enclosed bubbles of air and for occasional short lengths they may even be solid.

Celta

Celta was made by emulsifying air with the viscose before ripening; spinning was carried out as usual. Later similar yarns were made by incorporating sodium carbonate in the spinning solution; carbon dioxide bubbles were generated when the spinning solution passed into the acid coagulating bath and persisted in the yarn. In this case, though, there was some tendency for the carbon dioxide to be dissolved out by the water used in washing and finishing the rayon, so that the bubbles collapsed and even on drying the yarn they only partly refilled; probably it was much easier to spin uninterruptedly the viscose containing sodium carbonate which formed bubbles *after* its passage through the spinning orifice than it was to spin viscose containing bubbles of air, which would be expected to cause frequent breaks in spinning.

About 10 per cent of the section of a Celta fibre consisted of air space so that Celta threads were considerably thicker than ordinary viscose rayon of equal denier. Celta in fact had greater covering power than ordinary viscose rayon. Nevertheless the use of hollow filament yarn seems to have died out.

During the 1939–45 war a hollow fibre was introduced, but although lightness was once again the aim, it was directed not at better cover, but at buoyancy or floating power.

Bubblfil

Bubblfil consists of viscose filaments in which bubbles of air are entrapped. Each filament contains long air bubbles which are not continuous. Usually the bubbles are about $\frac{1}{4}$ in. long and there are about three of them to the inch. As a result, Bubblfil filaments and materials made from them are very light and have considerable buoyancy. They are made from the normal viscose spinning solution which is extruded through a single large spinneret orifice, and a small amount of air is blown into it at uniform intervals, so that, as the monofilament coagulates in the acid bath, bubbles of air are

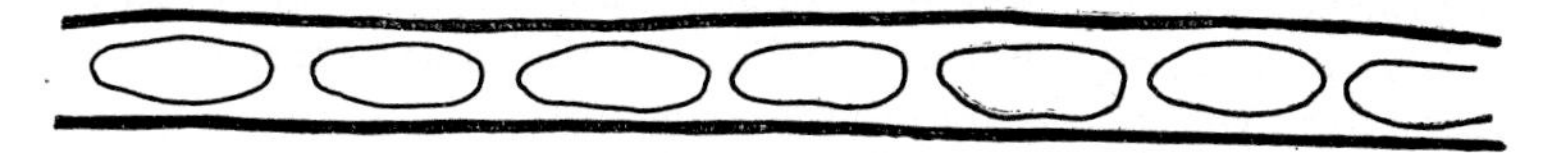

FIG. 80.—Sketch of Bubblfil.

trapped and enclosed in the filament. Normally, the bubbles are stream-lined, and resemble a string of beads in appearance, but the spacing can be altered to suit individual requirements. The material was introduced by du Pont de Nemours & Co. in 1942, and found very considerable application for war purposes, largely because of the shortage of kapok, which comes from Eastern sources that were then cut off. It was found that life-jackets made from Bubblfil were bulkier than those made from kapok, but they had good resistance to penetration of water, and so retained their buoyancy for a considerable time. Other uses to which Bubblfil was put included pontoons, rafts and as an insulating medium against temperature changes and shock in aviators' uniforms and in sleeping-bags. Under normal conditions—*i.e.*, when it was available—kapok would probably be preferred, to give buoyancy. Bubblfil has not been manufactured since 1943.

BASIFIED VISCOSE

Viscose will not ordinarily take a wool dye, and various products have appeared in which some basic constituent has been added to the viscose to confer on it an affinity for acid dyestuffs. Some of these are:

Rayolanda

This fibre was formerly but is no longer made by Courtaulds, Ltd. It was a viscose which incorporated a small proportion of a synthetic resin which conferred on it an affinity for acid dyes so that solid shades could be produced on wool/Rayolanda blends, with selected acid dyestuffs; it could also be dyed, because it was mainly a viscose, with the usual direct dyestuffs. When used as a fine (1·5 denier) filament with wool it gave a soft fullness, but in 4·5 denier filament it gave a springy handle. It was mothproof itself, but did not deter moth larvae from attacking wool with which it was blended.

Cisalpha and Lacisana

Both these fibres are Italian products, and have been given wool-dyeing properties by the inclusion of casein in the viscose solution. The former contained 4·5 per cent and the latter 3 per cent of casein on the weight of the cellulose. All these fibres, having been designed for mixture with wool, are used in staple form.

LANUSA

Lanusa was formerly made by the Badische Anilin & Soda Fabrik at Ludwigshafen. A viscose of very high degree of polymerisation was extruded through large orifices of about 0·8 mm. (ten times the usual diameter for viscose), and the issuing filaments were led through funnels and treated with water and dilute alkali before being carried forward to a second bath which contained dilute sulphuric acid and which completed the precipitation of the viscose rayon. The filaments (Fig. 81) had a nearly circular cross-section. This was partly due to the slowness of precipitation in the spinning bath and partly to applied stretch. Lanusa was sold as staple and had a good wool-like handle; its manufacture was discontinued in 1955. In some ways (p. 268) it anticipated the polynosic rayons.

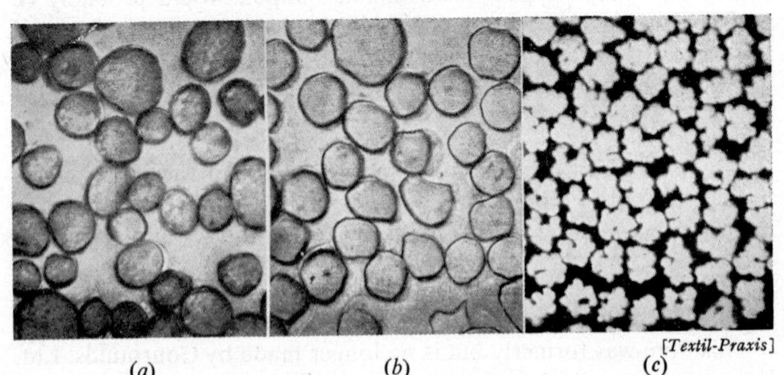

[*Textil-Praxis*]

(*a*) (*b*) (*c*)

FIG. 81.—Filament cross-sections of (*a*) wool, (*b*) Lanusa and (*c*) staple fibre viscose (× 300 approx.).

Cellophane

Cellophane is made from a viscose solution. The spinning solution is extruded through a narrow slot (instead of through a number of small orifices) as a film into a coagulating bath and is then purified, softened, and dried. In practice, the alkaline viscose solution is pumped into a rectangular casting head (the extrusion of the film is called " casting " instead of " spinning ") which has a narrow slit running along the bottom; the slit is positioned just beneath the surface of the acid-coagulating bath so that as the viscose is pumped through the slit it congeals to a film. The coagulated film is then drawn from the bath by a roller, which delivers the film to the processing baths. In these, the film is carried by rollers (Fig. 82) through a number of treating baths in which it is successively regenerated, washed, desulphurised, washed, bleached, washed, dyed

[*British Cellophane, Ltd.*]

FIG. 82.—Cellophane film passing through chemical baths.

It is carried through several baths in which it is successively regenerated, washed, desulphurised, washed, bleached, washed, dyed if desired, and finally softened with glycerine.

if desired, and finally (now transparent) softened with a little glycerine. The film is dried by passing over and under heated rollers and is finally wound on to a metal core and is then called the " mill-roll " (Fig. 83). Some customers buy the film in this form; others, *e.g.*,

[*British Cellophane, Ltd.*]

FIG. 83.—Cellophane being wound up as the mill-roll.

[*British Cellophane, Ltd.*]

FIG. 84.—Slitting machine.

For wrapping cigarettes and other small items, the customer requires the Cellophane cut into narrow bands. For this purpose the " mill-roll " is slit to produce a number of narrow rolls.

cigarette manufacturers, require it slit into a number of narrow bands and in a slitting machine (Fig. 84) the mill-roll is slit to produce a number of narrow rolls. Others buy it in cut sheets.

Corval

Corval is a cross-linked cellulosic fibre that is produced by Courtaulds (Alabama) and is designed for blending with the synthetic fibres to give increased loft, warmth and kindness of handle. Compared with viscose rayon it has a higher wet tenacity and a lower wet extensibility. The price (1962) in America is 37 cents per lb. It is made in bright and dull, regular and crimped, in several deniers up to 15. Physical properties are not very different from those of Fibro; some are listed in the table below.

Topel

This fibre is similar to Corval but has much less affinity for water, in which it swells only very little. It has good resistance to caustic alkali, a property useful in package dyeing. It is intended for blending with cotton, to which it is claimed to give improved appearance, handle and drape. In America, Topel costs 37 cents per lb., uncrimped in $1\frac{1}{2}$ or 3 denier, bright or dull, compared with 27 cents for ordinary rayon staple (1962). Physical properties of Topel are very similar to those of Corval. A later development is W-63 or Lirelle, stronger and still having a high wet modulus; 36 cents per lb. in 1965.

Avril

This is a cross-linked cellulosic fibre made by American Viscose Corporation. Its physical properties are rather different from those of Courtaulds' cross-linked fibres as the following comparison shows. Fibre characteristics quoted for Avril in 1964 are, however, much better than those shown in the Table.

	Avril	Corval	Topel
Tenacity (gm./denier):			
Dry	3·2	2·2	2·2
Wet	2·2	1·6	1·5
Ratio: wet/dry	69	73	68
Elongation (per cent):			
Dry	9·5	13	15
Wet	10·5	15	18

FURTHER READING

G. S. Ranshaw, " The Story of Rayon ". London (1949).
H. J. Hegan, " Talks on Rayon ", 1–22, The Cotton and Rayon Merchants Assn. (1944).
P. W. Frisk, " Review of Continuous Viscose Spinning ", *Rayon and Synthetic Textiles*, **30**, No. 9, 49 (1952).
" British Rayon Manual ", 15–25. Manchester (1947).
S. W. Barker and R. Alleston, *J. Text. Inst.*, **39**, P 1 (1948) (the development of continuous rayon spinning).
Industrial Rayon Corp., British Pat. 545,250.
V. Hottenroth (ripeness test), *Chem. Ztg.*, **39**, 119 (1915).
C. Dorée, " The Methods of Cellulose Chemistry ", 2nd Edn. London (1947).
" Talks on Rayon ", 20 (Rayolanda). Manchester (1944).
J. M. Matthews, " Textile Fibres ", 6th Edn. London (1954).
Courtaulds, Ltd., H. J. Hegan and F. Bayley, British Pat. 259,386 (preparation of fibres with included air).
M. R. Fox, " Vat Dyestuffs and Vat Dyeing ". London (1946).
Morton Sundour Fabrics, Ltd., R. S. E. Hannay and W. Kilby, British Pat. 620,584.
Standfast Dyers and Printers, Ltd., British Pat. 655,415.
(These last two references are for the Standfast process.)
J. Ardron, M. R. Fox and R. W. Speke, " The Continuous Dyeing of Vat Dyes ", *J. Soc. Dyers and Col.*, **68**, 249 (1952).
G. L. Boardman, " Continuous Piece Goods Dyeing with Vat Dyes ", *J. Soc. Dyers and Col.*, **66**, 397 (1950).
* J. Boulton, " The Application of Direct Dyes to Viscose Rayon Yarn and Staple ", *J. Soc. Dyers and Col.*, **67**, 522 (1951).
C. M. Whittaker and C. C. Wilcock, " Dyeing with Coal Tar Dyestuffs ", 5th Edn., 233–55. London (1949).
C. M. Whittaker, " Classification of Direct Dyestuffs ", *J. Soc. Dyers and Col.*, **58**, 253 (1942).
F. Lieseberg, " Spezial-Zellwolle für die Woll-Industrie ", *Textil-Praxis*, **8**, 360 (1949). (Description of Lanusa.)
W. A. Sisson and F. F. Morehead, " The Skin Effect in Crimped Rayon ", *Text. Res. J.*, **23**, 152–157 (1953).
E. S. Gilfillan and L. Linden, " Effects of Nuclear Radiation on the Strength of Yarn ", *Text. Res. J.*, **25**, 773–777 (1955).

* Boulton's paper includes a description of the classification of direct dyes which was recommended by the Society of Dyers and Colourists' Direct Cotton Dyes Committee. This was:

Class A: Dyes having good levelling or migration properties.
Class B: Dyes of poor levelling properties but which can be dyed uniformly by the controlled addition of salts to the dyebath.
Class C: Dyes of poor migration with which control by temperature as well as by salt is required.

CHAPTER 10

CUPRAMMONIUM RAYON

THE solubility of cellulose in aqueous solutions of ammonia which contained copper oxide was discovered by Schweizer in 1857 in Switzerland. Long before the practicability of making fibres from the solution of cellulose in cuprammonium was considered, the use of these solutions to give water-proofing and rot-proofing properties when applied to other textiles became important. The process was first patented in 1859 by Scoffern for proofing fabrics, and eventually became known as the " Willesden " finish. In 1881 Sir William Crookes suggested the production of cellulose filaments for electric bulbs from a solution of cellulose in cuprammonium; it was at about this date that Swan (p. 148) was working on the production of filaments from denitrated cellulose nitrate. In 1891, Fremery and Urban spun the first cuprammonium rayon in Germany, but it was not until 1901 when the " stretch-spinning " process was invented by Thiele that useful artificial silk was made. Thiele was engaged by J. P. Bemberg and silk was made by them in Barmen. The stretch-spinning process was successful and the filaments made were very fine and, for the times, strong. Bemberg started companies for the manufacture of cuprammonium rayon in England, America, Japan and Italy. The next major development was the introduction in 1940 of the continuous process; hitherto all yarn had been reeled.

At one time cuprammonium rayon was manufactured by British Bemberg Ltd. at Doncaster, but production was surprisingly discontinued. It is now made in Germany by J. P. Bemberg A.G. at Wuppertal-Barmen and by Farbenfabriken Bayer at Leverkusen, who call their continuous filament Cupresa and their staple Cuprama. Also in Italy by Bemberg S.p.a. at Milan, in America by the American Bemberg Corporation, and in Japan by the Asahi Chemical Industry Ltd. under the name Bemsilkie. The production of cuprammonium rayon in Japan has been given as about 48 million lb. for 1965 and it is quite likely that the German production is greater than this. Reports have also circulated of an increase in production by American Bemberg so probably the world production of cuprammonium rayon is rising fast, but even so it must be small compared with that of viscose or cellulose acetate rayon. The cuprammonium spinning process is still the only one which enables the commercial production of a regenerated cellulosic yarn as low in denier as 15.

Cuprammonium yarn is good, but it is not cheap. In 1968 100-denier Cupresa continuous filament on cone, for weft, cost 93*d.*/lb. in Britain. Staple fibre is much cheaper, roughly one-third the price of filament.

Chemical Constitution

Like viscose, cuprammonium consists of regenerated cellulose. The natural polymeric material, cellulose, which occurs so abundantly in nature, is dissolved, the solution extruded and the cellulose re-precipitated from solution. The chemical constitution of the cuprammonium rayon molecule (see p. 156) is therefore represented by the same formula as that already given for viscose rayon:

```
       ┌   CHOH—CHOH  ┐
       │   /       \  │
H —O—CH         CH—│OH
       │   \       /  │
       │    CH———O    │
       │    |         │
       └    CH₂OH     ┘n
```

It is an essential feature of the process for making cuprammonium rayon that the cuprammonium solution shall contain defined quantities of copper and ammonia, otherwise the cellulose will not dissolve properly in it. Suitable quantities are about 4 per cent copper and 29 per cent ammonia, and 9–10 per cent cellulose in the solution that is spun. The solution contains the cuprammonium ion which gives to it an intense blue colour.

Raw Material

Although at one time cotton linters were used exclusively for the cuprammonium process, the production of a purified wood pulp with a high α-cellulose content has led to its adoption in part. The wood cellulose is preferable in one respect, as it dissolves more quickly in the cuprammonium, but manufacturers still prefer cotton because it gives rayon of a better colour and strength, and despite the improvements that have been made in wood cellulose they only use it if linters are not available.

The α-cellulose to which reference is often made is that portion of a cellulose material that is undissolved after steeping for 30 minutes at 20° C. in 17·5–18·0 per cent caustic soda solution, and then, by addition of water to the latter, in a more dilute sodium hydroxide. β-cellulose is the material that dissolves in the α-test but is re-precipitated on acidification with dilute acetic acid. γ-cellulose is that which remains unprecipitated in the β-test. α-cellulose, the most insoluble form, is the most highly esteemed and the most useful for the manufacture of fibres. Severe previous process treatments invariably lower the α-cellulose content, and a high (90 per cent) α-cellulose content is an indication of careful pre-treatment—less degradation and depolymerisation.

Preparation. Cotton linters is the usual starting material, and is purified by kier-boiling at 150° C. with dilute caustic soda, and is then bleached with sodium hypochlorite. The purified cellulose is mixed with the necessary amounts of aqueous ammonia, basic copper sulphate and caustic soda, and is kneaded until it gives a clear blue solution. This is thinned until it contains 9–10 per cent cellulose, is de-aerated and filtered through nickel gauze. It is important to note that cuprammonium solution can be spun either when freshly prepared or after long storage, as it does not decompose —this is in striking contrast to the need for catching viscose just at the right time; there is no question of ripening with cuprammonium. The solution is pumped to the spinning room just when the spinners require it.

Spinning (Classical Method)

The de-aerated and filtered solution is metered by a pump into the nickel spinneret and extruded through large holes (0·8 mm. diameter) at the bottom of the spinneret, *A*, into a glass funnel, *B*, (Fig. 85) which is fed with running pure soft water at *C*. This water

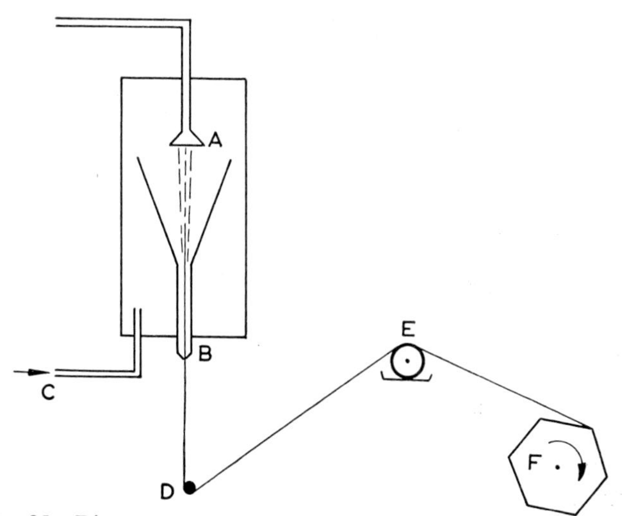

FIG. 85.—Diagrammatic representation of classical method of spinning cuprammonium rayon.

removes most of the ammonia and some of the copper, and precipitates the cellulose in a plastic form still containing some copper and a little ammonia. On their way from the nickel spinning nozzle, *A*, to the bottom of the glass funnel, *B*, the filaments undergo a considerable stretch caused by the pull of drawing them off, hence the name " stretch spinning process ". After having left the funnel, the filaments are drawn round a steel rod, *D*, to separate them from the greater part of the spinning water they are carrying, and then

[*J. P. Bemberg, A.G.*]

Fig. 86.—Reel spinning of cuprammonium rayon at the Bemberg factory in Germany.

round a roller, *E*, running in a small trough of acid, which completes the coagulation of the cellulose, converting the copper and the ammonia that were still in the yarn to copper and ammonium sulphates. In the classical method the yarn is wound up either in hanks, *F*, on reels or in a Topham Box as cakes. Either hank or cake has to be rinsed with pure water to remove the copper and ammonium sulphates and the free acid; the washed hanks or cakes are lubricated and dried in a continuous dryer. They are then backwound and the yarn is twisted and is ready for despatch to weavers or knitters. Fig. 86 shows the classical process in operation with the yarn being collected on reels at the Bemberg factory in Germany.

Spinning (Modern Continuous Method)

This method was developed during the 1939–45 war. As far as the acid rinse, the yarn is spun just as already described for the reel or classical method, but then instead of being collected on reel or in Topham Box, the yarn is led in the form of a sheet of 575 ends (from 575 spinnerets) through a washing range (Baths *F* and *G* in Fig. 87), and is then lubricated and dried on dryer *H*, then wound on

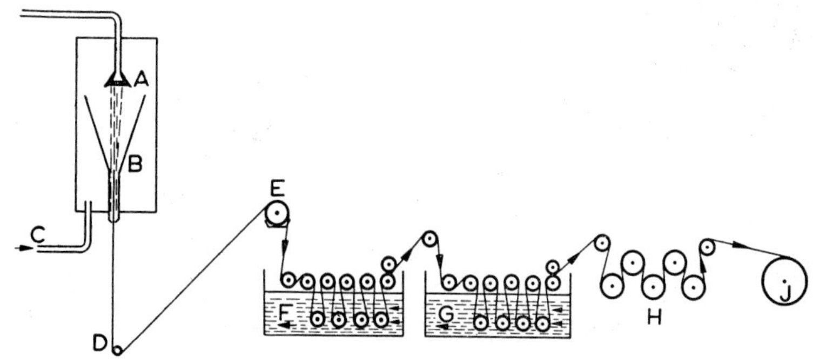

FIG. 87.—Diagrammatic representation of modern continuous method of spinning cuprammonium rayon.

to a beam, *J*. The beam can be used directly on warp knitting machines, or the yarn from four or five or six of them can be run on to a weaver's beam. This direct take-up on a beam (Fig. 88) represents a very great saving in textile winding and twisting operations, and consequently a big money saving. Additionally, time is saved because the transit of the yarn from spinneret to beam takes less than a minute, whereas the hank or cake washing operations necessary after the classical spinning process take up to 24 hours.

In both classical (reel) and modern (continuous) spinning methods, the precipitation of the cellulose is effected by dilution with water. In principle the method is simple: the cellulose is dissolved in cuprammonium solution, squeezed through orifices and the cellulose is precipitated in a plastic form containing copper and ammonia by dilution with water, and then coagulation is completed by treatment with acid, *i.e.*, the regeneration of the cellulose is completed.

Properties

As cuprammonium is a regenerated cellulose, it has properties that in many respects are similar to those of viscose. Perhaps the main points of difference are:

1. The fineness of filament, for cuprammonium yarns average about 1·2 filament denier—*e.g.*, a 75 denier yarn will contain

[*J. P. Bemberg, A.G.*]

FIG. 88.—Continuous spinning of cuprammonium rayon at the Bemberg factory in Germany.

about sixty filaments. Very much finer filaments, down to a filament denier of 0·4, are also spun for special purposes.

2. The strength of cuprammonium is rather higher at 2·3 grams per denier than that of viscose.

It will be appreciated that both these attributes are due to the stretch that is applied during spinning.

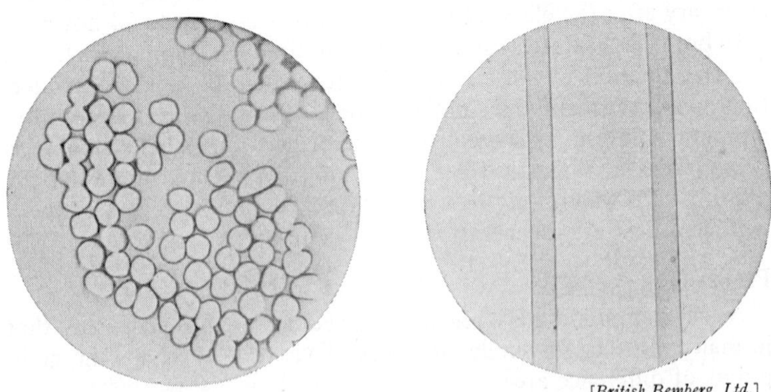

[*British Bemberg, Ltd.*]

FIGS. 89 and 90.—Photomicrographs of cuprammonium (×300). Note absence of structure of the longitudinal view. Filament denier 1·3.

Cuprammonium has a wet tenacity of about 1·2 grams per denier. The elongation is about 15 per cent dry and about 25 per cent wet. Moisture content is 11 per cent under standard conditions, similar, as would be expected, to that of viscose.

Cuprammonium, again like viscose, burns readily, chars at or below 180° C., and is degraded and weakened by exposure to sunlight. On ignition it leaves an ash which contains a trace of copper.

Under the microscope the longitudinal view is that of a featureless cylinder, and the cross-section is quite smooth and round (Figs. 89, 90).

In its chemical properties cuprammonium rayon resembles viscose rayon, and water swells it powerfully. Strong alkalis attack it, but it is stable to the action of weak alkalis. Oxidising agents attack it. The yarn may be bleached in hypochlorite solutions (mildly alkaline) or in hydrogen peroxide.

Dyeing

Direct dyestuffs are used for Bemberg. The fibre, being cellulosic, swells and is easily penetrated by dye-liquors; in addition, it has a high affinity for dyestuffs.

Uses

Bemberg yarn has found a great variety of uses. It is particularly suitable for apparel, and large numbers of ladies' stockings and undergarments are made from it. Mostly it is used in the better qualities, for sheer hose and for warp-knit fabric for underwear. The fineness of filament gives it great pliability and very good draping qualities. The filament denier of cuprammonium is about the same as that of real silk, and, as its lustre is also subdued, it makes a very good substitute. American Bemberg spin a heavy denier monofilament which is lustrous and is known as " Glitter ".

Cupresa

Cupresa is the name of a high-grade continuous filament yarn made by the cuprammonium process by Farbenfabriken Bayer at Dormagen. Cupresa CS. is a cuprammonium yarn specially made for stockings; it is made by the continuous process and is characterised by extremely good uniformity and level-dyeing properties. Cupresa is available in deniers of 25, 40, 60, 80 and higher and the commercial production of a 25 denier cellulosic fibre is worthy of note; the yarn is supplied in bright and matt and in a range of dope-dyed shades. Twist on the Continent is reckoned not in turns per inch but in turns per metre. Some Cupresa yarns are supplied untwisted, others in voile and crepe twists and others again with just sufficient twist for textile working. A twist of 150 turns per metre corresponds to nearly 4 turns per inch.

Cuprama

Farbenfabriken Bayer at Dormagen also make Cuprama, a cuprammonium staple fibre, and according to them " there is only one cuprammonium staple fibre ". Cuprama was first marketed in 1934 and has been very successful in combination with wool mixtures, for which its softness makes it very suitable. The yarn is made in a range of eighteen spun dyed colours, their fastness being similar to that of anthraquinone vat dyeings. The white yarn has the high affinity for dyestuffs which characterises cuprammonium fibres generally. The filament denier of standard Cuprama is $3\frac{3}{4}$ and it is made also in $2\frac{3}{4}$, and for special purposes in 6, 8 and 12 filament deniers. The high ratio of wet/dry strength of Cuprama which is 65–70 per cent is noteworthy; it is considerably higher than that of most cellulosic fibres.

Apart from its use in worsted spinning and in knitwear and tricots, Cuprama has been used considerably for spinning carpet yarns; high filament deniers are used and the variety intended for carpets, known as Cuprama TX, is made in 14, 22, 30 and 40 filament deniers with a staple length of 100 mm. It is cheap, springy, can be dyed in brilliant colours, and is said to be highly resistant to dust and dirt; in this last respect the higher the filament denier the greater is the resistance to soiling of any fibre, cuprammonium or otherwise.

FURTHER READING

M. H. Avram, " The Rayon Industry ", 523–543. London (1930).
J. T. Marsh and F. C. Wood, " An Introduction to the Chemistry of Cellulose ", 200–202. London (1945).
S. W. Barker and R. Alleston, *J. Text. Inst.*, **39**, P 3 (1948) (outline of continuous spinning machine).
R. Pummerer, " Chemische Textilfasern, Filme und Folien ". Stuttgart (1953).

CHAPTER 11

CELLULOSE ACETATE

CELLULOSE is chemically stable; its derivatives are not easily made, and it was not until 1869 that its acetate was prepared by Schutzenberger, who made it by heating cellulose with acetic anhydride in a sealed glass tube. In 1894 Cross and Bevan showed that the action proceeded readily at atmospheric pressure if either sulphuric acid or zinc chloride was present to act as a dehydrating catalyst. Using this method they obtained a cellulose triacetate, and found that it was soluble in chloroform.

It was discovered, in 1903, that if the cellulose triacetate is partly hydrolysed back to a stage half-way between the triacetate and diacetate, it loses its solubility in chloroform, but becomes soluble in acetone, which is a much more convenient solvent to use. If, however, cellulose is directly acetylated in the first place to the 2½-acetate stage it is not acetone-soluble (*cf.* p. 284).

In the 1914–1918 war aeroplanes had fabric wings, and it was found that if they were coated with a solution of cellulose acetate in acetone they tightened up and became impervious to air. The cellulose acetate was manufactured in Switzerland by the brothers Dreyfus, and as their supplies were inadequate to treat the aeroplanes of all the Allies they were asked to come to England to start a factory for the manufacture of cellulose acetate in large quantities. This they did at Spondon, near Derby.

When the war ended in 1918 there was a very large production of cellulose acetate coming from the Spondon factory, but no demand for it. After a period of intensive research carried out under the inspiring leadership of Dr. Henry Dreyfus, a method of converting it into artificial silk was discovered, and by 1921 this fibre was marketed as " Celanese ". In 1924 an allied company was also making and selling it in America. Since the 1957 merger of British Celanese with Courtaulds, ordinary acetate has not been quite so much to the fore in the U.K. Nevertheless, ordinary acetate has one of the most beautiful handles in the world of fibres. It is making great strides in Japan (in 1964 they made 70 million lb.) and, unexpectedly, in Chile. In America it is manufactured and sold as Celanese by the Celanese Corporation of America, as Estron by the Tennessee Eastman Co., as " acetate " (the generic term) by du Pont de Nemours and Co., and as Seraceta by the American Viscose

Corporation. Very large quantities of the order of 50 million lb. per year have also been made by the Rhodiaceta firms in Europe, notably in France and Germany. Acetate fibre is also produced under the name Aceta by Farbenfabriken Bayer: it was previously made under the same name by the former I.G. Farbenindustrie. One of the earliest and best spinners of acetate yarn was Lansil Ltd. of Lancaster; they still make it, but are no longer independent, having been taken over by Monsanto of America in 1962.

Chemical Constitution

The long-chain molecule which forms the basis of cellulose acetate is cellulose, the abundant natural polymer. It is the same basis as is used for viscose and cuprammonium yarns, but in the case of cellulose acetate most of the hydroxyl groups have been acetylated.

The unit from which cellulose is built up is the glucose molecule. Two stereo-isomeric forms are known, α- and β-glucose, but only the latter is a building-brick of cellulose. The constitution of glucose is:

```
            CH2OH                                   CH2OH
              |                                       |
     H        C——O      H                   H         C——O        OH
     |      /  |   \    |                   |       /  |   \      |
     C     H         C                      C     H          C
     |  \   OH  H   /  |                    |  \   OH  H   /  |
    OH    C——C        OH                   OH    C——C         H
          |    |                                 |    |
          H   OH                                 H   OH
         α-Glucose.                             β-Glucose.
```

Two molecules of β-glucose will combine to give one molecule of cellobiose:

```
                                              CH2OH
                                               |
        CHOH—CHOH                              CH——O
       /          \                           /      \
 HOCH              CHOH  +  HOCH                      CHOH  ⟶
       \          /              \                   /
        CH——O                      CHOH—CHOH
        |
        CH2OH
           Glucose.                         Glucose.
```

```
                                    CH2OH
                                    |
       CHOH—CHOH                    CH——O
      /         \                  /     \
HOCH             HC—O—CH                  CHOH + H2O
      \         /          \             /
       CH——O                CHOH—CHOH
       |
       CH2OH        Cellobiose.
```

Water is split off in the condensation of two molecules of glucose, and in an exactly similar way two molecules of cellobiose will combine, splitting off another molecule of water, and giving a longer chain. This process of condensation goes on continuously in Nature, and results in the formation of the long-chain polymer cellulose:

```
     ┌     CHOH—CHOH      ┐
     |    /         \     |
HO  —CH              HC—O—  H
     |    \         /     |
     |     CH——O          |
     |     |              |
     └     CH2OH          ┘n
```

It will be observed that glucose has the empirical formula $C_6H_{12}O_6$ and that it contains five hydroxyl groups, and, as would be expected, these can all be acetylated, and glucose penta-acetate is known. Cellulose, however, has the empiric formula $C_6H_{10}O_5$, and contains only three hydroxyl groups in each glucose residue, so that it is not possible to acetylate it beyond the triacetate stage. A triacetate is known, but no higher acetate, nor does it seem possible that a higher acetate could ever be made. The formula of cellulose may therefore be rather more informatively written as

$$[C_6H_7O_2(OH)_3]_n$$

Cellulose triacetate, which is the substance that was made by Cross and Bevan and that was later spun from chloroform solution by the Lustron Co., and which has recently (Chapter 12) been spun again commercially, has the constitution:

```
     ┌   OCOCH3   OCOCH3    ┐
     |   |        |         |
     |   CH———————CH        |
     |  /           \       |
HO  —CH              CH—O—  H
     |  \           /       |
     |   CH———————O         |
     |   |                  |
     └   CH2OCOCH3          ┘n
```

Whereas the empiric formula of cellulose is $C_6H_{10}O_5$, which corresponds to a (unit) molecular weight of 162, that of cellulose triacetate is

$$C_6H_7O_2(OCOCH_3)_3$$

which corresponds to a (unit) molecular weight 288. Therefore, when cellulose is converted into cellulose triacetate there is a gain in weight of 77·8 per cent. Conversely, when cellulose triacetate is converted back into cellulose, as it easily can be by saponification with alkali, there is a loss in weight of 43·7 per cent.

Assuming that an acetone-soluble cellulose acetate contains 2·3 acetyl groups per glucose residue, its formula will be:

$$C_6H_{7\cdot7}O_{2\cdot7}(OCOCH_3)_{2\cdot3}$$

with a (unit) molecular weight of 259. On complete saponification there will be a loss of about 37·5 per cent in weight.

Many other esters of cellulose have been made; of these, the mixed cellulose acetate-butyrate has been used to some extent, but for fibres cellulose acetate is supreme. Cellulose formate has also been made and spun into fibres. If, with a view to modifying the properties of cellulose acetate yarn or fabric, it is desired to insert other acid radicals, it is more convenient to do this by treating the yarn or fabric with a solution of the acid chloride in an inert solvent—*i.e.*, a solvent for the acid chloride but a non-solvent for the cellulose esters; this process is best conducted in the presence of an acid-acceptor such as pyridine to take up the hydrochloric acid formed and prevent it from damaging the cellulose acetate.

Although in the early days of rayon the acetone-solubility of the secondary acetate was the deciding factor in making it instead of the triacetate, the position is different to-day. There are solvents such as methylene chloride which are available for the triacetate and from which it is now spun (p. 243). But the secondary acetate should not be displaced by the triacetate, because the secondary is a better fibre, more kindly to the handle with higher moisture regain and stronger. Although the original adoption of the secondary acetate depended on its solubility characteristics, thirty years' experience has shown that it is a good fibre, particularly suitable for underwear and dresses.

Manufacture

Cellulose acetate rayon is made from cotton linters—*i.e.*, cotton of very short staple length which is removed from the cotton seeds, but which is unsuitable for spinning. The linters are purified by kier-boiling and are bleached. Wood pulp is also now being used increasingly for the manufacture of cellulose acetate, but in Russia cotton is still the main raw material for cellulose acetate.

The chemicals needed are: (1) acetic acid, which nowadays is frequently made by catalytic oxidation of alcohol, (2) acetic anhydride, which is made by dehydrating acetic acid at high temperatures to give ketene:

$$\underset{\text{Acetic acid.}}{CH_3COOH} \longrightarrow \underset{\text{Ketene.}}{CH_2{=}CO} + H_2O$$

and then passing the ketene into more glacial acetic acid with which it combines to form acetic anhydride:

$$\underset{\text{Acetic acid.}}{CH_3COOH} + \underset{\text{Ketene.}}{CH_2{=}CO} \longrightarrow \underset{\text{Acetic anhydride.}}{\begin{matrix} CH_3CO \\ CH_3CO \end{matrix}\!\!>\!O}$$

and (3) acetone. The cracking of petroleum yields large quantities of isopropyl alcohol, some of which is oxidised to acetone. In addition, sulphuric acid is needed, and a good supply of water. The water, and in fact all the chemicals used, must be substantially free from iron.

The stages in the manufacture of cellulose acetate rayon are as follows:

Cotton Purification. Cotton linters—fibres too short to be spun—are purchased in bales. The bales are broken and the linters kier-boiled under pressure for from four to ten hours with an alkaline liquor, which may be a solution of either sodium carbonate or caustic soda, or a mixture of the two. They are then rinsed, washed, bleached with sodium hypochlorite, washed and dried.

Pre-treatment. The purified cotton is steeped in glacial acetic acid to make it more reactive so that it will acetylate readily.

Acetylation. The pre-treated cotton is loaded with an excess of glacial acetic acid and acetic anhydride into a closed vessel fitted with a powerful stirrer. Suitable quantities are:

100 lb. purified linters (air-dry weight).
300 lb. acetic anhydride.
500 lb. glacial acetic acid.

These are thoroughly mixed together, but as yet no chemical reaction takes place.

When the mixing has been accomplished, 8–10 lb. of sulphuric acid dissolved in about eight times its own weight of glacial acetic acid is added. The sulphuric acid reacts with the acetic anhydride to form sulpho-acetic acid, which is the real acetylating agent.

Acetic anhydride and acetic acid would only acetylate the cellulose very slowly indeed—far too slowly for commercial practice—without the sulphuric acid. Sometimes the sulphuric acid is described as a catalyst, and in the sense that it promotes the reaction—*i.e.*, the esterification of the cellulose—the description is justified; on the other hand, the quantity of sulphuric acid used is considerable when compared with the weight of cellulose that is acetylated, and, furthermore, the sulphuric acid is consumed in the reaction; in these respects the sulphuric acid may more properly be regarded as one of the reagents than as the catalyst of the reaction.

The acetylation reaction is powerfully exothermic; it is desirable to keep the temperature low in order to avoid undue degradation of the cellulose—*i.e.*, to avoid, so far as possible, the breakdown of the long-chain cellulose molecules into shorter molecules of degraded cellulose. Consequently the acetylation vessels have to be cooled.

For the first hour of the reaction the temperature is kept below 20° C., and for the next seven to eight hours at 25–30° C. The mass becomes gelatinous and very viscous. At the end of about eight hours all the cellulose will have been converted into cellulose triacetate—*i.e.*, the cellulose will have been completely acetylated.

Samples are continually taken from the mass and examined, so that it can be known when this point has been reached. The indications that acetylation is complete are:

1. The fibres have all dissolved; a sample examined under a low-power microscope no longer shows (as it does in the early stages of the reaction) swollen but undissolved fibres.
2. The sample is completely soluble in chloroform, which is a good solvent for cellulose triacetate. (Note, however, that it is not a solvent, although it is a powerful swelling agent, for the secondary or commercial cellulose acetate.)

The degradation that the cellulose has undergone during acetylation can be assessed by measuring the viscosity of a solution of its ester. The cellulose triacetate which is formed by the direct acetylation of the cellulose, as described above, is known as "primary" acetate.

Hydrolysis. In order to convert the chloroform-soluble primary acetate into the acetone-soluble secondary acetate it is run, together with the excess acetic acid and anhydride, into water so that a 95 per cent solution of acetic acid results, and allowed to stand for twenty hours at a higher temperature. Acid hydrolysis takes place, and an acetone-soluble product results. Samples are removed and tested at intervals, so that the reaction can be stopped when the acetyl content has been reduced by the desired amount. When this is the case, the whole mixture is poured into excess of water, which so dilutes the acid that it will no longer hold the cellulose acetate in solution, and a precipitate of chalky flakes of cellulose acetate is formed. This is the secondary acetate. The degree of polymerisation is now about 350–400. The liquors which contain acetic acid are conducted to a recovery plant, where the acid may be extracted from the water with a solvent—*e.g.*, cresol. The secondary acetate is thoroughly washed, centrifuged and dried at a low temperature. The product from each batch is tested for viscosity, ash, and acetyl content, and every care is taken to blend different batches with each other. Blending—mixing of parts of one batch with parts of another—is carried out at every stage in order to ensure, as far as possible, that a uniform product results, and that finally the yarn may be mixed and woven without showing differential effects of lustre or dyeing affinity.

It is important to note that during the hydrolysis stage any mixed ester—*i.e.*, cellulose acetate sulphate (often called sulpho-acetate)—has the sulphate radicals removed. In the early days of making cellulose acetate rayon the importance of freeing it from combined sulphuric acid was not fully appreciated, and when rayon containing combined sulphuric acid was stored, after a time sulphuric acid was slowly liberated, and degraded the rayon. In addition, the presence of sulphuric acid in the rayon alters the dyeing affinity. Cellulose acetate rayon to-day is for all practical purposes free from combined sulphuric acid.

Preparation of Dope. The secondary acetate blended from many batches is mixed in a closed vessel with a powerful stirrer with about three times its weight of acetone. It dissolves slowly, and after twenty-four hours it will be completely dissolved. If a dull yarn is required, titanium oxide will be added at this stage, and if a spun black pigmented yarn is required, black pigment will be added. The product from several of these solution vessels is again blended in a mixing-tank, and the dope (the name used for the solution of the cellulose acetate in acetone) if it has not been pigmented is water-white and viscous. It is filtered and de-aerated, and then run into a feed-tank. The dope contains 25–35 per cent of cellulose acetate and has a high viscosity.

Spinning. Dope is fed from the feed-tank through pipes to the spinning cabinets (Fig. 91). A metering pump ensures that a constant flow (so many grams per minute) of dope is fed to the spinning jet, but between the pump and the jet is a candle-filter to give a final filtration and avoid trouble due to solid particles interfering with the smooth flow of dope through the jets. The spinneret consists of a metal plate through which a number of small holes, perhaps 0·03 mm. in diameter, have been drilled, usually concentrically. The number of holes in the jet determines the number of filaments in the yarn; if the number in a 150 denier yarn is sixty, the filament denier will be 2·5; if it is thirty the filament denier will be 5. The same total denier—*i.e.*, yarn denier of, say, 150—is often spun in a variety of different filament deniers; very fine filaments are desirable for some purposes, coarser for others. As the dope is squeezed out of the jets (Fig. 92) it emerges into the spinning cabinet and travels vertically down this a distance of from 2 to 5 metres to a feed-roller, from which it is guided on to a bobbin. If the bobbin is a cap-spinning bobbin, twist is inserted in the yarn as it is taken up; if, for example, the spinning speed is 300 metres per minute and the bobbin is rotating at 10,000 r.p.m. a twist of $\frac{10,000}{300 \times 39\cdot7} = 0\cdot76$ turn per inch will be inserted. Take-up speeds are normally between 200 and 400 metres per minute. Advantageously, a slight stretch will be imparted to the yarn in its draw-down from the jet; this imparts some degree of orientation to the molecules and yields filaments of strength greater than they would otherwise have. The diameter of the filament as taken up on the bobbin is therefore dependent on three factors: (1) the rate at which dope is fed by the pump, (2) the diameter of the jets, (3) the rate of draw-down. In the cabinet, hot air is fed in near the bottom at a temperature of 100° C.; this evaporates practically all the acetone in the dope emerging from the

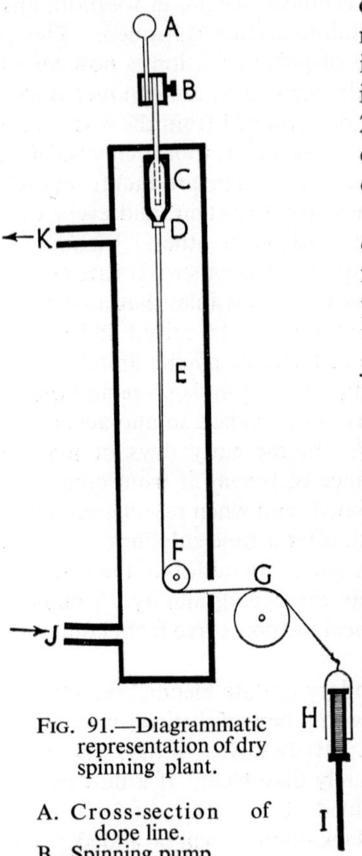

FIG. 91.—Diagrammatic representation of dry spinning plant.

A. Cross-section of dope line.
B. Spinning pump.
C. Filter candle.
D. Jet.
E. Spinning cabinet.
F. Guide roller.
G. Take-up roller.
H. Cap and bobbin.
I. Bobbin driving spindle.
J. Hot air inlet (100° C.).
K. Air and acetone vapour outlet (80° C.).

jets, and the acetone-laden air is withdrawn near the top of the cabinet and taken away to a recovery plant, where the acetone may be recovered either by (1) adsorption on active carbon, (2) scrubbing in water-towers, or (3) absorption in a concentrated solution of sodium bisulphite. Efficient recovery of both acetone and acetic acid from the earlier preparative processes is essential for the economical manufacture of cellulose acetate rayon, and in modern

[*American Viscose Corporation*]

FIG. 92.—Cellulose acetate spinning.
As the dope emerges from the spinnerets the warm air it encounters evaporates the acetone, leaving individual filaments.

plants the losses are very low indeed. Important factors in the spinning process are the temperature, moisture content and velocity of the air proceeding up the cabinet. Usually a trace of oil—perhaps 1 or 2 per cent—will be applied to the yarn before it goes on to the cap-spinning bobbin, in order to prevent damage to the filaments in winding, and also to prevent electrification.

When the yarn has been collected on the cap-spinning bobbin it is ready for textile use, and this method of spinning, known as " dry spinning ", has obvious and considerable advantages in economy of labour over the wet-spinning processes employed for viscose. Perhaps it is the elegance and simplicity of the dry-spinning process which have most largely contributed to the advancement of cellulose acetate fibres. A single workman may spin more than 200 lb. of acetate silk in his working day.

Subsequently the yarn may have more twist inserted in a separate operation before it is woven or knitted. Some makers supply the yarn on large packages; Farbenfabriken Bayer supply their acetate filament yarn on cones weighing up to 2 kg. and guaranteed knotless. The adoption of large packages is one of the signs of up-to-date processing; large packages mean less labour for package changing, and labour is so costly nowadays that it is well worth spending money on the development of large packages.

Properties

In considering the properties of cellulose acetate rayon, one has to bear in mind that the hydroxyl groups originally present in the cellulose have, for the most part, been esterified. Accordingly, the fibre is less hygroscopic and more water-repellent than viscose. At the same time, its "organic" nature has been increased, and it is correspondingly more prone to swell or dissolve in organic solvents. The esterified condition of the hydroxyl groups caused considerable dyeing difficulties when cellulose acetate rayon was first made, but these have long been largely overcome by the employment of special dyestuffs and the introduction of a method of dyeing from dispersion instead of solution of dyestuffs.

Tenacity. Cellulose acetate rayon has a tenacity of about 1·4 grams per denier and an elongation at break of about 25 per cent. Corresponding figures for wet yarn are about 0·9 gram per denier and 35 per cent. Its water-repellent properties ensure that there is less proportional decrease in tenacity on wetting than is the case with viscose. Up to about 5 per cent increase in length cellulose acetate has a high elasticity, but if it is stretched to higher elongations plastic flow or creep occurs, and the deformation is not fully recovered when the load is released. The stress–strain curves for wet and dry cellulose acetate yarns are shown in Fig. 93.

Resistance to Heat. Cellulose acetate is thermoplastic—*i.e.*, it softens on heating. It melts at about 230° C. with decomposition. If a very hot iron is used to iron fabrics made from it, "sticking" and eventually fusion occur. The acetate decomposes when melted in air, charring taking place. It will burn if ignited, but on account of the simultaneous melting, the spread of combustion is often slow. Cellulose acetate might reasonably be said to offer only a low fire risk, when in bulk. Unfortunately, when used for kiddies' clothing it is almost as dangerous as viscose or cotton. Research is urgently needed to find a way of making such fibres flameproof. The number of fatal accidents and serious injuries sustained by young children due to " clothes catching fire " is still shockingly high.

Lustre. The lustre is normally bright, but may be subdued by the incorporation of titanium dioxide in the dope. The lustre of bright cellulose acetate is greatly dulled by immersion in boiling water, and reappears on ironing, usually patchily. Care has to be taken not to dull it in dyeing, whilst for ordinary home washing warm water is safest.

Solubility. Cellulose acetate is readily soluble in some organic solvents—for example, acetone, methyl ethyl ketone, methyl acetate, ethyl lactate and dioxan—and is swollen by a large number of others,

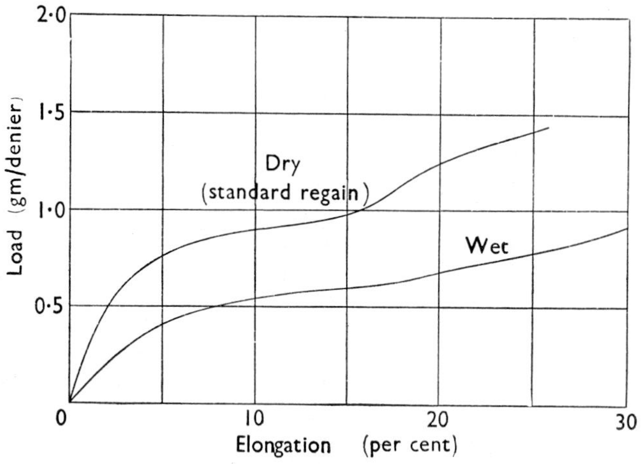

FIG. 93.—Stress–strain diagram of cellulose acetate yarn.

including chloroform, methylene chloride, ethylene chloride and Cellosolve. It is unattacked by ether, which is a suitable extractant for use with it.

Hygroscopicity. The moisture content under standard conditions is 6·5 per cent—considerably less than that of viscose.

Specific Gravity. The specific gravity is 1·32—very similar to that of wool.

Handle. The handle of cellulose acetate is particularly soft, and fabrics made from it drape well. These properties have very largely contributed to its successful development.

Electrical Properties. Cellulose acetate is an excellent insulator. It will readily develop static charges, and for some purposes it is desirable to apply an anti-static finish.

Biological Resistance. Organisms such as moths and mildew find no nutrient in cellulose acetate. Cases of damage are extremely rare, and are invariably due to the organisms feeding on the oil or finishing material applied to the fibre. Cellulose acetate will not produce dermatitis.

Chemical Resistance. Cold dilute acids do not affect it, but concentrated acids—*e.g.*, acetic and formic—attack it in the cold. Alkalis saponify it—*i.e.*, they remove acetyl groups—but dilute solutions up to *p*H 9·5 are safe.

Resistance to Light. Some tendering takes place when cellulose acetate is exposed to light, but it is usually not very serious; a loss of about 15 per cent in tenacity has been reported after 200 hours' exposure in a Fadeometer.

Morphology. Under the microscope the longitudinal view is that of a smooth featureless cylinder, whilst the cross-section is lobed as shown in Fig. 94.

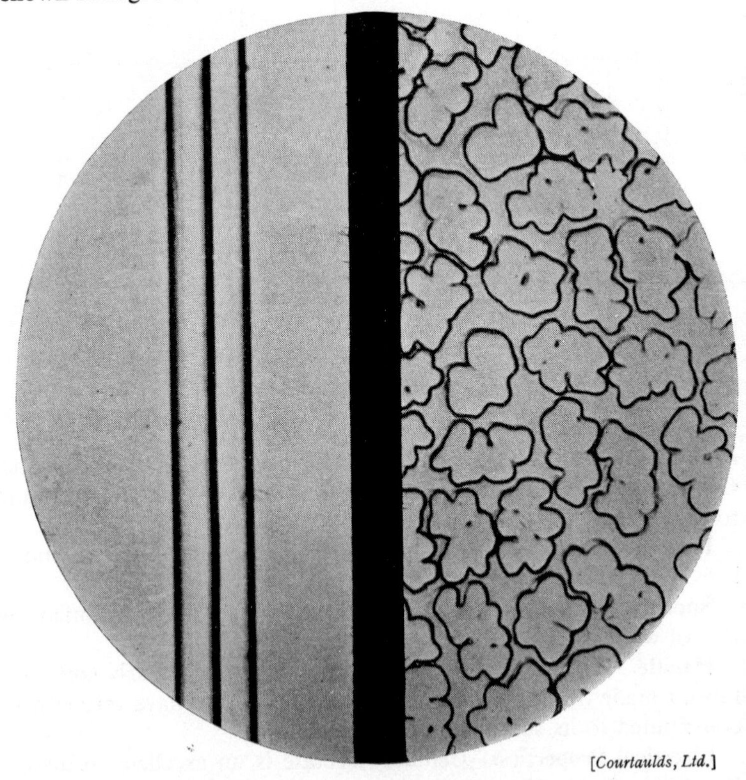

[*Courtaulds, Ltd.*]

Fig. 94.—Photomicrograph of cellulose acetate filaments (×500). Note the lobed outline and the absence of serrations.

Dyeing

In considering the dyeing of cellulose acetate four points have to be remembered.

1. Cellulose acetate, unlike the majority of older fibres, contains only very few reactive groups. Cotton and viscose have many free hydroxyl groups, silk and wool have free amino and carboxylic groups; but cellulose acetate has had all the free hydroxyl groups in the original cellulose esterified, although a small proportion of them have been regenerated during the hydrolysis of the primary (tri)-acetate to the secondary acetate. Even so, the number of free hydroxyl groups, which function as sites for chemical combination of dyestuff and fibre, is very small. Accordingly only a few of the acid, direct, and chrome dyestuffs show much affinity for cellulose acetate, and most of them only give slight staining. Nevertheless, a few individual dyewares which have been selected from repeated trials give reasonable dyeings, but for practically all purposes a new dyeing technique has had to be used.

2. Cellulose acetate is sensitive to boiling or nearly boiling liquors, in which either it is completely delustred or else its lustre is impaired. In practice, to avoid this trouble the dyeing temperature must be limited to 85° C., and a temperature of 75° C. is frequently used.

3. Cellulose acetate is saponified by caustic alkalis which split off the acetyl groups and regenerate cellulose. It is therefore impossible to dye it (except under very special conditions) with sulphur or vat colours. (*cf.* p. 765.)

4. Cellulose acetate, again because of its poverty of hydroxylic groups, swells only slightly when wet, so that dye-liquors do not penetrate it nearly so easily as they do viscose.

The dyeing methods that are, or have been, used with cellulose acetate are as follows:

Saponification. The cellulose acetate is partly saponified with caustic alkali to a weight loss of about 10–12 per cent. This regenerates more hydroxyl groups, and the fibre can then be dyed with direct cotton dyestuffs. Disadvantages are: (1) loss in weight, which to the manufacturer is serious, for there is less to sell, and (2) the softness of handle—one of the most attractive properties of cellulose acetate—is impaired by the saponification. The method is not often used, although it had some value in the early days, before other and more satisfactory dyeing methods had been evolved.

Dispersed Bases. These are complex dyestuffs, usually of the amino-anthraquinonoid type, which are truly insoluble in water, but can be made either from powder or paste into very finely divided dispersions in water. They are marketed under various trade names, of which the following may be noted:

Artisil Direct (Sandoz).
Cibacet (Ciba or Clayton Aniline Co.).
Duranol and Dispersol (I.C.I.).
Serisol (Yorkshire Dyeware Co.).
Setacyl Direct (Geigy).
S.R.A. (British Celanese Ltd.).
Supracet (L. B. Holliday).

Some are sold as pastes, others as powders. The dyestuff is first pasted with water and a wetting agent, and is added to the dye-bath, which already contains water and a wetting agent. The goods are entered cold and the temperature is raised to 70–85° C. The wetting agent may be a sulphated fatty alcohol, and it has to perform the vital duty of keeping the dyestuff well dispersed. Dyeing takes place, and it is believed to be simply a case of solution of the dispersed particles of dyestuff in the cellulose acetate fibres. These fibres function in just the same way as any other organic solvent such as ethyl acetate. The dyestuff has the choice of dissolving in either the water or the fibre, and as it is quite insoluble in the former, it chooses the latter. It is not always easy to get satisfactory penetration in the case of thick fabrics. Because of their water-insolubility, these dispersed bases give dyeings that are very fast to washing. Light fastness is usually good, although it varies from one individual member to another.

An example of the application of this type of dyestuff is as follows: 2 per cent (on the weight of the goods) of Artisil Direct Yellow GNP is pasted up and dispersed with 2 grams per litre soap and added to the dye-bath. Cellulose acetate fabric is entered at 35° C. and the temperature raised slowly to 70° C. The dyeing is complete after forty-five minutes. The fabric is washed off, hydro-extracted and finished.

Gas Fading of Dyestuffs. Some disperse dyestuffs, particularly blues, are subject to gas fading. By this is meant that fabrics dyed with them, and stored in gas-lit or gas-heated rooms, redden; it has been shown that this is caused by oxides of nitrogen which are present in the combustion products of coal gas. To some extent, the trouble has been overcome in two ways: (1) treatment of the dyed material with a protective agent such as diethanolamine or melamine, which absorbs or combines with the oxides of nitrogen; (2) the introduction to the market of some blue dyestuffs which are not reddened by oxides of nitrogen.

Subliming of Dyestuffs. One other defect of these dispersed dyestuffs is that they tend to sublime. A piece of dyed cellulose acetate fabric may discolour a piece of white cellulose acetate fabric stored in contact with it, because the dyestuff sublimes out of the coloured piece and dissolves in the white piece. It is a troublesome fault.

Dispersed Base Diazotised and Developed. In this method a dispersed base is applied as described above, and then diazotised and coupled with a hydroxy-compound such as β-naphthol or resorcin. According to the compound used for coupling, different shades can be obtained. For example, Artisil Scarlet R dyed direct gives a yellow, but coupled with β-naphthol it gives a scarlet, and coupled with resorcin an orange-brown shade. An example will illustrate the method of application:

Mix $\frac{1}{2}$ lb. Artisil Scarlet R with its own weight of hydrochloric acid (32° Tw.), disperse in $2\frac{1}{2}$ gallons of hot water and boil. Sieve into a dye-bath and make up to 200 gallons. Wet out goods (100 lb.), enter at 40° C. and raise to 65° C. in thirty minutes. Add 1 lb. sodium acetate slowly and dye at 65° C. until exhausted. Wash. To diazotise, treat for thirty minutes in the cold with $2\frac{1}{2}$ per cent sodium nitrite, and 7 per cent hydrochloric acid (32° Tw.) on the weight of the goods. Wash. Enter into the developing bath and treat for thirty minutes in the cold with 1 per cent β-naphthol and 1 per cent caustic soda (77° Tw.). Wash with dilute soap, rinse and dry. (Tw. or Twaddell is a measure of the specific gravity. If the Twaddell number is divided by 2 and the dividend is added to unity the result gives the specific gravity. For example, 32° Tw. = 1·16 S.G.; 77° Tw. = 1·385 S.G.)

Another favourite coupling agent is β-oxynaphthoic acid (more properly β-hydroxynaphthoic acid), which is very useful for blacks. The concentration of the diazotising bath is much higher than that needed for viscose. In the case of β-oxynaphthoic acid, this substance is dissolved by boiling in one-third of its own weight of soda ash (sodium carbonate), and the development bath, to which it is added, is made slightly acid with acetic acid. The diazotised material is entered cold, worked for fifteen minutes and the temperature slowly raised to 60° C., at which temperature development is completed in thirty minutes. If the bath is alkaline, coupling will not take place.

Cross Dyeing. Because cellulose acetate is unstained by many direct dyestuffs, and because the dispersed acetate colours will not dye viscose or cotton, two-colour and resist effects can be obtained on cellulose acetate/viscose (or cotton) fabrics from one dye-bath.

It is probable that this is at least partly why so little cellulose acetate is dyed in yarn (hank) form.

Water-soluble Dyestuffs. The dispersed colours do not give good penetration of heavy fabrics, but this can be obtained by the use of the Solacet colours, which are water-soluble dyestuffs that have been introduced for use with cellulose acetate by I.C.I., Ltd. Their degree of exhaustion can be controlled by the amount of Glauber's added to the bath. Their advantages are good penetration and levelling, but their wet fastness is inferior to that of dispersed colours.

Basic Dyestuffs. These are dyed with 1 to 2 per cent acetic acid. Goods are entered cold, the temperature raised to 70° C. and the dyestuff exhausts. The disadvantages are, as is usual with basics, the lack of light fastness and the fact that only few basic dyestuffs have a marked affinity for cellulose acetate. In the last few years new basic dyestuffs have appeared, and some of these although primarily intended for dyeing acrylic fibres, have secondary uses on acetate (*cf.* p. 761). It is odd that the fastness to light of basic dyes is much faster on acrylic than on other fibres. Nevertheless, the new basic colours are frequently used on acetate.

Dope Dyeing. Chromspun is the name given by the Tennessee Eastman Co. to their dope-dyed acetate; continuous filament is available, both regular and " lofted ", in twenty-five colours, and also staple in a good range—all very fast. A new colour, Diamond White, which contains a fluorescent agent, is brilliant. The lofted yarns (continuous filament) are made particularly for upholstery weaves and for draperies; they give an impression of weight and richness; their lustre is subdued and they appear to be admirable. Other manufacturers also make dope-dyed acetate.

Stripping. The " organic solvent " behaviour of cellulose acetate is well illustrated by the use of activated charcoal as a stripping agent, for this material adsorbs organic solvents readily. If dyed goods are heated in a bath containing 4 lb. soap and 4 lb. activated charcoal and ¾ pint Dispersol VL per 100 gallons of liquor at 85° C. they will usually be stripped. Slight reductions in shade, in cases where complete stripping is not required, can be obtained by heating in a bath at 70° C. with 1½ pints Dispersol VL alone per 100 gallons.

Cellulose acetate is a good white, and does not ordinarily require bleaching. Its thermoplasticity has enabled permanent embossed effects—*e.g.*, of a crêpe—to be applied to it by passing over heated engraved rollers.

Uses

The uses of cellulose acetate are manifold. Its soft and warm handle has made it popular for lingerie, dress-fabrics and all kinds of women's wear. For men's wear it has been used with some success for ties and dressing-gowns and to a limited extent for shirts, pyjamas, socks and underwear. It has proved very suitable for bathing-costumes.

Semi-soft collars are sometimes made by laminating a layer of fabric containing cellulose acetate (perhaps two ends cotton, two ends cellulose acetate in the warp) between two layers of cotton fabric. When heat is applied, fusion of the acetate takes place, and the composite structure acquires a permanent stiffening. The " Trubenising " process depends on this principle and on a similar application of it. A swelling agent may be applied before the fabric is heated, to assist in the fusion and make it more uniform.

Fabrics of normal width have been cut with heated blades into hat-bands, ribbons, *etc.*, the thermoplasticity of cellulose acetate ensuring a solid fused " selvedge ".

It has been used very widely as an insulator for electrical wiring and coils, and has found considerable application for military purposes.

Cellulose acetate is a versatile fibre, and apart from its specialised uses for obtaining cross-dyed effects, for Trubenising, for the preparation of fabrics resistant to mildew and bacterial attack, for insulation, *etc.*, it is to be found in nearly every field of textile usage for personal wear and for home-furnishing.

It is not suitable for handkerchiefs, on account of its low hygroscopicity, and its only moderate resistance to abrasion limits its use for fabrics that have to stand up to hard wear—*e.g.*, carpets and linings.

One interesting specific case of its employment was the incorporation of cellulose acetate fibres in the paper from which rationing coupons were made. The fibres had been dyed with fluorescent dyestuffs, and consequently when the coupons were examined in ultra-violet light, the fibres glowed, and this test was used to distinguish between genuine and forged coupons. An increasing use for acetate tow is in cigarette filters.

Staple Fibre Cellulose Acetate

This is discussed on p. 667.

Alon or Tohalon

This is a Japanese fibre, known in Japan as Alon; elsewhere as Tohalon to avoid confusion with the German acrylic Dralon. Alon is a remarkable fibre, it is cellulose acetate which has been made by acetylating viscose rayon fibre, *i.e.*, the acetylation process follows the spinning. The fibre marks a revolutionary change from the usual method which we have already considered of manufacturing cellulose acetate. Alon has the fundamental advantage that the spinning solution (for the viscose) is aqueous whereas ordinary acetate is spun from acetone solution, and water will always be cheaper than acetone. The manufacturers of Alon are the Toho Rayon Co. Ltd., of Tokyo. The research which led to the fibre was carried out at Kyoto University under the direction of Professor Sakurada.

Manufacture

The relevant British Patent is 750,702 (1956). From that and other published information the manufacturing process can be seen to be in principle as follows.

Wood pulp with a high α-cellulose content is converted into viscose rayon staple of the polynosic kind with a heart shaped cross-section and with high dry and wet strength; this rayon also has a higher than usual D.P. The rayon is usually in the form of staple with about 18 crimps per inch but tow is also used; in whatever form it is, the rayon is immersed at room temperature in a 15 per cent solution of sodium acetate in water, squeezed or centrifuged to 100 per cent liquor retention and then partially dried to 60 per cent retention.

The lower the moisture content the less wastage of acetic anhydride due to conversion to acetic acid. The sodium acetate, or some similar salt such as potassium acetate, disodium phosphate, zinc sulphate or ammonium oxalate, catalyses the subsequent acetylation process. The partially dried fibre is placed on a wire mesh conveyor belt and is led continuously into a chamber containing acetic acid vapour at 110° C. and thence into the acetylator in which vaporised acetic anhydride at 130° C. is circulated. The fibre takes several minutes to pass through the anhydride chamber; during this time acetic acid is continuously formed, so some of the mixed vapours of acid and anhydride are continually bled off and replaced by anhydride alone. When the reaction has proceeded far enough to give a product with a 50 per cent acetyl value (for comparison conventional secondary acetate has an acetyl value of 53 per cent) the fibre is washed with water, oiled, and dried in air at 50–70° C. The crimp originally in the viscose rayon is retained, not lost in processing. The economics of the process largely depend on recovering the acetic acid and reconverting it to anhydride without much loss and it is reported that this has been well worked out. The essential and novel feature of the process is the acetylation of the viscose staple without loss of fibrous form.

Properties

Because the vapour phase acetylation does not depolymerise the cellulose so much as the usual solution method, the D.P. of Alon is 340 compared with 380 for the viscose rayon from which it is made; it does not fall very much.

Mechanical. The strength of Alon 2·8 grams per denier is about twice as high as that of secondary cellulose acetate; so it is, wet, at 2·2 grams per denier. Knotted strength is 1·8 grams per denier. Elongation is 23 per cent, a little less than that of secondary acetate 27 per cent, but knotted elongation is better at 16 per cent against 10 per cent. Recovery from stretches up to 5 per cent is good, but not so good from higher stretches; it is slightly better than it is in ordinary acetate or triacetate but not so good as in Orlon. The outstanding feature of the mechanical properties is the very much higher strength of Alon compared with ordinary acetate and this derives mainly from the higher strength of the viscose rayon from which it is made. There is no loss of breaking load during acetylation, but because of the weight increase, the tenacity of Alon in grams per denier is only about two-thirds of that of the original viscose rayon.

Specific Gravity. Not very different at 1·34 from that of secondary acetate at 1·32. Both values are close to that of wool.

Moisture Regain. About 5 per cent compared with 6·5 per cent for ordinary acetate; the lower value is not advantageous for apparel uses.

Ironing. Like triacetate, Alon is more resistant to a hot iron than is secondary cellulose acetate. According to the makers it presents no problem of sticking or fusing in hot ironing, but in the author's experience it is difficult to press an Alon garment without glazing it. Alon yellows at 200° C.

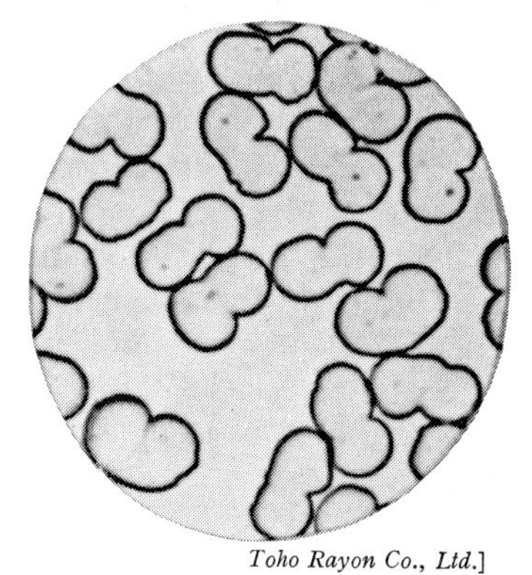

Toho Rayon Co., Ltd.]

FIG. 95.—Photomicrograph (× 1,200) of cross-section of Alon fibres.

Cross-Section. A little rounder (Fig. 95) than that of dry-spun acetate; the lobes are less pronounced.

Chemical Resistance. Not very different from ordinary acetate; good enough for ordinary purposes, but quite lacking the phenomenal resistance of such synthetics as Orlon.

Pilling. The makers claim that Alon shows little tendency to pill.

Dyeing

Similar to that of ordinary acetate. Disperse dyes are best for light colours, azoics for blacks, navys and maroons. Owing to its better heat resistance Alon can be dyed at higher temperatures than ordinary acetate; blended with Terylene it can be dyed solid. A resin finish is not usually necessary, but is sometimes applied; 2 or 3 per cent of urea-formaldehyde or melamine-formaldehyde or polyvinyl acetate to give a firmer handle.

Uses

Some Alon fabrics are very suitable for children's dresses and give pleasure and satisfaction; others are a little harsh to the hand. Clearly a lot depends on fabric construction. It would seem likely that whatever triacetate can do in the way of ease of care, pleat retention, and so on, Alon can do as well. Additionally, and very important, Alon has twice the strength of triacetate. It cannot hope to compete in strength and durability with nylon and Terylene, it cannot compete with the acrylics in chemical and microbiological resistance, it cannot compete in price with viscose rayon, but there seems to be no reason why it should not take most of the market that triacetate now has.

Price is clearly important; in 1961 Alon cost Yen 220 (about 4*s.* 3*d.*) per lb. in Japan and production capacity was 12 million lb. per year. Production costs will fall; already a new catalyst has been discovered which enables the acetylation process to be carried out in a few minutes instead of a few hours as was originally necessary. In 1961 ordinary acetate staple in Japan cost about 3*s.* 0*d.* per lb. so that Alon still has a good deal of price lag to catch up. Basically the Alon process would seem to have a bigger potential than conventional acetate processes. The only doubt concerns the handle; some of the Alon fabrics seen in the U.K. have not been as soft as they might be, certainly not as soft as secondary acetate.

FURTHER READING

M. H. Avram, " The Rayon Industry ", 559–580. London (1930).
H. Dreyfus, " The Birth, Development and Present Position of the Cellulose Acetate Artificial Silk Industry in this Country ", *J. Soc. Dyers and Col.*, **55**, 116 (1939).
G. H. Ellis (dyeing), *J. Soc. Dyers and Col.*, **40**, 285 (1924), and **57**, 353 (1941).
T. Takagi and J. B. Goldberg, " Alon ", *Modern Textiles Magazine* (April, 1960).
British Patent 750,702 (1956); U.S. Patent 2,780,511 (1957).

CHAPTER 12

CELLULOSE TRIACETATE

TRICEL, ARNEL

THE first cellulose acetate that was made was cellulose triacetate, but as it was not soluble in solvents that were then available and were safe to use (see p. 221), it was never made on a big scale in the early days of rayon. Small quantities of about 300 lb. per day of 150 denier cellulose triacetate yarn were in fact spun from a chloroform solution by the Lustron Company in America, starting in 1914 and continuing until 1924, but the process never reached any size. Nevertheless, considering the incomplete knowledge of the chemistry of the product that was then available, and considering also the frightening toxic hazards of using chloroform as a spinning solvent, this production, negligible as it seems to-day, must be rated as a stupendous achievement. Otherwise, until a few years ago all the enormous quantities of cellulose acetate rayon made have been of the secondary acetate.

Lately this position has changed and the manufacture of the triacetate has been carried out on a large scale: Tricel (British Celanese Ltd.) is made in the U.K.; Arnel is made by the Celanese Corporation of America and Trilan by Canadian Celanese Ltd. Courpleta, formerly made by Courtaulds, was discontinued after the 1957 merger with British Celanese. Its properties had been almost the same as those of Tricel. The two main reasons for the development of these fibres are:

1. Solvents for the triacetate which are easy and safe to handle have become available cheaply and in large quantities. Methylene dichloride, which is an excellent solvent for triacetate, but is not a solvent for secondary acetate, has been available cheaply since 1930.
2. The development of the synthetic fibres such as nylon, Orlon and Terylene has shown that there are uses to which a hydrophobic fibre can be put, and that there are many things that can be done with one of these that can never be done with the hydrophilic viscose and (secondary) cellulose acetate rayons and natural fibres.

It is mainly this new knowledge which had accrued that has driven the rayon manufacturers to make triacetate; the solvents had been available for long enough.

Chemical Constitution

The chemical constitution of the primary triacetate is discussed on p. 223.

It will be sufficient to note here that it is the product of complete acetylation of cellulose and that all of the three hydroxyl groups in each glucose residue are acetylated, whereas in the " secondary " cellulose acetate, with which we have been familiar for forty years, some hydrolysis has been carried out to reduce the number of hydroxyl groups from 3 to about 2·3 per glucose residue. The information already given on pp. 222–224 may be supplemented by reference to the formula:

```
                                                  OCOCH3
                                                    |
           H          OCOCH3                       CH2
           |            |                           |
 —O—┐      C————————————C      H      H             C——————————O           ┌———
    |     /|            |\     |      |            /|           \          |
    C    / OCOCH3  H     \     C      C           / H            \         C
    |   /  H                  /      |           /  OCOCH3  H      \       |
    H  \   |                 /       |          \          |       /       H
        \  C————————————O   /        └—O—┘       \ C————————————C /
           |                                       |            |
          CH2                                      H          OCOCH3
           |
         OCOCH3
```

In this the heavy bonds indicate the nearer edges of the planes in which the flat glucose residues lie. The acetyl groups stand away from these planes so that, in effect, the triacetate fibre molecule consists of a long core of some hundreds of glucose residues lined up serially, and bristling with acetyl side-chains. The core is the same as in viscose rayon, but viscose bristles with hydroxyl and not acetyl groups; the core in secondary acetate bristles with a mixture of acetyl and hydroxyl groups in the ratio of 23 : 7. The unit molecular weight of triacetate is 288 and its acetyl value " as acetic acid " is $3 \times 60/288$ or 62·5 per cent; in practice it varies from 61·5 to 62·5.

Manufacture

The raw material is either purified cotton linters or specially pure grades of wood pulp—in either case a pure form of cellulose. The reactivity of the cellulose is enhanced by pretreatment with acetic acid. Two methods of acetylation are available:

1. *Non-solvent process.* The activated cellulose is esterified with acetic anhydride in the presence of a non-solvent such as benzene, which preferably has a slight swelling action on the esterified

cellulose. An acid catalyst such as sulphuric acid, toluene sulphonic acid, or perchloric acid is used; this acid catalyst must subsequently be removed from the fibre by heating in a non-solvent medium with acetic acid, thus purifying the solid cellulose triacetate which is then dried.

2. *Solvent process.* The activated cellulose is esterified with acetic anhydride and acetic acid, using sulphuric acid as a catalyst. Alternatively, methylene chloride can be used instead of acetic acid. When acetylation is complete, all the cellulose fibre will have passed into solution. There is no need to hydrolyse any of the acetyl groups off, as in the preparation of secondary acetate (p. 226), but it is necessary to give a brief hydrolysis to remove combined sulphuric ester groups which, if left in, would make the product unstable. The cellulose triacetate is then precipitated into water, washed and dried.

The fibrous or flake cellulose triacetate is made into a 20 per cent solution in methylene chloride which may contain a small proportion of alcohol. Methylene chloride boils at 42° C. against acetone at 57° C., and as it is dearer, and as there are always some recovery losses, it is likely that cellulose triacetate will be more costly to produce than the normal acetate. Nevertheless, its lower boiling point should effect some economy (less heat) on the dry spinning process that follows. The dry-spun cellulose triacetate fibre is passed over a wick containing an anti-static agent and is collected on cap-spinning bobbins if it is in the continuous filament form. If required for staple, a number of ends are collected into a tow as they leave the spinneret, no twist is inserted, and the tow is crimped and cut to the desired length.

Wet Spinning. Cellulose triacetate is dissolved in glacial acetic acid and this is extruded into either water or dilute acetic acid. Arnel 60 was wet-spun, and it was considerably stronger than ordinary dry-spun Arnel. But it has been discontinued, and probably all of today's triacetate fibre is dry-spun.

Properties

Always the properties of a fibre depend on its constitution, and those of triacetate derive naturally from (1) the cellulosic backbone of its molecule, which gives it the very moderate strength of ordinary rayons and a similar extensibility, and from (2) the forest of acetyl groups which pretty well surround the cellulosic core and which have no liking for water molecules, and consequently make the fibre resistant to water and wet processing, give it a low moisture regain and water uptake and make it less easy to dye. But there is one other property of the fibre which could hardly have been foreseen, although it derives from its hydrophobic character, and that is that triacetate can be heat-set like the synthetics so that it will hold pleats that have been deliberately inserted even if subsequently washed, and so that it will resist subsequent creasing. Nowadays the importance that is attributed to thermal setting and dimensional stability is very great. Triacetate shares low production costs with the cellulosic fibres, and highly valued thermal-setting properties with the synthetics, but it also shares their less attractive features: the relatively low strength and durability of ordinary cellulose acetate, and the low moisture regain of the synthetics.

In detail the properties of triacetate are as follows:

Tenacity: 1·2 grams per denier dry (65 per cent R.H. and 70° F.) and 0·8 grams per denier wet. Loop and knot tenacity both 1·0 to 1·1 grams per denier.

Extensibility: 20–28 per cent (standard conditions) and 35–40 per cent (wet).

Initial modulus or load in grams per denier necessary to give a 1 per cent stretch (p. 309): 0·39, the same as ordinary acetate, which should contribute to a good soft handle.

Density: Tricel 1·32 gm./c.c. Ordinary acetate is usually quoted as 1·30, which is practically the same.

Melting point: 290–300° C., much higher than normal acetate (235° C.) both with decomposition. With both fibres there is some softening and sticking when ironed at much lower temperatures, but even so triacetate shows a marked advantage in safe ironing temperatures over ordinary acetate (see Fig. 96). It is noteworthy that the melting point is considerably higher than that of nylon and Terylene which is just over 250° C.

Morphology: Fibres show longitudinal striations, and the cross-section is bulbous (Fig. 97). The difference of appearance between triacetate and normal acetate (Fig. 94) is insufficient for positive identification.

Flammability: Shrinks and melts to a bead when ignited, but will burn and flame, especially if the fabric is of an open structure, such as a voile or ninon—about the same as normal acetate.

Moisture regain: Regain of triacetate at 65 per cent R.H. is 4·5 per cent. (compare nylon 4·2 and ordinary acetate 6·5). All of the natural fibres had high moisture regains; so had viscose and cuprammonium rayons, and ordinary acetate rayon was the first fibre to have a much lower regain. Then came nylon, much lower, and later Terylene and the acrylics with very much lower regains. The trend which one has hoped to see, and which is slowly coming,

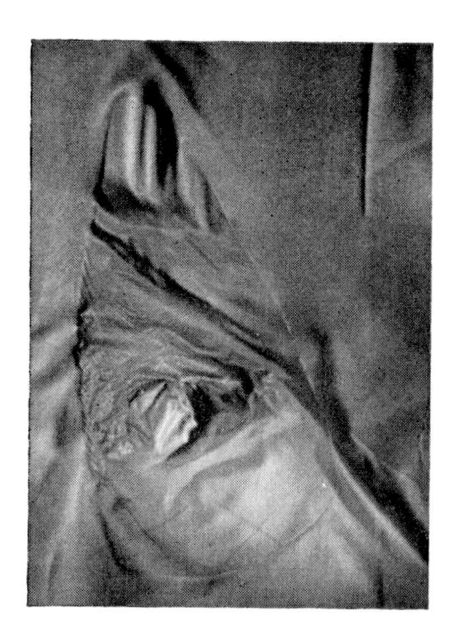

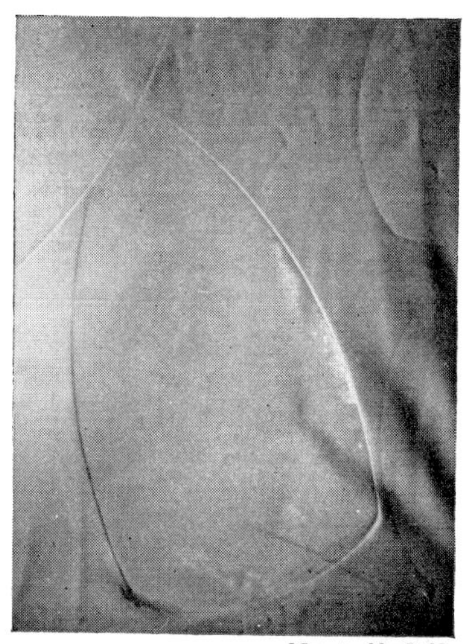

[*Courtaulds, Ltd.*]

FIG. 96.—The heat resistance of cellulose triacetate is superior to that of ordinary acetate. The same hot iron shrivels ordinary acetate (*left*) but not triacetate (*right*).

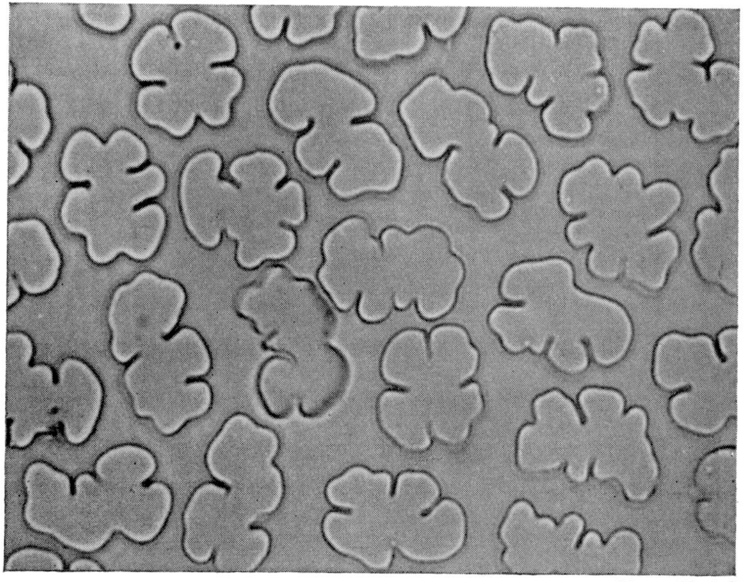

[*Courtaulds, Ltd.*]

FIG. 97.—Cross-section of filaments of triacetate (× 750).

is the synthesis of fibres with higher regains; in triacetate we have a reduction in the regain of a fibre chemically made from a natural polymeric material (cellulose) to approach that of nylon. It brings with it the advantage of a fairly high wet strength/dry strength ratio and also, because of the accompanying low water-imbibition, rapid drying after washing. It is important to note that when triacetate is dry-heated to 195° C. or heated to 130° C. in steam, a molecular rearrangement takes place that results in better crystallinity; the increased mobility of the molecules at the higher temperature enables them to sort themselves out and pack a little better; this rearrangement results in greater impregnability and a reduced accessibility to water. The moisture regain (65 per cent R.H. 70° F.) of *heat-set* Tricel fibres is given as 2·5–3·0, which is considerably lower than that of nylon.

Chemical resistance: Triacetate is resistant to boiling water, whereas secondary acetate is delustred by it. It is resistant to dilute alkalis such as might be met with in laundering, but is hydrolysed by hot strong alkali. It is resistant to dilute acids, but degraded by concentrated strong acids. It is stable to commercial bleaching procedures using peroxides, peracetic acid, chlorites and hypochlorites under mild laundry conditions. Sodium chlorite is an effective bleaching agent. It is soluble in methylene dichloride, chloroform, formic acid, glacial acetic acid, dioxan (slowly) and in cresol (slowly). It is swollen by acetone (also partly dissolved), ethylene dichloride and trichlorethylene, but unaffected by methylated spirits, benzene, toluene, carbon tetrachloride, perchlorethylene and most other solvents. Because it swells in trichlorethylene, this solvent should be avoided in dry-cleaning and perchlorethylene or white spirit used. The chemical resistance of triacetate is generally superior to that of secondary acetate.

Light: Resistance to ultra-violet light is similar to that of secondary acetate.

Biological resistance: Very good towards bacterial, fungal and insect attack.

Lustre: Bright, but can be subdued by the incorporation of titanium dioxide.

Electrical: Very high electrical resistance, which is only exceeded amongst textile materials by Terylene, polyolefines, Teflon and glass. The insulation resistance of triacetate is some fifty times greater than that of secondary acetate; this great superiority is attributable to its more hydrophobic character.

Handle: The handle of heat-set triacetate is much crisper than that of secondary acetate; this may be a good feature in taffetas and suitings, but it will not recommend the fibre for underwear. Handle and drape are two of the most valuable of the characteristics of secondary acetate.

Dyeing

Qualitatively, the dyeing behaviour of cellulose triacetate is similar to that of secondary acetate; quantitatively, the fibre is more hydrophobic and so more resistant to the entry of dye molecules from aqueous baths. In practice, this means that it has to be dyed at the boil instead of at 80° C.; it represents a stage of dye affinity intermediate between secondary cellulose acetate and the polyester and acrylic fibres. For example, a dyestuff which will give a full shade when dyed on secondary acetate at 70° C. and on Terylene at 120° C. will give a similar depth of shade when dyed on triacetate at 95° C.

Disperse dyestuffs are ordinarily used for triacetate, but they must be applied at a temperature not below 90° C. and preferably at 98–100° C. Arnel and Tricel are unaffected by mildly alkaline conditions up to *p*H 9·5 at 100° C; they are neither delustred nor saponified.

Dyestuffs of the Solacet class which are sometimes used for secondary acetate have practically no affinity for triacetate. When disperse dyestuffs are used, those that will not sublime at temperatures up to 200° C. must be used, because most commonly the dyed fabric will be subjected to a high-temperature heat treatment, and if a dyestuff that sublimes has been used, there will be a change of shade. Disperse dyestuffs give faster dyeings on triacetate than they do on secondary acetate, and this is in line with the usual experience that dyestuffs that have been difficult to get into a fibre are usually hard to get out, *i.e.*, they are fast. Correspondingly, the heat treatment of triacetate reduces its dye affinity, but if dyeing is done first and heat setting afterwards, one gets the relative ease of dyeing and the added fastness due to the dyestuff molecules becoming trapped in the fibre during the setting treatment. Triacetate materials show gas-fading with some dyes just like secondary acetate, but can be protected by the application of 1 per cent of inhibitor. Dispersed blacks are not very suitable for triacetate because of the prolonged time of application needed to build up the shade. Instead certain azoic combinations can be used, for example:

SRA Black IV is applied at 98–100° C. from a bath containing a sulphated fatty alcohol; surplus dye is washed off and then in a fresh bath β-oxynaphthoic acid is dyed on in a bath adjusted to *p*H 3·0 to 3·5 with formic acid. Surplus is washed off with cold water and then the dyestuff components on the fabric are coupled by diazotising with sodium nitrite and hydrochloric acid at room temperature. The fabric is finally washed with warm water and given a light scour.

Vat dyes are not much used on triacetate; they give only indifferent fastness to light, and they may change in shade when the dyed fabric is steam-press pleated.

The use of carriers has been advocated with Tricel; there is probably something to be said for their use in heavy shades, but whether this outweighs the difficulty of their ultimate complete removal from dyed fabric is questionable. Tripropyl phosphate and diethyl phthalate have been recommended; they swell the fibre and are adsorbed on to its surface, so providing a solvent for the dye at the fibre-liquor interface. Diphenyl has been recommended for mixtures of Arnel and Dacron.

Experience that has been gained with the dyeing of synthetic fibres generally seems to point to a well-founded preference in the industry to use high temperatures, almost always in closed pressure machinery and to avoid the use of carriers, which, apart from technical difficulties of securing uniformity of application, may be toxic and leave a residual smell that is not easy to eliminate. Generally this is true enough but in practice carriers are still used for dyeing triacetate.

Surface Saponification

As already indicated, cellulose acetate (the old Celanese) was at one time given a partial saponification (p. 233) in order to make it receptive to direct dyes, but this was in the 'twenties before better ways of dyeing it had been discovered. The best way to get a reasonably good affinity for direct dyes without excessive loss of weight was to restrict the saponification to the surface of the fibre. This was done by treating with a strong saponifying agent for a short time. The dyeing of triacetate has revived interest in the process; what used to be done for early Celanese is now done for Tricel. Now it is called " S " finishing, and the advantages claimed for it relate more to finishing than to dyeing; they are:

1. Improved anti-static properties. Cellulose has much more moisture affinity than triacetate, and the surface saponification will increase the moisture content of the fibre surface and so reduce the tendency to electrify.
2. Improved resistance to glazing during ironing. This, for

the same reason: cellulose does not soften or melt when it is heated, whereas triacetate does so.

3. The handle is improved and the saponified fibre is more receptive to soft finishes, again doubtless because of the higher moisture content.

4. The gas-fading of sensitive dyes on triacetate to which reference has just been made is reduced by the surface saponification.

5. Rubbing fastness, both wet and dry, is improved.

6. Resistance to hot air treatments is improved.

A suitable recipe for " S " finish on the winch is to treat the triacetate fabric for 1 hr. at 90° C. (or 1½ hr. at 80° C.) in a bath containing 0·35 per cent caustic soda flakes and 0·05 per cent sulphonated fatty alcohol surfactant with a liquor-to-goods ratio of about 35 : 1. If the treatment is done on the jig a lower liquor ratio is possible with a more concentrated bath and more severe treatment, *e.g.*, 2 hr. at 95° C. in a bath containing 0·7 per cent caustic soda and 0·2 per cent surfactant.

The " S " treatment is given to all-Tricel fabrics and to mixtures of Tricel with nylon or polyester, but not with wool or cellulose mixtures. Tricel/acrylic mixtures are sensitive, and if they are treated the temperature must not exceed 50° C., otherwise the handle will be impaired.

Heat-Setting

When triacetate is heated to 170° C. and above, the molecules become sufficiently mobile to rearrange themselves and pack better; they line up better and the crystallinity of the fibre is increased. The fibre becomes more impenetrable; its moisture regain falls from about 4·5 to about 2·5, the amount of water that it will hold after centrifuging falls from 16 to 10 per cent (for comparison, secondary acetate about 26 per cent), its affinity for dyestuffs is reduced and the fastness to washing of dyestuffs previously introduced is increased. Furthermore, pleats that have been deliberately imposed prior to or simultaneously with the heat treatment become permanent, and very durable pleated and moiré and embossed effects can be easily obtained. In this respect triacetate possesses one of the outstanding advantages of Terylene and some of the other synthetics, and, having been made from cellulose, it is very much cheaper. Dimensional stability and pleat-retention is one of the outstanding properties of triacetate fibre. There are machines available which will insert parallel pleats continuously throughout the length of the piece and will set them with thermostatically controlled rollers at the same time. But the bulk of the pleating trade is in " Sun-ray " pleating, a relatively simple operation in which cut skirt lengths are placed between radially pleated formers made of stiff manila paper, the whole then being folded like a concertina, tied and set.

There are two ways of setting the fabric: either dry heat treatment, already discussed, at temperatures of 170–220° C. or steaming with the fabric in roll form on a perforated tube at 110–130° C. Sun-ray pleating is usually carried out by an exposure of 10 min. to steam at 10–15 lb./in.^{2} pressure. Even with continuous machine pleating the machine-pleated fabric may advantageously be given a steam treatment to enhance the durability of the pleats. Blends of triacetate and cotton or wool which are to have permanent pleats set in them must contain not less than 50 per cent triacetate. Higher setting temperatures 220–240° C. for 10–20 sec. have been recommended for Arnel, but they may involve a loss of 10–20 per cent in strength. In practice, an exposure of 25 seconds at 200° C. is satisfactory for average fabrics. It is worth noting that whereas nylon and Terylene have been subjected to a stretching process during manufacture and consequently shrink when heat-set, cellulose triacetate, not having been stretched, does not shrink when heat-set.

Uses

The uses of a fibre are determined by its properties; triacetate has been used for durably pleated taffeta skirts and for blouses and dresses. Blends of staple triacetate and wool have been used for tropical suitings. The relatively low cost of fibres derived from cellulose may allow the dimensionally stable garment to enter a lower price field than hitherto. The handle of acetate is traditionally good, but heat-setting of the triacetate may affect it adversely, so that a finishing " breaking " treatment is necessary; any suggestion of filament fusion is bound to affect the handle deleteriously. In comparing the potential uses of the fibre with the applications of nylon, it has to be remembered that it lacks the strength and durability of nylon, and could not be safely used for the light and sheer fabric constructions in which nylon is familiar. The lightest denier in which triacetate continuous filament is sold is 55 and the heaviest is 300, a range more like that of the rayons than the synthetic fibres. Staple fibre is marketed in a range of staple lengths from $1\frac{7}{16}$ in. for blending with cotton, to 4 in. for the worsted system and 6 in. for the flax system. Nearly all the triacetate made is used for ladies' dress goods, an application in which it has been very successful. Much of it is 100 per cent continuous-filament woven fabric. In staple-fibre form triacetate is blended with wool, nylon and viscose. High bulk Tricel is used in knitwear. Tricelon is a continuous-filament blend of Tricel and Celon (nylon 6).

Identification

The chemical similarity of triacetate and ordinary secondary acetate calls for a note on distinguishing between the two. Probably the most straightforward method of distinguishing between the two is to treat the fibre with methylene dichloride, which dissolves triacetate but only swells secondary acetate. Conversely 80 acetone/20 water only swells triacetate but dissolves secondary acetate.

FURTHER READING

J. Boulton, " Courpleta—The Dyeing and Other Properties of Cellulose Triacetate Yarn and Staple ", *J. Soc. Dyers and Col.*, **71**, 451–464 (1955).
A. Mellor and H. C. Olpin, " The Dyeing and Finishing of Cellulose Triacetate Yarns and Fabrics ", *J. Soc. Dyers and Col.*, **71**, 817–829 (1955).
F. Fortess, " Dyeing, Finishing and Heat-treating Arnel triacetate ", *Amer. Dyestuff Reporter*, **44**, P 524–P 537 (1955).
" Arnel ", Technical Bulletin TD-12A, Celanese Corporation of America.
A. Murray and D. M. Byrne, " Colour Selection in the Dyeing of Triacetate ". *Text. Manfr.*, 342–347 (August 1968).

CHAPTER 13

HIGH-TENACITY CELLULOSIC FIBRES

TENASCO, CORDURA, DURAFIL, FORTISAN

TENASCO, Cordura, Durafil and Fortisan comprise a group of highly oriented yarns with a cellulose framework. Two early discoveries that did much to improve the tenacity of viscose rayon were:

1. In 1912 Napper (B.P. 406 (1911)) suggested the addition of about 1 per cent zinc sulphate to the spinning bath. It improved the strength and softness of the yarn.
2. In 1914 Wilson used two godets instead of one, stretched the just-spun yarn between them and took the tenacity up from 1·4 to 2·0 grams per denier thereby.

These two developments did not give high-tenacity (as we think of it to-day) yarn but they paved the way for it.

In the early nineteen-twenties Lilienfeldt introduced the process of spinning into sulphuric acid of parchmentising strength (65 per cent), and stretching the yarn as it is spun. It is then in a very plastic condition as a xanthosulphuric acid cellulose ester and, whilst it is so, it is stretched to more than twice its length and then immediately quenched in cold water. This was really a direct development of Wilson's earlier discovery. Lilienfeldt yarn aroused great interest but it was probably not made in any great quantity until Courtaulds adopted the process for making " old " Durafil. Even so, Lilienfeldt viscose yarn was the first that could fairly be called high-tenacity rayon.

Next came Weissenberg's process of stretching cellulose acetate in dioxan/water mixtures. This gave a high-tenacity cellulose acetate yarn. The process was developed by British Celanese in the early nineteen-thirties until it was replaced by the author's discovery of steam stretching. This process would give cellulose acetate yarn of 4–6 grams per denier and when this stretched yarn was saponified the resulting cellulose had a tenacity of 7–10 grams per denier. This was the first very high-tenacity yarn: in 1933 tenacities of 9–10 grams per denier were frequently obtained on experimental lots and pilot plant production was good enough to allow anything below 7 grams per denier to be rejected.

In 1935 the Tenasco process was discovered; this gave viscose yarn of 3·5 grams per denier; it depended on a big increase in the zinc sulphate content of the bath to 4 per cent and on a stretch being applied in hot water or dilute acid, continuously with the spinning. It was this discovery that started the replacement of cotton by high-tenacity viscose for tyre cord.

In 1950 du Pont discovered that certain quaternary compounds could be used to improve the tenacity of viscose and this discovery has been the basis of the manufacture of the " super " yarns that have ousted cotton from the tyre market.

It is a simple story with the benefit of hindsight. Those who have made many thousands of the many millions of trials that have laid it bare can view it with mixed feelings. With this simplified understanding of the way high-tenacity fibres have grown, we can consider some of the most famous of them in more detail.

TENASCO

It is normal practice when spinning viscose rayon to apply a stretch to the freshly formed filaments between two rotating godet wheels which revolve at different speeds. This method of Double Godet Spinning as it is called was devised in 1914 by L. P. Wilson, of Courtaulds, Ltd., and has been universally adopted. The stretch that it gives imparts a certain degree of orientation to the yarn, and increases its tenacity and reduces its extensibility. It was found that by interposing a water-bath at about 90° C. between the stretching godets, a still more substantial increase in tenacity was obtained. Subsequently this process was adopted for the production of Tenasco.

Tenasco appears to be spun from a normal viscose solution. It is spun into a special coagulating bath, which contains more zinc sulphate than usual, and also is at a higher temperature than would be used for normal viscose rayon. Some of the sodium cellulose xanthate is converted into zinc cellulose xanthate; probably the whole of the skin of each filament is converted into a mixed sodium zinc cellulose xanthate. The filaments are given a maximum stretch through hot water or dilute acid, and collected as usual in cake form in a centrifugal box.

" Maximum stretch " means that stretch which if slightly increased would cause the filaments to break. It might be thought that if, during manufacture, the filaments are stretched to just short of breaking point, the filaments when dry would have a very low extensibility—*i.e.*, they would have a very short break. In practice it is found that the yarn contracts after leaving the stretching godets and entering the

Topham Box, this contraction being that which normally accompanies the completion of the regeneration of cellulose. Later the *continuous* spinning process was used with Tenasco with the following advantages over Topham Box spinning:

1. The yarn is given complete and even treatment along the whole of its length.
2. The yarn is not handled and is therefore less likely to be " hairy " or otherwise damaged.
3. Larger packages can be spun.
4. The yarn as it comes from the machine is already pre-stretched to the required low elongation.

Using the method described, Tenasco yarn, with a tenacity of 3·3 grams per denier and 16–17 per cent extension, can be obtained. If the yarn is intended for tyre-cord where a low elongation is necessary the yarn may be subjected to a further stretching operation, known as pre-stretching. Whereas normal Tenasco has a tenacity of about 3·3 grams per denier and about 17 per cent extension, pre-stretched Tenasco has a tenacity of 3·6–3·7 grams per denier and an extension of only 9 per cent.

The Tenasco process gives best results on a moderately fine filament denier, and for this reason 2½ deniers per filament has been chosen as standard for Tenasco yarns.

X-ray diagrams, as well as birefringence studies, show that Tenasco is highly oriented. In the wet state Tenasco breaks at 2·3 grams per denier and 23 per cent elongation. Knotted it breaks at 2·1 grams per denier and 10 per cent elongation.

Under the microscope Tenasco has a cross-section not very different from that of viscose rayon (Fig. 98); similarly the longitudinal view is striated.

Tenasco requires one and a half times as much dyestuff as does normal viscose to give the same depth of shade.

Uses. The main use of Tenasco has been in tyre cords, which are usually made from yarn of very heavy denier, such as 1,100 or 1,650. It has been claimed by tyre-makers that the use of high-tenacity rayon in tyre cords leads to a greater tyre life, especially if the conditions of use have been severe—*e.g.*, long journeys, high speeds, heavy loads, bad roads, hot climates. Tenasco has also found employment in aeroplane tyres, which come in for particularly rough usage during the short time they are in use. It has been found that a tyre made from cotton frequently heats up to 130° C., whereas tyres made from Tenasco keep relatively cool, under similar conditions, at temperatures not above 100° C. An additional point in favour of Tenasco over cotton for tyre cords is that the high temperatures attained cause the fibres to dry out, and whereas cotton becomes a little weaker on thorough drying, Tenasco becomes quite considerably stronger.

Tenasco production increased slowly from 1936, but with the 1939–45 war came the need for new materials. The success

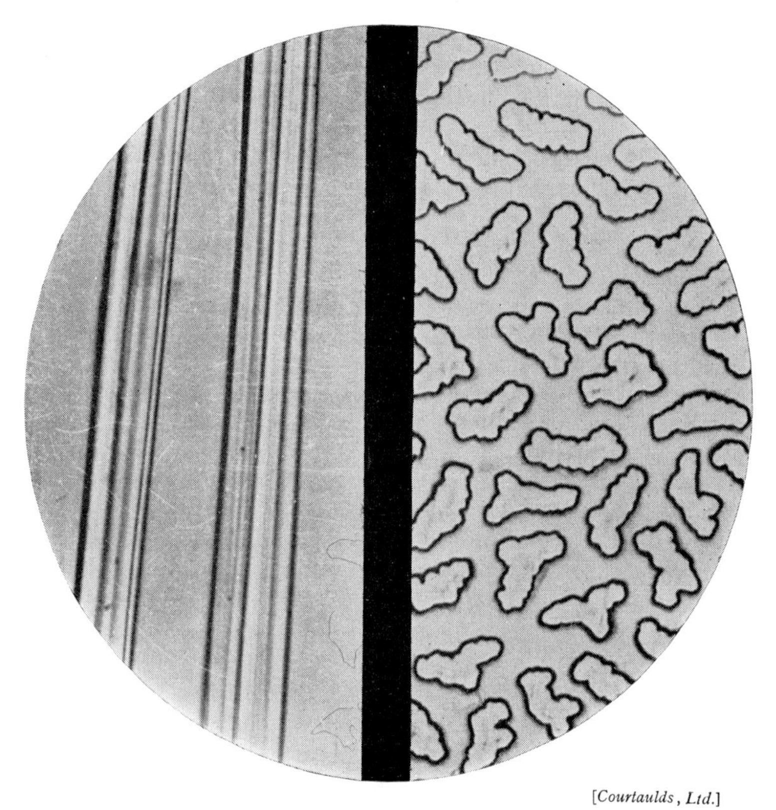

[*Courtaulds, Ltd.*]

FIG. 98.—Photomicrograph of Tenasco ($\times 500$).

achieved by Tenasco in tyres for trucks and aircraft led to a rapid increase in production. Furthermore, shortly after Japan entered the war, natural rubber became short and had to be replaced by synthetic. Tyres made from synthetic rubber generated much more heat and Tenasco proved to be much more suitable than cotton in synthetic tyres. Production rose rapidly after 1941 and continued to increase even after the end of the war.

Tenasco yarn or cords have to be " bound " to rubber in tyre manufacture; they are too smooth to adhere to rubber as well as, say, cotton, which has naturally a hairy surface. Tenasco yarns are therefore preheated in an aqueous mixture of resorcinol-formaldehyde and latex or some similar mixture which provides the necessary bonding between the smooth Tenasco filaments and the rubber.

Tenasco is also used as woven fabric to reinforce colliery and other conveyor belts, for belts for power transmission and for carrier ropes for paper drying, as well as for universal joint discs, link belts, hosepipes and liner fabrics which last are used in the rubber industry for preventing unvulcanised sheets of rubber from sticking together.

For ordinary textile purposes, Tenasco is inclined to crease more than viscose rayon, but this also applies to any highly oriented fibre.

CORDURA

The pneumatic tyre depends for its strength on a carcase of fabric made from heavy cord; without this fabric foundation to reinforce the rubber, pneumatic tyres would not be possible. Tyre cord constitutes the biggest individual use for a fibre, and in some years this has taken 500 million lb. Until the nineteen-thirties cotton was used exclusively for tyres and it was only when high tenacity rayons become available that cotton began to be displaced.

The first ton of " high tenacity " yarn with a strength of about 2·25 grams per denier was sent to tyre-makers in 1933; the following year du Pont started plant production of a tyre yarn under the name Cordura. They were followed within the next few years by other viscose manufacturers. The first tyre-cord yarn supplied was 275 denier 120 filaments, which was of equivalent counts to the cotton yarn that had hitherto been used in tyre manufacture, but as the construction of tyre-cord was simplified, a 1,100 denier 480 filaments yarn was introduced in 1939. Heavy denier yarn permits economies in processing in that breakages are less frequent and twisting is simpler. Deniers as high as 4,400 have been used, but the bulk of the trade is in 1,100 denier. In 1940 cotton still had 97 per cent of the tyre industry, high-tenacity rayon only 3 per cent; in 1950 cotton only retained 42 per cent, high-tenacity rayon had 57 per cent and the odd 1 per cent was nylon. By 1960, practically no cotton was used but nylon had taken a large slice of the market from high-tenacity rayon, and by 1968 nylon had taken most of the tyre-cord market.

Manufacture. The process is similar in principle to that for regular viscose; modifications that bring about the high tenacity are:

1. Retaining a higher than usual degree of polymerisation in the cellulose.

2. Better filtration—more frequent change of filter cloths.

3. Stretching the fibre before it is set, usually as much as 150 per cent at maximum stretch (*i.e.*, just short of breaking).

4. The presence of more zinc than usual in the coagulating bath; the high concentration of zinc ions causes coagulation of the xanthate on the skin, and this delays regeneration of the cellulose so giving time for the stretching to be carried out.

5. Because colour and dyeability are unimportant in tyre yarn, removal of sulphur from the yarn is unnecessary; in fact if the sulphur is left in it may help the bonding of fibre to rubber. Acid and salt contaminants are, of course, washed out of the yarn.

Properties. Tenacity is about 3 grams per denier and elongation about 13 per cent. Wet tenacity is about 1·9 grams per denier and elongation about 18 per cent. Typical stress–strain curves for high tenacity and regular rayons are shown in Fig. 99.

Uses. Almost entirely for tyre manufacture. Collectively for all the high tenacity rayons this constitutes the biggest single industrial use for man-made fibres. Production of high-tenacity yarn has

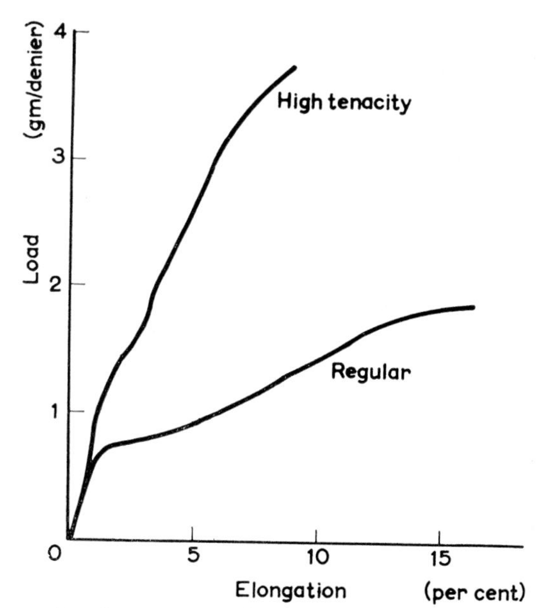

FIG. 99.—Stress–strain curves of high tenacity and regular viscose rayons.

grown enormously; in America it came to about 10 million lb. in 1940, to 308 million lb. in 1950 and to between 400 and 500 million lb. in 1960. For comparison, Japanese output in 1960 was 58 million lb., although in 1955 it had been only 21 million lb.

Other minor applications for Cordura are in the production of laminated and reinforced paper, clothes lines, rubber hose, and fabric for conveyor belts. In Japan its equivalent is used for bicycle tyres, upholstery, curtains and quite recently in the manufacture of mail bags. About one-quarter of the world production of viscose rayon is high tenacity (1965).

Super Fibres for Tyre Cord

The use of nylon for tyre cord has provided a great stimulus to the high tenacity viscose manufacturers, apprehensive that they may lose too much of the tyre cord market, to improve their products. This they have done and have produced a series of " Super " yarns, *e.g.*, Super Cordura, Super Tenasco and Suprenka, which have possessed remarkably good physical characteristics and have enabled the viscose manufacturers, five of whom have banded together into a mutual-help organisation called Tyrex Inc., to fight back.

These " Super " yarns made by different manufacturers seem to have the common feature that they are made by the use of additives to the spinning bath; these additives are often quaternary ammonium compounds, sometimes polyethylene oxides, but always they have the effect of retarding regeneration of the cellulose from the xanthate and of increasing the tenacity of the fibre. They have become important; the mechanism by which they work has not been elucidated, but their use is becoming widespread. The following extract from an important du Pont patent illustrates their use.

Quaternary Retardants. In U.S. Patent 2,536,014 (1950) N. L. Cox (assignor to du Pont) describes the use of quaternary, but not surface active (because their carbon chains are too short), ammonium compounds such as benzyl trimethylammonium hydroxide. These can be employed either as a constituent of the viscose solution itself or in the bath. Their use enables the viscose solution to be spun unripened or " green ", it slows down the rate of precipitation in the spinning process and gives a smooth round-to-oval section fibre with good strength and, particularly, good wet strength. The following example illustrates the way in which the quaternaries can be used and the effects that they have.

Two viscose solutions both containing 7 per cent cellulose and 6 per cent caustic soda, and one (but not the other) containing 1·4 per cent benzyl trimethylammonium hydroxide are prepared. They are spun unripened into a bath containing 8 per cent sulphuric acid, 14 per cent sodium sulphate, 15 per cent zinc sulphate and the viscose that does not contain the quaternary is also ripened and spun as usual as a control. The properties of the fibres so obtained are:

Viscose.	Tenacity (gm./denier).		Elongation (per cent).		Gel swelling factor.
	Dry.	Wet.	Dry.	Wet.	
No quaternary, ripened, *i.e.*, orthodox	3·7	2·2	9	20	3·2
No quaternary, unripened .	3·6	2·0	9	19	3·1
Quaternary, 1·4 per cent unripened	4·0	2·8	9	19	2·6

The improvement in those physical properties tabulated is not phenomenal, but it is accompanied by less swelling in water, lower imbibition of water and the skin of the fibre is increased to 60–100 per cent of the whole, *i.e.*, some of the fibres spun with the quaternaries are practically all skin.

Gel Swelling Factor. The gel swelling factor referred to in the

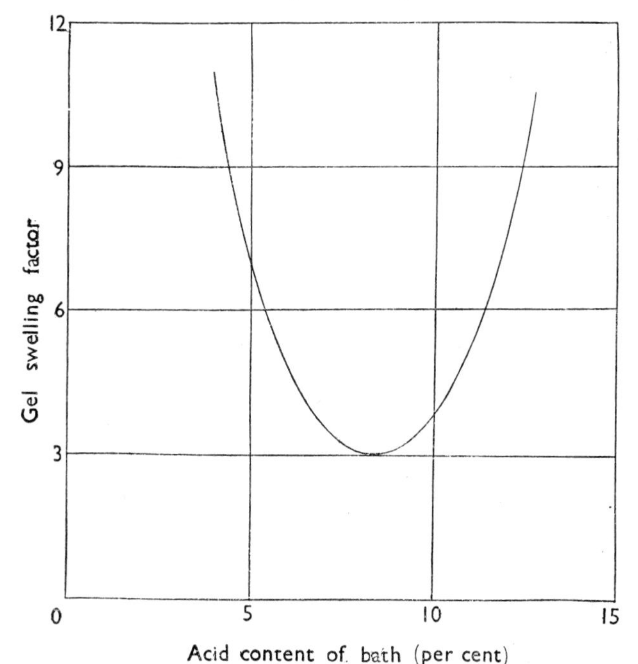

FIG. 100.—Variation of Gel swelling factor of viscose rayon with concentration of acid in the spinning bath.

table is determined as follows: the gel thread as it is spun is collected in a monolayer on a bobbin, the sample is centrifuged for 1 min. at 1,400 r.p.m., cut off and weighed. The sample is then washed free of acid, dried at 105° C. and reweighed. The ratio of the two weights—*i.e.*, wt. of gel/wt. of cellulose—is termed the " gel swelling factor "; the variation in gel swelling factor with acid content of the spinning bath is shown in Fig. 100. Usually the factor has a minimum near to the value of 3, but it depends on the viscose; that represented in Fig. 100 does have a minimum value of 3; in the du Pont patent example just discussed, rather lower values of gel swelling are obtainable. It is common to spin viscose yarn at or near the point of minimum gel swelling.

Spinning Speed. The modern tendency to spin fibres at high speeds has dictated the use of very acid spinning baths, but it has always been evident that such conditions do not make for the best fibre and to-day, to get still better viscose rayon, lower speeds have to be used; we are reverting to the lower speeds that were orthodox twenty years ago.

Super-Cordura

Du Pont's " super " rayon Type 272 has the following physical characteristics:

Tenacity:	
Dry	5·0 grams per denier
Wet	3·5 grams per denier
Ratio wet/dry	70 per cent
Elongation at break:	
Dry	10 per cent
Wet	20 per cent
Cross-section	Almost circular, smooth edge

This is an all-skin fibre, with smaller crystallites than ordinary rayon has. It is now believed that the main difference between skin and core ordinary rayon is in the size of the crystallites, skin having them smaller than core. Looked at from another viewpoint this is the equivalent of greater regularity of structure, so it is not surprising that an all-skin fibre should have great strength and resistance to abrasion and should be suitable for tyres.

All of the high tenacity yarns that are used for tyres are continuous filament.

Durafil

" Old " Durafil was a Lilienfeldt type highly stretched continuous filament rayon of extremely fine filament, about 0·3 denier, all "skin". Its dry strength was 5·6 grams per denier and wet about 3·9 grams per denier; elongation at break was about 7 per cent dry, 8 per cent wet. It has been discontinued and the same name Durafil given to a quite different fibre. This practice may not give rise to any commercial uncertainty but it can be confusing to the student. He should remember that "new" Durafil has nothing whatever to do with "old" Durafil.

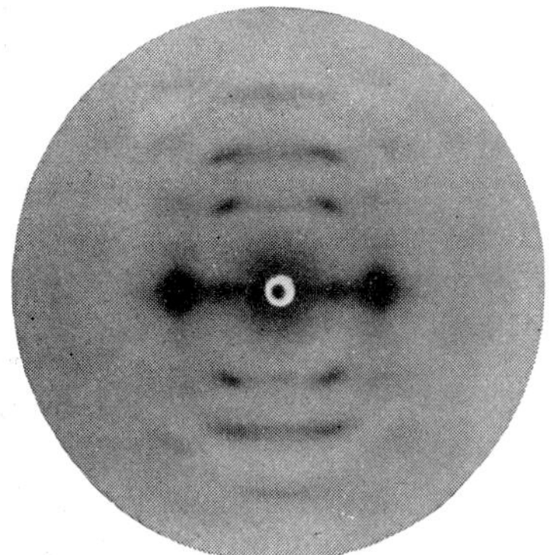

[*Professor W. T. Astbury*]

FIG. 101.—X-ray diagram of Lilienfeldt type old Durafil.

[*Professor W. T. Astbury*]

FIG. 102.—X-ray diagram of normal viscose rayon.

New Durafil is a viscose staple fibre in which the process of regeneration of the cellulose has been altered by the action of modifiers such as amines and polyethylene oxides in the viscose (B.P. 808,838). Durafil has an all-skin circular cross-section; filament deniers range from 1 to 4½; it has a high D.P. The crystalline regions are smaller in Durafil than in Fibro and total crystallinity is lower: 40 per cent against 47 per cent. Properties of the new Durafil compared with Fibro are as follows:

	Durafil.	Fibro.
Tenacity (gm./denier):		
Dry	4·0	2·5
Wet	3·0	1·4
Ratio wet/dry	75	56
Breaking elongation (per cent):		
Dry	30	18
Wet	35	22
Initial wet modulus (elongation (per cent) under load of 0·5 gm./denier)	13	11
Recovery (per cent) from 5 per cent extension	42	45
Water imbibition (per cent)	70	105
Moisture regain	13·7	13·0

A strong fibre with a high wet strength, poor recovery characteristics, and easily extensible. What has it to offer? Mainly, good flexibility and resistance to abrasion and good resistance to tear. It is a fibre that will stand up well to hard wear. It is excellent for pocketing; pockets made from it do not go into holes and it is proving much better and cheaper than the 85/15 Fibro/nylon pocketing formerly used. Shoe lining fabrics, shoe laces, industrial overalls, upholstery tapes and belts represent other uses. In method of manufacture, structure and physical properties the new Durafil is akin to the Super viscose tyre yarns. In its low water imbibition of 70 per cent (Fibro 105) and its high ratio of wet/dry tenacities it has something in common with the polynosic rayons, because like them it has a very fine crystalline structure. But there are important differences: tyre yarns are continuous filament whereas new Durafil is staple; polynosic fibres have excellent recovery characteristics whereas new Durafil has not. Furthermore both tyre cord fibres and polynosic fibres have fairly low elongations at break (about 10 per cent) whereas new Durafil stretches 30 per cent before it breaks; it is this high elongation which together with the high strength is responsible for the high work of rupture: new Durafil stands up to snags well; it is difficult to tear and extremely hard wearing. What it lacks is dimensional stability.

Fortisan

This fibre is made by stretching cellulose acetate in steam under pressure and then saponifying the stretched yarn with alkali to give regenerated cellulose. Fortisan yarn has an extension of about 8 per cent, which is very low, and precludes the use of the fibre for many purposes.

When cellulose acetate is saponified to cellulose there is a loss in weight of the order of 38 per cent, so that material made in this way is bound to be expensive.

Manufacture. Fortisan is made in two stages. The actual conditions can be varied considerably, but the following may be taken as typical for a small unit, the layout of which is shown in Fig. 103.

1. Cellulose acetate yarn, say ten ends of 150 denier, are run together through a guide and round a pair of rollers with a linear speed of 20 metres per minute. From the rollers the yarn passes into a chamber of compressed air at 32½ lb/sq. in. pressure containing a similar set of rollers. Thence it passes

through an orifice of 0·03 in. diameter into a tube, *D* in Fig. 103, which is fed with wet steam through jets *E*; there is a valve release so that the steam can be maintained at 33 lb./sq. in. pressure. Tube *D* may be conveniently about 6 ft. long; the yarn passes out of it through an orifice *F* of 0·015 in. diameter into the atmosphere and passes around the rollers *G* which are rotating at a linear speed of 200 metres per minute, so that the yarn is stretched to ten times its original length in making the

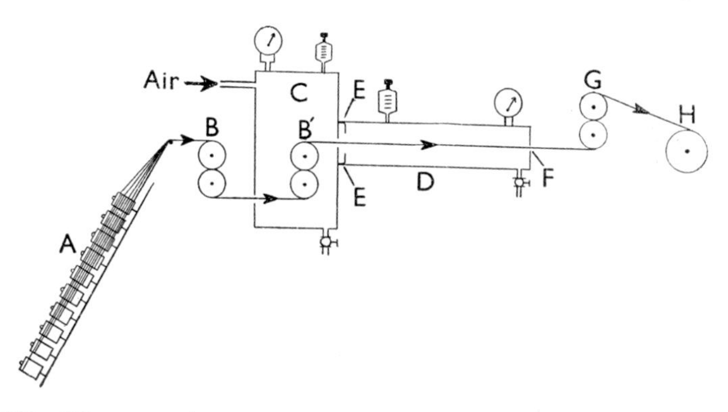

FIG. 103.—Diagrammatic representation of arrangement of apparatus for stretching cellulose acetate yarn to ten times its original length. The rollers on the right revolve ten times as fast as those on the left.

A. Let-off creel containing 150-denier 26-filament yarn on bobbins. B, B′ Rollers (rubber covered)—20 metres per minute. C. Compressed air chamber at 32½ lb/sq. in. pressure. D. Stretching tube. E. Steam (wet) jets. F. Outlet orifice. G. Rollers—200 metres per minute. H. Take-up bobbin for 150-denier 260-filament yarn.

passage through the tube and correspondingly its denier is reduced from (150 × 10) to 150. The yarn is wound on to a plastic bobbin with a perforated barrel. Whereas the initial yarn may have been 150 denier 26 filaments, the final yarn is 150 denier 260 filaments.

2. The stretched cellulose acetate yarn, wound on the plastic bobbins, is then saponified with a solution of caustic soda, say 1 per cent, and sodium acetate, say 15 per cent, in a pressure machine, the soda solution being pumped from a perforated spindle through the yarn, still on the plastic bobbin, first one way from inside to outside, and then the reverse. Finally the yarn is oiled, dried and wound. Because during saponification there is a loss in weight of 38 per cent, the package of yarn collapses and may be difficult to unwind. A yarn which is 150

denier 260 filaments before saponification will be 93 denier 260 filaments after saponification. The saponified yarn is known as Fortisan.

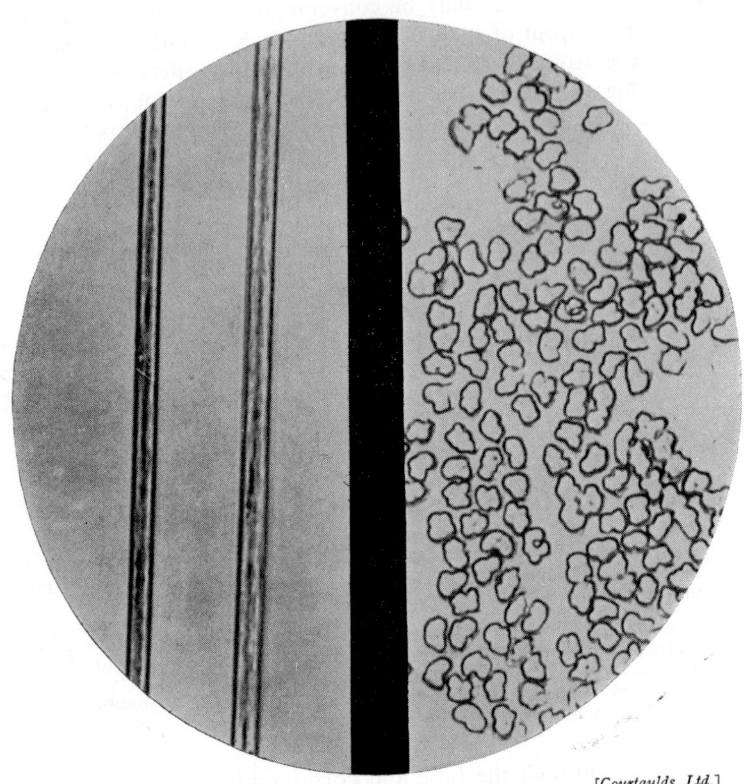

[*Courtaulds, Ltd.*]

FIG. 104.—Photomicrograph of Fortisan (×500).

Characteristics. Average characteristics are as follows:

Stage of manufacture.	Tenacity (gm/denier).	Extension (per cent).
Initial cellulose acetate yarn . . .	1·25	22
After stretching (× 10)	4·5	6
After saponification, *i.e.*, after conversion of stretched cellulose acetate to cellulose .	7	8

Even in 1935, yarn of this strength was being made on pilot plant scale. Fortisan is the most highly oriented and the strongest of any of the cellulosic fibres that have been made even on a small manufacturing scale.

FORTISAN 36

The Celanese Corporation of America in 1955 announced Fortisan 36. It seems to be no more than regular Fortisan in heavy deniers: 270–1,600.

Properties. Some of its characteristics are as follows:

Tenacity, dry:	8·5 grams per denier
wet:	5·5 grams per denier
Elongation, dry:	6·2 per cent
wet:	6·2 per cent
Loop strength dry:	4·2 grams per denier
Specific gravity:	1·52
Moisture regain at 65 per cent R.H. and 70° F.:	9·6 per cent
Cross-section:	Nearly round
Filament denier:	Low, usually about 1
Dyeability:	Not very good, more difficult than ordinary cellulose
Chemical and biological resistance:	Not very good, similar to cotton and rayon.

Uses. At first, Fortisan 36 was produced in 800 denier 800 fils and 1,600 denier 1,600 fils continuous filament yarn. It is intended for industrial uses, very probably for tyre-cords and certainly for transmission belting, fire hose, conveyor belts and carpet backing. It commands the high price (1962) of $2.40 in 60 denier 80 fils.

PROPERTIES OF ORIENTED CELLULOSIC FIBRES

They have high strength, and usually only a low extensibility. They crease badly. They have a lower moisture regain than viscose rayon, and swell less in water. When dyed they exhibit only a low colour value—so great is the surface of the very fine stretched fibres. They have a low affinity for dyestuff, and deny ingress to the dye liquors. Sometimes matters can be improved a little by pre-swelling the fibres in caustic alkali before dyeing. These properties are more pronounced in Fortisan than in Tenasco.

Uses. Mainly, they are used in the undyed state for purposes where strength is the paramount pre-requisite. By far their most important use has been in tyre-cords. For this purpose they are used in deniers of 1,100 and 1,650. During the 1939–45 war Fortisan was used extensively for parachute fabrics and cords. In its present state it is unlikely to be used for clothing, but for purposes which demand only very high strength it is useful.

FURTHER READING

Courtaulds, Ltd. and J. H. Givens, British Pat. 467,500 (1937) (discovery of Tenasco).

L. Rose, " High Tenacity Viscose Rayon ", *J. Soc. Dyers and Col.*, **61**, 113 (1945).

A. Mellor, " Celanese-Fortisan Dyeing and Printing ", *J. Soc. Dyers and Col.*, **62**, 168 (1946).

C. C. Wilcock, " Dyeing of Tenasco and Durafil ", *Dyer*, **93**, 127 (1945).

J. V. Sherman and S. L. Sherman, " The New Fibres ", 279–285 (Fortisan). New York (1946).

" Tenasco " (an outline of the development, manufacture, properties and uses of high tenacity viscose rayon produced by Courtaulds, Ltd.) (1952).

J. K. Berry, " The Chemical and Physical Properties of Modern Textile Fibres ", *J. Roy. Soc. Arts*, **94**, 403–17 (1946).

J. K. Berry, " The Performance of Rayons in Engineering Applications ", *J. Text. Inst.*, **40**, P 662–72 (1949).

British Celanese Ltd., H. Dreyfus, R. W. Moncrieff and F. B. Hill, British Pat. 438,584 (1934), (discovery of stretching cellulose acetate in steam).

British Celanese Ltd., R. W. Moncrieff and F. B. Hill, British Pat. 443,773 (1934).

J. D. Griffiths, " Modified Viscose Rayon Fibres ", *J. Text Inst.*, **52**, 575–591 (1961).

C. E. Coke, " Dynamic Changes in Viscose Fibres ", *Canadian Text. J.*, **76** (6), 41–51 (1959).

K. Weissenberg and B. Rabinowitsch, U.S. Patent 2,142,389 (1939) (stretching acetate in dioxan). This patent was applied for in 1930 and the corresponding German patent application was in 1929.

CHAPTER 14

POLYNOSIC RAYONS

VINCEL, ZANTREL

ORDINARY viscose staple has several advantages over cotton: it can be spun very uniform in diameter and therefore in denier or counts, of any desired lustre, and most important of all it can be made much more cheaply from wood-pulp than cotton which has to be grown and picked and ginned. The attribute which first popularised viscose rayon was its continuity of filament and its resemblance in this respect to real silk. This, added to the brightness of the early rayons, gave something novel and showy that sold easily. Fashion changed, people grew tired of the high lustre and nowadays much more viscose rayon is used as staple than as continuous filament. In this form, its competitor is cotton and although viscose staple has some important advantages it is no less at a disadvantage to cotton in several ways, notably: strength, especially when wet, is low; elongation is high and as elastic recovery is not very good, the rayon suffers from dimensional instability; it is too floppy for some purposes, not stiff or crisp enough; it absorbs too much water and swells much more than cotton when wet; it will not withstand the mercerising process. A great deal has been done to remedy these defects: the tenacity has been increased by new methods of spinning, the ratio of wet/dry strength has been greatly improved, and dimensional stability has come with the resin finishes. Even so, the best of the viscose rayons is not the equal of cotton in firmness, crispness and retention of strength on wetting; and it still requires the extra process of resin-finishing to give it dimensional stability. Two reasons underlie these several inferiorities of rayon:

1. The D.P. of cotton is 2,000–10,000, that of viscose rayon is about 250–270; the " molecular sliver " of cotton is very long " staple ", that of viscose is short " staple".

2. Cotton has a microfibrillar structure; if a fibre is broken by bruising or hammering it splits into thinner fibrils, and these in turn into microfibrils; the microscope and the electron microscope show that however far cotton is ground down the structure is always fibrillar; viscose rayon is not, each filament is solid and even when it has been bruised, shows little sign of a microfibrillar structure. Cotton has grown slowly, the structure of its fibre is beautiful and

wonderfully designed; viscose rayon has been squeezed as a solution through a tiny hole and very quickly coagulated into a solid filament. The slow natural growth has given a fineness of structure that the usual viscose process cannot match. Microfibrillarity is usually made evident by treating the fibre with 70 per cent nitric acid.

The polynosic fibres are a new kind of viscose rayon; in their manufacture steps are taken to maintain a higher D.P. by reducing the severity of the chemical processing so that a value of 500–700 is retained, and also to obtain a microfibrillar structure by precipitating the viscose gently and slowly, in particular by carrying out the stretching before the xanthate has been re-converted to cellulose. The fibres so made are much more like cotton; the degree of their disadvantage has been very greatly reduced, whilst they still retain their advantages of uniformity of size, length, lustre and potential cheapness. They represent a very fair attempt to get the best of both worlds: native cotton and viscose staple.

Development—The Influence of Lanusa

Much of the spadework which fostered the growth of the polynosic rayons was carried out by the Badische Anilin und Soda-Fabrik, A.G. at Ludwigshafen-am-Rhein. In 1933 German Patent 738,486 (U.S. equivalent is 2,155,934 of 1936) described the coagulation of high-viscosity viscose, first in a bath which contained only water or only a very little acid and salt, followed by further treatment in a second bath which contained a moderate amount of salt and acid. It was in this second bath that the cellulose was regenerated from the xanthate. This patent covered the first main principle of polynosic manufacture: retardation of the cellulose regeneration. In 1935, German Patent 746,991 described the use of alkali-cellulose which had not undergone ageing, this leading to the retention of a higher D.P. Both of these principles were incorporated in the manufacture of a fibre known as Lanusa which was made until 1955 when it was discontinued. Lanusa was not called a polynosic fibre, the word was then unknown, nor was it so highly stretched as are to-day's polynosics; it was in fact intended to be wool-like rather than like cotton. Because it was not stretched so much during the spinning process it had a lower tenacity, 2·25 grams per denier dry and 1·7 grams per denier wet, and a higher extension 15 per cent dry and 17 per cent wet than the modern polynosics. Nevertheless it was their precursor; it had the following features which they now have; round cross-section (see Fig. 81); high ratio, about 75 per cent, of wet/dry strength; low water imbibition of 80 per cent; low alkali solubility. All these derived from the high

D.P. (500 in manufacture) brought about by the omission of ageing, and from the fibrillar structure brought about by slowness of regeneration of the cellulose. These are the two fundamental principles on which production of the polynosic fibres is based.

Tachikawa's Development Work

Next came the important work of S. Tachikawa in Japan, described in U.S. Patent 2,732,279 applied for in 1951, granted in 1956. The main features of this patent specification, which described the preparation of viscose fibres that we now call polynosic, were:

1. Ageing of the alkali cellulose was eliminated.
2. The cellulose xanthate was dissolved in water instead of in (3 per cent) dilute alkali.
3. Ripening of the cellulose xanthate solution was eliminated.
4. The spinning bath consisted of acid of only a very low concentration and little or sometimes no salt.

A comparison is shown below of the stages for manufacturing ordinary viscose rayon and Tachikawa's rayon; this rayon has since been produced in Japan under the name Toramomen, and more recently Tufcel.

Operation.	Standard viscose.	Toramomen.
Steeping and pressing.	Soak cellulose in 17·5 per cent NaOH for 2 hr. Press to get moist mass of soda-cellulose.	As standard.
Shredding.	Soda-cellulose is mechanically disintegrated in 2–3 hr. into fine crumbs.	As standard, taking care temperature does not exceed 20° C. and time 2 hr.
Ageing.	During ageing from 1–3 days at 20° C. D.P. falls from 800 to 350.	Omitted.
Churning (xanthation).	CS_2 added in deficiency, only 70 per cent of theory. The crumbs dissolve in 3 hr. to orange gelatinous sodium cellulose xanthate.	Theoretical quantity CS_2 added at 15–20° C. over $2\frac{1}{2}$ hr.; then temp. raised to 25° C. for 1 hr.
Mixing (solution).	Sodium cellulose xanthate is stirred with 3 per cent NaOH for 4–5 hr. to give a solution containing 7·5 per cent cellulose (combined as xanthate) and 6·5 per cent NaOH.	Sodium cellulose xanthate is stirred with water to give a solution containing 6 per cent cellulose (combined as xanthate) and 2·8 per cent NaOH.
Ripening.	The solution is stored for 2–5 days at 10–18° C. The viscosity falls (degradation) and then rises (partial regeneration of cellulose).	Omitted.

Operation.	Standard viscose.	Toramomen.
Spinning.	Extrude into bath containing 10 per cent H_2SO_4, about 18 per cent $Na_2SO_4,10H_2O$, 1 per cent $ZnSO_4$ at 40–55° C. Stretching conditions vary enormously in practice, but for this comparison the standard viscose also undergoes the three times stretch.	Extrude into bath containing 1 per cent H_2SO_4 at 25° C. Then stretch by passing round the stepped godet (Fig. 105) to three times the spun length.

It will be seen that the Toramomen process is shorter in time and gentler in chemical action than is the standard. Elimination of the ageing and ripening stages preserves the D.P.; the more dilute spinning bath ensures slower regeneration of the cellulose and permits more of the stretching and its accompanying orientation of

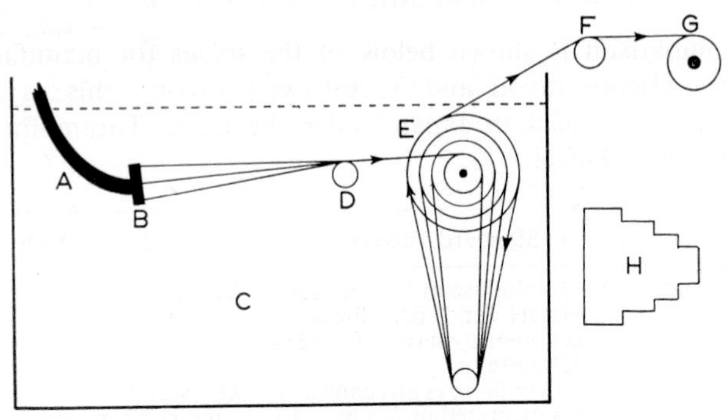

FIG. 105.—Spinning bath used by Tachikawa in making prototype polynosic rayon.

A. Viscose feed. B. Spinning nozzle. C. Bath of 1 per cent sulphuric acid. D. Guide. E. 4-step godet. F. Guide. G. Take-up bobbin. Another view of the 4-step godet is shown separately in H.

the fibrous molecules to take place before regeneration is complete. This device is said to give a higher crystallinity and smaller crystals; it gives better orientation and greater uniformity of orientation. It is significant that the stretching of Toramomen is carried out in stages, first to 100 per cent, second to 150 per cent, third and last to 200 per cent, so that the final length is three times the initial. Stretching in steps may give opportunity for better orientation than does single step stretching.

Properties. The effect that these changes in method have on the properties of the fibre is very considerable indeed; a comparison is shown below of the properties of the fibres prepared as indicated in the above table (the "Toramomen" procedure is taken from U.S. Patent 2,732,279).

	Standard viscose.	Toramomen.
Cross-section	Serrated	Round, smooth
Micro-structure	None	Fibrillar
Dry tenacity (gm/denier)	2·7	3·3
Wet tenacity (gm/denier)	1·6	2·5
Ratio wet/dry (per cent)	57	76
D.P.	270	500
Dry elongation (per cent)	17	9
Wet elongation (per cent)	22	12
Water of imbibition (per cent)	100	66
Increase in diameter on wetting (per cent)	26	15

Strength is higher, elongation at break lower, the ratio of wet strength to dry strength is much higher, there is less swelling in water, the D.P. is higher and there is microfibrillar structure in the Toramomen that is never found in standard viscose. Fig. 106 is an

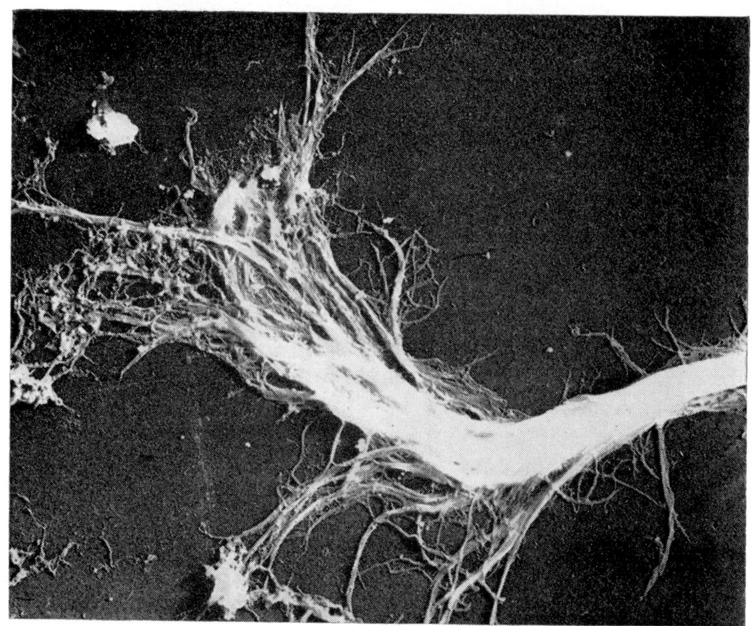

[*Japan Electron Optics Laboratory Co., Ltd., and Fisher Scientific Company*]

FIG. 106.—Electron micrograph of microfibril of Tufcel polynosic rayon (× 20,000). The microfibril which constitutes the whole picture was invisible to the naked eye.

electron micrograph of a Japanese polynosic rayon; it shows the microfibrillarity well.

Commercial Development

The Japanese polynosic fibres are Toramomen and Tufcel. In Europe, Fabelta of Belgium, the Société Chimiotex of Switzerland and Comptoir des Textiles Artificiels of France are working on the fibre under licence. Z54 (Fig. 107) is a name given to the pilot production of all the European licencees working together. In America, Hartford Fibers Company produce the polynosic and call

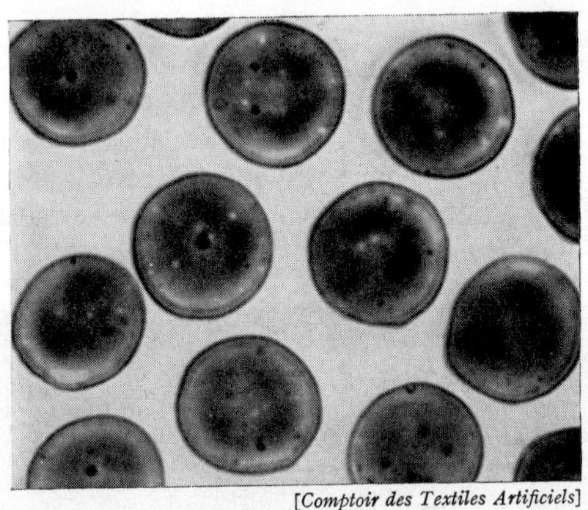

[*Comptoir des Textiles Artificiels*]

FIG. 107.—Photomicrograph of cross-section of polynosic fibre Z54 (× 1,500).

it Zantrel; it seems to be well established and has lately been taken over by American Enka.

There are a few instances, confusing to students, of a good name for a discarded fibre having been given to a new fibre. " Toramomen " has been made since 1944 but it was an orthodox high-tenacity rayon staple; only recently has the name been transferred to the high wet modulus or polynosic fibre. The French fibre " Meryl " was originally a high-tenacity Lilienfeld-type viscose but the name is now used for products made from the polynosic fibre Z54.

Commercial manufacturing methods follow closely the technique of Tachikawa. There are minor variations, for example the possible use of sodium dihydrogen phosphate or acetic acid as a weak acid for the spinning bath. These weak acids will neutralise the alkali in the extruded viscose but slowly so that there is time for the filaments to be stretched between the coagulation and the regeneration.

Properties. The properties of the commercial fibres show some improvement over those of the prototype: the following comparison is informative:

	Polynosic.		Ordinary viscose rayon.	Cotton.
	Zantrel.	Z54.		
Dry tenacity (gm/denier)	3·9	3·4	2·5	3·2
Wet tenacity (gm/denier)	2·9	2·7	1·4	3·5
Ratio wet/dry	0·74	0·79	0·56	1·1
Dry elongation (per cent)	11	10	18	9
Wet elongation (per cent)	13	12	25	10
Initial wet modulus (elongation (per cent) at 0·5 gm/denier)	3·1	3·1	11	3
Increase in diameter on wetting (per cent)	11·5	12	26	13
Imbibition of water (per cent)	60	66	100	50
Cross-section	Round	Round	Serrated	Flat, twisted
Microfibrillar?	Yes	Yes	No	Yes
D.P.	500	500	250	5,000
Recovery from stretch (wet)	Good	Good	Poor	Good

Of particular importance in the polynosic fibres is their high initial wet modulus; they elongate only a little for a small (usually 0·5 gram per denier) load. This means that the characteristic stress/strain curve (Fig. 108 shows that of Zantrel, one of the polynosics) for the wet fibre is fairly steep in the region of the origin. For example, Zantrel has an initial wet modulus (elongation (per cent) under a load of 0·5 gram per denier) of 3·1, whereas ordinary rayon has one as high as 11. This means that for such tensions as might be expected to occur during washing, the polynosic fibre will stretch much less than ordinary rayon. Added to this, the elastic recovery from strain of the wet polynosic fibres is excellent; from a 5-times repeated 3 per cent stretch wet Zantrel shows 96 per cent elastic recovery. It is these last two features, the high wet modulus and the good elastic recovery that give to the polynosics their excellent dimensional stability, especially when being washed and which set them apart from the ordinary run of rayons.

It can be said that there are four main ways in which the polynosics differ from ordinary rayon:

1. Dimensional stability in fabric form.
2. Ability to withstand mercerising. Their alkali solubility is much less than that of ordinary rayon.

3. Crisper, loftier handle more like cotton than rayon. Lustre is also more like that of sea-island cotton, sometimes like that of spun silk.

4. Swelling in water and imbibition of water are much lower.

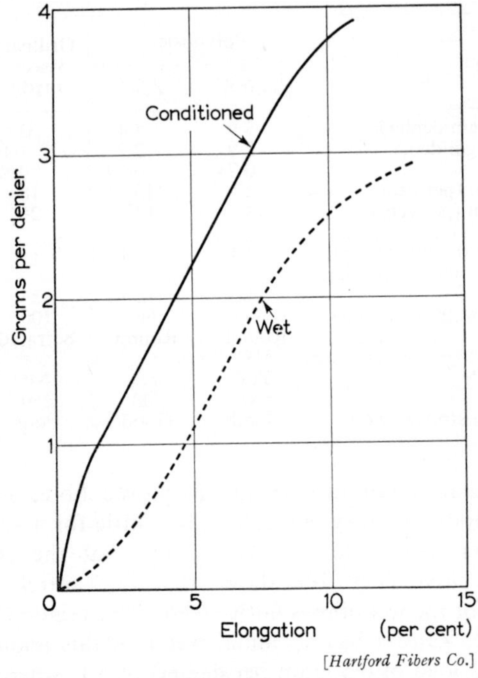

FIG. 108.—Characteristic curve for Zantrel polynosic fibre. The steepness of the curve for the fibre in the wet state is remarkable.

Definition

Polynosic fibres have been defined (F.T.C.) as follows: "A manufactured cellulosic fiber with a fine and stable microfibrillar structure which is resistant to the action of 8 per cent sodium hydroxide solution down to 0° C., which structure results in a minimum wet strength of 2·2 grams per denier and a wet elongation of less than 3·5 per cent at a stress of 0·5 grams per denier."

It is involved but it covers the main points: microfibrillarity and high initial wet modulus. The word polynosic has no connection with, or similarity of meaning to, such words as polymer, polyamide, polyester, polyvinyl and so on. It is said to mean multifibrillar.

Wet Processing

The greater wet strength possessed by the polynosic fibres, their lower degree of swelling in water and their resistance to caustic soda make it possible to wet process fabrics made from these fibres by methods that are used for cotton fabrics, and that are not as a rule suitable for rayon.

High wet strength and toughness open up the use of the continuous processes for the new fabrics, and their alkali resistance makes possible either mercerising or vat dyeing. Hartford Fibers Co. have recommended that 100 per cent Zantrel fabrics should be wet processed on a continuous cotton range which includes the following stages:

1. Singe.
2. Desize and scour; desizing can be done enzymatically followed by a scour; alternatively the fabric can be treated with caustic soda and then given an acid scour.
3. Bleach; sodium chlorite or peracetic acid is suitable; hydrogen peroxide is a good bleach, but care must be taken that the goods have not become contaminated with iron. Sodium hypochlorite is not a suitable bleach for Zantrel.
4. Continuous vat dye.
5. Resin treat if desired.
6. Wash.
7. Compressive shrink.

If instead of being all-Zantrel the fabric is a cotton–Zantrel mixture it can be caustic treated before the bleach stage and mercerised after bleach and before dyeing.

Dyeing

The polynosic fibres, because by comparison with ordinary viscose rayon they have higher D.P., lower swelling in water and hence lower accessibility to water, therefore have a lower affinity for dyestuffs. They dye more in the way that cotton does than viscose. The makers of Zantrel say that it can be dyed with vats, azoics, indigosols, reactives (Procion type), sulphur colours and some directs but these last are the least preferred. Vat dyes on Zantrel give fastness to light and washing as good as they give on cotton. Resin finishes can be applied with an application similar to that for cotton, which is lower than the application to ordinary rayon; these finishes do not weaken the polynosic fabric like they do cotton.

Uses

Some of the early Zantrel fabrics are really beautiful: their lustre is subdued and non-metallic like that of spun silk; they are uniform; they handle well and are strong; they have character. So far their use has been in dress-fabrics and shirtings, the traditional " departments " of the old " better end " cotton manufacturers. For fabrics such as taffetas, poplins, Bedford cords, shirtings and madapollams they should be excellent. Cotton and rayon have found a way into most textile applications; the polynosics should do the same. In particular their dimensional stability should fit them for curtains and other hangings.

Perhaps in the future we shall see cellulosic fibres with still higher D.P.s; here we go from the usual 250 to 500; when we go to 1,000 still better and tougher fibres will result. Although so far the polynosics are in staple form, they should be equally useful in continuous filament.

VINCEL OR MOYNEL

Vincel is a high wet modulus fibre produced by Courtaulds; in the U.S.A. it is known as Moynel. It is a polynosic fibre but is made by a process which is said to be different from Tachikawa's; details of the process have not been published, but the fibre is undoubtedly similar to the Japanese polynosic fibres in most respects, notably: D.P. is about twice that of normal rayon; cross-section is circular; crystallinity is high (55–60 per cent); water imbibition is low at 66 per cent; tenacity is 3·5 grams per denier dry, and 2·5 grams per denier wet, with a wet/dry ratio of over 70 per cent; breaking extension is low at 8 per cent dry, 10 per cent wet; initial wet modulus is comparable to that of Z54; Vincel's dyeing behaviour is more like that of cotton than of ordinary viscose; resistance to caustic soda is good and Vincel/cotton mixtures can be mercerised; handle resembles that of cotton and lustre that of mercerised cotton. Because the water accessibility is lower than that of normal viscose, the regain is also lower at 11·8 per cent against 13·0 per cent for Fibro. This lower regain is a characteristic of all polynosic fibres. Vincel's first successful application was as a 33 Vincel/67 Terylene blend which was used for raincoats and waterproof sportswear. It was sold (early 1962) in the U.K. at 30*d.* per lb. and late in 1965 was reduced to 27*d.* It does not wear so well as cotton.

SM27

SM27 is a very similar fibre to Vincel; it is made by Courtaulds (Alabama). Although the process by which it, too, is made is said to differ from Tachikawa's the properties of the fibre are much the same. Some comparative figures that have been published are:

	SM27.	Fibro.	Cotton.
Tenacity (gm/denier):			
Dry	3·2	2·5	2·5–3
Wet	2·3	1·4	2·5–3
Ratio wet/dry (per cent)	71	56	100
Elongation at break, dry (per cent) . .	7	18	6
Load (gm/denier) required to give 5 per cent extension wet	1·2	0·1	1·2

The figures for cotton are rather different from those given on p. 273—the characteristics vary considerably according to the quality of the cotton. The specimen used for this comparison was evidently not of such high quality. There is not a great difference between Vincel and SM27, but what advantage there is seems to lie with the former.

FURTHER READING

O. Eisenhut, H. Rein and E. Kaupp, German Patent 738,486 (1933) and the corresponding U.S. Patent 2,155,934 (1936).
O. Eisenhut, H. Schmidt and H. Rein, German Patent 746,991 (1935).
S. Tachikawa, *Rayon Synth. Text.*, **32**, 3, 31–33 (1951).
S. Tachikawa, U.S. Patent 2,732,279 (1956).
C. E. Coke, *Canad. Text. J.*, **76**, 6, 41–51 (1959).
Hartford Fibers Company, " Zantrel " brochure (1961).
J D. Griffiths, " Modified Viscose Rayon Fabrics ", *J. Text. Inst.*, **52**, P 575–593 (1961) (SC28).
G. V. Lund and J. Wharton, *Text. Research J.*, **30**, 692–697 (1960) (SM27).
" Polynosics: How it all Started", *Skinner's Record*, 673, 675–6, August (1964).
R. W. Moncrieff, " Vincel: Europe's major polynosic ". *Skinner's Record*, 743–4, September (1964).
H. W. Best-Gordon, " Second look at Vincel ", *Skinner's Record*, 862, 864–5, October (1964) and 1044, November (1964).
R. W. Moncrieff, " The New Rayon Fibres ", *Dyer and Textile Printer*, **127**, 557 (1962).
R. W. Moncrieff, " A Milestone in Rayon Manufacture ", *Textile Weekly*, 535, 539 (9 March 1962).

CHAPTER 15

CHEMICALLY MODIFIED CELLULOSIC FIBRES

THERE are a good many processes to which cotton and viscose rayon fibres are put which modify them chemically. For example, although the theory of dyeing has not yet been completely elucidated, there is no doubt at all that one of the several mechanisms involved in the union of cellulosic fibre and dyestuff is chemical in nature; indeed in the special case of the Procion dyestuffs (p. 203) it is known that this dyestuff reacts chemically with the hydroxyl groups on the cellulose. There are, too, various processes of applying functional finishes, such as flameproofing and waterproofing, in which the finish is bound chemically, although as a rule only very loosely to the fibre. Similarly the mercerising of cotton modifies the chemical and physical properties of the fibre. But in such processes as those mentioned, the change in chemical structure of the fibre is largely empirical and usually indeterminate and sometimes impermanent, and it is not with these slightly adventitious changes that this chapter is concerned, but with two quite well understood and major chemical modifications that can be made to cellulosic fibres, without loss of the fibre form; these two chemical modifications are respectively cyanoethylation and acetylation. Each has been applied both to the cotton fibre and to viscose rayon fibre.

CYANOETHYLATION

The cyanoethyl group is $-CH_2CH_2CN$; the formula of acrylonitrile is $CH_2{:}CHCN$ and the reaction of acrylonitrile with a compound that carries a hydroxyl group results in the formation of a cyanoethyl group, thus:

$$R-OH + CH_2{:}CHCN \longrightarrow R-OCH_2CH_2CN$$

The cellulose molecule carries plenty of hydroxyl groups and these react fairly easily with acrylonitrile, provided that the cellulose has been pre-treated with alkali. This reaction was discovered by the I.G. Farbenindustrie and was protected in French Patent 830,863 (1938). Many other mono-olefinic compounds such as acids, esters and ketones will react similarly, *e.g.*, acrolein, $CH_2{:}CH{\cdot}CHO$, and methyl vinyl ketone, $CH_2{:}CHCOCH_3$, and at first attention seems to have been directed mainly to these as reactants with cellulose, but later the use of acrylonitrile has overshadowed the other ethylenic compounds.

Increase in Weight

When cellulose reacts with acrylonitrile there is an increase in weight. The reaction can go to completion, as shown in the equation below:

```
         OH   OH
         |    |
         CH — CH
        /       \
   — CH          CH—O— + 3CH2:CHCN ——→
        \       /
         CH — O
         |
         CH2OH
       Cellulose                       Acrylonitrile
         162.                        3 × 53 = 159.

        OCH2CH2CN        OCH2CH2CN
         |                |
         CH ————————————— CH
        /                   \
   — CH                       CH—O—
        \                   /
         CH ————————————— O
         |
         CH2OCH2CH2CN
   Completely cyanoethylated cellulose
                 321.
```

When this happens the fibre gains in weight by $\frac{159}{162}$ or 98 per cent and the cyanoethylated cellulose has a nitrogen content of $\frac{42}{321}$ or 13·1 per cent.

If the reaction was allowed to go as far as adding acrylonitrile on to two of the three hydroxyl groups on each glucose residue to give the di(cyanoethyl ether) of cellulose this would show a gain in weight over the initial cellulose of $\frac{106}{162}$ or 65 per cent and would have a nitrogen content of $\frac{28}{268}$ or 10·4 per cent.

If the reaction was restricted to the formation of the monocyanoethyl ether of cellulose then the gain in weight would be $\frac{53}{162}$ or 33 per cent with a nitrogen content of $\frac{14}{215}$ or 6·5 per cent.

Although the gain in weight is attractive from the standpoint that fibres are sold by weight and there would be more to sell, it is unfortunately not accompanied by a gain in strength; if, therefore, the cellulose were to be completely cyanoethylated (98 per cent gain in weight) the tenacity of the product would only be about half that of the original cellulose, even if there was no loss in breaking load. In fact, there is a loss in breaking load, and in practice the cyano-ethylation of cotton is usually restricted so that the product has a nitrogen content of only about 3 per cent, *i.e.*, one hydroxyl group on every alternate glucose residue is etherified. If the reaction is taken so far as to give 6 per cent nitrogen content there is an undesirable loss in breaking load.

Cyanoethylation of Cotton

The advantage of treating the native cotton fibre and maintaining its fibrous form instead of a solution of cellulose is that the cotton fibre is wonderfully built up and has a fine internal structure which is itself fibrillar and helical, a structure which no chemist can ever hope to match. In the regeneration of cellulose from cotton (cuprammonium is the only yarn which is still to-day made mainly from expensive cotton instead of cheap timber) the fibre is dissolved and finally a solid cuprammonium fibre, which has none of the fine structure of the original cotton, is spun. Those processes, such as cyanoethylation and acetylation, which can be carried out on cotton without destruction of the fibre form do have the merit of preserving something good of nature. But, although there is this in their favour, and although cyanoethylation and acetylation of cotton are processes that are fashionable to-day, there is no need to be over-enthusiastic about either of them, for two reasons:

1. They do not have the academic merit of being the first chemical modification of cotton. John Mercer can be credited with that, when in 1844 he first mercerised cotton by treatment under tension with alkali, thus increasing the strength, dye-affinity and lustre of the cotton.
2. Cotton is unlikely to be used in the future in such enormous quantities as it has been in the past; it is too expensive a crop in labour to compete against viscose staple derived from wood pulp. Consequently, chemical modifications to cellulose are more likely to have the chance of large-scale development if carried out on viscose rayon instead of on cotton. (The Toho Rayon Company's production of 16 tons per day of acetate from viscose staple (p. 240) is in line with this view.)

Nevertheless, at present cyanoethylation of cotton has awakened a great deal of interest, partly because the supporters of cotton hope to see its decline postponed by the acquisition of new properties and partly by the desire of the manufacturers of acrylonitrile to find a new outlet for it.

Method. A process for the continuous cyanoethylation of cotton has been developed by the Institute of Textile Technology, Monsanto, and American Cyanamid. The plant for the process is illustrated schematically in Fig. 109. Cotton fabric is first run through a

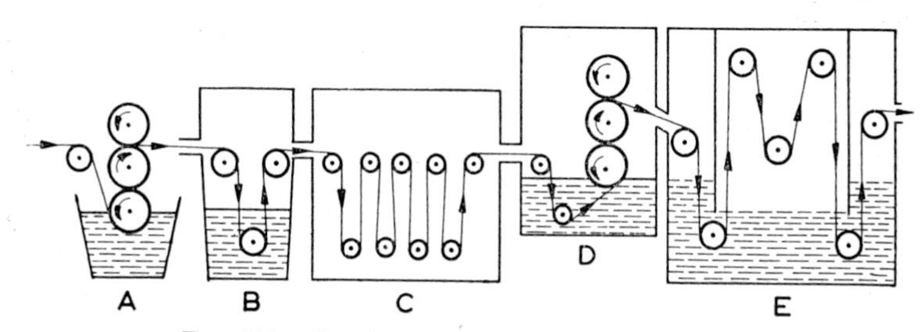

FIG. 109.—Continuous cyanoethylation of cotton.
A. Caustic pad.
B. Cold acrylonitrile.
C. Reaction oven.
D. Dilute acid dip.
E. Steam chamber to remove acrylonitrile from fabric.

caustic-soda bath, *A*, to swell the fibres, and then fed to an acrylonitrile bath, *B*, in which it has a cold dip; from there it passes to an oven, *C*, held at 72–76° C., just below the boiling point of acrylonitrile and in the oven at the high temperature the reaction between the cotton and the acrylonitrile is induced and accelerated. The fabric has a cold dip in dilute phosphoric acid, *D*, and next the excess acrylonitrile is removed in a steam chamber, *E*, in which the steam and acrylonitrile condense together and pass to a recovery plant. The fabric passes forward to a water washing and drying range. Cost is estimated to be about 1*s*. per lb. of finished product, if produced on a scale of 10 million lb. per year. The cost is quite high and is partly due to the conversion of some of the acrylonitrile to β-oxydipropionitrile:

$$2CH_2{:}CHCN + H_2O \longrightarrow O\begin{matrix} \diagup CH_2CH_2CN \\ \diagdown CH_2CH_2CN \end{matrix}$$

Acrylonitrile. β-Oxydipropionitrile.

A batch process has also been worked out, the cotton being treated in kiers first with caustic soda, then with acrylonitrile and finally with washing water. For large-scale production a continuous process is to be preferred, although for small lots a batch process is often good enough and less demanding in capital outlay for plant.

A third method which has been developed at the Southern Regional Research Laboratory, U.S.A., has been that of cyanoethylation of cotton fabric wrapped round a perforated beam similar in mechanical principles to the pressure dyeing methods recently developed (p. 730). The fabric is wrapped slackly round the beam and pretreated with a 2 per cent solution of caustic soda and then with commercial acrylonitrile saturated with water (about 3 per cent water is soluble in acrylonitrile at 20° C.). The fabric is treated with the acrylonitrile for from 1 to 3 hours at a temperature between 55° and 67° C.; time of treatment depends on the construction of the fabric. There was a marked tendency for local overheating to occur, due apparently to an insufficient flow of liquid through the fabric. Careful measurements of temperature throughout the piece of fabric enable the method to be controlled so that reasonably uniformly treated fabric could be made. The degree of cyanoethylation corresponded to a nitrogen content of 3 per cent, and the purpose of the work was to obtain sufficient fabric for wearing trials and assessment of the value of the cloth.

Properties. The enhancement of properties that cotton has gained from cyanoethylation are:

1. An undoubted and considerable increase in resistance to mildew and bacteria. Samples of cyanoethylated cotton that are buried in soil with untreated controls are apparently unattacked when the controls have rotted away. But the degree of microbiological resistance that cyanoethylation can confer on cotton cannot equal the almost complete immunity of the synthetics. Practically, the advantage of greater resistance to mildew might be no more than that washing left lying damp would be less subject to mildew attack.

2. A considerable improvement in resistance to heat. Whereas cotton loses 60 per cent of its strength if maintained for 15 hours in air at 160° C., cyanoethylated cotton (with a 3 per cent nitrogen content) loses only 15 per cent of its strength. If the cotton has been cyanoethylated so far as to have a 5 per cent nitrogen content the loss in strength is still lower, but there are reasons (one being loss of strength itself) which militate against the desirability of cyanoethylating cotton so far as this.

3. Increased receptivity towards dyestuffs, including the acid dyestuffs used for wool. If it became possible to get solid shades with one dyestuff on cotton/wool blends, that would indeed be useful, but so far it is not possible. Usually, less dyestuff, *e.g.*, direct colour, is needed for cyanoethylated cotton than for

untreated cotton to obtain the same depth of shade. Fastness of dyestuff to washing is about the same, but fastness to light is variable —the cyanoethylated cotton has better fastness than untreated cotton with vat colours, but poorer with direct colours.

4. The abrasion resistance of cyanoethylated cotton (3 per cent nitrogen) is about 40 per cent better than that of untreated cotton. Clearly, even with this improvement the treated cotton fibre is not in the same class as the nylons for resistance to abrasion.

Uses. After more than a year of field testing, cyanoethylated cotton, known in America as Azoton, has " in general, lived up to expectations ". Promising uses are said to be in fishing nets and twines, laundry press fabrics, buff wheel fabrics, high-temperature conveyor belts, acid filter cloth, awning fabric, reinforcing twine in rubber goods, and tobacco shade cloth. But the extra cost of the process making dear cotton dearer will reduce its promise for many applications, and even the most enthusiastic supporter of Azoton would have to admit that the synthetic fibres could do many of the suggested jobs better. To the looker-on it seems that laundry press covers and shoe-linings and shoe-welt threads (which commonly rot because of bacterial attack) might provide useful outlets, an advantage of much longer life than normal cotton will give offsetting the extra cost of the cyanoethylated material.

Cyanoethylation of Viscose Rayon

A method developed at Naniwa University in Japan is as follows: Viscose rayon is immersed in potassium cyanide (about 10 per cent), dried at room temperature and then hung in a desiccator filled with acrylonitrile vapour at 40° C. for 10 hours. The treated fibres were a little stronger than untreated (but if the treatment was carried too far they became much weaker) and were considerably swollen by acetone, whereas the untreated viscose is inert to acetone.

Prospects

To summarise, the process will not be particularly easy or cheap to apply; already a good deal of difficulty has been experienced in getting *uniformity* of treatment, particularly on fabric. Given that these difficulties will be overcome in time, all that we shall get for our trouble will be a little better wear, considerable improvement in microbiological resistance and marginal advantages in dyeing and heat resistance. There is nothing here to warrant making the cotton, already an expensive fibre, still more expensive. But there is no reason why the process should not eventually, when it has been cheapened and made more reliable, be used on viscose staple rayon, and that is probably the only part of its future which is important. Recently Russian workers have described the grafting of acrylonitrile on to cellulose and this may prove useful.

Acetylation

The chemistry of the acetylation process has already been discussed on pp. 223–225; irrespective of whether the reaction is carried out in solution or with retention of fibre form, structural changes are the same and so are stoichiometric relations.

Partial Acetylation of Cotton

It is found that if one-third of the available hydroxyl groups in cotton corresponding to the formation of the monoacetate are acetylated, then the product behaves almost like the nearly completely acetylated product in dyeing, water retentivity and rot resistance. Oddly, if cellulose is completely acetylated to the triacetate and then deacetylated to the monoacetate, the product is very much like ordinary cotton in respect of dyeing, water retentivity and rot resistance. The monoacetate made by progressive acetylation is rot-resistant, the monoacetate made by regressive acetylation is not rot-resistant. The explanation is probably that some of the hydroxyl groups are much more accessible either to chemicals or to organisms than others, that when cotton is monoacetylated, the easily attacked hydroxyl groups are protected and the bacteria are impotent; if, however, the cotton is triacetylated and partly deacetylated, it is the accessible groups that are hydrolysed by the alkali and so made available again to the bacteria. In the mass production of cellulose acetate for solution spinning, solubility properties are a determining factor in respect of degree of acetylation; in those processes now being discussed in which fibre shape is retained, solubility is unimportant, indeed undesirable. If, therefore, the desired change in physical properties of a cotton fibre can be brought about by acetylating to the monoacetate stage, there is not much point in going further.

Methods

There are two processes, one a batch process, the other continuous, that have been worked out by the U.S.A. Agricultural Research Service, and which have been adopted for commercial operation.

Batch process. The cotton, whether yarn or fabric, is desized and boiled for 2 hours in 2 per cent caustic soda solution to purify it; it is given an acid rinse and dried. Loose stock is not given this preliminary purification. The cotton is activated by soaking in glacial acetic acid at room temperature for 1 hour (or overnight if more convenient) and then the acid is squeezed from the cotton.

Esterification is effected by treating with:

85 per cent acetic acid
15 per cent acetic anhydride
0·15 per cent perchloric acid as catalyst

using eight times as much of this mixture as the weight of the cotton. Treatment is carried out at 20° C., this temperature being maintained by cooling. Time needed for the reaction varies according to the physical form of the cotton, but is about 1–1½ hours; the operation can be carried out on a stainless steel jig fitted with a hood and with some good means of temperature control. Finally, the cotton is washed and dried. The solution is analysed, made up to strength and re-used. Cost of *materials* only is 40 cents (2*s.* 10*d.*) for each pound weight of cotton treated so that the process is expensive. Commercial exploitation will doubtless lead to great economies in respect of recovery of unused chemicals and thereby to a very considerable reduction in price.

Continuous process. The continuous process is obviously most suitable for fabric. The separate stages are:

Presoaking in glacial acetic acid for 2 minutes at 82° C. to activate the cotton (the time is reduced from 1 hour to 2 minutes by the change in temperature from room in the batch process to 82° C.).

Cooling: By passing around a cooling can through which cold water is circulated.

Catalyst addition: Catalyst is added separately and before the acetylating materials. The fabric is kept in a 1–5 per cent solution of perchloric acid in acetic acid for 1 minute.

Acetylation: Treatment with 40 per cent acetic anhydride in acetic acid for 3 minutes at 20° C.

Washing: The fabric is washed and dried. The machine itself

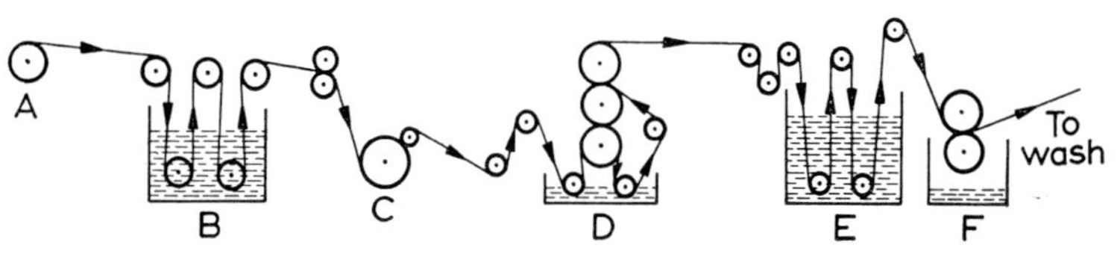

FIG. 110.—Continuous (partial) acetylation of cotton.

A. Untreated fabric (normal regain).
B. Pre-treatment in acetic acid.
C. Cooling can.
D. Catalyst addition.
E. Acetylation reaction.
F. Squeeze to remove surplus anhydride.

(Fig. 110) is about 35 feet long and will treat about 5 yards fabric per minute, more or less according as it is light or heavy. Cost so far is of the order of 33 cents per pound of fabric or warp treated, but can almost certainly be greatly reduced.

Improved Properties

There can be little doubt that rot-resistance is the main advantage to be derived from the acetylation of cotton and the improvement is very great. Whereas untreated cotton fabric lost all its strength after soil burial for 14 days, (mono)acetylated cotton retained all its strength after 6 months. Significantly monoacetylated cotton made regressively (via the triacetate) lost all its strength after 14 days' burial—it has no useful rot-resistance. However, the directly made monoacetate has excellent rot-resistance. For comparison, cottons treated with copper-containing preservatives lost 30 per cent of their strength in 10 weeks, so that acetylation is much more effective (and of course more costly) than conventional rot-proofing.

Heat-resistance is distinctly better, too, and acetylated cotton will withstand high temperatures for about ten times as long as untreated cotton. Laundry press pad covers made from acetylated cotton last four or five times as long as from untreated cotton.

Moisture regain is reduced from about 9 per cent to 3·5 per cent. Chemical resistance to acids is a little better than usual, and resistance to weathering is still under test but is giving promising results.

Uses

Too early to be sure, but flags, fishing nets, shoe-linings and tents seem to be indicated. Shoe-linings are very susceptible to bacterial attack and rotting and the shoe manufacturers generally seem unwilling to pay for the synthetics; they may perhaps find what they want in acetylated cotton (or acetylated viscose).

Uniformity

A good test for uniformity is to dye the monoacetylated cotton with a mixture of a yellow dispersed (acetate) dyestuff and a direct blue cotton colour—a uniform green coloration indicates uniform partial acetylation; yellow patches indicate local over-acetylation and blue patches local under-acetylation.

In thinking of the cost of this process it has to be remembered that conversion of cellulose to its monoacetate involves a weight gain of $\frac{42}{162}$ or 25·9 per cent because the molecular weight of the glucose residue repeat in cellulose, $C_6H_7O_2(OH)_3$, is 162 and that of the glucose monoacetate residue, $C_6H_7O_2(OH)_2(OCOCH_3)$, is 204.

Acetylation of Viscose Rayon

A method of continuously acetylating viscose rayon has been worked out at Kyoto University (Ichiro Sakurada *et al.*, *J. Soc. Textile & Cellulose Ind.*, Japan. **9**, 273–276 (1953) English Summary). It consists of pre-treating the viscose rayon in dilute (0·1 per cent) sulphuric acid to make it reactive, then pressing out as much liquor as possible, packing the fibres in a 6-inch tube quite loosely, so that each gram occupies 25 c.c., and passing through the tube air saturated with acetic anhydride, at 30° C. for 20 hours at such a rate that for every gram of fibres in the tube, 0·42 gram acetic anhydride is passed in each hour. At the end of the treatment the acetylation is 99 per cent complete—to all intents and purposes the triacetate is formed. The method is elegant, and the only danger that one could anticipate would be that of fibre swelling in the water carried from the pre-treatment and the acetic acid and anhydride formed on the fibre.

It should be added that there appear to be two separate aims involved in the acetylation of cotton and viscose rayon, viz.:

(*a*) Partial acetylation of cotton to obtain rot-resistance, treatment being taken only as far as the monoacetate.

(*b*) Complete or nearly complete acetylation of viscose rayon to obtain an acetate staple which will have the properties of the new triacetate fibres (pp. 241, 244).

Process (*a*) has been toyed with for some time, and if it ever reaches big production it is likely that acetylation will be carried out by the use of gaseous ketene, the " anhydride " of acetic anhydride.

Process (*b*) is a more recent development, one that seems to be going ahead in Japan and that has already resulted in the manufacture of Alon (p. 237).

FURTHER READING

I.G. Farbenindustrie, French Patent 830,863 (1938).

J. Compton, " New Textile Fibres with the Structural Elements of Natural Cellulosic Fibres ", *Amer. Dyestuff Rep.*, **43**, 103–111 (1954).

E. M. Buras, *et al.*, " Practical Partial Acetylation of Cotton ", *Amer. Dyestuff Rep.*, **43**, 203–208 (1954).

Anon., " Cyanoethylation Moves Forward ", *Chem. Engng News*, **34**, 5058 (1956).

L. H. Greathouse, *et al.*, " Cyanoethylation of Cotton Fabric ", *Industr. Engng. Chem.*, **48**, 1263–1267 (1956).

R. M. Livshits and Z. A. Rogovin, " Synthesis of Graft Co-polymers of Cellulose and Carbon Chain Polymers with the Use of Manganese Pyrophosphate ". *Khim. Volokna*, 38–40 (1963). Reviewed by R. W. Moncrieff in *Textile Weekly*, 765–7, 1st May (1964).

CHAPTER 16

ALGINATE FIBRES

SEAWEED has been used for many purposes. Dulse, a fleshy reddish-purple seaweed with deeply divided fronds, is used in Scotland and Ireland as a foodstuff, as an ingredient of stews, and, when it has been dried, as a substitute for chewing tobacco. Irish moss is also used extensively for jellies. In Scotland, cattle can sometimes be seen browsing on the kelp on the sea-shore. From immemorial times seaweed has been burnt and the ashes used as a source of potash and iodine, although these materials are now obtained from other sources. The Latin word for seaweed is *alga*, and it is from this that the word " alginate " is derived. One of the chief constituents of seaweed is alginic acid, and as this substance is a linear polymer it is a valuable potential source of fibres.

Textile Fibres from Seaweed. Some kinds of seaweed, especially the *Laminariae*, are strong and tough and quite obviously possess a fibrous structure; they must long have appealed to fibre technicians as potential sources of a fibre. The attractiveness which seaweed owes to its apparent fibrous structure must have been reinforced by its abundance; enormous quantities of weed have long been waiting to be used by anyone who could make a fibre from them. In fact their utilisation for fibre purposes waited on the isolation of alginic acid from seaweed. The fibrosity of seaweed derives from its content of long-chain molecules of alginic acid or of its salts; it was unlikely, therefore, that fibres could be made from seaweed until the existence of alginic acid had been recognised and until it had been isolated.

E. C. C. Stanford working in Sussex was the first to isolate alginic acid, probably about 1860. He observed that brown seaweeds could be dissolved in aqueous sodium carbonate and that when such a solution of seaweed was acidified, a transparent jelly-like substance was precipitated. This substance was an acid and Stanford named it alginic acid. It occurs in the combined form as the sodium and calcium salts in all brown seaweeds, but not in green or red seaweeds. As, however, the great bulk of seaweeds are of the brown kind, the potential supplies of alginic acid are almost illimitable.

Stanford was quick to realise the potentialities of alginic acid; the thickening and suspending powers of solutions of its sodium salt excited his interest and he set about the development of alginates with great enthusiasm. The first step was to acquire a large supply of seaweed and he leased the rights of the Isle of Tiree from the Duke of Argyll. In the next twenty years he built and ran factories in the West Highlands of Scotland, for the treatment of seaweed and the development of the products therefrom. It was a remarkable achievement to keep the factories going for so long as he did, and Stanford was undoubtedly in advance of his time, but eventually his schemes petered out, probably mainly because of the difficulty of disposing of his products.

The first alginate fibre was a coarse monofilament which was manufactured during the 1939–45 war; it consisted of chromium alginate, which is green in colour, and it was used for camouflage purposes. Pioneering work was carried out by Speakman and his school at Leeds University and in 1939 and 1940 the production of good multi-filament textile fibres was announced. These were practically as strong when dry as viscose rayon and were attractive in appearance. They consisted of calcium alginate and were characterised by the possession of two unusual properties:

1. They were flameproof.
2. They dissolved when washed in soap and water.

The first of these properties is an extremely valuable feature; the second precludes the use of calcium alginate fibres for those traditional textile purposes where the flameproof qualities would be most valuable, such as for kiddies' frocks. To a limited extent a virtue has been made of this defect, and the solubility of calcium alginate fibres in dilute alkaline (*e.g.*, soap) solutions has led to their being used for special purposes.

Chemical Structure

Alginic acid, which is one of the main constituents of seaweed, is a polyuronic acid, actually a polymer of *d*-mannuronic acid. There still exists a little doubt as to its exact chemical constitution, but it is thought to be as follows:

```
  ┌       COOH                                      ┐
  │        |                                        │
  │        CH———O              CHOH—CHOH            │
  │       /       \            /        \           │
 H│—O—CH           HC—O—CH              HC—│OH
  │       \       /            \        /           │
  │        CHOH—CHOH            CH———O              │
  │                             |                   │
  └                            COOH                 ┘n
```

The molecular weight, determined from the viscosity of sodium alginate, has been estimated at 15,000; osmotic pressure determinations have suggested much higher values. The alginic acid will be seen to consist of very long-chain molecules with reactive side-chains (the carboxylic groups). Solutions of sodium alginate (in which the carboxylic groups have been converted to —COONa groups) are readily soluble in water and, being very viscous, are highly suitable for spinning. They can readily be converted to alginic acid by spinning into an acid-bath or to calcium alginate by spinning into a coagulating bath containing calcium salts. For example, alginic acid filaments have been prepared by extruding sodium alginate into a coagulating bath of normal sulphuric acid saturated with sodium sulphate, 2·5 per cent olive oil, and 1 per cent Fixanol (as an emulsifier). The filaments so obtained are fairly satisfactory as regards strength, giving up to 1·2 grams per denier, but unfortunately they will dissolve in a few seconds in a solution containing 0·2 per cent soap and 0·2 per cent soda ash, so that they will not stand up to an ordinary scour or wash. In this respect, calcium alginate is not a great deal better, for filaments of calcium alginate disintegrate after treatment for a few minutes in a similar solution of soap and soda at room temperature. However, they are somewhat better than, and are preferred to, alginic acid filaments, so that the sodium alginate is ordinarily extruded into a solution of calcium salts. The calcium alginate can be spun as fine as 2-filament denier, and is reasonably strong with a tenacity up to 2 grams per denier, but, because of its susceptibility to the action of weak soap and alkali, it is clearly unsuitable for the great majority of textile purposes.

Manufacture

Seaweed—usually *Laminariae*—is collected, dried and milled. It can be stored safely in the form of a dry powder, but would soon rot from bacterial attack if not dried. The powdered weed is treated with a solution of sodium carbonate and caustic soda which converts all the alginate in the weed to sodium alginate; the pigment and the cellulose which are present in the weed are of course not dissolved. The viscous brownish solution of sodium alginate is suitably purified by sedimentation and is bleached and sterilised by the addition of sodium hypochlorite. The alginic acid is extracted from this solution by a series of chemical reactions and then purified and dried. Care is taken to avoid excessive depolymerisation. The alginic acid so isolated is reacted in the solid state with sodium carbonate, and the resulting sodium alginate is used for spinning. The grade used for fibre spinning has a viscosity of about 60–80 centistokes in 1 per cent solution at 25° C.; lower grades (lower viscosities) give poor fibres. Much higher grades are more difficult to dissolve and filter but they might well give a better fibre than any yet made from seaweed.

Spinning. An 8 to 9 per cent solution of sodium alginate is made, and is sterilised by the addition of a bactericide. It is filtered and spun on a viscose spinning machine into a coagulating bath which contains normal calcium chloride solution, 0·02 normal hydrochloric acid solution, and a small quantity of a cationic surface active agent. As the sodium alginate issues from the jet it is precipitated in filament form as calcium alginate. The filaments are drawn together, washed, lubricated and dried and wound.

The function of the hydrochloric acid is to prevent closure of the jets by the growth of calcium alginate on their sides; that of the cationic agent is to prevent filament adhesion.

The calcium alginate yarn is collected in a form ready for use. It can be stored for long periods without deterioration. The waste made during spinning is collected, is leached with acid to convert it to alginic acid and then re-converted to sodium alginate which can be used again.

Properties

Under the microscope the longitudinal view of alginate fibres is featureless or slightly striated; the cross-section is irregular but fairly well rounded, and the edges are serrated (p. 25).

Calcium alginate yarns have a dry strength comparable with that of viscose rayon, but their wet strength is low; their extensibility is sufficiently high to meet most textile requirements. The variation in physical properties with moisture content can be seen from the following table:

Atmosphere.	Tenacity (gm/denier).	Elongation at break (per cent).
Dry	2·20	10
65 per cent R.H.	1·14	14
100 per cent R.H. (saturated)	0·29	26

The low wet tenacity of calcium alginate is undoubtedly disadvantageous, but that of aluminium alginate is very much better.

The metal content of alginates is high; calcium alginate contains about 10 per cent calcium, and this brings with it the high specific gravity of 1·75.

One unique property of fibres made from metal alginates is their flameproofness. The other unique property of calcium alginate fibres is that they will dissolve in slightly alkaline soap solutions.

Whilst this precludes their use for dresses and garments generally that will be home washed or laundered (they can be dry-cleaned safely) it has opened up new uses which are discussed below.

Dyeing

Alginate fibres can be readily dyed with basic dyestuffs for which they have a remarkably high affinity; this high affinity is due to their possession of a multiplicity of carboxyl groups. Cibalan colours have a good affinity for alginate and give better fastness on it than do the basics, but there is really very little call for dyed alginate fibre.

Uses

The only use of consequence is for sock separation, when socks are knitted in a " string " with a few rows of stitches in calcium alginate rayon between the socks (Fig. 111). After separation by cutting through the courses of alginate yarn, the toes are linked in the normal manner and the remaining alginate courses at the welts are dissolved in the subsequent scouring process, leaving a perfectly finished edge to the welt. This process is not only easier than the old method of using a draw thread which had to be removed, but also gives a better finished welt. Alginate fibre is made by Courtaulds at Coventry, mainly as a convenience for some of their big viscose rayon

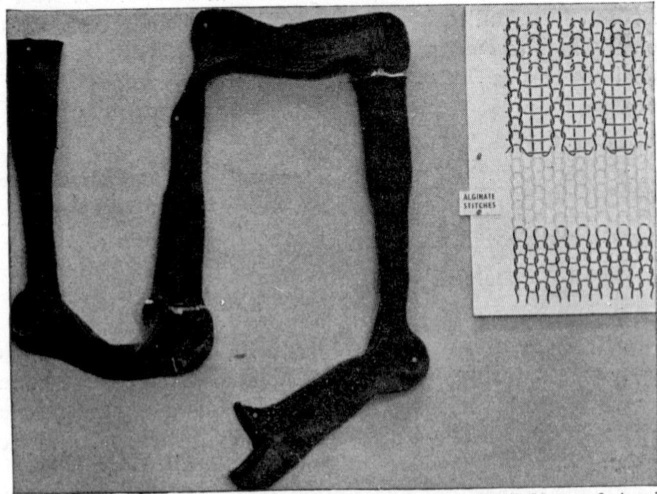

[*Textile Mercury & Argus*]

FIG. 111.—Woollen socks with a few rows of alginate rayon between the socks. On finishing the socks separate.

customers, who use it for sock separation. Not much of it is made. However, it does seem to be making some progress in the most progressive textile factories. Fig. 112 shows Helanca (p. 631) socks

[*Trigg Knit*]

FIG. 112.—Snipping the alginate draw thread to separate Helanca socks.

being separated by snipping an alginate draw thread at the Trigg Knit factory in Cadiz, Kentucky.

SODIUM ALGINATE FOR PRINTING

Aside from its use as raw material for a fibre, alginic acid in the form of its sodium salt, finds another important textile use. This is as a thickening agent for printing pastes; the dyestuff solutions have to be thickened for printing, to prevent them spreading beyond the confines of the desired pattern and sodium alginate—sold as "Manutex"—is recommended for printing viscose rayon fabric with Procion dyestuffs (p. 203), a striking tribute to the chemical inaccessibility of the hydroxyl groups in the alginate molecule. The Procion dyestuffs react chemically easily enough with the hydroxyl groups in the viscose, but they cannot master those in the alginate radical.

FURTHER READING

J. B. Speakman and N. H. Chamberlain, *J. Soc. Dyers and Col.*, **60**, 264 (1944).
J. B. Speakman, British Pat. 545,872 and 572,798 (formaldehyde treatment).
"Fibre Science", 48. Manchester (1949).
C. K. Tseng, "Alginate Fibres Compared", *Textile World*, **95**, 113 (December 1945).
E. G. V. Percival and A. G. Ross, *Nature*, **162**, 895 (1948) (the identity of sea-weed cellulose with land plant cellulose).
L. A. Bashford, M. D. Eastham, J. P. Hilton, W. S. Holden, L. Horton, R. S. Thomas and F. N. Woodward, "Alginate Hessian", *J. Soc. Dyers and Col.*, **73,** 203 (1957).

CHAPTER 17

CASEIN FIBRES

LANITAL, ARALAC, FIBROLANE, MERINOVA

BECAUSE wool has proved so eminently suitable for clothing, very many attempts have been made to produce an artificial fibre with similar properties. When we remember that wool and other hair-fibres, such as cashmere, mohair and alpaca, are Nature's provision to protect animals from rain, cold, and excessive heat from the sun, as well as to conserve the natural body-heat of the animal, it is not surprising that they should have proved the clothing fibres *par excellence*. Their characteristics which are most valued are warmth and softness of handle. Probably, in part, the warmth of wool is due to the imbricated (arranged like tiles on a roof) scale structure which forms the cuticle of the wool fibre, and it seems at present unlikely that it will be found possible to reproduce such a scale structure in any artificial fibre. In other respects, though, a fairly close approach has been made to the physical properties of wool, and nowhere nearer than in the protein fibres. It has to be remembered, too, that wool when used for human clothing has two outstanding defects: it shrinks on washing, and is very subject to attack by the larvae of moths and beetles; casein fibres are free from the defect of shrinkage, but some of them are subject to moth attack.

Casein is the protein which occurs naturally in milk, and, since wool also is a protein, it was perhaps natural that man, looking round for a substitute for wool, should have selected milk, an abundant natural protein, as the raw material. As to whether it is ethically justifiable to convert a valuable foodstuff into a fibre is open to question. If there is abundant food available, no objection to this course can be taken, but inasmuch as there is usually a shortage of food in the world, it may prove that this factor will retard the development of regenerated protein fibres like casein.

LANITAL AND ARALAC

Todtenhaupt in 1904 disclosed a method of making casein filaments, but his filaments were brittle and lacked sufficient resistance to water to withstand wet processing. His and the early attempts of others to make a commercially acceptable fibre were unsuccessful, and it was not until 1935 that the problem was really solved. An Italian, by name Ferretti, carried out a series of researches in 1924–1935 and succeeded in making pliable fibres with certain wool-like characteristics. The Italian rayon producers, Snia Viscosa, purchased Ferretti's patents and undertook large-scale production of casein fibre from milk. This fibre they called Lanital (*lana* is Latin for wool), and in 1937, 1,200 tons of this fibre was made.

In the U.S.A. the Atlantic Research Associates, Inc., carried out research independently, and in 1939 undertook production of a casein fibre, to which they gave the name " Aralac ". The company which manufactured this material was called Aralac Inc., and in 1943 the production was about 5,000 tons. However, in 1948, Aralac Inc. sold their entire plant and property to the Virginia-Carolina Chemical Corporation, who used it to make " Vicara " fibre from corn protein. Aralac has not been manufactured since then and Vicara itself was discontinued in 1957.

Chemical Structure

The protein in casein is not very different chemically from that in wool. Ultimate chemical analyses of Aralac and wool have given figures such as the following:

Element.	Aralac.	Wool.
Carbon	53·0 per cent.	49·2 per cent.
Hydrogen	7·5 ,,	7·6 ,,
Oxygen	23·0 ,,	23·7 ,,
Nitrogen	15·0 ,,	15·9 ,,
Sulphur	0·7 ,,	3·6 ,,
Phosphorus	0·8 ,,	—
	100·0 ,,	100·0 ,,

Perhaps the outstanding difference is the low sulphur content of the Aralac. The cystine cross-linkages which make such a difference to the properties of wool are not found in casein fibres. Casein is therefore weaker and less resistant to chemical attack than is wool. One may hope that further research will provide methods of introducing cross-linkages between the casein molecules, and this would doubtless greatly improve the properties of the fibres.

The proteins that occur in milk are " globulins ", and lack the straight chain structure of the linear polymers characteristic of practically all other fibres. Probably, during manufacture of the fibres the globulins are uncurled and take on a more linear form. Unless some change such as this does take place it would appear to be useless to hope that really strong fibres could ever be made from globular proteins. Doubtless, during stretching some degree of orientation is obtained.

Manufacture

The methods of manufacture of both Lanital and Aralac are similar. First the milk is skimmed to remove the cream, which is the most valuable part of the milk considered as a food but the least valuable part for fibre manufacture. Then the skimmed milk is heated to 40° C. and acid is added to coagulate the protein, which separates as a curd. The " whey "—*i.e.*, the liquid part—is separated, the curd is washed to remove acid and salts, most of the water is removed mechanically and the curd is dried. The curd consists of casein; 100 lb. of milk yields 3 lb. casein, from which about the same weight of fibre can be produced.

Casein is usually prepared in this way at the dairies; it would clearly be uneconomic to transport for long distances 100 lb. of skimmed milk from which only 3 lb. casein could be obtained. Argentina is the largest exporter of casein.

When the casein reaches the fibre factory it is carefully blended. Blending is an art practised in the preparation of more fibres than one. If different batches of raw material have even slight inequalities it is essential to mix them thoroughly, if there is to be any hope of securing a uniform fibre. After blending, the casein is dispersed in caustic soda solution. Ten kg. of casein is dispersed in 50 litres of water containing 0·27 kg. of caustic soda—*i.e.*, 2·7 per cent of caustic soda on the weight of the casein is used. The solution is clarified and extruded through jets into a coagulating bath and collected as " tow ". The coagulating bath consists of 100 parts by weight of water, 2 parts sulphuric acid, 5 parts formaldehyde and 20 parts glucose. The fibre is collected as " tow "—more or less similar to the slubbing of wool, a mass of thousands of parallel filaments in rope form (see Chapter 1). The tow is then hardened; this is essential to obtain a fibre which will withstand wetting; the hardening is performed by treating with formaldehyde, which may react with the free amino-groups of the casein in the following way:

$$RNH_2 + OCH_2 \longrightarrow RN{:}CH_2 + H_2O$$

or which by taking one hydrogen atom from each of two amino groups, may form a cross-linkage.

Aralac, but, so far as is known, not Lanital, was then given an acetylation process which gave greater stability to boiling water. This process was known as " Aratherming ". It has been stated that Lanital fabric had when wet a cheesy smell, but that Aralac, on account of the acetylation it had undergone, was free from this defect.

The tow is subsequently cut into staple, usually similar in length to that of wool fibres, and is made into top form for blending with wool, or is blended with it during the carding operation. Sometimes crimp is inserted in casein fibre to increase its resemblance to wool.

Properties

The outstanding properties of casein fibre are its warmth and softness of handle, which make it peculiarly suitable for mixing with

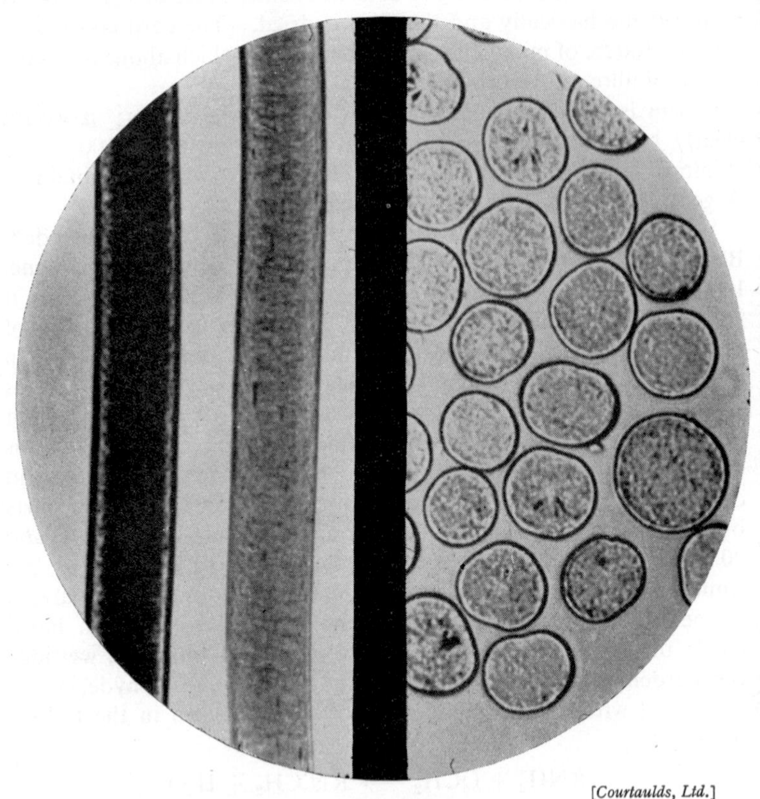

[*Courtaulds, Ltd.*]

FIG. 113.—Photomicrograph of milk casein filaments (×500). One of those shown in the longitudinal view has been dyed. Note the pitting.

wool. Casein fibre will not felt; but, rather curiously, the admixture of a small proportion of casein fibre with wool increases the feltability of the latter fibre.

The tenacity of casein fibres is about 0·8–1·0 gram per denier, and the elongation at break is about 15 per cent. When wet, the tenacity is very low; this is the main defect of casein fibres, but the elongation at break is correspondingly higher.

The specific gravity of casein fibre is 1·29—very similar to that of wool. Under the microscope the longitudinal view shows faint striations, and the cross-section (Fig. 113) is nearly circular, and exhibits pitting, which enables it to be recognised easily in wool/casein mixtures.

Casein fibres are sensitive to alkali, being gelatinised by strongly alkaline conditions, but they are stable to mild buffered alkalis such as sodium bicarbonate and disodium phosphate which may be used in scouring baths. The fibres can be carbonised safely in 2 per cent sulphuric acid in the cold.

Casein fibre is attacked by the larvae of clothes moths and carpet beetles, and is subject to mildew. In this respect it is similar to the natural hair fibres and different from the true synthetics, nylon, Vinyon and Saran.

Hydrogen peroxide provides the best bleach, but acts only slowly, and long treatment is necessary. At high temperatures it turns casein fibres yellow.

Solvents normally used in dry-cleaning have no effect on casein fibres and can be safely used.

Water absorption is similar to that for wool—about 14 per cent under standard conditions. The fibres wet out very easily. After casein has been wet, care must be taken to dry it at a temperature below 100° C., otherwise embrittlement occurs.

The stability of casein fibres to light is similar to that of wool.

Dyeing

Casein can be dyed in a manner similar to wool; it has a rather greater affinity for dyestuff. The reader will note that those fibres which have a high affinity for moisture—*e.g.*, viscose and casein—are easily dyed, whereas those which have a low affinity for water—such as cellulose acetate and nylon and, to take an extreme case, Vinyon or Saran—are more difficult to dye. Highly oriented fibres have closely packed molecules which prevent free ingress of the dyestuff molecules and, indeed, of the water in which they are dissolved, whereas fibres which have only a low degree of molecular orientation, as casein fibres have, are readily penetrated and dyed.

Acid colours on casein give only poor wet fastness, but metalliferous dyes—*e.g.*, Neolans and chrome colours—give a reasonable degree of wet fastness, not very different from that obtainable on wool. Mordant dyestuffs are, however, not altogether satisfactory, as the chroming process generally causes excessive shrinkage of the fibre. The Carbolan colours give strong shades from neutral baths, and the dyeings are of good fastness.

Casein may also be dyed with direct colours, basic colours, and the dispersed acetate colours, but with each of these wet fastness is very poor.

Uses

Casein fibre is almost invariably used in admixture with wool. Its properties of warmth and soft handle make it very suitable for this purpose, and the main inducement to the manufacturers to use it is one of low cost. Casein can be spun in fine filaments which approximate in diameter to high-grade wools. When spun with a filament diameter of 20 μ, it is equivalent to 70 s wool, one of the finest qualities in normal use and one that is largely used for baby wool, in which great softness of handle is a desirable feature, indeed an essential characteristic. If spun with a fibre diameter of 30 μ, casein fibres are equivalent to 50 s wool—a quality associated with a good crossbred. Casein fibre, because it felts well when mixed with wool, has been used in the hat trade.

Casein fibres are normally sold as loose staple fibre. They have been used in mixtures with rayon staple and with cotton in addition to wool.

In general, the multitudinous uses to which wool is put may, in a large measure, be shared by wool/casein mixtures. Typical individual uses to which they have already been put are in the manufacture of ties, socks, dressing-gowns, sweaters and hats. It may be added that 100 per cent casein fibre has been used for filling pillows and quilted goods.

Fibrolane BX

This is a casein fibre made in Britain by Courtaulds Ltd., by whom it is supplied as continuous tow or as staple or combed tops; it is similar to wool in some of its properties. The original " Fibrolane ", known as Fibrolane A, was very similar to Lanital, but is no longer made. A newer product, Fibrolane BX, which has been insolubilised with formaldehyde, and which thereby has improved properties, is the Fibrolane that is mostly used to-day.

Manufacture

Its raw material is casein which is dispersed in an alkaline medium, extruded into an acid bath, and collected as tow as previously described, but in addition it is stretched and hardened before insolubilising. These treatments strengthen the fibres and impart to them a good extensibility and resistance to acid dye liquors. In addition to its normal form, it is available in a number of Duracol (spun-dyed) shades, and also matt. The fibre is a good white and may be used in the production of a complete colour range, including white or pastel-coloured finished goods.

Dyeing

The dyeing of Fibrolane blends should be carried out in such a way as to preserve the characteristic properties of the fibre. Unless care is exercised, the fibre can be seriously damaged. The main points to remember are as follows:

1. Fibrolane BX has a lower wet strength than wool, and whereas the wet strength of the latter is about 70 per cent of its dry strength, that of Fibrolane BX is only about half its dry strength. Even so, this represents an advance, because the wet strength of the old Lanital was only about a third of its dry strength.
2. The greater affinity of Fibrolane BX than wool for most dyes.
3. The lower fastness to washing of dyeings on Fibrolane BX compared to that of the same dyes on wool. It is usually the same with dyeings—when the dyestuff is very easy to apply it is also easy to remove, and washing fastness is poor. Light fastness, however, is at least as good on Fibrolane BX as on wool.

A study of the wet processing conditions necessary to maintain the handle and to give satisfactory dyeing fastness on Fibrolane blends has shown that the following conditions are desirable:

1. Maintenance of the dyebath at about *p*H 4 as far as possible (alkalinity would be undesirable). If it is necessary to use a higher *p*H, the treated blend should be washed off and given a final treatment at *p*H 4. For example, blends of Fibrolane BX with cellulosic fibres are best processed at *p*H 6 in acetic acid–acetate baths, the fibre being subsequently returned to its optimum condition by being given an acid rinse.
2. The use of buffered solutions of acetic, formic or lactic acids, *e.g.*, a lactic acid–acetate solution.
3. The use of acid dyes with neutral-dyeing properties so that strongly acid baths are not required.
4. Dyeing at the boil should be for only a sufficient time to secure levelness of shade and penetration, in practice no more than about 30 minutes.
5. Fibrolane blends should not be subjected to heavy mechanical pressures in scouring or dyeing machines, nor through squeeze rollers.

Uses

Fibrolane BX can be used in a wide range of textiles, including floor coverings and felts of various kinds. The chief apparel uses are suitings, dresswear and velour coatings; the fibres being blended with either Fibro or wool, or possibly forming a three-fibre blend with both. Fibrolane BX has a more limited use in hosiery yarns as blends with Fibro or as a three-fibre blend with wool and Fibro.

Non-shrink blends with wool

The softness of handle of Fibrolane BX is comparable with that of a fine-quality wool; this property makes the fibre particularly attractive for blending with wool. Suitable blends are 30–50 parts Fibrolane BX and 70–50 parts 56 s quality wool. In many cases wool blends are given a shrink-resistant treatment, but the best way of conferring stability on Fibrolane BX/wool blends for knitting yarns is to use shrink-resistant wool, *e.g.*, dry-chlorinated combed tops. An interesting application for Fibrolane BX/wool blends is for felts; the blends actually felt faster than all-wool because the smooth Fibrolane BX fibre " lubricates " the migratory movements of the scaly wool fibre. It is possible, however, to make a felt from wool, Fibrolane BX and cellulose acetate, the last of which retards felting, which will felt at the same rate as wool.

FIBROLANE BC

The restrictions and special conditions that have to be observed in the dyeing of Fibrolane BX do not apply to Fibrolane BC, which is a later development in the series of regenerated protein fibres. Like Fibrolane BX, it is made from milk casein by dispersion, extrusion, stretching and hardening, but in addition the fibre is rendered exceptionally stable by a process of chrome tanning. Fibrolane BC is consequently pale blue-green in colour due to the presence of chromium salt; this colour has a light fastness equivalent to S.D.C.6 standard, and remains unaffected by most normal treatments met with in dyeing and finishing. (The colour, however, changes shade and becomes less fast when subjected to bleaching liquors containing hydrogen peroxide. No attempt, therefore, should be made to bleach materials containing Fibrolane BC.)

Fibrolane BC has an affinity for all classes of dye normally used in the dyeing trades. The *p*H range over which it can be treated without risk of causing damage to the fibre is similar to the stability range for wool. It has been found that the wet fastness of dyes on Fibrolane BC is markedly superior to the fastness previously obtained on regenerated protein fibres, and this is of particular interest in the case of levelling acid dyes, many of which can now be used to produce an acceptable standard of fastness to washing. A high standard of milling fastness can be obtained with aggregated and pre-metallised acid dyes. The physical condition of the fibre when wet is such that shrinkage and milling rates of wool blends are scarcely affected. Chrome dyes can be applied for the highest standards of wet fastness, and it will be found in practice that for the " afterchrome " process considerably less dichromate will be required to produce the necessary chelation.

The end uses of blends of Fibrolane BC are governed by the natural colour of the fibre. Pale mode shades and bright tones are not obtainable, but a wide range of medium shades is possible. Fibrolane BC is eminently suited to the woollen and worsted trades—it is useful for blending with cotton or Fibro. Fibrolane BC can be used in the production of evening wear or for any purpose where medium or darker shades predominate. Because of the chrome treatment it has undergone, the fibre possesses a measure of flame resistance.

The improved stability of Fibrolane BC is due to the chromium cross-linkages, which exercise somewhat the same function as the cystine cross-linkages in wool; that the chromium colours the fibre, somewhat reduces its field of application.

MERINOVA

Merinova is a casein fibre made by Snia Viscosa at Milan. It is a creamy near-white fibre with a soft but rather scroopy handle; it is supplied only as staple, either 100 per cent Merinova or in top form blended with other fibres. Annual production is about 8 million lb.

Manufacture

Milk is the raw material; it is heated and passed through a separator which removes the cream, which is used for butter-making. The skim milk is chilled and then acidified to precipitate the casein. This is washed, filtered, dried and different lots are blended. It is ground to a fine powder and dissolved in " chemicals " (probably soda); the solution is ripened and spun into a coagulating bath (probably sulphuric acid). The filaments are washed and cut. The fibre thus obtained is treated in autoclaves with formaldehyde in order to give it wet strength and some measure of resistance to weak alkali. The fibre is dried, opened and then baled in presses.

It is made in the following sizes:

3 denier $1\frac{9}{16}$ in., $2\frac{1}{2}$ in., 3 in.
5 ,, $1\frac{1}{16}$ in., $1\frac{9}{16}$ in., $2\frac{1}{2}$ in., 4 in., 6 in.
9 ,, 4 in.
18 ,, 6 in.
27 ,, 6 in., 14 in.

Properties

The fibre has the characteristic section of casein fibres, *i.e.*, it is nearly round and pitted. Its tenacity is 1·1 grams per denier dry and 0·6 gram per denier wet; breaking elongations are 50 per cent dry and 60 per cent wet, these figures not being very different from those for wool. It is susceptible to alkali but resistant to boiling for one and a half hours in 4 per cent sulphuric acid so that it will withstand the usual wool dyeing process. It is mothproof and shrinkproof; its specific gravity and moisture regain are not very different from those of wool. The mothproof quality derives from the formaldehyde treatment.

Uses

Primarily Merinova is suitable for blending with wool; it can be spun on the worsted system. Before use the moisture content of Merinova, which is sent out very dry, should be brought up to 20 per cent. Blended tops can be used; alternatively the combed tops of wool and Merinova can be blended in the mixing gill and recombed.

It can also be blended with cotton and with viscose rayon staple, and spun on the cotton system.

Merinova/wool blends are used for knitting yarns and hosiery, for blankets and for hat felts. Its admixture improves the handle of viscose rayon staple. Mixtures of 50 Merinova/50 nylon and 50 Merinova/50 Perlon have been used for men's socks.

The fine deniers (3 and 5) are recommended for felts, the 5 denier for apparel fabrics, the 9 denier for blankets, the 18 denier for quilting, and for mattresses the 27 denier 14 in. type is recommended.

Dyeing

Merinova is supplied in a range of twenty dope-dyed colours with good fastness to light and washing, *etc.* Uniformity of colour is good.

Merinova supplied in natural shade can be processed like wool provided that care is taken

1. to avoid the use of sodium carbonate;
2. to avoid drying at temperatures higher than 70° C.

It can be peroxide bleached, dyed with acid colours and chrome colours; it can be given an anti-crease resin finish; if necessary it can be carbonised with wool. Merinova alone does not felt as does wool, the reason being that it lacks the scale structure, and the directional friction effect of the wool fibres. If blended with wool, Merinova helps felting probably because its inert smooth fibres smooth the passage of, and make easier the movement of the wool fibre.

The recommended dyeing procedure for wool/Merinova blends is very similar to that for dyeing wool itself, *viz.*:

> Wet out the material; prepare the dyebath with 10 per cent Glauber's salts, 1–2 per cent acetic acid and the required quantity of dyestuff. Start dyeing at room temperature and raise the temperature to the boil over a period of thirty minutes. Turn off the steam and after fifteen minutes add 2 per cent formic acid; continue dyeing for another thirty minutes near boiling point.

The point of adding the formic acid is to ensure levelling; initially the Merinova is likely to absorb more dye than the wool, but when the formic acid is added some of the dyestuff leaves the Merinova to re-enter the dyebath and is absorbed by the wool.

In conclusion it is likely that Merinova will find its most useful application as a diluent for wool, the advantage being one of lower cost to produce wool-like fabrics in a price-range lower than that of all-wool fabrics.

FURTHER READING

A. Ferretti, British Pat. 483,731, 483,807–483,810.
A. G. Arend, *Fibres*, **7**, 194 (1946). Casein fibres in the production of felt hats.
A. E. Brown, W. G. Gordon, E. C. Gall, and R. W. Jackson, *Ind. Eng. Chem.*, **36**, 1171 (1944) (dealing with acetylation of casein fibres).
J. M. Matthews, " Textile Fibres ", 6th Edn., London (1954).
E. Sutermeister and F. L. Browne, " Casein and its Industrial Application ", 215–218. Reinhold Publishing Corporation : New York (1939).
J. B. Speakman, " The Chemistry of Wool and Related Fibres ", *J. Text. Inst.*, **32**, No. 7 (1941).
This does not deal with casein fibres but students of regenerated protein fibres are advised to read it, as many considerations are common to all protein fibres, including, of course, wool.
R. C. Cheetham, " The Dyeing of a Blend of Wool and Fibrolane for the Hand Knitting Trade ", *J. Soc. Dyers and Col.*, **69**, 76 (1953).
R. L. Wormell, " Milk Casein and Peanut Protein Fibres " *J. Text. Inst.*, **44**, P 258 (1953).
Canadian Chem. Processing, 35, No. 10, 776 1951).
Textile World, No. 9, 170 (1951).

CHAPTER 18

ARDIL, VICARA, SOYBEAN

THE natural protein fibres, wool and silk, have such superlative properties that many attempts have been made to manufacture regenerated protein fibres in the hope of reproducing some of the properties of wool and silk. The way has been hard and some fibres, excellent in many ways, that have reached commercial use, have been discontinued: Aralac, one of the casein fibres, has already been mentioned; in addition there have been Ardil, Vicara and Soybean. In one sense they are now of only historical interest: in another they have been the pioneers of those man-made, probably synthetic, protein fibres that will one day surely come, and in that sense they have played a part in the advance of fibre science. They are of interest to fibre students and even although they are no longer commercially available, a short description of each is desirable.

ARDIL

Ardil fibre is made from groundnuts; these contain about 25 per cent protein. In an attempt to help some of the less prosperous parts of the Empire, where groundnuts were grown to superfluity, I.C.I. embarked on their utilisation. Fibre was first spun at Ardeer in 1938 and after World War II was manufactured at Dumfries until 1957 when the fibre was discontinued.

Manufacture

The nuts grow under the soil and are usually hand pulled; they were shelled and the reddish brown skin was removed, so that it would not colour the fibre. The blanched nuts were ground to meal and this was extracted with solvents to remove the oil, a valuable by-product. The oil-free meal was dissolved in dilute alkali (some of the carbohydrates do not dissolve), and when the solution was acidified with sulphur dioxide to bring it to *p*H 4·5, the protein was precipitated. The precipitate when washed and dried was known as Ardein; it was a creamy-white powder. It was dissolved in dilute caustic soda, and the resulting viscous solution was spun into a bath containing 2 per cent sulphuric acid and 15 per cent sodium sulphate. The fibre was hardened by treatment with formaldehyde. Latterly it was spun as tow and cut into staple fibre.

Properties

Ardil's outstanding property was its handle; it had the nearest to that of wool which the author has ever known. Moisture content was high, about 14 per cent, a good feature in an apparel fibre. Tenacity was about 0·8 gram per denier, elongation at break about 50 per cent.

Dyeing. Ardil dyed like wool; its affinity for dyestuffs was usually greater than that of wool.

Uses

The fibre was used almost exclusively in blends, usually 50/50 with wool, sometimes with cotton or rayon. It had a negligible fire risk—rather less than that of wool. Sports shirts, pyjamas and dress fabrics were main outlets; a 50 Ardil/50 merino wool blend made wonderful sweaters; a 40 Ardil/60 viscose blend was used for carpets. The price of Ardil was about one-half of that of wool and it was an ideal diluent for it. Despite considerable publicity the consumption of Ardil never grew beyond 3 million lb. a year, and the manufacture of such a small quantity was unprofitable.

Fibrolane C was a somewhat similar peanut fibre that was at one time made by Courtaulds.

VICARA

Vicara was made from the maize protein, zein, by the Virginia–Carolina Chemical Corporation from 1948 until 1957.

Manufacture

Zein was extracted with 70 per cent *iso*propyl alcohol from corn meal; the alcohol was evaporated leaving the zein as a yellow powder. It was dissolved in caustic soda solution, filtered, de-aerated and ripened by storage; during ripening the globular protein molecules (globulins) uncoiled and straightened out. The uncoiling took place because the caustic soda converted the acid —COOH group on the protein molecules to sodium carboxylate —COONa groups; whereas the former attracted and held to them the free amino groups on other parts of the long molecule in so-called salt linkages, the latter had no attraction for the free amino groups and, as they were freed, the globulin molecules could disentangle and become linear. Secondary valency forces could now develop between the now-linear molecules and bind them together; their effective molecular weight increased (several molecules bound together behaved as one giant) and the viscosity of the solution increased. Then the solution was spun into an acid formaldehyde bath. The formaldehyde cross-linked the fibre and it was then stretched; then another hardening bath and washing, crimping, drying and cutting followed.

Properties

Tensile strength of Vicara was about 1·2 grams per denier dry, 0·75 wet; elongation at break 32 per cent dry and 35 per cent wet (for comparison wool 30 per cent dry, 70 per cent wet); moisture regain about 10 per cent. It was not so near as Ardil to wool in handle, but it was a good fibre and it blended well with other fibres. Resistance to washing was excellent and its wet strength finally overcame the bogey of wet weakness which had characterised the early regenerated proteins. Vicara was stronger and more stable than Ardil but less wool-like. It could be dyed like wool, even with vats.

Uses

Mostly in blends: it added moisture compatibility to nylon, loftiness and resilience to cotton, and suppleness to rayon. It was described as " the fibre that improves the blend " and was used in all kinds of clothing. In price it was about half of that of cross-bred wool.

SOYBEAN

For short periods several manufacturers have made fibre from soybean protein. At one time the Ford Motor Co. made it and used it for car upholstery; it was once made in Japan under the name Silkool. The attraction of soybean as raw material for a fibre lies in its high protein content (35 per cent) compared with groundnuts (25 per cent) and maize (10 per cent). Furthermore the beans which look like light brown peas grow prolifically in the East and in America.

Manufacture

The beans were flaked and their fat extracted by hexane. The residue was treated with a dilute solution of sodium sulphite; the protein dissolved in it. It was next precipitated with acid at *p*H 4·5. It was spun from caustic soda solution much as the other regenerated proteins, including the stretching and hardening processes.

Properties

Tenacity was 0·8 gram per denier dry but only 0·25 wet, a serious defect. Elongation at break was about 50 per cent; moisture regain was 11 per cent, lower than might have been expected. The fibre appears to have had no advantage over Ardil; its main attraction lay in the abundance and cheapness of the material from which it was made.

Protein Fibres and Body Comfort

Through long experience wool has been found to be the best all-round fibre for apparel and it is the regenerated protein fibres that have approached most closely of all fibres in physical and chemical properties to wool. No textile subject could be of more importance than the relation between fibre properties and comfort of clothing. How far can we trace this relationship? In trying to do so we can compare some of the properties of wool with the similar properties of some of the regenerated proteins, of some of the cellulosic, and some of the true synthetic fibres. Comfort is not at present capable of absolute definition in quantitative terms, but there do undoubtedly exist a number of well-recognised trends and we can use these as a basis for comparison of fibres. Yarn and fabric construction play a large part in the handle, feel and comfort of a garment, but assuming that they are kept suitable and the same, it is still a matter of everyday experience that the fibre itself can cause tremendous differences; there is no cotton shirt that will feel like a wool shirt, no viscose rayon dress that will drape and handle like one of acetate rayon, no other fibre that will give stockings as sheer and long-lasting as the nylons, and so on. The following properties of fibres are, amongst others, important in their contribution to comfort.

1. *Low Stiffness for Small Stretches or Bends.* This results in the fibres being soft to a light touch as when a garment is worn. It is measured as the "initial modulus" or load in grams per denier necessary to give a 1 per cent stretch. Comparative figures for some of the fibres are given in the following table:

Fibre.	Initial modulus.	Fibre.	Initial modulus.	Fibre.	Initial modulus.
Vicara	0·26–0·32	Nylon	0·20	Cotton	0·57–1·12
Wool	0·18–0·24	Dacron staple	0·20	Viscose	0·54
		Dacron continuous filament	0·86	Cellulose acetate	0·39
		Orlon staple	0·34		
		Orlon continuous filament	0·86		
		Ulstron	0·90		

Those fibres that are initially soft to the touch are wool, Vicara, nylon, Dacron staple and Orlon staple. It is interesting to note, too, how much lower is the value for cellulose acetate than for viscose rayon. The soft handle of cellulose acetate has been well recognised for these last thirty years.

2. *Compliance.* This is the *reduction* in stiffness for moderate stretches or bends, which enables fibres to stretch say from 5 to 10 per cent without a considerable increase in the load; this can be seen as a flattening of the stress–strain curves for moderate extensions. Compare the stress–strain curve on p. 231 for the compliant fibre cellulose acetate with that on p. 184 for the non-compliant viscose rayon. If the fibre easily, *i.e.*, without the application of much load, adjusts itself to the application of new forces, as it should do when a garment is worn, it is pliable or compliant. A high compliance ratio is desirable; it not only increases comfort of wear, but also confers good draping qualities. Comparative figures for different fibres are given in the following table:

Fibre.	Compliance ratio.	Fibre.	Compliance ratio.	Fibre.	Compliance ratio.
Wool	1·0–1·7	Nylon	−0·19	Cotton	0·21–0·04
		Dacron staple	1·05	Viscose	0·68
		Dacron continuous filament	−0·05	Cellulose acetate	0·91
		Orlon staple	0·68		
		Orlon continuous filament	0·11		

Those fibres which have high compliance ratios are wool, Dacron staple and cellulose acetate. It is interesting to note that the manufacturers of Dacron and Orlon have deliberately increased the compliance ratio of these fibres in the staple form, where wool-like properties are required. The negative values for nylon (continuous filament and staple have the same physical properties) and for Dacron continuous filament call for explanation. They indicate that these fibres become somewhat *more* stiff (so that the reduction in stiffness is negative) as they are stretched in the 5 to 10 per cent range; they tend to work harden. This can be seen in the increased gradient of their stress–strain curves in this region. Compliant fibres such as wool behave quite oppositely; their elongation can be increased without very much increase in load in this region.

3. *Frictional Coefficient of the Fibre.* All of the fibres listed in the above tables are smooth except cotton (even this is smooth when mercerised) and wool which has different coefficients of friction in the with-scale and anti-scale directions. This property, known

as the directional frictional effect or D.F.E., is peculiar to the natural hair fibres; it is largely responsible for their unique handle, and it will be a long time before the fibre synthesists can copy it.

4. *Resilience.* This determines the recovery of a garment from distortion; a resilient fibre will prevent baggy stockings, sagging frocks and badly creased garments. A high tendency to recover completely and quickly from deformation is desirable; the garment then keeps its shape. Comparative recovery figures from a *brief* in time) 2 per cent stretch are:

Fibre.	Recovery (per cent).	Fibre.	Recovery (per cent).	Fibre.	Recovery (per cent).
Wool .	99	Nylon . .	100	Cotton .	74
		Dacron staple	High	Viscose .	82
		Orlon staple .	100	Cellulose acetate .	94
		Acrilan .	80		
		Ulstron .	91		

The superiority of the protein fibre and of nylon and Orlon over the cellulosic fibres is evident. It is, of course, important that this resiliency should also be present when the fibres are wet; in this, Ulstron, nylon, Dacron and Orlon have a considerable advantage over the other fibres because of their great resistance to moisture; whilst this is beneficial in respect of retention of resiliency it brings with it other qualities detrimental to comfort.

5. *Loftiness.* A fabric should feel full and light, not heavy for its thickness. Two factors determine this; one is permanent crimp, which gives bulkiness, and the other is a low specific gravity of the fibre, which gives lightness. In respect of crimp, wool excels because it has a natural and permanent crimp; most of the synthetics, *e.g.*, nylon and Dacron, will take a permanent crimp and in the staple form they are sold crimped; viscose staple is usually crimped. Specific gravities can be compared in the table on p. 828. The cellulosic fibres cluster round the 1·5 mark, the protein fibres round 1·3, Dacron is fairly high at 1·38, nylon and the acrylates are all good in that they have much lower values, the polyolefines lowest of all.

6. *Moisture Absorption.* A high moisture absorbency is good because it means that perspiration can be taken up easily; in this respect wool excels, and even when it has absorbed some 40 per cent of its own weight of moisture, it still does not feel noticeably damp. Undoubtedly this, its hydrophilic nature, is due to its abundant content of amino and carboxylic groups. Under standard con-

ditions of temperature and humidity the natural moisture regains of some of the fibres are given in the following table:

Fibre.	Regain (per cent).	Fibre.	Regain (per cent).	Fibre.	Regain (per cent).
Wool .	16	Nylon .	4·0	Cotton .	8·5
		Dacron .	0·4	Viscose .	12·0
		Orlon .	0·9	Cellulose acetate .	6·5
		Ulstron .	0·2		

Wool has the highest and most desirable value, next come the cellulosic fibres and lastly the synthetics.

7. *Heat of Wetting*. Another factor which contributes to comfort is the " heat of wetting " of a fibre; wool and protein fibres generally give out heat when they are wetted and this is thought to help to keep the wearer of a garment warm. Some comparative figures are:

Fibre.	Heat of wetting (calories/gm.).	Fibre.	Heat of wetting (calories/gm.).	Fibre.	Heat of wetting (calories/gm.).
Ardil .	27·4	Nylon	7·6	Cotton	11·0
Wool .	27·0	Dacron	1·4	Viscose	25·0

The synthetics seem to be at a considerable disadvantage in this respect.

8. *Thermal Conductivity*. A fibre is not in itself either " warm " or " cold ", but it slows down to a greater or lesser degree changes of temperature. If the fibre is a very poor conductor of heat then a garment made from it will not transmit the natural warmth of the body to the atmosphere; the body warmth builds up and we feel warm. If we touch a fabric with a low thermal conductivity it does not conduct away the warmth of the finger and we say that it feels warm. Polyesters have low thermal conductivity and feel warm; so more than anything does polyvinyl chloride (P.V.C.) which is the best heat insulator of all fibres. The traditional warmth of wool is largely due to its complex surface scale structure (which no artificial fibre can reproduce) which entraps countless little pockets of air which reduce the thermal conductivity and contribute to the warmth of wool.

9. *Static*. Most fibres, even wool, will generate static when dry but they lose this faculty when they are moist because then the electric charges can leak away. Those fibres that have low moisture

regains are the most likely to generate static under normal conditions of wear. Nobody wants a garment to crackle as it is put on, or to attract to itself dust and dirt particles, as will those that electrify readily. Their low regain makes nylon, Dacron, and the acrylates particularly liable to static and cellulose acetate more liable to it than viscose, whilst wool and the other protein fibres are correspondingly less liable to it. A point of special interest is that P.V.C. fibres generate negative static electricity, whereas other fibres generate a positive static charge under the influence of friction (p. 482).

10. *Filament Cross-section.* The shape of the filament may play a bigger part in the handle of fabrics than is often realised. Wool and hair fibres have a round section and so have cuprammonium, the polynosics, nylon, Terylene and Acrilan; cotton, which has not such a good handle as wool, is flattish in section, viscose rayon is more highly serrated and rougher than the soft-handling cellulose acetate. Some of the new vinyl and acrylic fibres have flat and dog-bone sections. Will trilobal Antron be more comfortable than regular round section nylon?

Comfort, like flavour, is an experience that is caused by the integration in the brain of impulses passed up the nerves from a variety of peripheral receptors. For example, taste, smell, smoothness, consistency and even colour (through associations) all contribute to flavour: we know at once if we like a flavour or dislike it; the response is instantaneous and unmistakable, but if we try to analyse the liking or dislike, to break it down into factors, we are soon in difficulties. A person may like raspberry and dislike nutmeg; there is no doubt about these reactions, but analysis of the component perceptions which the brain builds up to give a reaction of pleasure, indifference, or displeasure is often beyond us. The wearer of a vest or a suit knows unambiguously whether it is comfortable or uncomfortable, but the causes underlying the comfort or discomfort will be many; probably at least a dozen different physical properties of the fibre make their contribution to or away from comfort. At present a full analysis of comfort cannot be made, but so far as fibres are concerned there are trends that are evident, and to some of the qualities we can already assign figures; fuller and more quantitative treatment will doubtless be possible as knowledge grows. Furthermore, the better does our knowledge become of the relations between physical properties and comfort, the more possible will it be to set out a specification for really comfortable fibres, and then to make them. This is a promising field for research.

FURTHER READING

C. E. A. Winslow and L. P. Herrington, " Temperature and Human Life ", Princeton (1949).

L. H. Newburgh, " Physiology of Heat Regulation and the Science of Clothing ", Philadelphia (1949).

A. D. B. Cassie, " Physical Properties of Fibres and Textile Performance ", *J. Text. Inst.*, **37**, P 154 (1946).

C. W. Hock, A. M. Sookne and M. Harris, " Thermal Properties of Moist Fabrics ", *J. Res. Nat. Bur. Stds.*, **32**, 229 (1944).

J. S. Gillespie, " From the Cornfield to the ' Haute Couture ' ", *Chemurgic Digest* (March 1955).

H. S. Jenkins, " Proteins as Fiber-forming Materials ", Paper presented at the A.C.S. Symposium on Fibrous Proteins, New York, Sept. 16th, 1954.

D. Traill, " Some trials by ingenious inquisitive persons; regenerated protein fibres ", *J. Soc. Dyers and Col.*, **67**, 257–270 (1951).

K. W. L. Kenchington, " Clothing and Climate ", *Texture*, **3**, 11–17 (1956).

W.I.R.A., " The Warmth of Clothing ", *Wool Research*, **2**, 43–70 (1955).

R. W. Moncrieff, " Wool: Molecular Basis of Unique Fibre Qualities ", *Text. Manufacturer*, 140–1, April (1963).

R. W. Moncrieff, " Clothing and Body Moisture Losses ", *Man-Made Textiles*, 75–7, November (1963).

PART 3

SYNTHETIC FIBRES

CHAPTER 19

NYLON

THE word *nylon* is generic; it is spelt without a capital letter and can be used in the same sense as " glass " to indicate a group of similar materials. An exception to this is provided by some European countries, notably France, where the word Nylon is a registered trade mark of Société Rhodiaceta.

Discovery of Nylon. Nylon was a product of the genius of Wallace H. Carothers. This man was a brilliant organic chemist at Harvard University, and forsook the academic life to undertake fundamental research work with the enormous American chemical combine of E. I. du Pont de Nemours & Co., an organisation similar to, but vastly bigger than, I.C.I. in this country. Probably the reasons which induced Carothers to enter industry were the greater facilities, more numerous assistants, greater laboratory space, and the unlimited apparatus and equipment which a large industrial organisation can provide, but which in the 'twenties were not so readily available at a university. Carothers tackled many chemical problems, made many chemical discoveries and it became quite a problem for the du Pont management to find commercial applications for these discoveries. In the first place, the programme that Carothers undertook was one of fundamental research, *without regard for any immediate commercial objective.* Carothers' initial interest was only general polymer research. The fact that some of the newly synthesised polymers were fibre-forming was a surprising and important discovery, but not the objective of the original fundamental research programme. This provides an outstanding example of the good and useful results that can accrue from fundamental research. It was solely with the extension of knowledge of polymer structures as his goal that Carothers started his work in the field of fundamental polymer research—work that led not only to nylon but to neoprene (a synthetic rubber with excellent chemical resistance), to macrocyclic compounds such as the synthetic musks, and to melt spinning itself.

Some two years after he started his work Carothers directed his attention particularly to the possibility of finding any product that might be used as a synthetic fibre. He found promise in some polymers known as polyesters, and finally achieved success with other polymers known as polyamides. In 1928, when Carothers embarked on his researches, polymers were not a favourite subject of study. Chemically, they were infusible, unreactive and insoluble —quite different from the substances the chemist more usually handled. These polymeric substances were intractable, and usually if an experiment produced a substance that was infusible and insoluble it was said to have " failed " and the product was consigned to the waste-bin.

Preparation of Polyesters. Carothers appreciated that these " failures " were much more closely allied to those materials, such as wood, rubber, cotton, wool and so on, that we encounter in our daily lives than are the soluble, small-molecule compounds which the chemist delights to manipulate in his laboratory. Carothers was not only a philosopher, but also an experimentalist, and when given a free hand and practically unlimited resources by his employers, he deliberately set out to make these chemically intractable polymers. He reasoned that just as ethyl alcohol and acetic acid will react with each other to form ethyl acetate, an ester, thus:

$$\underset{\text{Ethyl alcohol.}}{C_2H_5OH} + \underset{\text{Acetic acid.}}{CH_3COOH} \longrightarrow \underset{\text{Ethyl acetate.}}{CH_3COOC_2H_5} + H_2O$$

so, in a similar way, would a di-alcohol, usually known as a glycol, react with a diacid. Thus hexamethylene glycol would react with adipic acid to form an ester in this way:

$$HO(CH_2)_6OH + COOH(CH_2)_4COOH \longrightarrow HO(CH_2)_6OCO(CH_2)_4COOH + H_2O$$

Then two molecules of the ester would react to form a molecule twice as long, in this way:

$$HO(CH_2)_6OCO(CH_2)_4COOH + HO(CH_2)_6OCO(CH_2)_4COOH \longrightarrow HO(CH_2)_6OCO(CH_2)_4COO(CH_2)_6OCO(CH_2)_4COOH + H_2O$$

Next, two molecules of this ester would react to give a molecule twice as long, and so on. By building up these condensation products from molecules that had reactive or *functional* groups at either end of the molecule, Carothers prepared some very long molecules indeed. When these were made from glycols and acids they were called polyesters—*i.e.*, polymeric esters—and it was one of these polymeric esters that gave the first promise to Carothers of being a useful fibre. A chemist—his name has never been publicised, but one of Carothers' assistants—dipped a glass rod into a still containing one of these polymeric esters. Doubtless to his considerable astonishment and excitement, the molten material which adhered to the rod as he withdrew it stretched out into a long filament which solidified. Perhaps to his still greater surprise, this filament, even after it was cold, could be stretched by hand to several times its original length, but, unlike rubber, did not, on release, return to its original length. It is easy to imagine the excitement that must have prevailed in the laboratory on that day, to think of the gratification and satisfaction it must have afforded Carothers to see his reasoning from first principles so convincingly substantiated. Carothers had thought: Take molecules that can react at both ends, react them and long molecules will result. If the molecules are very long in relation to their other dimensions, they will exhibit fibre-forming properties.

The Introduction of Polyamides. Even so, there was still much hard work to be done. The fibres made from the polyesters were not altogether satisfactory, they were not very strong, and so great were the difficulties confronting the experimentalists that at one time the idea of dropping the work was seriously entertained. Then Carothers with rare insight realised that the solution of the problem lay in the substitution of polyamides for polyesters; these were made, and immediately showed such a big improvement that there could then be no possibility of dropping the work. These polyamides (polymeric amides) could be made from diamines and diacids in this way:

$NH_2(CH_2)_6NH_2 + COOH(CH_2)_4COOH \longrightarrow$
Hexamethylene diamine. Adipic acid.

$NH_2(CH_2)_6NHCO(CH_2)_4COOH + H_2O$

and then two molecules of this condensate would react to give:

$$NH_2(CH_2)_6NHCO(CH_2)_4CONH(CH_2)_6NHCO(CH_2)_4COOH$$

and this long molecule would then react with itself, eventually forming a long polymer. Another way in which the polyamides could be made was by starting with a molecule which at one end had an amino-group and at the other end a carboxylic acid group. This could react with itself in this way:

$NH_2(CH_2)_5COOH + NH_2(CH_2)_5COOH \longrightarrow$
ϵ-Amino caproic acid.

$NH_2(CH_2)_5CONH(CH_2)_5COOH$

Two molecules of this condensate would then react to produce:

$$NH_2(CH_2)_5CONH(CH_2)_5CONH(CH_2)_5CONH(CH_2)_5COOH$$

this would react with itself, and so both methods gave very similar polyamides, which were fibre-forming. The first method was chosen for use in preference to the second simply because it was easier and cheaper to make diamines and dicarboxylic acids than it was to make amino-acids.

The result of the research was that strong fibres could be produced

—fibres which were different from any other fibres, in that they were truly synthetic. The long-chain polymeric molecules had been built up in the laboratory by man from short, simple molecules. In no case previously had a fibre been so made. True, artificial fibres such as nitrocellulose, viscose and cellulose acetate had been made, but all these had used the original polymer which Nature had provided in the form of the cellulose molecule in the wood or cotton used as the raw material. Carothers synthesised the polymer—his nylon was the first true synthetic fibre—his achievement was outstanding.

By 1938 the du Pont Company were making nylon at a small pilot plant at Wilmington in U.S.A. Since then other large plants have been built in America. In this country the fibre is spun by British Nylon Spinners, Ltd., at Coventry, at Pontypool, and at Gloucester; and by other manufacturers throughout the world.

The history of nylon is one of the great romances of science; the success of nylon has exceeded all expectations.

Chemical Structure

Normal nylon, which is made from adipic acid,

$$COOH(CH_2)_4COOH$$

and hexamethylene diamine, $NH_2(CH_2)_6NH_2$, is referred to as " 66 " nylon because each of the raw materials contains 6 carbon atoms. A similar polyamide can be made from sebacic acid, $COOH(CH_2)_8COOH$, and hexamethylene diamine and by the same reasoning this is known as the " 610 " polymer. Another polymer known as nylon 6 and made from caprolactam

```
CH2—CH2—CH2—CH2—CH2
|                |
CO——————————————NH
```

which reacts as if it were ε-amino-caproic acid, $COOH(CH_2)_5NH_2$ has also been made on a very large scale (p. 354).

The " 66 " type is preferred to the 610 polymer, partly for reasons of superiority of the fibre—it has, for example, a melting point higher than that of the " 610 " polymer, and also because organic compounds containing 6 carbon atoms in a straight line are conveniently and cheaply derived from benzene or one of its substitution products such as phenol. There is no such cheap raw material for sebacic acid—it can be made from castor oil, which is probably not available in unlimited supply, or it can be synthesised rather deviously from one of the 6 carbon atom compounds. The manufacturing chemist finds it much cheaper and more convenient to make organic compounds which contain six carbon atoms than any other number.

Stabilisation. It should be noted particularly that the molecules of nylon are long and straight, that there are no side-chains or cross-linkages. Those polymers which are spun as nylon have molecular weight averages of the order of 12,000–20,000. If the molecular weight is below 6,000 it is unlikely that the polymer will form fibres at all; the fibres that are formed with low molecular weight (say 6,000–10,000) are weak and brittle; then, as the degree of polymerisation and the molecular weight increase, the fibres become strong. Correspondingly, the molecular weight must not be allowed to become too high; if it is well over 20,000 the polymer becomes difficult to melt or to dissolve. Therefore, the process of polymerisation must not be allowed to go on indefinitely, but must be stopped at a given molecular weight. This is achieved by a process known as " stabilisation ". If, instead of taking exactly equivalent quantities of adipic acid and hexamethylene diamine, an excess of, say, 2 per cent of the former is taken, the time will soon develop when all the long polymeric molecules have carboxylic groups at both ends, instead of a carboxylic group at one end and an amino-group at the other. When this happens, it is impossible for polymerisation to proceed further, and the polymer is " stabilised ". Perhaps it is advisable to elaborate this explanation of stabilisation a little. It will be appreciated that if, instead of taking equimolecular quantities of diamine and diacid, one takes 1 equivalent of diamine to 2 of acid, then the reaction will go as far as:

$$NH_2RNH_2 + 2COOHR'COOH \longrightarrow COOHR'CONHRNHCOR'COOH$$

It can go no farther, because all the end-groups are now acidic, and there are no free amino-groups for them to react with. Next, suppose that, instead of using a ratio of 1 : 2 of diamine to acid, we use one of 1 : 1·5 or 2 : 3. Then the reaction is as follows:

$$2NH_2RNH_2 + 3COOHR'COOH \longrightarrow COOHR'CONHRNHCOR'CONHRNHCOR'COOH$$

This time a longer polymer has resulted, and as the ratio of diamine : diacid approaches unity, the length of polymer formed, before stabilisation prevents longer chain formation, will become greater. A ratio of 1 diamine : 1·02 diacid gives a polymer with a stabilised molecular weight of about 12,000 and one which is suitable for nylon preparation. Stabilisation (this method is used on production) can also be effected by using exactly equimolecular quantities of diamine and diacid and adding perhaps 1 per cent of a mono-functional reagent such as acetic acid; this has the same effect as the excess of diacid; when the length of molecular chain becomes sufficiently great all the amino-groups are prevented from further participation in polymerisation because they are "protected" by acetyl groups. Accordingly, polymerisation can be stopped at a predetermined stage by using excess of one of the reactants, or by introducing a mono-functional reagent. As a corollary it will be appreciated how important it is that the reacting materials, usually adipic acid and hexamethylene diamine, should be very pure; the presence in them of a small quantity of impurity may be quite sufficient to prevent the polymer growing long enough to give high tenacity fibres.

It is sometimes said that nylon is made from coal, air and water; whilst this statement is perhaps a little misleading, it may be justified as follows. Phenol is a coal-tar derivative, the hydrogen necessary for the reduction processes can be obtained from water, and the ammonia from the nitrogen in the air and from hydrogen. All the primary materials necessary for the manufacture of nylon are available in this country. But most of the world's nylon is made from oil.

Manufacture

It is possible to synthesise adipic acid and hexamethylene diamine in more ways than one, but the method which was at first used on production started with phenol. Phenol is made usually by sulphonation from benzene, and benzene, in turn, is made by the distillation of coal tar or alternatively from petroleum.

The phenol is reduced by passing its vapour together with hydrogen gas over a catalyst. Alternatively, it may be reduced in the liquid

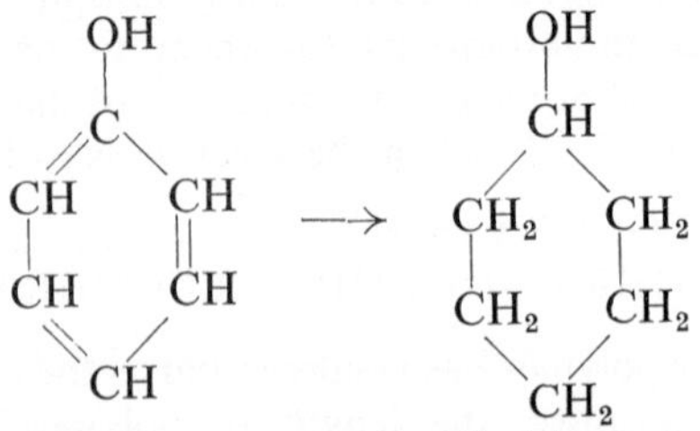

```
        OH                        OH
        |                         |
        C                         CH
     //   \                      /  \
   CH      CH       --->      CH2    CH2
   |       ||                  |      |
   CH      CH                 CH2    CH2
     \\   /                      \  /
       CH                        CH2
     Phenol.                  Cyclohexanol.
```

state in an autoclave, but the vapour-phase method has the advantage that it can be carried out continuously. The product is *cyclo*hexanol.

The *cyclo*hexanol is oxidised with concentrated nitric acid to adipic acid, the ring being broken.

$$\begin{array}{c} \text{CHOH} \\ / \quad \backslash \\ \text{CH}_2 \quad \text{CH}_2 \\ | \qquad\quad | \\ \text{CH}_2 \quad \text{CH}_2 \\ \backslash \quad / \\ \text{CH}_2 \end{array} + 2O_2 \longrightarrow \underset{\text{Adipic acid.}}{COOHCH_2CH_2CH_2CH_2COOH} + H_2O$$

Adipic acid is one of the compounds required for making nylon; the other is hexamethylene diamine, which can be made from adipic acid as follows. The adipic acid is caused to react with ammonia to give the amide:

$$COOH(CH_2)_4COOH + 2NH_3 \longrightarrow \underset{\text{Adipamide.}}{CONH_2(CH_2)_4CONH_2} + 2H_2O$$

The amide is dehydrated over a suitable catalyst to give the corresponding nitrile:

$$CONH_2(CH_2)_4CONH_2 \longrightarrow \underset{\text{Adiponitrile.}}{CN(CH_2)_4CN} + 2H_2O$$

The nitrile is reduced with hydrogen in the presence of a cobalt or nickel catalyst in an autoclave:

$$CN(CH_2)_4CN + 4H_2 \longrightarrow \underset{\text{Hexamethylene diamine.}}{NH_2CH_2(CH_2)_4CH_2NH_2}$$

The hexamethylene diamine and the adipic acid are then dissolved separately in methanol and on mixing the solutions a precipitate of " nylon salt " or hexamethylene diammonium adipate,

$$NH_2(CH_2)_6NH_2COOH(CH_2)_4COOH$$

is thrown down. The salt may be purified at this stage.

Modern Modification. Nowadays, the usual route for the first stages is no longer benzene ⟶ phenol ⟶ *cyclo*hexanol but benzene ⟶ *cyclo*hexane ⟶ *cyclo*hexanol. Air is passed through *cyclo*hexane at 120–150° C. under a pressure of 4 atmospheres to keep the system liquid; cobalt naphthenate is present as a catalyst and the *cyclo*hexane is converted to a mixture of *cyclo*hexanol and *cyclo*hexanone. This mixture is oxidised to adipic acid with nitric acid in the presence of a copper–vanadium catalyst. Molten adipic acid and preheated ammonia are fed to a boron phosphate catalyst at 360° C. and converted to adiponitrile in a yield of 90 per cent. Sometimes instead of the boron phosphate which requires frequent renewal, a molybdate/phosphate mixture mounted on silica is used as a catalyst; the yield is rather lower but the catalyst has a longer life. Reduction of the adipic nitrile to the diamine is, as before, by hydrogenation. The modern method just described is really very similar to the classical method; surprisingly so after an interval of 20 years.

Alternative Syntheses of Nylon Salt. The considerable increase in the production of nylon has brought about a shortage of raw materials, and in particular of benzene. The supply of this chemical from the coke ovens is limited, and even although the petroleum companies are doubling their production, benzene is still not so freely available that it is cheap. During the 1939–45 war benzene was obtainable at 15 cents (13*d.*) a gallon (the American gallon is about five-sixths of the Imperial gallon), whereas late in 1956 its price was about 36 cents (31*d.*). Early in 1962 it was 31 cents (27*d.*). This considerable increase in price and shortage of raw material directed attention to other ways of synthesising nylon salt. Two of these syntheses are believed to have been used on production; they are as follows.

From Oat Hulls and Corncobs. Various cereal products, notably husks and bran, can be used to make furfural and this chemical is made in large quantities by the Quaker Oats Company. The furfural is converted by the du Pont Company to furan which is hydrogenated to tetrahydrofuran:

Cereal Waste $\longrightarrow$ HC═CH, HC═C—CHO, ring closed by O (Furfural.) $\longrightarrow$ HC═CH, HC═CH, ring closed by O (Furan.) $\longrightarrow$ H_2C—CH_2, H_2C—CH_2, ring closed by O (Tetrahydrofuran.)

Furfural. Furan. Tetrahydrofuran.

The tetrahydrofuran is treated with hydrochloric acid which converts it to 1 : 4-dichlorbutane, which in turn is treated with sodium cyanide to give 1 : 4-dicyanobutane, which is adiponitrile. This is reduced by hydrogenation to hexamethylene diamine. The adiponitrile could also be hydrolysed to give adipic acid; in practice this is not done because it is cheaper to make adipic acid from phenol. Only the hexamethylene diamine half of the nylon salt is made from furfural, and it is combined with adipic acid made from benzene to give nylon salt.

$$\text{Tetrahydrofuran} \xrightarrow{2HCl} ClH_2C{-}CH_2{-}CH_2{-}CH_2Cl + H_2O \xrightarrow{NaCN} NC{-}H_2C{-}CH_2{-}CH_2{-}CH_2{-}CN$$

Tetrahydrofuran. 1 : 4-Dichlorbutane. Adiponitrile.

Adiponitrile $\xrightarrow{\text{Alkali}}$ $COOH(CH_2)_4COOH$ (Adipic acid.)

Adiponitrile $\xrightarrow{H_2}$ $NH_2(CH_2)_6NH_2$ (Hexamethylene diamine.)

From Butadiene. Butadiene has been made in enormous quantities for synthetic rubber; it is prepared from petroleum. The butadiene is converted to dichlorbutene by treatment with chlorine gas at 70° C. without a catalyst, yield 75 per cent. The dichlorbutene is treated with hydrocyanic acid in an aqueous medium at *p*H5 with calcium carbonate present as an acceptor base to pick up the hydrochloric acid released; the presence of some cuprous chloride speeds up the reaction. The 1,4-dicyanobutene that is formed is extracted with an organic solvent and then freed from the solvent by fractional distillation, yield 95 per cent. The dicyanobutene is reduced with hydrogen at 25 atmospheres pressure and 100° C. in the presence of palladium to give a 96 per cent yield of adiponitrile. The reactions may be represented:

$$\underset{\text{Butadiene.}}{CH_2{:}CH{\cdot}CH{:}CH_2} \xrightarrow{Cl_2} \underset{\text{Dichlorbutene.}}{ClCH{:}CH{\cdot}CH_2{\cdot}CH_2Cl} \xrightarrow{HCN}$$

$$\underset{\text{Dicyanobutene.}}{NC{\cdot}CH{:}CH{\cdot}CH_2{\cdot}CH_2CN} \xrightarrow{H_2} \underset{\text{Adiponitrile.}}{NC{\cdot}CH_2{\cdot}CH_2{\cdot}CH_2{\cdot}CH_2{\cdot}CN}$$

Of the three methods, the first from benzene doubtless holds, and will continue to hold, pride of place. As regards the second from furfural, any operation based on an agricultural commodity is subject to price fluctuations, but in Jan. 1957 the price of furfural was 12 cents (10*d.*) per lb. and in Jan. 1962 it was 11½ cents.

The third method depends entirely on the supply of butadiene, which is currently priced at about 10 cents (10*d.*) per lb. Economically, therefore, there is a good deal to be said for the two alternative methods. In the end butadiene or some similar 4-carbon compound such as butanediol will probably win.

There is no difficulty nowadays in making chemicals on a large scale; there are usually several possible routes and which one will be used depends on the prices and availability of the raw materials.

Polymerisation. Sufficient stabiliser is added—acetic acid is used for this—and the salt is melted under an atmosphere of nitrogen. Air has to be excluded, or the salt would discolour, and eventually the polymer would char; nitrogen is the most convenient gas to use to provide an inert atmosphere, but hydrogen may be used instead, or, again, a vacuum may be used. It is important only that air is excluded. Water that is split off from the nylon "salt" as polymerisation proceeds is allowed to escape. If matt nylon is required, an aqueous suspension of titanium dioxide is added to the reaction mass during the course of polymerisation; a quantity of 0·3 per cent pigment on the weight of the polymer is sufficient. A temperature of 280° C. and a time of four hours are suitable conditions for polymerisation. It is quite possible to polymerise nylon in another way—*viz.*, in solution in phenol or *m*-cresol—but on the manufacturing scale it is unlikely that any solvent is used. The molten polymer maintained at a temperature of 285–290° C. is extruded through a slot on to a wheel, in ribbon form several inches wide; this ribbon is quenched with cold water as soon as it solidifies, in order to reduce the size of the crystals.

Spinning. Nylon is melt spun. The ribbon of nylon is broken into nylon chips. These are fed through a hopper, *A* (Fig. 114), into a spinning vessel, *B*. In this they fall on to an electrically heated grid, *C*, which has a mesh too small to pass the chips until they have melted. As it melts, the molten nylon passes into the pool *D*; it is desirable to keep this pool small in order to reduce the risk of decomposition and discoloration to which nylon in the molten state is subject. There is therefore a float control on the pool level which controls the electrical heating of the grid—if the pool level rises, the current heating the grid is automatically reduced, and vice versa. In this way the level of the pool of molten nylon may be kept practically constant. A nitrogen atmosphere is maintained over the pool of polymer. The melt (288° C.) is metered by a pump *F* through a filter (not shown) to the orifices *E*. The filter may consist of several layers of metal gauze, the first relatively coarse, the next fine and ultimately very fine, about 300 mesh, and between the gauze layers there may be layers of sand graded in particle size; as the melt passes through the sand, impurities are removed. The orifices are of about 0·010 in. diameter and are drilled countersunk in a steel plate 0·25 in. thick and about 2 or 3 in. diameter. The melt solidifies immediately it issues from the jets and the so-formed filaments pass through a cooling chamber in which a cold air current *G* is swept across them. If one of the orifices has been too large the filament issuing therefrom will be heavier than the others and will bow more in the airstream than its companions, and if it touches one of them it will fuse to it; it is therefore essential for this reason, as well as for uniformity of denier, to have all the spinning orifices equal in size.

The spinning speed is about 1,200 metres per minute. As the yarn emerges from the cooling chamber it is at a temperature of about 70° C. It is next run through a steam chamber *H*, to wet it before it is wound; if it were wound without this treatment it would later extend a little in length on the package as it picked up moisture from the air to gain equilibrium and would slough off the package. If, however, the yarn is wetted by the steam before winding, this trouble is eliminated. A very slight twist which facilitates subsequent handling may be inserted during spinning.

Melt spinning has two big advantages over solution spinning:

1. It avoids the need for a solvent recovery plant and the losses which always occur during recovery, however well it is done;
2. The high spinning speeds that are possible.

Steam Spinning. Some depolymerisation always takes place in the melt and is known to be determined by the amount of water present; in order to reduce it, the polymer flake is dried before it is used for spinning, but there is always a little moisture present, and as this is variable the degree of depolymerisation that takes place is also variable. A method has been patented (Brit. Pat. 653,757 of 1951) in which a known amount of moisture vapour, usually 0·16 per cent, is added to the nylon in the spinning melt chamber so as to maintain a constant quantity of moisture; some depolymerisation occurs, but the amount that does is controlled and constant.

A process has been developed commercially by von Kohorn International Corporation in which the molten pool of polymer waiting to be extruded is maintained under a steam instead of a nitrogen atmosphere. It is said that this reduces the depolymerisation of the nylon with the resultant formation of a little monomer which normally occurs. (This steam spinning process will probably be extended to the polyester Terylene.)

Cold Drawing. The filaments as they are first spun are not very strong; their tenacity is 1·0–1·3 grams per denier, and they are dull. They are, however, cold drawn by stretching about 400 per cent, giving bright filaments with a tenacity of 5·8 and elongation of 17 per cent. The exact degree of stretch varies with the yarn being spun; fine filaments cannot be drawn down so much as coarse filaments. The reason for this difference is not really established, but is probably bound up with the partial orientation of the molecules on the outer parts of the filaments due to being drawn over the sides of the orifice; the finer the filament, the greater the proportion of " skin " to bulk, and consequently the better is its orientation; if fine filaments are already (before having been cold-drawn) a little better oriented than coarse filaments, then they will not be capable of being cold-drawn so much as the coarse filaments because the stage at which orientation is almost perfect will be reached earlier.

The stretching operation is simple and is illustrated in Fig. 115. Nylon yarn is pulled off the bobbin *L*, which is the primary spinning package; it passes round guides M, N, between a pair of nip rollers O, which determine its initial speed, and goes over a deflector P and then two or three times round roller Q which has a linear speed about five times that of rollers O; then it passes through a guide R, and on to a take-up bobbin S, twist being inserted as the yarn is taken up. The cold-draw, or degree of stretch, is equal to the ratio of the linear speeds of rollers Q and O. When a length of undrawn

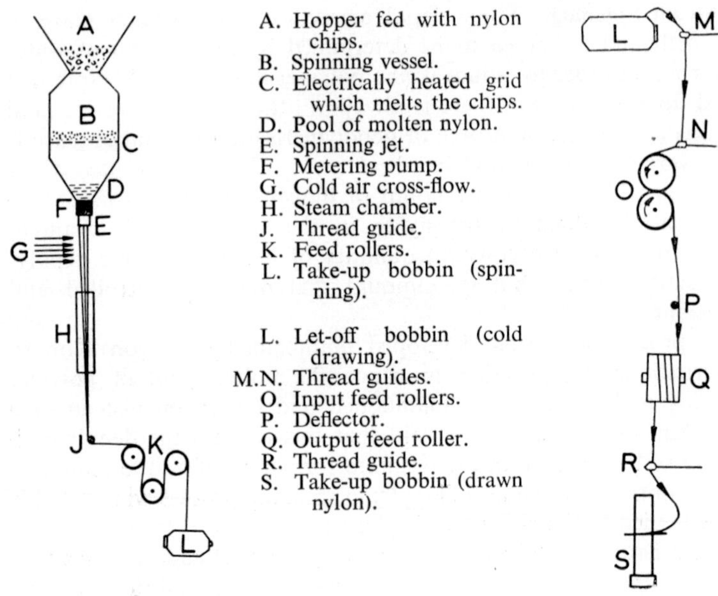

FIG. 114.—Diagrammatic representation of nylon spinning plant.

FIG. 115.—Diagrammatic representation of cold drawing of nylon.

nylon is held one end in each hand and pulled out, the sensation is rather like that of pulling out a telescope; photographs that have been taken during the stretching operation show that the nylon filament necks-down very quickly and not gradually along the free length of the yarn. When the yarn is cold-drawn and the filaments fine down they become lustrous and strong. This is due to the orientation of the molecules. As spun, the molecules are arranged in a random manner and are probably folded, and as they are cold-drawn the folds are removed as shown on p. 93. The final crinkle in the stretched nylon corresponds to the elongation at break. When a filament of nylon has been stretched almost to breaking point, the molecular crinkle has been straightened out. Further stretching causes breakage of the filament. This may be due either to breakage of the molecules or to overcoming the forces (Van der Waals) that hold the molecules together. Breakage of a fibre consists either in breaking the molecules, or in tearing them separate from each other, whichever is easier; the longer the molecules the less likely they are to be torn apart (p. 68).

During the orientation process oil may be applied to the yarn; any

[*British Nylon Spinners, Ltd.*]

Fig. 116.—Nylon yarn being tested for denier.

A known length is reeled off the package and weighed. In the illustration, lengths are being reeled off ten packages simultaneously.

high-grade vegetable oil is suitable. The oil is applied simply to facilitate textile processing—it has nothing to do with the orientation. Afterwards the yarn is twisted and is ready for use for textile purposes. Fig. 116 shows nylon yarn being tested for denier.

Properties

Nylon fibres are produced with a range of properties; some which are intended primarily for industrial application have a very high tenacity, others intended for apparel have lower but still high tenacities; where tenacity is very high, extensibility is lower. Depending on type, the tenacity and elongation at break range from 8·8 grams per denier and 18 per cent to 4·3 grams per denier and 45 per cent. The wet strength is also very high—about 80–90 per cent of the dry strength, a valuable feature. Another most valuable

[*British Nylon Spinners, Ltd.*]

FIG. 117.—Throwing or uptwisting nylon yarn from bobbin to cheese.

feature is that even when knotted the strength of nylon is very high. A half-hitch in a nylon yarn reduces the tenacity by only 15 per cent. Nylon has good flexing qualities and good resistance to abrasion—some four or five times that of wool. A stress–strain diagram for a nylon yarn compared with one for Dacron is shown in Fig. 146 (p. 407).

Elasticity. Nylon is fortunate in having not only a reasonably high elongation of 22 per cent at break, but also a high elasticity. If, for example, it is stretched 8 per cent, it completely recovers its original length when the tension is released, so that for an extension of 8 per cent it is 100 per cent elastic. For an extension of 16 per cent it is 91 per cent elastic. This property of high elasticity is extremely desirable, but it can also lead to trouble in processing. If nylon is wound on to cones, under conditions that would be suitable for a worsted yarn, it stretches slightly during winding, and after being wound on to the cone, the yarn, because of its high elasticity, will attempt to recover its original length and will contract, so that the " former " on which the yarn is wound will be crushed. Special winding conditions, particularly a low winding tension, are essential for nylon.

Specific Gravity. The specific gravity of nylon is 1·14, which is lower than that of most fibres (viscose 1·52, cellulose acetate 1·30), and it contributes to its potentiality of making light-weight fabrics with a good cover, although it is somewhat offset by the symmetrical cross-section of the fibres which permits very close packing.

Melting Point. Nylon melts at 263° C. under nitrogen and at about 250° C. in air. This is a little on the low side for a textile fibre, and, if a very hot iron is used, may lead to sticking, or even to fusion when nylon garments are ironed. If the iron is hotter than 180° C. sticking begins, and at 230° C. damage occurs. Nylon turns slightly yellow when heated in air to 150° C. for five hours, but is a little better in this respect than silk or wool, although not so good as the cellulose fibres. When ignited, it does not burn well, but melts to a glassy globule, rather like a borax bead; it will not support combustion, and if a quantity of nylon is ignited, the burning parts melt and drop away from the bulk; the fire risk is very low.

Stability. Chemically, it is extremely stable. Solvents which are normally used in dry-cleaning are quite harmless to nylon. Dilute acids do not affect it seriously, but it is hydrolysed, by boiling for several hours with concentrated hydrochloric acid, into adipic acid and hexamethylene diammonium hydrochloride, and this process is used to recover the original components from nylon waste, so that they can be used again for polymerisation. The stability of nylon to alkalis is most remarkable; treatment with a 10 per cent solution of caustic soda at 85° C. for ten hours reduces the strength of nylon by only 5 per cent. The only common solvents in which it is soluble are formic acid, cresol and phenol. The degree of polymerisation of nylon is ordinarily controlled by measuring the viscosity of its solution in *m*-cresol; the longer the polymeric chains, the higher is the viscosity.

Biological Properties. Nylon will not support mildew or bacteria, nor is it eaten by moth larvae, although they may bite their way through it, if imprisoned. It is quite harmless to the skin; there is no danger of dermatitis developing when nylon is worn.

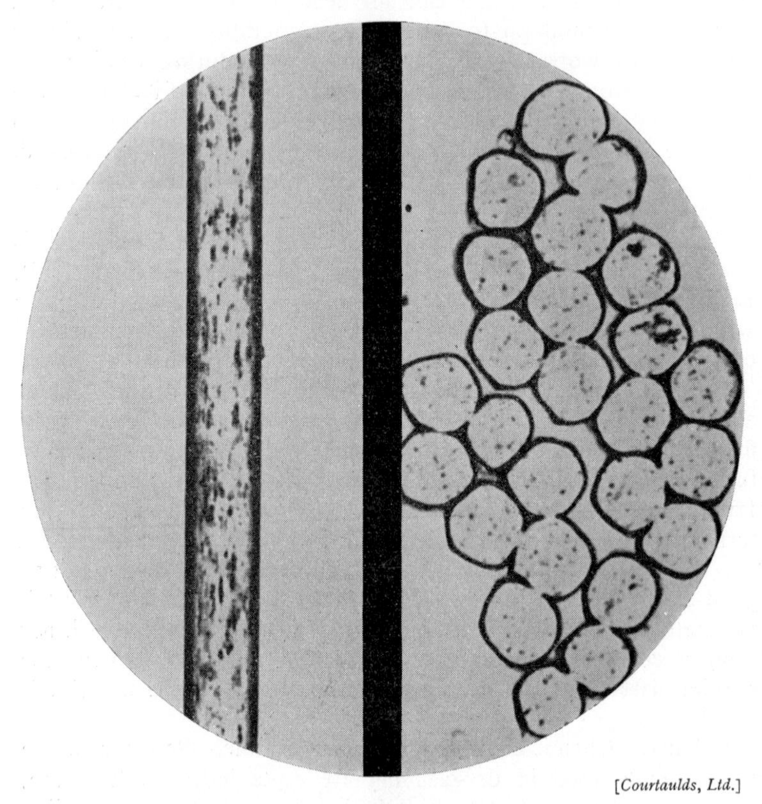

[*Courtaulds, Ltd.*]

FIG. 118.—Photomicrograph of nylon (×500).
The " pitting " is due to delustring pigment.

Moisture Absorption. The moisture absorption of nylon is low, only about 4·2 per cent water being in equilibrium with the fibre under standard conditions (65 per cent relative humidity and 70° F.).

Morphology. Under the microscope the longitudinal view is nearly featureless, almost like that of a smooth cylinder, whilst the cross-section is almost circular (Fig. 118).

Lustre. Before it is cold-drawn nylon is dull and semi-opaque, but on orientation the lustre is greatly enhanced, and normal nylon is clear. When pearl or matt yarns are required, titanium dioxide is added to the polymerisation mixture. In a similar way inert pigments can be added to produce spun-coloured yarns, although this is not ordinarily done.

Effect of Light. Nylon, like other textile fibres, is degraded by the action of light, but to an extent much less than real silk. On a comparative test over a period of sixteen weeks during which real silk lost 85 per cent of its tenacity, bright nylon lost only 23 per cent and semi-dull nylon 50 per cent. Under similar conditions cotton lost only 18 per cent of its strength.

Electrical Properties. Nylon is subject to static electricity because it is such a good insulator. This trouble can be reduced by working it in a moist atmosphere and also by the application of an " anti-static " finish. Static eliminators which ionise the air surrounding the fibre are useful; the Shirley Static Eliminator has already found considerable use. (Figs. 198, 199, p. 626).

Exposure to Heat and Moisture. Exposure to heat and moisture will give a permanent set to nylon—*i.e.*, it is given a shape and dimensions which will be permanent until the nylon is taken to a higher temperature than that at which it was set. If, for example, nylon hose are set in saturated steam under 25 lb. pressure, they will retain their shape on subsequent wearing and washing even in hot water.

Nuclear Radiation. There has been some suggestion that certain plastics when exposed to nuclear radiation derive an improvement in properties, and it was thought that fibres might be similarly benefited. The underlying idea has been that when the fibre, nylon for example, was bombarded with nuclear particles, some of them would hit and dislodge atoms from the nylon molecular chains and that the vacancies so formed would be immediately filled by atoms from adjoining chains, thus forming cross-linkages, which would give greater strength, better elastic properties and probably better chemical and biological resistance to the fibres. It was recognised that irradiation by nuclear particles would eventually degrade the fibre chemically, breaking the long polymer chains and also some of the postulated cross-linkages. The point at issue was whether the initial effect of irradiation would be to strengthen the fibre due to the beneficial effect of cross-linking being greater than the adverse depolymerisation effect, and whether, in fact, there was a *most favourable dose* of radiation which would give the greatest gain in strength. Experimental work on the subject is in its infancy, but so far no improvement has been found, and wherever the dose has been sufficiently severe to change the properties of the fibre, the change has always been for the worse. When a 70 denier 34 filament nylon yarn was exposed to γ-radiation from radioisotope cobalt 60 for 42 hours it lost about half its strength. When it was exposed to neutron irradiation in an atomic pile, the strength was reduced by a little more than one-half. For comparison, viscose rayon lost half its strength on γ-irradiation and was completely degraded by neutron irradiation. The effects of γ-irradiation and of neutrons on the characteristic curve of nylon are shown in Fig. 119.

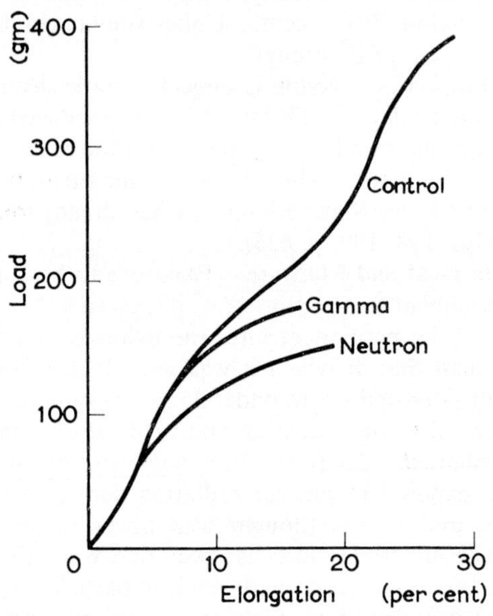

FIG. 119.—Effect of nuclear radiation on stress–strain curve of dry nylon.

Surface grafting. Use has been made of radiation to graft various materials on to nylon. Atoms are knocked out of the nylon polymer and the grafting material fills the vacancies. The work seems to have been originated by Korshak in Russia and to have been developed in other countries. The U.K. Atomic Energy group have irradiated nylon with cobalt 60 and then covered the irradiated nylon with an aqueous solution of monomeric methyl methacrylate. The methacrylate combines with the nylon even to the extent that it can give an increase in weight of 10 per cent. Because the fibre surface consisting of the methacrylate graft is more hydrophilic, its electrical resistance is considerably lower than that of untreated nylon, but hardly low enough to prevent the incidence of static which was one of the aims of the research. The solubility of the nylon in hot formic acid was reduced by the treatment and no complete solvent for the grafted nylon could be found (*J. Soc. Dyers and Col.*, **76**, 342–344 (1960)). Atomic Energy of Canada Ltd. have immersed nylon in styrene and then bombarded it with γ-rays from cobalt 60. Polystyrene was thereby grafted on to the surface of the nylon and its weathering qualities thereby improved (*Dyer*, 719 (3 November, 1961)). It has been reported that if irradiated nylon is treated with vinyl pyridine and then with ammonium sulphate a quaternary nitrogen compound becomes grafted on to the nylon and that this has sufficient hydrophilic powers to prevent static formation under some conditions.

But all of these processes are in a very early stage; they are interesting academically but it has to be remembered that the irradiation is likely to reduce the fibre strength by polymer degradation simultaneously with the desired surface reaction.

Mechanism of cross-linking on irradiation. In a detailed study of the effects of ionising radiation on natural and synthetic high polymers, F. A. Bovey (Vol. 1 of *Polymer Reviews*, London) writes that cross-linking occurs with nylon, polyethylene and polypropylene fibres whereas chain scission (degradation) is the main effect on P.V.C., Teflon and cellulose. However, it is not always easy to interpret the experimental results; a discussion of the difficulties involved is to be found in Bovey's book. One point of special interest concerns cross-linking: this reaction is attributed to the loss of hydrogen atoms at adjacent sites on neighbouring chains; it seems probable that when a hydrogen atom is initially ejected it picks up another hydrogen atom from a neighbouring chain forming a hydrogen molecule. If it did not do this, then the points of unsaturation, formed randomly, would not occur often enough to account for the observed degree of cross-linking.

Dyeing

Nylon during its manufacture has been oriented by stretching and unless this has been done very uniformly there will be differences of orientation which may lead to differences in appearance in the dyed nylon fabric. Two factors which can influence the dyeing characteristics under given bath conditions are:

1. The degree of orientation; the higher this is, the lower is the affinity of nylon for the dyestuff. It has been found that dyeings with Durazol Fast Blue 2R accentuate such differences very strongly and this dyestuff can be used to assess the uniformity of orientation.

2. The concentration of amino groups in the terminal positions of the molecular chains; the dyeing properties of nylon can be modified by the method that is adopted for stabilising the polymer. The acidic nature of the polymer stabilising agent used in nylon manufacture causes a fundamental difficulty in dyeing nylon with acid dyes, one due to the poverty of the fibre in the way of free amino end-groups (p. 46). These amino groups amount to only 0·04 milliequivalents per gram which is insufficient to build up heavy shades by combination with simple acid dyestuffs. Furthermore, if two or more dyestuffs are used there will be competition between them for the available sites of which there is a shortage, and one dyestuff may have a preferential affinity over another. An extreme (not a typical) instance of this preferential affinity is to be found with a mixture of Solway Blue BN (C.I.1054) and Naphthalene Red J (C.I.176): the sites available are occupied preferentially by the red dye. Indeed, if the fibre is first dyed with Solway Blue BN and then dyed with Naphthalene Red J, 95 per cent of the blue is turned out of the fibre and displaced by red. Phenomena of this sort, even although active to only a small degree, cause trouble in the dyehouse. More complex dyestuffs which are adsorbed on to the fibre from neutral solutions and do not rely on chemical combination with the fibre are less susceptible to these troubles than are the simple acid dyes.

When fabric is dyed it can either be " held " smooth on a jig during dyeing to prevent the formation of creases, or, alternatively, it can be " pre-set "—*e.g.*, on a roll in live steam at 15–25 lb/sq. in. and then subsequently dyed in rope-form on the winch. Because it has been pre-set at a higher temperature, no permanent creases will develop during treatment in rope form at a lower temperature. In order to minimise the tendency to form permanent creases dyeing is best carried out at a temperature not higher than 85° C.

Nylon can be dyed satisfactorily with silk and wool colours, but has only a low affinity for direct cotton, sulphur and vat colours.

Acetate dyestuffs are often recommended for nylon, mainly because they give uniformity of shade and because they exhaust well, but their light fastness is inferior to that given by the neutral dyeing acid colours. It is not always easy to get solid shades on wool/nylon mixtures, and only some dyewares are suitable when solid shades are required on mixtures. A tremendous amount of work on the selection of colours suitable for nylon has been carried out by the dyestuff manufacturers, from whom advice on this matter is always obtainable, and should be sought when required. Because of the low affinity of nylon for water, it is not easy to obtain good penetration of the fibres, or even of fabric by dyestuff solutions, and in this respect the dispersed acetate colours give the best penetration. The dispersed acetate rayon dyestuffs—*e.g.*, Duranol and Dispersol (I.C.I.), Supracet (L. B. Holliday), and Serisol (Yorkshire Dyeware Co.) are pasted, suitably with Dispersol VL, diluted with cold water and stirred until a good suspension is obtained. This is added to the dye-bath, and the dyeing carried out at 85° C. for forty-five minutes. Sometimes a dyestuff will give a shade on nylon different from that on cellulose acetate—*e.g.*, Dispersol Fast Orange A gives an orange shade on acetate rayon but a pure red shade on nylon. Nylon carpeting is being dyed with disperse dyes, suitably in the presence of borax at pH 8·5–9.

Logwood, a natural dyestuff, is important for dyeing blacks on nylon. It is interesting to note that it is still the standard black on real silk, as it far surpasses in richness of hue any black obtainable with other dyestuffs. Hematine, which contains the essential colouring matter of logwood, and is the best of all for black nylon hose, is supplied by the West Indies Chemical Works, Ltd., who recommend the following method for its application to nylon:

The scoured nylon is treated in a solution containing 8–10 grams unoxidised hematine crystals per litre, made faintly acid with acetic acid. Dyeing is started at 60° C., the bath raised to 90–95° C. in fifteen minutes, and continued for half to three-quarters of an hour at that temperature. The nylon is then washed with cold water until the wash-water is colourless. This is most important, as any unfixed hematine carried into the chrome bath causes the dyeings to rub, and may also cause brown stains. If the water is alkaline it is advisable to correct the alkalinity with acetic acid. Fixation is carried out by treatment in a solution containing 3·5–4·5 grams of sodium or potassium bichromate per litre with the addition of 5 per cent acetic acid (80 per cent) on the weight of nylon. The material is entered at 60° C., raised to 90–95° C. in ten minutes and treated at 90–95° C. for twenty to thirty minutes. (If 15 denier monofil nylon hose are being dyed, chroming should be carried out for a longer period owing to the poorer penetration into the monofil, in order to ensure complete fixation of the colouring matter.) This is followed by thorough washing in hot water and soaping, the severity of the soaping depending on the purpose for which the material is to be used. Standing baths may be employed.

The black produced is of very good fastness to light (6 or better, according to the depth of shade) and of excellent fastness to washing.

The dyeings will also withstand degumming with natural silk. A similar method to that described above can be used for dyeing nylon/cellulose acetate mixtures, provided that the temperature is not allowed to rise above 80° C.

Blacks on nylon can also be obtained by using the diazotised and developed acetate rayon blacks such as Dispersol Diazo Black 2BS. Twice the quantity of sodium nitrite and hydrochloric acid normally used for acetate rayon is necessary when processing nylon. The affinity of the nylon fibre for dispersed acetate dyestuffs is only about half that of acetate rayon for the same dyestuffs. Nylon is less subject to gas-fume fading than is acetate rayon.

Reactive dyes. The Procinyl dyes are insoluble disperse dyestuffs which are applied to nylon in the way usual for disperse dyestuffs, but which additionally combine with the nylon fibre if it is subsequently treated with alkali. Consequently the dyeing is carried out in two stages:

1. At 95° C. from a bath acidified with acetic acid to *p*H 3·5 for 40 min; the dye is least reactive in an acid medium and it is just adsorbed on the fibre like any other disperse dye with good levelling and penetration properties.
2. At 95° C. from the same bath made alkaline to *p*H 10·5 with soda ash, for 60 min.

Washfastness of Procinyls on nylon is better than usual; so is the cover up of inequalities deriving from the drawing process in the nylon. The dyeings with Procinyl dyes are much better than usual in respect of freedom from heat sublimation. In their early days the advantages that Procinyl dyes offer for nylon are less striking than those that Procion dyes have for viscose.

Wool dyestuffs which have been recommended for dyeing nylon in medium shades are Lissamine Fast Yellow 2GS, Azo Geranine 2GS and Solway Blue BS, dyed with 2 per cent formic acid on the weight of the nylon. Chrome dyestuffs can be used, but not so satisfactorily as on wool, because of the difficulty of chroming the dyestuff satisfactorily on nylon, twice the normal (*i.e.*, used for wool) amount of sodium bichromate being necessary as well as a longer time of processing.

Basic dyestuffs can be used for nylon, but ordinarily give only poor fastness to light and to washing. Manufacturers are very cognisant of any dyeing deficiencies in their fibres and will go to a lot of trouble to put them right. Thus I.C.I. in Dutch Patent (19)67-12505 state that the affinity of nylon 66 for basic dyes is increased if there is included in the monomer (nylon salt) 0·5–5 molar per cent of disodium 9,9, bis-β-carbamethyl-fluorene-2,7-disulphonate, a body with the structure

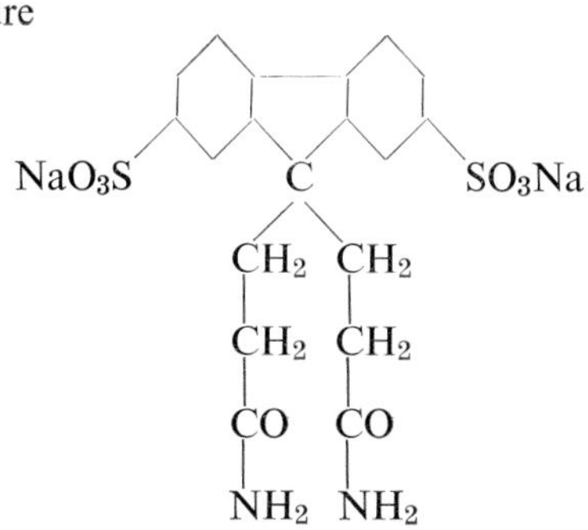

NaO_3S C SO_3Na

CH_2 CH_2

CH_2 CH_2

CO CO

NH_2 NH_2

An example is given in which 2 molar per cent of this substance is used and the polymerisation is carried out at 220° C. for 3 hr. The sulphonic acid groups could well be expected to provide suitable sites for basic dyes, but the function of the carbamethyl groups is less easy to define. Du Pont make an Antron Type 826 nylon which can be dyed with basic or cationic dyes.

Pigment padding (p. 198) gives excellent level dyeings of good fastness and depth of colour on nylon.

Cake Dyeing of Nylon. Continuous filament nylon is not easily dyed on conventional hank-dyeing machines, partly because of the fineness of denier in which it is usually spun, but mainly because of the tendency of nylon to float; its specific gravity is only 1·14 and a good deal of air is usually included between the very fine filaments. If floating does take place, the hanks become distorted and are liable to be very difficult to unwind so that wastage is high and the process is uneconomic. This directs attention to the possibility of package dyeing, and for the reason that nylon shrinks in hot liquors such as are used for dyeing, no former or centre can be used and a centreless cake or " mock cake " has had to be adopted. Packages of this type weighing 1 lb. are now dyed on normal Longclose (see also p. 713) and on Courtaulds' cake-dyeing machines in bulk lots. The packages must be of uniform size and density; a good range of shades can be obtained with good wash fastness and reasonable light fastness. The process is believed to be used only on yarn containing about 5 t.p.i. twist; if there was no twist the package would become soft and impenetrable.

USES

The uses of a fibre are determined by its properties.

The high tenacity of nylon has made it of paramount importance for parachute fabric, cords and harness, and during the war most of the fibre produced was used for such purposes. It is particularly suitable for glider tow-ropes. Again because of its high tenacity, and partly because of its good flexibility it has been used successfully for sewing threads.

The fact that it can be pre-set by steaming has been of use in the manufacture of ladies' hose. The stockings are " set " on shapes, and thereafter maintain their shape during wear. The high elasticity

[*British Nylon Spinners, Ltd.*]

FIG. 120.—A 1-in.-diameter nylon rope lifts $5\frac{1}{4}$ tons with safety.

of nylon ensures that the stockings will not " bag " at the knees or ankles, because after each flexing due to the wearer walking they recover their original shape. The high tenacity of nylon permits very sheer stockings, made of 30 denier, to have satisfactory strength for wear. In addition to the normal multifil yarns, a 15 denier monofil nylon has also been used extensively for stockings. High tenacity, too, has made nylon useful for ropes (Fig. 120); these can be much lighter and easier to handle than those made from natural fibres. If nylon yarn is twisted to 10 or more t.p.i. it is lively, inclined to snarl, and before being knitted or woven should be " set " in steam. That nylon can be set is a great advantage, not only in that it reduces the liveliness of twisted yarns, but also, and mainly, in that it gives a large measure of dimensional stability to fabric, and enables permanent pleats and creases to be " set " in garments. Most women who have used nylon for underwear and dresses find the property of shape retention of the garments valuable. Furthermore, if the garments are washed very carefully so that no compression creases are introduced, ironing can be dispensed with; normally, however, a light ironing greatly improves the finish and is desirable. Seersucker effects can be produced in nylon fabrics by

[*Northide, Ltd.*]

FIG. 121.—P.V.C. coated nylon sheets in use.

utilising the different shrinkage potentials of differently set (heat-set) yarns.

The high tenacity and resistance to abrasion of nylon have been of value in allowing light weight and even sheer garments to be made. Furthermore these properties have led to the use of nylon in the fabric foundation of motor and aeroplane tyres, the life of the tyres being considerably increased by the inclusion of nylon; some difficulty was at first experienced in bonding nylon to rubber, but this has been overcome by the introduction of new bonding agents such as Vulcabond TX, and in America about 250 million pounds per annum of nylon is used in tyre cords. Tarpaulins in large numbers have been made from nylon coated with P.V.C. (Fig. 121), and they stand up well to adverse conditions and have a long life, and as they weigh only half as much as usual they are easier to handle.

Nylon's high tenacity has made possible its use in a fine bolting cloth made with a leno weave; such cloths are used for sifting flour and they last at least twice as long as those made from real silk, although they cost no more.

[*British Nylon Spinners, Ltd.*]

FIG. 122.—Nylon belts support a great drop forge hammer; they have to withstand sudden terrific shocks.

The durability of nylon, combined with its strength, has made nylon very suitable for drop forge belting (Fig. 122), transmission belting (Fig. 123), and for universal joint discs and link belts. Men's socks made from nylon almost never need darning. Typewriter ribbons have been made from nylon; they are stronger and thinner than usual and give better definition and last longer, especially in electric typewriters where they have to withstand the constant hammering of the types, although perhaps the ink does not spread so

[*British Nylon Spinners, Ltd.*]

FIG. 123.—Nylon belt on a generator which supplies power to a quarry. Variable loads test the belt but it has a long life.

[*British Nylon Spinners, Ltd.*]

FIG. 124.—Nylon filter sacks in use at Wedgwoods.

well during rest periods from unworn to worn parts as it does with cotton. Durability has, too, led to nylon's use for spindle bands on textile (staple) spinning machines where it gives greater productivity, because there are fewer stoppages due to breaks; furthermore, starting loads are reduced to a minimum and a product with better uniformity of twist is obtained. Another textile use for nylon, also deriving from its durability, is as doup heald cords for leno weaving. Luggage which has to be durable is made in lightweight nylon canvas. Nylon has been used in the making of belts for telephone linesmen; telephone switchboard cords have an exceptionally long life; their resistance to abrasion enables the cords to withstand the severe rubbing which occurs each time a plug is connected or disconnected. A nylon/cotton mixture fabric has been used for fire hoses; not only is it more durable but also easy to handle. Furs (Furleen) made from nylon have proved popular and this application rests largely on durability.

Resistance to water and especially to sea water, combined with the strength and durability of nylon, has led to its use for ropes and cordage and for gill netting which is relatively translucent so that more fish are caught; furthermore, the nets are rotproof. Seine netting, used by our smaller fishing boats, is also made of nylon. Dracones, huge containers which can be filled with oil, to transport it cheaply, and towed along rivers, are made from coated nylon. The pottery industry (Fig. 124) has provided a large outlet for nylon, due mainly to its property of water-resistance and to the retention of a very high proportion of its high strength when wetted; filter sacks are filled with a slurry of clay and are squeezed in presses until the water filters through the fabric and leaves a cake of solid clay; when the filter sacks have been clothed with a 7 oz. (weight per yard) nylon fabric instead of a 12 oz. cotton fabric, the initial cost has only been 50 per cent greater, but the water extraction is quicker and better and the clay comes away more easily from the smooth surface of the nylon; moreover the nylon fabric has lasted three or four times as long as the cotton, partly because of its mildew resistance. Nets are used to support tobacco in conditioning-rooms and nylon netting holds taut through high temperatures and extreme humidity—nets made from natural fibres would soon sag. Nylon dough provers used in bakeries are easy to clean as well as durable and non-toxic. Nylon laundry bags last for about a thousand washes, compared with about a hundred for cotton bags, and as they pick up less moisture, there is less deadweight.

Chemical resistance has led to nylon's use for filter fabrics, especially for oils. Because of the ease with which they can be cleaned, nylon fabrics are used for fruit crushing; the pressed fruit residues are easily removed from the smooth nylon fabrics, which can, too, easily be sterilised, a treatment which would soon ruin the usual woollen fabric.

Biological resistance has led to the development of nylon insect screens. Mildew resistance has contributed to the use already mentioned of filter sacks in the pottery industry.

There are many other uses to which nylon has been put, but of all its applications the two in which it has proved pre-eminent have been for ladies' stockings and for tricot. A use which is consuming increasing quantities of nylon is for carpets; its durability gives long life despite hard wear and its chemical resistance makes it difficult to stain and easy to clean.

Monofilament Nylon. The most important application of monofil nylon has been for ladies' stockings for which 15 denier monofil has proved very suitable. Recently, still lighter monofils have been introduced; a 12 denier monofil is made in Japan and du Pont make a 7 denier monofil (one pound would reach from London to Scotland). Monofilament nylon, spun as a bristle, has been used very extensively for tooth-brushes, which have a life several times as long as those made from natural bristles, for clothes-brushes, hair-brushes and bottle-brushes. One case which illustrates the value of the stability of nylon to alkalis is to be found at the fellmonger's. When sheep have been slaughtered for mutton, the skins are removed from the carcases and are painted on the inside with a solution of sodium sulphide and lime and left overnight. Next day the wool can easily be pulled off the skin, and is known as " slipe " wool. When brushes of natural bristle were used by the fellmongers they had a useful life of only a few weeks because the strongly alkaline solution of sodium sulphide rapidly attacked them, but nylon brushes have shown no sign of deterioration after six months use. Paint-brushes have to be tapered, as of course natural bristles are, and it has been possible to make tapered nylon brushes by spinning a bristle with a rapid variation in denier, thick and thin places alternating, and then, of course, cutting at the thinnest places. The taper is retained in the cold-drawing process. Nylon, in fact, has brought the first outstanding improvement to the brush industry for a long time. Owing to its greater rigidity and lower moisture absorption, " 610 " nylon may be preferred to " 66 ", for use as bristles. Incredible as it may seem, there is a good chance that nylon in powder form may be used in the toothpaste that goes on the nylon tooth-brush; very finely divided nylon 66, with particle diameter between 1 and 10 μ and known as Nylasint, has the property of being a good polisher without scratching and it has shown promise in field tests as a toothpaste ingredient.

Nylon 610 made from hexamethylene diamine and sebacic acid has a melting point of 214° C. compared with 263° C. for the 66 polymer, and a moisture regain of 2·6 per cent compared with 4·2 per cent for 66 nylon. The resilience of 610 is greater (more like that of wool) than that of 66, and furthermore, the resilience is increased by orientation much more than that of 66. The improvement in resilience from 66 to 610 is not sufficient to justify making large quantities of the fibre; small quantities have been made for some years for bristles and also for plastics.

Staple Fibre Nylon. This is discussed on p. 668.

Antron and Similar Yarns

A good deal of new information about the spinning of nylon in sections other than circular is given in U.S. Patent 2,945,739 (1960) by D. J. Lehmicke (assigned to du Pont). The examples given in

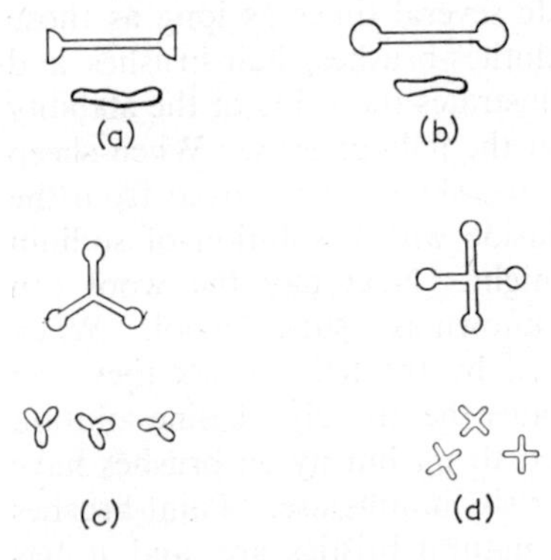

FIG. 125.—Slots with abrupt terminal expansions used as spinning orifices: (*a*) straight slot; (*b*) dog bone; (*c*) triangular; (*d*) cruciform. In each case the approximate shape of the spun filament is shown below the jet. The trilobal filaments made with orifices (*c*) are the nearest to Antron. As an indication of the magnification in the figure, the cross (*d*) is in reality $1\frac{1}{4}$ mm. across but the slots are drawn wider than to scale.

[*After Lehmicke (du Pont de Nemours & Co.)*]

the patent relate to both nylon and Dacron-type polymers, and they describe the use of orifices of special shapes through which the molten polymer can be spun. Some of the orifices are slots with abrupt terminal expansions (Fig. 125 (*a*)), some of them are dogbone (Fig. 125 (*b*)), some are Y-shape (Fig. 125 (*c*)) and some cruciform (Fig. 125 (*d*)). In each of these Figures, there is included an indication of the cross-sectional shape of the extruded and quenched filament.

The degree of modification of the filament section may be described numerically as $\frac{N}{n}$, when N is the diameter of the smallest circle that will enclose the filament and n is that of the greatest circle which the filament will include. The " filament ratio ", as $\frac{N}{n}$ is called, would be unity for a round filament and would be 5·0 for a ribbon-like filament of width five times its thickness (Fig. 126).

A cruciform jet may be so made that the arms of the cross are 40 mils (1 mil ≡ 0·001 inch) long and 2 mils wide, with an abrupt circular expansion 5 mils in diameter at the end of each arm, so that the total length of the cross is 50 mils. The average filament ratio

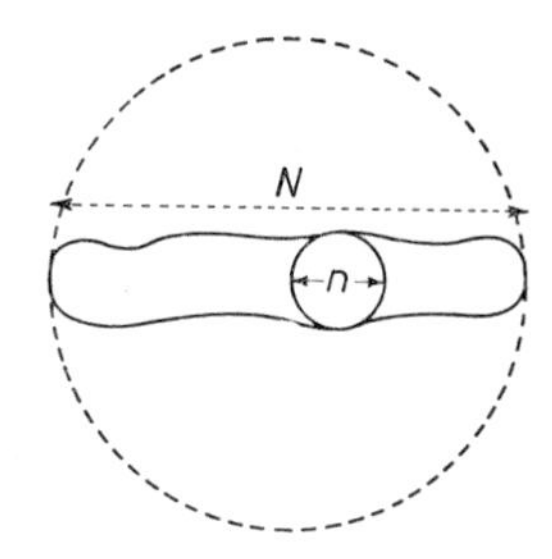

FIG. 126.—The " filament ratio " of the flat filament shown is 5, because $N/n = 5$. The " filament ratio " is a measure of the departure from circularity of any filament.

of the yarn spun from such a jet is 3·35. It is then cold drawn to 1·95 times its extruded length and its filament ratio is then 3·53. This is really very unexpected: that the departure of the filament cross-section from circularity should *increase* when the filament is stretched.

If the spinneret is deprived of its abrupt extensions so that it consists only of two intersecting slots each 40 mils long and 2 mils wide, it is impossible to spin yarn for long continuous periods because solid particles in the molten polymer choke the jets.

If the slot width is increased to 3 mils, still without the terminal expansions, spinning is possible but the filament ratio of the drawn yarn is only 1·9. It is clear enough that a high ratio of slot length to width is necessary, and the terminal expansions not only serve their obvious purpose of depositing heavy masses of polymer at places as remote as possible from the centre of the filament (so avoiding roundness of filament) but also afford an escape route for solid particles that contaminate the melt and that have passed the filters in their route to the jet.

" The essence of this invention is the combination in a spinneret orifice of a slot and at least one abrupt expansion in each tip of each slot. These abrupt expansions appear to have two functions. They provide a reservoir of polymeric material which stiffens the extruded shape, holding the shape and resisting surface tension forces until the filament solidifies. They also provide channels through which pieces of sand, foreign material and polymer gel which might otherwise obstruct the slots, are extruded."

The polymer has to set before surface tension pulls it into a round section and if necessary the solidification of the polymer to a filament can be speeded up by cooling the spinneret and filter pack to bring the polymer nearer to the solidification temperature at extrusion, or by cooling to a lower temperature than usual the cooling air that blows across the filaments just after they have been extruded; alternatively, the volume of this cooling air can be increased to accelerate the solidification of the polymer. The really essential condition is that the polymer shall be of high viscosity; nylon polymer " of at least 40 and desirably above 50 relative viscosity (measured on the spun yarn) " is preferred for the most highly modified filaments (*cf.* p. 40).

The manufacture of ordinary round hole spinnerets was once looked on almost as a magic art. Manufacture of these fancy jets must be extremely difficult even with modern techniques; it has to be remembered that the cross with arms, abrupt expansions and all is only $1\frac{1}{4}$ mm. across and the slots are only $\frac{1}{20}$ mm. across. Furthermore it is still important to have the holes as nearly alike as possible; if any hole is too small, the filament that issues from it will be too small and will probably break in the subsequent drawing operation. The slots are punched as " dimples ", these are sanded off so that the slots appear and the abrupt expansions are then drilled at the ends of the slots.

Properties of Non-Circular Filaments

These are what we should expect; yarn made from them has a better cover than round filament yarn. Fabrics made from this yarn have a firmer crisper handle than the very smooth handle of fabrics made from round filament nylon. The triangular sections of a melt-spun fibre shown in Fig. 127 are not greatly different from those of boiled-off real silk.

Effect on Pilling

The claim is made that filaments spun in ribbon shape do not pill, or at least not nearly so badly as do round filaments. Nylon and other high tenacity fibres in staple form that pill are a nuisance because the fibres are so strong that the pills do not drop off as they sometimes do from wool, but become almost permanent fixtures. It is claimed that the tendency to pill falls as the filament ratio increases. Some results are shown in the following table; the

" pilling " of the nylon staple fabric was produced by carefully measured brushing and sponging.

Filament denier.	Filament ratio.	Pills per sq. foot fabric.
3	1 (round)	290
3	2·75 (flat)	60
4·5	2·07 (flat)	110
4·5	3·54 (flatter)	35
4·5	4·64 (flattest)	0

The figures are very convincing but the author has seen so many remedies for pilling, which seemed all right in the laboratory, fail when a woman wore a pullover, that it is difficult to be enthusiastic.

One point that the patentees bring out that is undoubtedly true is that fine filaments pill worse than coarse ones.

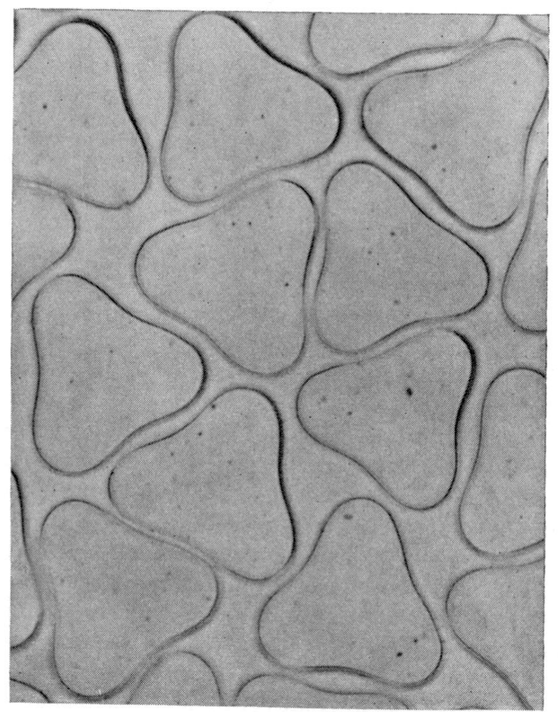

[*E. I. du Pont de Nemours & Co.*]

FIG. 127.—Triangular sections of a melt-spun fibre (Antron 520 denier, 34 fils) which will yield fabrics with a firm crisp handle.

Antron

Trilobal yarn made by a similar process, described in U.S. Patent 2,939,201, has been marketed by du Pont under the name Antron. The cross-sections which are described in the specified patent resemble those of Fig. 128. The characteristics associated with Antron are drier handle reminiscent of the scroop of real silk, increased covering power, greater bulk, and a three-dimensional highlight effect. Because of the greater surface area of the filaments, Antron needs from 10 per cent (for light shades) to 25 per cent (for dark shades) more dyestuff than round nylon to give equality of shade. Doubtless for the same reason, fastness to washing at

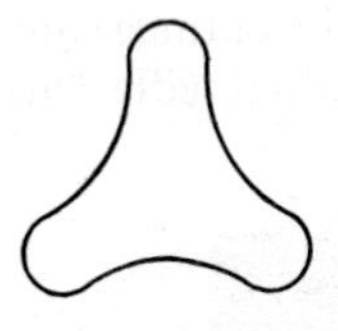
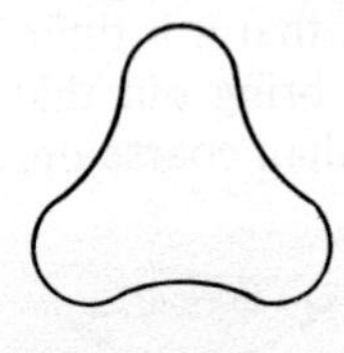

FIG. 128.—Two trilobal filament cross-sections described in U.S. Patent 2,939,201. Magnified several hundred times.

[*E. I. du Pont de Nemours & Co.*

temperatures higher than 60° C. is not so good on Antron as on round nylon.

Antron 24 is a textured Antron yarn which bulks on hot wet finishing and is supplied in heavy deniers for upholstery. Du Pont 501 carpet nylon staple is another variety; although not specifically called Antron, it has the same trilobal section filaments which are beautifully shown in Fig. 129. It is claimed that the carpets made from yarn of this type are more resistant to soiling than those made from ordinary circular filaments. Soiling is measured by the practical method of putting a piece of carpet on the floor and having an electric counter to count the number of times it is trodden on; then after say 8,000 treads the carpet is examined. Ideally it should be compared with a control carpet that has had just the same treatment; in the present case this would have been a similar carpet made from round nylon and subjected to 8,000 treads. Du Pont actually compared (U.S. Patent 2,939,201) the trodden special nylon carpet with untrodden (that is unsoiled) round nylon carpet, and it is impossible to say from this comparison how much (if any) improvement the novel cross-section has given in resistance to soiling. Du Pont also state that nylon yarn composed of filaments of this near-triangular section shows attractive lustre highlights which look well not only on the staple in the pile of a carpet but also in the continuous filament of a stocking.

These results are a little unexpected; the symmetrical trilobal filaments would, one might have thought, have been more inclined to enclose particles of dirt or soil between them than would circular section filaments; apparently they do not do this but the reason why is not clear. Furthermore lustre highlights have usually in the past been associated with unwanted flat cross-sections caused temporarily, perhaps only momentarily, by some irregularity in the spinning conditions—furthermore the highlights of "shiners" as they used to be called were considered an unmitigated nuisance. In

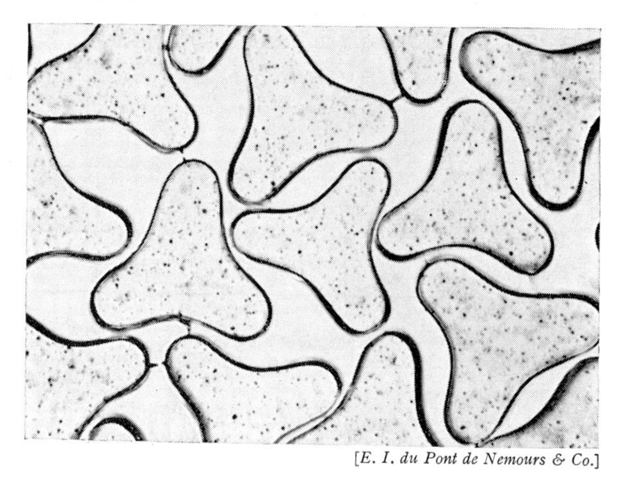

[*E. I. du Pont de Nemours & Co.*]

FIG. 129.—Photomicrograph of cross-section of trilobal nylon T-501. Magnification × 500.

the case of Antron the highlights are sufficiently uniform and appear with sufficient reliability and regularity to give a very attractive effect. But it is certain that very careful control will be necessary not only in spinning, but also in the subsequent textile processing to maintain uniformity of tension and so avoid irregularities of lustre. The production of Antron is a magnificent achievement and has brought the man-made fibre industry closer to the handle and appearance of real silk than ever before. And all this at the very reasonable cost of 10 cents per lb. more than for standard round nylon staple.

Cadon

This is also a multilobal nylon, one made by Chemstrand. It, too, has a silk-like handle and a built-in sparkle. The usual good qualities of nylon such as strength and durability are retained, as indeed they are in Antron. In light deniers, *e.g.*, 70, Cadon is used for printed taffetas for lingerie and blouses, the multilobal filaments assist in giving clear crisp prints. Heavy deniers are used for upholstery and carpets. Chemstrand's "Cumuloft" carpet nylon is a continuous filament textured multilobal nylon yarn and seems to have made a very successful start. The pile recovery of carpets made from this yarn is very good, as can be seen from Fig. 130. The 5 lb./sq. in.

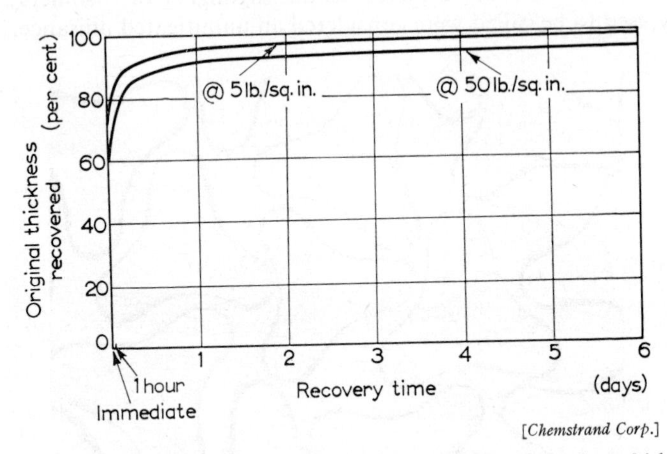

[*Chemstrand Corp.*]

FIG. 130.—Crush recovery of Cumuloft multilobal nylon carpet from 24 hours compression under 5 lb./sq. in. and 50 lb./sq. in.

load was to simulate the effect of light furniture imprints and the 50 lb./sq. in. load that of heavy furniture; if the resilience was assisted by subsequent vacuum cleaning then the recovery from the lighter load was 100 per cent after 24 hr.

FURTHER READING

U.S. Pat. (du Pont) 2,130,523, 2,130,947, 2,130,948, 2,163,636.
"Nylon Yarn : Properties and Processing", British Nylon Spinners, Ltd. (1944).
J. M. Matthews, "Textile Fibres", 6th Edn., London (1954).
Elastic nylon. British Pat. 582,518, 582,520, 582,522, 583,014.
G. Loasby, *J. Text. Inst.* March 1943.
Optical properties. A. Frey-Wyssling, *Nature*, **145**, 82 (1940).
Coffman, Berchet, Peterson and Spanagel (an account of some of the early work on polyamides), *J. Polymer Science*, **2**, 305 (1947).
W. H. Carothers and J. W. Hill, *J. Am. Chem. Soc.*, **54**, 1559–87 (1932). (Three papers.)
These papers describe fibre-forming polyesters, polyamides and polyanhydrides, which were the forerunners of nylon. These papers are classics.
"Cake Dyeing for Nylon", *Skinner's Silk and Rayon Record*, **26**, 1235 (1952).
K. H. Inderforth, "Nylon Technology", London (1953).

The West Indies Chemical Works Ltd., Manchester 2, " The Dyeing of Nylon with Hematine Crystals ".

H. C. H. Talliss, " Developments in Nylon 66 and 610 Fibres ", *The Textile Institute Report of* 41*st Annual Conference*, 45–58 (1956).

J. E. Waltz and G. A. Taylor, " Determination of Molecular Weight of Nylon, " *Anal. Chem.*, **19**, 448 (1947).

" Procinyl Dyes for Synthetic Fibres ", I.C.I. Technical Leaflet (1959).

" Borax in Dyebaths for Nylon Carpeting ", *International Dyer*, 233, August (1965).

R. W. Moncrieff, " Covering that Lorry ", *Commercial Vehicles*, 28, 29, 31, August (1964).

R. W. Moncrieff, " Improving the Thermal Stability of Nylon Yarn ", *Text. Manufacturer*, 340–1, August (1963).

CHAPTER 20

NYLON 6, NYLON 7, NYLON 11

PERLON, CAPROLAN, KAPRON, ENANT, RILSAN

THE linear polymer made from the α,ω-six carbon amino-acid is known as nylon 6. The acid is 6-aminocaproic acid:

$$NH_2(CH_2)_5COOH$$

and the polymer, nylon 6, is

$$H[-NH(CH_2)_5CO-]_nOH$$

where n is about 200.

In 1899, the acid was prepared by S. Gabriel and T. A. Maass; they heated it and found that a tenacious hard material resulted. This was a lump of nylon 6, but they were too far ahead of their time and there was nothing to suggest to them that the material might have some practical applications.

Carothers investigated the fibre-forming possibilities of the polymer of amino-caproic acid as early as 1930; he had some difficulty with it and it was not until 1932 that he published a description of fibre-forming nylon 6 polymers. Carothers and du Pont a little later selected nylon 66 (from hexamethylene diamine and adipic acid) as the most promising fibre and nylon 6 was neglected for a while in America. But in Europe work on it was pushed forward and in 1937 Paul Schlack of I.G. Farbenindustrie at Berlin-Lichtenberg polymerised *caprolactam* to obtain a polyamide that was similar in many ways to du Pont's nylon 66. The polymer, nylon 6, was spun into a fibre in several of the I.G. factories and was called Perlon L; all these factories were situated in what is now the Eastern Zone of Germany and after the war they were all dismantled and their technical staffs dispersed. But a few of the people re-assembled at I.G.'s Bobingen factory in Western Germany and by 1948 were again producing Perlon. Because Bobingen has been merged with Farbwerke Hoechst A.G., much of Germany's nylon 6 production is now called Perlon Hoechst. Another large part of it is made by Bayer and is known as Bayer-Perlon. Strictly, the word " Perlon " is the trademark of the nylon 6 fibre produced by members of the Perlon Warenzeichenverband.

Incidentally, it may be noted that during the war small quantities of nylon 66 were made in Germany under the name Perlon T, and furthermore that a polyurethane fibre (p. 441) was made and called Perlon U. But for a long time now, Perlon—the word without a suffix—has been used to denote nylon 6.

All over the world nylon 6 is made: Allied Chemical in America call it Caprolan, in Russia it is called Kapron, in Japan it is called Amilan (Toyo) or often just nylon. These names represent the products of but three of the many, perhaps fifty, manufacturers who make nylon 6.

Growth of Nylon 6

In America du Pont and Chemstrand have concentrated on nylon 66; so have B.N.S. in Britain. The enormous production of this fibre has overshadowed the world production of nylon 6. Furthermore, whereas nylon 66 melts at about 260° C., nylon 6 melts at about 215–220° C., which puts it at a quite serious disadvantage. Why, then, was nylon 6 manufactured at all? Firstly, those organisations in Germany which had developed the fibre would naturally tend to produce it, rather than switch to du Pont's nylon 66. Secondly, the process for nylon 6 is in some respects simpler than that for nylon 66; Schlack's discovery that nylon 6 could be made simply by heating caprolactam made a big difference because caprolactam can itself be made from *cyclo*hexanone and hydroxylamine which are both easy to make. Thirdly, there has always been a recognition of the superior resistance to light degradation of nylon 6. Fourthly, there has been a growing appreciation that its lower melting point may not make nylon 6 less suitable than nylon 66 for tyre cords; there is no likelihood of the temperature of a tyre exceeding 130° C. in use, and at temperatures such as this the resistance of nylon 6 to degradation is actually superior to that of nylon 66. Even in America, birthplace of nylon 66 (and indeed of all nylon), nylon 6 is advancing; at the end of 1961 five manufacturers there were making a total of about 50 million lb. a year, about 12 per cent of the U.S. production of nylon 66. Yet the fibre was not made in the U.S.A. at all before 1954.

A new factor which may push nylon 6 in future to the disadvantage of nylon 66 is the discovery by Snia Viscosa of a new and cheaper method of making caprolactam from toluene; this process has already been licensed to some spinners.

Nylon 6 has demonstrable advantages over nylon 66 in respect of dyeability, elastic recovery, fatigue resistance and thermal stability. In the U.K. British Enkalon now produce it in Northern Ireland, Courtaulds have built a 20 million lb. a year plant at Spondon, and I.C.I. have in production a 30 million lb. plant at Gloucester. It was at one time reported that I.C.I. were to use a Soviet process for their caprolactam but it appears now that both the Soviet and I.C.I. are to use a Swiss (Emser Werke/Inventa) process.

Manufacture

The reactions by which nylon 6 has been made are as follows:

From Benzene through Phenol. Coal is the raw material, tar is obtained from it and on distillation one of the fractions obtained is benzene. The benzene is chlorinated to give monochlorbenzene:

$$C_6H_6 + Cl_2 \longrightarrow C_6H_5Cl + HCl$$

Benzene. Chlorbenzene.

The chlorbenzene is treated with caustic soda to yield sodium phenate:

$$C_6H_5Cl + 2NaOH \longrightarrow C_6H_5ONa + NaCl$$

Sodium phenate.

and on acidification, phenol results

$$C_6H_5ONa + HCl \longrightarrow C_6H_5OH + NaCl$$

Phenol.

The phenol is reduced with hydrogen under pressure in an autoclave using nickel as a catalyst to give *cyclo*hexanol

$$C_6H_5OH + 3H_2 \longrightarrow \begin{array}{c} OH \\ | \\ CH \\ / \quad \backslash \\ CH_2 \quad CH_2 \\ | \qquad | \\ CH_2 \quad CH_2 \\ \backslash \quad / \\ CH_2 \end{array}$$

Phenol. *Cyclo*hexanol.

The *cyclo*hexanol is purified by distillation and then partially dehydrogenated using copper as a catalyst to the ketone *cyclo*-hexanone

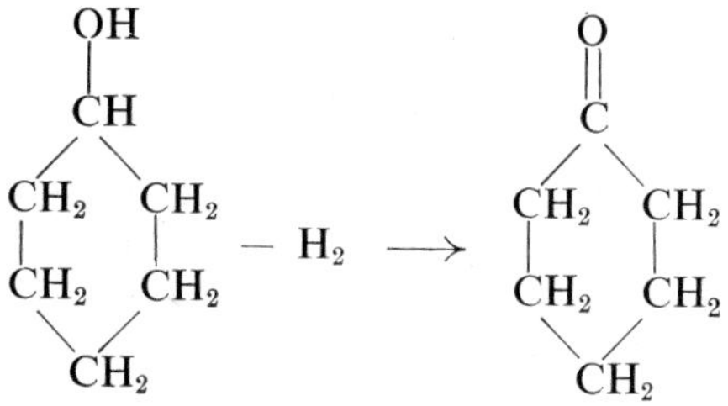

$$\begin{array}{c}\text{OH}\\ |\\ \text{CH}\\ / \quad \backslash\\ \text{CH}_2 \quad \text{CH}_2\\ | \qquad |\\ \text{CH}_2 \quad \text{CH}_2\\ \backslash \quad /\\ \text{CH}_2\end{array} - H_2 \longrightarrow \begin{array}{c}\text{O}\\ \|\\ \text{C}\\ / \quad \backslash\\ \text{CH}_2 \quad \text{CH}_2\\ | \qquad |\\ \text{CH}_2 \quad \text{CH}_2\\ \backslash \quad /\\ \text{CH}_2\end{array}$$

*Cyclo*hexanone.

Hydroxylamine which is produced from ammoniacal liquor, also derived from the coal tar, is then reacted with the *cyclo*hexanone to yield *cyclo*hexanone oxime. The hydroxylamine is used as the sulphate ($NH_2OH{\cdot}H_2SO_4$) in aqueous solution at 20° C. As it reacts with the *cyclo*hexanone, sulphuric acid is liberated and ammonia is passed in to neutralise the acid, the temperature rising to 90° C. After settling, the organic phase is crude *cyclo*hexanone oxime.

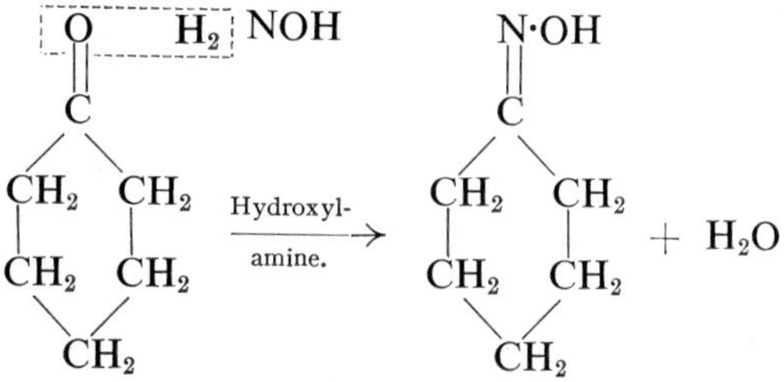

$$\begin{array}{c}\boxed{\text{O} \qquad \text{H}_2}\ \text{NOH}\\ \|\\ \text{C}\\ / \quad \backslash\\ \text{CH}_2 \quad \text{CH}_2\\ | \qquad |\\ \text{CH}_2 \quad \text{CH}_2\\ \backslash \quad /\\ \text{CH}_2\end{array} \xrightarrow[\text{amine.}]{\text{Hydroxyl-}} \begin{array}{c}\text{N}{\cdot}\text{OH}\\ \|\\ \text{C}\\ / \quad \backslash\\ \text{CH}_2 \quad \text{CH}_2\\ | \qquad |\\ \text{CH}_2 \quad \text{CH}_2\\ \backslash \quad /\\ \text{CH}_2\end{array} + H_2O$$

*Cyclo*hexanone. *Cyclo*hexanone oxime.

The oxime is then caused to undergo the Beckmann transformation by treatment with sulphuric acid which converts it into caprolactam:

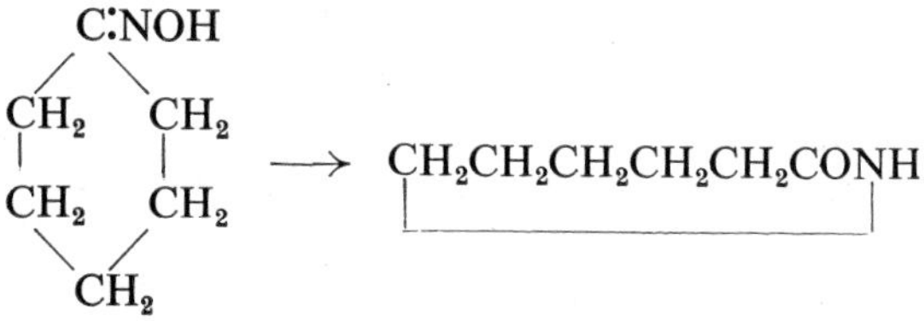

$$\begin{array}{c}\text{C:NOH}\\ / \quad \backslash\\ \text{CH}_2 \quad \text{CH}_2\\ | \qquad |\\ \text{CH}_2 \quad \text{CH}_2\\ \backslash \quad /\\ \text{CH}_2\end{array} \longrightarrow \underbrace{\text{CH}_2\text{CH}_2\text{CH}_2\text{CH}_2\text{CH}_2\text{CONH}}$$

*Cyclo*hexanone oxime. Caprolactam.

When the rearrangement is complete, the sulphuric acid is washed out, ammonia is added to it and the resulting ammonium sulphate is sold as a fertiliser.

The impure lactam is purified by distillation and is allowed to crystallise.

Synthesis from Aniline. A synthesis that has been used in Germany and in Russia is of caprolactam from aniline; it has the advantage that it does not require the use of hydroxylamine. Aniline is hydrogenated at 280° C. and 100 atmospheres pressure in the presence of cobalt and calcium oxides which act as catalysts; there is a 93 per cent yield of *cyclo*hexylamine. Treatment with hydrogen peroxide converts this to *cyclo*hexanone oxime, first as an addition compound with the peroxide; the oxime is liberated from this by treatment with a 2 per cent solution of ammonium tungstate in water. Then the oxime undergoes the Beckmann re-arrangement with conc. sulphuric acid at 130° C. The main stages of the synthesis may be represented:

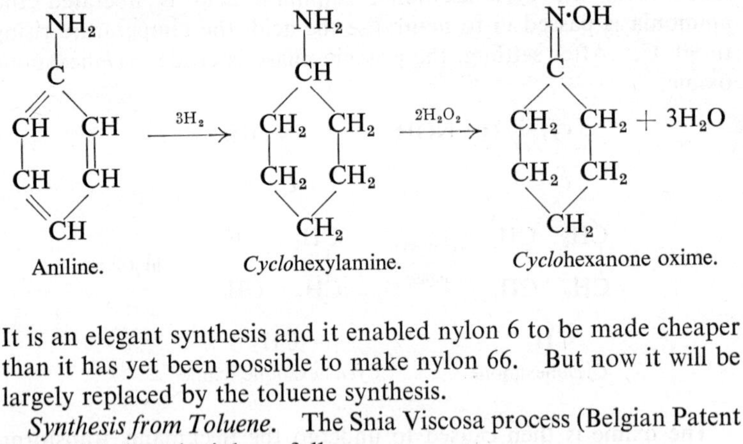

$$
\begin{array}{ccccc}
\begin{array}{c} NH_2 \\ | \\ C \\ CH \quad CH \\ CH \quad CH \\ CH \end{array}
& \xrightarrow{3H_2} &
\begin{array}{c} NH_2 \\ | \\ CH \\ CH_2 \quad CH_2 \\ CH_2 \quad CH_2 \\ CH_2 \end{array}
& \xrightarrow{2H_2O_2} &
\begin{array}{c} N{\cdot}OH \\ \| \\ C \\ CH_2 \quad CH_2 \\ CH_2 \quad CH_2 \\ CH_2 \end{array} + 3H_2O
\end{array}
$$

Aniline. *Cyclo*hexylamine. *Cyclo*hexanone oxime.

It is an elegant synthesis and it enabled nylon 6 to be made cheaper than it has yet been possible to make nylon 66. But now it will be largely replaced by the toluene synthesis.

Synthesis from Toluene. The Snia Viscosa process (Belgian Patent 582,793) is in principle as follows:

The starting material is toluene, which is a product of tar distillation and which nowadays is also one of the products of an oil refinery—a petrochemical. The toluene is oxidised to benzoic acid; this is hydrogenated over a platinum or palladium catalyst to hexahydrobenzoic acid. This reduced acid is then reacted with nitrosyl sulphuric acid ($HNO_2 + SO_3$ or $HOSO_2ONO$) at 60° C. in the presence of oleum (sulphuric acid enriched with SO_3); it gives a 90 per cent yield of caprolactam. The first stages of the synthesis are orthodox:

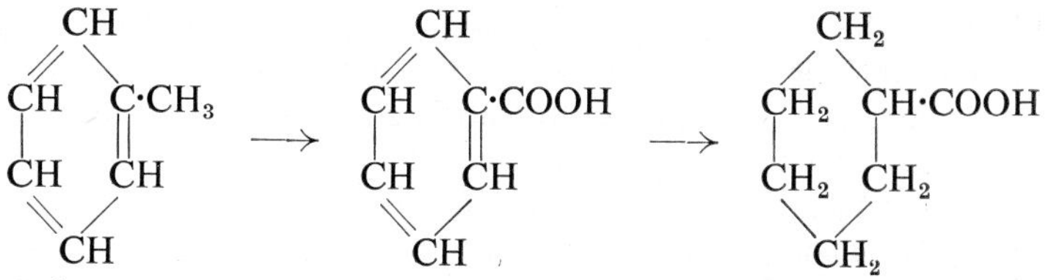

```
      CH                      CH                       CH2
    //  \                   //  \                    /    \
  CH    C·CH3             CH    C·COOH             CH2    CH·COOH
  |     ||        ——>     |     ||        ——>      |      |
  CH    CH                CH    CH                 CH2    CH2
    \\  /                   \\  /                    \    /
      CH                      CH                       CH2
   Toluene.              Benzoic acid.        Hexahydrobenzoic acid.
```

The conversion of hexahydrobenzoic acid to caprolactam is novel; it can be written:

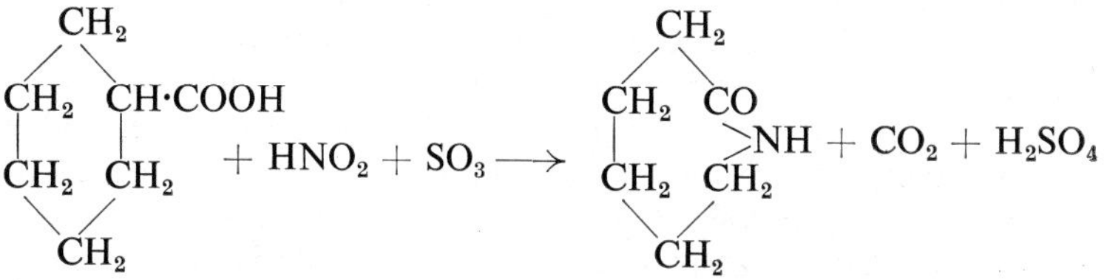

```
      CH2                                        CH2
    /    \                                     /    \
  CH2    CH·COOH                             CH2    CO
  |      |                                   |        >NH + CO2 + H2SO4
  CH2    CH2    + HNO2 + SO3 ——>             CH2    CH2
    \    /                                     \    /
      CH2                                        CH2
                                             Caprolactam.
```

Courtaulds making the toluene in their petrochemicals plant at Spondon are preparing to use this process, so are Allied Chemical in America. It is a process which could bring the price of nylon 6 down to be competitive on a weight basis with "super" high tenacity cellulosic fibres, and consequently very much cheaper than them on a performance basis in tyre cord. Only 1·1 lb. caprolactam is needed in practice to produce 1 lb. nylon 6.

Syntheses from Cyclo*hexane*. *Cyclo*hexane is easily obtainable by catalytic pressure hydrogenation of benzene. In the Toray process which is used in Japan the action of light and hydrogen chloride is used to convert *cyclo*hexane and nitrosyl chloride into *cyclo*hexanone oxime hydrochloride, thus:

$$C_6H_{12} + NOCl \xrightarrow[HCl]{light} C_6H_{10}NOH{\cdot}2HCl$$

Then the *cyclo*hexanone oxime is caused to undergo the Beckman rearrangement to yield caprolactam:

```
                H2SO4
C6H10NOH.2HCl  ————>  CH2CH2CH2CH2CH2CO + 2HCl
                      |________NH_______|
```

The yield in the first stage of the process is 86 per cent by weight; the second stage is commonly practised in other syntheses and gives a high yield. The light which is used to promote the first stage must not contain any component of wavelength less than 3,650 Å, otherwise some tarry material, which is useless and wasteful, is produced.

Light of this wavelength is ultra-violet; nothing of wavelength shorter than 3,930 Å (violet) is visible. The Toray process is known as the PNC (short for photonitrosation of *cyclo*hexane) process.

Polymerisation. Two alternative methods are used:

1. The lactam is liquefied, filtered and heated in an autoclave under high pressure and about 200 of the small monomeric molecules unite to give one large polymeric molecule of Perlon $H[-NH-(CH_2)_5CO-]_{200}OH$. Although the number 200 is assigned to this molecule this at the best represents a number average and actually the polymer will consist of molecules of different lengths. The polymer, Perlon, is known as " nylon 6 " because each repeating unit contains six carbon atoms.
2. The lactam has 10 per cent of its weight of water added, and the polymerisation is carried out at a high temperature with a controlled escape of steam. This process takes longer than the first but is easier to control in that there is less chance of locally overheating and so spoiling the molten material during the early stages of polymerisation. The caprolactam, it will be noted, behaves on being heated as if it were ε-aminocaproic acid and this is the fundamental basis of the process.

Whichever method of polymerisation is used, some unchanged monomer remains and this is washed out with water in an extractor, otherwise it would weaken and spoil the final fibre.

Spinning. The washed and dried polymer is melted to a clear liquid. It is one of the advantages of nylon 6 that because its melting point is lower than that of nylon 66 it can be melt-spun at a lower temperature. It is metered through pumps to the spinning orifices and is spun straight into atmosphere; the spinning speed can be as high as 1,000 metres/min. The polymer freezes at once in the cold of the atmosphere and the fibres which result from the freezing are passed round two rollers; the first applies water and a wetting agent and the second an oil–water emulsion. This conditions the yarn.

The method used for nylon 66 of passing the yarn as it is spun through a steam chamber is not suitable for Perlon (nylon 6) because there is such a relatively high concentration of monomer in the Perlon that the filaments of the yarn would become sticky if steamed and liable to adhere either to the walls of the vessel or to each other. Next, the yarn is stretched to about five times its original length to orient the constituent linear molecules and thereby to make the filaments strong, supple and unshrinkable (*i.e.*, the stretch is irreversible). The yarn is washed again with water to remove any low polymer, dried and cone wound. If staple fibre is required instead of continuous filament, then instead of being cone wound the yarn is crimped, cut and baled. A part of the process—from the molten lactam to the staple—is shown in Fig. 131.

Continuous polymerisation processes have been developed in recent years: molten caprolactam containing titanium dioxide delustrant and some acetic acid as stabiliser is passed through tubes about 50–100 feet long heated at 260° C. and the polymer that emerges continuously is ready to be spun. The practical difficulty is to secure uniformity of heating; the advantages are higher and cheaper production.

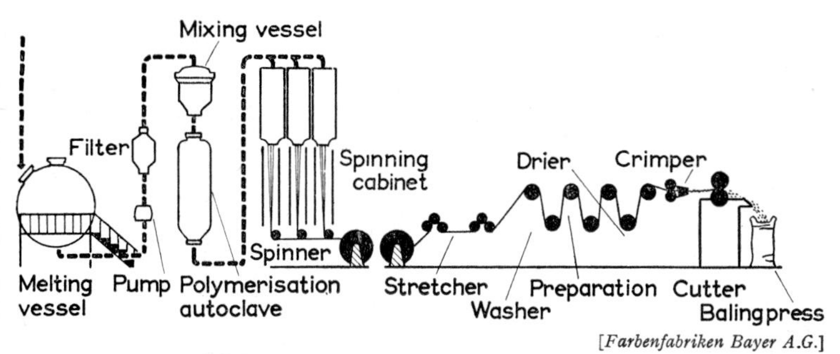

[*Farbenfabriken Bayer A.G.*]

FIG. 131.—Perlon manufacture from molten lactam to staple fibre.

Vickers-Zimmer Process. This is the Consumer–Synthesist Process, the do-it-yourself job (*cf.* p. 64). There is a variety of processes available for nylon 6. If the spun yarn is to be used for industrial or technical purposes, such as tyre cord or carpet yarn, then the process can be shortened by feeding the polymer straight to spinning without intermediate formation and isolation of "chip".

As an example, the process for the production of nylon 6 carpet yarn consists of the following stages:

(*a*) Solid caprolactam in the form of powder or flakes is melted in the melter, 1 (Fig. 132).

(*b*) The molten caprolactam is fed batchwise into the mixer, 2, where the catalyst and the stabiliser are mixed in.

(*c*) The mix is drained through a filter into an intermediate vessel, 3, to a dosing and mixing system, 4, where the delustrant, normally titanium dioxide, is added.

(*d*) In vessel 5 polymerisation takes place continuously under pressure, the polymer being raw nylon, 6.

(*e*) The polymer passes to vessel 6, where polymerisation is continued, but under atmospheric pressure. This is an equalisation stage and improves the uniformity of the product.

(*f*) The molten polymer is transferred continuously to the vacuum stage in vessels 7 and 8 to remove by distillation any incompletely polymerised and consequently low-molecular-weight polymer.

(*g*) The melt from the vacuum stage enters the spinning plant through a distribution system, 9, and is spun directly.

The elimination of the chip isolation cuts down the cost of running the process. The capital cost for a plant including buildings which

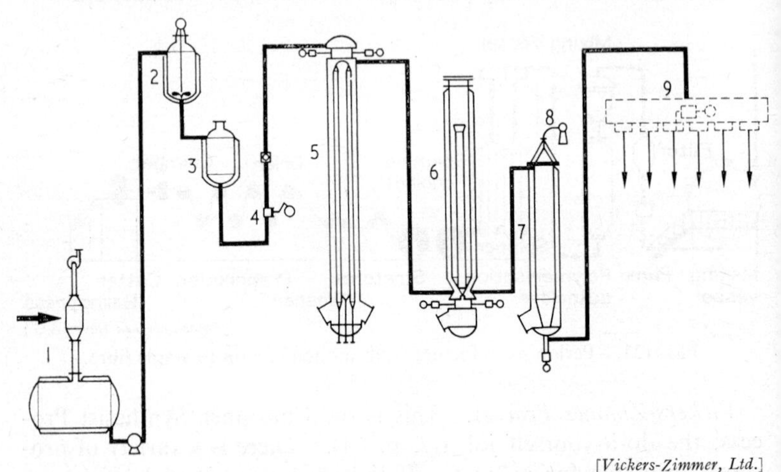

[*Vickers-Zimmer, Ltd.*]

FIG. 132.—Vickers-Zimmer plant flow sheet for conversion of caprolactam into unpurified spinning melt.

working continuously could produce 4·75 million lb. of carpet yarn per year, *i.e.*, roughly six tons a day, would be £1·6 million. The cost of producing the yarn would be somewhat as follows:

	(*d.*/lb.).
Depreciation	8·0
Maintenance	8·0
Personnel	7·6
Raw materials	31·9
Power, oil, water, nitrogen, hydrogen . .	5·5
	61·0

The biggest single item is naturally the caprolactam, which accounts for more than 97 per cent of the raw material costs: it is costed at 30·5*d.*/lb. Its polymerisation and conversion into carpet yarn costs as much again, so that the final cost is 61*d.*/lb. The personnel item (7·6*d.*/lb. of product) includes labour for the production of yarn only, and does not include labour for any textile processes that may follow the spinning operation.

The specification that the caprolactam which is used for the synthesis should meet is:

Melting point	68·5° C.
Permanganate number	More than 950
Volatiles	Less than 1 c.c. $N/10$ HCl per 20 gm. lactam
Moisture	Less than 0·15 per cent
Iron content	Less than 10 p.p.m.

The specification that the product, the polycaprolactam or nylon 6, should meet is:

Relative viscosity	2·5–3·2
Moisture	Less than 0·07 per cent
Extractables	Less than 3 per cent

The process described above consists essentially of continuous polymerisation at elevated pressure, of vacuum treatment to remove impurities and of direct spinning.

Vickers-Zimmer Process with Chip Isolation. But if the yarn to be spun is intended for some demanding end-use, such as continuous filament for textile processing, then it will be preferable to use a modified process, one in which the polymer is isolated intermediately as chip which can be extracted to remove unchanged monomer and oligomers, *i.e.*, low polymers consisting of only a few molecules of monomer strung together and still sufficiently soluble to be extractable. The chip can thus not only be thoroughly purified but can also be completely dried. Essentially, this whole process consists of the continuous polymerisation of caprolactam at atmospheric pressure, chip production, extraction and drying to provide a product suitable for spinning. The stages in the process are:

(*a*) Solid caprolactam powder or flakes is melted in the melter, 1 (Fig. 133).

(*b*) The melt is fed batchwise to the mixer, 2, where catalyst and stabiliser are mixed in.

(*c*) The mix is drained through a filter into an intermediate vessel, 3, to a dosing and mixing system, 4, where delustrant is added.

(*N.B.* So far the process is just the same as in that described above, but now comes a difference.)

(*d*) The lactam is passed continuously to reactor 5, where it polymerises continuously at *atmospheric* pressure under a blanket of nitrogen to nylon 6. The polymerisation is stimulated by heat and by the presence of the catalyst which was mixed in earlier.

(*e*) The melt is pumped to a chip-spinning head, 6, and is pumped through jets, from which it emerges like spaghetti. The " spaghetti " is cooled in a cooling vat, 7, and is then cut to chip of uniform length in a cutter, 8.

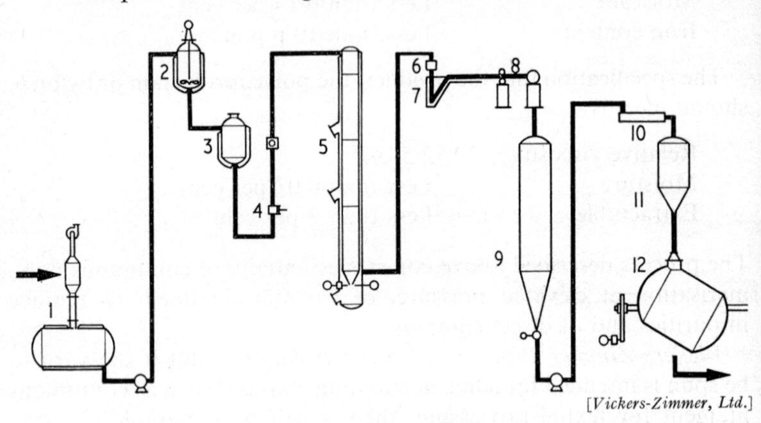

FIG. 133.—Vickers-Zimmer plant flow sheet for conversion of caprolactam into purified chip polymer.

(*f*) The chip falls into an extractor, 9, where any unchanged monomer and oligomers are washed out with demineralised water.

(*g*) The washed chip passes through a separator, 10, into a drip pan, 11, and is dried under vacuum in the tilted drum drier, 12. It is then ready for spinning.

The specification for the raw material, the caprolactam, should be the same as adduced earlier (p. 363). The product, the polymer, should have these properties:

Relative viscosity	2·2–2·6
Moisture	Less than 0·08 per cent
Extractables	Less than 0·7 per cent
Chip dimensions	Diameter, 1·5–3 mm.
	Length, 2–4 mm.

It will be appreciated that isolation of the polymer in the solid chip form at an intermediate stage increases the cost of the process, compared with that described earlier.

Properties

The main difference between this polymer (nylon 6) and nylon 66 is that it has a much lower melting point; this is a serious disadvantage, as any garments made from it must be ironed with considerable care. Stockings, of course, do not need ironing, and for some other garments Bayer's have put the question " to iron or not ". The truth of the matter is that all woven goods are much better for a light ironing, and most knitted goods, except hose, are so too, and that although some people maintain that ironing is not necessary with some of the synthetic fibres, its omission leaves some crumpling, even if it is only very light, and a lack of smartness.

What has Perlon got to offset this quite serious disadvantage of low melting point? There are two things that merit consideration:

1. The synthesis of caprolactam is easier than that of hexamethylene diamine used in nylon 66; in particular the high pressure catalytic reduction of nitrile to diamine is avoided; probably it is cheaper to make Perlon than nylon 66.
2. The affinity of Perlon for acid dyestuffs seems to be greater than that of nylon 66, due to the greater number of amino end-groups in Perlon than in nylon molecules; the more amino groups there are, the better the affinity of the fibre for acid dyestuffs.

Strength and Extensibility. The strength of nylon 6 can be varied at will up to about 8 grams per denier. Such very high strength is characteristic of yarn for industrial use, with an elongation at break of 16–20 per cent. In yarn for apparel uses where softness of handle is important, representative figures are 5 grams per denier tenacity and 30 per cent elongation. The industrial yarns have been subjected to greater stretch in the drawing process than the apparel yarns.

Specific Gravity. The fibre is light with a specific gravity of 1·14 (wool is 1·31, nylon is 1·14, most of the acrylics are 1·14–1·19).

Moisture Regain. Moisture regain is about 4 per cent, very similar to that of nylon and because nylon with its relatively high moisture regain is probably the best yet of the synthetic fibres for purposes of apparel, Perlon, too, stands high in the list in this respect.

Swelling. Swelling is low; if Perlon is steeped in water and then centrifuged its volume has increased by 13–14 per cent compared with 40–45 per cent for cotton and 80–110 per cent for viscose rayon. It dries quickly.

Light Fastness. Similar to that of the natural fibres in strength retention when exposed to sunlight.

Biological resistance. Very good. When wool, cotton, real silk and Perlon were buried in soil, the wool and cotton were badly attacked after 10 days (Fig. 134) and rotten after a month (Fig. 135); real silk had lost half its strength after 1 month and all of it after 6 months, but Perlon retained more than 95 per cent of its strength after 6 months' burial—it was substantially unattacked (Fig. 136). No one wants to bury their clothing, but tests such as this do guarantee that Perlon clothing, normally used, will be absolutely immune to micro-biological attack.

Heat. Perlon can be heat set and thereby made dimensionally stable by exposure to a temperature of about 150° C. It will withstand a temperature of 100° C. for long periods without undergoing any change in shade. It can be ironed at 130–150° C., softens at 170–180° C. and melts at 215° C.

Chemical resistance. It is resistant to most organic chemicals such as benzene, chloroform, acetone, esters and ethers, but is dissolved by phenol, cresol and strong acids. Resistance to alkalis is generally good, and to acids fair; for example 10 per cent mineral acids at room temperature attack it slowly, 10 per cent acetic acid does not attack it, and glacial causes only slight damage, but does cause some shrinkage, again at room temperatures. Formic acid is a solvent for Perlon. Hot acids attack Perlon badly. Nylon 6 is not attacked by pure trichlorethylene or perchlorethylene (dry cleaning agents), but both these substances tend to decompose under the action of light or catalytically when distilled in the presence of metallic chlorides. This decomposition is accompanied by the liberation of hydrochloric acid and if the cleaning agent is strongly contaminated with hydrochloric acid it will degrade and weaken the fabric. If the dry cleaning machines are made from stainless (non-corrodible) material, there will be no formation of metallic chloride, no catalytic liberation of hydrochloric acid and no damage to the nylon 6 fibre. If the cleaning machines are not made of stainless material then perchlorethylene and trichlorethylene should not be used as cleaning agents for nylon 6.

Dyeing

Perlon L has a higher affinity for acid dyestuffs than has nylon 66; this is attributable to its greater number of free amino groups on the ends of the linear polymer molecules which can combine with acid dyestuffs and in effect act as sites for acid dye molecules. The more of these amino acid sites there are, the more easily will acid dyestuffs

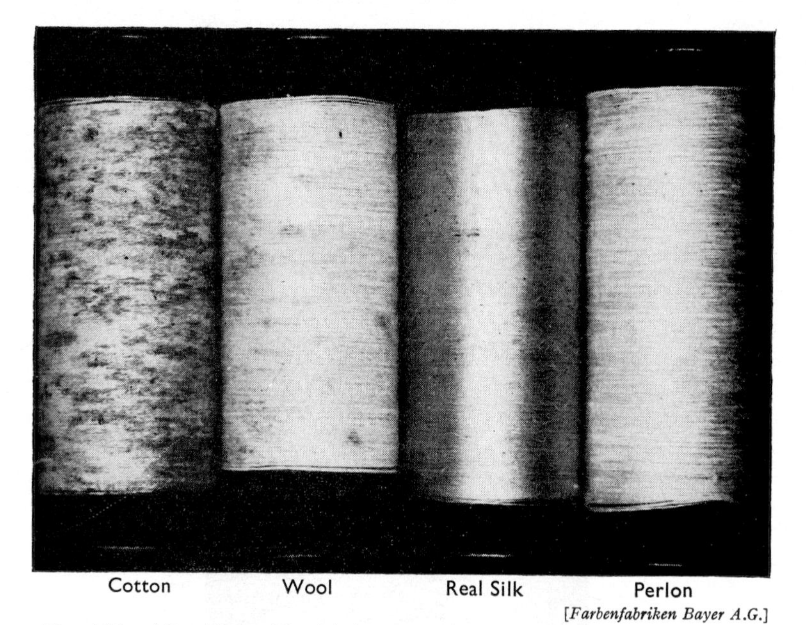

[*Farbenfabriken Bayer A.G.*]

FIG. 134.—After 10 days' burial the cotton is badly rotted, and the wool has been attacked, but the real silk and Perlon are all right.

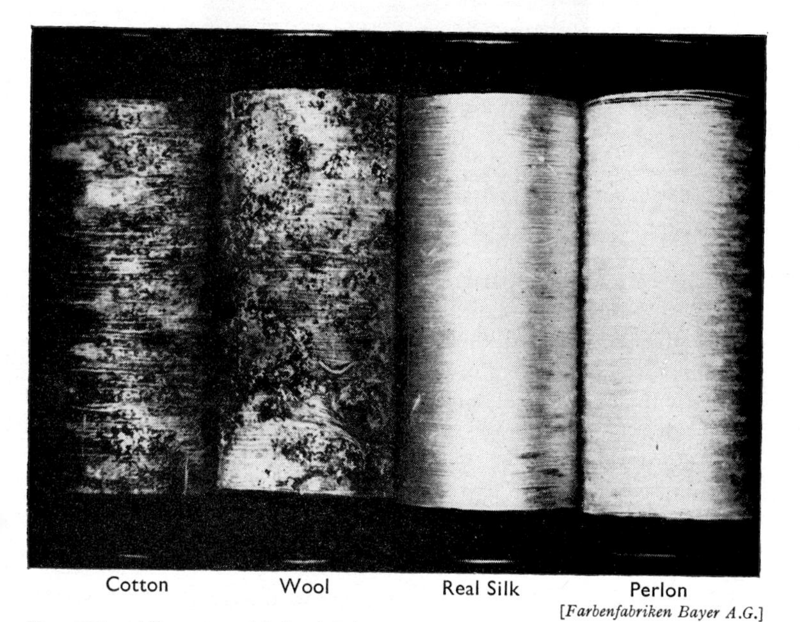

[*Farbenfabriken Bayer A.G.*]

FIG. 135.—After a month's burial the cotton and wool are quite rotten, and the real silk has been attacked but the Perlon is still all right.

dye on to the fibre. Whereas nylon (p. 47) has about 0·04 milliequivalents of free amine groups per gram of fibre, Perlon has about 0·098 milliequivalents. For comparison, the natural protein fibres silk and wool have corresponding values of about 0·20 and 0·80

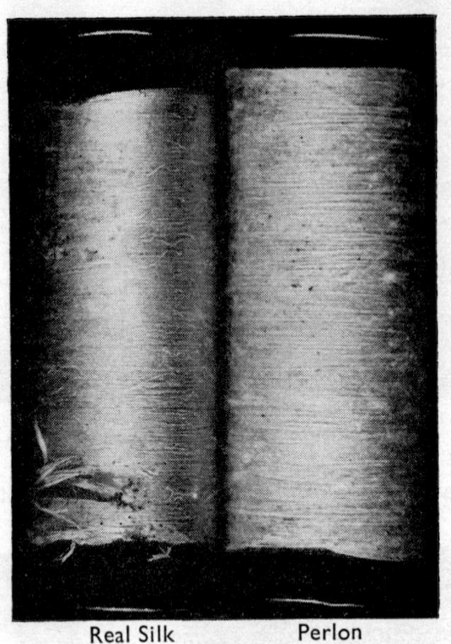

[*Farbenfabriken Bayer A.G.*]

FIG. 136.—After 6 months the cotton and wool are completely rotted away and the silk is rotten, but the Perlon is still all right.

respectively, and they, particularly wool, dye easily enough with acid dyes. Probably a value of about 0·1 milliequivalents per gram free amine groups will give an adequate number of dye sites at equilibrium; it should therefore be, as it is, easier to build up heavy shades with acid dyestuffs on Perlon than it is on nylon. So marked is the advantage of Perlon in this respect that Allied Chemical and Dye Corporation have publicised their caprolactam polymer Caprolan as the " deep dye nylon ".

Uses

Perlon L—or " Perlon " as it is now universally and simply called—has been made in great quantities and has found many uses. Some of these are as follows:

Tyre cord. This, the most eagerly sought fibre outlet of all, is falling to nylon. At first the higher melting point of nylon 66 led people to forget about the possibilities of nylon 6 for tyres. But

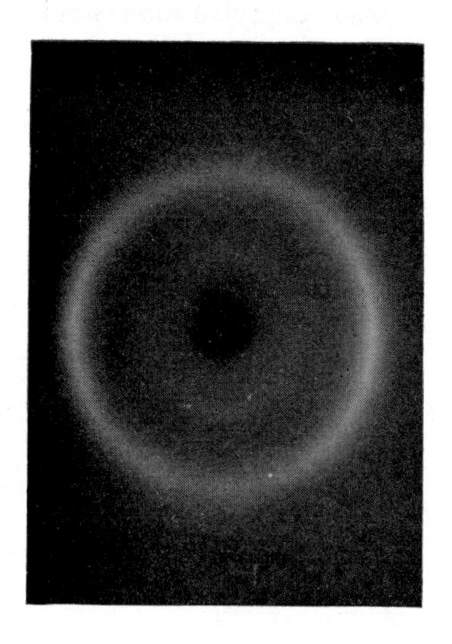

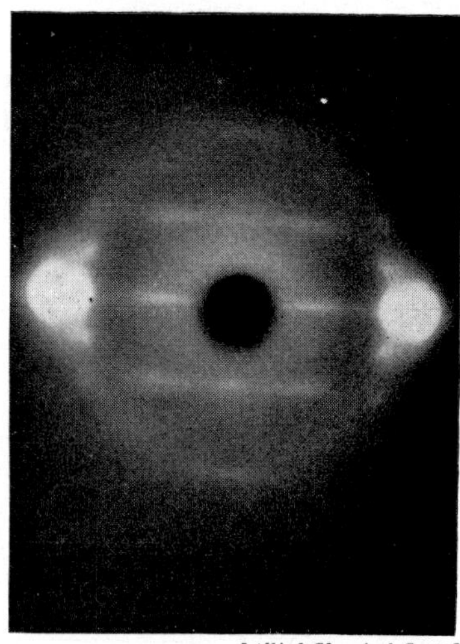

[*Allied Chemical Corp.*]

FIG. 137.—X-ray diffraction diagrams of nylon 6 tyre yarn; *left*, undrawn; *right*, drawn.

this outlook has quite changed. Firestone makes nylon 6 for its tyres at Hopewell, Virginia. Allied Chemical make Caprolan for tyre fabrics; their nylon that is intended for this purpose is a little different from their deep dye Caprolan; it is more highly drawn, stronger and perhaps different in the end groups it possesses. The

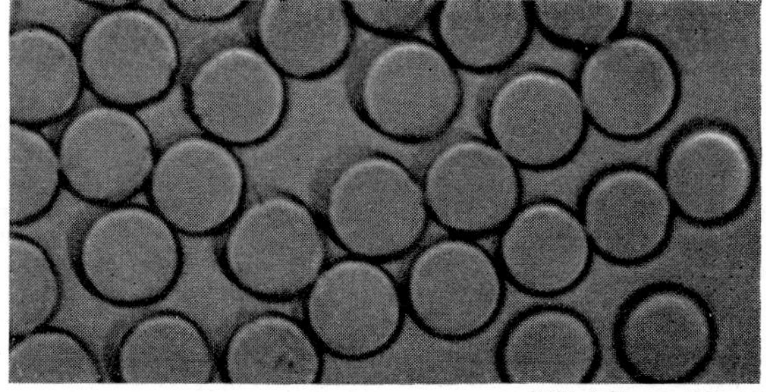

[*Allied Chemical Corp.*]

FIG. 138.—Photomicrograph showing cross-section of nylon 6 drawn tyre yarn. Magnification × 875.

high orientation of Caprolan tyre yarn can be judged from Fig. 137 which shows X-ray diffraction diagrams of undrawn and drawn Caprolan tyre yarn. Cross-section of the drawn tyre yarn is shown in Fig. 138. Those factors which have made Caprolan successful in tyre yarn are:

1. Good resistance to heat. High temperatures weaken it (Fig. 139) but its thermal stability is rather better than that of

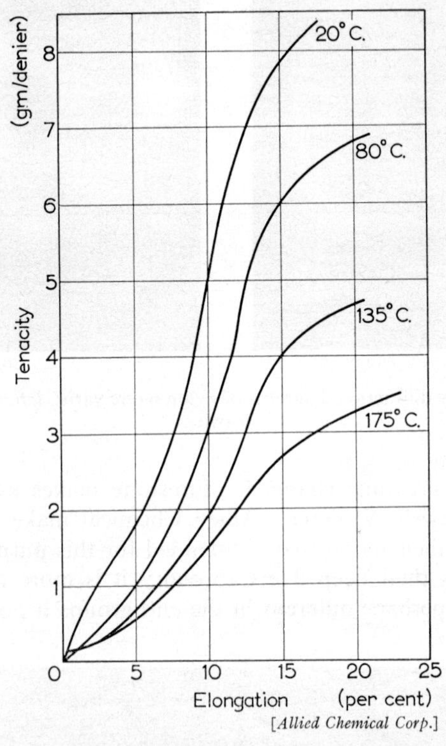

FIG. 139.—Stress–strain curves (Instron setting gauge 10 in., rate 100 per cent extension per minute) for nylon 6 Caprolan tyre yarn at different temperatures. The curves were determined with the yarn actually at the indicated temperatures. As tyres rarely run hotter than 130° C., it can be seen that a high proportion of the original strength will be retained under working conditions.

nylon 66, despite its lower melting point as can be seen from Fig. 140.

2. Good adhesion to rubber; when nylon 66 cord is pulled out of a tyre it comes out fairly easily but Caprolan cord requires

more force to pull it out and when it does come out appreciable quantities of rubber adhere to the cords.

3. Good resistance to flex fatigue; this varies with the rubber stock used but is usually rather better than the resistance of nylon 66 under similar conditions.

The main difficulty that nylon tyres, whether 6 or 66, have had to contend with has been " flat-spotting ". When, after running, a

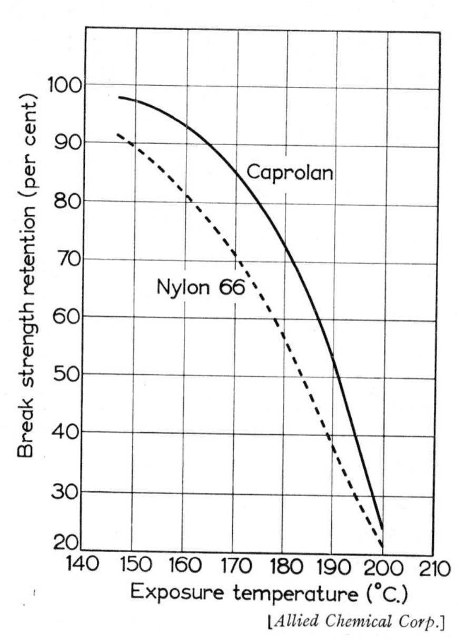

[*Allied Chemical Corp.*]

FIG. 140.—Nylon 6 and nylon 66 tyre yarns were heated to the " exposure temperature " for 77 hours in air, then cooled, conditioned and tested. The percentage of the strength retained is shown above. It can be seen that after being heated to temperatures at which nylon 6 weakens, it regains most of its strength on cooling, provided that the "exposure temperature" does not exceed 150° C. So, of course, does nylon 66 and the difference between the two nylons is only small.

car stands still the tyres, hot from the running, develop flat spots where they touch the road; when the car is re-started these flat spots persist for a little and cause the car to run unevenly, sometimes even to thump, for a few hundred yards. The answer to this seems to lie in improving the elasticity of the nylon when hot. Much has already been done by various stretching and annealing treatments to which the nylon is subjected, but the complete answer to the problem may lie in some chemical modification. Possibly the introduction of cross-linking, such as occurs in the wool molecule, into the nylon 6 polymer would be a fruitful approach; possibly the introduction of silicon atoms into the carbon backbone on the lines of already reported Russian work would be useful.

The battle between nylon and high tenacity viscose rayon for the tyre market began in 1947 when nylon first appeared as a tyre cord material for truck tyres. In the U.K., nylon's progress has not been fast for tyres, and to get a tyre with a nylon carcase still means a special order. But in America more than half the replacement tyres are built on nylon; tyres fitted to new cars are still mainly built on rayon. Nylon's share in that country of the car and truck tyre market has grown and rayon's has slumped.

Year.	Million lb. of tyre cord.	
	Nylon.	Rayon.
1955	50	400
1957	85	295
1959	125	290
1963	200	240*

* Estimated

In 1958 five leading rayon makers (American Viscose, American Enka, Courtaulds (Canada), Beaunit and Industrial Rayon) banded themselves together in a non-profit making organisation known as Tyrex Inc. to promote high tenacity rayon for tyres and their powerful influence was felt in this country. But irrespective of what influence there may be, the better fibre will win the field; so far nylon is doing well and it is highly significant that du Pont who make both viscose rayon and nylon are supporting nylon for this use. Prices have come down on both sides.

Although Allied Chemical's tyre yarn is nylon 6, all of du Pont's has been nylon 66. Du Pont's price for 840 denier nylon tyre cord has been as shown below. Some figures for super rayon are included for comparison.

Year.	840 denier nylon (price per lb.).	1,650 denier super rayon (price per lb.).
1958	$1·20	$0·55
1959 (Dec.)	$0·97	$0·50
1962 (April)	$0·92	$0·51
1967	$0·82	$0·47

Because nylon is stronger, less nylon can be used than rayon in car tyres, and a 1 : 1·7 ratio is used in calculating relative costs. Rayon at 47 cents is equivalent on a cost basis to nylon at 80 cents. Of late a good deal has been heard about polyesters for tyres. Polyester was never successful in replacing nylon stockings, because the knees became baggy; it seems likely the imperfect elasticity may make it unsuitable for tyres. But some manufacturers appear to be satisfied that polyester cord will be suitable for tyres (see p. 426). In 1967 polyester tyre cord was \$0·90 per lb. in the U.S.A. compared with nylon cord at \$0·82. Other fibres are being tried; polypropylene will be cheap and is very strong, but its liability to oxidative breakdown will be against it. The likelihood seems to be that nylon 6 will eventually get the bulk of the tyre cord trade and will keep it for a very long time. Several important manufacturers who have hitherto made only nylon 66, notably du Pont and British Nylon Spinners, are preparing to make nylon 6. The battle between nylon and super rayon has swayed back and forth with nylon gradually gaining ground, but late in 1961 this trend was reversed, and in the last quarter of that year super rayon pulled nylon back considerably in the tyre market. Early in 1962 British Nylon Spinners introduced a Type 900 specially for tyres, in 840 denier, with the very high tenacity of 8·8 grams per denier and 15 per cent extension at break.

It cannot be without significance that the U.S. production of high-tenacity rayon fell very greatly in 1967. From 1960 to 1965 it had been within the limits of 246–279 million lb., but in 1966 it fell to 236 and in 1967 to 161 million lb.; over the same period the U.S. production of synthetics went steadily ahead, every single year breaking the previous record.

Fishing lines. Monofilaments are used in diameters from 0·10 mm. to 1·50 mm. and with tensile strengths of 0·5 to 60 Kg.; they seem to behave rather better than these figures suggest, *e.g.*, a 12 Kg. carp was caught on a 0·20 mm. line (tensile strength rather less than 2 Kg.). Round, smooth, cross-sections and freedom from knots make unreeling smooth; strength is very useful; the line is almost invisible and so more deadly; it will not rot even if left wet.

Tow ropes. Bayer Perlon tow ropes have been used in many difficult salvage operations; one German tug has used a 4-in. diameter Perlon for four years and it is still serviceable. It is stronger than manila of similar size, and easier to work than thicker manila ropes; it does not get much heavier when wet, does not stiffen when exposed to temperatures below freezing point; it is resistant to bacteria and sea-water and can be stored wet; it has a long life. When used for towing its high extensibility gives an even pull because in the slacker moments the rope recovers its original length and it will absorb sudden shocks in its high extensibility. The tow is made from 20 denier filament Bayer Perlon, bright undyed, also from spun black. The low specific gravity and high strength of Perlon are advantageous; whereas a manila rope of 3-in. diameter weighs 8 lb/yd. and has a tensile strength of 30 tons and a low extensibility, a Perlon 3-in. rope weighs only 6 lb/yd. and has a tensile strength of 160 tons, and an extensibility of 30 per cent. Ropes made with Perlon are as white as the cotton rope material in use on luxury yachts.

Hose manufacture. Always a major market for polyamide yarns, the hosiery trade uses Perlon continuous filament mainly in 10, 15 and 20 deniers monofil and untwisted. The welt or reinforcement areas may be made from Perlon 30, 45 and 60 deniers multifilament. Perlon stockings have the good fit, sheerness, freedom from sag, and quick washing and drying properties.

Woven fabrics. Perlon continuous filament has been used for light fabrics such as 100 per cent Perlon georgettes. Other outlets which by now have become almost traditional for most of the synthetic fibres are: upholstery materials and carpets (made from the spun staple fibre), swim-wear, uniform cloths, protective clothing, flags and bunting, and filter cloths.

Knitting. Perlon monofil yarns in 10, 15 and 20 denier and multifilaments in 30, 45 and 60 deniers are used for tulles, blouse fabrics and shirtings.

Fibre. A point of special interest is that Bayer Perlon fibre is the first dope-dyed polyamide fibre. The colours are fast and stand up to steaming and permanent pleating treatments. For nearly all colours it is cheaper to buy the dope-dyed yarn than to buy white yarn and dye it fast; furthermore, the fastness of the dope-dyed material is nearly always superior. Shades available are peach, sky blue, grey, brown and black. The fibre is available in a wide range of deniers and staple lengths for spinning either alone or as a blend on the cotton, woollen, worsted and schappe systems. For carpet yarn spinning the heavy filament deniers of 12, 20 and 30 are available all in staple lengths of 60, 80, 100 and 120 mm.

Essentially the same fibre is made by the Industrial Rayon Corp. at Covington, by American Enka, by Allied Chemical and Dye Corp. at Hopewell; it is made by a great many firms on the Continent and is known as Perlon and Phrilon in Germany, Enkalon in Holland, Grilon and Mirlon in Switzerland and Capron (or Kapron) in the U.S.S.R. Société Rhodiaceta at Lyons make a very large quantity; like most manufacturers they produce a high-tenacity quality mainly for industrial use, and a normal (regular) quality for apparel. There is a considerable difference between them in physical properties as their characteristic curves in Fig. 141 show.

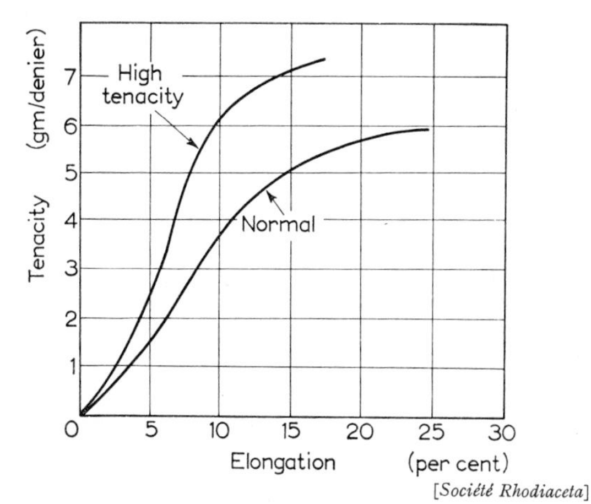

FIG. 141.—French nylon. Rhodiaceta make nylon 6 in high tenacity and normal kinds. This figure shows the characteristic curves of the two fibres.

BICOMPONENT NYLONS

Bicomponent or conjugate filaments, in connection with viscose rayon, have already been described on p. 183, and as shown in Fig. 71 two different viscoses are fed simultaneously through the same jet. A similar device has been used to make Orlon Sayelle, and the bicomponent structure of its filaments is shown in Fig. 183 on p. 516. The same principle, that of making a bicomponent filament from two different polymers, has also been applied to nylon. Whereas viscose rayon and acrylic fibres are spun from polymer solutions, the nylon is spun from molten polymers. The apparatus is accordingly a little different; one type of equipment which has been used successfully in Russia is shown in Fig. 142. The top part of the illustration shows the actual bicomponent thread formation, and the lower part shows the separate operation of hot-drawing to gain strength; the bobbin 4, which is the take-up package for the first operation, becomes the bobbin 8, which is the let-off package for the second operation.

One of the bicomponent fibres that the Russians describe is of nylon 6 and the equivalent of Terylene (polyethylene terephthalate), the two molten polymers converging on the same jet and a bicomponent filament emerging. Spinning conditions were: temperature 285° C. stretching at 90–130° C. at a speed of 37 m./min. with a stretch ratio of 4–5·5, followed by relaxation for 2 min. at 160–210° C. There can, however, be some difficulty with separation of the two components of the fibre in later textile processing. If this happens, then the crimp which it is hoped to obtain in finishing will be only poor.

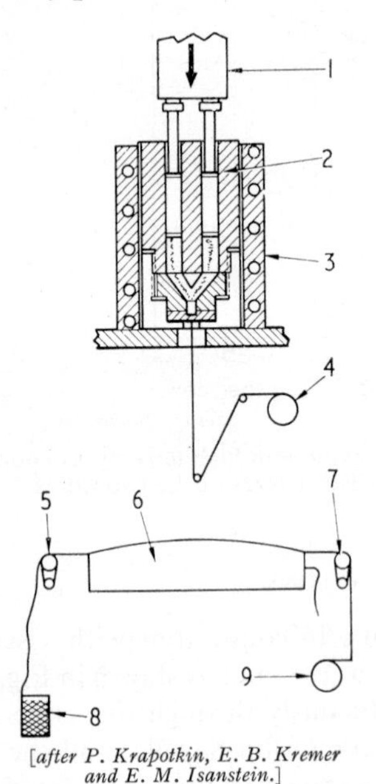

[*after P. Krapotkin, E. B. Kremer and E. M. Isanstein.*]

1. Press.
2. Spinning block with filter and spinnerette.
3. Electric heater.
4. Spinning take-up bobbin.
5. Thread guide and let-off control.
6. Hot iron.
7. Draw roller.
8. Bobbin 4 used as let-off in drawing operation.
9. Take-up of drawn thread.

FIG. 142.—Laboratory apparatus for spinning bicomponent filaments from two molten polymers.

The Dutch organisation N.V. Onderzoekingsinstituut have claimed, in B.P. 981003 (1965), that if two not very dissimilar polyamides are used as the two components, then separation does not subsequently occur. If, for example, the two components are: (1) nylon 6, and (2) a co-polymer of caprolactam and of the salt of hexamethylene diamine and terephthalic acid, then there is no separation, and subsequent crimping of the yarn made from the bicomponent threads is good. It will be seen that in this example molten nylon 6 constitutes one feed to the jet and molten co-polymer of the monomers of nylon 6 and nylon 6-T constitutes the other. Quite small differences between the two components will be enough to yield differential shrinking, *i.e.*, crimpability on wetting and drying, and the smaller these differences, the better will be the cohesiveness and strength of the filaments.

" Source " the new Allied Chemical fibre is a nylon filament with much finer polyester fibrils embedded in it and running along the length of the fibre. It is not a conjugate fibre and is not made by the method described above, but it is in the widest sense of the term a bicomponent fibre. It consists of 70 per cent nylon, 30 per cent polyester. It is strong with a tenacity of 8 gm./den.; it has a moisture content of 2·7 per cent, intermediate between nylon and polyester. To begin with it is being made into carpet yarns of 1,100 and 2,200 denier.

ENANT

It has been reported that nylon 7 called Enant is being developed in Russia and it has been claimed to be more stable to heat and ultra-violet light than nylon 6. Whereas nylon 6 (Capron) is made by polymerising caprolactam, the lactam of caproic acid, nylon 7 (Enant) is made from the lactam of heptanoic (or œnanthic or enanthic) acid:

$$ {}_{n}(CH)_6CONH \longrightarrow (-CH_2CH_2CH_2CH_2CH_2CH_2CONH-)_n $$

Lactam of enanthic acid. Enant.

The method that has been used to synthesise the lactam of heptanoic acid is of considerable interest, as it enables this intermediate to be made from abundant and cheap raw materials. These are ethylene and carbon tetrachloride, which will " telomerise " (R. M. Joyce, W. E. Hanford and J. Harmon, *J. Amer. Chem. Soc.*, **70**, 2529 (1948)) under the catalytic influence of benzoyl peroxide at pressures within the range 50–15,000 lb./in.^{2}. The telomerisation may be represented thus:

$$ nC_2H_4 + ClCCl_3 \longrightarrow Cl(C_2H_4)_nCCl_3 $$

When the pressure is about 1,500 lb./in.^{2}, the reaction products are those where $n = 1, 2, 3$ or 4. These differ in boiling point quite a lot and can be separated by fractional distillation. The product where $n = 3$ is $Cl(CH_2)_6CCl_3$, which boils at 143° C. at 20 mm. Hg. This αααω-tetrachloroheptane can be hydrolysed by aqueous sulphuric acid to yield ω-chloroheptanoic acid:

$$ Cl(CH_2)_6CCl_3 + 2H_2O \xrightarrow{H_2SO_4} Cl(CH_2)_6COOH + 3HCl $$

The ω-chloroheptanoic acid has been studied by A. N. Nesmeyanov, R. K. Freidlina and L. I. Zakharkin at the U.S.S.R. Academy of Sciences, Moscow (*Quart. Rev. Chem. Soc.*, **10**, 330–70 (1956)), and apparently they have found that it reacts with anhydrous ammonia at 230–250° C. (or its ethyl ester with alcoholic ammonia at 120–140° C.) to give a good yield of the lactam of heptanoic acid:

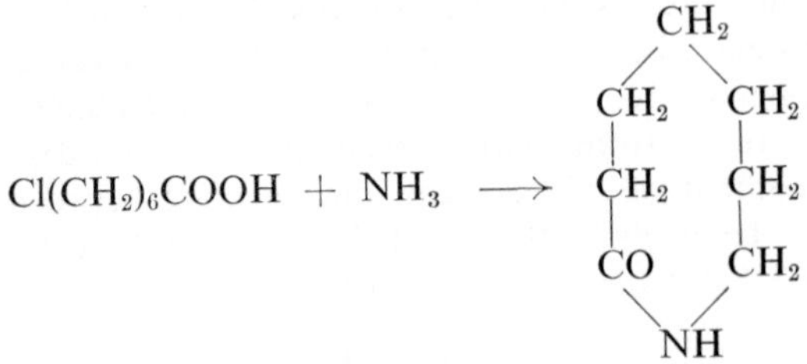

$$Cl(CH_2)_6COOH + NH_3 \longrightarrow \begin{array}{ccc} & CH_2 & \\ \diagup & & \diagdown \\ CH_2 & & CH_2 \\ | & & | \\ CH_2 & & CH_2 \\ | & & | \\ CO & & CH_2 \\ \diagdown & & \diagup \\ & NH & \end{array}$$

and this on polymerisation will give the polylactam of heptanoic (œnanthic) acid which is Enant.

Telomerisation. The word is used to describe a process in which polymers of *low* molecular weight are formed in a polymerising system by chain transfer from the solvent or added retarding agent.

In the ethylene–carbon tetrachloride system the ethylene if alone would polymerise almost indefinitely, but the carbon tetrachloride splits into –Cl and $-CCl_3$ radicals which add on to the ends of the polyethylene chain and stop further polymerisation. If the amount of carbon tetrachloride is considerable, the polymerisation will be stopped at an early stage and only two or three or four ethylene groups will add together.

RILSAN

Just as Perlon L is nylon 6, so Rilsan is nylon 11. The fibre has been developed in France, and the same material as a plastic under the name Rilsanite. Rilsan has two quite serious disadvantages compared with Perlon L, namely:

1. Its raw material is castor oil which being an agricultural product is more subject to fluctuation in availability and price than are materials made from coal or oil.

2. Its melting point is lower at 186–187° C. That of Perlon L is quite low enough (220° C.).

Quite clearly Rilsan must possess some advantages to offset these disadvantages. Possibly it is intended for use as bristles, and in that event its lower moisture regain than Perlon would be advantageous.

Manufacture

Castor oil is heated at 300° C. under reduced pressure and undecylenic acid can be steam distilled off. Hydrogen bromide is added on by treatment with a solution of it in the presence of a peroxy catalyst, to give ω-bromo-undecanoic acid and this is treated with ammonia to convert it to ω-amino-undecanoic acid. The acid is polymerised at a temperature of 215° C. to give the polymer. The stages may be represented:

$$\text{Castor Oil} \xrightarrow[\text{reduced pressure}]{300°\text{C.}} CH_2{:}CH(CH_2)_8COOH$$
Undecylenic acid.

$$\xrightarrow[+\text{ catalyst.}]{\text{HBr}} CH_2BrCH_2(CH_2)_8COOH$$
ω-Bromo-undecanoic acid.

$$\xrightarrow{NH_3} NH_2CH_2(CH_2)_9COOH$$
ω-Amino-undecanoic acid.

$$\xrightarrow{\text{Heat}} H[-NH(CH_2)_{10}CO-]_nOH$$
Rilsan.

The polymer is melt-spun into fibres.

Properties

Similar to Perlon except for a lower melting point and a lower moisture regain with accompanying greater difficulty of dyeing.

FURTHER READING

Combined Intelligence Objectives Sub-Committee Report, File No. XXXIII—50, H.M.S.O.
S. Gabriel and T. A. Maass, " Über ε-Amidocapronsaure ", *Ber.*, **32**, 1266–1272 (1899).
P. Schlack, U.S. Patent 2,241,321 (1941).
W. H. Carothers and J. W. Hill, *J. Am. Chem. Soc.*, **54**, 1266–1269 (1932).
Modern Textiles Mag., **40**, No. 6, 6 (1959), for nylon 4.
R. W. Moncrieff, "Fibres of the Future?, Nylons 1, 2 and 3", *The Dyer*, **129**, 221 (1963).
" Soviet Work on Nylon 6 Stabilisation ", *Skinner's Record*, 187, 189, March (1965) and 361–2, April (1965).
R. Graf *et al.* (Dimethyl Nylon 3), *Angewandte Chemie*, Internat. Edn., **1**, 481–8, (1962).

CHAPTER 21

AROMATIC POLYAMIDES

NYLON 6-T, NOMEX

DOZENS of nylons or polyamides which contain aromatic groups have been made experimentally, and a few of them have got as far as being extensively tested. But there are only two of them that need to be discussed in detail: nylon 6-T and Nomex.

NYLON 6-T

Reference has already been made (p. 376) to this polymer. It is half-aliphatic, half-aromatic, made by condensing hexamethylene diamine with terephthalic acid, one component from nylon 66 the other from Terylene. Structurally its synthesis may be represented as

$$n\mathrm{NH_2(CH_2)_6NH_2} + n\mathrm{HOCO}\!-\!\mathrm{C_6H_4}\!-\!\mathrm{COOH} \longrightarrow \left[-\mathrm{NH(CH_2)_6NHCO}\!-\!\mathrm{C_6H_4}\!-\!\mathrm{CO}-\right]_n + 2n\mathrm{H_2O}$$

One part of each recurring unit is aliphatic, the other is aromatic. Its outstanding property is its high melting point of around 370° C., compared with 263° C. for nylon 66. Its specific gravity of 1·21 is, as would be expected, intermediate between that of nylon 66 (1·14) and Terylene (1·38). Moisture regain at 4·5 per cent is the same as that of nylon 66; its tenacity is a shade lower and its extensibility a little higher than the corresponding values for nylon 66, but the differences are already small and could probably be much diminished by altering the stretching conditions. The biggest difference lies in the thermal resistance of nylon 6-T; its strength is unaffected after being kept at 185° C. for 5 hr. (nylon 66 is reduced by half) and even after 5 hr. at 220° C. it retains 60 per cent of its strength (nylon 66 has lost all its strength under these conditions). It also has better dimensional stability and is much less subject to flat-spotting in tyres than is nylon 66. These, then, are the differences from nylon 66: heavier, much more resistant to high temperatures, less tendency to creep and consequent flat-spotting. All told, a useful group of properties. Furthermore, hexamethylene diamine and terephthalic acid are already both made in enormous quantities and are relatively cheap. The properties described above are those of fibre which has been obtained by wet-spinning a solution of polymer in cold concentrated sulphuric acid.

Preparation

The formation of the polymer and its spinning have been described by Cipriani in U.S. Patent 3,227,793 (1966), as follows:

To a solution of 0·88 per cent by weight of hexamethylene diamine in 3 l. water there is added 2·5 gm. magnesia (an acid acceptor) and also a solution of 1·65 per cent by weight of terephthaloyl chloride in 3 l. xylene. The two immiscible parts are agitated for 30 min. The precipitate which forms is polyhexamethylene terephthalamide (inherent viscosity 1·3 measured in conc. sulphuric acid). It is filtered, washed and dried. The polymer is then dissolved at room temperature in 96 per cent sulphuric acid to yield a spinning solution of 14·9 per cent polymer. This is extruded through a jet of 0·1 mm. diameter into an open trough containing 47·5 per cent by weight sulphuric acid at 50° C. It is wound up at 54 m./min., washed and dried. The filaments are 3·2 denier, their tenacity is 2·0 gm./den. and their extension at break is 92 per cent. They can be drawn up to 500 per cent increase in length either hot or cold. If they are suitably drawn one can obtain yarn with a tenacity of 4·0 gm./den. and 18 per cent extension at break. This, or something very like it, represents the best way of making nylon 6-T.

Nomex

Nomex is a du Pont fibre; it is nylonic in type, a polyamide, but it stands out from the large number of nylons that have been made by virtue of one invaluable property. For practical purposes it is flameproof. If a single thread of Nomex is lit with a match it flames brightly for a few seconds, but then goes out. A yarn represents a very severe test, because there is only a little fibre and there is a lot of air surrounding it. If a flame is held underneath a piece of Nomex fabric, the fabric hardens as it starts to melt, discolours and chars. The fire risk is negligible, and this is of great value not only for space suits and so on but also for nursery nightwear. The " proof " is intrinsic in the fibre and is retained however many times a garment is washed or laundered or dry-cleaned; as long as there is any fabric left, it will be flameproof, and just as flameproof as when it was new. It is a stupendous achievement on the part of du Pont to have made such a fibre. The fibre is commercial and is being made and sold and can be bought by anybody.

Constitution

The constitution of Nomex has not been divulged by the makers. They do not now rely altogether on patent protection, nor on the advantage of a flying start (let those who can, catch up with us), but on secrecy. When Carothers, working for du Pont, made the first highly polymeric amides and esters his work, and that of his colleagues, was published in the *Journal of the American Chemical Society* and in various patents which described in detail the preparative methods for and the physical properties of a large number of new fibre-forming polymers. It was the first real synthesis of fibres, it had cost a fortune, but everything (or at any rate a very great deal including the underlying principles) was published. All could see, all could read of the work. It was much the same when C.P.A. made Terylene and du Pont made Orlon. Information, new knowledge was given freely to the world. The masters could afford to do it; it would take any copyists some time to catch up, and by that time the masters would have advanced further. There was, too, the protection afforded by good patent specifications. It must be added that patents often divulge more than they protect.

Nevertheless, it is fairly widely believed today that Nomex is *N,N′-m*-phenylene-bis-*m*-aminobenzoamide or more simply poly-(*m*-phenylene isophthalamide). It is apparently a condensation product of *m*-phenylene diamine and isophthalic acid and has the structure

$$\left[-NH-C_6H_4-NHCO-C_6H_4-CO-\right]_n$$

Most likely the condensation is brought about by interaction of the diamine with the diacid chloride, thus:

$$n\,NH_2-C_6H_4-NH_2 + n\,ClOC-C_6H_4-COCl \longrightarrow \left[-NH-C_6H_4-NHCO-C_6H_4-CO-\right]_n + 2n\,HCl$$

The product is an all-aromatic polyamide and the first to have been used commercially for fibres.

Preparation

If, as seems to be likely, Nomex consists of poly(*m*-phenylene isophthalamide) it will be of interest to look at a way in which this polymer has been made and spun into fibres. These processes have been described in a du Pont patent U.S. 3,006,899 (1961). The examples given therein indicate alternative possibilities, but one that looks as likely as any for practical use is as follows:

> A solution of 5·41 parts *m*-phenylene diamine and 10·6 parts sodium carbonate (acid acceptor) in 150 parts of water is vigorously blended with another solution containing 10·26 parts of isophthaloyl chloride in 155 parts tetrahydrofuran (a very useful solvent). Agitation is carried out for 10 min., by which time condensation polymerisation is complete. The polymer formed thus:
>
> $$Cl-\overset{\overset{O}{\|}}{C}-\langle\text{ring}\rangle-\overset{\overset{O}{\|}}{C}Cl + H_2N-\langle\text{ring}\rangle-NH_2 \longrightarrow -\overset{\overset{O}{\|}}{C}-\langle\text{ring}\rangle-CONH-\langle\text{ring}\rangle-NH- \quad +2HCl$$
>
> is filtered off, washed and dried. The yield is 94 per cent of theory. This completes the formation of the polymer. Note how the use of the chloride of the diacid accelerates the polymerisation.
>
> A sample of this polymer with an inherent viscosity of 1·65 is dissolved in dimethyl formamide to give a solution containing 20 per cent polymer and 4·5 per cent lithium chloride which is added to improve the solvent power. This solution is dry spun through orifices of 0·13 mm. diameter into hot air at 200–210° C. and wound up at a speed of 125 m./min. It spins well. Then the wound-up fibre is extracted with cold water for 64 hr. and drawn 5·5 times its original length in steam at 56 lb./in.^{2} pressure.
>
> The so-obtained yarn had a tenacity of 3·6 gm./den. and an elongation of 23 per cent at break. The fibre stick temperature

(*cf.* p. 131) is 305° C. This may be Nomex, and the method of making it is probably very similar to that just described.

Properties

First, the high thermal resistance. With a melting point of 371° C. and a stick temperature of over 300° C. the fibre breaks new ground among the synthetics. It is this single property that is at the heart of the fibre. Towards ionising radiation Nomex fibre shows better resistance than does nylon 66, and under some conditions of exposure that will completely degrade nylon 66, so that it has no measurable strength left, Nomex will retain 70 per cent of its original strength. To β-, γ- and X-rays Nomex shows good resistance. The protection that a compact aromatic group on either side of it might confer on the amide linkage may be responsible for this unusual resistance. Nevertheless, such aromatic groups give no protection against ultra-violet light, which degrades Nomex just as it does nylon 66.

Chemical resistance is good, although not nearly so good as that of Teflon. Nomex is not significantly affected by dilute acids or alkalis, but it is attacked by such agents when they are concentrated, and especially when they are hot. It has good resistance to most organic solvents, such as phenol, formic acid (both are solvents for ordinary nylon) and methanol.

Dimensional stability is such that in dry air at 260° C. (melting point of ordinary nylon) Nomex shrinks only about 2 per cent. In boiling water, too, it shrinks about 2 per cent. Strength is good; its tenacity is about 5·3 gm./den. at 65 per cent R.H. and 70° F., but it is only 4·1 gm./den. wet. Elongation is normal at 22 per cent conditioned and 16 per cent wet. The stress–strain curves at different temperatures are of interest in view of the fibre's thermal resistance; they are shown in Fig. 143 recorded after the fibre has been heated for 5 min. at the indicated temperature. It is, too, of very considerable interest that even after having been kept at a temperature of 300° C. (at which ordinary nylon and Terylene would be liquid) for as long as a week, Nomex retains half of its original strength. There is nothing marginal about the special qualities of Nomex; they stand out boldly.

The specific gravity of the fibre is 1·38, which is the same as that of Terylene and very different indeed from ordinary nylon, which is about 1·14. The compactness and solidity of molecular structure which are responsible for the fibre's stability to heat is reflected in the higher density of the fibre. The constituent molecules are so firmly held that they cannot oscillate wildly and wrench themselves free until a much higher temperature than usual is reached.

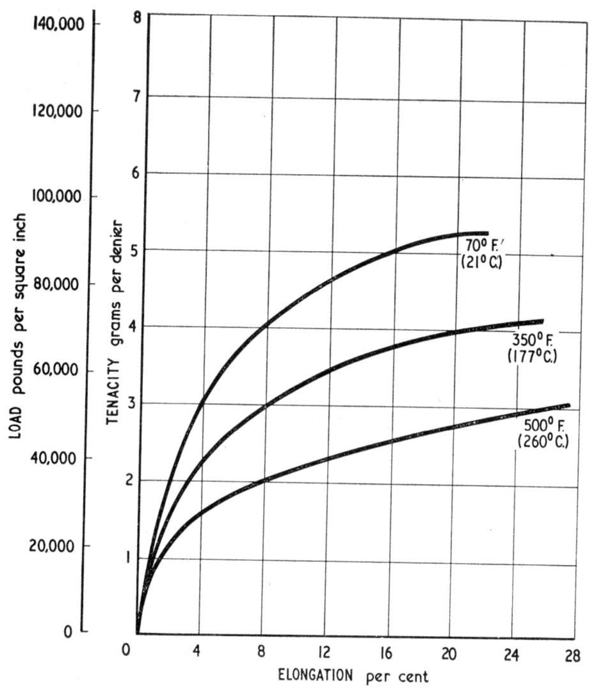

[*E.I. du Pont de Nemours & Co., Inc.*]

FIG. 143.—Characteristic curves of Nomex after the fibre has been held at the indicated temperature for 5 minutes.

Uses

First and foremost must be those applications where the flame-proof property is important: for space suits and, if fabrics of suitable construction can be made, for nursery wear. Nomex has been used for racing drivers' overalls; if the driver crashes and the fuel in his car ignites he can survive only for 20–30 seconds before he breathes in the flames and is thereby killed; in those 20 seconds Nomex gives much better protection than treated cotton or glass fibre, which latter conducts heat too readily. Nomex is still not nearly so good as an aluminium-foil-covered asbestos suit, but it is much more comfortable and practical to wear. It is also essential to wear two sets of Nomex underwear under the Nomex suit. Such underwear is available in America at around $20 a suit. It is necessary that any Nomex suit be stitched with Nomex thread; English Sewing Ltd. make one which they call "Arklex ". Other uses for Nomex include foundry workers' clothing, gloves, protective footwear, etc. The fibre is not easy to dye, and at the time of writing is only available in its natural off-white (a light biscuit colour) and in jungle green. Laundry presses and ironing-machine covers have received a lot of publicity; the Nomex press cover has a much longer life than nylon, for example, or cotton. At a temperature of 170° C. Nomex covers outlast the nylon covers that they replace by eight times; at 230° C. they outlast cotton by twenty times. The way in which the press cover of Nomex is used over a felt of Nomex is shown in Fig. 144. Apart from the long life, fabric glazing of the clothes being pressed is reduced.

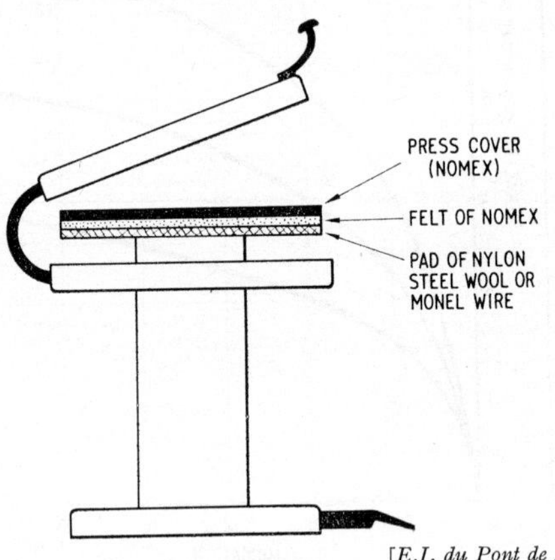

[*E.I. du Pont de Nemours & Co., Inc.*]

FIG. 144.—In the steam presses used by laundries and dry-cleaners Nomex covers last longer than most. The way in which they are used is shown in this illustration.

High-temperature gas filtration is another special use. Gases and air which contain dust are customarily filtered through fabric; this fabric is often in the form of bags which are internally supported by frames from collapsing; the air passes from the outside to the inside of the bag and leaves its dust on the outside of the bag, from which it has to be shaken off every few hours, but the air that gets inside the bag and escapes from its open end is clean. Nomex in the form of continuous-filament yarn, or staple fibre (for both are made, as well as tow) can be converted into conventional filter fabrics. One of the troubles of dust collection is, as a rule, condensation of water on the fabric bag; if Nomex is used the air can be heated to 125° C. or need not be cooled (dilution cooling is costly) lower than that if it is already hot; it can be filtered hot. Moisture condensation ordinarily causes bags to blind, and if there is an acid gas, such as sulphur trioxide, in the air that is being filtered, then this will form a corrosive acid which will be deposited on the surface of the bag. Nomex has good chemical resistance and dimensional stability. It does, however, remain that both Teflon (p. 572) and glass-fibre fabrics (p. 577) can be used at temperatures up to 260° C., whereas the regular use of Nomex above about 230° C. is not recommended. Other uses for Nomex are as dryer felts for paper machines, leader tapes for textile finishing machinery, coated fabrics and conveyor belts.

One of the first uses found for Nomex was for the filtration of hot fumes emerging from steel furnaces and other metal smelters. Such plants work continuously, and the temperature of the effluent may reach 320° C. at peaks, but Nomex filter bags can sometimes last for years in such conditions.

Availability

(1) Crimped staple fibre as 2 denier in lengths of $1\frac{1}{2}$, 2 and 3 inches.
(2) Tow as a 120,000 denier bundle.
(3) Continuous filament as 200 and 1,200 denier.

The fibre is made at Richmond, Virginia, and a second factory is being built.

The Choice of Aromatic Components

There must have been some hard thinking before Nomex was made. When once the decision had been reached, to try and emulate the success of the polyester Terylene by using aromatic groups in a nylon, it must have been a matter of much interest to look at the likely candidates for inclusion. Considerations that would seem to be important were doubtless of cost and availability of materials, with probably a preference to keep the monomers as simple as could be.

A dibasic aromatic acid to take the place of the adipic acid which is ordinarily used in nylon 66 was evidently needed. The simplest dibasic aromatic acid is phthalic $COOH(C_6H_4)COOH$, but there are three isomers of it:

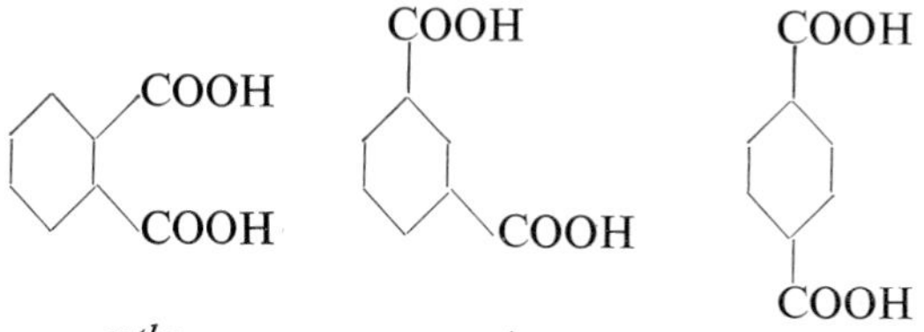

COOH COOH COOH COOH COOH COOH

ortho
decomposes
at 191° C.

meta
m.p. 346° C.

para
sublimes at about 300° C.
without melting.

The *para* acid is usually called terephthalic acid. The use of " tere " meaning " para " is unusual in the nomenclature of organic chemistry. The *meta* acid is called isophthalic. The three acids are made by the oxidation of: naphthalene for *ortho*, *m*-xylene for *meta* and *p*-xylene for *para*. Naphthalene is the cheapest source and *ortho* (ordinary) phthalic acid is the cheapest acid. But *o*-phthalic acid has its two carboxyl groups so close together that they tend to react and form a ring which prohibits the formation of a linear polymer. If an attempt is made to condense *o*-phthalic acid with a primary diamine NH_2RNH_2 no linear polymerisation takes place. Possibly the acid, or its derivative the chloride, and the diamine combine thus:

$$C_6H_4\begin{matrix}COCl\\ \\COCl\end{matrix} + H_2NRNH_2 \longrightarrow C_6H_4\begin{matrix}CO\\ \\CO\end{matrix}NRNH_2 + 2HCl$$

and even if this product were to combine with another molecule of the diacid chloride it would only yield:

$$C_6H_4\begin{matrix}CO\\ \\CO\end{matrix}NRN\begin{matrix}CO\\ \\CO\end{matrix}C_6H_4$$

which would be incapable of linear or indeed of any further polymerisation. The *ortho* acid is the only one of the three which forms either an anhydride or an imide.

Piperazine polyamide. But, unexpectedly, if *o*-phthalic acid is reacted not with a primary but with a secondary diamine such as piperazine, then linear polymerisation can take place, and strong film-forming materials are formed thus:

$$n\ C_6H_4\begin{matrix}COCl\\ \\COCl\end{matrix} + n\,HN\begin{matrix}CH_2-CH_2\\ \\CH_2-CH_2\end{matrix}NH \longrightarrow$$

$$\left[-CO-C_6H_4-CO-N\begin{matrix}CH_2-CH_2\\ \\CH_2-CH_2\end{matrix}N-\right]_n + 2nHCl$$

Such polymers have remarkably high softening temperatures (glass transition), as high as 325° C.; they are a little higher still at 340° C. if isophthalic acid is used and 350° C. if terephthalic acid is used instead of *ortho*. There is not much difference in these polymers of piperazine with the three different phthalic acids; all seem to have wonderfully good thermal resistance. Difficulties may crop up, but at present these piperazine phthalamides (even the *o*-phthalamide) look like fibres of the future. Under the simplest set of conditions, *i.e.*, reaction with a primary diamine, *o*-phthalic acid undergoes ring closure and will not yield linear fibre-forming polyamides, but it does very well with secondary amines.

Katz has described in U.S. Patent 2,949,440 (1960) how to make a polyamide from piperazine or one of its derivatives, such as dimethyl piperazine, and phthalic acid. He puts into an ice-jacketed (everything must be kept cool for acid chloride reactions) Waring Blender jar, 24 parts of an alkaline aqueous solution of dimethyl piperazine and 47·7 parts of a chloroform solution of phthaloyl chloride. The alkaline solution has been previously made of:

4 parts caustic soda (acid acceptor).
21 parts 5 per cent solution sodium lauryl sulphate (an anionic surface-active agent to aid the mixing of the two phases: aqueous and chloroform).
4·5 parts dimethyl piperazine (one of the monomers).
150 parts water.

The chloroform solution consists of:

10·2 parts phthaloyl chloride (apparently the *ortho* isomer).
37·5 parts chloroform (organic diluent).

When the two solutions, the watery one of dimethyl piperazine and the chloroform one of phthaloyl chloride, were mixed in the blender they condensed at once, and polymerisation (with the release of hydrogen chloride picked up by the caustic soda) was rapid and apparently immediate. In such reactions with an acid chloride the reactivity is usually very great, so that the action, whatever it is, takes place almost at once. When the polymerisation had taken place the chloroform was boiled away and the polymer filtered off, washed and dried in the form of a white powder; the yield was 80 per cent of theory. The polymer had an inherent viscosity (p. 41) of 1·66 measured in *m*-cresol. The stick temperature (p. 131) of the polymer was 350° C.—very high. The polymer was soluble in chloroform, hot dimethyl formamide, acetic acid, and 1,1,2-trichlorethane. In order to make a spinning solution it was dissolved in a 12 methyl alcohol/88 chloroform (by weight) mixture to give a 20 per cent solids solution. This was extruded through orifices of 0·1 mm. diameter into air at 75° C.: the chloroform and methanol evaporated and left the solid fibre. This was drawn to 1·6 times its original length over a pin heated to 90° C.; it had

1·1 gm./den. tenacity and 15 per cent extension at break, with good recovery from elongation. The preparation is described in detail as being instructive. The tenacity of the yarn is rather low, but it could doubtless be increased by one of the following devices: (1) use of piperazine instead of dimethyl piperazine, and (2) a high degree of stretch. But it is probably true that one cannot expect such high tenacities when *o*-phthalic acid is used as when the *meta* or *para* acid is used.

Nylon 6-T and Homologues. Isophthalic and terephthalic acids are not subject to the tendency to ring closure. Either of them will react with primary diamines to give linear polyamides; the polyamide made from hexamethylene diamine and terephthalic acid is the comparatively well-known nylon 6-T already discussed, which has a melting point of about 370° C., with decomposition unfortunately. The corresponding polyamide made from hexamethylene diamine and isophthalic acid melts as low as 198° C. Similarly, tetramethylene diamine condensed with terephthalic acid yields a polyamide "melting" at 430° C., but with isophthalic acid the polyamide melts at 250° C. It is a general rule that terephthalic acid yields much higher melting polyamides than does isophthalic acid. There is a partial exception to this rule, when the acids are condensed not with a primary diamine but with the secondary diamine piperazine. But piperazine polymers are abnormal; they are interesting, they may be of importance industrially in the future, but at present their interest is mainly academic.

Structural isomerism. Some of those polyamides that have just been discussed are grouped in the table on page 391.

Side chains. A quite typical effect of methyl substitution in aliphatic diamines has been described by two of du Pont's workers. Their findings are interesting although not surprising: substitution in the aliphatic diamine causes the polymeric terephthalamide to have a significantly lower melting point. Some of the polymers that were made by Shashoua and Eareckson of du Pont illustrate this:

Melting points (° C.) of polyterephthalamides from	
hexamethylene diamine	371
N,N'-dimethyl hexamethylene diamine $CH_3NH(CH_2)_6NHCH_3$	260
N,N'-diethyl hexamethylene diamine	182
ethylene diamine	455
N,N'-dibutyl ethylene diamine $C_4H_9NHCH_2CH_2NHC_4H_9$	190

Nylons Containing Phthalic Acid

id monomer derivative uch as the loride may be used).	Basic monomer.	Structure of molecular repeat.	Melting point (° C.).
thalic acid	Hexamethylene diamine	Ring closure; no linear polymerisation	—
hthalic acid sophthalic)	Hexamethylene diamine	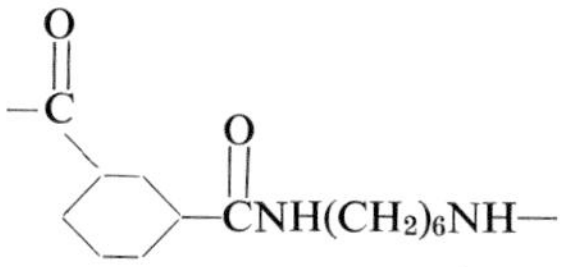 O ‖ —C O ‖ $CNH(CH_2)_6NH$—	198
thalic acid erephthalic)	Hexamethylene diamine	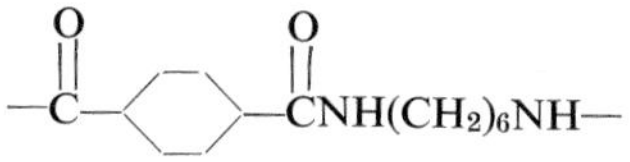 O ‖ —C— O ‖ —$CNH(CH_2)_6NH$— (Nylon 6-T)	370
thalic acid	Tetramethylene diamine	Ring closure	—
hthalic acid sophthalic)	Tetramethylene diamine	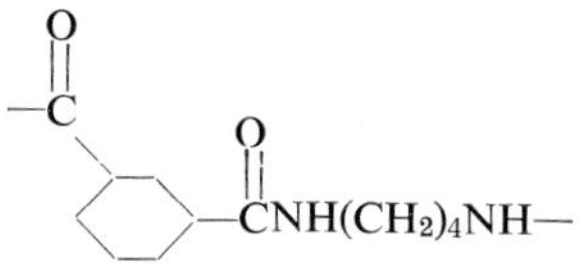 O ‖ —C O ‖ —$CNH(CH_2)_4NH$—	250
thalic acid erephthalic)	Tetramethylene diamine	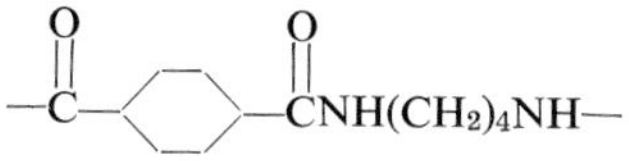 O ‖ —C— O ‖ —$CNH(CH_2)_4NH$—	430
			Softening point (° C.).
thalic acid	Piperazine	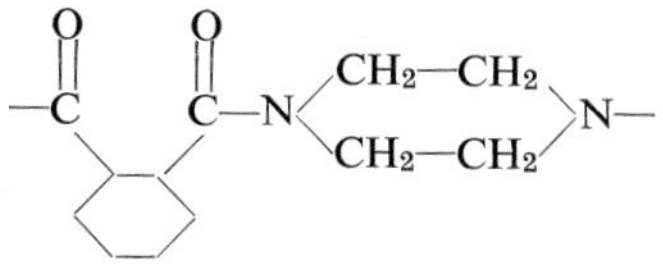 O ‖ —C O ‖ C—N<(CH_2—CH_2)(CH_2—CH_2)>N—	325
hthalic acid sophthalic)	Piperazine	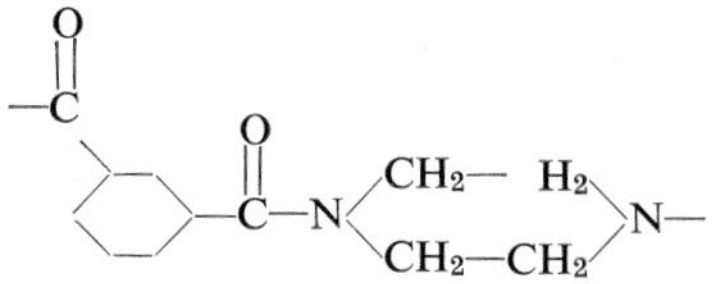 O ‖ —C O ‖ —C—N<(CH_2— H_2)(CH_2—CH_2)>N—	340
thalic acid erephthalic)	Piperazine	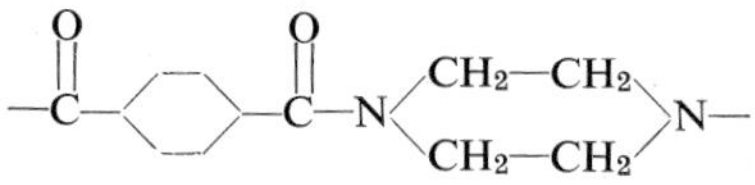 O ‖ —C— O ‖ —C—N<(CH_2—CH_2)(CH_2—CH_2)>N—	350

Very very clearly, quite inevitably and understandably, side chains are *de trop* in a linear polymer; they stick out awkwardly, preventing the orderly arrangement of the molecules; a high degree of order or orientation cannot be achieved, and the polymers are never very stable and will melt at temperatures much lower than otherwise.

Nylon MXD-6. Another aromatic diamine which has been used in the preparation of aromatic nylons is *m*-xylylene diamine

$$C_6H_4(CH_2NH_2)_2 \quad (1,3\text{-}bis(CH_2NH_2)\text{ benzene ring})$$

This compound is equivalent to *m*-phenylene diamine, but with the interposition of two methylene groups which might be expected to increase the reactivity of the amino groups compared with those of phenylene diamine. These side chains might well be expected also to reduce the melting point of any polymer made, because they stick out from the aromatic ring and make close packing and a compact structure less likely. They introduce an aliphatic quality into the polymer. In fact, the polymer formed from this amine and adipic acid is poly-*m*-xylylene adipamide, and it has the structure

$$\left[-HN-CH_2-C_6H_4-CH_2NHCO(CH_2)_4CO- \right]_n$$

It melts at 243° C., which is rather lower than does nylon 6; its glass transition temperature is 90° C. (nylon 66 is 60° C.), and it might therefore be expected to have rather better thermal resistance at relatively low temperatures. The polymer has been spun from the melt followed by drawing at 100° C. into fibres. It is known as Nylon MXD-6 (the nomenclature is logical, *cf.* nylon 66 and nylon 6-T), and it has been tested extensively by the Celanese Corporation of America: its physico-mechanical properties are not very different from those of nylon 66, it is not quite so extensible but not much inferior in this respect. According to Sprague and Singleton, the fibre is " unacceptable due to excessive moisture sensitivity ". The wet fibre loses a lot of strength at temperatures of 40–60° C.

Phenylene Diamine Adipamide. A simple condensation of *m*-phenylene diamine with the chlorides of adipic (C6), suberic (C8) and sebacic (C10) acids has been described by Kwolek and Morgan in U.S. Patent 3,287,323 (to du Pont). Equivalent quantities of the acid chloride, *e.g.* adipyl chloride and *m*-phenylene diamine, are mixed in an acid-accepting diluent such as *N*-methyl pyrrolidone or dimethylacetamide and agitated. Oxygen (air) must be absent, and the temperature must be kept within the range −20° C. to +25° C. A polyamide *m*-phenylene diadipamide is quickly formed; it has an inherent viscosity of 0·8 measured at a concentration of 0·5 gm. polymer per 100 gm. *m*-cresol.

Wholly Aromatic Polyamides.

In the preceding paragraphs there have been discussed polyamides made from an aromatic acid (usually phthalic) and aliphatic diamines, on the one hand, and on the other, those made from an aliphatic acid and aromatic diamines (usually phenylene). In either case considerable elevation of the melting point and of the glass transition temperature has been achieved; either way the thermal properties of the fibres have been much improved. If the possibility of melt-spinning has had to be retained, then substituent groups which will bring down the melting point have had to be introduced, or structural concessions made, *e.g.*, the use of isophthalic instead of terephthalic acid, of octamethylene diamine instead of hexamethylene diamine. A similar effect can also be obtained, as has been seen by the use of methyl or dimethyl piperazine instead of unsubstituted piperazine or of alkyl substitution in hexamethylene diamine. Side chains in a polymer nearly always lower the melting or softening point. The reader must rationalise wool for himself; it is loaded with side chains, but does not melt, although it chars with some approach to melting at a low temperature.

Aromatic Diamines. The other aromatic component in addition to an acid of all-aromatic polyamides is an aromatic diamine; the simplest is phenylene diamine, which exists in its three isomers:

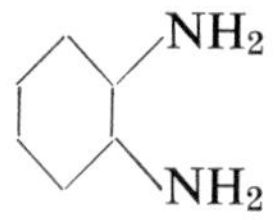

NH_2 NH_2 | NH_2 NH_2 | NH_2 NH_2

ortho m.p. 102–104° C. | *meta* m.p. 63° C. | *para* m.p. 147° C.

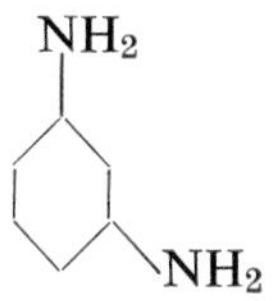

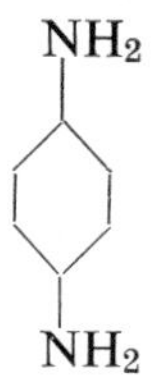

All of them darken on exposure to air, and the *meta* compound is particularly unstable. All of them are much less reactive than the aliphatic diamines, such as hexamethylene diamine, and because of this low reactivity it is impossible to make linear polyamides from them by high-temperature condensation with adipic and similar aliphatic diacids. But if the acid is used in the form of the acid chloride, then its reactivity is enormously increased and is great enough to react with the phenylene diamines in the cold.

It will be observed as we progress that *m*-phenylene diamine is made more use of than its *ortho* and *para* isomers, if we are to judge from the patent specifications. This is logical enough, because it should be the most reactive. If we consider the three diamines in the light of Lapworth's theory of alternating polarities and we start by considering the N atom in the 1 position as electro-negative we find that this induces

polarities in the ring carbon atoms, as shown above, and that these also induce polarities, as shown below, in the three diamines:

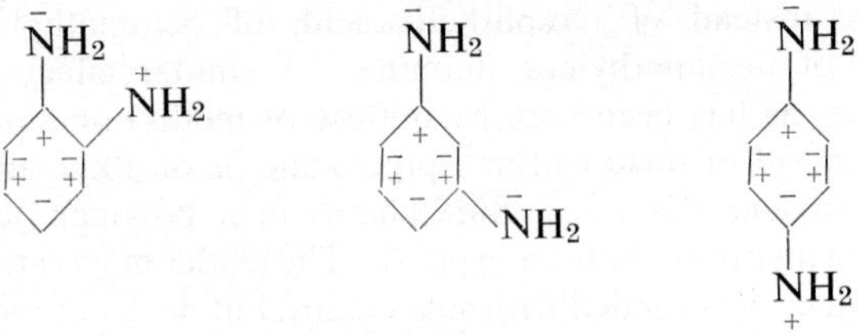

Only in the *meta* isomer is the electro-negative character of the amino N atom reinforced by that in position 1; in the *ortho* and *para* arrangements it is opposed. A fair inference is that the amino groups in the *meta* compound will be more reactive than those in the *ortho* or *para* compounds.

One example of the condensation of isophthalyl chloride and *m*-phenylene diamine has already been given (p. 383), that taken from U.S. Patent 3,006,899, and this one probably represents the stage when all-aromatic polyamides became commercial; it represents a way in which Nomex could be and probably is formed. Patent specifications are designed to cover as wide a field as they can without thereby becoming meaningless and valueless; often at the time specifications are filed there is remaining uncertainty about the details; sometimes matter which seems to be extraneous is included. But usually the principles that are necessary shine through the cloud of irrelevancies, and U.S. Patent 3,006,899 is taken to show anyone how to make Nomex.

FURTHER READING

Nomex Nylon. Bulletin N-201 issued by du Pont's Textile Fibers Division (October 1966): " Properties of Nomex High Temperature Resistant Nylon Fibre ".

H. W. Hill, S. L. Kwolek and P. W. Morgan, " Polyamides from Reaction of Aromatic Diacid Halide Dissolved in Cyclic Non-aromatic Oxygenated Organic Solvent and an Aromatic Diamine ", U.S. Patent 3,006,899 to du Pont (1961).

B. S. Sprague and R. W. Singleton, " Fibers from Aromatic Polyamides ", *Text Res. J.*, **35** (11), 999–1008 (1965).

V. E. Shashoua and W. M. Eareckson, " Polyterephthalamides from Short Chain Aliphatic Primary and Secondary Diamines ", *J. Polymer Sci.*, **40**, 343–358 (1959).

C. Cipriani, " Spinning of a Poly(polymethylene) Terephthalamide ", U.S. Patent 3,227,793 (1966) (assigned to Celanese Corp.).

S. L. Kwolek and P. W. Morgan, " Process for the Production of a Highly Orientable Crystallizable Filament-forming Polyamide ", U.S. Patent 3,287,323 (1967) (assigned to du Pont).

M. Katz, " Preparation of Piperazine Phthalamide Polymers ", U.S. Patent 2,949,440 (1960) (assigned to du Pont).

CHAPTER 22

POLYUREAS

URYLON was a synthetic polyurea fibre that was developed in Japan by Toyo Koatsu Industries Inc., who are primarily a fertiliser firm. Looking for an outlet for urea, of which they make large quantities, they embarked in 1950 on research into the development of a polyurea fibre. The research proved to be expensive and in all it cost more than £1·5 million. By 1954 Toyo Koatsu had developed a laboratory equipment which would make 20 lb. of fibre a day, and by 1958 they had a pilot plant producing 1 ton a day. But the expected large-scale manufacture never materialised. The cardinal difficulty was the high cost of one of the monomers, nonamethylene diamine. This was never overcome, and in 1963 the manufacture of the fibre was discontinued altogether. Toyo Koatsu is continuing to protect its patent rights, and in October 1968 it merged with the Mitsui Chemical Industry, and the joint company is known as Mitsui-Toyo. Whether Urylon will ever be spun again must be regarded as more than problematical. However, the fibre had some points of interest, and a little information about it is given below.

Chemical Nature

Polyurea fibre is characterised by the recurring group —NHCONH— as an integral part of the main polymer chain, just as nylon is similarly characterised by possession of the —CONH— group. Urylon was synthesised by the condensation polymerisation of urea with a diamine; the particular diamine used being nonamethylene diamine. The condensation may be expressed:

$$\underset{\text{Nonamethylene diamine.}}{nNH_2(CH_2)_9NH_2} + \underset{\text{Urea.}}{nNH_2CONH_2} \longrightarrow \underset{\text{Polyurea.}}{[-(CH_2)_9NHCONH-]_n} + 2nNH_3$$

It might be thought that the first advantage of such a fibre-forming polymer would be one of low cost, inasmuch as urea, which is very cheap (about 4*d*. per lb.), is one of the monomeric compounds. But although urea is used in equimolecular weights with the expensive diamine, its molecular weight is low so that in absolute weight the proportion of it that is used is not very great and this reduces the cost advantage. In parts by weight:

158 Nonamethylene diamine + 60 Urea $\longrightarrow$
184 Polyurea + 34 Ammonia.

Manufacture of Polymer

Urea is manufactured in enormous quantities; the most common process is to react carbon dioxide (either from heating mineral carbonates or from the combustion of coke) with ammonia (from atmospheric nitrogen). Reaction takes place at high temperatures and pressures to give ammonium carbamate which ultimately decomposes to urea and water

$$2NH_3 + CO_2 \longrightarrow NH_2COONH_4 \longrightarrow NH_2CONH_2 + H_2O$$

Nonamethylene diamine always presented a problem, at least to make it cheaply. It can be made from the corresponding C_9 dibasic acid, just as hexamethylene diamine is made from the C_6 dibasic acid (adipic) in the nylon synthesis. So, the problem resolved itself into one of making azelaic, the C_9 acid, cheaply. The raw material used by Toyo Koatsu was rice bran oil which contains the unsaturated oleic and linoleic acids. When the oil is treated with ozone the molecules of the unsaturated acids split at the point of unsaturation thus:

$$\underset{\text{Oleic acid.}}{CH_3(CH_2)_7CH{:}CH(CH_2)_7COOH} + 2O_2 \longrightarrow$$

$$\underset{\text{Pelargonic acid.}}{CH_3(CH_2)_7COOH} + \underset{\text{Azelaic acid.}}{COOH(CH_2)_7COOH}$$

$$\underset{\text{Linoleic acid.}}{CH_3(CH_2)_5CH{:}CHCH{:}CH(CH_2)_7COOH} + 2O_2 \longrightarrow$$

$$\underset{\text{Caproic acid.}}{CH_3(CH_2)_5COOH} + \underset{\text{Azelaic acid.}}{COOH(CH_2)_7COOH}$$

The azelaic acid is separated from the reaction mixture by distillation and is purified by recrystallisation. Its conversion into nonamethylene diamine is exactly analogous to the nylon synthesis of hexamethylene diamine from adipic acid, and comprises the following steps:

$$\underset{\text{Azelaic acid.}}{HOOC(CH_2)_7COOH} \xrightarrow{NH_3} \underset{\text{Ammonium azelate.}}{NH_4OOC(CH_2)_7COONH_4} \xrightarrow{-H_2O}$$

$$\underset{\text{Azelaic dinitrile.}}{NC(CH_2)_7CN} \xrightarrow{H_2} \underset{\text{Nonamethylene diamine.}}{NH_2CH_2(CH_2)_7CH_2NH_2}$$

The diamine must be very pure for use in polymerisation; this is achieved partly by purification at the acid and nitrile stages by distillation, and finally by distillation of the diamine.

The purified diamine is reacted with urea to give the polyurea.

New Improved Process for Dinitrile. Azelaic dinitrile was always the problem component for Urylon. It is still of interest that the makers described in B.P. 889,360 (1962) an improved method for making this substance. It consisted of the vapour-phase reaction of azelaic acid with ammonia in the presence of such a catalyst as molybdate/phosphate or vanadate/phosphate. It was so arranged that the acid was fed into a vaporising zone at a temperature of about 370° C. so that part only of the acid vaporised in a limited time, leaving the remainder in the liquid phase. The acid vapour reacted with the ammonia, and the catalyst and the high temperature dehydrated the so-formed salt to the dinitrile. The process is well conceived and seems to depend on two factors: (1) a highly active catalyst, preferably vanadate/phosphate mounted on silica, and (2) reduction to a minimum of the time that the azelaic acid is kept at a high temperature, so reducing thermal degradation.

Spinning

The polymer melted at 235–240° C. It contained very little low molecular weight material and could be spun directly. It was extruded by gear pumps through spinnerets into air, and the filaments were collected on bobbins. Thereafter they were hot-drawn to improve the orientation and so give better tenacity and lustre. In addition to continuous filament, staple fibre was manufactured.

Properties

The properties of Urylon were not very different from those of nylon and Terylene. Its dyeing affinity was rather better than that of either of those two. It was lighter in weight (S.G. 1·07), and this should have contributed to good cover.

If some foreign manufacturer should wish to develop the fibre Toyo Koatsu would license its manufacture; otherwise Urylon will be no more than a name in fibre history.

Russian Polyurea Fibres

According to the Russian technical press, extensive trials made on polyurea fibres had shown that their thermal stability was unsatisfactory and had thrown doubt on their commercial value. Investigation had shown that the instability was caused by the terminal urea groups, $-NHCONH_2$, on the fibres and that if these were blocked or protected in some way, the thermal stability of the fibre was much improved. The inclusion in the monomer mix of small proportions of monoamines, such as octadecylamine, will stabilise the polymer and will replace some of the unstable terminal urea groups by stable unreactive hydrocarbon groups, thus:

$$H_2N(CH_2)_9NH_2 + NH_2C_{18}H_{37} \longrightarrow H_2N(CH_2)_9NHC_{18}H_{37} + NH_3$$

FURTHER READING

du Pont de Nemours, U.S. Patent 2,190,770 (1940).
Y. Iwakura, *J. Chem. Soc. Japan*, **78**, 1416 (1957).
Toyo Koatsu Industries Inc. " Urylon " (1959).
R. W. Moncrieff, " Russian Polyurea Fibres ", *Man-Made Textiles*, 57, June (1965).
Toyo Koatsu Industries Inc., B.P. 889,360 (1962).
R. W. Moncrieff, " New Japanese Nitrile Synthesis ", *Text. Manufacturer*, 293, 295 (July 1962).

CHAPTER 23

POLYESTERS

TERYLENE, DACRON, KODEL, VYCRON

TERYLENE is a synthetic linear polymeric fibre which was invented and developed in this country by chemists (J. R. Whinfield and J. T. Dickson) of the Calico Printers' Association. It is a direct development of the work carried out by W. H. Carothers on polyesters. Whereas Carothers found that the polyamides were more suitable than the polyesters for making fibres, the C.P.A. have made new polyesters with improved properties. A plant for the manufacture of Terylene is operated by I.C.I. at Wilton, Yorks. The du Pont Company in America purchased Whinfield and Dickson's U.S. patent application from C.P.A. Ltd., and received the subsequently issued U.S. Pat. 2,465,319. The polymer has been made and the fibre spun from it in a plant at Kinston, North Carolina, which was opened in March, 1953. The name that has been given to it in America is "Dacron". Terylene and Dacron are the same chemically; the fibre and yarn made in the U.K. are called Terylene, those made in the U.S.A. are called Dacron. Since those early days it has done well and is now made throughout the world.

Chemical Nature

Terylene is a polymeric ester; an ester is formed by reacting an acid with an alcohol, in the case of Terylene the acid is terephthalic acid and the alcohol is ethylene glycol. It will be seen that both acid and alcohol are bifunctional, the acid containing two –COOH or carboxylic acid groups and the glycol two hydroxyl groups. If the acid and alcohol contained only one functional group, the reaction would stop at the monomeric stage, thus:

$$C_6H_5\text{—}COOH + HOC_2H_5 \longrightarrow C_6H_5\text{—}CO{\cdot}OC_2H_5 + H_2O$$

Benzoic acid. Ethyl alcohol. Ethyl benzoate.

The ethyl benzoate has no sufficiently reactive end-groups and the reaction cannot go further. Even if one of the components, say the alcohol, was bifunctional but not the other, still only a simple monomeric ester can be formed, thus:

$$C_6H_5\text{—}COOH + HOC_2H_4OH \longrightarrow C_6H_5\text{—}CO{\cdot}OC_2H_4OH + H_2O$$

Benzoic acid. Ethylene glycol. Hydroxyethyl benzoate.

A little ethylene dibenzoate

$$C_6H_5\text{—}CO{\cdot}OC_2H_4O{\cdot}CO\text{—}C_6H_5$$

would also be formed. But if *both* the acid and the alcohol are bifunctional then the reaction can proceed indefinitely to form a polymer, thus:

$$HOOC\text{—}C_6H_4\text{—}COOH + HOC_2H_4OH \longrightarrow HOOC\text{—}C_6H_4\text{—}CO{\cdot}OC_2H_4OH + H_2O$$

Terephthalic acid. Ethylene glycol. Hydroxyethyl terephthalate.

$$2\ HOOC\text{—}C_6H_4\text{—}CO{\cdot}OC_2H_4OH \longrightarrow$$

$$HOOC\text{—}C_6H_4\text{—}CO{\cdot}OC_2H_4O{\cdot}CO\text{—}C_6H_4\text{—}CO{\cdot}OC_2H_4OH + H_2O$$

Dimer of hydroxyethyl terephthalate.

The product, the dimer, still contains reactive groups at either end of its molecule, so that polymerisation can proceed, and a high polymer, containing, say, eighty benzene nuclei, will eventually be formed and will yield good fibres. The American fibre Dacron is made as just outlined from the acid, and the complete reaction can be written:

$$nHOOC\text{—}C_6H_4\text{—}COOH + nHO(CH_2)_2OH \longrightarrow$$

$$HO\left[\text{—}OC\text{—}C_6H_4\text{—}CO{\cdot}O(CH_2)_2O\text{—}\right]_n H + (2n - 1)\ H_2O$$

Dacron.

The British fibre Terylene is made by polymerising the dimethyl ester of terephthalic acid with ethylene glycol, and the complete reaction can be expressed:

$$nCH_3O{\cdot}OC\text{—}C_6H_4\text{—}CO{\cdot}OCH_3 + nHO(CH_2)_2OH \longrightarrow$$

$$CH_3O\left[\text{—}OC\text{—}C_6H_4\text{—}CO{\cdot}O(CH_2)_2O\text{—}\right]_n H + (2n - 1)\ CH_3OH$$

Terylene.

The product is essentially the same as Dacron, the only difference being that one of the end groups may be an ester group in Terylene, whereas it would be a carboxyl group in Dacron. Probably the use of the ester instead of the acid is preferred as purity of reactants is essential, and it is easier to purify dimethyl terephthalate than terephthalic acid itself by distillation at a fairly low temperature.

Manufacture

The schematic flowsheet for the manufacture of Terylene is shown in Fig. 145. The raw material for British Terylene is oil which comes from the Middle East, and from which *p*-xylene and ethylene glycol are made at Wilton. In America the raw material is American oil.

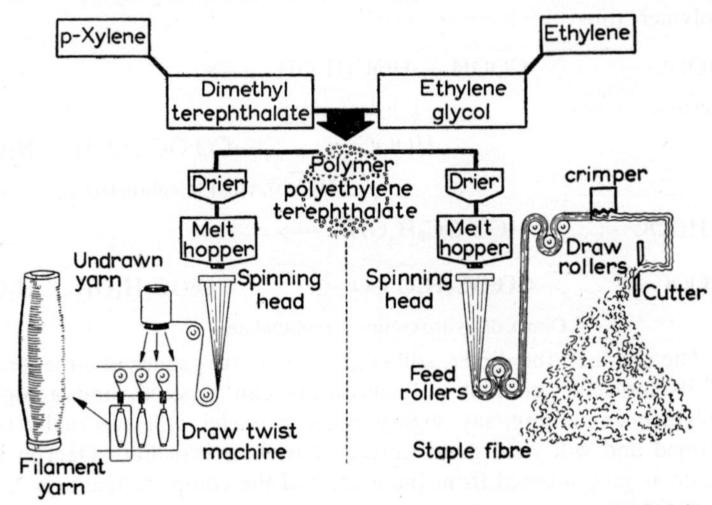

[*I.C.I., Ltd.*]

FIG. 145.—Flowsheet for manufacture of Terylene.

The oil is cracked to give ethylene, which is catalytically oxidised with air to ethylene oxide, which is hydrated to ethylene glycol.

$$\begin{matrix} CH_2 \\ \| \\ CH_2 \end{matrix} \xrightarrow{\text{Oxidation}} \begin{matrix} CH_2 \\ | \\ CH_2 \end{matrix}\!\!>\!O \xrightarrow{\text{Hydration}} \begin{matrix} CH_2OH \\ | \\ CH_2OH \end{matrix}$$

Ethylene (from a "cracking" plant). Ethylene oxide. Ethylene glycol.

Terephthalic acid is made from *para*-xylene, which must be free from the *ortho* and *meta* isomers. The *p*-xylene comes from the C_8 fraction of the naphtha that is distilled from the petroleum. It cannot be separated from the *ortho* and *meta* isomers by distillation, because the boiling points of all three are very close together; it is in practice separated by crystallisation: *p*-xylene freezes at 13° C., *m*-xylene at $-48°$ C. and *o*-xylene at $-25°$ C. Nitric acid oxidation of the *p*-xylene is used by du Pont; at a temperature of 220° C. and 30 atmospheres pressure to maintain liquid conditions, the yield is 80–90 per cent of terephthalic acid. An alternative which is probably used in the U.K. is to oxidise the *p*-xylene with air at 200° C. using cobalt toluate as a catalyst, first to toluic acid which is esterified to methyl toluate and this on further oxidation gives monomethyl terephthalate. This or the terephthalic acid from the nitric acid process is converted into dimethyl terephthalate. Either terephthalic acid or its ester can be used to synthesise Terylene or Dacron. The reactions involved in the first of these two processes are:

$$\text{Petroleum} \longrightarrow \underset{p\text{-Xylene.}}{CH_3\text{-}C_6H_4\text{-}CH_3} \xrightarrow{HNO_3} \underset{\text{Terephthalic acid.}}{COOH\text{-}C_6H_4\text{-}COOH} \xrightarrow{CH_3OH} \underset{\text{Dimethyl terephthalate.}}{COOCH_3\text{-}C_6H_4\text{-}COOCH_3}$$

The terephthalic acid or its ester and ethylene glycol are polymerised *in vacuo* at a high temperature. The polymer is extruded in the form of a ribbon from the autoclave (pressure vessel in which the polymerisation took place) on to a casting wheel. The ribbon of polymer solidifies on the wheel and is then cut into chips (little cubes of about 4 mm. sides) for easy handling, and these are conveyed by suction to the spinning building. The polymer chips are dried to remove residual moisture and are then put into hopper reservoirs ready for melting. The fibre is spun from the molten polymer—a very advantageous feature—through a spinneret with circular holes; the individual filaments solidify almost instantaneously and are drawn together and wound on to cylinders as undrawn yarn. This yarn is taken to draw-twist machines, where it is hot-stretched to about five times its original length and correspondingly to about one-fifth of its original denier. This yarn is then supplied to customers, who often re-twist it before use, and set or stabilise the twist by heating it in an oven. Two kinds of Terylene filament yarn are made, (*a*) yarns of normal strength, which are used mainly for clothing, (*b*) high-tenacity yarn for industrial uses; the higher tenacity of the latter is accompanied by a lower extensibility, and these changes are brought about by " drawing " or stretching to a greater extent than in the normal yarn. Very coarse filament yarns, owing to their poor thermal conductivity, have to be drawn or stretched cold instead of hot; fine filament yarns are always hot-stretched; the undrawn yarn can be stretched or drawn cold, but less easily and less uniformly than when hot. Accordingly, except for very coarse filament yarns, the drawing process is carried out hot.

Staple Fibre Production. Staple fibre is made in just the same way until the spinning operation; a great number (very many times more than when filament yarn is being spun) of filaments are spun and are brought together to form a thick tow. This tow is drawn, crimped mechanically to reproduce in part the crimp of wool, and the crimp is set or stabilised by heating; the tow is cut into specified lengths of a few inches according to the textile process for which it is intended and is baled. It is made in deniers and staple lengths suitable for spinning on the worsted, woollen, cotton and flax systems. Normal yarn is produced both bright and dull, the reduction of lustre being achieved in the usual way by the inclusion of a little titanium dioxide pigment, but the high-tenacity yarn, for industrial use, is produced only bright.

The flowsheet for the manufacture of continuous filament and staple fibre Terylene is shown in Fig. 145. The same fibre is produced in Canada under the name Terylene and under licence in Germany, where it is called Diolen and Trevira, in Italy (Terital), in France (Tergal), in Holland (Terlenka) and in Japan (Tetoron). World production of polyester fibre in 1968 was about 1,500 million lb. per annum.

Vickers-Zimmer Process. If a consumer uses two tons a day or 1·7 million lb. a year of polyester it may be worth his while to make his own. Vickers-Zimmer supply the plant, get it going and leave technical staff working on it until it is running smoothly. The cost of installing a plant is not so colossal as might be thought. For example, a plant capable of making 18 tons a day (14 million lb. a year) of polyester staple fibre would cost not much more than £3 million including buildings. Production of staple fibre would cost around 34–36*d.*/lb., made up as follows:

	Pence per lb. of fibre.
Amortisation (10 per cent of plant cost per year)	5·4
Maintenance	0·9
Personnel	3·4
Raw materials	22·7
Power, steam, water, nitrogen, etc.	3·4
Research and development	0·9
	36·7
Less value of recovered glycol and methanol	2·4
	34·3

In this costing the biggest item is that of the raw materials. The price of dimethyl terephthalate is taken as 14·5*d.*/lb. and that of ethylene glycol as 7·3*d.*/lb. This plant does not provide for intermediate isolation of polymer chip; the polyester is fed directly from the chemical plant to the spinning machine. The process is continuous; it represents an advance over the batch process, and the various stages that are carried out continuously are as follows:

(1) Dimethyl terephthalate is fed into a hopper.
(2) It passes continuously to a melter.
(3) Ethylene glycol is fed into a storage tank.
(4) The products from (2) and (3) are run through a dosing system into a reaction vessel where ester interchange takes place.
(5) The mix from (4) is fed into a stirring vessel at high temperature and reduced pressure, and thence into reactors at high temperature under vacuum where the condensation is completed.
(6) The product in the melt form is continuously discharged into the spinning machine.

Plants are made in various other sizes, yielding from 1,400 to 29,000 lb. per day of spinnable melt. In other designs the polyester is isolated as chip (about 3·5 mm. long by 2·5 mm. diameter) before being spun.

Properties

The early polyesters that Carothers had made suffered, as fibre-forming materials, from two defects: (1) they were too readily hydrolysed and lacked the chemical stability characteristic of nylon, (2) their melting points were so low that ironing troubles would inevitably have resulted when fabrics made from them were laundered. The discoverers of Terylene have shown that the inclusion of an aromatic (benzene) nucleus in the chain increases the chemical stability and raises the melting point. Terylene melts at 249° C. It will, however, inevitably make the fibre more inflammable. It has a density higher than that of nylon, and its handle may be preferred.

In other respects the yarn bears a close resemblance to nylon. It is strong, resistant to bacteria, mildew and moths, and can be " set " —*i.e.*, made dimensionally stable—at high temperatures. It is not severely degraded by the action of light; in this respect it is superior to nylon. It has only a low absorption of moisture, and has a wet strength nearly as high as its dry strength. Detailed properties are as follows:

Tenacity and Elongation. The tenacity and elongation at break of both Terylene and Dacron can be varied over a considerable range, according to the degree of drawing that is applied to the undrawn yarn. High tenacity, normal tenacity and staple fibre are all made from the same polymer, but they are drawn to different extents after spinning; the higher the draw or stretch the higher the tenacity and the lower the elongation of the resulting yarn, thus high-tenacity yarn is drawn to a greater extent (possibly six or seven times the undrawn length) than normal-tenacity yarn (probably about five times the undrawn length). The makers of Terylene say that " tenacity can be varied within the range 4·5 to 7·5 grams per denier, with corresponding extension at break from 25 to 7·5 per cent ", and du Pont say that " Dacron is produced with a wide range of properties. Depending on type, the tenacity and elongation at break is from 4·0 grams per denier and 40 per cent to 6·9 grams per denier and 11 per cent."

Typical figures are:

	High tenacity Terylene or Dacron.	Normal Terylene or Dacron.	Staple fibre Terylene or Dacron.
Tenacity, wet or dry (gm/denier)	6–7	4·5–5·5	3·5–4
Elongation at break (per cent) .	12·5–7·5	25–15	40–25

The lower tenacity and the higher extensibility of the product intended for staple should be noted; it is a general practice for fibre manufacturers to stretch the fibres for staple less than the continuous filament; this gives the staple lower strength but higher extensibility. Stress–strain curves are shown in Fig. 146 for the normal " medium-tenacity " Terylene and for high-tenacity Terylene and curves for nylon, viscose rayon and cellulose acetate rayon are included for comparison.

Moisture Regain. Under normal conditions the moisture regain of Terylene and Dacron is only 0·4 per cent (nylon is 4 per cent, Orlon is 1–2 per cent). Even at 100 per cent R.H. their moisture regain is only 0·6–0·8 per cent. Consequently, even when thoroughly wet their tenacity and elongation at break are not significantly different from those of the dry fibre. The low moisture regain of Terylene and Dacron is, however, conducive to static troubles in winding and weaving, although these may be mitigated by the application of a hydrophilic sizing material. Furthermore, garments made from them may, during wear, develop electric charges which attract to the fabric particles of " soil " (dirt, swarf, dust) flying about in the air, so that the cuffs of shirts, for example, become soiled quickly and are not easily laundered clean.

Elasticity. Dacron and Terylene fibres recover well from stretch, and similarly from compression, bending and shear, and it is to this good recovery that their power of rapidly losing wrinkles and creases is due. Yarn samples that were loaded rapidly and the load

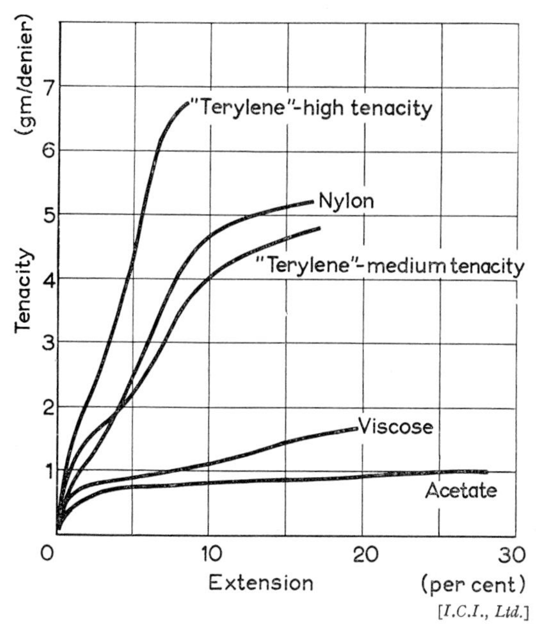

[*I.C.I., Ltd.*]

FIG. 146.—Comparison of stress–strain curves of Terylene (high tenacity and medium tenacity) with those of other fibres.

released immediately gave the following recovery after only one minute:

From 2 per cent stretch . . .	97 per cent recovery.
From 4 per cent stretch . . .	90 per cent recovery.
From 8 per cent stretch . . .	80 per cent recovery.

With respect to elasticity, Terylene is much inferior to nylon.

Modulus of Stretch. The initial modulus (resistance to early or initial stretch) of Dacron is high. A load of 0·9 gram per denier stretched Dacron filaments by only 1 per cent and a load of 1·75 grams per denier (which would break a good acetate rayon) stretched Dacron by only 2 per cent. Terylene has similar properties; it can be seen from Fig. 146 that in order to stretch the various fibres up to 2 per cent, a much greater load is required for medium- and high-tenacity Terylene than for the other fibres. This high modulus of stretch is a good feature; it means that under the comparatively low tension of winding, *etc.*, the yarns will hardly suffer any appreciable stretch.

Microscopical Appearance. Terylene and Dacron fibres are circular in section and often contain pigment (see Fig. 148). The longitudinal view (not shown) is smooth and uniform without striations.

Density. Terylene and Dacron fibres have a specific gravity of 1·38 (compare nylon 1·14, Vinyon 1·37).

Chemical Resistance. Resistance to weak acids is good even at boiling temperature; to strong acids, even hydrofluoric acid, it is good in the cold. Resistance to weak alkalis is good; resistance to strong alkalis is less good. Resistance to oxidising agents such as bleaches is good. The examples given in the following table are illustrative:

Chemical.	Temperature.	Concentration (per cent).	Time.	Effect on strength.
Hydrochloric acid	Room	18	3 weeks	None
,,	75° C.	18	4½ days	Appreciable
,,	Boiling	10	3 days	Degraded
Nitric acid	Room	40	3 weeks	Moderate
Sulphuric acid	Room	37	6 weeks	None
,,	,,	50	3 weeks	Moderate
,,	75° C.	37	2 weeks	Appreciable
Caustic soda	Room	10	3 days	Moderate
Sodium hypochlorite	70° C.	2½	4 hours	None

N.B.—" None " means not more than 5 per cent loss in strength, " moderate " means from 6 to 30 per cent loss in strength, " appreciable " means from 31 to 70 per cent loss in strength and " degraded " means more than 70 per cent loss in strength.

That the resistance to sulphuric acid is excellent can be seen from Fig. 147; even after three days' immersion at 60° C. the tenacity of Terylene is substantially unchanged, although it falls fairly rapidly at higher temperatures. Susceptibility to degradation by alkali is Terylene's chemical weakness.

Most alcohols, ketones, soaps, detergents and dry cleaning agents have no chemical action on Dacron or Terylene.

Solubility. The following substances will dissolve Dacron and Terylene: hot *meta*-cresol, trifluoroacetic acid, *ortho*-chlorphenol, a mixture of 7 parts trichlorophenol and 10 parts (by weight) of phenol, a mixture of 2 parts of tetrachloroethane and 3 parts (by weight) of phenol.

Action of Swelling Agents. The following solutions are swelling agents, *i.e.*, they increase the diameter of the fibre immersed in them: 2 per cent solutions in water of benzoic acid, salicylic acid, phenol, and *meta*-cresol; 0·5 per cent dispersions in water of monochlorbenzene, *p*-dichlorbenzene, tetrahydronaphthalene, methyl benzoate, methyl salicylate; 0·3 per cent dispersions in water of *ortho*-phenylphenol, *para*-phenylphenol.

Melting Point. Terylene and Dacron melt very similarly to nylon. Melting point of fibres such as these which decompose when they

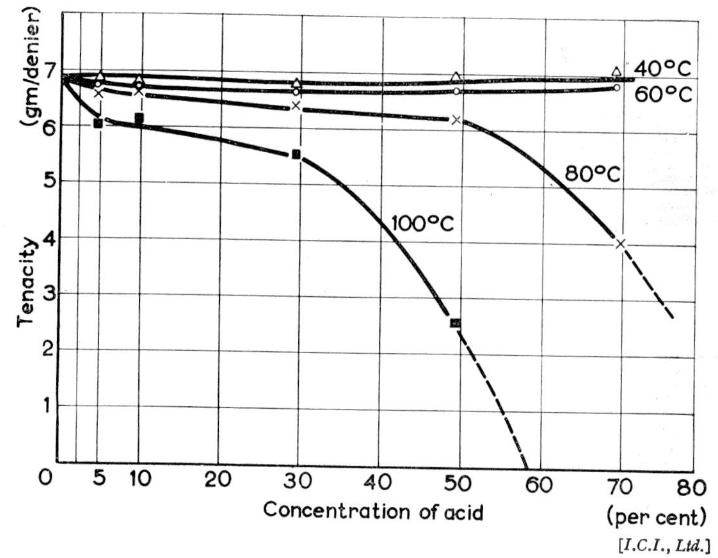

[*I.C.I., Ltd.*]

FIG. 147.—Resistance of Terylene to sulphuric acid for 72 hours at concentrations of 5–70 per cent.

melt in air depends a good deal on the method of measurement; an average melting point in air for all three is about 250°C. On an electrically heated copper block nylon melts at about 263° C. and Terylene rather lower at 249° C., but practically there is very little difference in performance. They withstand high temperatures well; Dacron after eleven days' exposure at 175° C. showed only slight yellowing and still had a tenacity of about 1 gram per denier.

Ironing Temperature. The recommended ironing temperature is 135° C. Sticking of the iron on the fibre takes place at 205° C.

Electrical Properties. Terylene, partly because of its low moisture regain, has good dielectric properties. So have most of the synthetic fibres, and the advantage that Terylene has is that its electrical and

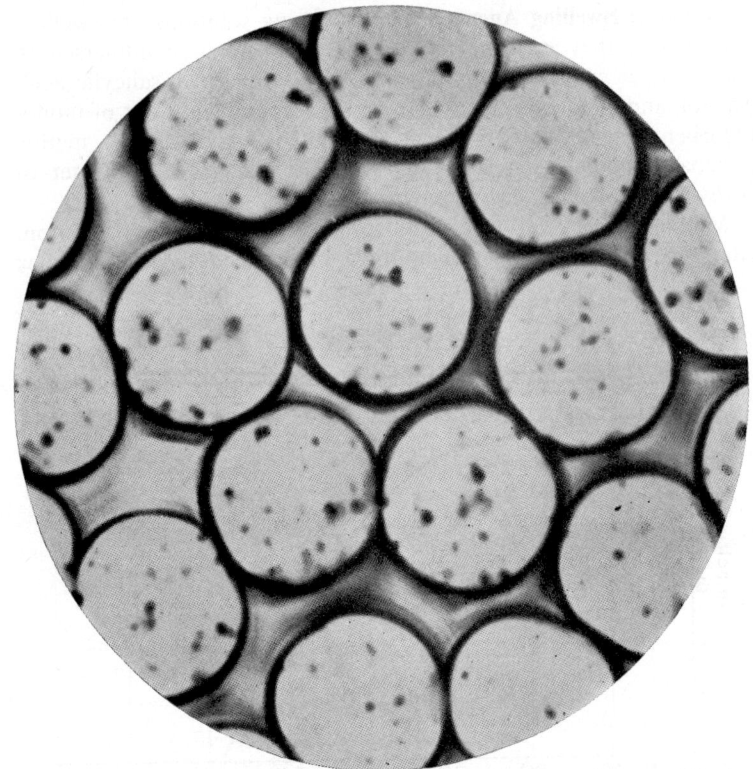

[*E. I. du Pont de Nemours & Co., Inc.*]

FIG. 148.—Cross-section of Dacron polyester fibre (× 1,500).

The fibres have the round section characteristic of melt spun fibres. Particles of pigment can be seen.

mechanical properties are maintained at higher temperatures up to 180° C. It has been found, too, that a smooth Terylene surface is " non-tracking ", *i.e.*, an accidental flash-over does not form a carbonised, and hence conductive, track.

Biological Resistance. Resistance is good. No insect has nourished itself on the fibre, although some have cut their way through fabric in which they have become entrapped. Moulds, mildew and fungi may grow on finishes that have been applied to Terylene or Dacron, but they do not attack the fibre.

Light Resistance. Like all other textile fibres, Terylene and Dacron lose strength when exposed for long periods to light, apparently due to absorption of light in the near ultra-violet range (300–330 μ), *i.e.*, " light " just too high in frequency to be visible.

Compared with other textile fibres resistance to full sunshine is fairly good; behind glass, which cuts out the ultra-violet, Terylene and Dacron are better than most fibres and are quite suitable for curtains.

Nuclear Radiation. When Terylene is irradiated with γ-rays from cobalt 60 it suffers a marked decrease in tensile strength and elongation. Little or no hydrogen is evolved and there appears to be no cross-linking induced.

Shrinkage. Terylene yarn as manufactured shrinks by approximately 7 per cent when immersed in an unrestrained state in boiling water; the high-tenacity yarn shrinks rather more. Dacron yarn shrinks similarly, as much as 10–14 per cent after 70 minutes' immersion. Many organic liquids such as acetone, chloroform and trichlorethylene will also shrink the yarn or fabric at their boiling points. If, however, the fabric has been previously "heat-set" it will not shrink in either boiling water or boiling dry-cleaning solvent; to get the best out of Terylene and Dacron for clothing it is imperative that they should have been "heat-set".

Flammability. Terylene and Dacron will burn, but the burning is usually accompanied by local melting of the fabric, which drops away, so that as a rule the fire does not spread. If, however, the Dacron is mixed with another fibre that will support it, it will burn. Rather oddly, a combination fabric of Dacron and Fiberglas may burn when ignited, although neither the Dacron nor the Fiberglas would support combustion alone.

Heat-setting

Terylene and Dacron fabrics can be dimensionally stabilised by heat-setting so as to be completely undisturbed by subsequent treatments such as washing or dry cleaning at lower temperatures. Pleats that are inserted at the setting temperature are in for the life of the garment, whether on a frock or as a crease in trousers; furthermore, any creases or crumpling that may occur during later washing processes are never severe and can be removed by a very light ironing.

Heat-setting consists essentially of exposing the fabric while under dimensional control—usually held out at full width—to a temperature 30–40° C. higher than any likely to be encountered in its subsequent life; accordingly, for fabric which may be ironed, 220–230° C. is a suitable setting temperature for material intended for clothing. If Terylene fabric is not heat-set, very troublesome creases which are difficult to remove may be "set" by hot domestic washing, but if the fabric has been pre-set even at a temperature as low as 130° C. such washing creases will be fewer and less severe and easy to remove. The setting process stiffens fabric, which is undesirable, but the stiffening is lost as the fabric is subsequently treated mechanically, *e.g.*, by dyeing on a winch; it is accordingly better to heat-set the fabric at as early a stage as possible, preferably before dyeing, and sometimes even before the preliminary scour, so that as much of the processing as possible will be utilised to break down the stiffness caused by heat-setting of the fabric. If heat-setting is done before dyeing, care must be taken to see that the heating is uniform otherwise the affinity for dyestuffs may be irregular and unlevel dyeing will result; heat-setting at temperatures from 100–160° C. causes a progressive diminution in affinity for dyestuff, but from 160–220° C. there is a progressive increase, so that fabric which has been set at 220° C. has about the same affinity as unset fabric for dyestuffs. But the importance of *uniformity* of heat-setting before dyeing will be clear. If, on the other hand, heat-setting is done after dyeing, there is a danger of dyestuff volatilising from the fabric and giving patchiness. It is probably better to heat-set before dyeing.

Heat-setting of Terylene and Dacron can also be carried out in steam under pressure, but this method is undesirable because it causes some hydrolysis of the ester groups in the polymer chain, and the fibre is partly depolymerised, with a resultant loss in tenacity. Dry heat-setting is the best process, and the dimensional stability which is thereby conferred on Terylene and Dacron fabrics is one of their greatest assets. The degree of heat-set of polyester fibres has been correlated with their iodine absorption by the Russians; the method is outlined in *International Dyer*, 19 February 1965.

Bleaching

Terylene does not usually need bleaching, but its natural colour is ivory, and not dead white. If bleaching is necessary for special end uses, then boiling sodium chlorite, brought to *p*H 2–3 with nitric acid, is the most effective.

Progress has recently been made with optical bleaching, and Uvitex RE (Ciba) has been found to be very suitable for Terylene. It is insoluble and is applied as a dispersion in water at the boil on a winch or jig. When fabric treated with the optical bleach is subsequently washed in warm to hot water (45° C.) the bleach is fast, but a little of it is lost if the fabric is washed in boiling water. There are many optical bleaches available, but they vary considerably in their fastness to washing; in this respect Uvitex RE is satisfactory.

Dyeing

Terylene and Dacron have offered considerable difficulties in dyeing. These have been due to the lack of hydrophilic properties and of reactive groups such as those that abound in the cellulosic fibres and in the protein fibres. For similar reasons cellulose acetate, when it was first made, presented the same difficulty, and it was eventually overcome by the introduction of the dispersed dyes, which are applied from an aqueous dispersion, not from solution, and which, having a greater affinity for the organic fibre than for the water of the dyebath, eventually migrate to the fibre and form a solid solution in it. This technique has been applied to most of the synthetic fibres as they have come along, and notably to nylon. It was the obvious line of approach to dyeing Terylene, but it did not prove to be very easily applicable to this fibre. The reason for

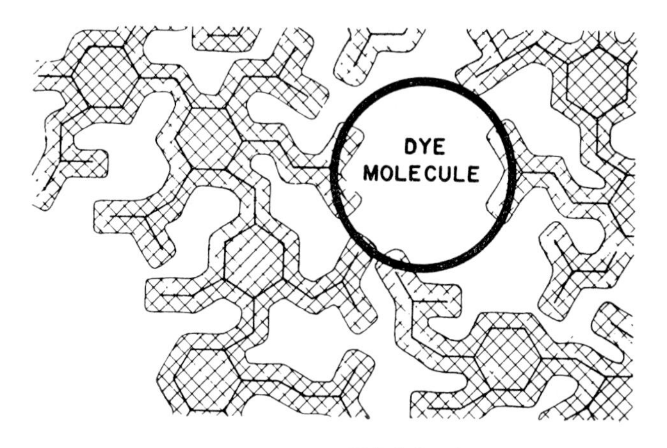

[*E. I. du Pont de Nemours & Co., Inc.*]

FIG. 149.—Cellulose acetate chain and dye molecule ($\times$ 10^8).

this is that the polymer chains in Terylene and Dacron are much more closely packed than in cellulose acetate. Fig. 149 is a representation of cellulose acetate fibres based upon information gained from X-ray diffraction studies; it represents a magnification of about 100 million times; the dye molecule is drawn approximately to scale and it can be seen that there are gaps in the chains which will accommodate the dye molecule, particularly so because the chains are fairly free to change their positions. Fig. 150 is a similar representation of Dacron fibres, similarly magnified; the long-chain molecules can be seen to be neatly packed in a zig-zag pattern, and there are no gaps which will suitably accommodate the dyestuff molecule, and furthermore the chain molecules are very reluctant to change their positions, *i.e.*, to lose their orientation. The result is that the dyestuff particles cannot penetrate the fibre easily and in fact the process is so slow that it takes a matter of days or even weeks to dye a Terylene fibre at a dyeing temperature of 85° C. This difficulty has been overcome in four ways:

1. *Selection of Dyestuffs.* The only dyestuffs with measurable affinity for Terylene under ordinary conditions are those which may be applied from aqueous dispersion. Some of these will give light to medium shades in about 1½ hours at the boil; it is necessary to use a temperature of practically 100° C., as the affinity even of the selected dyestuffs is low at 85° C. Six selected dyestuffs for use with Terylene are Dispersol Fast Yellow G, Dispersol Fast Orange G, Duranol Red GN, Duranol Red 2B, Duranol Violet 2R and Duranol Blue GN. Even if only pastel shades are required it is still necessary

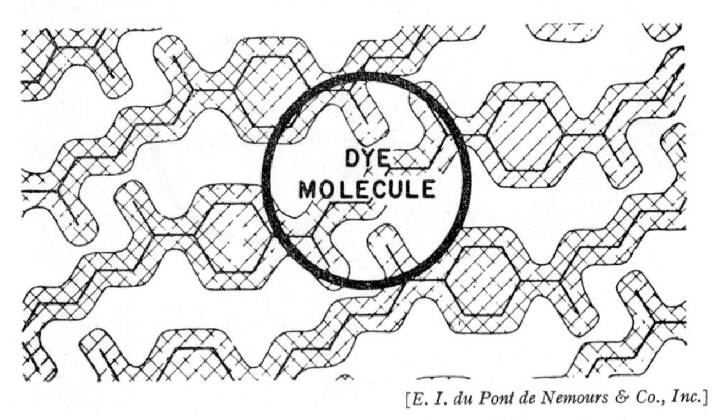

[*E. I. du Pont de Nemours & Co., Inc.*]

FIG. 150.—Dacron polyester chains and dye molecule.

to use a temperature of 100° C., otherwise, if a lower temperature is used, most of the dyestuff will be on the surface of the fibre and rubbing fastness will be very poor.

When azo dyestuffs are used it is essential to use as the phenolic component (azoic dyeing consists of the combination of a phenol and a primary amine by a process known as diazotisation; both components may be colourless but the compound is highly coloured) one which has a small molecule, so that it can get into the Terylene molecule. If one of the usual complex phenolics is used and coupled with a base, the Terylene is not dyed. If, instead, a phenol such as β-oxynaphthoic acid (Brentosyn BB) is used with one of the usual bases, such as Bentamine Fast Red RL Base, good dyeings can be effected at 100° C. The essential point is to use selected dyestuffs with small molecules which have a better chance than most of penetrating the fibre; even then a high dyeing temperature (100° C.) is necessary, and only light and medium shades can be obtained.

2. *High-temperature Dyeing.* By using a higher temperature, *e.g.*, 100° C., the Terylene or Dacron molecules are more free to move and the dyestuff molecules can penetrate faster and pale shades can be dyed fairly satisfactorily in a reasonable time at this temperature; even so, penetration is poor and most of the dyestuff is located on the surface of the fibres. If, however, the temperature is taken still higher, say to 120° C., the chain molecules are much freer to move and the dyestuff can penetrate the fibres well, so that good medium and heavy shades can be obtained within a reasonable dyeing time of say one hour. This method, however, involves the use of pressure equipment (water or aqueous dyebaths cannot be maintained at a temperature higher than 100° C. at atmospheric pressure), and whilst it is admirable for dyeing loose fibre or yarn, it has, until recently, been unsuitable for fabric because no suitable machines have been available—the jig and the winch do not lend themselves to use under pressure. The advantages to be gained from dyeing some of the synthetic fibres in fabric form at temperatures above 100° C. have attracted attention, and have encouraged the development of several pressure fabric dyeing machines; examples of them are described on pp. 730–740. *Selection* of dyestuffs must be made for high-temperature dyeing because some decompose at 120° C., *e.g.*, some blues will redden. See also p. 759.

3. *Carrier Dyeing.* By using a carrier; this in effect is a swelling agent. A small quantity of it is included in the dyebath, and it swells the fibres, so enabling the chain molecules to move about more easily and increasing the distance between them; the dimensions of the chain molecules themselves are not changed (they do not swell), the molecules move farther apart so that the diameter of the fibre increases and so that there are bigger spaces to accommodate the dyestuff molecules. The " carriers " that are used are the known swelling agents for polyester fibres, notably *p*-phenylphenol and *o*-phenylphenol, diphenyl, benzoic acid and salicylic acid. In practice diphenyl which is supplied in an easily emulsifiable form as " Tumescal D " and *o*-phenylphenol (Tumescal OP) have been mainly used for dyeing Terylene, and *p*-phenylphenol has proved to be the best carrier for printing it.

The advantages of using a carrier in aqueous dyeing are:

(*a*) Dyeing is much more rapidly carried out.

(*b*) The dyestuff penetrates the fibre. Fig. 151 shows sections of polyester fibre dyed without carrier, (*a*), and with carrier, (*b*). The improved penetration gives better fastness, especially to rubbing.

(*c*) Improvement in levelling, *i.e.*, transfer of dye during application.

(*d*) *Fabric* can be dyed by this method as no pressure is necessary.

Disadvantages are:

(*a*) The extra cost is considerable; it may be about 1¼*d*. per lb. of fibre if *o*-phenylphenol (this substance is marketed as the sodium salt under the names "Tumescal OP" and "Dowicide A") is used, and as much as 7*d*. per lb. if benzoic acid is used.

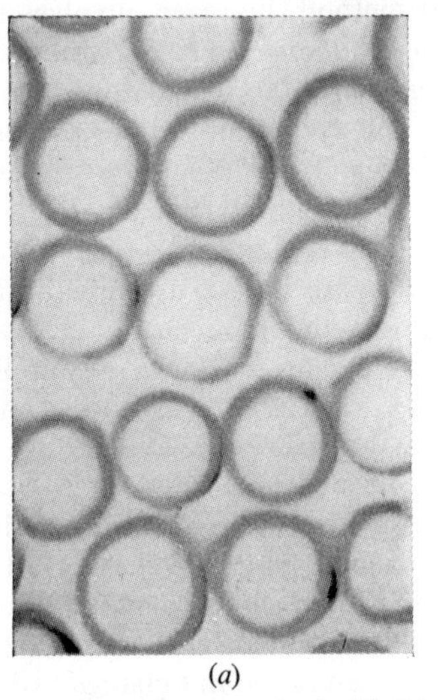

(*a*)

(*b*) [*I.C.I., Ltd.*]

FIG. 151.—Terylene dyed with Dispersol Fast Scarlet B 150 Powder Fine (*a*) for 4 days without carrier, (*b*) for 1 hour with carrier.

(*b*) The goods are inclined to retain the odour of some of the carriers and need very thorough washing to remove it; diphenyl is particularly odorous.

(*c*) The light fastness of the dyeing is badly impaired by *p*-phenylphenol and to some extent by *o*-phenylphenol and salicylic acid. Monochlorbenzene is good in this respect but impossibly toxic. Benzoic acid is good but expensive.

Terylene has been well dyed in heavy shades using diphenyl as the carrier at 85° C. and at 100° C.; *o*-phenylphenol does not give quite such good results. The optimum concentration for benzoic acid is 20 grams per litre (2 per cent), whilst that for *o*-phenylphenol, which is used as the sodium salt, is 3 grams per litre (0·3 per cent).

In all cases where a carrier is used it should be added slowly during the dyeing process; if added all at once the dyestuff may go on to the goods too quickly.

The carrier method is probably less to be preferred than the pressure (high temperature) method for staple fibre or for yarn, but the carrier method has the outstanding advantage that it can be used for dyeing fabric, *e.g.*, on an open jig.

4. *Thermosol Method of Dyeing.* This method of dyeing consists essentially of the following stages:

(*a*) The Dacron is thoroughly scoured as a preparatory step, a wetting agent being used to ensure complete removal of warp size, *etc.* This scouring is most suitably carried out in a beck.

(*b*) The Dacron is padded in a dispersion of an "acetate" colour; the "Latyl" class of colours are very suitable. A thickening agent such as salt-free CMC (carboxymethyl cellulose, sodium salt) and an organic solvent such as Cellosolve to help in spreading the dispersion on the fabric should also be present.

(*c*) The Dacron fabric is dried in a flue dryer at about 70° C. with good air circulation (steam-heated cylinders are unsatisfactory because they cause excessive marking off).

(*d*) The fabric is passed through an oven at 175–200° C., the time of passage being about one minute. The dyestuff which has been applied to the fabric in (*b*) and dried on the surface in (*c*), dissolves in and penetrates throughout the fabric in (*d*).

(*e*) The fabric is given a good scour to remove traces of loose colour and is then finished and dried normally.

Whilst primarily intended for fabric, the Thermosol process can be used for dyeing staple, tow and top of Dacron, *e.g.*, staple can be immersed in a suitable dye dispersion, the excess removed by centrifuging and the colour developed in a belt conveyor, rawstock dryer. This development requires up to ten minutes because thicker layers are being heated than when fabric is dried.

The Thermosol process has also been used on a semi-commercial scale for printing Dacron.

Some vat pigments can also be used on the Thermosol process. A diagrammatic representation of the Thermosol equipment is shown in Fig. 152.

Fastness. Always those fibres which are the most difficult to dye give the fastest dyeings. It is no easy matter to introduce dyestuff molecules into the ordered arrangement of chain molecules (Fig. 150) in Dacron and Terylene, but when once they are in, it is equally difficult to get them out and ordinarily they are in "for keeps". Washing fastness is excellent; light fastness is very good provided that a carrier has not been used; if it has, light fastness is fair to good.

Reduction Clear. The insoluble (in water) dyestuffs that are used on Terylene are inclined to aggregate at 100° C., so that they may be deposited on the surface of the fibre in relatively large particles; this deposition leads to poor rubbing fastness. The deposit can be removed by a Reduction Clearing Treatment. This consists of treating with an alkaline sodium hydrosulphite bath at 70° C. for about 10 minutes; this removes the dyestuff that lies on the surface of the fibre, but not that which has penetrated into the fibre.

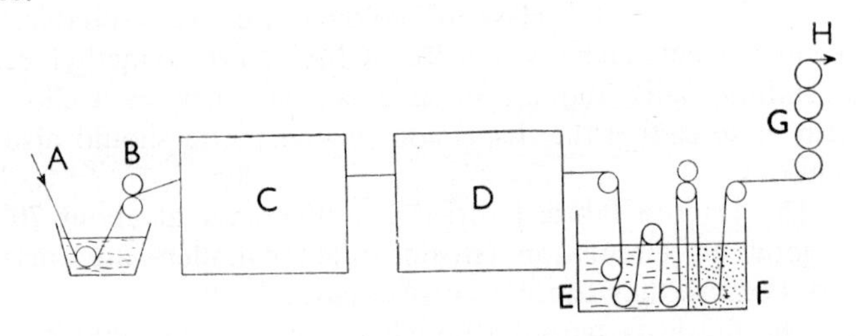

FIG. 152.—Diagrammatic representation of Thermosol equipment.

A. Undyed fabric. B. Padding mangle. C. Hot air dryer at 70° C. D. Oven at 175–200° C. E. Soaper. F. Rinse. G. Drying cylinders. H. Dyed fabric.

Terylene Mixtures. The extreme resistance of Terylene or Dacron to ordinary dyeing methods makes the dyeing of Terylene unions with other fibres difficult. It is still practically impossible to get solid shades on nylon/Terylene or on acetate/Terylene mixtures. Wool/Terylene also is very difficult, the Terylene component always being light in shade, although good results have been claimed for the use of dispersed dyes for the Dacron and pre-metallised acid dyes for the wool, dyeing being carried out in an acid bath containing *o*-phenylphenol.

Pigment padding (p. 198) gives level dyeing of good depth of shade and excellent fastness on Terylene.

Uses

First of all, the properties imparted to fabrics by the polyester fibre are resistance to and recovery from creasing or wrinkling when wet or dry; high resistance to stretch in the filament yarns but not in the staple; high abrasion resistance; good texture and appearance; resistance to heat ageing; good chemical resistance and good resistance, behind glass, to sunlight.

Perhaps the most valued of the distinctive properties of Terylene is its ability to be set in pleats and creases, an example of its outstandingly useful property of dimensional stability. Dresses and skirts ranging from haute couture models to those that sell in the multiple shops have pleats that are really fast, and furthermore, the garments retain their shape—because they have been heat-set—during wear; the strength and resistance to abrasion are complementary properties which give the garments long life. Tropical suitings have been made in 100 per cent Terylene, and these also possess shape and crease retentive properties. These properties are also to be found in Terylene 55 per cent/wool 45 per cent blends which have been used as suitings.

Curtains in Terylene net are popular, and because their resistance behind glass to sunlight is good and because Terylene can be dyed fast they should have a long life; many of them are to be found in a real leno weave, which allies stability of fabric construction to stability of material.

It is frequently claimed for light garments, *e.g.*, blouses and lingerie, made from synthetic fibres that they need no ironing. This may be true if they are washed very carefully so that no compression is applied whilst wet and warm, but in practice washing usually introduces a little crumpling; there are no sharp creases, but nevertheless, such garments, including those made from Terylene, look all the better for a light ironing. The warm handle of Terylene makes it suitable for lingerie, and it can be washed almost indefinitely; blouses which drape like a heavy silk but can be washed like cotton are made from Terylene. Whereas nylon is subject to a disadvantage that after long wear it " yellows " or " greys ", its pristine whiteness is quite lost, Terylene does not suffer from this defect, so that the useful life of Terylene garments will not be cut short.

In the staple form Terylene has been used for socks which are fairly comfortable for town wear and which practically eliminate the housewife's darning chores. A " fiberfill " is made from Dacron for filling pillows; it is fluffy, resilient from crushing, non-allergic and easy to wash and dry. Compared with wool, Terylene (or Dacron) has three defects as an apparel fibre:

1. It becomes saturated with perspiration at a moisture content below 1 per cent, whereas wool will absorb 30 per cent moisture and still feel dry to the touch.
2. It " wicks " so that polyester fabric transfers rain or snow

quickly from the outside to the interior of a garment; wool does not do this.

3. It has a very low heat of wetting (see p. 312).

It must be added, however, that many people who have worn Terylene underclothes say that the lowness of its moisture regain is not disadvantageous, and that the moisture travels through the garment doubtless by wicking and then evaporates; they maintain, too, that there is no chilliness when cooling off after being hot. Something has to be allowed for personal preferences and for conditions of wear. As a tie fabric, Terylene is superb, and the ties can be washed without trouble.

Industrially, the biggest use of Terylene fabrics is in the laundry (Fig. 153), where its heat resistance, strength and hard-wearing qualities have established its use for calendar sheeting, packing flannel, laundry blankets, knitted padding, polished head press covers and laundry and dye bags (Fig. 154); it lasts a long time. The hosiery dyeing industry is using Terylene dye bags to dye nylon stockings; the resistance of Terylene to dyestuffs which will dye nylon here becomes a virtue.

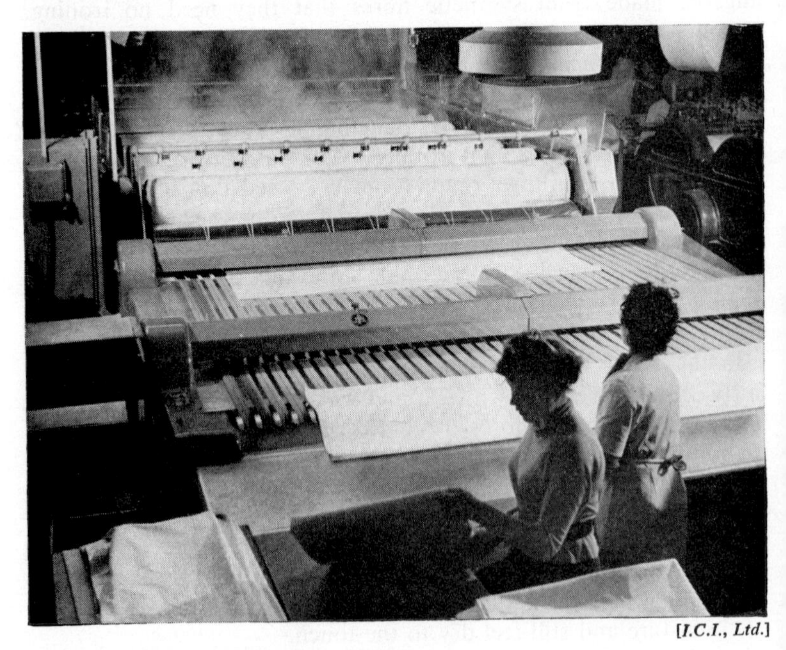

[*I.C.I., Ltd.*]

FIG. 153.—Terylene fabrics in use in a modern laundry.

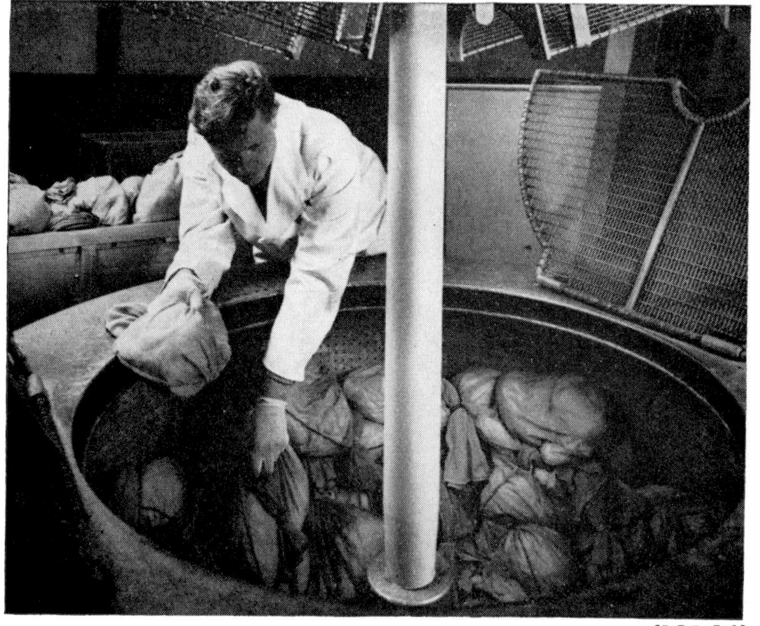

[*I.C.I., Ltd.*]

FIG. 154.—Terylene dye bags being loaded into a centrifuge.

Terylene has also been used (100 per cent) as a woven felt in paper-making to convey the paper web over heated cylinders in the last stages of drying; the requirements for a paper machine felt are that it shall be resistant to a temperature of 120° C. and moisture, resistant to acid which arises from the use of aluminium sulphate as a filler for the paper, as well as to microbiological attack when the machine is left standing, and shall be resistant to the abrasion caused by the continuous flexing and rubbing of the felt at high speeds; in one instance a Terylene felt lasted for two and a half years, whereas the average life of the felts formerly used was only about six months.

Conveyor belts in chemical factories are traditionally made of rubber with a cotton reinforcement, but if the rubber becomes punctured acid can rapidly destroy the cotton; in one chemical factory cotton reinforced belts lasted about ten weeks, whereas a (100 per cent) Terylene-reinforced belt has lasted about a year. It might have been thought that glass-fibre reinforcement would have been ideal for this purpose; it was, in fact, tried, and gave a life of twenty weeks before breaking down due to failure of the bond

FIG. 155.—Terylene in use in a hydrated lime elevator belt; its good heat resistance gives it a long life. [*I.C.I., Ltd.*]

between the glass fibre and the rubber covering. Those properties that make Terylene especially good are its chemical resistance, strength and resistance to abrasion. Fig. 155 shows its use in a hydrated lime elevator belt.

Most of the synthetic fibres have good dielectric properties, but none of them (apart from the very expensive Teflon) has such good resistance to high temperatures as Terylene, and since electric gear is inclined to run hot, this resistance to heat is valuable. Hitherto insulants have mainly been required to withstand 105° C., but there is a trend towards higher temperature motors, and it is likely that the standard British general purpose motors of up to 50 h.p. will in future incorporate class E (suitable for use up to 120° C.) insulation materials. Terylene is included in the I.E.C. (International Electrical Technical Commission) classification as a class E insulator. Terylene electrical fabrics are heat-set at 220° C. on a hot-air pin stenter, to obtain adequate dimensional stability to the varnishing temperatures. The fabrics can be hot slit to form straight or bias cut tapes. The use of Terylene tape is illustrated in Fig. 156. Terylene filament yarns are also used for lapping bare wires and Terylene braided sheaths for covering battery leads.

Fire hose is in the main of two kinds: unlined and lined. Unlined hose is made from cotton or some similar fibre that swells when it is wetted so that all the interstices between the threads are filled up and water does not pass through them; such a hose could not be made from synthetic fibres because there would be practically no swelling and water would leak along the length of the hose. Lined hose is lined with rubber or P.V.C. and can be made with almost any fibre as a covering. Terylene (high-tenacity type) has been found to be very suitable; its high strength makes lighter structures possible, so that the hose can be more easily carried, its low elongation and high modulus mean that the hose does not stretch much, and because of its low moisture absorption, used hose dries quickly, and even if neglected will not rot. Usually Terylene is used in the weft with a cotton warp, but 100 per cent Terylene hose are also being developed. Large quantities of $1\frac{1}{2}$-in.-diameter hose reinforced with Terylene have been supplied to Canada for forest-fire fighting—the required burst pressure was best obtained with Terylene.

Terylene has been used for ropes, fish-netting, sailcloth (Fig. 157) and similar purposes. The attributes that make it good for these purposes are strength, flexibility, low stretch (high tenacity type), durability and resistance to sea water, sunlight, heat and micro-organisms. Splicing with Terylene needs special attention, all splices should have at least four tucks and be firmly whipped—

[*I.C.I., Ltd.*]

FIG. 156.—G.E.C. 500 h.p. motor insulated in part with bias cut varnished Terylene tape.

[*I.C.I., Ltd.*]

FIG. 157.—Terylene sails and ropes. The majority of sails are made of Terylene nowadays.

Terylene will slip far more easily than manila. Terylene ropes sometimes give more than fifteen times the life of an ordinary rope; they comprise tug and mooring ropes, yacht cordage and mast support ropes. Tests were carried out by a lighterage company operating on the Thames to compare 2-in.-diameter ropes made from sisal which is ordinarily used, nylon and Terylene. Results were as shown in the following table:

Fibre used for rope.	Life in days.	Cost in pence per day of life.
Sisal	6	194
Nylon	65	320
Terylene	137	152

The longer life of Terylene compared with nylon used in ropes seems to be due to its considerably lower stretch, with consequent low inter-strand friction.

So far as concerns Terylene for fishing-nets, it is excellent, easy to handle, light, tough and long-lasting. The trouble is that nets are often lost accidentally, and it is more expensive to lose a Terylene net than one of cotton; it is no good having a net with a very long expectation of life if it is going to be lost in its first few trips. Sail-cloth made from Terylene has proved popular with yachtsmen (Fig. 157), and some of the best American yachts have raced with Dacron sails; present opinion in this country seems to be that the Terylene sail should be at least 90 per cent as heavy as the cotton sail, otherwise in very strong winds it may suffer some permanent elongation; one would expect this view to be modified with more experience so that advantage can be taken of the very high strength of Terylene. Terylene awnings, hatch and lifeboat covers are suitable for use on oil-burning ships, where ordinary canvas is rapidly degraded by the acidic diesel fumes. Terylene fabric has also been used successfully as a basis for tarpaulins—lighter and longer lasting than the usual.

The chemical resistance of Terylene towards acids makes it useful for protective clothing and it is used on acid-handling plants. A special use is for anode bags in tin-nickel plating where the electrolyte is hydrofluoric acid, which no fibres other than Teflon and Terylene will withstand. The microbiological resistance of Terylene has enabled it to be used for some water filtration work, notably in the Thames valley, where cotton and wool had been unsatisfactory.

Blends of Terylene or Dacron staple with wool and with viscose rayon are considered to be amongst the most important uses of these polyester fibres. These are discussed on pp. 670–673.

The biggest single industrial use for man-made fibres is in tyres for cars and lorries. This was the preserve of cotton, then of high-tenacity rayon, then nylon took a big share and now polyester fibres are apparently with some confidence being used in tyres. The only real trouble with nylon is that of flat-spotting; if the car stands in the showroom for a week or two, flats develop where the tyres touch the floor, although in fact they are soon lost when the car is on the road. Terylene is less subject to creep and to this defect. Apparently the main difficulty with Terylene has been that of bonding it to rubber; a two-stage dip has been necessary and expensive. I.C.I. have announced that a product they call Pexul will do the job with one dip. This will help; so presumably will the reduction that we have already seen in Terylene prices. The way things have shaped and are shaping in the chemical industry, polyester is going to be cheaper than nylon. But it may still be that for a time polyester tyres will be premium tyres, dearer than those with a cotton or a nylon carcase. What will decide is whether polyester makes better tyres than nylon. It may not do. However, Monsanto (the old Chemstrand) are reported to be building a polyester tyre yarn plant at Decatur in Alabama. The total U.S. market for tyre yarn was, in 1968, about 450 million lb. a year, and Monsanto expect it to increase to 700 million lb. Evidently they have great confidence in polyester for tyres. In 1968–69 the consumption of fibres for tyre cord was probably approximately (in million lb.):

	High tenacity cellulose.	Nylon.
In U.S.A.	150	350
In U.K. and Western Europe .	220	120

Du Pont have made a special tyre fibre, Nylon 44, in America; it is claimed to be less subject to flat-spotting than the usual nylon. A plant to produce 100 million lb. per annum is being built.

Terylene–Cotton Blends

Considerable success has been achieved by blends of 2 parts Terylene staple to 1 part Egyptian cotton. The cotton gives an opacity which is desirable in shirts and blouses; it also gives the fabric greater moisture absorbency and improves the handle.

Dacron Type 62

Dacron Type 62 differs from standard Dacron in two important respects:

1. Its filament cross-section is nearly triangular (Fig. 158) instead of round.
2. It can be dyed not only with disperse dyes but also with basic colours, resulting in easier dyeing and particularly in brighter colours.

Cross-section. The section is more triangular than is that of trilobal Antron nylon. The method used is doubtless the same in principle: the use of a specially shaped jet, possibly triangular or perhaps intermediate between those shown in Figs. 125 (*c*) and 159.

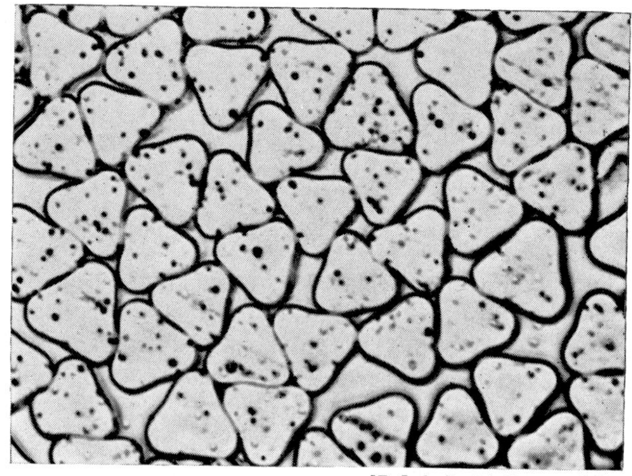

[*E. I. du Pont de Nemours & Co.*]

FIG. 158.—Photomicrograph of cross-section of Dacron Type 62 fibres. Magnified several hundred diameters, probably 15 denier × 250.

It must not be forgotten that the drawing process may modify the sectional shape of the spun filament. The triangular cross-section gives an improvement in aesthetic qualities (better handle and appearance) over circular section Dacron.

[*After Holland*]

FIG. 159.—*Left:* Design of jet described in U.S. Pat. 2,939,202 (M. C. Holland, assignor to du Pont) for spinning polyester filaments with convex triangular shape. *Right:* Cross-section of polyester filament spun from such a jet. Both diagrams represent a magnification of the order of 100 times.

Affinity for Basic Colours· Ordinary Dacron can be dyed satisfactorily only with disperse colours; there are no suitable sites in the polyester molecule to adsorb acid or basic dyestuffs. Dacron Type 62 is different, it can be well dyed with basic dyestuffs; it must have some very convenient acid groups in its molecule to which the basic dyestuffs may join. Perhaps the patent literature can give us the principle, if not the detail, of how these acid groups got there.

Du Pont in Belgian Patent 549,179 (1959) have described polymers which consist essentially of polyethylene terephthalate but which include in their chain about 2 or 3 molar per cent of sodium-3,5-di-(carbomethoxy)-benzene sulphonate

$$\mathrm{CH_3OC(=O)}-\underset{\mathrm{SO_3Na}}{\mathrm{C_6H_3}}-\mathrm{C(=O)OCH_3}$$

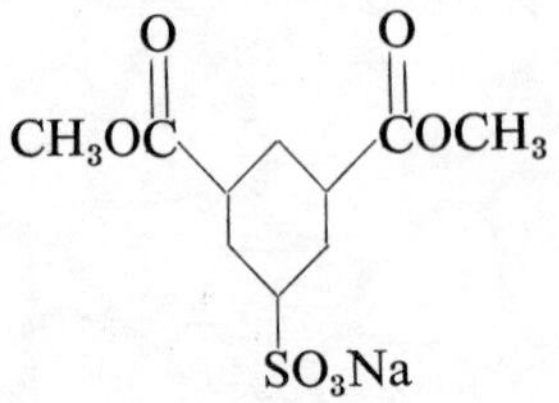

The sodium atom can be replaced by other atoms. The inclusion in the polymer of this sulphonic acid grouping by copolymerisation enhances the affinity of the Dacron for basic dyestuffs.

Alternatively to the acid specified above, which is sulphoisophthalic acid, there can be used the corresponding 2,5 acid, that is sulphoterephthalic acid instead of the 3,5 acid. In effect the sulphoterephthalic acid takes the place of a small part of the terephthalic acid in the condensation polymerisation. Both are used as their methyl esters. In a typical example (Belgian 549,179) the following are mixed (all parts by weight):

1·6 Potassium-2,5-di(carbomethoxy)benzene sulphonate
49 Dimethyl terephthalate
34·5 Ethylene glycol
0·022 Manganese acetate
0·015 Antimony trioxide

The corresponding molecular weights for the major components are:

Potassium-2,5-di(carbomethoxy)benzene sulphonate, 312
Dimethyl terephthalate, 194
Ethylene glycol, 62

so that the molar percentage of the sulphonate of the total sulphonate and terephthalate esters is:

$$\frac{1{\cdot}6}{312} \div \left(\frac{1{\cdot}6}{312} + \frac{49}{194}\right) \times 100 = 1{\cdot}99 \text{ (\textit{i.e.}, 2 molar per cent)}$$

The quantity of ethylene glycol present is more than double that which is theoretically required; one presumes that the excess is used to give good heat conductance throughout the mix in the early stages (dimethyl terephthalate, m.p. 140° C.) of the polymerisation and is boiled off in the later stages. The mixture is heated for 3 hr. to a maximum of 220° C., methanol being split off and lost; the pressure is reduced to 1 mm. Hg and the temperature raised to 275° C. for 5 hr. At this stage the polymer has an intrinsic viscosity of 0·53. The actual spinning of a polymer very similar to this, spinning into trilobal filaments, is described in U.S. Patent 2,939,202. As a delustrant, 0·3 per cent titanium dioxide is added. The polymer is melt-spun at 296° C. through triangular orifices (side of the triangle, 12 mils). The filaments immediately after extrusion are quenched in cross-flowing air at 70° C. and with a velocity of 65 ft./min. They are collected at 1,500 yd./min. as 152 denier 50 fils yarn. This is hot drawn at 107° C., using a $\frac{1}{4}$ in. diameter ceramic snubbing pin, a draw ratio of 2·14 and a wind-up speed of 600 yd./min. to give a 70 denier 50 fils yarn with a tenacity of 3 grams per denier and an extension at break of 24 per cent.

The yarn so made had a good affinity for basic dyes which will do no more than stain ordinary Terylene or Dacron. For example, treatment at 125° C. with 3 per cent (on the weight of the fibre) Genacryl Blue 60 in water under pressure coloured it a fast level mid-blue. Ordinary Dacron absorbs practically no dyestuff under such conditions.

Because of the triangular section soiling was reduced, cover improved, and lustre highlights and surface sheen were good. One or two slight liberties have been taken in combining examples from two different patent specifications, but not enough to invalidate any principle. One has to admit that to be able to make such a polyester, with built-in sulphonate groups, and with a triangular filament section not unlike that of real silk, is indicative of an advanced mastery of the art of making fibres. Forty years ago nylon and Terylene were unthought of, the science of fibre structure was being born, and in those days to spin cellulose acetate faster than 60 metres per minute was asking for trouble. The fibre scientists have done well in the last few years. To spin at a mile a minute through jets of intricate shape is one of their finest achievements.

Fortrel

Fortrel is a polyester made by Celanese Fibers Company, a joint subsidiary of American Celanese and I.C.I.; it appears to be another polyethylene terephthalate, similar to the original Terylene and Dacron.

KODEL

Kodel is a polyester fibre that has been developed by Eastman Chemical Products Inc., a subsidiary of Eastman Kodak Company; it was introduced in the autumn of 1958 and has been well tested and has gained a large measure of acceptance. It is different chemically from the other polyester fibres on the market; it can be distinguished chemically from the others in the following way.

Boil a sample of the polyester fibre for 10 min. in a 10 per cent solution of hydrazine in butanol. Remove the specimen and allow the solution to cool. If on cooling a precipitate forms, the polyester is not Kodel; if no precipitate forms, the polyester is Kodel. Kodel is not soluble in boiling 10 per cent hydrazine in butanol whereas the other polyester fibres are, and that is the reason underlying the test.

Chemical Structure

The chemical structure of Kodel has not been disclosed by the maker, but examination of the patent literature (U.S. Patent 2,901,466 of 1959 is relevant) suggests that it is a polymer of 1,4-*cyclo*hexane-dimethanol and terephthalic acid. (Disclosure made late in 1962 confirmed this suggestion.)

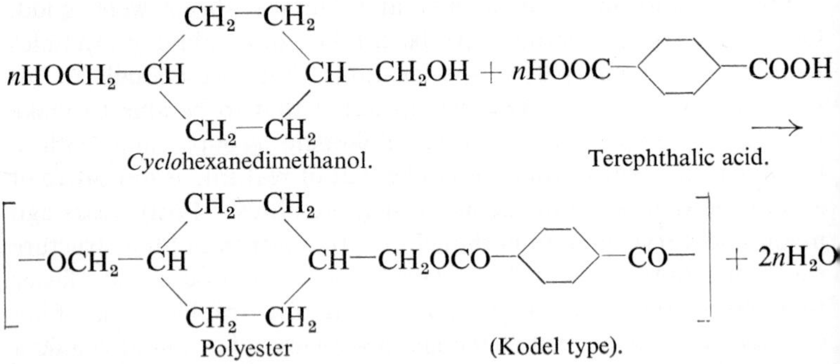

$$n\mathrm{HOCH_2{-}CH}\begin{matrix}\diagup\mathrm{CH_2{-}CH_2}\diagdown\\ \diagdown\mathrm{CH_2{-}CH_2}\diagup\end{matrix}\mathrm{CH{-}CH_2OH} + n\mathrm{HOOC{-}C_6H_4{-}COOH} \longrightarrow$$

*Cyclo*hexanedimethanol. Terephthalic acid.

$$\left[\mathrm{{-}OCH_2{-}CH}\begin{matrix}\diagup\mathrm{CH_2{-}CH_2}\diagdown\\ \diagdown\mathrm{CH_2{-}CH_2}\diagup\end{matrix}\mathrm{CH{-}CH_2OCO{-}C_6H_4{-}CO{-}}\right] + 2n\mathrm{H_2O}$$

Polyester (Kodel type).

1,4-*Cyclo*hexanedimethanol exists in two stereoisomeric forms, *cis* and *trans*. If the *trans* isomer is used exclusively in the formation of the polymer, the melting point of the latter is as high as 320° C.; if all *cis* isomer is used, the melting point is 260° C. In practice a mixture of the *cis* and *trans* isomers of *cyclo*hexanedimethanol is used and these two give a polymer with terephthalic acid having a melting point of about 290° C. The actual melting point of Kodel is 290–295° C. and it is now known to be a straight condensation polymer of terephthalic acid and mixed *cis* and *trans* 1,4-*cyclo*-

hexanedimethanol. It is surprising that the polymer does have such a high melting point; a very similar polymer from *p*-xylene diol

$HOCH_2-C_6H_4-CH_2OH$ and terephthalic acid melts as low as

253° C. A melting point of 290° C. is really higher than is needed for most domestic purposes; it is 25° C. higher than nylon or Terylene. This opens up the possibility of co-polymerising a third component and still retaining a melting point as high as that of nylon. The third component might be:

1. A replacement of part (say one-quarter) of the terephthalic acid by a commercial mixture of isomers of sebacic acid, which would cheapen the polymer; or
2. A part replacement of the *cyclo*hexanedimethanol by an amine such as hexamethylene diamine; this would increase the affinity of the polymer for acid dyes. The product would then be a polyester–polyamide.

Manufacture

Dimethyl terephthalate and *cyclo*hexanedimethanol are reacted together at 190–200° C. using titanium *iso*butoxide as an ester exchange catalyst. The methanol distils off rapidly and the temperature is raised to 270° C. and finally, under vacuum, to 300° C. or just over and kept there for an hour. The polymer is an opaque polyester. Fibres are melt-spun through round holes of 0·3 mm. diameter into air at room temperature and wound up at about 500 metres per minute. They are hot-drawn by stretching to 4½ times their original length by passing round a roller heated to 120° C., the yarn being pulled off the roller nearly five times as fast as it is fed on to it.

Properties

In a general way, the properties of Kodel are similar to those of the other polyesters but there are some significant differences. Like the other polyesters the fibre is made in several modifications, the two most important being standard Kodel and Kodel HM.

Tenacity and Elongation. The characteristic curves are shown in Fig. 160 which should be compared with Fig. 146. Tenacity and extension are both lower than of Terylene or nylon. The characteristics wet are almost the same as dry.

Moisture Regain. The same as Terylene, *i.e.*, 0·4 per cent.

Elasticity. Recovery from stretch is excellent, practically as good as nylon, better than any other fibre except wool. Will this give Kodel a chance in the stocking industry or the tyre market?

Cross-Section. Round.

Specific Gravity. 1·22, intermediate between nylon and Terylene, very different from all the other polyesters which are about 1·38.

Chemical Resistance. Excellent against acids and alkalis, perhaps not quite so phenomenal as that of Terylene against acids. For example, after 6 weeks in 10 per cent hydrochloric acid at 50° C.,

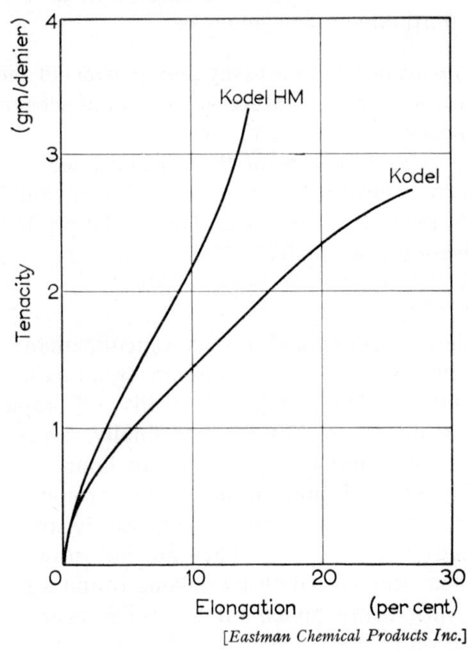

Fig. 160.—Stress–strain diagram for Kodel HM and standard Kodel.

Kodel retained only about 75 per cent of its strength. Trichlorethylene and methylene chloride shrink it and must be avoided in cleaning; perchlorethylene is considered safe but may induce very slight shrinkage. Swollen by phenol, acetone, ethyl acetate and toluene; dissolved by 60 phenol/40 tetrachlorethane (wt/wt) at 100° C.

Thermal. Kodel melts at 290–295° C., wonderfully high; hydrogen bonding between the molecules must be very strong. Safe ironing temperature is 215° C. The fibres will burn slowly but in practice the fire risk of Kodel (provided that it is undecorated with other fibres) is negligible, as the burning fabric melts and drops off. This behaviour is common to nylon and Terylene as well; what is not always appreciated is that if a dress made mainly with one of these fibres is decorated, *e.g.*, with panels of rayon or cotton, then the whole dress, polyester and all, may burn like a torch if ignited.

Boiling water. Fibre shrinkage of the standard quality is only 1 per cent—a very good feature. Kodel HM has a little higher shrinkage at 2 per cent.

Heat setting. Fabric should be set after dyeing, at 205° C. A special Kodel Type II for use in wool blends can be set at 160° C.

Good recovery from stretch and very high melting point distinguish Kodel.

Dyeing

Kodel S is available solution-dyed in black, navy and brown. Other colours will doubtless be forthcoming.

The standard fibre is a good white and does not alone require a bleach. If it is blended with a fibre that does require bleaching either hydrogen peroxide or sodium hypochlorite is suitable.

Disperse dyes are usual and can be applied either by high temperature (120° C.) pressure dyeing or with the use of carriers at 98° C. The phenylphenols are unsuitable carriers; they cause too rapid a strike; emulsified butyl benzoate and methyl salicylate are preferred. Azoic dyestuffs can also be used.

Uses

Ease of care is the property that is emphasised. There may be advantages in handle and crease-resistance compared with other polyesters. Kodel is produced in staple and tow. It can be used 100 per cent. Type II is particularly suitable for blending with wool because it can be set at a relatively low temperature; Type HM is recommended for use with cotton and high wet modulus rayons such as Zantrel. Its excellent recovery from stretch may open big markets for the fibre. The chances are that Kodel will be used to begin with in those applications that Terylene has been successful in—except for such as require very great strength, as do some industrial uses. If Kodel has the edge in shape and size retention that will help it a lot. It is not easy to see just where its higher melting point is going to be useful; more useful it would be to sacrifice a little (still keeping it above 250° C.) and include some nitrogenous material in the polymer to enable the fibre to be dyed with acid dyestuffs.

VYCRON

Vycron is manufactured by Beaunit Mills Inc. (North American Rayon Corporation) at Elizabethton, Tenn. Production of staple fibre and tow (200,000 denier) may be in the region of 12 million lb. per year; production started in 1958.

Chemical Constitution

It is a polyester differing from other polyesters in its total chemical constitution; it is advertised as being produced from Vitel, a new polyester resin made by Goodyear. Filament cross-section is roundish but not nearly so regular as Dacron; in fact the published photomicrographs suggest that the fibre has been spun from solution rather than from the melt. The manufacturers have not disclosed the precise chemical structure of the fibre but it appears to be a modified polyethylene terephthalate. In Japan, Toyo Spinning Co., by agreement with Goodyear, are to make 10 million lb. of Vitel per year.

Properties

Mechanical. Tenacity is high at 5·6 grams per denier combined with 35 per cent elongation at break; taken together rather better than Dacron. Modulus of elasticity is 0·5 grams per denier per cent, *i.e.*, a load of 0·5 gram per denier must be applied to give a 1 per cent stretch. Recovery from a 2 per cent extension is very good, from a 5 per cent extension not so good.

Moisture Regain. 0·4 per cent at 65 per cent R.H., 0·6 per cent at 80 per cent R.H., 0·8 per cent at 95 per cent R.H. About the same as Dacron.

Specific Gravity. 1·38, the same as Dacron, although at one time the figure given to Vycron was 1·36.

Heat Resistance. Melts at 235° C., intermediate between nylon 6 and nylon 66; not so good as the latter, nor so good as Dacron. Safe ironing temperature 175° C. Vycron will burn slowly but melts and drops off like nylon or Dacron.

Resistance to Chemicals. Good, not affected by common solvents or cleaning agents, or chemicals likely to be met.

Weathering and Light Resistance. Good.

Biological Resistance. Unaffected.

Dyeing

The affinity of Vycron for dispersed acetate colours and developed azoics is reported to be slightly greater than that of some polyesters.

Nevertheless in medium and dark shades it is necessary to use either high temperature pressure dyeing or else carriers such as *o*-phenylphenol, trichlorbenzene and methyl salicylate.

Finishing

Vycron can be heat set at 185° C. for 45 sec. Heat setting and singeing are recommended to keep down the pilling. Singeing must be carried out after dyeing, otherwise the melt balls formed would dye more heavily. Creases can be inserted (and will be retentive) at temperatures lower than the setting temperature; it is easier to insert low temperature creases in Vycron than in some synthetic fibres.

Uses

Good fibre to fibre cohesion is claimed so that staple yarns spun from it are stronger than those spun from other polyesters. The fibre was about 12 per cent cheaper than Dacron but this advantage has now (1965) disappeared. It is claimed that pilling does not normally occur and if this is borne out in practice it will be of great help. The fibre is intended for blends with cotton and wool and rayon, *e.g.*, a tweed made from 74 Vycron/26 Bemberg, summer trousers from 50 Vycron/50 cotton. It is marketed in the following forms:

Continuous filament for industrial use, 420 denier and higher in 1·5 and 3 filament denier.
Staple in 1·5, 3·0 and 4·5 denier.
Crimped tow for Pacific Converters in $1\frac{1}{2}$ and 3 filament denier; total denier 200,000 for industrial and apparel fabrics.
Tow for Direct Spinners. (This direct spun yarn has a high shrinkage, 12 per cent in width of yarn used as weft, from loom to finished fabric.)

It remains to be seen how it goes in blouses, skirts, underwear, dresses, suits, raincoats and so on, but the manufacturers are progressive people and it is unlikely they would have touched it unless it had something useful to offer.

Grilene

Grilon (p. 374) is a nylon, but Grilene is a newer fibre and is a polyester. At first it was different chemically from the other polyesters that we know in that it had some ether groups in its molecular chain; it was therefore properly described as a polyester–polyether.

It is manufactured by Grilon S.A. (Emser Werke AG) at Dormat-Ems GR Switzerland. The monomers that were co-polymerised to yield Grilene polymer were terephthalic acid, glycol, and *p*-hydroxybenzoic acid. The first stage of manufacture was the formation of an ether by condensing the glycol and *p*-hydroxybenzoic acid; the second stage was one of esterification with the terephthalic acid. But a change has come over the scene. After supplying the polyester–polyether with its splendid dyeing properties (dye at the boil without carriers) for a year or so the makers are now supplying a straight polyester apparently basically the same as Terylene, and have regrettably given up the polyester–polyether. Thus, the big advantage of improving the dyeing by the presence of ether groups has been sacrificed. The homopolymer is similar in dyeing properties to other homopolymeric polyesters. The physical properties of Grilene fibre are not very different from those of Terylene. Grilene is made in the following four different types, all staple fibre, not continuous filament:

T-type—a high tenacity fibre for technical and industrial uses, made in deniers of 2·6 and 6 and in various staple lengths.

W-type—a lower tenacity fibre with a much higher extension, in effect a wool-type intended for blending with wool, especially in the worsted and carpet fields. It is made in staple length roughly from 1½ to 9 in. and in various deniers.

AP-type—anti-pilling with a significantly lower tenacity intended for the wool, rayon and linen fields, especially for fabrics of an open, loose structure. Its resistance to abrasion and to flexing is lower than of the other types; its characteristics more nearly approach those of the natural fibres than do the other types. It lacks the superlative strength of say the T-type, but it is a kindlier fibre. Staple lengths and deniers are similar to those of W-type.

B-type—a cotton type made in short staple (roughly 1–1½ in.) and fine deniers of 1·2 and 1·5 and obviously intended for blending with cotton.

A trilobal cross-section is also available in the W-type and so is spun black in W- and AP-types. Top and tow are both available for the worsted and wool trade.

The differences in the physical properties of the four types of Grilene fibre are clearly seen in the stress–strain curves in Fig. 161.

Their other properties are typical of polyester fibres, *e.g.*, wet strength is 98 per cent of dry strength, moisture content (regain) is about 0·4 per cent, elastic recovery is 95–100 per cent from a 2 per cent elongation and 80–90 per cent from a 5 per cent elongation, shrinkage in boiling water is 1–2 per cent, specific gravity is 1·38, melting point is 250° C. Chemical resistance, too, is good. As with all polyesters, there is some liability to hydrolysis, especially with hot mineral acids. But, all told, the resistance is good, *e.g.*, blends of Grilene/cotton can be scoured, or mercerised, or vat dyed, under the conditions normally used for cotton, without damage. Furthermore, the reduction clearing with hydrosulphite in order to

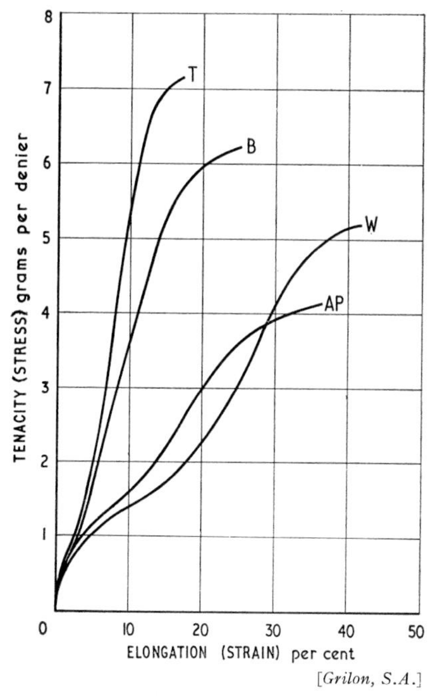

[*Grilon, S.A.*]

Fig. 161.—Stress–strain curves of the four types of Grilene.

increase the fastness of dyes has no effect on the physical properties of Grilene. Resistance to light and weathering is good, better than that of natural fibres or of polyamide. Resistance to heat is important; in hot dry air temperatures up to 175° C. is safe.

Dyeing can be carried out with all the dyestuffs recommended for polyester fibres. Dyeing can be carried out at the near-boil if carriers are used. Pressure dyeing is best done at 125–130° C.; if the Thermosol process (p. 417) is used the temperature should be 190–210° C. Finally, heat setting is suitably carried out at 180–195° C. for 45 sec. The fibre will shrink about 6 per cent if exposed to a temperature of 180° C. for 15 min. Grilene is reported to be making rapid progress and to have been widely accepted by the trade.

A-Tell

A new polyester known as A-Tell has been introduced by the Nippon Rayon Co., Japan, thus making still one more original contribution to the world of fibres. The new fibre is made from *p*-hydroxybenzoic acid and ethylene oxide; this latter is equivalent to ethylene glycol from which the elements of water have been removed. Perhaps the first reaction is that of ester formation:

$$HO-C_6H_4-COOH + \underset{\diagdown O \diagup}{CH_2-CH_2} \longrightarrow$$

$$HO-C_6H_4-COOCH_2CH_2OH$$

followed by condensation to yield a polyether–polyester

$$-O-C_6H_4-COOCH_2CH_2O-C_6H_4-COOCH_2CH_2-$$

Polyethyleneoxybenzoate.

but the information available is too scanty to be sure. It might be that there is more acid present than this equation would indicate. The little that is known about it is very reminiscent of that polyester–polyether which was the original Grilene.

The Japanese claim that the fibre is like natural silk (*cf.* du Pont and Quiana; it is odd that the fibre synthesists are searching again for Artificial Silk). A-Tell is said to have a similar specific gravity to natural silk; this is 1·33 raw and 1·25 boiled off, *i.e.* with the gum removed; one might have guessed that it was the value of 1·25 that was referred to, but one would have been wrong, for A-Tell's specific gravity is 1·34. Its tenacity varies according to manufacturing conditions from 4 to 5·3 and is the same wet as dry. Elongation is 15–25 per cent, moisture regain 0·4–0·5 per cent, and melting point 222° C. It is available in deniers of from 30 to 90, but only in small quantities because production only started in May 1968; by the end of 1969 it should be about 5 tons per day. In Japan the price of A-Tell is $4·46 per lb., the equivalent of 37*s.*/lb. Doubtless it will fall, but even so it is hard to see how any fibre with a moisture regain of only 0·5 per cent can be much like real silk. One of the makers, asked if A-Tell resembled Quiana, has been reported (July 1968) as having said that no sample of Quiana had yet reached Japan.

The handle of A-Tell is said not to be waxy. Perhaps one would agree that most synthetic fibres do have a slight smoothness, slickness and waxiness compared with natural silk. Even if one falls short, it is good to have a worthwhile target. Real silk and wool are still the best targets for manufacturers of apparel fibres.

FURTHER READING

J. R. Whinfield, *Nature*, **158**, 930 (1946).
G. Cook, *Silk J. and Rayon World*, **270**, 28 (1946).
J. R. Whinfield and J. T. Dickson, British Pat. 578,079 (1941).
Silk and Rayon, **1946**, 1380.
" Papers Presented at the Technical Conference on Dyeing of Orlon Acrylic Fiber and Dacron Polyester Fiber. Wilmington, Aug. 5, 1952 ", E. I. du Pont de Nemours & Co.
J. R. Whinfield, " The Development of Terylene ", *Text Research J.*, **23**, 289 (1953).
A. Pace, U.S. Pat. 2,578,899 assigned to du Pont (superstretching).
I.C.I. Ltd., " Dyestuffs for ' Terylene ' Polyester Fibre ".
I.C.I. Ltd., " The Dyeing of ' Terylene ' Polyester Fibre ".
Anon., " Kodel ", *Modern Text. Mag.*, **40**, 43–45 (Feb., 1959).
I.C.I. Ltd., Dyehouse leaflet 819 (Grilene).
R. W. Moncrieff, " Grilene—Properties and Colouration ", *International Dyer*, 379, 381–2, 3 September (1965).
R. W. Moncrieff, " Polyester Cords for Premium Tyres ", *Text. Manufacturer*, 51, 52 (Feb., 1965).

CHAPTER 24

SNAP-BACK FIBRES

LYCRA, VYRENE, SPANZELLE

THE fibres discussed in this chapter are those which resemble rubber in that they have a high extensibility and highly retractive forces which derive from their chemical nature. Some stretch nylon *yarns* such as Helanca have high stretch and recovery but these properties are due to special configurations of their filaments brought about by various devices such as special forms of crimping and twisting; the individual *filaments* of such yarns are not elastic in the sense that rubber is.

Fibres which have an extension at break in excess of 200 per cent, and have also the property of rapid recovery when the tension is released, are known as elastomers. Undrawn nylon has an extensibility of several hundred per cent, but practically no recovery and is not an elastomer. First attempts to make true rubber-like fibres centred on the modification of nylon; these were not completely successful and this line of attack has been abandoned. The polyurethane fibres have been more satisfactory and some of them are now on the market. In Britain, Courtaulds (Elastomeric Fibres Ltd. of Coventry) make Spanzelle, a multi-filament yarn in which the filaments have been fused together. Vyrene is made by the Dunlop Rubber Co.; du Pont are already making Lycra at Maydown in Northern Ireland at the end of 1969; Glospan is made by the Globe Elastic Thread Co. Ltd., and it also is a fused multifil yarn and so is Blue C Elura made by Polythane Fibres Ltd. Snap-back fibres originated in the U.S.A.; the first two were du Pont's Lycra and the U.S. Rubber Co's Vyrene. Also in the U.S.A. is Monsanto's Blue C Elura, another fused multifil. Glospan is made there by the Globe Manufacturing Co. and Spandelle by Firestone, who were associated with Courtaulds in the manufacture of Spanzelle. The fibres are all very similar, despite the ramifications of the commercial interests; they are an American development, and those that are made in Britain have been made with American help. In 1968 similar fibres were also being made in West Germany, the Netherlands and Japan, again with American help. The fibres are known generically as spandex fibres. Those synthetic fibres that have already been discussed—*i.e.*, the polyamides, polyureas and polyesters—are all condensation polymers. Those organic fibres still to be discussed are all vinyl-type addition polymers. The polyurethanes do not fit at all satisfyingly into either group. So, perhaps, here between the two main groups is as good a place as any to deal with them.

Modified Nylon

Although the elastic nylons have been abandoned, they were of considerable academic interest. The general idea was to make use of the high extensibility of undrawn nylon, but by the inclusion of bulky side groups to reduce hydrogen bonding and make it impossible for the nylon molecules to fit together nicely and find restful positions; instead they were bulky, ungainly, and when stretched were always in a state of strain so that when tension was released they reverted to their original positions and the nylon retracted.

Nylon 610 made from hexamethylene diamine and sebacic acid was the basis, but a part of the hexamethylene diamine was replaced by a substituted diamine carrying butyl or *iso*butyl groups, which are very bulky. The product had a tenacity in excess of 1 gram per denier and an extensibility of up to 400 per cent and a rapid recovery of 95–99 per cent; it was certainly an elastomer. Another method that was used was the after-treatment of nylon 66 with formaldehyde and methyl alcohol; this introduced methylene cross-linkages and methoxymethyl side groups. These bulky side groups and the cross-linkages both contributed to retraction from stretch and the product was rubbery, but it was not very stable.

Such fibres were never quite right; they were either unstable chemically or showed excessive stress decay; the first meant that garments made from them would not withstand repeated hot water washing, the second that they would gradually get bigger and sag. It is a pity that this line of attack was never quite successful; the nylon manufacturers poured millions into it. If it ever is completed, the products will probably be much stronger than the polyurethane spandex fibres.

Polyurethane Fibre—Perlon U

Modern spandex fibres are polyurethanes, but the first polyurethane fibre to be spun (first spun in Germany during World War II) was not an elastic yarn but an ordinary " hard " yarn, rather like nylon but not so good. It seems apposite to describe it briefly here, because it had some chemical features in common with the new spandex fibres.

It was made by reacting 1,4-butanediol with hexamethylene di-*iso*cyanate at 195° C., the two adding together to give the polyurethane:

$$n\ HO(CH_2)_4OH + n\ OCN(CH_2)_6NCO \longrightarrow$$

Butanediol. Hexamethylene di-*iso*cyanate.

$$[-O(CH_2)_4OCONH(CH_2)_6NHCO-]_n$$

Polyurethane.

The polymer was melt spun and the resulting fibre was known as Perlon U. It melted at 175–180° C. which is rather low for apparel uses, it was stiffer than nylon and was used for bristles, artificial horsehair, filter cloths and the like. It should not be confused with Perlon L (now known familiarly as " Perlon ") which has made great strides and is a nylon 6, nor with Perlon T which was a rather unsatisfactory German war-time nylon 66. Perlon U was entirely a polyurethane, the spandex fibres are block polymers which contain urethane groups—there the similarity ends, Perlon U was not an elastic fibre.

Synthesis of Spandex Fibres

Lycra. Spandex fibres are polyurethanes; the exact compositions of the two fibres that have been marketed have not yet been disclosed, but their general type can be inferred from a study of the patent specifications. In U.S. Patent 2,692,873 (1954) du Pont describe elastomers made by copolymerising polyethylene glycol (molecular weight 750–10,000) and tolylene-2,4-di-*iso*cyanate in the presence of water and of a small quantity of an acid chloride. The glycol, the di-*iso*cyanate and the acid chloride are heated at 50–100° C. for 2 hr. to form a " prepolymer " (low molecular weight linear polymer). Then water is added and heating continued. The product is cured at 140° C. and 200 atmospheres pressure to introduce the retractile cross-linkages. It has a tensile strength of 2,000 lb./in.^{2} (less than 0·2 gram per denier) and an elongation at break of 500 per cent. It has to be remembered that a tenacity of 0·2 gram per denier in a fibre which has an extension of 500 per cent means that the tenacity of the stretched fibre just before it does break is 1·2 gram per denier. Later patents have described methods of improving the strength of polyurethane fibres but as it is not known which of them have been put to use, they will not be discussed here.

Vyrene. The U.S. Rubber Co. have described some of their polyurethanes in U.S. Patent 2,751,363 (1956). According to this a mixture of ethylene and propylene glycols is esterified with adipic acid using excess of the glycols so that there results a polyester with terminal hydroxyl (not carboxylic acid) end groups; this is made as a low polymer with a molecular weight of about 2,000. This polyester is reacted at about 120° C. with *p,p′*-diphenylmethane di-*iso*cyanate using 2 moles of the latter to one of the polyester. This gives a liquid " polyester–di-*iso*cyanate " intermediate which is in fact a linear polyurethane having terminal *iso*cyanate groups.

$$7\,HOCH_2CH_2OH + 3\,HO\underset{\displaystyle CH_3}{\underset{|}{C}}HCH_2OH + 9\,COOH(CH_2)_4COOH \longrightarrow$$

Ethylene glycol. Propylene glycol. Adipic acid.

$$HOCH_2CH_2OCO(CH_2)_4COO\underset{\displaystyle CH_3}{\underset{|}{C}}HCH_2OCO(CH_2)_4CO \ldots OCH_2CH_2OH$$

Polyester of molecular weight 1,652 (or shortly HOROH).

$$OCN-C_6H_4-CH_2-C_6H_4-NCO +$$

Diphenylmethane di-*iso*cyanate.

$$HOROH + OCN-C_6H_4-CH_2-C_6H_4-NCO \longrightarrow$$

Polyester. Diphenylmethane di-*iso*cyanate.

$$OCN-C_6H_4-CH_2-C_6H_4-NHCOOROCONH-C_6H_4-CH_2-C_6H_4-NCO$$

Polyurethane (so-called polyester–di-*iso*cyanate).

This liquid intermediate is then reached with a small quantity of water which converts some, but not all, of the *iso*cyanate groups to amine groups:

$$OCN-C_6H_4-R' + H_2O \longrightarrow H_2N-C_6H_4-R' + CO_2$$

*Iso*cyanate. Amine.

(only a fragment of the molecule is depicted).

The so-formed amine groups are very reactive and react eagerly with unchanged *iso*cyanate groups to form urea bonds:

$$-C_6H_4-NCO + H_2N-C_6H_4- \longrightarrow -C_6H_4-NHCONH-C_6H_4-$$

*Iso*cyanate. Amine. Urea.

This reaction gives a cross-linked product which is no longer liquid but is a gum which resembles uncured sheet raw rubber. The gum is " cured " (this increases the degree of polymerisation) by heating under pressure at 130° C. for 1 hr. There is nothing very simple about this, the final polymer contains urethane, urea and ester linkages whilst the chain includes ethylene, propylene, tetramethylene and diphenylmethane groups. U.S. Rubber Co. define one of their polyurethane polymers as " an elastomeric polymeric cross-linked chain-extended di-*iso*cyanate-modified polyester or polyester–amide ".

Despite the complexities of the above two syntheses it is possible to see that each consists of four essential steps and that there is really a very close parallelism between the syntheses of Lycra and Vyrene. Perhaps the following table may clarify this:

Stage of synthesis.	Essential feature of stage.	In detail in the synthesis of:	
		Lycra.	Vyrene.
1	Preparation of a low linear polymer with terminal hydroxyl groups.	The low polymer is polyethylene glycol.	The low polymer is a polyester made with excess of ethylene and propylene glycols (excess so as to ensure terminal hydroxyl groups).
2	The low polymer is reacted with excess of a di-*iso*cyanate to give a polyurethane which has terminal *iso*cyanate groups.	Tolylene-2,4-di-*iso*-cyanate is used.	Diphenylmethane di-*iso*cyanate is used.
3	Water, in deficiency, is added to convert some of the terminal *iso*cyanate groups to amine groups.	At this stage a linear prepolymer with terminal *iso*cyanate and amine groups.	At this stage a linear polymer with terminal *iso*cyanate and amine groups.
4	The linear polymer is cured by heating. Amine and *iso*-cyanate groups react to give urea cross-linkages. The abundance of cross-linkages give the snap-back property.	The resulting cross-linked polymer is Lycra. It contains ether (from the glycol) urethane and urea linkages and probably others.	The resulting cross-linked polymer is Vyrene. It contains ester (from the first stage) urethane and urea linkages and probably others.

The Federal Trade Commission defines spandex yarns as " a manufactured fiber in which the fiber-forming substance is a long chain synthetic polymer comprised of at least 85 per cent of a segmented polyurethane". It seems to be clear that they are essentially polyurethanes but that in between the urethane groups there are long chains which may be polyglycols, polyesters or polyamides or copolymers of them.

Properties of Spandex Fibres

The stretch before break of Lycra yarn is from 520 to 610 per cent (rubber 760 per cent); Lycra can be stretched to six or seven times its original length, rubber to $8\frac{1}{2}$ times. The breaking strength of Lycra is about 0·7 gram per denier equivalent to about 4·5 grams per breaking denier; that of rubber is 0·25 gram per denier equivalent to 2·3 grams per breaking denier. At a given stretch the recovery of Lycra is inferior to that of rubber; recovery from a 50 per cent elongation is only 93·5–96 per cent; it might be thought that this was a serious defect. Lycra has an advantage over rubber in being white and dyeable. It has good resistance to chemicals; it is degraded and yellowed by hypochlorites but will withstand swimming pool concentrations (0·5 p.p.m. available chlorine maximum); it withstands the action of perspiration and of sun-tan and other cosmetic oils (rubber is not very good against oil). Lycra can be washed repeatedly in washing machines at 60° C. and tumble-dried at 80° C., but bleaches that contain chlorine must be avoided as they degrade and yellow the fibre.

Lycra yarn is spun as multifilament but the filaments are joined together into what is in effect a monofil. Fig. 162 illustrates the filament adhesion. Vyrene is a monofil. Deniers available range from 70 to 560, 420 being the most popular. Moisture regain is 0·3 per cent, specific gravity about unity, melting point is 250° C. but sticking occurs at 175° C. The fibre is soluble in boiling dimethyl formamide. Sometimes Rubber Gauge is used instead of Denier to express the size of a spandex thread. An approximate relation between them is: Rubber Gauge $= \frac{2350}{\sqrt{\text{Denier}}}$, *e.g.*, 100 Rubber Thread Gauge = 560 denier, 200 Rubber Thread Gauge = 140 denier. This is for cut rubber. For extruded rubber the gauge number is about 10 per cent less, *e.g.*, 180 extruded Rubber Thread Gauge = 140 denier.

Dyeing

Direct dyes as a rule have no affinity for Lycra and are not used. Disperse dyes have good affinity, so have acid and basic dyes.

Lycra is usually dyed in combination with nylon and disperse dyes are the most suitable. Wash fastness of colour is usually only fair, and light fastness is fair to poor. Demands on light fastness are probably not great as the fibre is used mainly for girdles and similar garments, but poor light fastness might be a serious defect in swimming costumes. Blacks can be obtained with good washing fastness and fairly good light fastness, by dyeing with an acid dyestuff, *e.g.*, Pontachrome Black TA, and after-chroming. Generally,

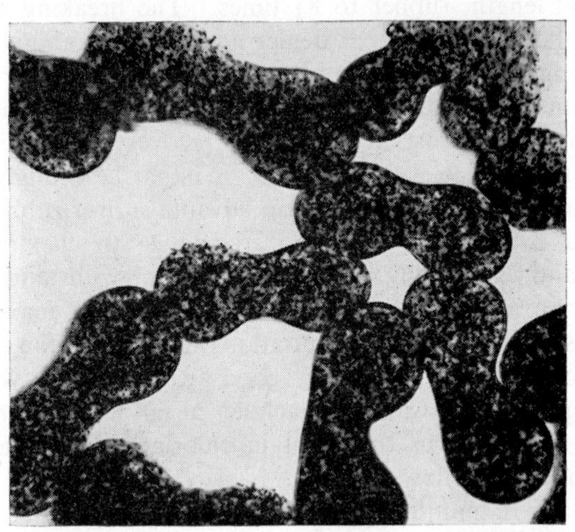

[*E.I. du Pont de Nemours & Co.*]

FIG. 162.—Photomicrograph showing cross-section of Lycra filaments. Adhesion between them can be seen. Magnification × 110—the filaments are very big.

it will not be practicable to get an exact match between the Lycra and the nylon with which it is knitted, but at least the Lycra will get somewhere near the nylon in colour, which rubber would not do.

The popularity of spandex yarns and fabrics has brought with it much investigation into their dyeing properties. J. Ehlert of B.A.S.F. has recommended the use of a range of Celliton dyes on Lycra (the dyeing properties of other spandex fibres may be rather different). He has also used Vialon Fast Dyes and Indanthren Dyes. Light fastness with Indanthren colours is about 5–6 (out of a possible 8) and with Celliton and Vialon dyes about 4. Elastomeric fibres are inclined (like cellulose acetate) to yellow under the influence of nitric oxides or gas fumes, and an inhibitor such as GFN Inhibitor should be used in a concentration of 4 gm./l. for

30 min. at 40° C. on yarns that have been dyed with Cellitons or disperse colours generally. The inhibitor prevents or greatly reduces the yellowing. Wash fastness on spandex yarns that have been dyed with disperse dyes is only moderate, about 3 (out of 5), but is improved half a point by after-treatment with tannin–tartar emetic.

Spandex yarns are nearly white, so that they do not need to be bleached for their own sake, but if they have been blended with other fibres bleaching may be necessary. Blends with polyamide or acrylic fibres may be given a mild reduction bleach, the bleach including some hydrosulphite, and finished with an optical brightening agent. Blends of spandex with cellulosic fibres are suitably bleached in hydrogen peroxide. Chlorine-containing bleaches may yellow spandex fibres and are better avoided. But it still remains that for foundation garments, which take most of the spandex that is made, light fastness is unimportant. Swim-suits are a very different matter, and there light fastness is very important indeed.

Uses

Mainly for belts, girdles, corsets, brassières, garters, surgical stockings and sock tops. Being stronger and lighter than rubber, garments with the same holding power can be made lighter in spandex fibres than in Lastex (covered rubber) fibre, which is what they are supplanting. Elastomeric fibres are usually blended with other synthetic fibres.

All elastic garments are made mainly of some ordinary " hard " fibre such as nylon and the elastic fibre whether it is rubber or spandex constitutes only a small proportion of the garment. Knitting or other textile process of manufacture is carried out with the Lycra under controlled tension.

Covered Yarns. All rubber elastic yarns are covered with a hard or rigid fibre such as rayon or nylon. Usually two layers of the rayon or nylon are wound on to the elastomer yarn in opposite directions while the latter is moving through the covering machine under controlled tension (stretched to four or five times its relaxed length); the two wrappings give a balanced structure with a minimum of twist liveliness. In the covering machine, the elastomer passes through the hollow spindles which carry the covering yarns and rotate at high speed. However, Lycra can also be used bare, giving lighter fabrics and eliminating the costly covering process. Vyrene, a monofilament can also be used uncovered, but most of it is supplied nylon-covered. Some velvets, taffetas and broadcloths which incorporate Vyrene look just like orthodox high quality fabrics, but when their capacity for being stretched is put to the test, the fabrics are seen to be most unorthodox: they are marvellous.

In considering the potentialities of spandex yarns, there was in 1961 a great shortage of them and du Pont planned a sixfold increase in Lycra output. Venturesome prophecies have been made (not by the two manufacturers of spandex yarns) that spandex fibres will completely replace rubber elastic in foundation garments in a year or two. It is well to remember though that the price of natural rubber is about 2*s.* per lb., and that the elastic recovery of rubber yarns is better than of spandex. Nature's polymers are usually better than those we make ourselves and the development of the spandex yarns may well spur the rubber technologists to new achievements; in the past they have not had very much competition from other snap-back fibres.

FURTHER READING

Modern Textiles Mag., **40**, 38 (December 1959).

E. O. Langerak, L. J. Prucino and W. R. Remington, U.S. Patent (to du Pont) 2,692,873 (1954).

CHAPTER 25

VINYON AND VINYON HH

It has long been known that vinyl chloride, which is a colourless liquid, would polymerise readily and that the polymer could be drawn out into long filaments. Naturally, this attracted much attention, and many chemists tried to make useful fibres from polymeric vinyl chloride and acetate too, but the fibres they made were too weak to be useful. In 1933 E. W. Rugeley, T. A. Feild and J. F. Conlon tackled the problem again, this time with success. They obtained useful fibres, to which the name "Vinyon" was given, and assigned the patent rights of the process to the Carbide and Carbon Chemicals Corporation in America. This firm made the polymer in the form of a white, fluffy powder, but did not spin it. The spinning was performed by the American Viscose Corporation, who in 1939 converted the polymeric powder into textile filaments. It is interesting to note that Vinyon, which is like nylon in that it is a true synthetic fibre, made its appearance at almost the same time. For thousands of years Man pursued textile crafts without producing a synthetic fibre; then two such fibres, quite distinct from each other, appeared in a period of two years.

Chemical Structure

Vinyon is a co-polymer of vinyl chloride (88 per cent) and vinyl acetate (12 per cent) It is called a co-polymer because the two are polymerised together. Either vinyl chloride or vinyl acetate will polymerise separately, but the polymers that result have their limitations for forming fibres. Polyvinyl chloride alone is a hard, tough, water-white resin which must be plasticised for extrusion purposes because it has a low decomposition temperature. Polyvinyl acetate is a clear resin of good acetone solubility, but one which softens at temperatures not much higher than room. Mixtures of polyvinyl chloride and polyvinyl acetate have not displayed any very promising properties, but resins (co-polymers) produced by the simultaneous polymerisation of mixtures of monomeric vinyl chloride and monomeric vinyl acetate have quite different and very useful properties. The function of the vinyl acetate is to plasticise the vinyl chloride internally—*i.e.*, in the same molecule. The reaction may be represented:

$$CH_2{:}CHCl + CH_2{:}CHCl + CH_2{:}CHOCOCH_3 + CH_2{:}CHCl \longrightarrow$$

Vinyl chloride. Vinyl acetate. Vinyl chloride.

$$-CH_2CHClCH_2CHClCH_2CH(OCOCH_3)CH_2CHCl-$$

Vinyon polymer.

The most suitable polymers for making Vinyon fibres are those that have molecular weights in the range of 10,000–28,000. If the molecular weight is lower than 10,000 the fibres are weak, for very long molecules are necessary to give *any* strong fibres, and if the molecular weight is higher than 28,000 the polymer is insoluble in the solvents, from solution in which it is usually spun.

Manufacture

The co-polymerisation of vinyl chloride and vinyl acetate is effected by heating, probably in the presence of a catalyst of the aluminium chloride, $AlCl_3$, or boron trifluoride, BF_3, type. The polymer is dissolved in a solvent, acetone, or its homologue methyl ethyl ketone (M.E.K.), and a solution which contains 23 per cent by weight of the co-polymer is made. This is filtered and de-aerated, and then extruded through fine orifices by the pressure of a pump. The "filaments" emerge into a counter-current of warm air which evaporates the acetone so that the filaments solidify and can be wound. The process of spinning is very similar to that used in the manufacture of cellulose acetate. The air from the spinning cabinets which contains acetone is drawn through ducts to a recovery plant, where it may be scrubbed in water and the solutions of acetone in water gradually concentrated. When it is spun, Vinyon is weak and has to be oriented by stretching in the same way as nylon. It is first twisted in the wet state, and is then given a stretch of about 800 per cent—*i.e.*, it is stretched to nine times its original length. The yarn before stretching has a tenacity of about 0·8 gram per denier and after stretching one of 3·4 grams per denier. If a dull yarn is required, a pigment such as titanium oxide may be incorporated in the dope of vinyl resin and acetone.

Properties

Vinyon yarn has a tenacity of about 3·4 grams per denier and an elongation at break of 18 per cent. It is as strong wet as dry, probably because it is very resistant to the action of water. The moisture content at standard conditions is less than 0·5 per cent. If this is compared with the value of 12 per cent for viscose rayon the difference will be very evident

Because Vinyon is a stretched yarn, the filaments are usually very fine—*e.g.*, a yarn of 40 denier may have twenty-eight filaments, *i.e.*, a filament denier of 1·43. Its specific gravity is 1·37, which is not very different from that of cellulose acetate (1·33) or wool (1·32), but is considerably lower than that of viscose (1·52) and higher than that of nylon (1·14). Vinyon is a non-conductor of electricity.

The fibre is thermoplastic at temperatures over 65–70° C. When heated to 150° C. it becomes tacky and begins to melt—a very serious defect. It is unattacked by bacteria, fungi and the larvae of moths and carpet beetles. Chemically it is very stable, and resists attack by even high concentrations of alkalis and mineral acids, although it is softened at high temperatures by acetic acid.

Under the microscope, the fibre resembles mercerised cotton, with a lumen-like channel appearing to run through the middle of the filament. This is an optical illusion, for the filament is solid, and is caused by the peculiar cross-section of the filaments, which is flat and slightly dumb-bell shaped (Fig. 163). The " lumen " does not show more than faintly in this picture.

Vinyon is satisfactorily stable to the action of sunlight.

Dyeing

Dyeing is not easy, owing to the very high resistance of the material to water, and special dyeing assistants which temporarily soften the surface of the filament, and so make it more penetrable, have been used. A series of dyestuffs has been marketed for dyeing Vinyon, under the name Calcovins. It may also be dyed satisfactorily with dispersed colours of the cellulose acetate type. Owing to the thermoplasticity of Vinyon, its maximum safe dyeing temperature is 60° C. The use of pigments in the dope might be expected to give good bright colours, and so avoid the dyeing difficulties, but the use of pigmented dope is not always attractive to the manufacturer, as he has to segregate carefully the different colours; pipe-lines that have been used for a red dope will take a long time to become completely free from red. Perhaps the best method is to spin three basic colours—a red, a blue and a yellow—and modify these by subsequent dyeing of the filaments to obtain the large range of colours that the market demands.

Vinyon, because of its very considerable content of chlorine, will not support combustion, although it will burn in a flame.

Uses

Vinyon has found very considerable application. Its superlative chemical resistance makes it very suitable for use in filter-pads and in protective clothing for chemical workers. Its resistance to water has enabled it to be used for fishing-lines and nets. In fabric form it effectively replaced No. 10 silk bolting cloth as a screen printing material when this became unavailable. Other uses include felts, sewing-threads and twines, and ladies' gloves which have been made from warp-knitted Vinyon fabric. Its great defect of having such a low melting point means that it cannot be used for materials and

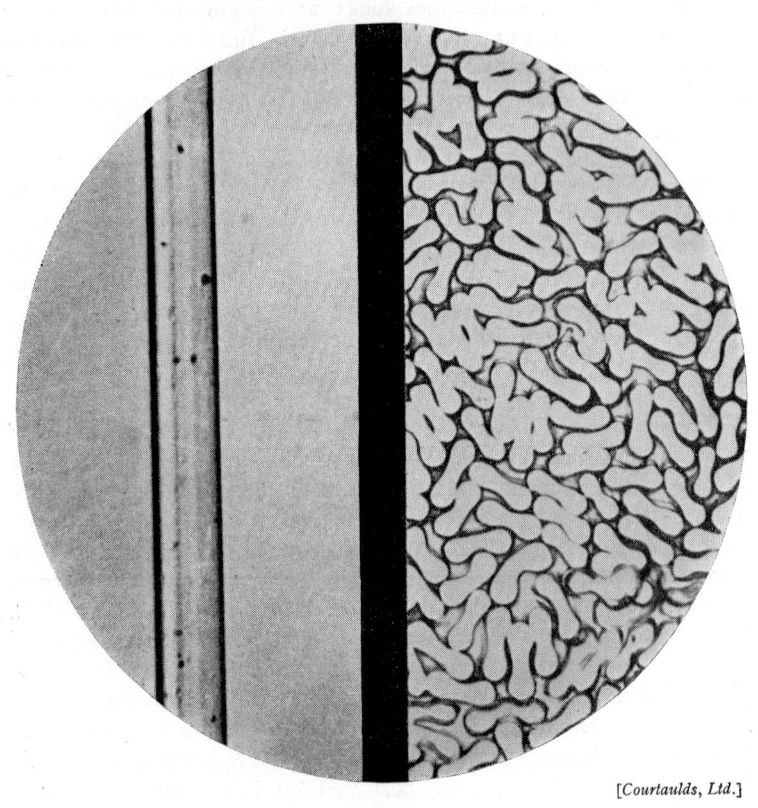

[*Courtaulds, Ltd.*]

FIG. 163.—Photomicrograph of Vinyon filaments ($\times 500$)

garments that normally are laundered, and apart from this its very poor absorption of moisture makes it unsuitable for underwear. Most important applications are felts, carpets and filter fabrics.

It has been used almost exclusively in continuous filament form, although a little staple fibre has been made. The unusual properties of Vinyon have made it possible to use it for purposes for which most textiles are unsuitable, but, on the other hand, they have prevented it being used for the more common purposes. Vinyon N and

Dynel, which are developments of Vinyon, have probably now taken over most of the uses that Vinyon promised to fill.

VINYON HH

Vinyon HH is a staple fibre that is spun by the American Viscose Corporation from a solution of co-polymerised vinyl chloride and vinyl acetate, *i.e.*, it is similar in chemical composition to the original Vinyon. It is made in staple form only. In 1962 Vinyon HH was still being made by the American Viscose Corporation and is sold at 80 cents per lb.

Manufacture

The co-polymer is dissolved in a solvent, probably acetone, and dry spun. The method of spinning is similar in principle to that used for cellulose acetate. The filaments are cut to staple and opened before they are shipped. Vinyon HH is made in the following sizes:

2 denier	$\frac{3}{4}$–5 in.
3 ,, ...	$\frac{1}{2}$–5 in.
5·5 ,, ...	$\frac{1}{4}$–5 in.

Properties

Vinyon HH has a low strength (0·6–0·8 gram per denier) and a high extensibility. It has exceptional water resistance and is practically as strong when wet as dry; it absorbs less than 0·1 per cent moisture. It will not grow bacteria, moulds or fungi and will not support combustion; it has the high dielectric strength of 650 volts per mil. Specific gravity is about 1·35.

Chemical Resistance. Its chemical resistance is good; it is practically unaffected at room temperatures by concentrated acids such as sulphuric, nitric, hydrochloric, hydrofluoric and aqua regia; it is resistant to 30 per cent caustic soda or caustic potash solutions and is unaffected by salt solutions, cuprammonium solutions, alcohols, glycols, paraffins, petrol and mineral oils. At higher temperatures, mineral acids may char and embrittle the fibre. Examples of its resistance to sulphuric acid are given below; it can be seen that, as the temperature is raised, the concentration that the fibre will withstand falls.

Temperature (° C.).	Maximum concentration of sulphuric acid which has no noticeable chemical action on fibre (per cent).
25	95
58	80
70	75
80	65

The fibre is, however, dissolved by ketones and softened or partly dissolved by esters, ethers, aromatic hydrocarbons, and some amines and chlorinated hydrocarbons.

Effect of Heat. The most important characteristic of Vinyon HH staple fibre is its ability to soften, shrink and bond to other fibres when heated under pressure or in the presence of some solvents. Temperatures at which the fibre changes character are as follows:

Softening temperature	52–60° C.
Shrinkage ,,	60–66° C.
Tacky ,,	85–102° C.
Melting point	135–149° C.

Dyeing

Vinyon HH presents considerable difficulties; it cannot be dyed deep shades with both wash and light fastness. The best method is to apply an acetate dyestuff in a bath containing 5 per cent of a swelling agent such as *o*-hydroxydiphenyl or dibutyl phthalate, and to dye at 55° C. for three-quarters of an hour.

Uses

Spinning with Other Fibres. The low strength and high extensibility of Vinyon HH have made it a not very satisfactory fibre to spin into yarn. All-Vinyon HH gives static trouble; this can be reduced by the application of a suitable finish and the use of a static eliminator, and 100 per cent Vinyon HH has been spun experimentally to 20 s cotton counts. Blending with 25 or 50 per cent cotton or rayon is advisable for spinning on the cotton system, for which 3 denier 1½ in. staple is preferred. Mixtures of Vinyon HH and wool can be spun more satisfactorily on the woollen system.

Bonding. The ability of Vinyon HH to shrink when heated in combination with other fibres is made use of in the manufacture of rubber-coated elastic fabric and of embossed carpets. The shrinkage of Vinyon HH fibres at different temperatures is as follows:

Temperature (° C.).	Shrinkage (per cent). U.S.T.
60	Nil
71	18
74	27
79	45
85	50
91	55
100	60
110	65
121	70
132	Melting range

The main use of Vinyon HH is for bonding and heat-sealing. The ability of Vinyon HH to bond to other fibres is used in making pressed felts, bonded fabrics and heat-sealable papers.

The temperature at which the bonding is carried out is important; in batch processing it should preferably not exceed the "tacky" temperature, otherwise the fabric will be unduly stiffened. In continuous bonding processes where the fabric is exposed to heat and pressure for a very short time it may be necessary to use a temperature almost as high as the melting point of Vinyon HH.

FURTHER READING

U.S. Pat. 2,161,766 (Carbide & Carbon Chemicals) 1937.
Silk and Rayon, **18**, 261 (1944).
F. Bonnet, *Amer. Dyestuff Reporter*, **1940**, 547; *Ind. Eng. Chem.*, **32**, 1564 (1940).
Rayon Textile Monthly (Elastic Vinyon). Nov. 1942.
J. Woodruff, "The Dyeing of Vinyon", *Amer. Dyestuff Reporter*, Apr. 22, 1946.

CHAPTER 26

DYNEL

DYNEL, a modacrylic fibre, is made in staple fibre and in tow by the Union Carbide Chemicals Company, a division of the Union Carbide Corporation. Dynel was introduced in 1951 and has been very successful; it is considerably lower in price than most of the synthetics. Reference must be made, however briefly, to Vinyon N, now obsolete, but made by Union Carbide before they made Dynel. It was in continuous filament form, whereas Dynel is staple; it was dry spun from acetone, whereas Dynel is wet spun. Large-scale production of Vinyon N continuous filament was stopped in 1954; thereafter only relatively small quantities were made to satisfy existing customers who still wanted to use it.

Synthesis of Acrylonitrile

The fibres that will be discussed in this and in some of the following chapters are largely based on acrylonitrile. This chemical has been made in enormous quantities for fibres; it is also used for the cyanoethylation of cellulose (p. 278). It has become an important chemical in the fibre world. There are four methods by which it has mainly been made.

From Ethylene Oxide. The classical method but not much used nowadays. Ethylene oxide is made by oxidising ethylene obtained from oil-cracking plants; it is reacted with hydrocyanic acid. The hydrocyanic acid comes from the catalytic partial oxidation of methane in the presence of ammonia. The reaction which yields the acrylonitrile may be written:

$$\begin{matrix} CH_2 \diagdown \\ | \quad\quad O \\ CH_2 \diagup \end{matrix} + HCN \longrightarrow \begin{matrix} CHCN \\ \| \\ CH_2 \end{matrix} + H_2O$$

From Acetylene. Most of the world's acrylonitrile is made from acetylene nowadays. Acetylene is traditionally and still largely made by the action of water on calcium carbide, but more and more is coming from petroleum distillation. Calcium carbide needs a lot of electrical heat energy to make it although the raw materials, lime and coke, are cheap enough; not much can be done about the cost of the electricity except to make the carbide in Canada and

Norway where hydroelectric power is cheap. Some acetylene is made at petrochemical factories; for example, the French acrylic fibre Crylor is made from acrylonitrile that in turn is made from acetylene obtained by the partial oxidation of methane at 1,200° C. The acetylene whether from carbide or from methane is reacted with hydrocyanic acid thus:

$$C_2H_2 + HCN \longrightarrow CH_2{=}CHCN$$

The addition process takes place in the presence of cuprous chloride as a catalyst in aqueous solution with ammonium chloride added to increase the solubility of the cuprous chloride. Reaction conditions are 85° C. at one atmosphere pressure, with about 10 moles of acetylene present for each mole of hydrocyanic acid, so as to convert completely the hydrocyanic acid. The yield of acrylonitrile based on the acetylene is about 80 per cent; based on the hydrocyanic acid it is 95 per cent.

Sohio Process. The starting materials are propylene and ammonia. The propylene is oxidised to acrolein, ammonia adds on to it, the complex is dehydrated and dehydrogenated to acrylonitrile.

$$\underset{\text{Propylene.}}{\begin{array}{c}CH_3\\|\\CH\\||\\CH_2\end{array}} \longrightarrow \underset{\text{Acrolein.}}{\begin{array}{c}CHO\\|\\CH\\||\\CH_2\end{array}} \longrightarrow \underset{\text{Ammonia complex.}}{\begin{array}{c}HOCH{\cdot}NH_2\\|\\CH\\||\\CH_2\end{array}} \longrightarrow \underset{\text{Acrylonitrile.}}{\begin{array}{c}C{\vdots}N\\|\\CH\\||\\CH_2\end{array}} + H_2O + H_2$$

The overall reaction can be written:

$$CH_2{=}CHCH_3 + NH_3 + 3/2\ O_2 \longrightarrow CH_2{=}CHCN + 3H_2O$$

It is believed that du Pont in America and Asahi and Mitsubishi in Japan all use acrylonitrile that has been made from propylene. This chemical route has very obvious advantages to those fibre manufacturers who include polypropylene in their range.

Acetaldehyde. The reaction is simple. Hydrocyanic acid is added on to acetaldehyde to give the cyanhydrin, and the elements of water are removed to yield acrylonitrile:

$$CH_3CHO + HCN \longrightarrow CH_3CH\begin{matrix}\diagup OH\\ \diagdown CN\end{matrix} \longrightarrow CH_2{=}CHCN + H_2O$$

In America alone, production of acrylonitrile is about 250 million lb. a year, of which 70 per cent is used in fibres. In July, 1961, the price of acrylonitrile fell to $14\frac{1}{2}$ cents per lb. but has since risen. In late 1969 it is once again $14\frac{1}{2}$ cents.

Manufacture of Vinyl Chloride

The best method is the addition of hydrogen chloride to acetylene in the presence of a catalyst of mercuric chloride mounted on charcoal:

$$HC\vdots CH + HCl \longrightarrow CH_2\colon CHCl$$

An alternative method is to react ethylene with chlorine to give ethylene dichloride and to heat this to 550° C. and 4 atmospheres pressure when it breaks down:

$$CH_2Cl{\cdot}CH_2Cl \longrightarrow CH_2\colon CHCl + HCl$$

but methods that produce hydrogen chloride are not looked on favourably as there is a glut of it and it is almost impossible to sell the hydrogen chloride. Production of vinyl chloride is very great, *i.e.*, $1\frac{1}{2}$ thousand million lb. a year in America alone. Price in June, 1961, fell to $7\frac{1}{2}$ cents per lb. and later in 1969 was about 8 cents per lb.

Chemical Nature of Dynel

The two chemicals just described are co-polymerised in the ratio 40 acrylonitrile/60 vinyl chloride, and the co-polymer is dissolved to a 21 per cent solution in acetone.

Manufacture

The acetone solution is de-aerated, filtered and then extruded by the usual wet spinning technique into a water bath where it coagulates to form continuous strands. After spinning it is stretched hot (not cold like nylon but hot like Terylene) as much as 1,300 per cent and then annealed by heat treatment; it is crimped and cut to staple; alternatively it is supplied as tow; this tow is then cut into staple lengths and crimped. The Dynel flowsheet is shown in Fig. 164.

Properties

Dynel is a light-cream fibre; it can be bleached nearly white. Its filament cross-section is irregular and ribbon-shaped, much more like that of cotton than wool. Tenacity ranges from 2·5 to 3·5 grams per denier, depending on filament denier, and the elongation varies correspondingly from 42 to 30 per cent. The characteristic curve is shown in Fig. 165. The moisture regain of Dynel is less than 0·4 per cent under standard conditions, and consequently its tenacity and elongation are almost the same when wet as in the dry state. The specific gravity is 1·31.

Chemical Resistance. Perhaps the outstanding characteristic of

Dynel is its extremely good chemical resistance, a factor of great importance in the selection of fabrics for industrial and military use. Dynel has very good resistance to long exposure to weak or moderately strong solutions of inorganic alkalis, acids and salts,

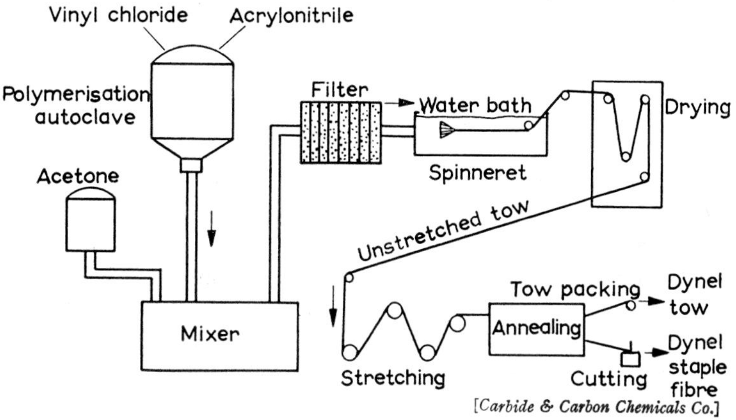

[*Carbide & Carbon Chemicals Co.*]

FIG. 164.—Schematic representation of the stages of Dynel manufacture.

and generally it has outstanding resistance to organic chemicals. In some tests carried out by an independent organisation, a woven all-Dynel fabric was immersed in various chemical reagents and the loss in strength noted after twenty hours' immersion. Some of the results are given in the table on p. 460.

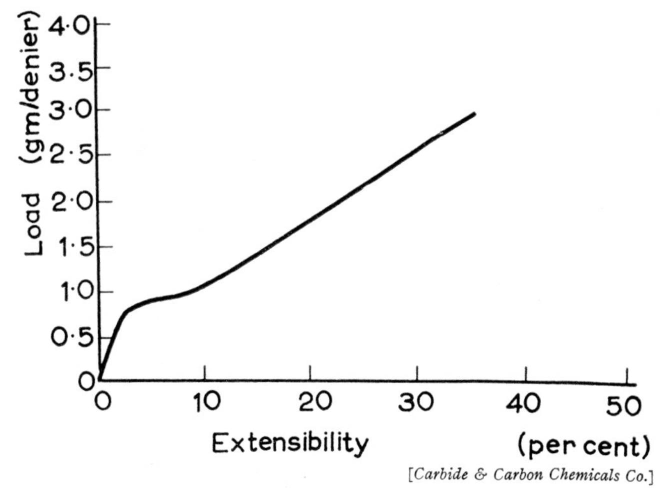

[*Carbide & Carbon Chemicals Co.*]

FIG. 165.—Stress–strain curve of 3-denier Dynel.

Acetone is the best solvent; *cyclo*hexanone and dimethyl formamide also have some solvent action. Acetic anhydride, acetaldehyde, aniline, ethylene dichloride and methyl ethyl ketone all plasticise or swell Dynel.

Biological Resistance. Dynel is resistant to clothes moths' larvae and to carpet beetles, also to mildew and fungus. Dynel fabric

Chemical.	Concentration (per cent).	Temperature (° C.).	Effect on appearance of fabric.	Shrinkage in area (per cent).	Loss in strength (per cent).
Sulphuric acid	70	50	None	None	6
,,	25	100	Stained brown	None	12
,,	5	100	None	None	0
Hydrochloric acid	38	50	None	None	0
,,	38	100	Stained brown	10	12
Hydrofluoric acid	50	20	None	None	0
Nitric acid	20	100	Slight bleaching	5	9
Phosphoric acid	85	100	Slight darkening	None	0
Acetic acid	75	100	Shrunk, curled, stained brown	15	14
Formic acid	25	50	None	None	0
Phenol	5	20	Stiffened and shrunk	22	28
Caustic soda	50	50	None	None	5
,,	25	100	Stained dark brown	5	17
Cuprammonium	5	20	None	None	0
Zinc chloride	50	100	Slight darkening	None	0
Hydrogen peroxide	90	20	Slight bleaching	2·5	7
Acetone	10	50	None	None	0
,,	100	20	Dissolved	—	—
*cyclo*Hexanone	5	50	Slight stiffening	None	33
Ethyl acetate	100	50	Bleached white	None	13
Formaldehyde	40	50	Slight yellow staining	None	0
Perchlorethylene	100	50	None	None	0
Toluene	100	50	Slight bleaching	None	0

buried in soil in tropical conditions was unchanged after six months, whereas a heavy cotton duck disintegrated after ten days.

Flammability. Dynel will burn in a flame, but if the flame is removed it is self-extinguishing: it does not drip beads of molten fabric. When tested according to ASTM D 626-41 T, the length of char on a blanket was $3\frac{1}{4}$ in. There is little or no risk of fire with Dynel.

Dimensional Stability. Dynel is resistant to water and is non-felting and non-shrinking below the boil. At the boil, Dynel will shrink progressively with increase of time. If the temperature is taken higher, shrinkage is faster. Dynel fabrics can be dimensionally stabilised by dyeing at the boil; thereafter they are stable in boiling water and up to dry heat temperatures of 115° C.

Resistance to Ironing. Dynel is susceptible to stiffening and shrinking if an ordinarily hot iron set at the " rayon " temperature is used. It has to be ironed with the lowest iron setting and a dry cotton cover over the fabric. This is a marked defect in a fibre, but blends of Dynel and wool or Dynel and rayon behave much better.

High-bulk Dynel. Type 63 Dynel (regular Dynel is called Type 60) is a high-bulk fibre; it can be shrunk up to 35 per cent of its length by the controlled application of dry heat. If Type 63 is blended with worsted and then woven into a pile fabric with regular Dynel, the pile can be sheared to a uniform height; then, if the fabric is exposed to dry heat, differential shrinkage of the Type 63 and the regular Dynels will give a fabric with two lengths of pile (*cf.*, Verel, p. 493). The Japanese equivalent of Type 63 is Kanekalon Hibulkee.

Dyeing

There are three difficulties which present themselves in dyeing Dynel. These are:

1. Its low affinity for water. This is partly overcome by the common use of dispersed acetate dyes and mainly by the use of a high temperature, not lower than 96° C.
2. Its liability to delustre or " blush " at temperatures of 96° C. and over. This is overcome by the incorporation of sodium sulphate in the dyebath; this prevents blushing. If the dyeing process which is being used does not permit the addition of sodium sulphate, the fabric will become delustred, but after the dyeing is finished the lustre can be restored by treatment of the dyed fabric with a 30 per cent solution of salt. This salt probably brings out by a process of osmosis water that has been trapped in the hot fibre. But the salt treatment can now be omitted. It has been superseded by a method of drying at 120° C. to relustre the Dynel.
3. Its thermoplastic nature at temperatures over 82° C. If creases get into fabric at temperatures higher than this they will be permanent. Accordingly, when socks are dyed at 96° C., in order to avoid "crows' feet" in them it is essential that the socks be permitted to float freely during dyeing and whilst they are cooled down to say 50–55° C.

Because of these difficulties, there have in the past been several special methods developed for dyeing Dynel. Many of these involved the presence of the cuprous ion which greatly increased the affinity of the fibre for the dye. Combined with this there were some ingenious expedients to supply a full range of colours. Almost all of these techniques have now disappeared, and in particular the use of cuprous ion is no longer necessary.

Broadly, there are two classes of dye used nowadays: disperse or " acetate ", and the new basic or cationic colours. With these, it is possible to get a fastness to light of at least 5–6 (out of 8 points) and a maximum rating for wet fastness. The old trouble of poor fastness to light on modacrylic fibres has been very largely overcome.

1. **Disperse Dyeing.** Dynel can be dyed with almost the entire range of disperse dyestuffs. They level well, a feature of this class of dyes, and they have good fastness to washing and light. The general process may be summarised thus:

(*a*) Scour if necessary in 0·5 per cent (all on the weight of material) of non-ionic detergent and 1 per cent of sodium tri-polyphosphate. Rinse.

(*b*) Dye at a temperature of from 80° C. to 105° C. Because the disperse dyes level well, the dyebath can be raised quickly from the starting temperature to the dyeing temperature. Hold at the dye temperature for 60 min. for light shades, or for 90 min. for dark shades. Correction to shade if necessary (*i.e.*, addition of more dyestuff) should be run for at least 30 min. at dyeing temperature. (Temperatures over 100° C. involve the use of pressure machines. But they are often not necessary and even when they might seem to be, they can usually be avoided by using a carrier.)

(*c*) Rinse, hydro-extract and relustre by drying at 120–130° C. for 3–10 min. The time necessary can best be determined by observation in the dyehouse. Another way of relustring is to expose to steam heat at 10 lb./in.^{2} for 10 min.

But if package dyeing is the method used, all hot (100° C.) baths should be cooled to 75° C. before the circulating pump is stopped. If the circulation is stopped with the bath at the boil the packages will sag, because the fibre is very thermoplastic at 100° C. Typical disperse colours that are suitable are Cibacete, Setacyl and Eastone.

2. **Cationic Dyeing.** Typical classes of cationic dyes are the Basacryls, Sevrons, Astrazones and Deorlenes. The process is:

(*a*) Scour at 70–75° C. for 20 min. in 0·5 per cent non-ionic detergent. Rinse.

(*b*) Dye in bath containing for a Green say

Basacryl Yellow 5RL	0·15 per cent
Basacryl Blue GL	0·03 per cent and
Sevron Blue 5G	0·3 per cent
Sodium acetate	1 per cent
Sodium chromate	0·1 per cent
Acetic acid to *p*H 5.	

Dye at 90° C., cooling slowly to 70° C.

(*c*) Rinse, hydro-extract and relustre by drying at 120° C. Care must be exercised not to put tension on fabric in this last high temperature stage, otherwise it will be starched.

Pile Fabrics. A great deal of Dynel goes into pile fabrics (fur coats and so on). In order not to damage the pile by heating to a temperature at which the fibre is thermoplastic, dyeing is done at a low temperature, *e.g.*, not above 82° C. It may be necessary to use a carrier to get the depth of shade at this temperature, and if it is so, then butyl benzoate (3 per cent on the weight of goods) is suitable.

Pile fabrics are often printed, and for this process either disperse or cationic dyes can be used. Processing details are available from the fibre manufacturer.

Dope Dyed. Dynel is also spun coloured in a range of shades.

Whiter Dynel. Type 80 includes a fluorescent agent, added in the dope stage, and is whiter than the regular fibre.

Boarding. Dynel socks are usually boarded to shape, after they have been air dried on internally heated forms (shapes), at a steam pressure of 5–6 lb. This ensures that they will be the right size and shape to begin with, and moreover that they will retain that size and shape throughout future laundering.

Uses

The chemical resistance of Dynel makes it very suitable for protective clothing, and this suitability is enhanced by its inability to support combustion. It has accordingly been used for clothing for chemical workers and miners; Fig. 166 illustrates two shirts, one of cotton (left) and one of Dynel (right), which have been placed in 60 per cent sulphuric acid, agitated for two minutes, then washed and dried. In one plant which processed aluminium chloride, zirconium and titanium tetrachlorides, workers often found their cotton garments in tatters before a single shift was over, but since having been given Dynel shirts and trousers they have worn these garments for almost a year. Because, too, of its resistance to chemicals, it has been used for chemical filter cloths, for acids and acid dye liquors. It has also been used for diaphragms in copper-plating plants; in this process a heavy cotton duck had a life of from four to six weeks, whilst Dynel fabric lasts from four to six months. For similar reasons it has found use in bags for collecting anode sludge, for holding water-softening chemicals and in dye and laundry nets.

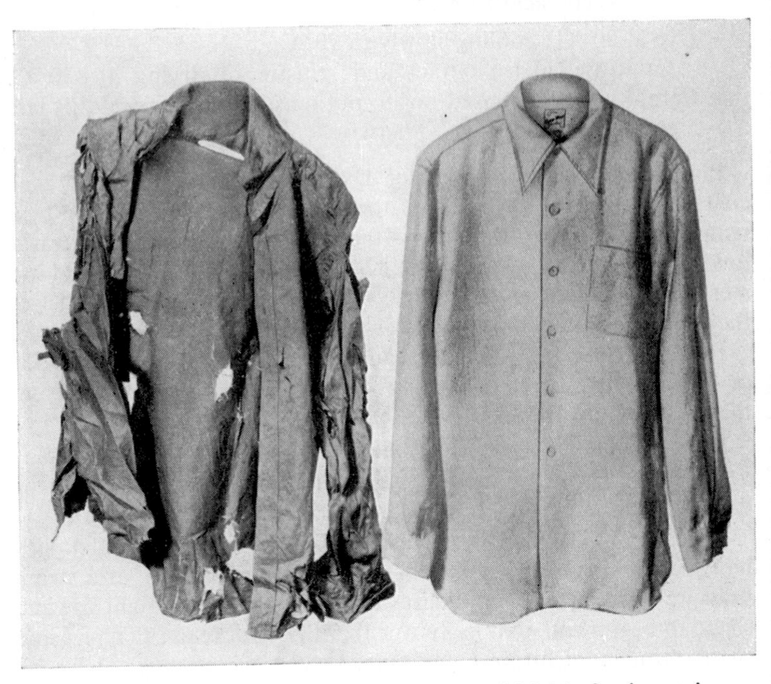

FIG. 166.—Two shirts of cotton (*left*) and Dynel (*right*) after immersion in acid.

A paint roller (instead of a brush) has found considerable popularity for home use, and is now being used industrially; Dynel is used for the roller cover, and provides good flow-off of paint, no matting or curling and easy cleaning.

One of the most successful outlets for Dynel has been in blankets made of all Dynel. These are easily washable, impervious to damage by disinfectants and easy to clean from stains; furthermore, they are shrinkproof and mothproof; owing to Dynel's very low moisture regain, blankets made from it can be air-dried in a few hours. Bedspreads and upholstery have provided other uses; the " United States " liner used 13½ miles of Dynel fabric for draperies because of its fire-resistance and mothproof and easy cleaning qualities. The fire-resistance of Dynel is responsible for its fastest expanding end-product, which is one of furnishing and drapery products for ships, hotels, aircraft and the G.L.C. Part of this expansion is attributable to a new American regulation stating that any ship carrying American passengers must be furnished with inherently non-flammable fibres, rather than with Proban-treated cotton. Its mothproof and non-allergenic properties have enabled Dynel to be used for filling cushions, pillows and eiderdowns; if these are covered with nylon fabric they can be washed and dried very quickly. Straw hats made of Dynel are light and keep their shape well when wet.

Pile fabrics suitable for coats are particularly good in Dynel; the pile is bouncy, and remains so even after washing; blends with Orlon and mohair are in current use; the production of high-quality simulated fur fabrics constitutes the largest single use for Dynel. For this application Type 183 Dynel is used; this is high shrink and gives a definite two-height-pile fur fabric (*cf.* pp. 461, 493). Rug manufacturers make rugs with Dynel pile up to 3 in. long; with most fibres there would be a serious risk of flame flash through the pile if it was accidentally ignited, but not with Dynel. Dynel is still used extensively for Wildemann Knitted scatter rugs. Large quantities, running into hundreds of tons of Dynel, have been sold for carpets, where the Dynel is blended with an acrylic fibre to prevent the acrylic fibre flashing in the event of a fire. As little as 30 per cent Dynel in the blend will considerably reduce the fire risk. The fibre is used industrially in the form of a batt for air filtration (Ulok filters), Dynel's rough surface and irregular cross-sectional outline enabling it to pick up a lot of dust. Pleat and crease retention is good in Dynel; the pleats are put in by steam pressing and are retained permanently. Other apparel uses are for underwear, sleeping garments, knitted blouses, men's socks, which wash easily, dry quickly and do not shrink. Tent fabrics, umbrella fabrics and awnings, and tape and braid for insulation are others.

The inclusion of Dynel in blends for suitings will give good shape retention and resistance to rain. Pressed creases stay in even after a wetting, and further pressing is unnecessary. A blend of 50 per cent rayon, 25 per cent cellulose acetate and 25 per cent Dynel is recommended for suitings.

There is a U.S. Marine Corps uniform jersey which contains 15 per cent Dynel, and a series of U.S. Army, Navy and Air Force fabrics containing up to 50 per cent Dynel; advantages expected to accrue from the inclusion of Dynel include better strength, durability, and resistance to creasing.

One odd application of Dynel is as hair and wigs for dolls; the hair can be washed, waved, brushed, combed, set and tinted.

As time has passed this use has grown; Dynel is used nowadays as hair and wigs, not only for toy dolls but also for the human variety. Continuous-filament Dynel is sold in 24 and 40 denier for ladies' hair-pieces, falls and chignons; the lustre of the Dynel closely resembles that of human hair, whereas similar " hair " made from nylon and most fibres is inclined to look artificial. Furthermore, Dynel is inherently flameproof, whereas nylon, if accidentally ignited, could melt and burn the skin. The Dynel hair-piece is easily managed and can be washed and dried at home. It is not a mammoth outlet, but it is one that is not inconsiderable and one for which Dynel is uniquely suitable. Type 150 uncrimped tow is used, and " hair " is the main end use for this particular type of Dynel.

FURTHER READING

C. R. Denbridge, *Fibres*, **9**, 335 (1948). (This paper deals with Vinyon N.)

T. A. Feild, " The Dyeing of Dynel and Related Products ", *Amer. Dyestuff Reporter*, **41**, 475 (1952).

CHAPTER 27

SARAN

VELON, PERMALON, TYGAN

VINYLIDENE CHLORIDE is a clear, colourless liquid boiling at 32° C. It was discovered in 1840, and it was noticed by many chemists that it would easily polymerise. More recently, systematic work on its polymerisation was undertaken by the Dow Chemical Co. in U.S.A., and in 1940 they introduced a material called Saran. This was made from vinylidene chloride with the admixture of a little vinyl chloride. The product was spun by other firms, notably (1) Pierce Plastics, Inc., who called their product Permalon and (2) Firestone Industrial Products Co., who called their product Velon. Permalon, Velon and Tygan are trade names given by different spinners and weavers to fibres of this type that they have spun or woven.

Saran is spun in the United Kingdom by BX Plastics from polymer made by British Geon, an associate of the Distillers Co. Ltd. The spun monofilament is woven into Tygan fabrics by Fothergill and Harvey Ltd. A similar fibre is also spun in Japan; Asahi-Dow Ltd. (trade name of fibre Saran) and Kuahi Kasei (trade name Krehalon) each make about 1,500 tons a year. In France a similar fibre known as Clorene is made by La Société Rhovyl.

Chemical Structure

Vinylidene chloride has the formula $CH_2{:}CCl_2$; it is unsymmetrical dichlorethylene. When heated in the presence of a catalyst it polymerises rapidly in this way:

$$CH_2{:}CCl_2 + CH_2{:}CCl_2 + CH_2{:}CCl_2 \longrightarrow -CH_2CCl_2CH_2CCl_2CH_2CCl_2-$$

The monomeric vinylidene chloride contains a double bond and is unsaturated, but the polymeric form contains only single bonds and is saturated. The polymerisation reaction may be regarded as proceeding in the following way. First the double bonds open, leaving unused single valencies available:

$$CH_2{:}CCl_2 + CH_2{:}CCl_2 \longrightarrow -CH_2{\cdot}CCl_2- + -CH_2CCl_2-$$

and then these single bonds mutually saturate each other:

$$-CH_2CCl_2- + -CH_2CCl_2- \longrightarrow -CH_2CCl_2{\cdot}CH_2CCl_2-$$

A co-polymer from vinylidene chloride and vinyl chloride would be formed in the following way:

$$\underset{\text{Vinylidene chloride.}}{CH_2{:}CCl_2} + \underset{\text{Vinyl chloride.}}{CH_2{:}CHCl} + \underset{\text{Vinylidene chloride.}}{CH_2{:}CCl_2} \longrightarrow \underset{\text{Co-polymer.}}{-CH_2CCl_2CH_2CHClCH_2CCl_2-}$$

and Saran is such a co-polymer—*i.e.*, it is made by polymerising vinylidene chloride and vinyl chloride together. The molecular weight of Saran is about 20,000, so that it consists of molecules approximately eighty times as long as that shown just above. Saran is a linear polymer, in just the same way as the other fibres are.

Manufacture

The raw materials are ethylene, $CH_2{:}CH_2$, which is obtained by cracking natural petroleum, and chlorine, made by electrolysing sea-water, so that we might expect the material to be reasonable in price. When ethylene is treated with chlorine, trichlorethane, $CH_2ClCHCl_2$, is formed by addition and substitution:

$$CH_2{:}CH_2 + 2Cl_2 \longrightarrow \underset{\text{Trichlorethane.}}{CH_2ClCHCl_2} + HCl$$

Then the trichlorethane is treated with lime to give monomeric (unpolymerised) vinylidene chloride:

$$\underset{\text{Trichlorethane.}}{2CH_2ClCHCl_2} + \underset{\text{Lime.}}{Ca(OH)_2} \longrightarrow \underset{\text{Vinylidene chloride.}}{CH_2{:}CCl_2} + CaCl_2 + 2H_2O$$

The function of the lime is to abstract the elements of hydrochloric acid, HCl, from the trichlorethane.

The vinylidene chloride is mixed with some vinyl chloride and polymerisation induced by means of heat and a catalyst. The mixture actually used for the co-polymerisation is believed to consist of 85 parts (by weight) vinylidene chloride, 13 parts vinyl chloride and 2 parts acrylonitrile (vinyl cyanide).

The co-polymer is extruded through fine orifices at about 180° C. and solidifies immediately it leaves the spinning jet and meets the cold air. Note that Saran, like nylon, is spun from the melt, and not from a solution. The filaments are immediately quenched to cool them rapidly; this converts the crystalline structure of the filaments into an amorphous state. Next, the filaments are stretched in a manner analogous to the way in which nylon filaments are cold-drawn, and are then wound on to bobbins.

The melt—*i.e.*, the polymer from which the filaments are spun—may have been impregnated either with pigment to give a coloured thread, or with titanium dioxide to give a matt or pearl filament.

A new note is struck by Clorene monofil which is available not only in circular but also in flat or elliptical cross-section, *e.g.*, with major and minor axes of 0·20 × 0·50 mm., 0·25 × 0·75 mm. and 0·35 × 0·75 mm. The flat section gives greater cover and approaches the use of grass, raffia and split cane.

Properties

Vinylidene chloride polymers are of a pale gold or straw colour, and this militates to some extent against the purity and brightness of some pastel shades that are applied to them. Under the microscope the outer surface appears smooth and almost structureless and the cross-section of the fibres nearly round. Saran is practically unaffected by light.

The tenacity is about 2·4 grams per denier and the elongation at break from 15 to 25 per cent. Like many other bodies which contain a high percentage of chlorine, such as carbon tetrachloride, which is used in fire-extinguishers, Saran fibres will not support combustion; if lighted, they are self-extinguishing. This is a very valuable feature, as it means that there is no fire-hazard attached to the use of such fibres. Like Vinyon, they are extremely water-resistant, showing an absorption of less than 0·1 per cent. Chemically they are very stable, and are unaffected by acids and most alkalis, except ammonium hydroxide. They are unaffected by common dry-cleaning agents, but are soluble in some oxygen-containing solvents, notably *cyclo*hexanone and dioxan. Saran is relatively heavy, specific gravity 1·68–1·75, which is not a good feature in a textile material. The fibres are reasonable in price.

It might well be thought that there will be a considerable future for vinylidene fibres. First of all, they are made from materials—petroleum and sea-water—that are abundant and cheap; and, secondly, they can be spun from the melt, which is much more economical than spinning from a solvent. The initial cost of a solvent recovery plant is avoided, and as solvent recovery is never perfect—*i.e.*, there is always some loss of solvent—this running expense, too, is avoided.

Dyeing

The natural resistance of Saran to water makes it difficult to dye. The methods normally employed for Vinyon may be taken as the most suitable for Saran. In the case of fibres, like Vinyon and Saran, which are extremely water-resistant, the normal methods of dyeing are not very suitable, and the inclusion of a pigment in the polymer before it is spun is a method which may be preferred. In this case very bright and fast colours are produced. This method of pigmentation of spinning material has of course been used for the commoner artificial fibres, such as viscose and cellulose acetate, but as these fibres lend themselves to dyeing by normal methods, dope-pigmentation is not likely to be used so extensively with them. In the case of the vinyl and vinylidene fibres, pigmentation may be the best solution to the problem of coloration; certainly some very fine effects have already been obtained. Tygan is available in a good range of colours.

Uses

As is always the case, the uses to which a fibre can advantageously be put are determined by its properties. Its resistance to bacterial and insect attack makes Saran very suitable for insect screens, its resistance to chemical damage has enabled it to be used successfully for filter cloths, its resistance to water has made it suitable for fishing leaders. In addition, it has been used extensively for upholstery and for many narrow fabrics such as belts, suspenders and braids. Its main use has undoubtedly been for the upholstery of motor-cars in which it has largely displaced leather cloth. It has been used for upholstery of cafés and bars, and also for public transport. Its main advantages for these purposes are that it is hygienic, it can be washed, it does not stain, it wears well and does not fade. The B.B.C. have used it to line the walls of studios; a cinema circuit has adopted it to cover the walls of cinemas, so that dirt and cigarette smoke can be washed off them. It has been used for office furniture. Some municipal authorities have adopted Tygan for use on their deck-chairs and its bright pigments make a brave splash of colour on the promenades and round the bandstands; it does not fade nor stain from spilt drinks which can just be wiped off (Fig. 167); nor does it smudge from hair-grease. The author left a Tygan deck-chair exposed to all weathers in the garden for six years and it seemed to be unaffected; the fabric costs about twice as much as a good canvas. In Japan the monofilament yarn finds the usual uses, 120 denier for filter cloths and fishing-nets, 400 denier for mosquito-nets and deniers up to 1,000 for car fabrics and a 3,000 denier monofil ribbon for outdoor seating; additionally, multi-filament yarns, *e.g.*, 110 denier 10 filaments, are made and these in deniers up to 1,000 are also used for fishing-nets and a large proportion of the Japanese " vinylidene " is consumed in this form.

There also is made a crimped " vinylidene " staple which is used for rugs and carpets and which more recently has been mixed with wool and rayon staple.

FIG. 167(*a*).—Tygan deck-chair fabric with drops of (coloured) water standing on it. Tygan is impervious to water.

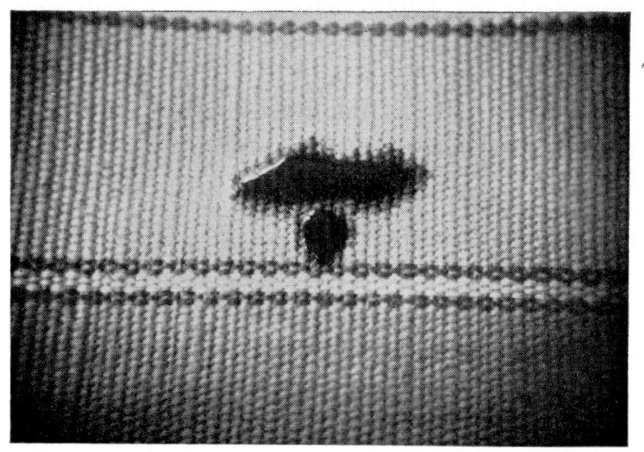

FIG. 167(*b*).—Cotton deck-chair canvas with drops of (coloured) water spreading over the fabric. Cotton absorbs the water.

It has been used for radio baffles, for golf and tennis bags and in admixture with cotton for shoe fabrics. Its chemical resistance is so excellent that steel pipe internally lined with Saran is now being made and sold; for this purpose the plastic is used, not the fibre. Velon plastic also serves a number of other uses.

The defects of Saran are its excessively low ironing temperature, which prohibits its use for garments that are normally laundered, its inability to absorb any appreciable amount of moisture, which renders it unsuitable for underwear, and finally its yellowish colour. It has to be remembered that a very high percentage of textile material is white—*e.g.*, more white baby wool is sold than all the other colours put together. Saran will not give a white, nor will it give really *bright* pastel pinks and blues, for these can be obtained only by dyeing (or pigmenting) on a good white ground. It must be accepted that Saran is a fibre for special uses; not for apparel, but for transport seating, for out-of-doors and that kind of use. That Saran has lost some of the deck-chair market to polyethylene is unfortunate. The polyethylene fabric is cheaper than Saran, but it is unconscionably slippery to sit on, which Saran is not. Polyethylene has a remarkably low frictional coefficient but this is one of the applications where it can be a disadvantage. In the U.K. Saran seems to be having a little set-back; elsewhere in the world it is doing famously. In Japan production amounts to 7 million lb. per year and about one-third of it is used for fish netting; some of it is used for blinds on their prosperous railways. The price of vinylidene chloride is low (14½ cents per lb. in America) and there seems to be plenty of scope for the fibre to develop; in Britain its wider use in public transport might be advantageous to everybody.

FURTHER READING

H. P. Staudinger, *Chemistry and Industry*, 685 (1947).
Rowland, *Rayon Text. Monthly*, **27**, 540 (1946).
W. C. Goggin and R. D. Lowry, *Ind. Eng. Chem.*, **34**, 327 (1942).
" Saran ", The Dow Chemical Co. (1942).
" Clorene ", La Société Rhovyl.

CHAPTER 28

POLYVINYL CHLORIDE

PE CE, PCU, RHOVYL

PE CE fibres were developed in Germany, and achieved some importance for the 1939–45 war effort.

Chemical Nature

Pe Ce fibres consist of chlorinated polyvinyl chloride (the other fibres discussed in this chapter have not been after-chlorinated). The vinyl chloride is first polymerised according to the following equation:

$$CH_2{:}CHCl + CH_2{:}CHCl + \ldots . + CH_2{:}CHCl \longrightarrow$$
$$—CH_2CHClCH_2CHCl \ldots . CH_2CHCl—$$

It will be realised that the chlorine content of this polymer is $\frac{35{\cdot}5}{62{\cdot}5}$, or 56·8 per cent. This material is then chlorinated until it has a chlorine content of 64 per cent, and this chlorinated product is the material from which the fibres are made. In ultimate composition it may not be very different from Saran, which, being polymerised from 13 per cent vinyl chloride, 85 per cent vinylidene chloride and 2 per cent acrylonitrile has a chlorine content of

$$\frac{(0{\cdot}13 \times 35{\cdot}5) + (0{\cdot}85 \times 71)}{(0{\cdot}13 \times 62{\cdot}5) + (0{\cdot}85 \times 97) + (0{\cdot}02 \times 53)}$$

or 70 per cent. The purpose of the chlorination of polyvinyl chloride is that of obtaining a product soluble in acetone, which is perhaps the most convenient of all solvents normally used for spinning.

Manufacture

Preparation of Polyvinyl Chloride. The monomeric vinyl chloride is polymerised in emulsion form in autoclaves at a pressure of 50 atmospheres and a temperature of 65° C. The polymer is spray-dried at 130° C.

Chlorination of P.V.C. The polyvinyl chloride (P.V.C.) is made into a 25 per cent solution in tetrachlorethane and chlorinated for thirty hours at 80° C. in a vessel cooled by a water-jacket. Some manufacturers use a catalyst, but the Germans claimed that the process could be carried out without a catalyst. After chlorination, the polymer is spray-dried, the solvent being removed by vacuum and recovered. The Pe Ce polymer has a softening point (it does not melt sharply) of 80–85° C.

Spinning. A 28 per cent solution of Pe Ce polymer is made in dry (*i.e.*, nearly 100 per cent) acetone and, after filtration, is extruded through spinnerets into cold water. The spinnerets are made of tantalum, and usually have 120 holes of 0·08 mm. diameter. When the dope is extruded into water, the acetone dissolves in the water, and the flow is arranged so that a 4 per cent solution of acetone is obtained and sent to recovery. Before winding, the yarn is stretched to three times its length. The yarn, as wound, still contains 7 per cent acetone, which is subsequently lost by evaporation.

Properties

The yarn has a tenacity of 1·8–2·2 grams per denier and an elongation at break of 40 per cent. Its softening temperature of 100° C. (or possibly less) renders impracticable its use for textile materials that have to be laundered. The fibre has good non-flam properties and good chemical stability.

Staple Fibre

A tow of 20,000 denier is collected, and then run straight to the crimper and cutter.

Uses

The main use for Pe Ce fibre appears to have been on the secret list of the Luftwaffe, who took considerable quantities of it. It has been hazarded that it was for the construction of fire-resistant clothing for airmen. Other uses to which Pe Ce fibre was put included filter-pads for the chemical industry, fishing-nets and mosquito-nets. Owing to its low softening temperature, it is unsuitable for electrical insulation. Its potential uses are probably similar to those of Vinyon and Saran.

PCU Fibre

PCU is a post-war development of the Badische Anilin & Soda Fabrik branch of the I.G. Farben. It is a polyvinyl chloride which has *not* been after-chlorinated. The theoretical chlorine content for such a substance is 56·8 per cent; in fact the " found " chlorine content of PCU fibre is 53·1 per cent. This compares with a " found " value of about 62–64 per cent for the original Pe Ce fibre.

Manufacture

The raw materials are acetylene and hydrochloric acid, which combine to give vinyl chloride

$$CH{:}CH + HCl \longrightarrow CH_2{:}CHCl$$

This is polymerised to give the long-chain polyvinyl chloride

$$[-CH_2{\cdot}CHCl{\cdot}CH_2{\cdot}CHCl{\cdot}CH_2{\cdot}CHCl-]_n$$

which is dissolved in a solvent—probably an acetone mixture—and is spun and then cut into staple form.

Properties

PCU fibre has a tenacity of 3·5–3·8 grams per denier and an extension of 25–28 per cent; these characteristics are practically the same whether the fibre is wet or dry. Specific gravity is 1·39, very similar to that of Terylene (1·38) and of Vinyon (1·37). The cross-section is irregular, almost circular, strongly indented.

Resistance is excellent to rotting and mildew, to water and to most inorganic acids, alkalis, salts and some oxidising agents and organic acids, *etc.* It is safe in paraffin, the lower alcohols, glycol, glycerin, carbon tetrachloride and formaldehyde, but is attacked by benzaldehyde, chloroform, trichlorethylene, benzene, acetone and methyl acetate.

It is non-flam, but is thermoplastic and shrinks considerably at 75–80° C.

Uses

It is used for filter cloths, diaphragms and packings industrially. Other uses are for fishing gear, ropes, sailcloth and swimsuits.

RHOVYL, FIBRAVYL, THERMOVYL, ISOVYL

These four fibres are manufactured by Société Rhovyl in France and by Deutsche Rhodiaceta AG. in Germany. They are pure polyvinyl chloride that has not been after-chlorinated as had Pe Ce. They are spun from solution, the solvent used being a mixture of carbon disulphide and acetone.

Rhovyl is the continuous filament yarn, marketed in deniers ranging from 75 to 1600. It is highly oriented, and has a breaking load of 3 grams per denier and an elongation at break of 12 per cent, these characteristics being the same wet as dry. Specific gravity is 1·4. It is impervious to water, resistant to some chemicals, rot-resistant, non-flam, and is not degraded unduly by light. Its defect is that it has a very low softening temperature, and at temperatures higher than 78° C. it contracts; at 100° C. it may shrink to half its length. Filter cloths and non-flam furnishings are probably its chief outlets. Although general chemical resistance is good there are exceptions;

it must not, for example, be cleaned with benzene, nor with trichlorethylene, but is safe in petrol, white spirit or carbon tetrachloride. Alcohol and ether have no effect on the fibres, but they are swelled by toluene, carbon bisulphide, acetone, ethyl acetate, chloroform and nitrobenzene. They are exceptionally resistant to caustic soda and to nitric and sulphuric acids.

Fibravyl is Rhovyl in staple form. Its characteristics are the same as those of Rhovyl. It is marketed with a filament denier of 1·8 or 3; it is intended for spinning on wool, cotton, schappe and linen systems. Standard staple fibre lengths are 65 mm. (2½ in.) and 90 mm. (3½ in.).

Thermovyl, like Fibravyl, is a cut fibre intended for spinning, but it differs from both Fibravyl and Rhovyl in that it has been deliberately disoriented by heat treatment. This disorientation has been accompanied by shrinkage, so that the fibre has a much lower tenacity—0·9 grams per denier—and the extremely high extensibility of 150–180 per cent. In its case also these physical characteristics are the same wet as dry. Its specific gravity at 1·35 is rather lower than that of Rhovyl and Fibravyl. The usual filament denier is much higher at 8, another indication of the shrinkage to which it has been submitted in manufacture, but it is also supplied in 3½, 5 and 15 filament deniers. It is stable in resistance to shrinkage up to 100° C., *i.e.*, it does not undergo further shrinkage at this temperature. It appears to be essentially Rhovyl that has been pre-shrunk by heat treatment and then cut to staple fibre. It is matt in appearance, another result of the shrinkage, whereas Rhovyl and Fibravyl are bright.

Isovyl. This is similar to Thermovyl in being weak and disoriented. It is a large-diameter fibre of the kemp type and is made in filament deniers of 30, 15 and 7½; the usual staple length is 80 mm. (3¼ in.). It is mainly used as a blend with Thermovyl for the manufacture of felts and paddings, which in turn are used for bedding and also for insulating panels against both heat and sound. Resistance to water, moisture, perspiration, combined with good thermal and acoustic insulating properties, makes the fibre suitable and highly esteemed for such uses; the felts are attractive in appearance, resistant to crushing, light and flimsy and soft to the touch; they look as if they would flare into flame at the touch of a light but in fact they are quite non-flam, a property common to all fabrications of the P.V.C. fibres. Isovyl differs from the other Rhovyl yarns mainly in the smaller degree of stretch which it has undergone during manufacture.

Retractyl 15 *and* 30. Retractyl is a similar P.V.C. fibre, made in staple form only, with a molecular orientation intermediate between those of Fibravyl and Thermovyl. This molecular state is shown in the degree of shrinkage which each of the fibres undergoes at 100° C.

Fibravyl contracts	55	per cent	of	its	length
Retractyl 30 contracts	30	,,	,,	,,	,,	,,
Retractyl 15 contracts	15	,,	,,	,,	,,	,,
Thermovyl contracts	0	,,	,,	,,	,,	,,

Both Retractyl 15 and Retractyl 30 are supplied crimped, the former in 4 and 7½ deniers, the latter in 3 and 6 deniers. In other respects the properties of Retractyl 15 and Retractyl 30 are similar to those of the other members of the P.V.C. group.

Crinovyl. This is a P.V.C. bristle or monofil used mainly for sweeping brushes, and supplied cut to length (not continuous).

Tevilon. This is a P.V.C. fibre of the Rhovyl type which is being spun in Japan by the Teikoku Rayon Co. Ltd. There are now five manufacturers making P.V.C. fibre in Japan; it is gaining ground largely because the monomer is available in larger quantities and at lower cost than the raw material of any other Japanese synthetic fibres. In 1959 Japanese production of P.V.C. fibre was about 8 million lb. and in 1966 it was about 17 million lb. Some of this production is spun from solution, but most of it from the melt. Solution spinning has the advantage that finer deniers are obtainable, and the fine yarns are often converted into bulky stretch yarns to be used in sweaters and underwear. Solution spun fine denier staple is used for blending with cotton and wool; these natural fibres are very fine and will blend better with fine than with coarse synthetic fibres. In Japan, fish-netting is a very important outlet for P.V.C.

Khlorin. A P.V.C. fibre made in the U.S.S.R. is called Khlorin.

Uses of Polyvinyl Chloride Fibres

Technical applications such as filter cloths, wadding and braid, protective clothing, flying-suits (non-flam) for airmen, canvas, blinds, curtains, fairings (resistance to light), fishing-nets (resistance to water) are one main class of application for these polyvinyl chloride fibres. So far as filtration is concerned, the shrinkage that Rhovyl undergoes on being heated enables a fabric of very close construction to be made, one which will trap the finest particles in either the liquid or gas which is being filtered. Static, which is usually nothing but a source of trouble in textile processes and products, exercises a useful function when Rhovyl fabrics are used to filter gases. The Rhovyl, owing to its very high dielectric constant and its water repellency, develops a considerable static charge which attracts to the fabric fine particles of dust, thus freeing the gas from them, in addition, of course, to the normal filtration effect of the fabric.

The other main application of these fibres is for furnishings, where their moisture and rot-resistance, allied with their good resistance to light, are of value. Rhovyl and fibres made from it have become important in the mosquito-netting industry. The decoration of cinemas, theatres, liners and aircraft are outlets where the non-flam properties have proved useful. The softening and shrinkage of Rhovyl and Fibravyl enable fabrics made from them to be moulded to shape for such garments as brassières.

Fabrics made from Thermovyl will withstand repeated washings, but must not be ironed.

The contraction that Fibravyl undergoes on heating enables it to be used in wool mixtures, which do not then need to be milled. The inclusion of some Fibravyl in the yarns for a worsted coating allows shrinkage to take place (thus closing up the fabric in the desired way) during finishing without the necessity for milling. Wool and Fibravyl blankets can, too, be produced without the need for milling.

Thermovyl lends itself to blending with nylon, and blends of 82 Thermovyl/18 nylon and 67 Thermovyl/33 nylon are referred to as Rhovylon.

Dyeing and Finishing of Polyvinyl Chloride Fibres

Thermovyl and Isovyl can be dyed at temperatures of 80–95° C. They are first of all de-sized with a weak solution of caustic soda (5 grams per litre) and a wetting agent. Acetate colours are very suitable; dyeing should be started at 65° C. and the temperature increased to 95° C. Naphthol colours can also be used.

Rhovyl and Fibravyl must be dyed at temperatures not higher than 65–70° C., and this restricts the range of shades available with acetate colours to light ones. The use of a swelling agent enables darker colours to be obtained. Indigosol dyestuffs can also be used on Rhovyl, and light shades can be obtained without exceeding 60° C.

Mixtures can be dyed at higher temperatures, *e.g.*, a fabric with a Rhovyl warp and a non-vinyl weft (or even a Thermovyl weft) can be dyed at higher temperatures on the jigger if care is taken to prevent warp shrinkage by not allowing the unrolling cylinder to get a lead on the enrolling cylinder. Dark shades can be obtained in this way. Methods of printing these polyvinyl chloride fibres have also been developed. Blacks are best obtained by the use of mass-dyed yarn. When Rhovyl, Fibravyl or Retractyl have been included in a fabric,

[*Société Rhovyl*]

FIG. 168.—Check velvet made by using the shrinkage properties of Fibravyl and the double cloth construction.

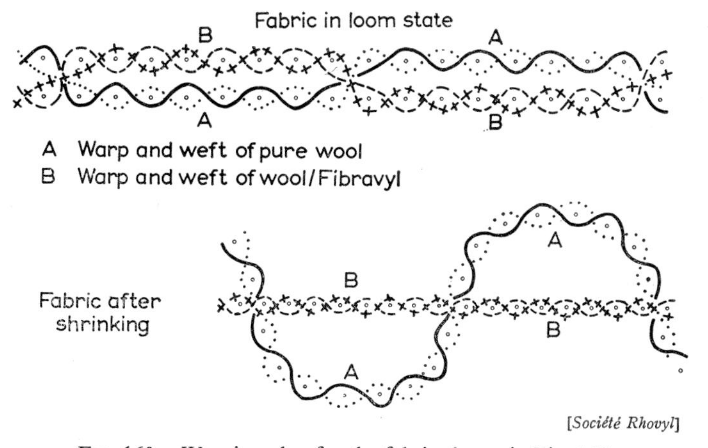

[*Société Rhovyl*]

FIG. 169.—Weaving plan for the fabric shown in Fig. 168.

with the intention of shrinking during finishing to obtain a more compact fabric, *e.g.*, collar velvet by shrinking a velvet with a Rhovyl or Fibravyl ground, the shrinkage is brought about by heat. The heat treatment may be dry or wet, *e.g.*, the fabrics can be finished dry on the stenter and hot air circulated; alternatively, they can be steamed on a cylinder drier, or again they can be processed in open width in hot liquors in jigs.

Fig. 168 shows a most attractive check velvet made with alternate groups of A. wool B. wool/Fibravyl in both warp and weft. The weaving plan (Fig. 169) shows the fabric construction in the loom stage at the top and after shrinking to produce the cockled check at the bottom.

Shaping, moulding and embossing can be carried out on P.V.C. fabrics; two ways are used:

1. Flexible tubes are covered with cloth and then the cloth sleeves are shrunk on to the tubes; when only unidirectional shrinkage is required the fabric is best made from a Rhovyl warp which shrinks and a Thermovyl weft which does not shrink.
2. The P.V.C. cloth is first shrunk to make it stable to changes in humidity and is then pressed with heat on to shapes; in this way are made loud-speaker coverings, parts of orthopædic apparatus, and breast-supporter gussets.

Amination of P.V.C. Fibres

The good features of P.V.C. fibres are their strength and chemical resistance. Their bad features are their low melting points and their lack of affinity for dyestuffs. It is true that black is spun-dyed and that other shades are obtainable by using disperse dyes, but the affinity of P.V.C. for these is much less than the affinity for them of cellulose acetate. For acid dyes which are amongst the most useful in the textile world, P.V.C. fibres have no affinity at all; it would clearly be very advantageous if they did have some such affinity. The obvious way to confer it on them is to introduce into the P.V.C. some basic groups, such as $-NH_2$, to give the acid dye molecules something to hold on to.

One interesting approach to this aim has been described in Japanese Patent 9,597 (1962): the P.V.C. was irradiated in the presence of ammonia; the radiation, the nuclear particles, knocked some of the chlorine atoms out of the P.V.C., the ammonia seized its opportunity and amino groups filled the vacant spaces, yielding a product which had a nitrogen content of up to 1·5 per cent.

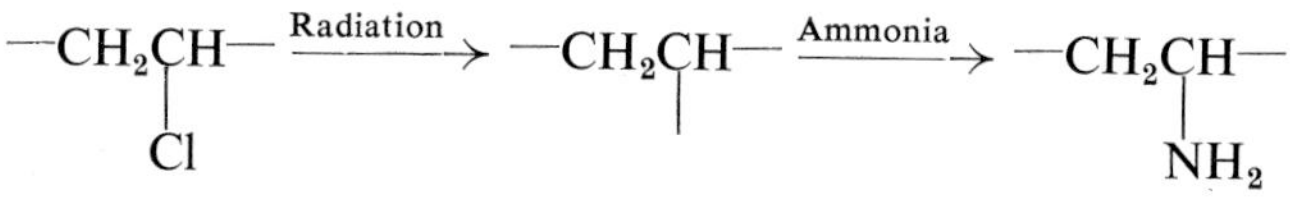

$$-CH_2\underset{\displaystyle Cl}{\underset{|}{C}}H- \xrightarrow{\text{Radiation}} -CH_2\underset{|}{C}H- \xrightarrow{\text{Ammonia}} -CH_2\underset{\displaystyle NH_2}{\underset{|}{C}}H-$$

What happened to the other H atom of the ammonia? Probably it combined with the ejected Cl atom; perhaps it seized another vacant space thus:

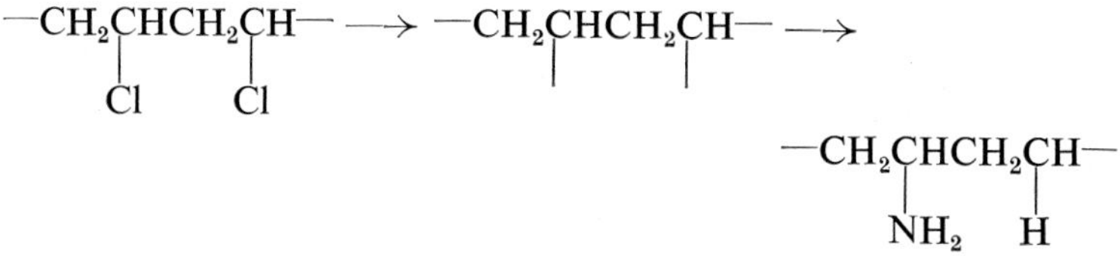

$$-CH_2\underset{\displaystyle Cl}{\underset{|}{C}}HCH_2\underset{\displaystyle Cl}{\underset{|}{C}}H- \longrightarrow -CH_2\underset{|}{C}HCH_2\underset{|}{C}H- \longrightarrow$$

$$-CH_2\underset{\displaystyle NH_2}{\underset{|}{C}}HCH_2\underset{\displaystyle H}{\underset{|}{C}}H-$$

The possibilities of inducing chemical reaction, particularly substitution, by first bombarding with nuclear particles the substance in which it is desired to substitute, are considerable. The problems of controlling such substitution are likely to be even more considerable.

A more controllable approach to the substitution of amino groups has been reported by Vaiman and Fikhman from Russia. Their starting material was some domestic P.V.C. which had a chlorine content of 56·6 per cent and a characteristic viscosity when measured in cyclohexanone solution of 0·89. The theoretical chlorine content for pure P.V.C. $-CH_2CHCl-$ is 56·75 per cent (*cf.* p. 473), but in practice it is sure to be slightly reduced by the presence of the terminal groups, which do not contain chlorine, on the long linear molecules. It is clear that the P.V.C. used was essentially pure and had not been after-chlorinated or similarly treated.

The amination was carried out by treating the P.V.C. in an autoclave at either 80° C. or 120° C. for lengths of time up to 20 hr. The reaction medium was water or dimethyl formamide or cyclohexanone, and the quantity of P.V.C. in the reaction mixture was 5 per cent. Three different aminating agents were used: ammonia, methylamine and dimethylamine. The reaction products after autoclaving were precipitated with methanol and were washed with 3 per cent acetic acid and then with water until there was no ammonia or amine in the wash water, and finally they were dried at 60° C. The products, the samples of aminated P.V.C., were tested for their nitrogen and chlorine contents, their characteristic viscosity to reveal polymeric degradation if any and for the percentage of the dye Acid Scarlet that they would absorb. Some of the samples were unfortunately discoloured by the treatment, and the extent of this was also observed. Some of the findings are assembled in the following table:

Aminating agent.	Diluent.	Temp. and time.	N content (%).	Cl content (%).	Viscosity.	Acid dye (%) absorbed by product.	Colour before dyeing.
None (untreated)	—	—	0	56·6	0·89	0	White
Ammonia	Water	120° C. 5 hr.	1·35	52·8	0·81	4·85	Cream
Ammonia	Dimethyl formamide	80° C. 5 hr.	1·45	53·0	0·80	5·0	Yellow
Methylamine	Dimethyl formamide	120° C. 5 hr.	3·00	38·0	Low	7·2	Brown

All the trials in which dimethylamine was used as the aminating agent gave a seriously discoloured product. The trials with ammonia and methylamine often gave a considerable affinity for acid dyes. A pick-up of 5 per cent of dye on the weight of the fibre is enough for a heavy shade. Under selected conditions there was no serious depolymerisation or weakening. The process is one therefore which might well have an industrial potential in addition to its academic interest. The discoloration is troublesome, but can be minimised by blending the treated fibre with untreated, so diluting the adverse discoloration.

Therapeutic Properties of Polyvinyl Chloride

When polyvinyl chloride fabrics are worn against the skin, negative static electricity is generated; this is of the opposite charge from that generated by all other fibres. A good deal of evidence has been accumulated that these negative electric charges alleviate the symptoms of rheumatism and kindred complaints. On the continent the use of Rhovyl underwear to afford such treatment is well known, but in the U.K. it is not so well known. A review of the subject by Froger appeared in the *Textile Research Journal* for February, 1962.

Since this was written there has been developed a yarn known as Thermolactyl, an 85/15 blend of Rhovyl and Courtelle. Garments made of it are obtainable from Damart Thermawear (Bradford) Ltd., Bingley, Yorks.

FURTHER READING

H. Rein, " Die Pe Ce-Faser, ihre Eigenschaften, Verarbeitung und Ausrüstung ", *Melliand Textilberichte*, **22**, 5 (1941).

" Synthetic Fibre Developments in Germany ", File XXXIII—50, B.I.O.S. H.M.S.O. (1946).

British Pat. 600,490 (solvent spinning of Rhovyl).
G. Mouchiroud, " The Blending of Polyvinyl Chloride Fibres with Other Fibres ", *J. Text. Inst.*, **43**, P 466 (1952).
F. Lieseberg, " PCU-Faser im Rahmen der Polyvinylchlorid-Fasern ", *Melliand Textilberichte*. **32**, 169 (1951).
Dyeing, " Les progrés récents dans la teinture et l'apprêt des fibres en chlorure de polyvinyle *Teintex* ", 671 (Nov. 1953); also *Teinture et Nettoyage*, 47–53 (Feb., 1956).
L. Hochstaedter, " Rhovyl–Polyvinyl Chloride Fibres ", *Text. Research J.*, **28,** No. 1, 78–85 (1958).
" Rhovyl Bulked by the Helanca Process ", *Hosiery Trade J.* (Jan., 1960).
E. Ya. Vaiman and V. D. Fikhman, " Amination ", *Karbotsepnye Volokna*, 68–77 (1966).
R. W. Moncrieff, " Amination of P.V.C. Fibres ", *Textile Weekly*, **67**, 661, 663 (1967).

CHAPTER 29

VINYLON AND KURALON

VINYLON is the name for a fibre that has been developed and is now being produced in Japan. It owes its development to the Synthetic Fibre Manufacturers group of the Japanese Synthetic Textile Association. The fibre is already of considerable importance and may well grow in importance. Output of Vinylon has been:

1952	3,000 tons
1955	6,000 tons
1960	23,000 tons
1964	40,000 tons
1966	58,000 tons

Almost all of this has been staple; filament production still only is small. " Kuralon " is the trade-name adopted by the Kurashiki Rayon Co. (who spin the fibre) for the overseas market in order to avoid confusion with American Vinyon. In 1965 Communist China ordered from Japan a plant to make 30 m. lb. per annum of Vinylon.

Chemical Nature

Vinylon is a polyvinyl alcohol which has been treated with formaldehyde to give water-resistance. A portion of its long-chain molecule can be represented as follows:

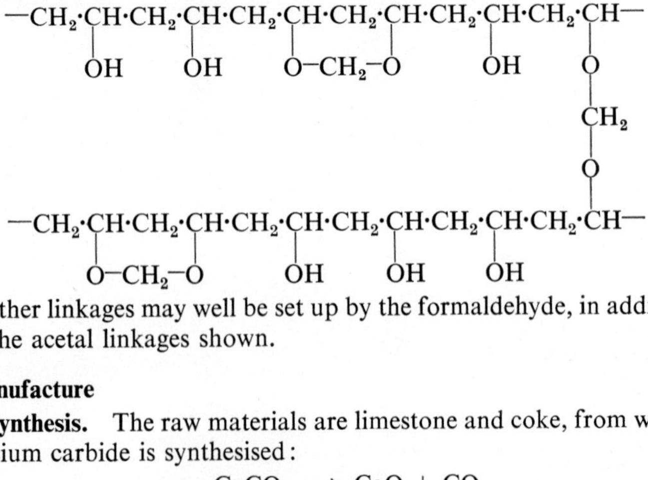

```
—CH₂·CH·CH₂·CH·CH₂·CH·CH₂·CH·CH₂·CH·CH₂·CH—
      |       |       |       |       |       |
      OH      OH      O—CH₂—O         OH      O
                                              |
                                              CH₂
                                              |
                                              O
                                              |
—CH₂·CH·CH₂·CH·CH₂·CH·CH₂·CH·CH₂·CH·CH₂·CH—
      |       |       |       |       |
      O—CH₂—O         OH      OH      OH
```

Ether linkages may well be set up by the formaldehyde, in addition to the acetal linkages shown.

Manufacture

Synthesis. The raw materials are limestone and coke, from which calcium carbide is synthesised:

$$CaCO_3 \longrightarrow CaO + CO_2$$
$$3C + CaO \longrightarrow CaC_2 + CO$$

Acetylene is generated by the action of water on the calcium carbide:

$$CaC_2 + 2H_2O \longrightarrow Ca(OH)_2 + C_2H_2$$

Part of the acetylene is converted into acetic acid by hydration and oxidation:

$$C_2H_2 + H_2O + O \longrightarrow CH_3COOH$$

The acetic acid is reacted with the other part of the acetylene, using zinc acetate as a catalyst, so that vinyl acetate is formed:

$$CH_3COOH + C_2H_2 \xrightarrow{\text{Zn acetate}} \underset{\text{Vinyl acetate.}}{CH_3COOCH{:}CH_2}$$

The vinyl acetate is dissolved in methanol and, using a peroxide as a catalyst, is polymerised into polyvinyl acetate

$$nCH_3COOCH{:}CH_2 \longrightarrow \left(\begin{array}{c}-CH_2CH- \\ | \\ OCOCH_3\end{array}\right)_n$$

Caustic soda is added to the methanol solution; this saponifies the polyvinyl acetate to polyvinyl alcohol, which is precipitated:

$$\underset{\text{Polyvinyl acetate.}}{\left(\begin{array}{c}-CH_2CH- \\ | \\ OCOCH_3\end{array}\right)_n} + nNaOH \longrightarrow \underset{\text{Polyvinyl alcohol.}}{\left(\begin{array}{c}-CH_2CH- \\ | \\ OH\end{array}\right)_n} + nCH_3COONa$$

The polyvinyl alcohol (known as Poval) is pressed and dried.

Spinning. The Poval is dissolved in hot water to make a 15 per cent solution. This solution is spun into a coagulating bath of sodium sulphate in water; the fibre is subsequently given water resistance by heat treatment and treatment with formalin. The fibre is washed, a little oil applied and it is dried. Dry spinning is contemplated at a later stage of development.

Properties

Vinylon has a specific gravity of 1·26, which is fairly close to that of wool; tenacity has been stated to be 3·5–6·5 grams per denier and extensibility from 30 per cent to 15 per cent. Evidently the properties of Vinylon depend very greatly on the degree of orientation that the fibre has undergone; the higher the stretch the higher is the tenacity and the lower is the extensibility; the wet strength is 75 per cent of the dry strength. Vinylon has a regain of 5 per cent, illustrating the low moisture affinity of all vinyl fibres, but it is noteworthy that the figure of 5 per cent is much higher than that of other vinyl fibres, and this relatively high value is due to the high proportion of hydroxyl groups in the molecule; it would be expected

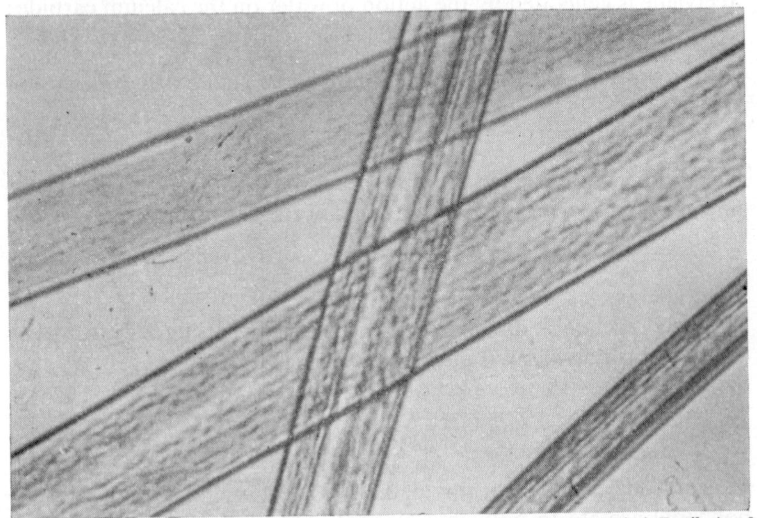

[*Japan Synthetic Textile Assn.*]

FIG. 170.—Longitudinal view of Vinylon staple fibre, 1·5 denier (× 600).

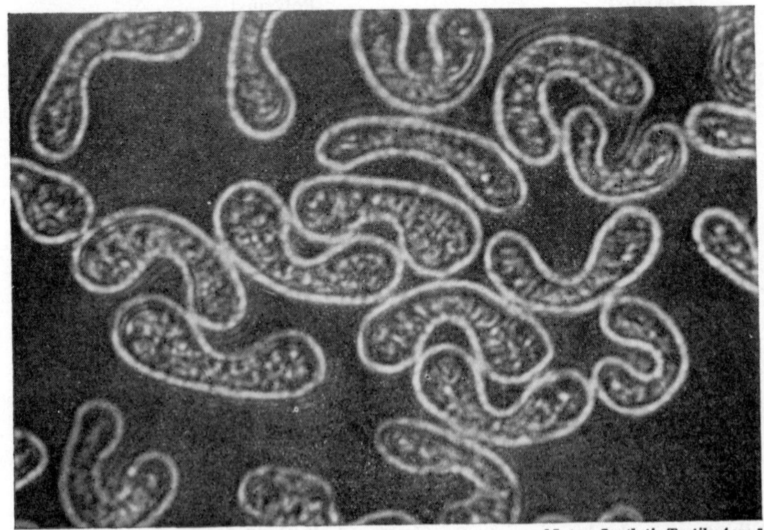

[*Japan Synthetic Textile Assn.*]

FIG. 171.—Cross-section of Vinylon staple fibre, 1·5 denier (× 1000).

that a polyvinyl alcohol fibre would have a greater affinity for water than would a polyvinyl chloride or polyvinyl acetate fibre. Vinylon softens at 200° C., and melts at 220° C., which is rather low for traditional textile purposes. Elastic recovery on release from a 5 per cent elongation is 50–60 per cent, which again is rather low for many textile purposes.

[*Kurashiki Rayon Co., Ltd.*]

FIG. 172.—Japanese fishermen using a Vinylon fishing-net to catch tuna. The net is held by the people in the large boat in the background and by those in the two smaller boats.

Chemically it has good resistance to both acids and alkalis, *e.g.*, it resists the action of 20 per cent sulphuric acid at 20° C. and that of 5 per cent sulphuric acid at 65° C. It will withstand the action of boiling dilute caustic soda. Vinylon dissolves in formic acid at 55° C. and also in hydrogen peroxide, in phenol and cresol (note that three of these four are also nylon solvents); it is resistant to most other solvents. Resistance to fungi, mildew and insects is good. Burial for six months in soil or steeping in sea-water is said not to have affected it. Its appearance under the microscope is shown in Figs. 170 and 171.

Uses

Vinylon, owing to its chemical resistance and its relatively low affinity for water, has found the following uses: school uniforms, raincoats, protective clothing, umbrellas, surgical sewing thread, filter cloths and fishing-nets (Fig. 172). It has also been used for suitings, linings, stockings, socks, gloves, hats and sewing thread. It is manufactured as a monofil for bristles, as continuous filament and mainly as tow and staple fibre; continuous filament production, inconsiderable until the end of 1953, has been increased and is used mainly for fishing-nets, tyre cord and tarpaulins. Some of the 55 Vinylon staple/45 cotton fabrics are really beautiful, reminiscent of sea-island cotton but even silkier. The fibre is hard-wearing; a jacket made in 1953 of Vinylon/rayon blend was used by the author for ten years.

Varieties of Vinylon

Kuralon and Cremona are spun by the Kurashiki Rayon Co., Mewlon by the Dai Nippon Spinning Co., and Kanebian by the Kanegafuchi Spinning Co. This last variety is coagulated by an aqueous solution of ammonium sulphate instead of sodium sulphate. Kuralon and Cremona are the same in substance; the name "Cremona" is used for Japanese home trade, and "Kuralon" for export.

VINAL FO

Vinal is the American name for Vinylon, really an F.T.C. approved name for polyvinyl alcohol fibres. The Air Reduction Co. in U.S.A. is interested in the production of two kinds of Vinal, both made by Kurashiki. The one known as Vinal 5F is a typical Vinylon and has characteristics similar to those already described; the other, Vinal FO, is made differently and has different properties. Instead of being spun into water, it is mixed with a partial solvent to reduce its melting point, is extruded by a semi-melt process and the

	Vinal 5F (normal Vinylon).	Vinal FO (semi-melt spun Vinylon).
Tenacity (gm./denier):		
Dry	5·0	7·2
Wet	4·0	7·0
Ratio wet/dry (per cent)	80	97
Elongation at break (per cent) dry	20	10
Stress needed (gm./denier) for 5 per cent strain	1·3	3·0
Elastic recovery (per cent) from 5 per cent stretch	55	75

fibre is hot drawn. No formalisation treatment is applied; evidently the polymer must have been water-insolubilised before the spinning mixture was prepared. A comparison of some of the physical properties of the product Vinal FO with those of standard Vinylon (Vinal 5F) is interesting (p. 488).

Clearly Vinal FO is stronger, less compatible with moisture, and tougher and is well suited for industrial uses, whereas the standard Vinylon is more suitable for apparel use. Its potentialities for tyre cord are being investigated by Goodyear.

Much work on polyvinyl alcohol fibres has been reported from Russia, mostly concerned with an analysis of optimum spinning conditions. A novel approach to the coloration of the fibres has been to acetalise the fibre with an aromatic amino-aldehyde instead of the more usual formaldehyde. The aldehyde group combines with the hydroxyl groups of the polyvinyl alcohol and the free amino group can be combined with a chromogenic amine such as *p*-nitraniline, resulting in a coloured fibre.

China, too, is engaging in the manufacture of polyvinyl alcohol fibre. In 1964 she arranged to buy from Japan (Nichibo Company) a plant capable of making 30 million lb. per annum of Vinylon. The cost was reported to be £10 million.

Water-Soluble Polyvinyl Alcohol

Although every effort is ordinarily made to improve the water-resistance of the fibre Vinylon, there is another fibre which is used only in the Japanese home trade which is water-soluble (*cf.* alginate fibres); this fibre, whilst consisting essentially of polyvinyl alcohol, has been subjected to a partial oxidation so that some of the carbinol groups have been oxidised to carboxyl groups, probably with breakage of main molecular chains. Although this yarn is not exported, polyvinyl alcohol in powder form is being exported for use as a sizing material.

FURTHER READING

Textile World, **101**, 123, 232, 234 (1951).

R. D. Wells and H. M. Morgan, " New Developments in Polyvinyl Alcohol Fibres ", *Text. Research J.*, **30**, 668–674 (1960).

R. W. Moncrieff, " China to Make Polyvinyl Alcohol Fibre ", *Text. Weekly*, 521, 19 March (1965).

K. E. Perepelkin and M. D. Perepelkina, " Synthetic Polyvinyl Alcohol Fibres ", *Tekstil'n Prom* **23** (6), 20–23 and (8), 27–31 (1963). Reviewed by R. W. Moncrieff in *International Dyer*, 231, 233, 7 February (1964) and in *Skinner's Record*, 876, 879, October (1963).

CHAPTER 30

VEREL, TEKLAN

VEREL (pronounced verEL) is a modified acrylic fibre (modacrylic, *i.e.*, between 35 and 85 per cent acrylonitrile) which in 1956 was introduced by Eastman Chemical Products Inc., of Kingsport, Tennessee, a subsidiary of the Eastman Kodak Company. Perhaps its outstanding characteristic amongst the acrylics is its good dyeability; other notable properties to which reference is made later are its relatively high moisture regain, its good strength and its non-flam nature.

Chemical Structure

No direct information other than that it is a " modified acrylic " fibre is available. The very much higher specific gravity of Verel (1·38) than of Orlon, and Acrilan (1·17) suggests that the degree of modification has been considerable. It is of interest that the original Vinyon co-polymer of 88 per cent vinyl chloride and 12 per cent vinyl acetate had a specific gravity of 1·37. High specific gravity and non-flam nature suggest that Verel contains a considerable proportion of combined chlorine. Its properties suggest a monomer composition similar to 60 acrylonitrile/40 vinylidene chloride; possibly with some amide (methyl acrylamide) to give the high regain.

Types

Verel is supplied only in staple, but in various filament deniers (2, 3, 5, 8, 12, 16, 20 and 24) and staple lengths, bright or dull. Two types, with different shrinkage characteristics, are made:

Regular Verel (conventional textile type) with a shrinkage in the spun yarn form of 1–3 per cent in boiling water and 2–5 per cent dry at 150° C.

Verel Type III (high shrinking fibre) with a spun yarn shrinkage of 28–32 per cent in boiling water and 35–40 per cent in air at 150° C.

Properties

Two of the properties of this modacrylic command attention, neither separately but taken together they are important: the tenacity of the fibre is 2·5–2·8 grams per denier and the moisture regain is 3·5–4·0. There are few synthetic fibres which have higher regain figures, although nylon has, and Verel is the first of the acrylics to have come along with a moisture regain of 4 per cent; this is practically the same as that of nylon and certainly makes for comfort in wear. Details, so far as they are available, of the properties of Verel are:

Tenacity: 2·5–2·8 grams per denier (standard conditions).
Extension: 33–35 per cent (standard conditions).

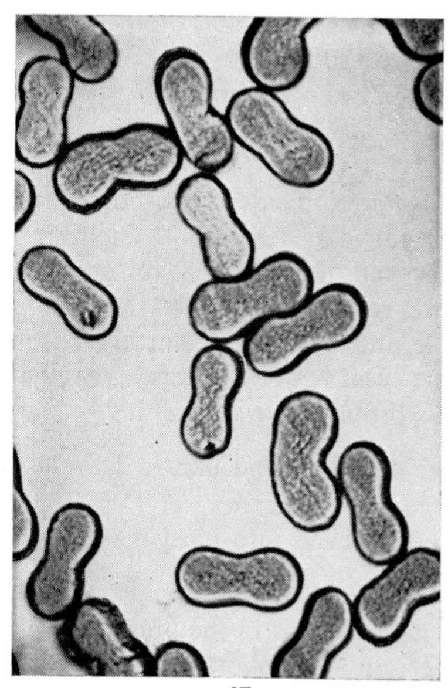

[*Tennessee Eastman Co.*]

FIG. 173.—Photomicrograph showing cross-section of Verel filaments (dyed). Magnification × 1,000.

Elasticity: 88 per cent recovery from 4 per cent extension, 55 per cent recovery from 10 per cent extension.

Specific gravity: 1·37, this is about the same as Terylene (1·38) and the original Vinyon (1·37), much higher than the acrylonitrile fibres such as Orlon (1·18) and Acrilan (1·17).

Morphology: Cross-section is pea-nut shaped, indicative of the fibre having been spun from solution, probably in acetone (Fig. 173).

Flammability: Fibre will char and melt but will not support combustion. Dull fibre is more flame-resistant than bright. Medium and heavy fabrics are usually self-extinguishing if ignited, very light fabrics may burn. All should be tested, nothing taken for granted.

Chemical resistance: Good; exposure to acids, common alkalis, bleaches and dry-cleaning solvents has little or no effect on the tenacity of Verel. Soluble in warm acetone.

Biological resistance: Good.

Weathering: No loss in tenacity is found in Verel fibres after 30 weeks' outside exposure but after 50 weeks' exposure there is a 45 per cent loss in strength.

Effect of heat: Not very resistant to heat; fabrics can be safely pressed and ironed at the low settings on the iron, but damage occurs at temperatures above 150° C.

Handle: Attractive and soft handle.

Dyeing

Most acrylic fibres are difficult to dye but Verel has an affinity for a surprisingly wide range of dyestuffs. The fibre is a good white so that bright shades can be obtained on it. Neutral dyeing pre-metallised dyes are recommended for the best light and washing fastness. Disperse and basic dyes can also be used to get bright shades and in some cases better fastness. Originally Verel was dyed at the boil but this brought two difficulties:

1. The fibre delustred and had to be relustred with salt solutions or otherwise.
2. Raw stock packed into hard masses that were difficult to open and dry.

It has since been found that in the presence of Verel Fiber Dyeing Assistant (a carrier, an alkyl phosphate), dyeing can be carried out at 70° C., and as this is below the temperature at which serious plastic deformation occurs, both troubles are largely obviated (if some delustring does take place, the fabric can be relustred by drying at 90–120° C.).

The lower dyeing temperature can be used for raw stock (loose fibre), yarn, piece goods or carpets, and it is particularly valuable for pile fabrics which are dyed at 80–85° C. and relustred at 105° C. Fig. 174 shows Verel carpets of identical construction dyed at 100° C. and at 70° C. and the superiority of the latter is evident. Verel is probably unique in its capacity for being dyed at low temperatures.

Uses

The uses are determined by the properties. Softness of handle, controlled shrinkage, high moisture regain (for a synthetic), good dyeability, low pilling, non-flam nature, and good chemical and weathering resistance are the significant ones. Pile fabrics have been successful. Natural furs can be simulated by making a fabric composed of both regular and high shrinkage (Type III) Verel. If the fabric is run through an oven at about 145° C., the

[*Tennessee Eastman Co.*]

FIG. 174.—Piece-dyed Verel carpets of identical construction. *Left:* dyed at 100° C.; *Right:* dyed at 70° C. The lower temperature gives better quality and texture.

Type III fibres contract to produce a ground structure, corresponding to the dense undercoat of a natural fur; the regular Verel is unaffected and forms a long pile corresponding to the long silky guard hairs of a natural fur. Carpets too have been found to be resistant to soiling (due to the smooth fibre cross-section), resistant to crush and to wear. It is of interest that Verel fur-type fabrics are suitable for making up into garments at home, whereas natural furs require an expert.

Teklan

Early in 1962 Courtaulds introduced B.H.S., a provisional name for a new modacrylic fibre, now called Teklan, made on pilot-plant scale at Coventry. It is a co-polymer made with " equal weight proportions of vinylidene chloride and acrylonitrile, together with small but critical quantities of other substances ".

Spinning. This co-polymer of vinylidene chloride and acrylonitrile apparently contains a small amount, less than 5 per cent, of some such additive as itaconic acid (used by the Russians in their Nitron) or of vinyl pyridine (used by Monsanto in their Acrilan) in order to enhance the polymer's affinity for dyes (B.P. 898,734 of 1962). The fibre is thus apparently made from two major and one minor monomers.

The polymer which is to be spun should have an intrinsic viscosity measured in dimethyl formamide of 1·4–1·5. Spinning is carried out from a 22 per cent solution of the polymer in acetone. After having been spun the fibre is immediately stretched, continuously with the spinning, to twice its length (100 per cent stretch) at room temperature. Tenacity is 1·5 gm./den., extensibility 50 per cent at this stage. Later, in order to secure good textile properties the yarn is stretched in saturated steam at 30 lb. pressure to 10 times its length (20 times the original spinning winding up speed). Tenacity is 4·3 gm./den., extensibility 11 per cent at this stage. The yarn is then passed, to relax it, through boiling water. Its final tenacity is 3·4 gm./den. and extensibility 14 per cent. Where does the flameproof quality come from? Presumably from the high chlorine content. A 50/50 co-polymer of vinylidene chloride and acrylonitrile should possess a chlorine content of $71/(98 \times 2)$ or 36 per cent, given that the two are present in equal *weight* proportions.

Properties. A comparison of what is known of Teklan with Dynel and Verel, the two other best known modacrylics, is interesting:

	Dynel.	Teklan.	Verel.
Identity of monomers	Acrylonitrile and vinyl chloride	Acrylonitrile and vinylidene chloride	Acrylonitrile and undisclosed component (might be vinylidene chloride)
Tenacity (gm./denier)	3	3·5–4	2·8
Extension at break (per cent) . .	30	18–20	33
Moisture regain (per cent) . . .	0·4	Low	4
Flameproof quality .	Self-extinguishing, no risk	Flame-resistant	Non-flam

Additionally Teklan has good abrasion resistance, low shrinkage in boiling water, is resistant to photodegradation and to microbiological attack. It is suggested that it be used for children's clothing because of its flame-resistant quality: it will be cheaper than nylon or Terylene.

Dyeing. The coloration of fibres of this type, *i.e.*, co-polymers of equal quantities of acrylonitrile and vinylidene chloride, has been described by Courtaulds in B.P. 940,372. In the preamble to this patent they state that although such fibres can be dyed with disperse (cellulose acetate type) dyes, deep shades are not obtainable in a reasonably short time of dyeing. They therefore suggest that the co-polymer (Teklan) be immersed in an aqueous solution containing 10 per cent of such a dye as Duranol Blue 2 GS in a closed dyeing vessel. Dyeing is started at 65° C. and the temperature is raised to 110° C. in 15 min. and dyeing continued at 110° C. for another 20 min. Then the temperature is reduced to below 100° C. and the fabric rinsed and dried. " A full blue shade of excellent fastness to washing and to rubbing was obtained." The novel feature is the use of a temperature above 100° C. Other colours are obtained similarly by the use of such dyes as Dispersol Fast Orange BS, Duranol Red 2 BS and Duranol Violet BRS. Other kinds of dye which can be similarly used at superatmospheric pressure on these Teklan fibres are azoics (p. 763) and 1 : 2 metallised dyes (p. 759), specific examples of both being given on the pages indicated.

The first British acrylic fibre, Courtelle, was a long time coming after the American Orlon and Acrilan, but when it did come it was good. The first British modacrylic, Teklan, has come a long time after Dynel and Verel. Its flameproof qualities are superb, they are perfection itself.

FURTHER READING

R. J. Fortune and V. G. Paul, " Recent Developments in Dyeing of Verel Modacrylic Fiber ", *Amer. Dyestuff Reporter*, **50**, No. 9, 41–44 (1961).

R. O. Rutley, " Modacrylic Fibres: Properties, Treatment and Features ", *Text. Manufacturer*, 211–213 (May, 1968).

CHAPTER 31

ORLON, PAN, DRALON

ORLON was introduced and is made by E. I. du Pont de Nemours & Co., the original manufacturers of nylon; it is manufactured at their Camden, South Carolina, plant.

The early laboratory name for Orlon was Fiber A. Du Pont started research on it in the early 1940's, and by 1942 experimental samples had been offered to the U.S. Government for military applications. The fibre was in the pilot-plant stage in 1945; in 1948 the trade-mark Orlon was announced, and commercial production was started in 1950.

Chemical Structure

The fibre is a polymer of acrylonitrile formed by the reaction shown below. It should be noted that another name for acrylonitrile is vinyl cyanide. Just as the formula of vinyl chloride (the main constituent of Vinyon) is $CH_2{:}CHCl$, so that of acrylonitrile or vinyl cyanide is $CH_2{:}CHCN$. When this substance is polymerised the following reaction takes place:

$$CH_2{:}CHCN + CH_2{:}CHCN + CH_2{:}CHCN \longrightarrow -CH_2\underset{\displaystyle CN}{\underset{|}{C}}H CH_2\underset{\displaystyle CN}{\underset{|}{C}}H CH_2\underset{\displaystyle CN}{\underset{|}{C}}H-$$

As the reaction proceeds the length of chain increases, and ultimately a long chain polymer, which has the composition

$$\left(-CH_2\underset{\displaystyle CN}{\underset{|}{C}}H-\right)_n$$

results. The reaction by which it is made is one of *addition* polymerisation. When Orlon was introduced it was described as a polymer of acrylonitrile and it was then a straight polymer (a homopolymer). To-day (1968), staple fibre Orlon at least is a co-polymer of acrylonitrile (probably about 90 per cent) and of a minor constituent which has been co-polymerised with the acrylonitrile to increase dye receptivity (the dye affinity of staple fibre Type 42 is much higher than that of continuous filament Type 81). A variety of ethylenic compounds have been used with acrylonitrile in published patent examples; which one of them has been included in Orlon has not been disclosed; substances specified in patents include such as vinyl acetate, vinyl chloride, styrene, *iso*butylene, acrylic esters, acrylamide and other similar compounds. It is likely that Orlon Type 81 always has been a homopolymer of acrylonitrile.

Degree of Polymerisation. The value of n (as used in the above formula) in Orlon continuous filament polyacrylonitrile is about 2,000, corresponding to a molecular weight of about 100,000. An analysis of the relative abundance of molecules of different length (*cf.* p. 33) has been made by R. C. Houtz. The method that he used was to dissolve the polyacrylonitrile (Orlon) in dimethyl formamide and then to add a little heptane, which dissolved when the mixture was warmed to 60° C. This decreased the solvent power of the dimethyl formamide, and some of the polyacrylonitrile was precipitated; the first precipitate consists of the longest molecules—those that are least soluble. Then a little more heptane is added and a little more precipitate obtained, and this process is carried out ten times, giving ten fractions of the polyacrylonitrile, the first fraction containing the longest polymers and the last the shortest polymers. The fractions are redissolved and the viscosities of the solutions measured; thence the average molecular weight of each fraction was calculated. The results obtained by Houtz are given in the following table:

Fraction No.	Percentage heptane in dimethyl formamide.	Percentage weight of polyacrylonitrile precipitated.	Average molecular weight of poly-acrylonitrile.	Average degree of polymerisa-tion of poly-acrylonitrile.
1	22·3	6·6	135,000	2540
2	22·5	33·0	113,000	2140
3	22·8	5·2	131,000	2480
4	23·4	15·8	77,000	1460
5	24·2	6·9	66,000	1250
6	25·0	10·1	62,000	1170
7	25·7	5·9	69,000	1300
8	28·0	4·8	44,000	860
9	30·5	3·1	17,600	330
10	—	8·6	less than 17,600	less than 330
		100·0		

Fractionation is not perfect and one or two anomalies can be discerned, but the trends are unmistakable.

If we divide the percentage weight of polyacrylonitrile precipitated by the average molecular weight for that fraction, we arrive at the relative *numbers* of molecules in each fraction, and if we group these into high molecular weight (over 100,000), medium molecular weight (50,000 to 100,000), low molecular weight (below 50,000), we obtain the figures in the table on p. 498.

It can be seen that we have the makings of a graph similar to that shown on p. 33, wherein the percentage number of molecules is plotted against molecular weight, or, as Houtz himself wrote, " it would appear that the distribution follows the normal distribution curve, with some skewness toward the high-molecular-weight end ".

Fraction No.	Average molecular weight.	Relative *numbers* of molecules.	Relative *number* of molecules in groups.
1	135,000	49	381 high molecular weight
2	113,000	292	
3	131,000	40	
4	77,000	206	560 medium molecular weight
5	66,000	104	
6	62,000	164	
7	69,000	86	
8	44,000	108	291 low molecular weight
9	17,600	183	
10	Low	—	

N.B.—The figures in the third column have been multiplied by 10^6 to make them easier to read; as they are *relative*, this is of no consequence.

Suitability of Polyacrylonitrile for Fibres. It has long been known that acrylonitrile will polymerise to give high polymers, and that these polymers are characterised by insolubility and absence of swelling in most liquids. It was realised that the high polymeric nature of polyacrylonitrile would lend itself to the formation of good fibres; the difficulty was that there was no known suitable solvent for it, so that there was no means of making a solution and spinning it. Research work on this problem was in progress in the thirties in both America and Germany. The chemical structure of polyacrylonitrile suggests that there are no cross-linkages; why, then, should the polymer be so insoluble? The most likely reason was thought to be that there were strong secondary valence forces between adjacent chains, and that these were due to hydrogen bonds (see p. 96) existing between the α-hydrogen atom of one chain and the nitrile nitrogen of an adjacent chain. This may be represented thus (the hydrogen bonding shown dotted):

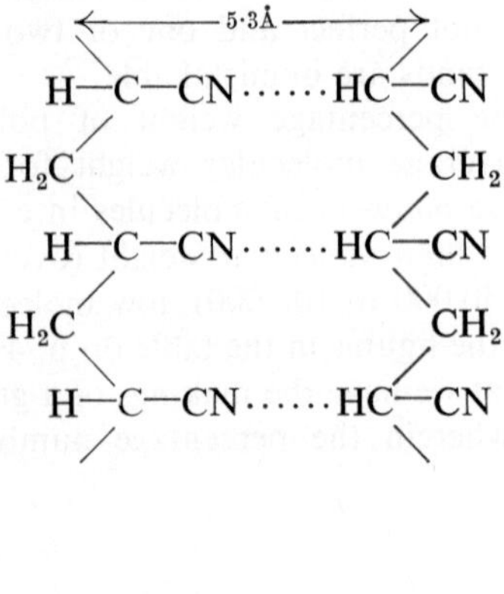

```
      <------- 5·3Å ------->
       \                 /
     H—C—CN······HC—CN
        /               \
   H2C                   CH2
        \               /
     H—C—CN······HC—CN
        /               \
   H2C                   CH2
        \               /
     H—C—CN······HC—CN
        /               \
```

The long molecular chains run vertically in this diagram, and the hydrogen bonds are shown horizontally. Furthermore, X-ray diffraction studies (see p. 79) have established the existence of a spacing of 5·3 Å, the distance between the polymer chains through the nitrile group.

It is likely, therefore, that in order to make the polymer soluble, a solvent should be used which will break the hydrogen bonds. Such a solvent is likely to be one which itself has a tendency to bond at those points in the polymer which are hydrogen-bonded, so breaking the hydrogen bond and associating itself (the solvent) as intimately as possible with the polymer; such intimate association becomes, or is the equivalent of, solution. Strongly polar solvents are indicated, and when several thousand organic liquids were tried as solvents for polyacrylonitrile it was found that quite a number of them, all very strongly polar, would in fact dissolve the polymer: they do this by breaking the hydrogen bonds in the polymer. Even so, very slight differences in constitution could mean the difference between solvent and non-solvent properties as the following table of solvents and closely related non-solvents for polyacrylonitrile reveals.

Solvents.	Non-solvents.
Dimethyl formamide $HC(=O)N(CH_3)_2$	Formamide $HC(=O)NH_2$
	Diethyl formamide $HC(=O)N(C_2H_5)_2$
Dimethyl sulphone $CH_3S(\rightarrow O)(\rightarrow O)CH_3$	Diethyl sulphone (swells) $C_2H_5S(\rightarrow O)(\rightarrow O)C_2H_5$
m-Nitrophenol $HO-C_6H_4-NO_2$ (meta)	*o*-Nitrophenol $HO-C_6H_4-NO_2$ (ortho)
p-Nitrophenol $HO-C_6H_4-NO_2$ (para)	
Succinonitrile $CN(CH_2)_2CN$	Acetonitrile CH_3CN (swells)
Adiponitrile $CN(CH_2)_4CN$	Suberonitrile $CN(CH_2)_6CN$

The discovery of suitable solvents for polyacrylonitrile made possible the manufacture of fibres from it.

Manufacture

No details of the manufacturing processes, neither of Orlon nor the other (more or less modified) acrylonitrile fibres, have yet been disclosed, but there are uniformities of procedure in the examples of patent specifications dealing with claimed improvements in properties which make it almost certain that manufacture runs along the following lines:

Forty parts (all parts by weight) ammonium persulphate catalyst and 80 parts sodium bisulphite as an activator are dissolved in 94 parts distilled water at 40° C. and then, over a period of 2 hours, 16 parts of a mixture of about 90 per cent acrylonitrile/10 per cent other ethylenic monomer are added slowly with stirring. The polymer—polyacrylonitrile modified with the other monomer—precipitates from the solution; it has a molecular weight of around 60,000. In principle the process is simple, the acrylonitrile is added to water containing a peroxy catalyst, it dissolves in the water, is polymerised and the polymer which is insoluble is precipitated as a slurry. (The solubility of acrylonitrile in water is about 7 per cent; sometimes a greater proportion than this is used and then some of it is in emulsion.) The precipitated polymer is filtered off, washed, dried and dissolved in a suitable spinning solvent which may be dimethyl formamide; the solution contains not less than 10 per cent and probably more like 20 per cent solute. It is heated and extruded into a heated spinning cell; a heated evaporative medium which may be air, nitrogen or steam, moves countercurrent to the path of travel of the freshly formed filaments, removes the solvent and carries it off to a recovery plant. The filaments are hot-stretched to several times their length, the yarn is heated either by passing it round a heated pin or through hot air or hot water; the hot stretching temperature can vary from 100° C. to 250° C., depending on the time of contact during heating.

Properties

The first Orlon that was spun was not a good colour, but for some years now the fibre has been a good white. Moisture regain is 1–2 per cent. Flammability is similar to that of rayon and cotton. The two main types of Orlon that have been marketed are continuous filament Type 81 and staple fibre Type 42 (which towards the end of 1953 superseded an earlier Type 41). The mechanical properties of Type 81 and Type 42 are very different and must be considered separately. More recently Type 81 continuous filament has been discontinued. Type 42 staple is the main product with Type 44 (dyeable with acid dyes) staple also important.

Continuous Filament Type 81. The dry strength is 5 grams per denier, and the wet strength 4·8 grams per denier. The high ratio of wet to dry strength indicates that the fibre is water-resistant, and in this respect similar, as it would be expected to be from a consideration of its constitution, to Vinyon and Saran. Its looped (half-hitch knotted) strength is 3·6 grams per denier; this is 72 per cent of its normal strength, and the high value is a good feature. The elongation at break is 17 per cent dry and 16 per cent wet; the elastic recovery is 85 per cent after a 4 per cent extension provided that the yarn is released instantaneously, but if it is held stretched for 100 seconds and is then released, recovery is only 66 per cent; in each case the recovery was measured one minute after the stretch was released.

The stretch resistance is good; quite high loads are necessary to produce even small elongations. This is illustrated in the table below:

Elongation (per cent).	Load (gm/denier).
0·5	0·4
1·0	0·8
1·5	1·1
2·0	1·3
2·5	1·5

A load which would break a cellulose acetate fibre will only stretch an Orlon fibre by 2½ per cent of its length. The stress–strain curve for continuous filament Orlon is shown in Fig. 175.

The work of rupture as determined from the area under the stress–strain curve is high, and this gives good toughness. Flex life is good.

Staple Fibre Type 42. This yarn is much less highly oriented and has a much lower tenacity of only about 2·3 grams per denier with an extensibility at break of 28 per cent. The stress–strain curve (Fig. 176) is generally similar in shape to that of the continuous filament yarn but it flattens very much more quickly. For assessing the value of staple Orlon, a point to be taken into consideration is that the strength and extensibility of the yarn are not very much greater than those of staple viscose rayon. The stretch resistance of Type 42 is much lower than that of Type 81—about 0·5 grams per denier instead of 0·8 grams per denier for a 1 per cent stretch. Whereas Type 81, the original continuous filament Orlon, was a homopolymer, Type 42 is a co-polymer with greatly reduced strength and chemical resistance.

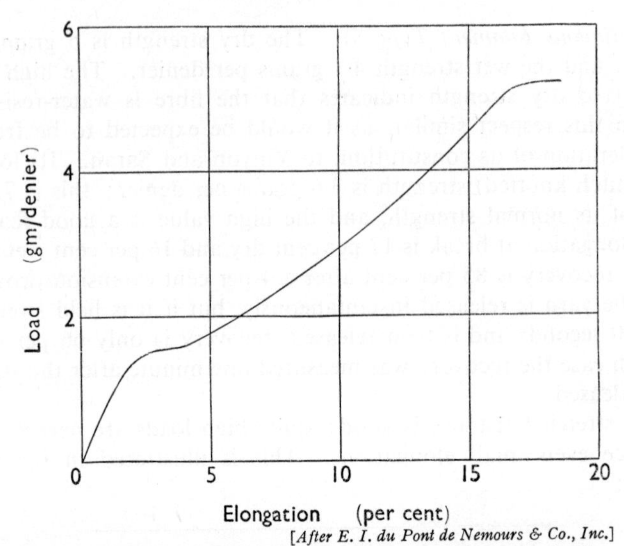

[*After E. I. du Pont de Nemours & Co., Inc.*]

FIG. 175.—Stress–strain diagram of Orlon Type 81 acrylic fibre.

The initial steep gradient is noteworthy; over this range the fibre obeys (nearly) Hooke's Law and its elasticity is high. The curve was recorded on an Inclined Plane Tester (*cf.* Fig. 6).

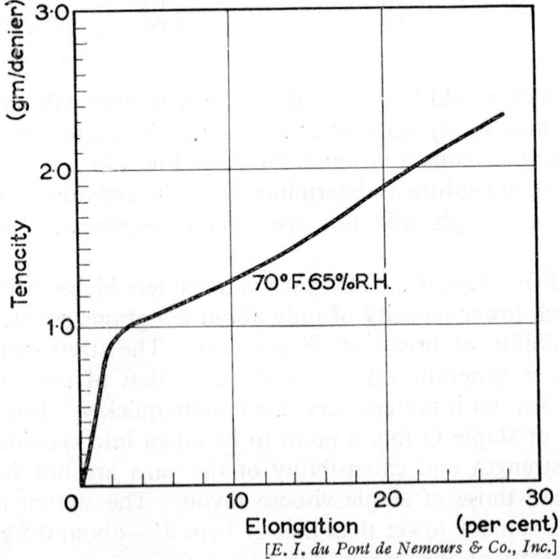

[*E. I. du Pont de Nemours & Co., Inc.*]

FIG. 176.—Typical stress–strain curve of 3-denier Orlon staple Type 42.

Single filaments of crimped tow tested on Instron with rate of extension of 60 per cent/min. 10 in. sample test length. Samples of tow produced during May, 1954, were prepared by boiling-off for 1 hour in 212° F. water containing a detergent, rinsing, and then drying and conditioning at 70° F. and 65 per cent R.H.

There are various other types made, *e.g.*, T37 and T39 for blending and T25 for spinning on the cotton system; all staple. Orlon 21 and Sayelle are apparently acrylics and have a polyacrylonitrile content of at least 95 per cent. Orlon 28 and Orlon 44 appear to be modacrylics (35–88 per cent polyacrylonitrile) with some basic constituent.

Chemical Resistance. Generally, Orlon has very good resistance to mineral acids and excellent resistance to common solvents (the difficulty experienced in finding a suitable solvent for it bears this out), oils, greases and neutral salts; its resistance to weak alkalis is fairly good, but strong alkalis, especially when hot, degrade it rapidly. The results of some typical immersions of Type 81 are given in the following table:

Chemical.	Temperature.	Concentration (per cent).	Time.	Effect on strength.
Aqua regia	Room	100	20 weeks	None
Aqua regia	75° C.	100	4 days	Moderate
Hydrochloric acid	Room	37	3 days	None
Hydrochloric acid	Room	37	5 weeks	Considerable
Nitric acid	Room	40	30 weeks	None
Nitric acid	75° C.	40	1 day	Considerable
Sulphuric acid	75° C.	25	7 days	None
Caustic soda	Room	5	5 weeks	None
Caustic soda	Room	5	15 weeks	Considerable
Caustic soda	Room	20	3 weeks	Moderate
Caustic soda	79° C.	5	1 day	Degraded
Caustic soda	100° C.	20	8 hours	Degraded
Common salt	Room	3	50 days	None
Potassium permanganate	Boiling	3	15 minutes	Considerable
Acetone	Room	100	3 days	None

N.B.—" None " means 10 per cent or less loss in tenacity, " moderate " means 11–30 per cent loss in tenacity, " considerable " means 31–75 per cent loss in tenacity and " degraded " means over 75 per cent loss in tenacity.

It can be seen that polyacrylonitrile is relatively easily saponified by hot aqueous alkali. These tests were made on Orlon Type 81, the original Orlon that was a homopolymer of acrylonitrile, and a necessary accompaniment of its excellent chemical resistance was its resistance to coloration. So great were the difficulties of dyeing it that towards the end of 1956 production was stopped. Orlon Type 42 is a co-polymer in which the acrylonitrile has been co-polymerised with another monomer to improve its dye receptivity, and, in fact, Orlon Type 42 is relatively easy to dye, but correspondingly its chemical resistance has been reduced, and probably that of Orlon Type 42 is not very different from the chemical resistance of Acrilan (p. 522), which is another easy-to-dye acrylic fibre.

Effect of Heat. Orlon is resistant to degradation even for long exposure times up to temperatures of 150° C. Even after two days' exposure at this temperature there is no loss in strength. Fabric can be set during finishing by high-temperature treatment. An ironing temperature of 160° C. is recommended; higher temperatures may cause yellowing (short contact ironing, for example, at 200° C. will cause yellowing), but the fabric does not stick until a temperature of 230° C. is reached. Provided that the iron is not too hot, Orlon fabric can be ironed repeatedly and effectively without discoloration.

The behaviour of Orlon on being subjected to more drastic heat treatment is interesting; it changes progressively to yellow, brown and black as heating is continued. Even after sixty hours heating at 200° C. the yarn, although quite black, retains more than half of its initial strength; furthermore, the black yarn is remarkably stable to further heating even in a Bunsen burner flame. Very little combustion takes place, the yarn merely glowing in the flame, and only about 30 per cent of the initial weight is lost. When polyacrylonitrile is dry heated the long molecular chains do not break down extensively, but some rearrangement of the molecule occurs, colour develops, hydrogen is lost, basic groups appear (the heated yarn will absorb acid) and the yarn becomes insoluble in some of the normally good solvents for acrylonitrile. It has been suggested that these changes could be accounted for by the following rearrangement:

```
   CH2   CH2                          CH    CH
  /   \ /   \                       //  \  /  \\
—CH    CH    CH—CH2—          —C      C      C—CH2—
 |     |     |          →      |      ||     |          + H2
 CN    CN    CN               HC      C      C—NH2
                                \\   /  \   //
                                  N       N
      Orlon.                  Dry overheated Orlon (black).
```

Nuclear Radiation. Orlon Type 81 is very much more resistant to nuclear radiation than most fibres. When it was exposed for 42 hours to γ-radiation from Cobalt 60, a radioisotope which is a powerful γ-emitter, the Orlon was not reduced in strength, whereas cotton, viscose rayon and nylon lost about half their strength when similarly treated. When exposed to neutron bombardment in an atomic pile for 26 hours so that each square centimetre was bombarded by $2{\cdot}3 \times 10^{17}$ neutrons, the Orlon lost about one-quarter of its strength, whereas nylon lost more than half its strength and cellulosic fibres lost practically all their strength on similar exposure. The neutron irradiation slightly flattened the characteristic curve of the Orlon (Fig. 177). The considerable superiority of Orlon Type 81 to nylon and to cellulosic fibres in respect of the way it withstands bombardment by nuclear radiation, may prove to be of considerable use, and it is all the more a matter for regret that the manufacture of Type 81 should have been discontinued. Cantrece, the more recent continuous filament is apparently a modified polyacrylonitrile and somewhat different from the original Type 81.

Physical. Orlon has excellent resistance to sunlight, and after one and a half years' outdoor exposure, test-pieces have retained 77 per cent of their strength. In resistance to abrasion it is inferior to nylon. Orlon shows dimensional stability when exposed to hot gases and liquids. It is also resistant to moulds, mildew and insects.

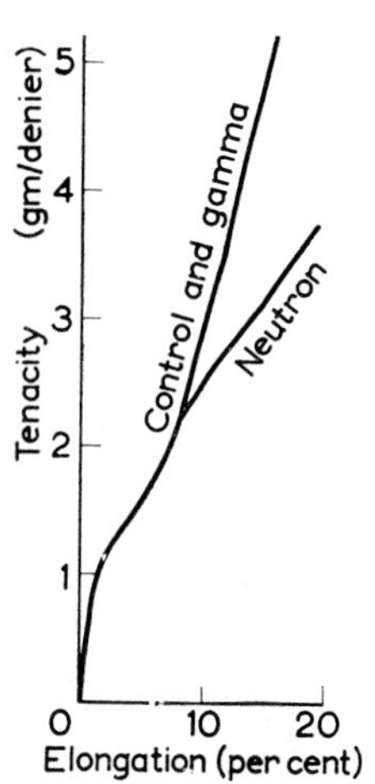

FIG. 177.—Effect of nuclear radiation on the stress–strain curve of Orlon Type 81.

The specific gravity is 1·17. This low value contributes to the high bulking power of the fibre. Fabrics made of Orlon have the feel and appearance of appreciably heavier fabrics made of other fibres of higher specific gravity. It is warm to the touch, drapes well and does not crease easily.

The filament cross-section is dumb-bell or "dog-bone" (Figs. 178, 179). Some of the filaments in the more recent continuous type approximate to a three-lobed structure, rather more like that of cellulose acetate (*cf.* p. 232). This tendency might be expected to reduce the incidence of "shiners" in fabric. The longitudinal views (Figs. 180, 181) are slightly striated.

Stress–Strain Curves for Continuous Filament and for Staple Fibre

It was noted above that whilst the tenacity of Orlon continuous filament is 4·5–5 grams per denier that of the staple is only 2·3 grams per denier. Simultaneously the elongation at break is very much higher in the staple than in the continuous filament. These differences are seen graphically in Fig. 182, which indicates the stress–strain relationships for both staple fibre and continuous filament Orlon. Those for natural silk and wool are included, and it can be seen that the continuous filament Orlon has a stress–strain curve not very different from that of natural silk, whilst that of the staple fibre approximates to that of wool. The fibres have been so designed as to make these approximations, which are in no way fortuitous; if a fibre which has a stress–strain diagram similar to that of staple fibre Orlon is stretched or drawn considerably and then re-tested, it will have a stress–strain diagram somewhat similar to that of continuous filament Orlon; the main difference between the

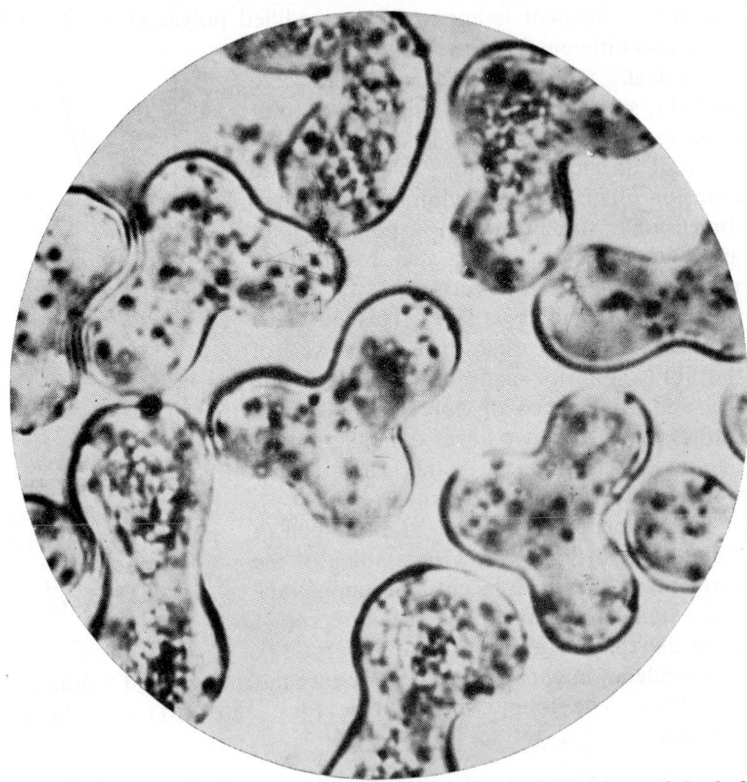

[*E. I. du Pont de Nemours & Co., Inc.*]

FIG. 178.—Photomicrograph showing cross-section of Orlon continuous filaments (×1500).

two fibres is that the continuous filament has been more highly stretched during manufacture than the staple fibre.

Dyeing

The low affinity of Orlon for water has made its dyeing difficult, and at first basic dyes provided the most practicable method, but

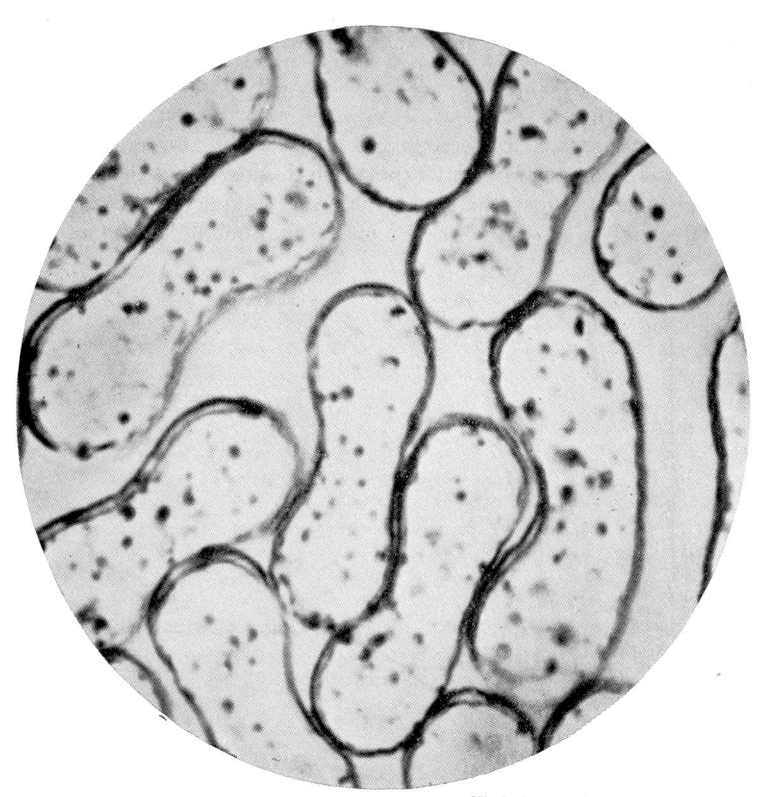

[E. I. du Pont de Nemours & Co., Inc.]

FIG. 179.—Photomicrograph showing cross-section of Orlon staple fibre ($\times$ 1500).

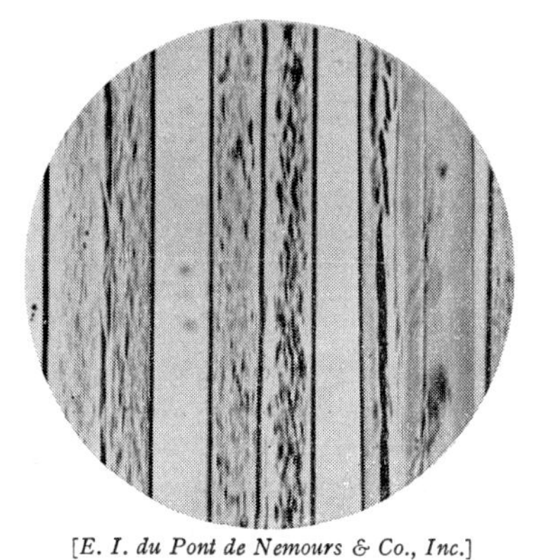

[E. I. du Pont de Nemours & Co., Inc.]

FIG. 180.—Photomicrograph showing longitudinal view of Orlon continuous filaments.

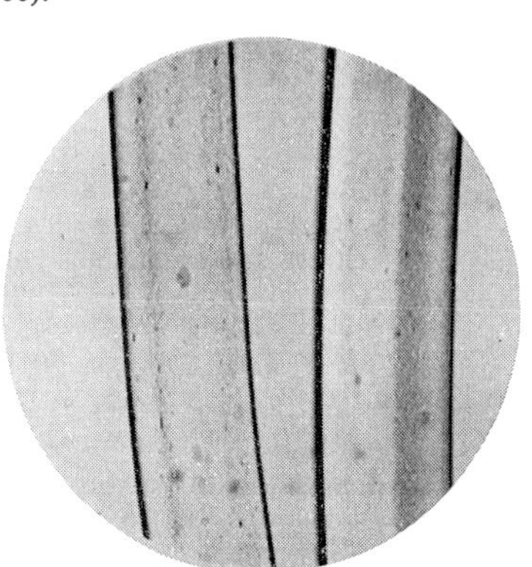

[E. I. du Pont de Nemours & Co., Inc.]

FIG. 181.—Photomicrograph showing longitudinal view of Orlon staple fibre.

their light fastness was as usual poor. Acid dyestuffs had no affinity for Orlon. However, the position was considerably ameliorated by the discovery of the cuprous ion process. Late in 1950 the Union Carbide and Carbon Corporation disclosed that the presence of copper salts helped the dyeing of Dynel (staple Vinyon N) with acid dyestuffs. It was soon found by du Pont de Nemours that

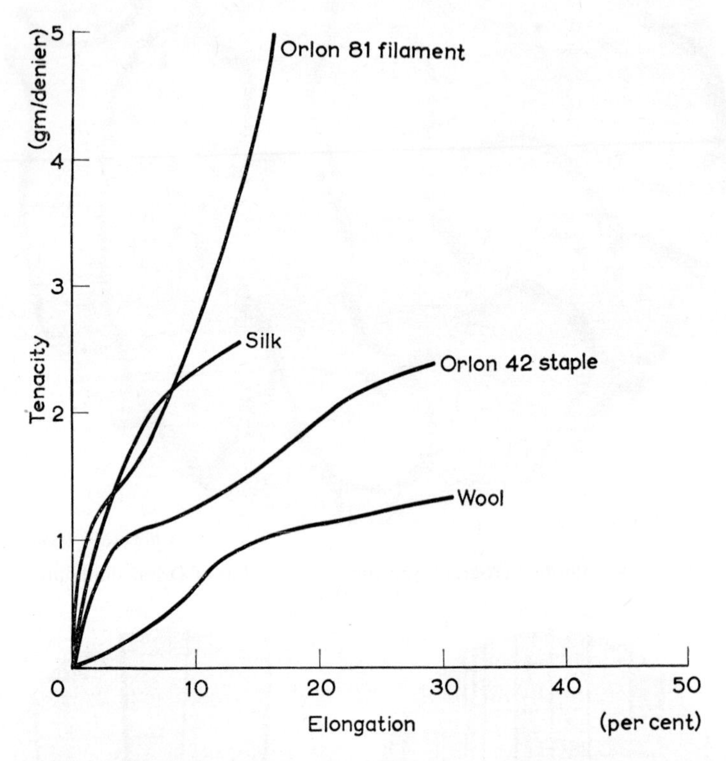

FIG. 182.—Stress–strain diagrams of Orlon continuous filament, silk, Orlon staple and wool. Orlon continuous filament is like silk; Orlon staple is like wool.

the presence of a copper salt, in the cuprous form, enabled Orlon to be dyed with acid and with some direct dyestuffs.

There are to-day five standard methods of dyeing Orlon, and two of these depend on the presence of the cuprous ion; they are described below. There are differences in dyeing behaviour which apply to all five of these methods, between continuous filament and staple. The filament yarn has been more highly stretched in manufacture than has the staple, and this has resulted in better packing of the chain molecules, with greater inaccessibility; there is still not one of the five methods that is entirely satisfactory on continuous filament, *i.e.*, Orlon Type 81. Most of the Orlon made to-day is staple fibre, and fabrics made from this material show up differences in affinity from thread to thread much less than do continuous filament yarns.

Fibres that are difficult to dye by ordinary methods, and continuous filament Orlon is certainly one such of these, can usually be dyed satisfactorily by one of the pigment padding methods, and in America particularly large quantities of Orlon have been dyed with Aridye pigment colours (see p. 794) to a good depth of shade and with excellent fastness. Probably this represents the best solution to date for the coloration problem of Orlon Type 81; such a pigment padding process can give good level dyeings even on fabric made from yarn that is not uniform.

The five following methods are used more on, and are more suitable for, staple fibre Orlon than for continuous filament Orlon Type 81. They are:

1. *Acid Dyes at the Boil with Cuprous Ion.* Orlon staple can be dyed in a full range of colours using acid dyes at the boil in the presence of cuprous ions. The copper plays a definite chemical part in the dyeing, the mechanism of which is probably as follows:

Cuprous ions add on to the nitrile groups in the Orlon and then combine with the dyestuff molecule; the copper which has an affinity for both the Orlon and the dyestuff joins the two together, although these two have no affinity for each other. There are several points that have to be watched to ensure good results; they are:

(*a*) the copper must be in the cuprous state; it is actually added to the dyebath in the cupric state as copper sulphate and is reduced *in situ* to the cuprous form;

(*b*) the temperature must not be below 100° C., otherwise the rate of absorption by the fibre of cuprous ion and of dyestuff is very slow;

(*c*) the *p*H value of the bath must be between 2 and 3; below *p*H 2 the absorption of copper by the fibres is poor, and above *p*H 3 coppered Orlon turns yellowish;

(*d*) the copper present must be in the correct proportion relative to the dyebath, *not* to the fibre;

(*e*) because the amount of dye absorbed by the fibre is proportional to the amount of copper absorbed, the concentration of cuprous ion in the bath has to be carefully controlled.

The copper sulphate in the dyebath is reduced to the cuprous form by hydroxylamine sulphate, and as both this substance and the cuprous copper itself are liable to atmospheric oxidation, some sort of control is necessary. The best method to do this is based on redox potential measurements. It is well known that an electrical potential difference is generated between certain electrodes when they are immersed in either an oxidising or a reducing agent. This potential difference can be measured easily in millivolts on a suitable potentiometer, or even on most *p*H meters. In the cuprous ion process the reducing agent is added to the dyebath at such a rate as to maintain this potential difference at about 80 millivolts. Hydroxylamine sulphate is an expensive chemical to use, and other cheaper reducing agents, notably zinc formaldehyde sulphoxylate, have been used. The use of metallic copper instead of hydroxylamine sulphate has been protected by Sandoz Ltd., and forms the basis of their Sandocryl process; as it is a milder reducing agent than hydroxylamine it has two advantages:

(1) It generates cuprous ions more slowly thus making it easier to produce level dyeings.
(2) It is less liable than hydroxylamine to reduce the dyestuff itself and thereby to dull some azo dyes.

The metallic copper is used either as strips or in mesh; it is unsuitable for some kinds of equipment and further experience is necessary to show the real value of the process.

As an example of the original process, Orlon staple fibre has been dyed with the assistance of cuprous ions in a circulating machine as follows: the Orlon is scoured with an acid detergent at the boil and is rinsed; then 10 per cent of copper sulphate on the weight of the Orlon is added and the liquor circulated at 75° C. for five minutes, and then 3 per cent of zinc formaldehyde sulphoxylate is added and the liquor re-circulated; then the dyestuffs are added, the *p*H value adjusted to 3·2 (just beyond the recommended range), and the liquor is circulated for from one and a half to two hours at the boil. The method allows a wide range of shades of good depth and fastness to be obtained on staple fibre Orlon; continuous filament Orlon, on account of its more highly oriented nature, is more difficult to dye, but even in this case the inclusion of a copper salt in the dyebath has been found advantageous.

2. *Acid Dyes above the Boil with Cuprous Ion.* This method

follows the lines of 1, but a temperature of 120° C. is used and this, of course, involves the use of pressure apparatus, and excludes certain dyeings. The higher temperature improves the exhaustion and general wet fastness, and allows a wider range of dyestuffs to be used. The improvement in exhaustion may reduce the cost of dyestuff to one-third of that necessary in dyeing at 100°. Furthermore, cheaper reducing agents than hydroxylamine sulphate are more easily used; the best is sodium bisulphite; it is true that at 120° C. this will pit stainless steel equipment, but this trouble can be eliminated by the use of sodium nitrate as a corrosion inhibitor. For heavy shades a carrier such as *o*-phenylphenol may be added (*cf.* p. 415). Typical of the dyestuffs that are used for this process are the following: Pontacyl Fast Black N2B Conc. 200 per cent; du Pont Quinoline Yellow PN Extra Conc. These two dyestuffs, the black shaded with the yellow, give a good black of superb fastness.

The monosulphonated acid dyes are more suitable than the more highly sulphonated dyes for cuprous ion dyeing, and at temperatures above the boil some acid milling dyes are also suitable. It has to be added that between 1953 and 1956 considerable progress was made in the application of disperse and basic dyes to Orlon Type 42 and they can be applied so easily and give such good fastness, that the cuprous ion method is unlikely to be widely used; its main application may become restricted to navies and blacks, particularly on loose fibre. There is, too, an unwelcome tendency for coppered Orlon to turn brown on being steamed.

3. *Indigo and Selected Vat Dyes at the Boil.* A new technique has recently been developed for piece-dyeing with indigo to give deep navy shades of good light, washing, perspiration and crocking fastness. A bath is made up with du Pont Indigo PLN Paste, it is reduced to the greenish-yellow leuco-compound, dyeing is carried out for two hours at 99° C., an oxidising agent is added, and the fabric rinsed and scoured. The key to this process is *p*H control. At a *p*H value of about 5, Orlon has a natural affinity for the *acid* leuco form of indigo (and some other vat dyes) that exist at that *p*H level. Some indigoid and thio-indigoid dyes such as Durindones can be applied to Orlon by dyeing as the alkaline leuco form at about 95° C. in the presence of Formosul at *p*H 10. The dyes are vatted in the usual way and the *p*H of the dyebath is reduced to the required level with sodium bicarbonate; finally the dyes are oxidised to the insoluble form by oxidation with sodium perborate or sodium percarbonate at 95° C. Excellent light and wet fastness results, but the dyestuffs are expensive, and as such superb fastness is unnecessary

for jumpers and pullovers, vats will not be widely used on staple Orlon Type 42.

4. *Basic Dyes Above the Boil.* A solution of the dye is made with acetic acid, and dyeing is carried out in a pressure machine at 110° C. Basic dyes have little affinity for Orlon below 85° C.; as the dyeing temperature is raised to 100° C. rapid absorption on to the fibre *surface* takes place, but penetration is still extremely slow, and well-dyed Orlon in which the dyestuff has penetrated into the fibre can be obtained in a reasonable time only at above the boil temperatures. The temperature should not exceed 110° C. for staple Type 42, otherwise fibre shrinkage will be excessive. Light pastel shades are difficult to get level mainly because of the speed at which basic dyestuffs rush on to the surface of Orlon, and it is advisable to employ a cationic retarding agent such as Lissolamine A to give light shades; pastels are best obtained with disperse dyes which will give level results. The main use of basics on Orlon is to give medium to heavy shades, and when dyed at 110° C., good penetration may be complete in 30 min. Basic dyes are often bright and sometimes they must be used for bright light shades because disperse dyes will not give the brightness; they have good washing fastness and as a group are characterised by poor light fastness, although selected members have light fastness as good as 4–5 and new basic (or cationic) dyestuffs have been produced with light fastness as good as 6. Du Pont's range of Sevron basic dyestuffs gives a full range of shades on Orlon Type 42. Continuous filament Type 81 is best dyed with basic colours at 120–125° C. Nowadays (1968) the new basic or cationic colours represent the best way of dyeing Orlon (see also p. 761).

5. *Disperse Dyes.* At 95–100° C. disperse dyestuffs dye only slowly, but they migrate well in the presence of a levelling agent such as Lissapol C; consequently, pastel shades can be obtained which level easily; whereas it is very difficult to get pastel shades level with basic dyes. It is difficult with disperse dyes to build up a good depth of shade; this is due to the inherent low affinity of the fibre for the dyestuff. This can be illustrated by comparing the Saturation Uptake of the three fibres Orlon, nylon and acetate for some disperse dyes; the saturation uptake is a figure obtained by dyeing the fibre with a large excess of dyestuff for a long time (until fibre is saturated with dyestuff), *e.g.*, for 48 hours with 100 per cent dyestuff on the weight of the fibre. Typical values are shown in the table on page 513.

Such figures help us to understand the difficulty in building up heavy shades on Orlon with disperse dyes. Nevertheless, by the selection of suitable individual dyestuffs, medium shades can be built up on Orlon Type 42 with extended times of dyeing. Light fastness of the disperse dyes on Orlon is good, usually 5–6, but fastness to steam pleating is less good, and whilst it is true that Orlon fabric is much less likely than Terylene to be used for pleated skirts (because it cannot be heat-set nearly so satisfactorily) it is a point that has to be watched; some disperse dyestuffs withstand steaming on Orlon better than others, and when necessary these individuals have to be selected.

Saturation uptake (%) of Disperse Dyes at 95° C.

Dye.	Orlon Type 42.	Nylon.	Acetate rayon.
Dispersol Fast Yellow G 300 .	1·4	4·8	7·4
Dispersol Fast Orange G 300 .	1·1	1·8	7·3
Duranol Red 2B 300 . . .	1·8	4·5	11·0
Duranol Blue Green B 300 .	1·0	9·5	10·8
Duranol Brilliant Blue CB 300 .	3·5	8·0	10·5

If disperse dyes are used at a temperature of 110° C. in pressure equipment then medium shades can be built up in a much shorter time than at 95–100° C. although the saturation uptake of Orlon being so low (not very much higher at 110° C. than the figures given for 95° C.) it is impossible to build up really deep shades except with selected individual dyestuffs that have fairly high saturation uptakes. The main advantage to be derived from high-temperature pressure dyeing is that of speed; a process that would require days at the boil can sometimes be completed in a commercially acceptable time of a few hours at 110° C. Blue acetate dyestuffs when dyed on Orlon are not subject to gas-fading; evidently fibre as well as dyestuff plays a part in gas-fading.

The position of the dyer in respect to the application of disperse colours to Orlon for medium shades has improved considerably in the last few years, and with experience and suitable selection of dyestuffs, light to medium shades can be obtained fairly satisfactorily at the boil. Simultaneously the use of Orlon has increased and it is estimated that 1 million lb. of it, in forms from loose fibre to garment, were dyed in the United Kingdom in 1956. These remarks apply only to staple Type 42; continuous filament Type 81 is still quite impossible to dye at the boil, without the use of special devices such as the cuprous ion technique already described, and even then not very satisfactorily.

Heat Setting

Orlon is not receptive to heat-setting treatment in the same way as are Terylene and nylon.

High Bulk Orlon

The high bulk yarns of Orlon are a comparatively recent introduction; they are designed to give a lofty, voluminous handle and appearance to garments such as sweaters and pullovers made from them. These yarns are made by combining fibres of Orlon of different shrinkage characteristics in the same yarn; one component has the normal very low residual shrinkage at temperatures near the boil, of Orlon Type 42, and the other component consists of fibres of Orlon which undergo the relatively high shrinkage of 16–20 per cent at the boil. About 60–70 per cent of the normal low shrinkage fibre will be combined with about 40–30 per cent high shrinkage fibre; a yarn so made when relaxed in steam or boiling water or hot air either in yarn or fabric form will bulk, because as the high shrinkage component shrinks, the low shrinkage fibres buckle and so increase the effective diameter and the loftiness of the yarn. Naturally, allowance has to be made for the finishing shrinkage when designing a fabric. When it is desired to preserve the lofty handle of high bulk yarns, dyeing should be done at temperatures not above 100° C. whenever possible.

Uses

Du Pont are now producing 100 million lb. of Orlon per year, and are planning to produce twice as much Orlon as nylon, so that considerable uses must have been found for the fibre. Some of these are as follows:

For Outdoor Furnishings. The excellent resistance of Orlon Type 81 to weathering has lent itself to its use for awnings, tents, tarpaulins, shades, car-tops and outdoor furniture. Even in hot climates, where the light, and particularly its ultra-violet component, is strong, Orlon awnings have behaved excellently; one, painted maroon and natural, that was exposed for two years in Florida retained 75 per cent of its strength, although unpainted it retained only 40 per cent of its initial strength; another, of rather heavier construction, that had been painted " Regular Green ", retained 90 per cent of its strength after two years (unpainted retained only 30 per cent). Orlon awnings will also resist industrial fumes such as contaminate the air in big towns; in Philadelphia the strength retention was even better than in Florida, the improvement being most marked in the case of the unpainted awnings, although the *absolute* performance of the painted awnings was always better than that of the unpainted in the same locality. So far as weathering is concerned, Orlon Type 81 is supreme amongst the established synthetics.

For Indoor Furnishings. Orlon's outstanding resistance to sunlight makes it especially suitable for curtains. It has not got the non-flam properties of glass or Dynel that are an additional attribute for curtainings, but it is no more inflammable than cotton or rayon. Orlon is more suitable than Dacron for pile rugs.

In Industry. Orlon Type 81 has been used for anode bags in the electro-plating industry. These bags collect any particles that may drop from the anode, so maintaining the clarity of the chemical solution in which the articles for plating are immersed; such clarity is essential for good plating, and because the solutions used are often strongly acid (sometimes alkaline), Orlon bags have a much longer life than those made of any natural fibre. Orlon has also found application for filter cloths and chemical protective clothing, but its considerably higher price than Dynel may make its development difficult in industry.

Type 42 *for Knitwear.* Orlon staple fibre Type 42 is extensively used in the U.S.A. and in the United Kingdom for knitted outerwear of 100 per cent Orlon and also for underwear of 100 per cent Orlon. Jersey knit dress fabrics of 80 per cent Orlon/20 per cent wool are much in demand. Fleece fabrics of 100 per cent Orlon or 65 per cent Orlon/35 per cent Dynel are produced for coats. For sweater knitting yarns, a high bulk yarn of two-thirds normal Orlon staple fibre Type 42 and one-third high-shrinkage Orlon Staple Fibre Type 42 gives a good combination of loftiness, handle and body. The stress–strain curve of Type 42 (Fig. 176) is not dissimilar from that of wool, and mixtures of the fibre with wool have been successful; The low specific gravity and the flattish cross-sections of Orlon Type 42 give a natural loft to blends of it with wool. Men's socks made from Orlon have achieved some success; 50/50 blends with wool have been sold as men's suitings and blends of two parts of Orlon to one part cotton have been used for a seersucker washable suiting. Other uses have included umbrella fabrics, dress-shirts and ties; but it is in knitted goods that staple Orlon has so far excelled; some of the dresses and costumes made from it have been superb. The continuous filament Type 81 has not had the same degree of success, mainly because of the difficulty it presents to the dyer, but for outdoor furnishings it is first class, but has to compete with the cheaper Saran.

Orlon Type 39 staple is intended for the woollen spinning system; it consists of coarse filaments of various lengths and deniers. In an average 2·4 filament denier it is recommended for soft baby blankets, in an average 6·5 filament for crisp blankets.

Orlon Type 37. A coarse staple for rugs and carpets.

Orlon Cantrece. Continuous filament Orlon.

Orlon Type 72. This is a fine filament (1·5 denier) fibre for use in cotton process spinning; it has a fluorescent agent built in to give good whiteness. The fineness of filament makes it comfortable to wear and, by offering a very large surface, accelerates the evaporation of perspiration.

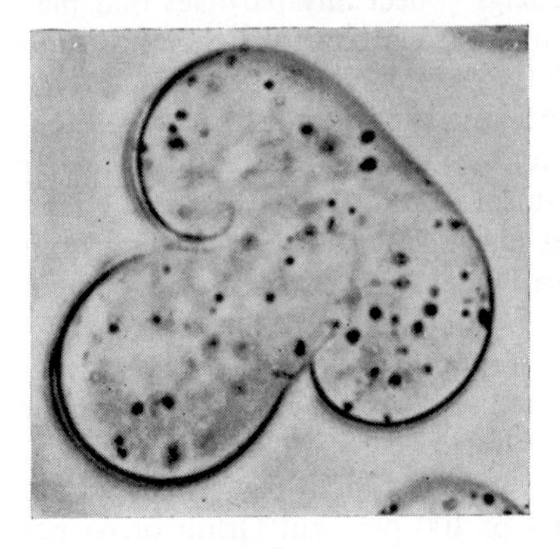

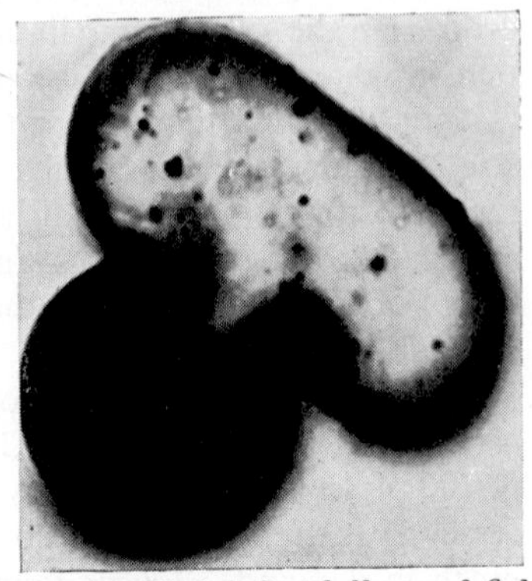

[*E. I. du Pont de Nemours & Co.*]

(*a*) (*b*)

FIG. 183.—Photomicrograph showing cross-section of Orlon Sayelle. *Left:* as spun. *Right:* stained. The two-component structure is clearly seen. Probably 6 denier magnified × 2,000.

Orlon Type 44 is a modification of ordinary Type 42 Orlon, one which has some affinity for acid dyestuffs. It is very widely used in the U.K. (1968).

ORLON SAYELLE

Orlon Sayelle is an achievement of chemical engineering. Two polymer solutions are fed to the spinning jet; the spun filament consists of two components one merging into the other. A cross-section of the filament as spun is shown in Fig. 183(*a*), and a similar filament after it has been stained in Fig. 183(*b*). The large flat part of the filament is ordinary Orlon, the highly stained lobe is a modified acrylic polymer, one which swells more, has a higher moisture regain and dyes deeper shades. When the fibre is wet this lobe swells more than the other and, if the fibre is then dried, it tends to curl and to give a spiral crimp which is claimed to bear some resemblance to that of wool. But the retractive and crimping forces in Orlon Sayelle are nothing like so strong as are those in wool: for example a sweater made from Orlon Sayelle must be tumble dried after it has been washed; it cannot be hung up to dry because the crimping power of the fibre is insufficient to curl the fibres against the weight of the wet garment.

The fibre has a moisture regain of 2·6 per cent (ordinary Orlon about 1·6), and has a much higher affinity for basic dyestuffs. It is made in two filament deniers, 3 and 6, in staple of mixed lengths suitable for worsted process spinning and also in tow of 420,000 denier for use on the Pacific Converter. It has the usual crimp impressed mechanically, but has the second crimp—the potential spiral form when boiled and dried—built in. In order to develop the spiral crimp, yarn or fabric or garment must be maintained in water at 98° C. or at the boil for 15 min.; lower temperatures will not do; usually such a hot wet treatment can be included in the dyeing or finishing treatments; then if the Sayelle is dried without tension the spiral crimp develops. It is useful for giving bulk in sweaters and similar garments. If the garment is washed repeatedly, the crimp will re-appear after each drying (provided it is either tumble dried or dried spread out loosely on a flat surface): the tendency to crimp and bulk reduces the liability of the fabric to glazing when ironed, a defect often to be found in fabrics made from synthetic fibres. Sayelle is used in large quantities in the U.K. (1968).

Pan and Dralon

Pan is a continuous filament yarn of the modified polyacrylonitrile type that is made at Frankfurt-am-Main by Cassella Farbewerke Mainkur A.G. for Farbenfabriken Bayer A.G. of Leverkusen. The nature of the monomer which is co-polymerised with acrylonitrile has not yet been published. The co-polymer is dry-spun from solution. Dyeing of Pan follows the general lines of dyeing the other acrylic fibres such as Orlon Type 42, although behaviour is not identical; it has, for example, been reported that disperse dyes have a much lower light fastness on Pan than on Orlon Type 42 or on Dralon.

Properties seem to be very similar to those of other acrylics, although the specific gravity is a little lower at 1·14. The uses to which Pan has been put include the following: underwear, swimsuits, pullovers, sportswear and marquisette curtains.

Dralon (staple fibre) is made by Farben Bayer A.G. at Dormagen near Leverkusen. It is based mainly on polyacrylonitrile modified with a small proportion of some other monomer to confer greater dye affinity on the co-polymer fibre-forming material. In Germany acrylonitrile is made from coal and chalk to give calcium carbide, thence acetylene and by addition of hydrocyanic acid, acrylonitrile.

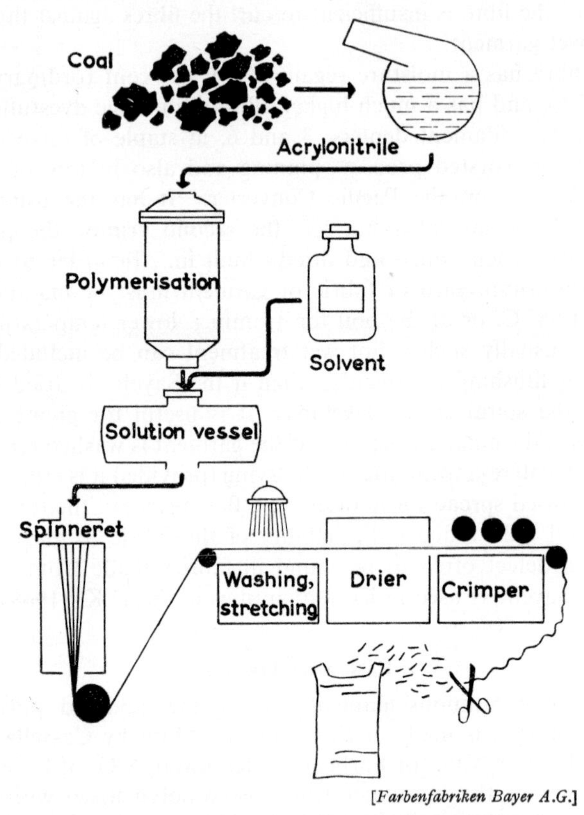

[*Farbenfabriken Bayer A.G.*]

FIG. 184.—Schematic flowsheet for manufacture of Dralon.

This is polymerised with another monomer dissolved in a solvent, possibly dimethyl formamide or ethylene carbonate and dry-spun; the fibre is stretched, dried, crimped and cut into staples. The operations are shown schematically in Fig. 184.

The fibre is made in filament deniers of 1·5, 2, 3, 6, 10 and 15; its tenacity ranges from 2·5 to 3·2 grams per denier dry, and the wet value is about 85 per cent of the dry. Elongation at break varies from 24 to 30 per cent wet or dry. Elasticity is good with immediate recovery of 95 per cent from a 2 per cent stretch, of 79 per cent from a 4 per cent stretch and of 63 per cent from an 8 per cent stretch. These figures are very similar to those that would obtain from wool under similar conditions and it is well known that one of the outstanding virtues of wool is its ability to recover from stretch, *i.e.*, to behave elastically. Specific gravity is 1·14 which is rather lower than most of the acrylics which range up to about 1·19 and is, of course, as the makers point out, very considerably lower than that of any natural fibre—Dralon and the other acrylics, too, are light fibres. Cross-section is dumb-bell shaped (*cf.* Dynel and Orlon Type 42) indicative of having been dry solvent spun. Dralon fibre is naturally a light cream colour. When heated, the fibre softens at 235° C. and melts with decomposition at about 300° C. Heated in air at 150° C. it shrinks only 1 per cent and its heat resistance is one of its good features for industrial use. In resistance to sunlight, ultra-violet and industrial contaminated atmospheres Dralon is better than any of the natural fibres. Chemical resistance is good, as it is with all of the acrylic fibres; it withstands 30 per cent sulphuric acid well and 60 per cent moderately, but is dissolved by concentrated; it withstands acetic and lactic acids even concentrated. Resistance to alkalis is not so good, but is good enough to be useful; for example 10 per cent caustic soda or potash at room temperature is withstood, but there is serious attack at 75° C.; furthermore, it withstands concentrated solutions of hexamethylene diamine, the nylon intermediate, and may be used for filtration of this compound. It is practically unaffected by bleaching agents and by most organic solvents.

Dralon can be dyed with wool colours using the cuprous ion technique in a full range of shades. Light to medium shades are obtainable with disperse colours.

Uses include suitings, gloves, underclothes, socks, furnishings such as moquettes and plush, handknitting yarns and sportswear. Sportswear or playwear or weekend wear is a common outlet for acrylic fibres. The garments are usually light, airy and loose-fitting, and of the type that once would have been all cotton; acrylics are the cheapest of the synthetics and they offer a good measure of dimensional stability, easy washing, quick drying and a product that looks a bit more novel than cotton. This applies not only to Dralon but generally to the spun acrylics. Ultimately, perhaps when the novelty has faded, acrylic fibres will have to sell lower than cotton to command an apparel market.

Not much Pan and Dralon have yet been seen in the United Kingdom, and the dyeing properties seem to be a little different from those of Orlon, and perhaps unfamiliar as yet to the home dyeing industry, but it is certain that with the impetus of German technology and post-war drive behind them, the German acrylics will be good. Dralon is very good in upholstery fabrics.

Other Acrylonitrile Fibres

The following fibres are made in Europe and are very similar to Orlon Type 42 and to Pan and Dralon.

Crylor	Rhodiaceta S.A.
N 53	N.V. Kunstzijd-Spinneri Nyma, Holland.
Redon	Phrix A.G.

The main differences between these yarns and Orlon is in their dyeing behaviour. Rate of dyeing is not sufficiently different from that of Orlon Type 42 to be troublesome, although most of the fibres dye rather more slowly. The maximum depth of colour that can be obtained with basic dyes applied at 100° C. for 1–2 hours varies considerably, the heaviest shades can be obtained on Orlon and the least heavy with Redon. Light fastness is best on Orlon and Dralon; variations are not greater than one point on the S.D.C. scale for basic dyes but may drop up to two points for disperse dyes especially on Pan, Redon and N 53. Disperse dyes which give pale shades of good light fastness on Orlon are inadequate on some of the other acrylonitrile fibres and probably the cuprous ion method will have to be more often used with these other fibres.

Redon. Made only in staple for use on all spinning systems. For the cotton spinning system, 1½ denier is recommended. Strength is good at 3·3 grams per denier and elongation at 30–40 per cent. Moisture regain is 1·3 per cent, specific gravity is 1·17. It is supplied spun-dyed in a range of colours. Disperse colours are recommended for dyeing. A Phrix fibre.

Dolan. Made by Süddeutsche Zellwolle A.G. in Bavaria and spun-dyed in 48 shades. It has a high tenacity of 4–5·5 grams per denier and an elongation of 18–22 per cent. Specific gravity is 1·14, lower than many acrylics. After 4 months burial in soil it is virtually unaffected; in a similar test all natural fibres are completely lost. An odd use has been reported: " Dolan fabrics have been used in fuel oil at 200° C. 6 to 10 times as long as cotton fabrics of the same texture ".

Crylor. Made by Rhodiaceta. Possesses all the usual good properties of acrylics. Is remarkable for its low specific gravity, only 1·12.

FURTHER READING

Silk and Rayon, **23**, 662 (1949).
British Pat. (du Pont) 579,887, 603,873 (I.C.I.), 583,939, 584,548.
R. J. Thomas and P. L. Meunier, *American Dyestuff Reporter*, 38, P 295 (Dec. 12, 1949).
" Papers Presented at the Technical Conference on Dyeing of Orlon Acrylic Fiber and Dacron Polyester Fiber ", Wilmington, Aug. 5, 1952, E. I. du Pont de Nemours & Co.
R. C. Houtz, " Orlon Acrylic Fibre, Chemistry and Properties ", *Text. Research J.*, **20**, 11 (1950).
J. B. Quig, " Orlon Acrylic Fiber ", *Papers of A.A.T.T.*, **4**, 61–70 (1949).
I. M. S. Walls, "The Dyeing of Orlon and Orlon Mixtures", *J. Soc. Dyers and Col.*, **72**, 261–266 (1956).
B. Kramrisch, " Recent Developments in the Dyeing of Acrylic Fibres ", *J. Soc. Dyers and Col.*, **77**, 237 (1961).

CHAPTER 32

ACRILAN AND COURTELLE

ACRILAN is a fibre that is made by the Chemstrand Corporation, Decatur, Alabama. This Corporation was formed in 1949 as an associate company of the Monsanto Chemical Company, who are one of the biggest producers of acrylonitrile, and of the American Viscose Corporation; in 1962, Monsanto owned it all. In 1968 they have given up the name Chemstrand, which is regrettable because it was an apt and euphonious name and had acquired respect as a result of the advances in fibre science and technology made by the Chemstrand Corporation. Now, instead, it is called Monsanto—Textiles Division.

Pilot-plant production of Acrilan started in November 1950, and the fibre was introduced to the public with high-pressure advertising and publicity in September, 1952. The large plant at Decatur, Alabama, started to produce in 1953, but some teething troubles were encountered. However, in 1954 an improved version of the fibre was marketed and has been well received. The Decatur plant, which cost $30 million, has a capacity of about 30 million lb. per annum.

A subsidiary is producing Acrilan at Coleraine in Northern Ireland. Capacity, there, is 25–30 m. lb. per annum.

Chemical Structure

Acrilan acrylic fibre is a co-polymer of acrylonitrile probably 85–90 per cent and of a minor constituent (15–10 per cent) of a mildly basic character. The basic constituent is important, because it confers on the polymer an affinity for acid dyestuffs and Acrilan stands out from the earlier acrylic fibres in having a much greater affinity for dyestuffs and consequently in being very much easier to dye—a great advantage.

The nature of the basic component has not been positively disclosed, although it has frequently been reported to be vinyl pyridine. Probably it was this compound in the first Acrilan, but whether it is so to-day is more doubtful. In the absence of definite information, some inferences may be drawn from published patent specifications.

One such (U.S. Pat. 2,744,086) describes the preparation of a fibre from the product resulting from the polymerisation of: 18 parts of a co-polymer of 95 parts acrylonitrile/5 parts vinyl acetate; and 2 parts of a polyester made by condensing azelaic acid with methyl diethanolamine. The polyester is graft polymerised on to the acrylonitrile–vinyl acetate co-polymer and this has the effect of increasing the dye affinity. Thus, in a test under standard conditions a co-polymer of 95 parts acrylonitrile and 5 parts vinyl acetate absorbed 10 per cent of a dyestuff from a dye-liquor, but the graft polymer (with the polyester grafted on) absorbed 99 per cent of the dyestuff under the same conditions; clearly this represents a considerable and important improvement.

In another specification (U.S. Pat. 2,749,325) the preparation of a fibre-forming co-polymer containing 6 per cent vinyl pyridine is described. Apparently a variety of vinyl pyridines can be used; the most usual is 2-vinyl pyridine,

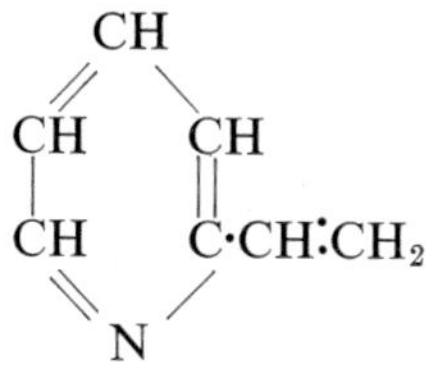

```
        CH
      //  \
    CH     CH
    |      ||
    CH     C·CH:CH2
      \\  /
        N
```

This co-polymer is made by co-polymerising two co-polymers, grafting one on to the other. These two are: (i) 1 part co-polymer of 50 parts acrylonitrile and 50 parts 2-vinyl pyridine; (ii) 7 parts co-polymer of 95 parts acrylonitrile and 5 parts vinyl acetate. This graft co-polymer dissolves in dimethyl acetamide to give a perfectly clear solution, whereas a dry blend of the two co-polymers gives a turbid solution.

These two examples indicate some of the possibilities that are being explored; it seems to be clear that complex and graft co-polymers can have a combination of properties not obtainable in any simple co-polymer of acrylonitrile and a vinyl compound. Polyvinyl acetate has been reported to constitute 10 per cent of Acrilan.

It is now thought that the original Acrilan, which is now often referred to as Acrilan C.I., was in fact a co-polymer of 88 per cent acrylonitrile, 6 per cent vinyl acetate and 6 per cent vinyl pyridine and that by virtue of the pyridine groups it had a good affinity for acid dyes and a poor one for basic dyes. However, protracted dyeing in acid baths caused deterioration of Acrilan C.I.

Acrilan 16, which has come more recently, is believed to be at least 95 per cent polyacrylonitrile, not quite a homopolymer, but not very far off. It carries acidic groups which permit easy dyeing with basic colours (*cf.* p. 761).

Manufacture

The raw materials are obtained from petroleum by cracking processes. Acetylene is formed and is reacted with hydrocyanic acid to give acrylonitrile, a liquid:

$$HC{\vdots}CH + HCN \longrightarrow CH_2{:}CHCN$$

The acrylonitrile is co-polymerised with about 12 per cent of other constituents, including the basic material, the resulting polymer being a white powder similar in appearance to talc. The polymer is dissolved, probably as a 20 per cent solution in dimethyl acetamide, and the solution is spun into a bath made up of dimethyl acetamide and water which precipitates the fibres, which are stretched 350 per cent and permanently crimped. The fibre is sold as tow with 72,000 filaments or more, depending on the filament denier, or is cut to staple of various lengths.

Wet Spinning. Wet spinning of modified acrylic fibres (Fig. 185 shows Acrilan after extrusion from the spinnerets) may be carried out into a variety of precipitating baths; some that have been described are:

1. A 20 per cent solution of polymer in ethylene carbonate is heated to 120° C. and extruded into a bath of 80 per cent dipropylene glycol and 20 per cent ethylene carbonate at 130° C. The yarn is washed with water at 80° C., stretched ten-fold at 150° C. and finally relaxed at 140° C.
2. A 20 per cent solution of polymer in dimethyl acetamide is spun into glycerol at 140° C. and wound on to a bobbin, which is washed free of glycerol with water and dried. The yarn is maintained under tension during its passage through the glycerol bath. Sometimes two glycerol baths may be used, the first at 120° C. and the second at 170° C.
3. A 20 per cent solution of polymer in dimethyl acetamide is spun into a 40 per cent solution of calcium chloride in water at 90° C.
4. An 18 per cent solution of polymer in dimethyl acetamide is spun into a mixture of 2 parts dimethyl acetamide and 1 part water; as the yarn emerges from the bath it is washed with water, which flows counter-currently into the spinning bath at such a rate that the bath composition is kept constant; dimethyl acetamide is continually being added by the polymer as it is extruded; water is being added from the wash.

Which of the various methods is used for any particular yarn has not yet been published; the processes are new and the one that offers the most advantages has not yet revealed itself.

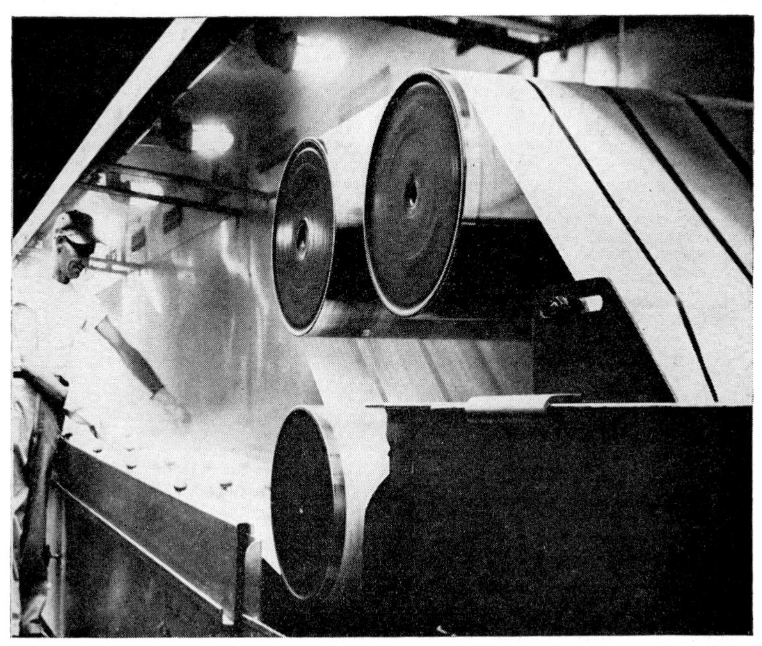

[*Chemstrand Corporation*]

FIG. 185.—Acrilan spinning: after the spinnerets.

Properties

Cross-section. Fig. 186 shows Acrilan of 3 denier filament cross-section (×1,000 magnification) and the particles of delustrant are easy to see; it will be noted that the cross-section is roughly round with occasional indentations and this is attributable to its having been wet-spun into an aqueous bath and not dry-spun. Fibres that have been dry-spun (from solution) usually have collapsed or dumb-bell sections; melt-spun fibres have round sections, and wet-spun fibres that have been highly stretched (cuprammonium, casein, polynosic) also approximate to round.

Density. The density of Acrilan is 1·17 gm./c.c. at 25° C.

Mechanical. The dry tensile strength is 2·5 grams per denier and extension at break is 35 per cent; corresponding figures for wet

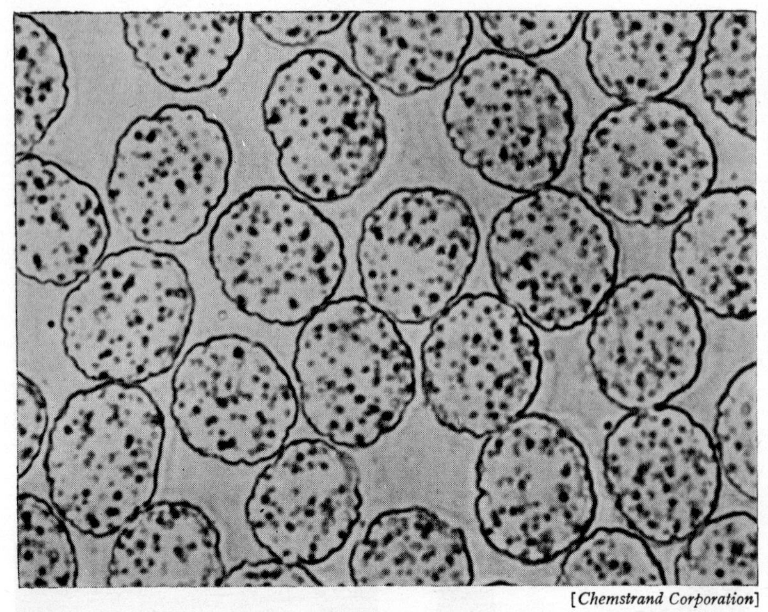

[*Chemstrand Corporation*]

FIG. 186.—Cross-section of Acrilan filaments (×1,000).

fibre are 2·0 grams per denier and 44 per cent. Representative stress–strain curves are shown in Fig. 187.

Heat. The fibre decomposes before it melts; under pressure it sticks to surfaces at about 245° C. Exposure to air at 150° C. for 20 hours causes less than 5 per cent loss in strength, measured when cool again. Prolonged exposure in air at elevated temperatures

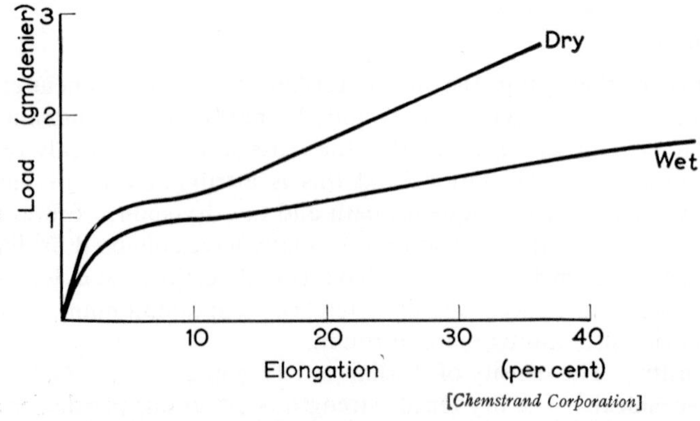

[*Chemstrand Corporation*]

FIG. 187.—Stress–strain curves, dry and wet, of Acrilan.

will cause some yellowing. Short exposures to temperatures used in drying fabrics will not cause discoloration. Fabrics of Acrilan are not readily ignited and do not burn rapidly; their rate of burning is less than that of cotton, viscose rayon or acetate rayon.

Moisture Regain. The regain at 70° F. and 65 per cent R.H. is 1·2 per cent when equilibrium is reached from the dry side, and 1·6 per cent when reached from the wet side. With most fibres it is assumed that the moisture content at standard conditions is independent of the path by which equilibrium has been reached, but it is likely that other fibres as well as Acrilan really do show a hysteresis effect in this way.

Abrasion Resistance. The abrasion resistance and durability of acrylic fibres generally are greatly inferior to those of nylon and Terylene. Nevertheless, Acrilan withstands wear better than wool, and Fig. 188 shows the decrease in pile height with traffic over them

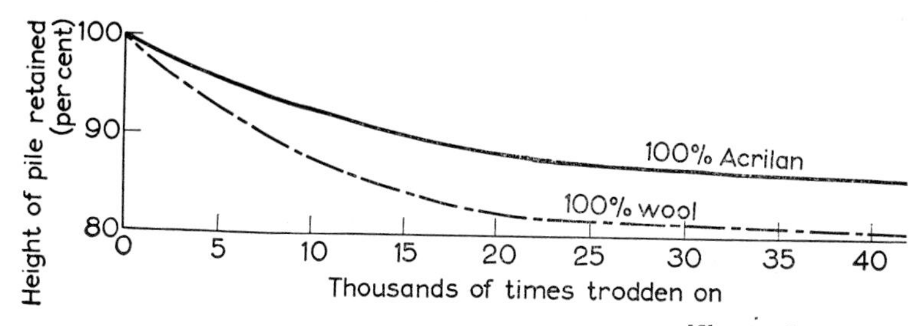

[*Chemstrand Corporation*]

FIG. 188.—Loss in height of carpet pile with wear; Acrilan compared with wool.

of two rugs, one of wool and the other of 15 denier filament Acrilan. It must be added, though, that heavy filament denier is advantageous towards wear resistance and it is for this reason that 15 denier and sometimes heavier deniers are used in carpets (compare Rhovyl and Verel), and that Acrilan of the same denier as the wool would not show the same advantage in durability. For many of the uses to which acrylics are put, notably sweater and knitted outerwear, supreme abrasion resistance is not really necessary. The person, certainly the woman, who wears his or her clothes till they are worn out is exceptional.

Chemical Resistance. Acrilan is insoluble in and unaffected by all common solvents. It has very good resistance to mineral acids and its resistance to weak alkalis is fairly good. The behaviour of Acrilan on extended exposure to some chemicals is shown in the table on p. 528 (*cf.* the behaviour of Orlon Type 81, p. 503).

By earlier standards chemical resistance is extremely good, by comparison with Orlon Type 81 it is not so good: the constitutional features that make Acrilan accessible to dyestuffs make it similarly accessible and susceptible to other chemicals.

Weathering. The resistance of acrylic fibres to sunlight and to weathering is very good; it is stated by the manufacturers that "chemical modifications used in order to improve the dye-ability of 'Acrilan' have no harmful effect on the sunlight durability".

Biological Resistance. Unaffected by mildew, moulds, and larvae of moths and carpet beetles.

Chemical.	Temperature.	Concentration (per cent).	Time.	Effect on strength.
Hydrochloric acid	Room	37	3 days	Degraded
Hydrochloric acid	Room	30	20 days	None
Nitric acid	Room	40	26 weeks	Considerable
Nitric acid	Room	10	26 weeks	Moderate
Sulphuric acid	Room	40	26 weeks	None
Sulphuric acid	Room	60	30 days	Moderate
Caustic soda	Room	20	10 days	None
Caustic soda	Room	10	26 weeks	Degraded
Caustic soda	100° C.	1	1 hour	Moderate
Potassium permanganate	Room	5	30 days	Degraded
Acetone	Room	100	30 days	None
Hydrogen peroxide	Room	30	10 days	None
Hydrogen peroxide	Room	30	26 weeks	Moderate
Sodium hydrochlorite	Room	5C1	10 days	None
Sodium hypochlorite	Room	5C1	26 weeks	Considerable
Calcium chloride	Room	Saturated	30 days	None
Stannous chloride	Room	Saturated	30 days	Considerable

N.B.—"None" means 10 per cent or less loss in tenacity; "moderate" means 11–30 per cent loss in tenacity; "considerable" means 31–75 per cent loss in tenacity; and "degraded" means over 75 per cent loss in tenacity.

Hi-Bulk Acrilan

By a modification of the manufacturing process, doubtless by increasing the degree of stretch, Acrilan can be given a high potential shrinkage in boiling water or steam. This type of Acrilan when blended with normal Acrilan gives the product which is called "Hi-Bulk", a blend containing 35–40 per cent high shrinkage fibres; it begins to shrink in water at 65° C. and shrinkage is complete (about 19 per cent in length) at 88° C. Fig. 207 shows Hi-Bulk Acrilan blended yarn before and after relaxation; yarns of this kind have been used successfully for outerwear knitted garments, particularly sweaters. The modification to process consists of stretching a large number of filaments in a rope-like form, allowing some (say 35–40 per cent) to relax, but not the others, then breaking or cutting all together and spinning into a high bulk staple yarn (p. 643).

Dyeing

Amongst the acrylic fibres, Acrilan has good dyeing properties; some of these depend on the basic constituent of the co-polymer of which it consists, and the co-polymer has been designed to dye as nearly as possible as wool dyes, in the expectation that a large proportion of Acrilan will be used in blends with wool. Most acid dyes that will dye wool, will dye Acrilan, although with differing fastness properties and with a rather different technique. The main classes of dyestuff used for dyeing Acrilan are:

Acid Dyes. It is essential to use about 5–8 per cent of sulphuric acid (on the weight of the goods) to get penetration of the dyestuff into the fibre. Acid metallised dyestuffs are commonly applied to wool with this quantity of acid and the technique can be applied directly to Acrilan. The neutral dyeing acid colours and the neutral dyeing premetallised dyes which are applied to wool with very much less acid or with acetic acid, can be applied to the surface of Acrilan in the same way, but then extra sulphuric acid must be added to get penetration into the Acrilan. Wool dyestuffs usually have better wash fastness and slightly inferior light fastness on Acrilan than on wool. Light fastness of acid dyes on Acrilan depends on the method of application and British dyers have had better results than American dyers.

Disperse Acetate Dyes. These can be applied to Acrilan to give shades from pastels to black at a temperature of 95–100° C. Uniform results and good fibre penetration can be obtained; washing fastness is better on Acrilan than on cellulose acetate, as would be expected—the more severe the conditions for dyeing, the harder it is to wash out the dye subsequently. Gas fading does not occur on Acrilan. Light fastness is nearly as good as on acetate rayon.

Chrome Dyestuffs. Chrome dyes are of importance, especially in dyeing 100 per cent Acrilan stock (loose fibres). They are dyed from a formic acid solution on to the surface of the fibre, and then sulphuric acid is used to give penetration into the fibre. Then the material is rinsed and chromed in a fresh bath using from 1–3 per cent sodium bichromate and 4 per cent formic acid (on the weight of the goods). Finally the material is washed and neutralised.

Chrome dyes are also used to dye the Acrilan component of Acrilan/rayon fabrics, and when this is done phosphoric acid replaces the sulphuric acid; it gives good penetration, but does not tender the cellulose fibres so much as would sulphuric acid.

Basic Dyestuffs. Older basic dyestuffs had very poor light fastness; some recently introduced have reasonably good light fastness.

Basics generally give a brightness of shade that cannot be matched by other types of dyestuffs. Acrilan is dyed with basic colours at *p*H 5·5–6·5 in the presence of Igepon T Gel and ammonium acetate; dyeing temperature is 85–96° C. Sometimes a disperse colour can be used to give depth of shade and a small amount of basic colour to give brightness—a more economical method than using all basic colours. A later recommendation is that basics should be applied to regular Acrilan with 3 per cent urea and a non-ionic product such as Emulsifier OC. On Acrilan 16 the basics should be applied from an acetic acid bath; this method is described on p. 761.

Vat Colours. Stabilised vat leuco esters can be applied under mildly acid conditions and in the presence of a retarding agent; sodium dichromate is used to oxidise to the insoluble vat form. Indigoid and thio-indigoid vats can be applied at 98–100° C., but anthraquinonoid vats (the fastest of all) must be applied in pressure machines at 110° C. and at a *p*H of 8·5–9·5. Fastness to light and washing is very good when these vats are used.

Generally, Acrilan is easy to dye with acid colours, disperse colours and basics; washing fastness does not give much trouble but light fastness does.

Acrilan/Wool Blends. Acid dyes will dye both fibres, but at different rates, and devices have to be adopted to get solidity of shade on the union. If neutral dyeing acid colours are used, the blend must be pretreated in 1 gram per litre sulphuric acid at 90–95° C. for 20 minutes and the acid washed off; then the colour is applied, and in the early stages it goes preferentially on to the Acrilan, but as the temperature is raised, the affinity of the dyestuff for the wool increases, catches up at 65° C. and then exceeds the affinity for the Acrilan, and by the time 95–99° C. has been reached, the shade on the wool will have caught up with that on the Acrilan. If the wool is too dark, acid can be added and dyestuff will then migrate from wool to Acrilan; with practice the system can be controlled to obtain solid shades on the union.

If the dyestuff chosen is one that is ordinarily applied to wool from an acid solution, then pre-treatment is as before, but the actual dyeing is done with 2 per cent of acid on the weight of the goods. As before, Acrilan has the higher affinity for the dyestuff at first, then the wool catches up at 65° C. and after that dyes faster than the Acrilan.

After dyeing, the goods are cooled slowly to prevent the formation of creases, neutralised with sodium bicarbonate, and given a light scour to remove surface colour. The essential point in dyeing Acrilan/wool blends is the pre-treatment with acid to allow the dyestuff into the Acrilan.

Acrilan/Rayon Blends. The simplest method is to use a disperse colour for the Acrilan and a direct dye with Glauber's salt for the rayon, all in one bath. The temperature is raised to 80° C. by which time the rayon is substantially dyed and then raised to 95° C. for 1–2½ hours to dye the Acrilan. The goods are cooled slowly to prevent the formation of cracks and creases, and given a scour.

Acrilan/Nylon Blends. These can be dyed with acid colours. The nylon will initially have the greater affinity for the dyestuff, but by the addition of sulphuric acid, that of the Acrilan can be increased and excellent union shades obtained. The goods are cooled slowly, neutralised and scoured.

Hi-Bulk. Yarn containing Hi-Bulk Acrilan should not be dyed by pressure on package equipment, as the pressure of the dye liquor being pumped through puts sufficient tension on the yarn to stretch it and diminish its bulkiness. It is important, too, with fabric or garment dyeing of Hi-Bulk to keep the temperature below 90° C. in order to prevent stringiness.

Finishing. Softness of handle of sweaters and jersey knit outerwear is improved by the addition of a cationic softening agent of which there are many on the market; the softening agent should be applied at *p*H 7·5–8. The cationic agent also prevents static; the anti-static agent applied to all Acrilan during manufacture may have been lost during the dyeing process. After hydro-extraction, sweaters are best tumbler dried to avoid the development of creases. Jersey knit fabric after extraction should be opened and, while stretched to full width, steamed from above and below and then passed through the rubber rollers of a steaming machine, which irons out all cracks and creases. Drying can be done on a loop drier.

Uses

In 100 per cent form Acrilan is used mainly in sweaters, jersey knit outerwear fabric and blankets. In blends with cotton it is used in work-clothes, where its chemical resistance is useful. In blends with rayon, Acrilan is used to give good resistance to creasing and for permanent pleating. With wool, Acrilan gives dimensional stability. Acrilan is easy to wash and quick to dry. Perhaps those qualities that have mainly led to the success of Acrilan are its warmth and softness (with cationic finish) and the cover and loft of the Hi-Bulk variety, all made more attractive to the manufacturer by the relative ease of coloration. Other qualities, such as strength and durability, although not approaching those of nylon, are more than adequate for the desired end uses and better than possessed by the rayons.

Acrilan cannot be heat-set in the way that nylon and Terylene can, but it can be durably pleated.

An interesting development that took place around 1968 was the blending of Acrilan with a small proportion of kempy wool; it gives to knitwear made from it a Shetland look.

COURTELLE

Courtelle is an acrylic fibre that is made at Grimsby by Courtaulds Ltd. It is pigment dulled. The standard material shrinks 1 per cent in boiling water, and a high shrinkage variety shrinks 15 per cent (for comparison Acrilan Hi-Bulk 19 per cent).

Properties

The properties of Courtelle suggest that it is very similar indeed to Acrilan; they are as follows:

Density: 1·17 grams per c.c.
Colour: Off-white.
Cross-section of filaments: Round.
Tenacity: 3·0–3·5 grams per denier (dry), 2·5–3·0 grams per denier (wet).
Extensibility: 30 per cent (both dry and wet).
Moisture regain: 2 per cent.
Water imbibition: 20 per cent.
Chemical resistance: Good to acids and oils, fair to alkalis.
Biological resistance: Excellent to fungi, mildew and moths.
Resistance to light: Good.
Resistance to abrasion: Much inferior to nylon and Terylene; better than viscose rayon.
Effect of heat: Faint coloration on heating to 110° C. for 16 hours, pale yellow after 1 hour at 150° C. It softens at about 160° C. and sticks at about 230° C. No true melting point (decomposes first). Inflammability rather less than that of rayon. The makers claim that it has a comfortable warm handle and good crease resistance.

Dyeing

It can be dyed with disperse and basic dyes (*cf.* p. 761). Selected dyestuffs will give bright and deep shades of good light and wash fastness. It does not felt (none of the synthetics do) and it can be used to give dimensional stability in blends with wool. It is supplied with a 0·5 per cent application of an anti-static finish and if this is removed by the dyeing process, some cationic agent must be applied to replace it. The fibre's low moisture regain and water imbibition give it quick drying properties; knitted garments require little or no ironing. Disperse dyestuffs give maximum brightness if applied under slightly acidic conditions. It is also made spun-dyed in a range of colours. The name given to it then is Neochrome, a good name formerly used by Harbens Ltd. for their spun-dyed viscose rayon.

Differences from Acrilan

It is stronger and has a lower extensibility wet (Courtelle 30 per cent, Acrilan 44 per cent). Moisture regain is a little higher at 2 per cent against 1·2–1·6 per cent, a good point so far as it goes. No claim is made that it can be dyed with acid dyes which are preferred for Acrilan and which are exceedingly useful for dyeing wool blends. The main differences between the two fibres seem to be the higher strength and lower dye affinity of Courtelle.

There is another difference which is probably more fundamental: one that has become evident with extended use. Courtelle is finding its way into quality, even couture, fabrics; some of the early dresses made from Courtelle were smart and kept their shape, and washed and drip-dried like new, so that experience gained with them has given designers and makers-up the confidence to put Courtelle into quality fabrics. Acrilan's future seems to be shaping rather differently, one sees it and its blends in very low priced goods and in very cheap shops as well as in other shops. Adaptability for common usage and low price are of course both really sound characteristics of a fibre and it would certainly run contrary to the general interest to try to restrict a fibre that could be used by all, to be the prerogative of the well-to-do. But the speed with which Courtaulds are building up their capacity (a 3-fold increase) to make much more Courtelle at Grimsby suggests that the experience gained with their early production built up a conviction of the fundamental goodness of the fibre. In mid-1968 Grimsby production of Courtelle was at the rate of 100 million lb. per annum. Smaller quantities are still made at Coventry. It is planned to increase the capacity at Grimsby to 185 million lb. per annum by 1970. It is perhaps slightly to be regretted that Courtelle is made from American raw material. But it is important that the fibre maker should be free to gather what materials he will to make the best product he can. Product is more important than process.

FURTHER READING

T. A. Feild, " The Dyeing of Dynel and Related Products ". *Amer. Dyestuff Reporter*, **41**, 475 (1952).
J. A. Woodruff, *ibid.*, **40**, 402 (1951).
A. B. Craig, U.S. Pat. 2,749,325 (assigned to the Chemstrand Corp.) (1956).
D. T. Mowrey and A. B. Craig, U.S. Pat. 2,744,086 (assigned to the Chemstrand Corp.) (1956).

CHAPTER 33

CRESLAN

CRESLAN is made by American Cyanamid Co., the world's biggest producers of the raw material acrylonitrile. Cyanamid's capacity for the nitrile is 100 million lb. a year and they probably use about a quarter of what they make in their Creslan production at Pensacola in Florida. Cyanamid's first acrylic fibre was called X-51 and Creslan eventually replaced this.

Chemical Nature

No information is available beyond that it is based mainly on acrylonitrile; it is doubtless a co-polymer possibly with acrylamide or a substitute acrylamide.

Properties

It may be of interest to compare the properties of Creslan with those of its forerunner, X-51.

Property.	X-51.	Creslan.
Availability	Continuous filament and staple (various deniers)	Staple and tow, various filament deniers (1½, 2, 3, 5) and high loft
Colour	Off-white; filament yellowish, staple blueish	Off-white
Cross-section	Round	Round
Specific gravity	1·17	1·17
Tenacity at 50 per cent R.H. and 73° F.	Filament 3·7 gm/denier Staple 2·5–3·5 gm/denier	2·7 gm/denier
Elongation (same conditions)	Filament 22 per cent Staple 20–35 per cent	33 per cent
Immediate recovery from stretch	69 per cent from 8 per cent stretch	40 per cent from 5 per cent stretch
Regain at 65 per cent and 70° F.	Probably under 2 per cent	1·5 per cent from dry side 2·1 per cent from wet side
Effect of heat	Tacky at 290° C. Shrivels lower	Sticks at 210° C.
Effect of boiling water	Probably slight	1 per cent shrinkage
Effect of sunlight	Slight, 60 per cent tenacity retained after one year's weathering	Negligible
Resistance to acids	Good including mineral acids	Good except to mineral acids

Property.	X-51.	Creslan.
Resistance to alkalis	Good to dilute, fair to concentrated	Fair to dilute, poor to concentrated
Resistance to cleaning solvents	Good	Good
Biological resistance	Good	Good

The outstanding differences seem to be that the new fibre is intended for staple spinning only, that its tenacity is rather lower than that of X-51, that its elasticity is much lower and that its chemical resistance, particularly to acids, is poorer. None of these differences is necessarily of much consequence; a yarn of 2½ grams per denier is strong enough for apparel uses and no one is going to put a sweater in mineral acid; the loss in elasticity is rather more to be deplored, but it may not be serious. However, the analysis, so far as it goes, reveals Creslan as being poorer rather than richer in desirable attributes than the discontinued X-51. Something must offset these disadvantages and apparently it is dyeability.

Dyeing

According to its makers, " Creslan is outstanding in the ease with which it can be dyed a full range of shades from light to dark, in blends or in the pure state, with a great variety of dyestuffs having good fastness. X-51 had a moderate affinity for acetate colours, Creslan had a moderate-to-good affinity; X-51 could be dyed satisfactorily with acid dyes only by using the cuprous ion method, Creslan has a good affinity for them without cuprous ion; both fibres have a good affinity for basic dyestuffs." Affinity, particularly for acid dyes, seems to be better with Creslan, and indeed it has a good affinity also for neutral milling, chrome, direct and developed dyes. Fastness, too, seems to be better on Creslan, but improved dyestuffs that have appeared since X-51 was dropped may have something to do with this; even so, the following table illustrates a general advantage of Creslan:

Type of dyestuff.	Fastness to					
	Light.		Washing.		Crocking (rubbing).	
	X-51.	Creslan.	X-51.	Creslan.	X-51.	Creslan.
Acetate .	Fair	Moderate–good	Fair	Moderate–good	Good	Moderate–good
Basic .	Fair	Moderate–good	Fair	Good	Good	Good
Acid .	Moderate	Good	Good	Good	Moderate	Good

Creslan shows a general superiority in fastness over the old X-51. Cibalan and similar 1 : 2 metal complex dyes have a good affinity for, and give good results on, Creslan. According to a recent announcement Creslan Type 58 staple gives a full range of shades with good fastness with a wider range of dyes than any other acrylic. There is no doubt that amongst the acrylic fibres dyeability has been greatly improved. The original Orlon was nearly undyeable; now there are several fibres that can be well dyed without resorting to heroic measures. Creslan Type 63 is readily dyeable filament yarn. Creslan Type 84, an acrylic staple intended for carpets, can be dyed with acid colours. Creslan Type 68 is dyeable with cationic dyes, and the two Types 84 and 68 can even be cross-dyed.

Uses

Those uses that acrylics generally fill are recommended for Creslan, notably: knitwear, sportswear, blankets, fleeces and simulated fur fabrics, dresses, men's and women's suitings and overcoatings, children's wear and industrial fabrics. It is claimed that Creslan can be durably pleated. The circular section will not help the cover; it is indicative of wet-spinning. High-loft (or high bulk) Creslan is similar in type to high bulk Orlon and Acrilan.

In Japan the same fibre is spun by Japan Exlan Ltd. and is sold as Exlan; it probably accounted for a considerable proportion of the 200 million lb. of acrylic fibres that were spun in Japan in 1968. In Japan, too, sweaters seem to be the most important use.

CHAPTER 34

ZEFRAN

ZEFRAN, made by the Dow Chemical Company, is outstanding amongst the American acrylic fibres in two respects. First of all it is made in very large quantities, probably inferior only to Orlon; secondly it differs fundamentally from other acrylic fibres in that it can be dyed with most of those classes of dye applicable to cellulosic fibres. Dow are unique amongst the large American acrylic fibre manufacturers in not making their own acrylonitrile. There is an excess capacity (450 million lb. a year) of nitrile plant in America and it is probably as cheap to buy it as to make it. Production of acrylonitrile runs at little more than half capacity.

Chemical Nature

The makers say, " Zefran staple fiber can be described as a nitrile alloy. It has been engineered to obtain a balance between a crystalline component, which provides strength and ease of care properties, with a dye-receptive component, which provides ease of dyeability." It was later described as " a graft co-polymer of dye-receptive groups on a backbone of heat-resistant polyacrylonitrile." It may be a co-polymer of polyacrylonitrile and poly-*N*-vinyl-2-pyrrolidone. The formula of the monomer of this last component is

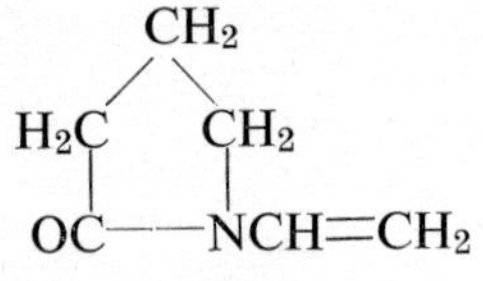

$$
\begin{array}{ccc}
 & CH_2 & \\
 \diagup & & \diagdown \\
H_2C & & CH_2 \\
| & & | \\
OC & - & NCH{=}CH_2
\end{array}
$$

and that of the polymer

$$
\left[
\begin{array}{ccccc}
 & & CH_2 & & \\
 & H_2C & & CH_2 & \\
 & | & & | & \\
 & OC & - & N & \\
 & & & | & \\
 & -CH & CH_2 & CH & -CH_2- \\
 & | & & & \\
OC & - N & & & \\
| & | & & & \\
H_2C & CH_2 & & & \\
 & CH_2 & & &
\end{array}
\right]_n
$$

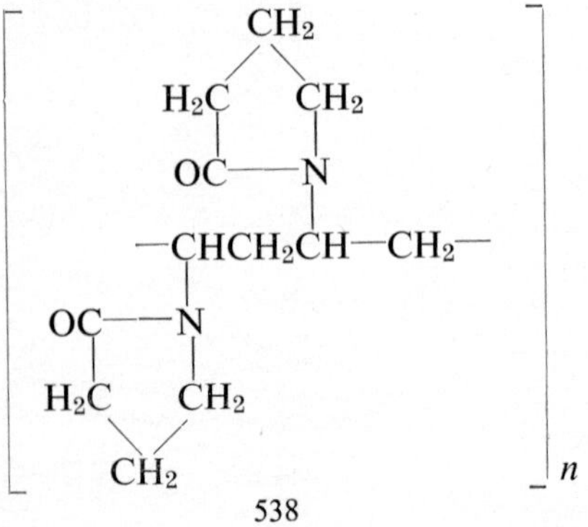

Properties

Under standard conditions (65 per cent R.H., 70° F.) tenacity is 3·5 grams per denier dry and 3·1 grams per denier wet; elongation is 33 per cent wet or dry. Loop tenacity is 1·4. Cross-section of filaments is round, indicative probably of a wet-spinning process; the fibre is a white colour, and moisture regain is 2·5 per cent, which is rather higher than most acrylics, which average about 1·7. Chemical resistance is good to solvents and to most acids, and moderate to alkalis. The fibre burns a little briskly. Specific gravity is 1·19. The yarn shrinks 5 per cent in boiling water.

Dyeing

Dyeability is good on standard equipment with a variety of dyestuffs applied at atmosphere temperatures. The best combination of colour yield and fastness is obtained with vats, naphthol colours, sulphur dyes, after-treated direct colours and neutral premetallised; the Cibalans and Irgalans are representative of the last group. The after-treatment for the direct colours consists of treatment with a resin/copper salt such as Gycofix 67 (Geigy) in dilute acetic acid. Blacks, navys, browns and greens can be obtained with sulphur colours; fastness is good. The neutral premetallised dyes are of special interest because of their light fastness and method of application which is essentially dyeing at 98° C. for 1 hr. Zefran is remarkable amongst the acrylics for its dye-receptivity and it can be dyed or printed in a full range of fast colours; the method used for its identification emphasises its affinity for dyestuffs, and is as follows:

Dye for 3 minutes at room temperature with 1 per cent (on weight of fibre) Fastusol Pink BBA Extra Conc. (Colour Index 353); if the fibre dyes pink it is Zefran—other fibres are unaffected or only slightly stained. The deep colour of the stain given by Shirlastain A (p. 822) is indicative of good dye receptivity.

Uses

Like other acrylics, Zefran is used for interlock and jersey knit goods, both 100 per cent Zefran and in blends. Woven fabrics are used for play clothes, suitings, skirts, work clothes and draperies.

CHAPTER 35

DARVAN OR TRAVIS

IN 1955, the B. F. Goodrich Chemical Co. announced a " dinitrile " fibre which they called Darlan. They developed it and made it on a pilot-plant scale at Avon Lake, Ohio; subsequently they changed its name to Darvan. The fibre was not particularly strong but it had a soft and attractive handle. In 1960 Goodrich sold the rights of Darvan to Celanese Fibers Co. and in 1961 Celanese agreed with Farbwerke Hoechst to build a jointly owned plant and produce Darvan in Germany. The reason underlying the decision to produce in Germany is that Europe is more wool-minded than America and that Darvan is said to be wool-like. The fibre would have been marketed in America under the name of Darvan, but in Europe the same fibre would be called Travis; it is also called, generically, a nytril fibre. This American invasion of the Common Market would have been of interest to the student of economics if it had materialised. In fact it seems doubtful if Darvan will be produced.

Chemical Nature

Darvan fibre is manufactured from a co-polymer of about equal parts of vinylidene dinitrile and vinyl acetate. Vinylidene dinitrile bears the same relation to acrylonitrile as vinylidene chloride bears to vinyl chloride.

Vinyl chloride	$CH_2{:}CHCl$ basis of Rhovyl; main constituent of Dynel
Vinylidene chloride	$CH_2{:}CCl_2$ main constituent of Saran, Tygan.
Vinyl cyanide, usually known as acrylonitrile	$CH_2{:}CHCN$ basis of Orlon, Acrilan.
Vinylidene dicyanide loosely termed vinylidene dinitrile	$CH_2{:}C(CN)_2$ main constituent of Darvan.

It has been found (and it is a matter for surprise) that when vinylidene cyanide and vinyl acetate are co-polymerised, the co-polymer consists of a molecular chain in which the alternation of the two vinyl compounds is nearly perfect. If it were perfect the structure would be

$$\left[\begin{array}{cccc} \mathrm{H} & \mathrm{CN} & \mathrm{H} & \mathrm{H} \\ | & | & | & | \\ -\mathrm{C}- & \mathrm{C}- & \mathrm{C}- & \mathrm{C}- \\ | & | & | & | \\ \mathrm{H} & \mathrm{CN} & \mathrm{H} & \mathrm{OCOCH_3} \end{array}\right]_n$$

and the nitrogen content would be $\frac{28}{164}$ or 17·1 per cent. It has been shown by H. Gilbert *et al.* (*J. Amer. Chem. Soc.*, **78**, 1669–1675, April 20, 1956) that, almost irrespective of the relative quantities of the two monomers that are used, the nitrogen content of the co-polymer lies between 16 and 20 per cent; if, for example, four times as much vinylidene cyanide is used as vinyl acetate the co-polymer contains about 19·9 per cent nitrogen, whereas if all the cyanide had engaged in the reaction, the nitrogen content would have been about 28 per cent. The spontaneous selection by the two monomers of the 1 : 1 ratio is well shown in Fig. 189, if the

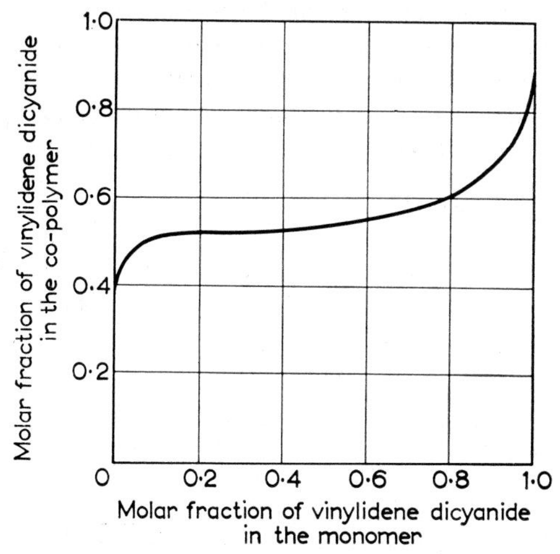

[*After Gilbert* et al.]

FIG. 189.—Curve showing that the 50/50 proportions of vinylidene dicyanide and vinyl acetate in the Darvan co-polymer are relatively undisturbed by quite wide departures from 50/50 proportions of the monomers.

molar fraction of either of the components is between 20 per cent and 80 per cent then the polymer contains very nearly 50/50 of the two components on a molar basis. Strict proof of the alternating effect is based on a method devised by Lewis *et al.*; in this it is shown that in certain pairs of monomers each monomer prefers to react with the opposite type radical rather than with itself. Why? The simplest and probably the truest explanation is that there is more room and less crowding that way. Imagine an escalator with every step occupied by equal numbers of fat men and small boys; the most convenient arrangement from the point of view of the men at least will be 1 : 1 alternation of men and boys. The Darvan polymer is unique because of its nearly perfect alternating structure, which transmits to it remarkable elastic properties and excellent resistance to sunlight and to heat ageing. The two nitrile groups on the one carbon atom of the vinylidene cyanide portion of the chain afford excellent sites for hydrogen bonding, and the close molecular packing which results from this may be responsible for the fibre's resistance to boiling water and to steam at 10 lb./in.^{2}.

Manufacture

Whilst details of the manufacturing processes for Darvan have not yet been released, something can be learnt from a study of U.S. Pat. 2,615,866 and others, assigned to B. F. Goodrich Chemical Co. There are several ways of making vinylidene dicyanide but the one most likely to be used is the reaction of malonitrile and formaldehyde in aqueous solution to give 1:1:3:3-tetracyanopropane:

$$
\begin{array}{c}\mathrm{CN}\\|\\\mathrm{H{-}C{-}}\\|\\\mathrm{CN}\end{array}
\boxed{\mathrm{H} + \begin{array}{c}\mathrm{H{-}C{-}H}\\\|\\\mathrm{O}\end{array} + \mathrm{H}}
\begin{array}{c}\mathrm{CN}\\|\\\mathrm{{-}C{-}H}\\|\\\mathrm{CN}\end{array}
\longrightarrow
\mathrm{H{-}}\begin{array}{ccccc}\mathrm{CN}&&\mathrm{H}&&\mathrm{CN}\\|&&|&&|\\\mathrm{C}&-&\mathrm{C}&-&\mathrm{C}\\|&&|&&|\\\mathrm{CN}&&\mathrm{H}&&\mathrm{CN}\end{array}\mathrm{{-}H}
$$

Malo-nitrile. Formaldehyde. Malo-nitrile. Tetracyanopropane.

which on heating decomposes to malonitrile (which is re-used) and vinylidene dicyanide.

$$
\mathrm{H{-}}\begin{array}{ccccc}\mathrm{CN}&&\mathrm{H}&&\mathrm{CN}\\|&&|&&|\\\mathrm{C}&-&\mathrm{C}&-&\mathrm{C}\\|&&|&&|\\\mathrm{CN}&&\mathrm{H}&&\mathrm{CN}\end{array}\mathrm{{-}H}
\xrightarrow{\text{Heat}}
\begin{array}{c}\mathrm{CN}\\|\\\mathrm{H{-}C{-}H}\\|\\\mathrm{CN}\end{array}
+
\begin{array}{ccc}\mathrm{H}&&\mathrm{CN}\\|&&|\\\mathrm{C}&=&\mathrm{C}\\|&&|\\\mathrm{H}&&\mathrm{CN}\end{array}
$$

Tetracyanopropane. Malonitrile. Vinylidene dicyanide.

Distillation at low pressure, 2 mm. Hg, separates the malonitrile which at this low pressure boils at 90° C. from the vinylidene cyanide which boils at 47° C.

The relative quantities of vinylidene dicyanide and vinyl acetate used for the polymerisation may be about 50 of the dicyanide to 50 of the acetate. Both monomers are dissolved in benzene and 0·2 parts of *o-o'*-dichlorobenzoyl peroxide is added to catalyse the polymerisation which takes place on warming the solution to 45° C. The co-polymer precipitates, as it is insoluble owing to its high molecular weight and is filtered off after the reaction has proceeded for two hours. A 12 per cent solution of the co-polymer in dimethyl formamide is prepared and is then spun into a water bath which precipitates filaments of the co-polymer, and these filaments are hot-stretched in a series of steps, crimped and cut to give Darvan.

It will have been noted that the polymerisation takes place at a fairly low temperature, and this is due to the influence of the peroxy-type catalyst; in fact, vinyl acetate reacts with vinylidene cyanide less readily than styrene and vinylidene chloride do, but much more readily than vinyl chloride does.

Properties

The manufacturers of Darvan claim that their fibre " combines the æsthetic qualities of natural fibres with (the) ease-of-care which is typical of the synthetics ". A detailed summary of the properties of the fibre will enable an assessment to be made of the basis of this claim.

Tenacity: 1·75 grams per denier dry (standard conditions)
1·5 grams per denier wet.

These values are in the usual rayon range. The ratio of 0·86 for wet/dry strength is low for a synthetic fibre, although it is far higher than that of any natural or regenerated fibre.

Elongation: 30 per cent, both dry and wet. This is a useful value. The characteristic curve is shown in Fig. 190, and it is apparent from this that Darvan has a very high compliance ratio.

Elasticity: The elastic behaviour is good, although it does not approach that of nylon (p. 331). When fibres are stretched at the rate of 10 per cent per minute, held at the stretch for 30 seconds and then released, the recovery from the extension is 70 per cent from a 3 per cent stretch, and 45 per cent from a 10 per cent stretch (even after contraction there is a residual 5·5 per cent stretch compared with the original length).

Specific gravity: 1·18, a low value which is useful in contributing good cover (*cf.* nylon, 1·14, Orlon, 1·18, Terylene, 1·38).

Moisture regain: Between 2 and 3 per cent under standard conditions (65 per cent R.H., 70° F.). This is an interesting value; it is lower than those of nylon, Verel and Vinylon, but considerably higher than other synthetic fibres; in the author's view it is a little on the low side for an apparel fibre. It is, too, so low that protection against static is necessary. When Darvan is examined under the microscope it can be seen that it carries a finish.

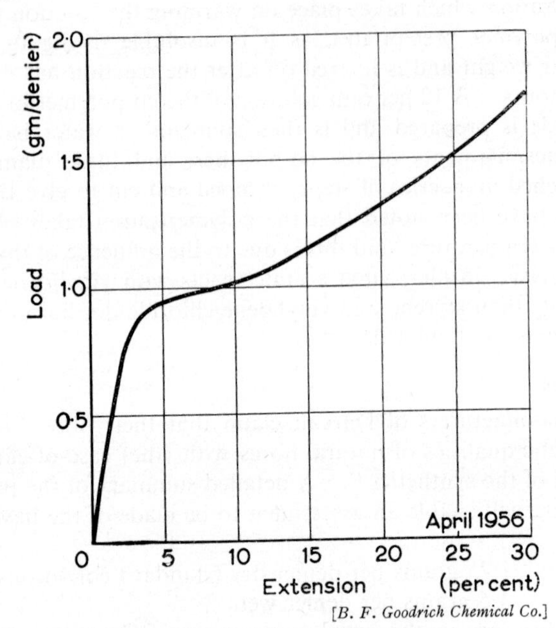

[*B. F. Goodrich Chemical Co.*]

Fig. 190—Characteristic curve of Darvan.

Light: Weathering seems to be extremely good; Darvan continuous filament yarns which were weathered in direct sunlight in Florida retained 88 per cent of their strength after two years' exposure. A fabric made from staple fibre Darvan was not measurably weakened after five months' exposure to direct sunlight in Arizona.

Biological resistance: Immune to clothes-moth and carpet-beetle larvae attack; no evidence of failure in standard tests of mildew resistance.

Chemical resistance: There is only 1 per cent shrinkage when the fibre is immersed in boiling water for 3 minutes and little if any more in steam at 10 lb./in.^{2}. This stability is attributable to the hydrogen bonding which takes place between the cyanide groups of the molecular chains; chemical resistance is, of course, very much more marked in some fibres such as Orlon and Dynel which consist mainly of polyacrylonitrile, but it has to be added that these fibres, having been stretched during manufacture, usually tend to shrink on steaming. The behaviour of Darvan suggests that it has not been stretched unduly during manufacture, and this view is reinforced by the comparatively low tenacity of the fibre. Resistance to chemicals is moderate; good by comparison with the natural fibres, poor compared with Dynel and Terylene and of course Teflon, which is much the best of all. Using the same arbitrary description of the effect on strength as that on p. 408, we have for Darvan:

Chemical.	Temperature (° C.).	Concentration (per cent).	Time.	Effect on strength.
Sulphuric acid	100	10	4 hours	Moderate
Caustic soda	43	0·5	8 hours	None
,, ,,	43	0·5	7 days	Moderate
,, ,,	43	0·5	10 days	Appreciable
,, ,,	Room	10	24 hours	Moderate
,, ,,	100	10	4 hours	Degraded

Darvan is insoluble in acetone and in methylene chloride.

Softening temperature: About 175° C.; it retains a significant portion of its strength at 160° C.

Flammability: Similar to cotton and viscose.

Morphology: Cross-section is shown in Fig. 191, it is rather flatter than desirable, probably indicative of having been spun from a not very concentrated solution.

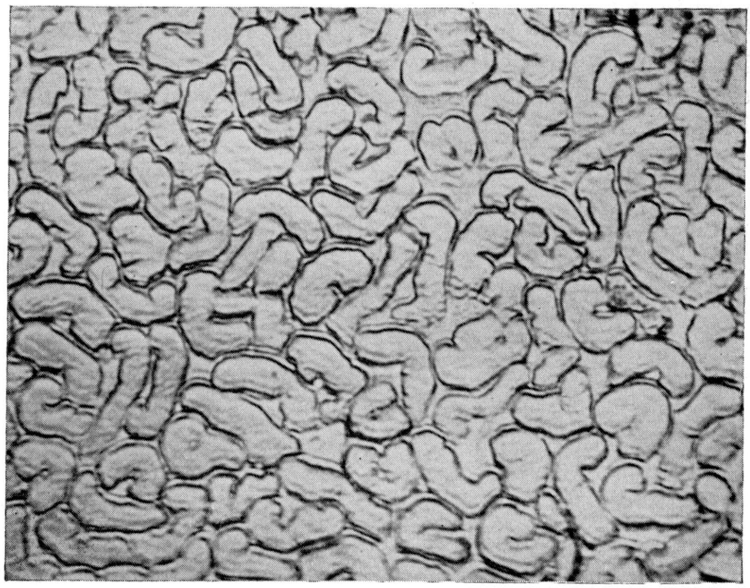

[*B. F. Goodrich Chemical Co.*]

FIG. 191.—Cross-section of Darvan (× 140).

Dyeing

Dyeing is the main trouble with the vinyl and vinylidene fibres; there is, of course, no gainsaying that devices have appeared whereby these difficulties have been surmounted, but the difficulties are not really removed. If, for example, some modification could be made to Orlon so that it could be dyed easily by traditional methods there is no doubt that the use of the fibre would very greatly increase. Darvan has a low dye affinity; it cannot be dyed with direct, acid, metallised or chrome dyestuffs. Dispersed dyestuffs at the boil will give pastel shades, and heavier shades such as bright red, maroon, navy and black can be obtained by azoic dyestuffs, of which β-oxy-naphthoic acid is one component (*cf.* p. 414). Carriers can be used to increase the depth of shade from disperse dyes, and the cuprous ion method can be used with acid dyestuffs; *o*-phenylphenol used at 4 gm./litre or methyl salicylate (oil of wintergreen) used at 8 gm./litre are the best carriers. Vigorous after-scouring is required to remove excess dyestuff or residual carriers. The picture is familiar in outline, but so far the detail has not been filled in. Apparently the fibre can be heat-set. It can be bleached with acid hypochlorite.

Uses

Darvan's outstanding properties are good handle, resistance to sunlight and relative freedom from pilling. Intended uses were in hand-knitting yarns, sweaters, women's pile coats and in blends for tropical suitings and men's shirtings. It will probably be made if it is made at all only in staple, not in continuous filament, and mostly in fine deniers.

FURTHER READING

H. Gilbert, *et al.*, *J. Amer. Chem. Soc.*, **78**, 1669–1675 (1956).
H. Gilbert and F. F. Miller (assignors to B. F. Goodrich Co.), U.S. Patent 2,615,866 (1952).
F. M. Lewis, *et al.*, *J. Amer. Chem. Soc.*, **70**, 1519 (1948).

CHAPTER 36

POLYETHYLENE

COURLENE, MARLEX

LIKE many other fibre-forming materials—for example, cellulose acetate and Saran—polyethylene was made originally as a plastic material, and only later, after it had established itself in the plastics field, was its use for fibres developed. Polyethylene resulted from research carried out by I.C.I. Ltd. on the effect of very high pressures on gases. The plastic made by I.C.I. Ltd. is given the name " Alkathene ".

Chemical Nature

When ethylene gas is subjected to high pressures and temperatures, it polymerises and forms a solid. The mechanism is one of addition polymerisation, very similar to that whereby Vinyon and Saran are formed. The reaction by which polythene is produced is as follows:

$$CH_2{:}CH_2 + CH_2{:}CH_2 + CH_2{:}CH_2 + CH_2{:}CH_2 \longrightarrow$$
$$-CH_2CH_2CH_2CH_2CH_2CH_2CH_2CH_2-$$

The higher the pressure to which it is subjected, the higher the molecular weight of the polymer.

Manufacture

Ethylene gas may be obtained by cracking petroleum or, alternatively, from alcohol. At one time alcohol was made by fermentation and ethylene was made by dehydrating the alcohol, but since the erection of oil cracking plants, ethylene has been obtained from petroleum. Indeed most of to-day's alcohol is also made from this ethylene and it is estimated that by 1964 there was only very little fermentation alcohol made; almost all of it coming from oil cracking. The ethylene is polymerised in autoclaves at 200° C. and 1,500 atmospheres pressure (about 10 tons per square inch) in the presence of a trace (0·01 per cent) of oxygen, which acts as a catalyst. The polymer (M.W. 15,000) is spun from the melt at a temperature of 300° C. through spinneret orifices of 0·1 mm. diameter into a current of cooling gas. After spinning, it is cold drawn to six times its original length. In addition to melt spinning, solvent spinning has been used but, owing to the relative insolubility of polyethylene, the solvents, of which benzene is one and xylene another, have to be used hot. Melt spinning would appear to be a proposition much more attractive than hot solvent spinning. If the polythene has a molecular weight of 6,000 it gives filaments with a tenacity of 0·5 grams per denier, but if the molecular weight is 21,000 the tenacity is as high as 3 grams per denier. The polyethylene yarn made by Courtaulds Ltd. is known as Courlene. Several American manufacturers make polyethylene yarns, notably Reeves Bros. Inc. (Reevon) and National Plastic Products Co. (Wynene I).

Properties and Uses

Courlene has a tenacity of 2–3 grams per denier and an elongation of 40 per cent. The outstanding property of polyethylene is its chemical stability, which has proved of great value when the material, as a plastic, has been used for applying protective coatings on other material. This coating can be carried out with a Schori spray-gun. The fibre is also very resistant to microbiological attack.

From the textile standpoint, the very low melting point of 110–120° C. militates against its use for normal textile purposes; it would be quite impossible to iron it with a normally " cool " iron. It does not possess the non-flam properties of Vinyon and Saran, so it is unlikely to be used for some of the purposes for which these particular fibres have been found suitable.

Polyethylene has a very low frictional coefficient, probably because of the lack of polar chemical groups in its molecule. This gives it a waxy handle which is very noticeable; the low frictional coefficient is responsible for some of the special uses to which polythene is put. Fabric made from 500 denier monofil has been used for car upholstery; the fabric is pre-shrunk (British Pat. 673,879) to avoid shrinkage during summer use. Courlene has been woven into fabric and used for protective clothing in industries where contact with corrosive chemicals is likely. Boiler-suits made from it have proved satisfactory in a sulphuric acid plant, as have also women's overalls in a factory where acid powders are handled. The woven structure allows the fabric to breathe, and overalls made from it are more comfortable to wear than similar garments made from plastic film. Protection against chemical splashing, dust and vapour is excellent. Courlene fabrics have also been used for low-temperature filtration and for clay and sewage filtration. The U.S. Rubber Co. make a fabric (Trilok) which consists of polyethylene yarn in the warp and conventional yarns, *e.g.*, rayon, in both warp and weft; the raw fabric is immersed in boiling water and the polyethylene yarn shrinks about 50 per cent, cockling the fabric, giving it a three-dimensional figure and a cushioning action.

Effect of Nuclear Radiation. Polyethylene is one of those fibres that become cross-linked when irradiated with cobalt-60. The effect of this cross-linking has been strikingly demonstrated in the following way:

Polyethylene normally softens at 110–115° C. and at that temperature can no longer support its own weight, but if it has previously

[*Minnesota Mining and Manufacturing Co.*]

FIG. 192.—Irradiation with cobalt-60 increases the melting point of polyethylene. The polyethylene bottle (*a*) which has not been irradiated melts in an oven at 135° C.; the bottle (*d*) which has had a big dose of radiation does not melt in the same oven; the others intermediately.

been irradiated with γ-radiation or with high speed electrons it becomes rubbery rather than liquid at 110–115° C.; the cross-links restrict molecular freedom. Fig. 192 shows four polyethylene bottles that have been in an oven at 135° C. for 15 min.; the one on the left (*a*) has not been irradiated and has melted to a shapeless blob; the one on the right (*d*) has had a high dose of radiation and is consequently able to withstand the temperature of 135° C.; the other two (*b*) and (*c*) have had smaller doses. The higher the dose of radiation, the better will the polyethylene subsequently withstand the effect of heat.

LOW-PRESSURE POLYETHYLENE—MARLEX 50

It has been found that in the presence of certain catalysts (pp. 110–114), ethylene can be polymerised at temperatures and pressures that are considerably lower than those first used. Various catalysts have been suggested, notably aluminium alkyls and chromium oxide mounted on silica-alumina. The Phillips Petroleum Co. of Bartlesville, Oklahoma, have used 2·5 per cent chromium oxide mounted on silica-alumina, to polymerise ethylene in a hydrocarbon solvent at a temperature of 130–150° C. and a pressure of 450 lb./in.^{2}. The very much lower pressure necessary in the presence of the catalyst means a big capital saving on the cost of the plant. Polymers with molecular weights of 40,000 and higher have been made. One of these is called Marlex 50.

Low-pressure polyethylene has a higher density (0·96) than ordinary high pressure polyethylene (0·92) and softens at about 125° C. instead of 105° C.; its chemical resistance is slightly better and its tenacity, when spun into a monofilament, is high at 4·4 grams per denier, whereas 2·7–3·0 grams per denier is good for conventional high-pressure polyethylene. It still remains that, with a melting point of about 125–130° C., Marlex 50 is unsuited for traditional textile purposes, but it does show a significant improvement in most ways over the older material, and in view of the more moderate conditions of polymerisation it seems likely to supersede the older high-pressure polyethylene. The makers of Marlex 50 have suggested its suitability for " automobile interiors, curtains, furniture covers, tarpaulins, filter cloth, rugs, rope and fish nets ".

Courlene X 3

Courlene X 3 is a polyethylene monofil that Courtaulds Ltd. have introduced with certain improved properties. Melting point is higher at 135° C. and it is more stable in boiling water—an important point for cleaning protective clothing. It is, however, still subject to a shrinkage of 12 per cent at 100° C., and allowance must be made (not easy) for this in the design of a garment. In strength (4 grams per denier) and abrasion resistance Courlene X 3 is superior to the original Courlene. Its specific gravity is 0·96.

The improved properties of low-pressure polyethylene are due to better molecular packing of the fibre molecules and to the consequently higher degree of crystallinity of the fibre. This brings with it a higher density, strength and melting point, as well as improved resistance to chemical attack. Low-pressure polyethylene is a linear polymer (see pp. 118–120).

Complementary Polyolefine–Nylon Blends

The reader will have gathered that the properties of polyethylene and nylon are very different. The question arises as to whether blends of the two would give the best or the worst of both. The test may not yet have been made on fibres, but it has certainly been made on plastics.

In B.P. 889,354 (1962) the Continental Can Company described some blends of nylon 6 and polyethylene as plastics. A comparison of the properties of the two components was as follows:

	Polyamide (*e.g.*, nylon 6).	Polyolefine (*e.g.*, low-density polythene).
Chemical inertness	Good	Good
Toughness and flexibility	Good	Good
Ease of fabrication	Temperature range for extrusion is narrow	Good
Cost	High	Low
Permeability to water	High	Low
Permeability to lower alcohols	High	Low
Permeability to hydrocarbons	Low	High
Permeability to esters, ketones, etc.	Low	High
Printability	Good	Poor, unless pre-treated by flaming or chemically
Discoloration on extrusion	Yellows	None

The two polymers have no common solvent, and the only way of making an intimate mixture is in the melt. Provided that a high pressure, preferably of 130 atmospheres, was maintained, that each of the two components was present to the extent of at least 5 per cent (this is to give impermeability, not uniformity) and that the melt was kneaded with a shearing action and with turbulence, suitably obtained by the use of mixing screws, then a homogeneous melt was obtained. A blend of 20 per cent nylon 6 and 80 per cent polyethylene was almost impervious to esters such as in oil of wintergreen and to hydrocarbons such as heptane. The use of similar blends for fibres is a development that we shall probably see.

Application to Wool

Plastic materials are sometimes applied to wool in order to cover its scale structure and thereby make the wool unshrinkable. There are various ways in which a measure of shrink-resistance can be obtained, and these have been discussed in the author's " Wool Shrinkage and its Prevention ". The application of a plastic to the fibre is not altogether good, because it alters the handle of the wool fibre. Warmth and softness of handle are the two most valuable attributes of wool; there is no fibre (excepting goat and rabbit hairs which are essentially the same) to compare with it. Perhaps the polyolefines with their low frictional coefficients and smoothness of handle are less unsuitable than most plastics to put on to wool. It has been shown by Baldwin, Barr and Speakman at Leeds that if a resin is to be applied to wool it is better to apply it in the form of a monomer and to polymerise it on the wool fibre, rather than to apply the already formed polymer to the wool, because much better uniformity of application and better fastness are obtained by the former method. In 1962, Wolfram and Speakman formed polymethylene on wool. They impregnated the wool with a 0·1 per cent solution of copper chloride catalyst dissolved in methyl alcohol, dried the wool with the catalyst in it and then put it in a 1·4 per cent solution of diazomethane in ether. After 20 min. treatment 3 per cent of polymethylene had formed on this fibre and after 2 hr. treatment 7 per cent. Whereas untreated wool shrunk 32 per cent in area under a standard milling treatment, wool with 3 per cent polymer shrunk only 10 per cent and with 7 per cent polymer only 3–4 per cent. The formation of polymethylene from diazomethane may be represented:

$$n\,CH_2\!\left\langle\begin{matrix}N\\ \|\\ N\end{matrix}\right. \longrightarrow [-CH_2-]_n + n\,N_2$$

The formula of polymethylene is essentially the same as that of polyethylene $[-CH_2CH_2-]_n$, but the former has the advantage as an additive to wool in that it can be made without the application of high temperatures and pressures. Perhaps this application points the way to the use of polymethylene as a fibre itself; it might well constitute an ideal form of polyethylene.

FURTHER READING

I.C.I. Ltd. *et al.* British Pat. 471,590 (1936) (conditions of polymerisation).
E. I. du Pont de Nemours & Co. British Pat. 598,464 (1944) (spinning from solvents).

" Fibre Science ", edited by J. M. Preston, 327–28 (1949).
British Nylon Spinners Ltd. and G. Loasby, British Pat. 565,282 (1944).
R. A. V. Raff and J. B. Allison, *Polyethylene*, Vol. XI of " High Polymers ", pp. 451–453, London, 1956.
A. Clark, J. P. Logan, R. L. Banks and W. C. Lanning, " Marlex Catalyst Systems ", *Industr. Engng Chem.*, **48**, 1152–1154 (1956).
R. V. Jones and P. J. Boeke, " Properties of Marlex 50 Ethylene Polymer " *Industr. Engng Chem.*, **48**, 1155–1161 (1956).
R. W. Moncrieff, " The Application of Plastic to Wool ", *Text. Manufacturer*, 533–4 (Dec. 1962).
R. W. Moncrieff, " Wool Shrinkage and its Prevention ", pp. 374–409, National Trade Press, London (1953).
L. J. Wolfram and J. B. Speakman, *J. Soc. Dyers and Col.*, **78**, 407 (1962).
G. C. Travers, " Polyolefine–Nylon Blends ", *Plastics*, 117 (Sept. 1962).

CHAPTER 37

POLYPROPYLENE

MERAKLON, ULSTRON

PROPYLENE is a by-product of the oil refineries; it comes from the cracking process and is produced in even greater quantity than is ethylene; it is accordingly very cheap and is much the cheapest of those substances that are within sight as a possible raw material for synthesising fibres. When propylene is polymerised under certain conditions it will give fibre-forming polymers, *i.e.*, polypropylene. The primary attraction of polypropylene fibres is that of low cost, and that is why big chemical manufacturers like Montecatini in Italy and I.C.I., whom Montecatini have licensed for the production of polypropylene fibre in the U.K., are spending so much effort on the production of good fibres from propylene. Montecatini call their polypropylene fibre Meraklon; I.C.I. call theirs Ulstron.

It is a long time since polyethylene appeared; it was extensively used as a plastic material in the early days of World War II, yet solid polypropylene was not made until 1954. Ethylene and propylene are not very remote chemically and ethylene is easy to convert into fibre-forming polyethylene by the application of very high pressures at high temperatures. Yet when the same processes were applied to propylene the so-formed polymer was not the expected fibre-forming solid, but was a grease. The reason was that whereas ethylene is a compact molecule and polymerises to regular linear molecules which can pack closely together, or crystallise, propylene can give polymers which are irregular in the arrangement of the methyl side-chains, thus:

$$nCH_3CH{:}CH_2 \longrightarrow -\underset{\substack{|\\CH_3}}{C}HCH_2\overset{\substack{CH_3\\|}}{C}HCH_2\underset{\substack{|\\CH_3}}{C}HCH_2\underset{\substack{|\\CH_3}}{C}HCH_2\overset{\substack{CH_3\\|}}{C}HCH_2-$$

Propylene. Atactic polypropylene.

The methyl groups can fall in random order on either side of the carbon backbone and so give long linear molecules which bulge here and there with the methyl groups, and because the bulges on different molecules occur at different intervals the molecules cannot pack snugly together; their assembly lacks cohesiveness and cannot crystallise; instead of a solid plastic it constitutes a viscous liquid or grease. This state of affairs could not obtain with ethylene because this olefine has no side-chains and so there was little opportunity for irregularity. Propylene with its methyl side-chains had every opportunity to form irregular or atactic polymers of the type shown above. It was only when methods of controlled propagation became available as the result of work by Ziegler and Natta that isotactic polypropylene could be made. In the isotactic polymer all the methyl groups are arranged in very orderly fashion and they all lie on the same side of the plane of the main carbon chain. Fig. 42 (p. 108) represents atactic polypropylene (not fibre-forming because it is too irregular in structure for the molecules to line up nicely) and Fig. 43 (p. 109) represents isotactic polypropylene which is fibre-forming. In both Figures the plane of reference is that in which the zig-zag main carbon chain lies. More simply, but not quite so informatively, we can represent atactic (not fibre-forming) polypropylene:

$$
\begin{array}{lllllll}
 & CH_3 & CH_3 & & CH_3 & & \\
 & | & | & & | & & \\
- & CHCH_2 & CHCH_2 & CHCH_2 & CHCH_2 & CHCH_2 & CHCH_2- \\
 & & & | & & | & | \\
 & & & CH_3 & & CH_3 & CH_3
\end{array}
$$

and isotactic (fibre-forming) polypropylene as:

$$
\begin{array}{lllllll}
 & CH_3 & CH_3 & CH_3 & CH_3 & CH_3 & CH_3 \\
 & | & | & | & | & | & | \\
- & CHCH_2 & CHCH_2 & CHCH_2 & CHCH_2 & CHCH_2 & CHCH_2-
\end{array}
$$

Isotactic polypropylene is stereoregular, atactic polypropylene is stereo-irregular.

Catalytic Polymerisation. Exactly what catalysts are used in the commercial manufacture of isotactic polypropylene is not known; nearly all the information available is what can be gleaned from a study of patent specifications, and it is evident from these that a large number of catalysts will control the propagation of propylene so effectively that an isotactic polymer results. The general considerations relating to controlled propagation in addition polymerisations have already been discussed (pp. 108–114) and these apply directly to the polymerisation of propylene.

Conditions which probably approximate to those that are used commercially are as follows:

Ziegler catalyst: titanium trichloride, $TiCl_3$ (or possibly the tetrachloride $TiCl_4$, which is reduced *in situ* to the lower chlorides).
Ziegler co-catalyst: aluminium triethyl, $Al(C_2H_5)_3$.
Liquid medium (solvent and diluent for propylene): heptane.
Pressure: 30 atmospheres; highly purified propylene is fed in repeatedly as pressure drops.
Temperature: 100° C.
Time of reaction: 8 hours.
Yield: solid polymer, 85–90 per cent, based on propylene used.

The crude solid is extracted with acetone to remove any low molecular weight material.

The molecular weight of the polymer is about 80,000; it is controlled by (*a*) the choice of polymerisation temperature and (*b*) additives which can terminate the reaction by combining chemically with the ends of the growing molecular chains. The tool used to measure molecular weight is intrinsic viscosity (p. 41) and it is related to the number average molecular weight (p. 47) by the relation:

$$\text{Intrinsic viscosity} = \text{Number average molecular wt.} \times 2{\cdot}5 \times 10^{-5}$$

Melting point affords a measure of control of tacticity. Atactic polypropylene is only a grease. Unstressed isotactic polypropylene will melt at 165° C. Under stress with improved crystallisation, the melting point may rise.

In the polymerisation process there must be a solid phase present which may either be the catalyst itself or some solid, for example silica (or even sodium chloride), on the surface of which the catalyst is adsorbed. This solid phase adsorbs the monomer molecules one by one and orients them in an activated form in relation to the end-group of the growing polymer chain; the growing polymer chain adds on monomers one by one and each one is fed to it in a special orientation by the catalyst. Monomer molecules enter one after another between the catalyst and the polymer chain, tagging on to the chain. The catalyst, in effect, acts as a template which determines the shape of the growing polymer chain by a series of repetitive actions carried out at a speed by comparison with which any man-made mass production repetitive process is a laboured crawl. The solid phase is essential to hold the monomer molecules in a defined position so that they all enter the chain, facing the same way. If the same point, the same molecule of catalyst, presents a thousand monomer molecules one after another, and one side of the monomer molecule will fit into the catalyst molecule like a ball into a socket, but not the other side which we can imagine as flat, then it is easy to understand why monomer molecule after molecule is presented in the same right hand (or left hand as the case may be, but always the same) way. Ball and socket, lock and key, or template, or whichever way you look at it, the solid geometry of the final polymer chain is determined by the solid geometry of the monomer and catalyst molecules, and the catalyst molecules are anchored to (or adsorbed on to) or form the surface of some really solid base. Could you unlock a door, even with the right key, if the screws had come loose and the lock was no longer firmly fixed to the door? The polypropylene fibre grows like a wool fibre does, *i.e.*, from the root end.

Spinning and Drawing. The polymer from which low molecular weight material has been dissolved out is melted, and spun by gear pump pressure through spinnerets. Melt spinning is usual, partly because it is cheap, mainly because polypropylene of high D.P. is difficult to dissolve. The polymer before spinning is about 50 per cent crystalline; after having been spun, the fibre has a 33 per cent crystalline content. It is stretched or drawn and this increases the crystallinity to 47 per cent; then it is annealed, *i.e.*, given a final heat treatment in which the higher temperature increases the mobility of the polymer molecules and gives them an opportunity to re-arrange themselves a bit and pack together better; this increases the crystallinity to 68 per cent. During the spinning the polymer becomes birefringent as would be expected from the orientation induced by the traction of the jet, and the degree of birefringence is increased by the drawing process, but the highest value obtained (see p. 72) is 0·035, whereas " theoretically " it could be 0·067. This discrepancy is interpreted as indicating that the molecules never open out completely but exist in a stable helical form.

Most of the polypropylene fibre that has been spun as yet (1968) is monofilament, but some multi-filament is already being made and it will doubtless increase in importance, relative to the monofilament. Ulstron, I.C.I.'s polypropylene, is also made as staple fibre.

Properties

Mechanical. Polypropylene is like all other highly stretched fibres in that its breaking load and breaking extension can be controlled, one at the expense of the other, by altering the degree of stretch to which the fibre is subjected in manufacture. Tenacity is about 8·5–9·0 grams per denier and extension at break about 17–20 per cent; in these respects the fibre is excellent. Knotted strength is high, a measure of the transverse strength of the fibre. Impact strength is also high; the fibre can absorb energy. Elastic recovery is as follows: if a 2 per cent extension is held for 30 sec. and then released there is 91 per cent immediate recovery and 9 per cent delayed recovery; if the 2 per cent stretch is held for 3 min. then on release the immediate recovery is 82 per cent and the delayed recovery 18 per cent; if it is held for 6 min. then recovery within 1 sec. is 45 per cent (for comparison Terylene will recover 85 per cent of its stretch under similar conditions), but within 100 sec. it is 92 per cent.

Static Electricity. Like other hydrophobic fibres, polypropylene will develop static charges; its tendency to do this is very great because of its negligible moisture regain.

Specific Gravity. Very low at 0·90 to 0·92 depending on the degree of tacticity; polypropylene is the lightest of all commercial fibres.

Chemical Resistance. Generally excellent. Acids and alkalis have very little effect on polypropylene; 1 week in 20 per cent caustic soda at 70° C. (destroys Terylene) or in conc. hydrochloric acid at 20° C. (destroys nylon) has practically no effect. 50 per cent nitric or sulphuric acid causes a little discoloration and on lengthy immersion of the order of days, some loss of strength. Polypropylene is insoluble in cold organic solvents, but will dissolve in hot decalin and tetralin or in boiling tetrachlorethane, whilst trichlorethylene at the boil causes heavy shrinkage. Strong oxidising agents such as hydrogen peroxide will attack the fibre. Liability to oxidative degradation may prove to be a serious defect in the fibre; there may be vulnerable spots in its structure which will be attacked by air.

Moisture Regain. Practically nil, less than 0·05 per cent. Tenacity and extension are the same " wet " as dry. The low compatibility for moisture will not enhance the value of the fibre for apparel.

Abrasion. Resistance to abrasion is good. The fabric does not " pill " and this is probably a reflection of its good resistance to abrasion. Wool/polypropylene blends will pill but if the pills are examined they are found to consist entirely of wool. Whether the blending of polypropylene with wool has any effect on the tendency of the wool to pill is not known; it is unlikely that it would reduce it, but not impossible that it might increase it.

Effect of Heat. Melting point is about 165° C., softening point about 155° C., *i.e.*, too low for safe ironing. For comparison Perlon is about 215° C. and nylon about 260° C. The highly oriented (but not the undrawn) fibre is very resistant to extreme cold, and keeps its strength down to −100° C., a property which might find special uses for it.

Effect of Light. Polypropylene may be sensitive to oxidation initiated by the action of light; apparently there may be a few weak spots in the fibre, even although the main polymer chains are unaffected. Its resistance can be improved by the application of antioxidants and radiation absorbents and some such stabiliser is included in Ulstron fibre.

Colour. Colourless.

Dyeing

Polypropylene presents a real problem to the dyer; there are no polar groups in the structure to which dye molecules are attracted. So far, the only satisfactory method of coloration has been the inclusion of pigment in the material to be spun, that is in the melt; this has given excellent results on polyethylene and will doubtless do the same on polypropylene. It does, of course, restrict the number of available shades, it is costly, it is only worth the spinner's while if he can get a really long run on each colour, it is useless for small quantities, and it means that the weaver or knitter must hold much larger stocks of raw cloth than he need do if he could send it to the dyer to be dyed in any required shade.

One method that has been tried to improve the dyeability of (uncoloured) polypropylene has been to graft side-chains of polymethyl methacrylate on to it; these contain polar groups and afford sites for dyestuff molecules to be adsorbed, and at the same time they give some moisture compatibility so that the moisture regain of the fibre is increased. All along the polypropylene chain there are tertiary carbon atoms (those to which the methyl side-chains are attached) and the single hydrogen atom that is attached to each of these has a little more reactivity than, for example, the hydrogen atoms in a typical methyl group. In particular the tertiary carbon atoms can be oxidised with oxygen or even air at temperatures of 70–80° C. and a pressure of about 3 atmospheres. When they have been oxidised the tertiary carbon atoms on the polymer backbone will react with monomeric methyl methacrylate which forms branches of its polymer on the polypropylene chain. The side-chains form a combined hydrophilic layer some 10–100 μ thick on the polypropylene surface. It is noteworthy that isotactic polymer can be graft-polymerised only on the surface, whereas atactic polypropylene, because of its greater accessibility, can be graft-polymerised throughout its mass. Another possibility that has received consideration is that of high energy irradiation, for example by bombardment with γ-rays from cobalt-60. However, these methods are a long way from fruition and at present they do not seem to be very practicable. It may be that the controlled propagation method that is used for isotactic polymerisation has an inherent disadvantage in that it cuts out the possibility of co-polymerisation of the propylene with some more polar monomer, and brings the need for two-stage polymerisation: (1) isotactic polymerisation of the propylene; (2) separate grafting of polar groups on to the backbone of the isotactic polypropylene. The difficulty will doubtless be overcome, but it looks as if it may be a serious one; a possible approach is that of block polymerisation, blocks of isotactic polypropylene alternating with blocks of polyvinyl acetate or some similar polar vinyl compound. Whatever method is adopted, either manufacturing costs, tenacity, or chemical resistance will have to pay for the improvement in dyeing affinity and moisture compatibility.

There are a very few disperse colours which will give pale to medium shades on Ulstron at 90° C. in 30 min. by orthodox dyeing methods. Dyeable Meraklon polypropylene has been reported by Montecatini's U.S. affiliate the Chemore Corp. So far there is no information respecting other changes in the properties of the fibre. This dye-receptive polypropylene was available early in 1962 in limited quantities at a premium of about 10 per cent. Although Montecatini has not revealed the nature of the chemical modification to the fibre it is perhaps relevant that two methods of improving the dyeability of polypropylene which Montecatini have patented are:

1. About 10 per cent of an epoxy resin is included in the spinning solution. It is claimed that the fibre spun from such solution will dye to deep shades that are fast to light, perspiration and washing.
2. A substituted pyridine is grafted on to the fibre to provide dyeing sites for acid dyes.

Many grafts on polypropylene have also been made by the Russians (see Further Reading).

Another method of tackling the problem is that of making dyes specifically for the purpose. To begin with, one must consider what is the new and outstanding characteristic of polypropylene which makes it more difficult than other fibres to dye. There are two such outstanding characteristics; they are:

1. The extremely high regularity, stereo-regularity in fact, of the polypropylene fibres, their immaculate perfection of molecular arrangement, that makes it very difficult for a dye molecule to get into the fibre. How can this apparently fundamental difficulty be overcome? The most likely method is by swelling the fibre.

2. The very hydrophobic nature of polypropylene, with no significant moisture regain, makes it less amenable than most to aqueous processing. The fibre and the dye should be compatible. Perhaps, therefore, dyes that have a minimum of polar solubilising groups, dyes that are very insoluble, might be the best for dyeing polypropylene. If they were dispersed (quite insoluble) in an aqueous medium, the polypropylene fibre or fabric put in such a bath might dissolve the dye in its fibre and thereby become dyed. Like calls to like. What appears to be an application of this point of view will be described.

Some water-insoluble disperse dyes which are called "plastosoluble" have been described in B.P. 1,045,431 by Aziendi Colori Nazionale Affini Acna (in the Montecatini group) as being particularly suitable for dyeing polypropylene. They have the general formula:

$$R-\left[NH-C_3N_3(X)(Y)\right]_n$$

when n is 1 or 2 and X and Y are alkoxy or alkylamino or similar groups, and R is a dye radicle free from water-solubilising groups. A suitable member has the formula:

$$C_6H_5-N{=}N-C_6H_4-NH-C_3N_3(NHC_4H_9)_2$$

If 2 gm. of such a dye is dissolved in a little acetone and this solution is dispersed in 2 l. water, then polypropylene fibre can be dyed in it using the dispersion in a dyebath. The dyebath is heated to 100° C. and 100 gm. polypropylene fabric is immersed in it and the whole kept at 90–95° C. for 30 min. The dyed fabric is removed and treated for 30 min. at 80° C. with 2 l. of a 5 gm./l. soap solution. The dried fabric shows good fastness to light and sublimation. The soaping is an integral part of the dyeing process. It was with disperse dyes that G. H. Ellis solved the problem of dyeing Celanese. Possibly Montecatini will succeed in dyeing their polypropylene by a very similar device making sure that the dye is just about as water-resistant as the fibre is.

Uses

For industrial uses, dyeing of a fibre is often unnecessary, and it may be that polypropylene fibres will find more use in industry than in the home for some time to come; the fibre is already being made into ropes and for this use its great strength, resistance to abrasion and its negligible affinity for water should stand it in good stead; ordinary Manila ropes swell and get very heavy and unwieldy when used in water; polypropylene ropes will not.

For a given breaking load, polypropylene and Terylene ropes are very similar in diameter, but the polypropylene rope will be perhaps 20 per cent lighter and will cost only half as much. Ropes that are lost overboard will float if made of polypropylene and there is a chance that they may be retrieved. Hitherto the fishermen have looked a bit askance at synthetic fibre ropes that cost three or four times as much as Manila, even if they did last ten times as long; too often ropes are lost so that their potential longevity is only of academic interest. Perhaps ropes that will float will have a real life that will be longer. Furthermore, the chance that a rope will foul the propeller of a boat is much less if the rope is a floater and not a sinker. Baler twine of great strength and durability is another new product that has been made from polypropylene; this product is made by Eastman Chemical Co. The high price of jute (it has risen by 50 per cent since 1962) is allowing polypropylene to nibble at its market. This nibbling may grow into something more serious.

Fishnets are already made of Ulstron. Lightness, negligible moisture regain, high strength, and probably a fairly low cost are favourable. The fibre is liable to a certain amount of creep under heavy loads, and this can affect mesh size. Large scale fishing trials are in progress.

One fantastic use that has been investigated for polypropylene fibre is as artificial seaweed. The idea apparently came from Denmark, and I.C.I. have taken out a licence and have carried out large-scale tests at Bournemouth, on the North East coast and at Dorset. The basic idea is that the fibre is made into fronds which are planted offshore to prevent beach erosion. The "seaweed" reduces the wave motion and allows sand which is carried in from offshore to drop to the beach. More sand is deposited on to the beach as a result of the presence of the seaweed; this is considered useful. It might be a more attractive proposition to import the giant Pacific kelp *macrocystis* and let it grow off our shores. In fact, this was suggested a long time ago so that the kelp could provide an abundant source of raw material for alginate extraction. The idea was turned down on the grounds that the kelp might grow too rampantly and choke our harbours. It might be worth reconsideration to-day. The artificial seaweed made by I.C.I. to-day is in the form of 8-ft.-high tufts, suitably held in place on the sea-bed. But what about the swimmers? It is bad enough to swim into a tangle of seaweed, but a tangle of high-tenacity polypropylene might be a death-trap.

Carpets are another outlet for which polypropylene is being got ready; it will be used as the pile. One of the factors that is important in the behaviour of a carpet is for how long a time the height of pile is maintained; it should not be quickly flattened nor worn away. Nylon pile carpets are very popular in America and have good pile length retention properties and it is of interest that a carpet pile made of 50/50 wool/polypropylene behaves nearly as well, as the following figures illustrate.

Pile fibre.	Percentage of pile height retained after		
	2,000 treads.	5,000 treads.	10,000 treads.
Nylon	82	76	73
50/50 wool/polypropylene	74	72	70

It has to be added that in America where the room atmosphere is often very dry, electrification is rather a nuisance. When one has walked about a room on a nylon carpet, one's shoes are covered with a fluff of fibres that are retained electrostatically; polypropylene would probably do the same thing even more, although if used in blends with wool the problem might be somewhat reduced.

Other uses envisaged for polypropylene are for pile and loosely constructed fabrics; velvets, voiles and so on: the high strength of polypropylene will enable it to be used in light fabric constructions.

The freedom of polypropylene fabrics from the pilling scourge has already been noted.

There is no doubt that the particular suitability of polypropylene as a moulding plastic has prepared the way for its use as a fibre. By virtue of its high fluidity it will mould at low pressures by extrusion, pressing, injection or compression and the moulded products have fine detail and a high finish. The higher melting point of polypropylene, 165° C., compared with 130° C. for Ziegler polyethylene is advantageous and permits sterilisation. In the U.S.A. alone production capacity (mainly for use as plastic) was 10,000 tons in 1958, 60,000 tons in 1960, and was expected to reach 250,000 tons by the end of 1962. In the U.K. polypropylene is made at Wilton by I.C.I., present (1961) capacity is 11,000 tons per year and it will probably be trebled shortly. Courtaulds make it as sheeting and call it " Propylex ". At Wilton, pilot-plant quantities of fibre are being produced, and it should soon go into big production. Montecatini, by far the biggest Italian chemical company, makes the fibre at Milan and calls it Meraklon; 1960 production was 5,000 tons, output for 1961 being 10,000 tons; in America under the name Novamont they plan to build a fibre plant at Neal, West Virginia, which should produce 12,000 tons of fibre a year starting in 1962 or 1963. North American Rayon Corporation (Beaunit) are already producing fibre; in 1962 staple was only 80 cents per lb. Japan will make 10,000 tons in 1963. There are great years ahead for polypropylene as a plastic and the immense capital investments now being made will find their reward. But as a fibre? For industrial purposes, yes. For tyres, the biggest single industrial use? Very doubtful with that low softening temperature, but undergoing intensive examination. For the home? Ironing temperature and dyeing are against it; possibly spun-dyed for carpets, much less likely for apparel.

Split Film. Fibre can also be made by spinning film and then splitting it, the advantage of this method being low cost. The Chevron Chemical Co. are building a factory at Dayton, Tenn., to exploit this process.

Production. In 1968–69 output of polypropylene fibre exceeds 500 million lb. per annum. The main producing countries are:

U.S.A. . . .	300 million lb.
U.K. . . .	120 million lb.
Japan . . .	50 million lb.
Italy . . .	50 million lb.

Trilobal Cross-section Fibre

Ordinarily, polypropylene fibres have cross-sections that are almost perfectly round. Melt-spun through a circular jet, just as nylon and Terylene are, polypropylene necessarily has, as they do, a round cross-section. Experimentally, polypropylene has been spun with a trilobal cross-section as described by du Pont in U.S. Patent 2,939,201 (1960). The molten polypropylene at 200° C. was extruded through spinnerets in which each hole was an equilateral triangle 9 mils long on each side and at each vertex of the triangle a square slot was cut running 3 mils in length from the vertex and being 3 mils wide. The yarn was wound up at 65 yards per minute and was hot-drawn at 124° C. to six times its length over a hot pin. Final tenacity was 4·6 grams per denier and elongation at break was 76 per cent. " The yarn of trilobal polypropylene filaments exhibited a unique luster highlight to the unaided eye, compared to the dull flat appearance of the round filament yarn." It may not be long before all the melt spinners abandon round-section fibres; they have the lowest area/volume ratio.

Crystal Reality

As a rule, the crystals which compose a part (often a large part) of synthetic fibres are too small to be seen with a microscope; their presence is inferred from the behaviour of the fibre towards X-rays. It is doubtful too if discrete crystals usually form, more likely there are parts of the molecular assembly that are so orderly that they warrant the description crystalline.

But with polyethylene and isotactic polypropylene discrete crystals can actually be observed under the microscope.

Marker *et al.* have studied the crystallisation of polypropylene (m.p. 171° C.) at different temperatures. The progress of crystallisation was followed by microscopy: using crossed polarisers the amorphous layer of the polypropylene appeared uniformly dark, and crystals appeared first as small spots of light and gradually grew in circular patterns, forming beautiful spherulites nearly 1 mm. in diameter. Apparently the crystallisation is started by traces of foreign matter. All of the crystals start to grow within about 2 min. of solidification from the melt; after about 2 min. no new ones start but those already started continue to grow (the temperature being kept at about 130° C.) until after 8 hours they begin to touch each other.

Polypropylenes from different manufacturers behaved differently;

the higher the molecular weight of an isotactic polymer the slower was the crystallisation and secondly the presence of some atactic (that is sterically irregular) polypropylene of high molecular weight in the nominally isotactic polymer slowed down crystallisation of the polymer.

The Vicat Softening Point

The Vicat Softening Point is often encountered in discussions on polyethylene and similar materials. It relates primarily to plastics and particularly to polyolefines. As these are fibre-forming materials the concept is useful to us, even although the actual test has to be made on a massive specimen and cannot be made on a fibre.

The Vicat Softening point is defined (American Society for Testing Materials, *i.e.*, A.S.T.M. Designation D1525/58T) as the temperature at which a flat-ended needle of 1 sq. mm. cross-section will penetrate a thermoplastic specimen to a depth of 1 mm. under a specified load (usually 1 kg. or 5 kg.).

Figures that come from different sources show some spread, but the following are typical.

	Vicat softening point (° C.) with	
	1 kg. load.	5 kg. load.
Old (low density) polyethylene	85–91	—
Ziegler polyethylene	121–124	82–88
Isotactic polypropylene	130–145	85–105

It is of interest to us that the softening point (Vicat) of polypropylene varies with the degree of isotacticity, the greater the tacticity the higher the softening point as the following figures illustrate.

Isotactic index (per cent).	Vicat softening point at 5 kg. load (° C.).
77	72
81	74
85	78
90	86
95	98
97	106

POLYSTYRENE

Well known and extremely successful as a plastic material, polystyrene is a doubtful qualifier for inclusion in a book on fibres. It has been made as a flat monofilament, particularly by National Plastic Products Co. Inc., of Odenton, Maryland; its application was for use in knitted pot cleaners but it has been replaced by polypropylene.

The chemical structure of styrene or phenylethylene

$$C_6H_5CH{:}CH_2$$

is not very encouraging for conversion into a fibre-forming polymer, because most of the substance (the phenyl group) will constitute the side-chain and will not contribute to strength. Nevertheless, isotactic polystyrenes that have been made have had useful properties including a sharp melting point (220° C.) and good chemical resistance. Furthermore, the phenyl group offers a heaven-sent opportunity to organic chemists to introduce almost any desired functional group to give dye receptivity.

Perhaps some of the intensive research currently carried out on polypropylene may awaken interest in fibres from isotactic polystyrene, fibres that melt rather higher than nylon 6. Perhaps more likely, styrene will play the humbler part of co-monomer with propylene to make polypropylene dye receptive.

In 1965 Toyo Rayon Co. started to make polystyrene fibre at Hiroshima and contemplate a production of 30 m. lb. per annum.

FURTHER READING

A. B. Thompson, " Fibres from Polypropylene ", *J. Roy. Inst. Chem.*, **85**, 293–300 (1961).

Anon., " Grafting Cures Polymer Ills ", *Chem. & Eng. News*, **36**, No. 32, 51 (1958).

L. Marker, *et al.*, " Kinetics of Crystallization of Isotactic Polypropylene between 120 and 160° C. ", *J. Polymer Sci.*, **38**, 33–43 (1959).

W. C. Teach and G. C. Kiessling, " Polystyrene " (Reinhold, New York, 1960).

Jung-Jui Wu, A. A. Konkin and Z. A. Rogovin, " Structures and Properties of Modified Polypropylene Fibres ", *Khim Volokna*, (6), 14–18 (1964). Reviewed by R. W. Moncrieff in *International Dyer*, **133**, 546–8, 550 (1965).

P. N. Hartley, " Coloration of Polypropylene (fibre contains nickel to make it dyeable) ", *International Dyer*, **134**, 541–3, (1965).

CHAPTER 38

TEFLON

POLYTETRAFLUOROETHYLENE was another of the fibre-forming materials to be made and used as a plastic before it was developed as a fibre. The plastic material is made and marketed as a granular white powder under the name " Fluon " by I.C.I. Ltd., but is not at present made into a fibre in the U.K. In America, where the material was developed, it is used as a plastic, and is also spun into a fibre under the name Teflon. Chemically it is a development of the fluorine–chlorine–carbon compounds which have been used as refrigerant liquids under the names " Freons " and " Arctons ", but it was not until 1941 that a polymeric form of tetrafluoroethylene was discovered. In 1954 the fibre was introduced by du Pont.

Chemical Nature

Tetrafluoroethylene has the composition $CF_2{=}CF_2$, and its polymer, which is the substance from which the fibre is prepared, is:

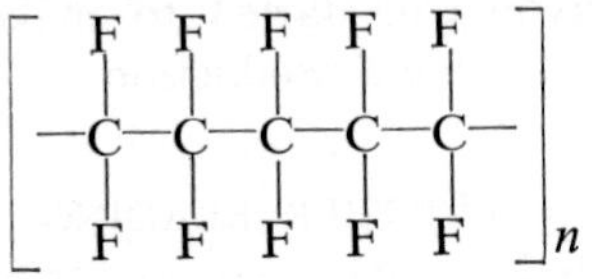

$$\left[-\underset{\substack{|\\ F}}{\overset{\substack{F\\ |}}{C}}-\underset{\substack{|\\ F}}{\overset{\substack{F\\ |}}{C}}-\underset{\substack{|\\ F}}{\overset{\substack{F\\ |}}{C}}-\underset{\substack{|\\ F}}{\overset{\substack{F\\ |}}{C}}-\underset{\substack{|\\ F}}{\overset{\substack{F\\ |}}{C}}- \right]_n$$

Neither the length of the chain nor the nature of the end groups on it is known. The molecular structure is significant; the fluorine atoms, themselves bound to carbon atoms, are unreactive and the voluminous fluorine atoms pack so densely round the chain of carbon atoms that the latter are quite inaccessible, and there is no common chemical whose molecules can pass through the defences of the fluorine atoms to attack the carbon chain, so that the chemical resistance of Teflon is extremely high. Secondly, the uniformity and symmetry of the molecular Teflon chains permit them to pack together in a highly organised way, so that the van der Waals' forces between the atoms of adjacent chains add up to produce substantial intermolecular attraction, and it is the sum of these forces, weak but very numerous, that hold together the molecules of Teflon so that it has a high melting point. The molecules are electrically neutral, and there are no strong polar forces such as are so important in holding together the molecules of nylon, Terylene and cellulosic fibres; the van der Waals' forces are relatively unimportant in such fibres which carry polar groups; in Teflon they are extremely important and it is only because the regularity of the Teflon molecules permits extraordinarily close packing that they are so strong in aggregate. The close packing is also responsible for the high specific gravity, 2·2, of Teflon.

Manufacture

The basic raw materials are fluorspar, CaF_2, from which anhydrous hydrofluoric acid is prepared, and chloroform. These two are reacted together to give difluoromonochloromethane, a gas which boils at −41° C.

$$CHCl_3 + 2HF \longrightarrow CHClF_2 + 2HCl$$

This gas is pyrolysed at 600–800° C. to give tetrafluoroethylene:

$$2CHClF_2 \longrightarrow C_2F_4 + 2HCl$$

The tetrafluoroethylene is scrubbed and fractionated to remove hydrochloric acid and various fluorocarbons which are formed during the pyrolysis. The purified tetrafluoroethylene is then polymerised in stainless-steel autoclaves in the presence of an aqueous solution of ammonium persulphate or other peroxy catalyst; the reaction is rapid and exothermic, and conditions have to be carefully controlled. The polytetrafluoroethylene produced is a granular white solid which is washed and dried hot.

Spinning. Polytetrafluoroethylene presented a great problem to the spinner. It is quite insoluble, so it could not be spun from a solvent, and it only melts at a very high temperature, 400° C., to a melt which is too viscous to spin and which decomposes fairly quickly at that temperature. This apparently insurmountable difficulty has been overcome to the great credit of the du Pont technical staff. The underlying principles of the spinning method appear to be two:

1. The formation of the polymer in such a form that it has a high length/width ratio so that these polymer particles will hold together in the form of a weak yarn. The difficulty to be overcome is to obtain a filament with sufficient strength to be handled before sintering; the necessary strength is at least 25 lb./in.^{2} and preferably more. The polymer is prepared under conditions such that the polymer particles have at least one dimension less than 0·1μ with a width of less than 0·07μ and a ratio of length/width of 5 or preferably higher. The dispersion should contain at least 5 per cent by weight of the polymer. If such a dispersion is allowed to settle, the ribbon-like particles settle preferentially, and if they are removed, the particles can be aligned more or less parallel by passing the dis-

persion through a capillary tube. Essentially, sub-microscopic particles of the polymer must have a fibrous form. These dimensions are ascertained by investigation with the electron-microscope; some of the particles can be 6μ long and sometimes the length/breadth ratio will equal 500. As can be seen from the method below, a colloidal dispersion can be made in large quantities with 15 per cent of polymer, of which some 30 per cent is in ribbon-like form.

2. The very rapid melting of the very tiny particles of polymer is so quick that there is no time for decomposition to take place. Chips or lumps melt so slowly that decomposition occurs before melting is complete. This incipient melting is done at about 385° C.

In practice, Teflon is spun from an aqueous colloidal dispersion, and as it is in this same medium that the polymerisation of the tetrafluoroethylene has taken place, the two processes, polymerisation and spinning, will be described together. Teflon spinning is sometimes described as emulsion spinning. This is not really correct, because the Teflon particles themselves are solid, not liquid, and in the form of a colloidal dispersion. The preparation of the polymer dispersion may be carried out as follows:

Into an autoclave fitted with a stirrer there is charged

2,000 parts distilled water—the dispersion medium

50 parts ammonium eicosafluoroundecanoate—the dispersing agent (eicosa means 20, the number of fluorine atoms)

1 part ammonium persulphate—the catalyst for the polymerisation.

100 parts paraffin wax as a " stabiliser ".

This mix is heated to 79° C. and tetrafluoroethylene gas charged in to 100 lb./in.^{2} pressure. The temperature is maintained for 4 hours during which time the tetrafluoroethylene polymerises.

$$n\ CF_2{=}CF_2 \longrightarrow -CF_2-CF_2-CF_2-CF_2-CF_2-CF_2-$$

Fluoroethylene. Polyfluoroethylene.

The mix now consists of a dispersion containing 15 per cent polymer, of which about 30 per cent of the particles are ribbon-like in form, some of them 5μ long but less than 0·07μ wide. It is interesting to observe here how the formation of a linear polymer induces a *long* shape, *i.e.*, a high length/breadth ratio. The dispersion is extruded through an orifice of 0·5 mm. diameter (very big!) into a 5 per cent solution of hydrochloric acid in water at 25° C. A filament is precipitated due to the breakdown of the colloidal dispersion by the acid, and the filament is dried. Considering that the particles of polytetrafluoroethylene are still discrete and disconnected it has a remarkably high tenacity of 1,000 lb./in.^{2} (0·04 gram per denier). The filament is then sintered on a metal surface, such as a roller, at 385° C. for a few seconds, rapidly quenched by passing into water at 25° C., and the resulting homogeneous filament is cold drawn to four times its initial length to give a strong clear filament with a tensile strength of 14,000 lb./in.^{2} (0·5 gram per denier) and 80 per cent elongation. The fibre which is marketed has a higher tenacity than this, actually 1·5 grams per denier.

The fundamental difficulty of the insolubility and relative infusibility of the polymer is overcome by polymerising under conditions where many of the particles are themselves fibre-shaped, although sub-microscopic, and by lining these up so that when the dispersion is broken the fibrous particles hold together firmly enough to allow the filament to be handled. Then the filament is heated high enough to start the fibrous particles melting so that they stick together to form a continuous and strong film. Nature has helped by precipitating the polymer particles in fibrous form, but the chemists and spinners had to observe that she did this, and it calls for a great deal of acuity to observe something that happens quite unexpectedly in the sub-microscopic region, and furthermore, they had to ascertain the quite narrow range of conditions under which she did it. Familiarity will blunt appreciation and different ways of reaching the same end have already been indicated, but the first spinning of polytetrafluoroethylene should always rank as a brilliant achievement.

At present Teflon is supplied in 400 denier 60 fils, and also as 4½ in. staple fibre of the same filament denier.

Properties

The properties of Teflon are in many respects unique amongst fibres.

Physical. Tenacity is moderate at about 1·5 grams per denier (50,000 lb./in.^{2}); elongation is medium at about 13 per cent. Water absorption is zero, so that " wet " characteristics are the same as dry. Pliability is good, and loop and knot tensile strengths are about 75 per cent of the straight strength. The cross-section is round.

Colour of the yarn is tan, but it can be bleached to white, without loss of strength, by treatment in boiling conc. sulphuric acid to which nitric acid is added dropwise. Frictional properties are uniquely low, the dynamic coefficient being 0·28 and the static 0·20; this gives the fibre a slick greasy handle quite unsuitable for apparel, but may lend itself to special uses, as it certainly has done in the plastic form. The low coefficients of friction are probably associated with the lack of polar groups in the molecules and with the extreme inertness of the fibre; the adhesive properties of the fibre are correspondingly poor.

Heat resistance is the best of all fibres; the fibre still has a measurable strength of 0·1 gram per denier at 310° C. (called the zero strength temperature) and gels at 327° C. and decomposes at 405° C.; it can be used at temperatures up to about 290° C., and at this temperature the weight loss is only 0·0002 per cent. At this temperature cellulosic fabrics would be charred and nylon and Terylene would be pools of molten polymer. Heat transfer is low and electrical properties are good—fabric made from the fibre does not "track" if accidentally arced. Teflon is non-flam.

Chemical. Chemical stability is quite unique; the fibre can be boiled in aqua regia or mineral acids or concentrated caustic soda solution without decomposition; in one test, fabric was treated in concentrated sulphuric acid for a day at 290° C., then in concentrated nitric acid for a day at 100° C., then in 50 per cent caustic soda for a day at 100° C., and finally in concentrated aqua regia for a day at 100° C.; its strength was unimpaired; it is resistant, too, to all the usual strong oxidising agents. There are no known solvents, except certain perfluorinated compounds at temperatures over 300° C. It is attacked and decomposed by fluorine gas at high temperature and pressure and by chlorine trifluoride. In effect it is chemically resistant and heat resistant up to temperatures higher than any other synthetic fibre will withstand at all.

Because of its resistance to water it is almost impossible to dye, although it can be stained with acetate dyestuffs.

Uses

The outstanding properties of PTFE (polytetrafluoroethylene) fibre are excellent temperature, chemical and solvent resistance, good mechanical and excellent electrical characteristics, very low frictional coefficients with slippery handle, no moisture regain and almost no dyeability. These at once rule out ordinary apparel uses, but look extremely promising for industrial uses. Add that the fibre is £5 or $12.00 per lb. (it was at one time $14.00 per lb.) and all but very special industrial uses will be ruled out, too, for the present at least. Such are:

Braided packing, which is used for chemical pump shafts; not only its chemical resistance but also its very low frictional coefficients should make it suitable. A braided packing material was made and was tested on a pump handling 102 per cent fuming nitric acid; after seven months it is still in good condition, whereas the best previous packing lasted two to three weeks. Tested on a centrifugal pump handling 40–78 per cent caustic soda at 165° C., it lasted eleven days, whereas the best previous was two to four days. Other tests gave striking results, and it is evident that Teflon will assist the chemical industry to solve troublesome sealing problems. The performance of the Teflon braid is improved if the braid is lubricated with a Teflon dispersion.

Filtration fabrics, for the liquid filtration of corrosive mixtures, behaved very satisfactorily and their freedom from any tendency to adhere (lack of adhesion) to the filter cake made them easy to clean. Teflon fabric has also been used successfully for the filtration of particles from hot corrosive gases.

Other applications have been for gasketing fabrics, laundry fabrics for press pad-covers and roll covers, electrical tapes and wire wraps for corrosive service, corrosion resistant cordages and anti-stick bandages. Some attention has been paid to a staple yarn to overcome in part at least the slick greasy handle which is disliked when Teflon is used for protective clothing; the usual process was almost impossible because card webs and slivers fell apart owing to their lack of cohesion. More successful has been a direct spinning system (in which the carding is eliminated) by randomly breaking the filament in a heavy continuous filament, low twist, yarn immediately drawing down to the required size and then twisting the fibres into a spun staple yarn; the resulting fabrics feel much less waxy than those made from filament Teflon. "Taslan" textured yarn of Teflon has also produced fabrics with a dry spun type handle without resorting to conventional spinning systems.

Teflon is non-toxic, but when heated above 200° C. toxic gases are given off, and proper care (good ventilation and no smoking) should be taken.

Japan. Early in 1965 it was announced that the Toyo Rayon Co. were producing a PTFE fibre in their Ehime factory at the rate of 40,000 lb. per annum and were planning to make 1 m. lb. per annum.

FURTHER READING

J. T. Rivers and R. L. Franklin, *Text. Res. J.*, **26**, 805–811 (1956).
K. L. Berry, U.S. Pat. 2,559,750 (1951); assigned to du Pont.
E. I. du Pont de Nemours & Co., Brit. Pat. 689,801 (1953); also of interest Brit. Pat. 686,438.

CHAPTER 39

GLASS

THE idea of making yarns and fabrics from glass is centuries old. Glass has unique properties, and it was thought that if they could be conferred on fabric a most attractive product would result. In 1841 a machine for spinning glass fibres was demonstrated at the British Association meeting in Manchester. Nothing came immediately of the project, but it is of additional interest because the machine incorporated a primitive form of spinneret which was afterwards used so effectively to spin rayon. Until comparatively recently no real success was achieved in making fabrics from glass.

In the 1914–18 War the Germans experienced a shortage of asbestos, and attempted to use glass in place of it. They made a primitive spinning machine in which a filament from a heat-softened glass rod was touched on to a rotating bicycle wheel and was wound on to it, until a skein of fibres as thick as a tyre had collected. The first apparatus was actually driven by a man pedalling the bicycle, and although some obvious mechanical improvements were made, notably the substitution of a mechanical drive for the man, no real textile developments ensued for some time.

In 1931 two American firms—the Owens Illinois Glass Co. and the Corning Glass Works—developed a method of spinning glass through fine orifices from the melt. In 1938 these two firms amalgamated and worked as the Owens-Corning Fiberglas Corporation. In this country, glass fibres, yarns and fabrics are produced by Fibreglass Ltd. at St. Helens and at Glasgow.

Very extensive uses have been found for the material. It is still unsuitable for apparel, but there are many purposes for which it has proved satisfactory. In the U.S. the production of textile glass fibre has risen from 60 million lb. in 1954 to 300 million lb. in 1967. This is a high rate of growth and is indicative of the true worth of glass fibres, not so much for apparel use, but certainly for their industrial applications.

Chemical Nature

The primary ingredients of glass are silica sand and limestone, other constituents, such as aluminium hydroxide, soda ash and borax, are often included. The quantities of ingredients taken vary according to the properties required in the glass; a different mixture will be used for glass required for electrical resistance than for one required for freedom from chemical attack. A typical glass may be made by melting sand, soda ash (sodium carbonate) and limestone together in a furnace. No definite chemical compounds are formed, and no crystallisation takes place. The atoms arrange themselves roughly in some sort of pattern, but there are none of the exact repeats which are typical of the crystalline state. Glass can be thought of as a supercooled liquid—a liquid so thick that it will not flow perceptibly. Glass fibres and fabrics consist of glass, and only of glass.

Manufacture

Continuous Filament. Batches of selected silica sand, limestone, soda ash and borax (or other ingredients) are prepared and fused in an electric furnace. The glass that results from the melting together of these materials is formed into marbles about $\frac{5}{8}$ in. diameter, weighing about 50 to the pound. These marbles or *cullet* are inspected to weed out any that contain imperfections. Those marbles that are passed as satisfactory are then re-melted in an electric platinum furnace and, when at the correct temperature, the glass is run out of about 100 orifices at the bottom of the furnaces (Fig. 193). The filaments are drawn together, lubricated and wound as a strand. One marble gives approximately 100 miles of filament. The winding speed is extremely high—about 2,000 metres per minute. After winding, the yarns can be put through twisting and other normal textile operations.

[*Owens-Corning Fiberglas Corp.*]

FIG. 193.—Continuous filament Fiberglas textile fibres being drawn from a small electric furnace in which glass marbles are melted.

Staple Fibre Process. Staple fibre yarns are used largely for chemical filterpads. For this reason they are made from a quality of glass specially mixed to resist chemical action. Marbles are heated in a furnace as before, and filaments emerge through fine holes at the bottom; but shortly after emergence the filaments are hit by a steam-jet, which breaks them into staple varying from 6 to 15 in. and draws them down on to a drum. They pass round this drum, and are guided on to cardboard tubes, on which they are collected as sliver. The sliver can be drawn and twisted into yarns.

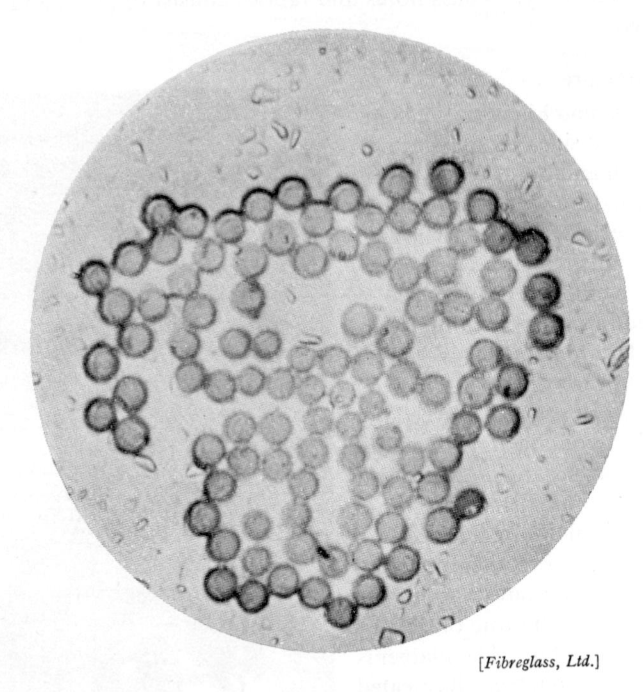

[*Fibreglass, Ltd.*]

FIG. 194.—Photomicrograph of cross-section of glass filaments (×1,000). Note the round section.

The Bat Wool Process. Molten glass is run from the very fine holes at the bottom of the furnace as before, and the streams of molten glass are attenuated into fine fibres by jets of high-pressure steam which have a high velocity. The fine fibres are allowed to form a fleecy, resilient mat, about 10 in. thick on a conveyor belt, along which the mat moves to be compressed to the required thickness and then cut to the required dimensions. The bat wool is used for insulation rather than for textile purposes.

Properties

The outstanding properties of glass fibres are their high strength (usually 6 gm./den.), resistance to heat, non-flammability, mildew resistance and chemical resistance. They lack resiliency and stretch; the finer the filaments—they are usually about 0·008 mm. diameter—the more pliable they are. Their elongation at break is only about 2 per cent, which is much too low for many textile uses. Glass

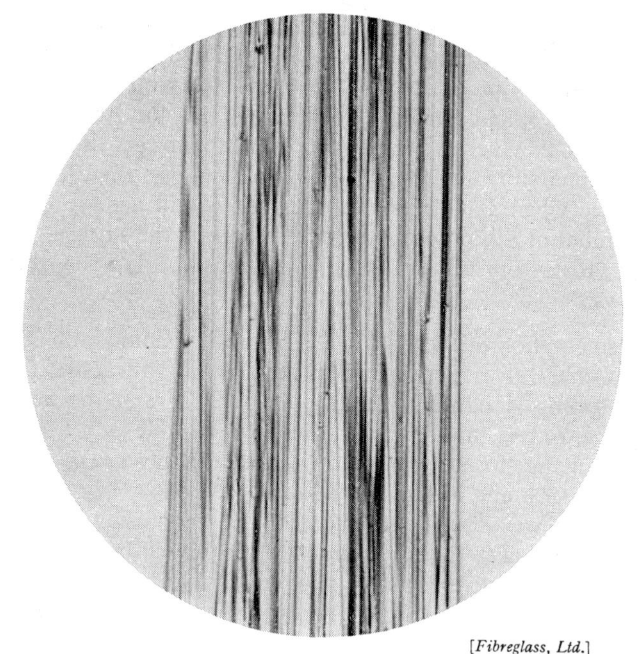

[*Fibreglass, Ltd.*]

FIG. 195.—Longitudinal view of a glass multi-filament yarn (×100).

fibres are heavy, with a specific gravity of about 2·5–2·7, which is similar to that of aluminium. The wool which is made from glass fibres, containing, as it does, large volumes of air, has a specific gravity of only 0·025. Glass will not soften below about 700° C. It has a negligible moisture content, and can be used under very wet conditions. The cross-section of glass filaments is shown in Fig. 194.

The high strength of glass is noteworthy. Fibres, as a rule, increase in tenacity as they are oriented, as the degree of crystallisation increases. Why, then, should glass, which is not at all crystalline, have a high tenacity? All other fibres are partly crystalline, and it is thought that when rupture of a fibre occurs, it is at an imperfection in the crystal structure, where adjacent crystals are not conveniently arranged. Orientation reduces the number of these imperfections, and so increases the strength. But glass, which has no crystalline part at all, but is completely amorphous, is free from the crystalline imperfections, and therefore very strong.

Dyeing

Glass lacks affinity for dyes. Attempts have been made to increase its affinity by etching the surface of the fibres, but without a great deal of success. Nevertheless, the writer once made net-bags of glass in which to bundle hanks of wool for dyeing in a pressure machine. It was anticipated that the glass bags would be unaffected by the wool dye, and so capable of repeated use, but it was found in practice that they became very badly stained—in fact dyed a pastel shade. Apparently the roughening or etching of the surface which takes place when glass materials are used is sufficient to enable a large number of dye particles to be adsorbed on the surface. There are two satisfactory ways in which glass can be coloured; they are as follows:

1. Pigmentation of the glass melt, before it is spun, with ceramic colours. This gives good fastness, but is not attractive from a manufacturer's standpoint, because if stocks are to be kept not unduly big, only a limited range of colours can be spun. Furthermore, the change from one colour to another involves very thorough and tedious cleaning of tanks, pipes, *etc.*

2. Coronising. The development of the Coronising process has lifted the role of glass fabrics from that of a safety (fireproof) measure to that of a decorative fabric of good performance. It gives a soft handle, dimensional stability and good coloration. Coronising consists of three stages:

(*a*) The glass fabric as received is very subject to slippage of one thread over another. It is therefore padded through a colloidal silica dispersion, such as Syton. This application of silica stops weave slippage, and also the padding ensures a flat uncreased fabric for the next step.

(*b*) The fabric, having been impregnated with silica, is heat-set. The fabric is passed through a hot oven and the fibre is crimped—this gives a soft fabric which is very crease-resistant. Most of the heat-setting of fabric is done for weavers by the manufacturers of the glass fibres. The heat treatments burn away any size that may be on the fabric, but the silica remains on the fabric, being more permanently attached than before the heat treatment.

(*c*) After leaving the heat-setting oven the fabric passes into an aqueous bath containing pigment for coloration and a resin which is usually a latex which binds the pigment to the fibre. One formula used is:

Pigment padding colours (*cf.* p. 794)	5 parts
Butadiene–acrylonitrile latex	10 parts
Water	85 parts

The fabric is then dried at 160° C. At this stage the binding agent—which is the latex—does not have good wash fastness. It must therefore be given another padding, this time in a dilute stearato chromic chloride aqueous bath, and a final drying at 120° C. The stearato chromic chloride is obtainable as a dispersion in *iso*propyl alcohol under the name Quilon.

Printing

Glass fabric can be screen-printed by a process very similar to Coronising. The first two steps, those of silica treatment and of heat-setting at 650° C. are as already described. A light coating of a cationic softening agent is then applied to serve as a lubricant and protective agent for the fibres. The next step is the screen printing. Here another latex preparation is used, thickened with sodium alginate to the proper consistency of a printing paste. To this the paste pigment is added; most of the water-dispersible pigments on the market are satisfactory. The pigment paste is applied through the screen by a sharp paddle. After printing, a five-minute cure at 160° C. is given, followed by padding in the stearato chromic chloride bath, and the fabric is dried for fifteen minutes at 120° C. Finally it is given a scour in a warm soap solution and is dried. Development work on the application of the process to roller printing is in hand. Roller printing is more useful when large quantities of material have to be printed to one design; screen printing is commonly used on fabrics other than glass, when fairly small lots to a design and colour are needed.

Printed glass fabrics are crease-resistant, water-repellent, of good handle and drape and of course they will not burn; they do not need to be ironed.

Another method that has been used to combine the colourings of normal textile materials with the peculiar advantages of glass, has been to laminate drapery fabrics in modern designs and colours on to a glass fibre-plastic base. Lamp-shades made in this way are washable and fire-resistant.

Uses

Wedding dresses have been made from glass fibres, and look very lovely, but the time has not yet come when glass fibres are suitable for use as apparel. They lack resistance to abrasion, for glass cuts glass, and as glass fabrics are flexed, the filaments roughen and break and the fabric becomes hairy; the resistance to abrasion is poor. Secondly, they lack moisture absorption and would be cold to the skin. Thirdly, they lack extensibility, which is an essential feature of fibres for apparel. It has been reported that there are no toxic hazards associated with the use of glass fibres.

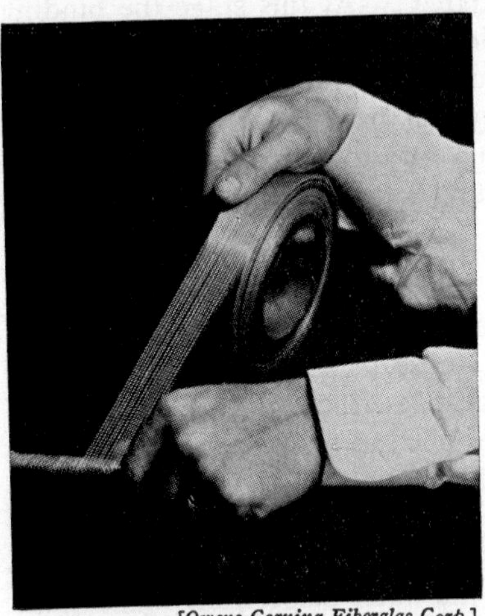

[*Owens-Corning Fiberglas Corp.*]

FIG. 196.—Fiberglas-reinforced acetate film barrier tape for electric cable construction.

Advantages are good insulation, high strength, resistance to moisture, and, because the film is very thin, low cost.

It will be appreciated that glass fibres are complementary to natural fibres, and not substitutes for them. Uses to which they have been put are as follows:

Fireproof fabrics for lampshades, awnings, screens, curtains and furniture covers, tablecloths, *etc.* Even if a lighted cigarette is dropped on one of these, no harm results. Their use will apply chiefly to continuous filament yarns and fabrics made from them. For the same reason, lifeboats of oil tankers are fitted with a glass canopy over them as protection against spouts of burning oil which rain down in the vicinity of a burning tanker.

Staple fibre glass, when woven into cloth, is largely used for filtration purposes. Most chemicals are safely manipulated in glass vessels, and all of these can safely be filtered through glass fabric. Lamp-wicks and ropes, unaffected by water, are also made from staple fibre glass.

Tapes and fabrics, braided tubular sleevings, and yarn itself are used for insulation purposes. Yarn is used for wrapping bare copper wire, and as braids for high tension cables. Glass fibre yarns are used to give strength to plastic insulating film (Fig. 196). When motors are overloaded, the insulation, if made of rubber, burns, but glass insulation gives no trouble in this respect. Similarly, motors

[*Owens-Corning Fiberglas Corp.*]

FIG. 197.—Coils, made with Fiberglas-insulated wire, being lashed with Fiberglas cord into stators of $\frac{1}{27}$ h.p. totally enclosed spinner bucket (Topham Box) motors.

Resistant to acid, moisture and overload conditions, so that there are fewer burn-outs and less interruption of production of rayon.

insulated with glass fibres are particularly suitable for work in very humid plants and in plants where chemical fumes exist (Fig. 197). Because glass provides a low-weight insulation, it has enabled such appliances as vacuum cleaners to be constructed lighter in weight.

For heat insulation, glass wool is ideal, *e.g.*, Cosywrap in walls and roofs of houses; it saves fuel costs. It can be used to lag hot pipes and cylinders. It has also found use for sound insulation.

Some new uses to which glass fibres have been put are as follows:

Fire Blanket. A blanket made of Fiberglas cloth coated with a synthetic rubber is used in industrial plants to smother fires.

Crash Bags. Bags made of Fiberglas cloth impregnated with a synthetic rubber and with Fiberglas interlining have been used by the New York Federal Reserve bank for the air transport of documents; they are designed to withstand the shock of an air crash and fire.

Protection against Radiation. A gown made of a lead-containing glass cloth is used by radiologists as protection against X-ray and β-radiation.

Packing Material for Stills. Glass fibres have been used as packing material in the reflux columns employed for the distillation of industrial alcohol.

Reinforcement of Plastics. The biggest single use for glass fibres is in reinforced plastics, mainly in flat and corrugated panelling. Applications are for boat hulls (great impact strength), sports car bodies, T.V. cabinets, pipes, tanks, aircraft parts, *etc.*

Paper Reinforcement. Glass fibre has been used as a scrim base for paper mailbags intended for one trip only.

Other Inorganic Fibres

Rocketry has demanded fibrous materials with exceedingly good resistance to high temperatures and with excellent insulating powers. There are several such inorganic materials; they are not really textile fibres, but they are fibrous. Some of the most important are:

Sil-Temp. This is a pure silica (99 per cent SiO_2) fabric made by Haveg Industries Inc. of Wilmington, Delaware. It is made by treating glass fabric (electrical grade) with acids (mainly HCl) to leach the cationic components out of the glass. What is left is practically pure silica; it is washed and then baked at 700° C. for a few minutes to shrink and finish it. Melting point is about 1,600° C. It is used as a reinforcement for plastics for missile nose cones; also for jet engine insulation blankets and for filtering dross or slag from molten metals. Price about £1–£2 per lb.

Refrasil. A silica fibre also made by the acid leaching of glass.

Fibrous Potassium Titanate. Made by du Pont. It has the elementary composition $K_2O.6TiO_2$. It is a fibrous material (fibre dimensions 1 mm. long $\times$ 0·001 mm. diameter) used for insulation at temperatures up to 1,200° C. Used also for gaskets, filtration. It is now in the development stage, but is available in lump form at about £5 per lb.

Fiberfrax. An aluminium silicate fibrous material made by the Carborundum Co. from white alumina ore (1 part), flint (1 part) and borax glass (0·03 part), melted together. It can be spun into a yarn as a blend of 85 per cent Fiberfrax and 15 per cent organic carrier yarn, and made into ropes. It will withstand temperatures of 1,200° C. (m.p. about 1,800° C.). Used for insulating blankets, gaskets, protective clothing, filter media for the filtration of hot exhaust gases from atomic energy reactors and other radioactive materials, and in the form of laminates with metal foils for insulating and fire wall materials for aircraft. Fibre diameter about 0·004 mm.

Quartz. This has been spun into vitreous silica fibres 0·020 mm. diameter by Bjorksten Research Laboratories. The fibres either matted or woven into textiles have good heat resistance (*Chem. Engng News*, **36**, No. 15, 41 (1958)).

FURTHER READING

J. M. Matthews, " Textile Fibres ", 6th Edn., London (1954).
Atkinson, *Trans. Amer. Inst. Elect. Eng.*, 1939 (58) (electrical applications).
Owens-Corning Fiberglas Corp., " Fiberglas, An Engineering Material ", " British Rayon Manual ", 39–40. Manchester (1947).
M. Steiner, " Preliminary Processing and Weaving ", *Textile Industries*, 109 (Jan. 1951).
R. F. Caroselli, " Dyeing, Printing and Finishing ", *ibid.*, 107 (Jan. 1951).
J. D. Nordahl, " Manufacturing Glass Fabrics ", *ibid.*, 95 (May 1950).
L. P. Biefeld and T. E. Philipps, " Sizes for Glass Textiles for Reinforcing Polyester Fabrics ", *Amer. Dyestuff Rep.*, Aug. 18, 1952.
" Fiberglas Bibliography ", 3rd Edn., Toledo (1950).
" Haveg Steps up Silica Activity ", *Chem. Engng News*, **39**, 30 (13 March 1961).
" Insulate with Titanate ", *Chem. Engng News*, **36**, No. 43, 54 (1958).
C. B. Woolworth, " Aluminium Silicate Ceramic Fibre ", *Materials in Design Engng*, **46**, No. 5, 124 (1957).
" Continuous Process for Drawing Quartz into Vitreous Silica Fibres ", *Chem. Engng News*, **36**, No. 15, 41 (1958).

CHAPTER 40

CARBON FIBRES

HYFIL, THORNEL

SOME of the first man-made fibres were of carbon. In 1850, Swan made carbon filaments out of paper (Bristol board, of which calling cards are made, was the best); later he used for his electric lamps cotton threads which had been treated with sulphuric acid and were mounted, after 1875, in glass vacuum bulbs. The heat generated by an electric current passing through the filament would burn it away in air; in a vacuum it does not burn the carbon but heats it so that it becomes incandescent. In 1880 the Edison incandescent carbon-filament lamp was patented. Carbonised bamboo provided the best carbon fibres, and lamps made from them would give several hundred hours of life. (To-day's lamps have a life of roughly 1,000 hours.) Then, in 1883, Swan made carbon filaments from extruded fibrous material. These two men, Edison and Swan, made electric lighting practical, and they both depended on carbon filaments, which was the best material then available. In fact, the use of carbon survived until about 1907 when tungsten filaments were introduced, and because they converted electric energy into light so efficiently they replaced carbon filaments and have continued to be used ever since. But apart from threads of gold and silver, carbon filaments were the first man-made fibres that found any real use and now, one hundred years later, and with some very special properties, they are the newest and the most recent development of all in the fibre field of activity.

All forms of carbon are insoluble, so that it was never possible to extrude a carbon filament from solution; neither will carbon melt, so that it has never been melt spun. Swan (p. 148) extruded solutions of cellulose nitrate in acetic acid through orifices and so obtained filaments; these were converted by denitration into cellulose fibres which were carbonised. Still to-day when a carbon fibre is needed the best starting material is an extruded organic material; one could presumably still use cellulose just as Swan did, but nowadays the preferred starting material is the polyacrylonitrile fibre Courtelle or the highly oriented cellulosic fibre Fortisan. Swan was never unduly enamoured of the textile employment of his filaments; he looked rather to their industrial development, and perhaps he looked better and farther than some of his contemporaries. Now, a century later, carbon fibres with unusual physical properties have been made and have found employment in the reinforcement of resins and plastics, which can then withstand the most testing of conditions, particularly in the aerospace field. It is the kind of advance that both Edison and Swan would have liked to have seen.

Development

The new high-performance carbon fibres have been developed both by the Royal Aircraft Establishment at Farnborough, Hants, and by Rolls-Royce at Derby. Teams of scientists from both centres have published work on the fibres: Rolls-Royce refer to theirs as Hyfil. The Shirley Institute, the Cotton Silk and Man-Made Fibres Research Association, at Didsbury, Manchester, have carried out work on the textile aspects of the fibres. A number of patents have been applied for by the Ministry of Aviation in connection with this work, and these have been hopefully assigned to the National Research Development Corporation for exploitation. It is envisaged that industrial concerns will produce the fibres commercially under licence.

Graphite Whiskers

Before considering the preparation and properties of the new carbon filaments it may be well to note that short graphite whiskers of great strength have already been known for some years. In fact, very fine filaments of very strong graphite have been observed by many workers. Usually these filaments have occurred by accident as deposits on various surfaces inside electric furnaces. They have varied in diameter from 0·001 to 200 μ and they have been up to 5 cm. long. Some of these short fibres have been found to be very strong, probably stronger than anything else, but they have been difficult to make in bulk and length. Bacon of Union Carbide grew (purposely not accidentally) graphite whiskers in an arc under a pressure of 92 atmospheres of argon and at a temperature of about 3,600° C. His whiskers were only a few centimetres long, and they varied from less than 1 μ up to 5 μ in diameter. They had tensile strengths up to 2,000 kgm./mm.^{2} with a maximum elastic strain of about 2 per cent. This corresponds to 4,400 lb./mm.^{2} or to 2,840,000 lb./in.^{2} and if the density of the graphite was about 2, then it follows (p. 12) that their tenacity was $\frac{2,840,000}{12,800 \times 2} = 111$ gm./den. Some of the illustrations in Bacon's paper (see Further Reading) are most interesting as showing natural filament formation, *i.e.*, the spontaneous growth of filaments, and are well worth looking at by any fibre scientist.

Whiskers have a near perfection of molecular structure, which the polycrystalline man-made carbon fibres have not yet been able to attain, and it is for this reason that the whiskers are so much stronger. As the graphite crystals grow they do so regularly, elegantly and by natural design. The result is the amazing tenacity of more than 100 gm./den. This compares with values of about 10 gm./den. for the strongest continuous-filament orthodox fibres (Fortisan and industrial nylon and polyester) yet made. It has been recognised for very many years that on the basis of the known strength of the C—C and C—O— chemical bonds a cellulose fibre with a tenacity of 50 gm./den. was theoretically possible. Nobody could ever make one much stronger than about 10 gm./den. and the discrepancy has always been attributed to faults in the crystal structure of the fibre. But the graphite whiskers apparently grow without faults and thereby attain enormous strengths.

Whiskers only a few centimetres long must clearly have only very limited uses, but there seems to be a market of some sort for them in America, where those of alumina and silicon carbide sell at the handsome price of around £400 per lb.

Carbon does not have a monopoly of whisker-forming, but the carbon whiskers are the best known.

What exactly is a whisker? According to Watt *et al.*, it may be roughly defined as a single crystal with a large but finite length : thickness ratio. Such crystals have very smooth surfaces and may possess very high strengths of the order of 10^6 lb./in.^{2}.

Preparation of High-strength Carbon Fibres

Some such organic fibre as Courtelle is the starting material; it is oxidised in air at 200–300° C., then carbonised at 1,000° C. and then converted into graphite by heating at temperatures up to 3,000° C.

Thus, A. F. Standage and R. Prescott of Rolls-Royce reported in 1966 that they had " prepared carbon fibres of high elastic modulus in continuous lengths and in relatively large quantities, by precisely controlled heating of commercially available textile fibres to 3,000° C. in both reactive and inert atmospheres ".

But since then R. Moreton, W. Watt and W. Johnstone of R.A.E. have shown that the maximum temperature can be varied between 1,500° and 3,000° C. and that useful products can be thereby obtained. The properties of the treated fibres can be controlled to a considerable extent by controlling the maximum temperature used in their preparation. It seems that when the final heating is done at 2,500–3,000° C. then the fibres have an elongation at break of about 0·5 per cent and a tensile strength of about 250,000 lb./in.^{2}, but that if the final heating is done at 1,600° C. the fibres may have an elongation at break of about 1·3 per cent and a tensile strength of about 440,000 lb./in.^{2}. Accordingly, control of the final heating temperature in their preparation can be used to develop a method of producing carbon fibres to any Young's modulus desired within the limits of 25–70 $\times$ 10^6 lb./in.^{2} and of specific gravity varying between 1·74 and 2·0.

Young's Modulus

At present the strong carbon fibres have engineering rather than textile applications. Their properties are expressed in terms of tensile strength in lb./in.^{2} and in terms of Young's modulus also in lb./in.^{2}. Those who have forgotten what they were taught about Young's modulus may find it convenient to be reminded.

If a force is applied to a solid the force is called the stress. If, as a result of the stress, the shape or size of the solid is changed, that change is called the strain. The strain, just like a strain in the ankle, is due to deformation of the solid (or the ankle). The ratio stress : strain is the modulus of elasticity; for some purposes volume changes of solids are considered, but for our purposes in the consideration of fibres we are concerned only with length changes. When only length changes are considered, then the ratio stress : strain is called the Young's modulus. If, in a fibre, a stress of 10,000 lb./in.^{2} causes an extension, a strain, of 8 per cent, then the Young's modulus of that fibre is:

$$\frac{\text{Stress}}{\text{Strain}} \text{ or } \frac{10{,}000 \text{ lb./in.}^2}{0{\cdot}08} = 125{,}000 \text{ lb./in.}^2$$

The strain is given as 0·08 because the increase in length was 8 per cent; the strain is the extension that the thread undergoes divided by the original length of the thread. If, therefore, the thread or filament is extended by 8 per cent, then that is equivalent to a strain of 8/100 or 0·08.

Stress in this connection is reckoned in units of force per cross-sectional area, not in units of weight as denier is. Therefore, two different kinds of fibres might have the same tensile strength in breaking load per cross-sectional area (lb./in.^{2}), but if their specific gravities were different they would have different tenacities expressed in gm./den. Imagine two fibres, one with a specific gravity of nearly unity (polypropylene for example) and another with a specific gravity of 2·5 (glass), both might have a tensile strength of 130,000 lb./in.^{2}, but if they did their tenacities would be 10 gm./den. and 4 gm./den., respectively. High specific gravity brings down the value in gm./den. compared with lb./in.^{2}.

(In this example the strength of glass fibre has been understated in order to clarify the principle involved; it is usually about 6 gm./den.)

According to a recent report by the R.A.E., the ratio of Young's modulus to specific gravity is remarkably similar for all normal structural materials. For example, steel, titanium, aluminium, magnesium and even glass and wood have values for this ratio, which is called the " specific modulus ", of about 4×10^6 lb./in.^{2}. If the specific modulus is high, then light, stiff structures are obtainable, so that what are needed to make lighter, stiffer aerospace structures are a high Young's modulus and a low specific gravity. It will be of interest to see how this ratio, the specific modulus, varies between some of the fibres discussed in the preceding parts of this book. We know that Young's modulus is given by the ratio stress : strain, and in order to make the estimates as comparable as possible we shall take both stress and strain at the breaking point. If, for example, a fibre (polyethylene, for example) needs a stress of 23,500 lb./in.^{2} to break it, and if it has an extension at break of 40 per cent (p. 548), then its Young's modulus is

$$\frac{23{,}500 \text{ lb./in.}^2}{0{\cdot}40} \text{ or } 58{,}800 \text{ lb./in.}^2$$

and because its specific gravity is 0·92 its specific modulus must be

$$\frac{58{,}800 \text{ lb./in.}^2}{0{\cdot}92} \text{ or } 64{,}000 \text{ lb./in.}^2.$$

This is a long way short of the R.A.E. value of 4×10^6 lb./in.^{2} for structural materials, and this presumably indicates that polyethylene fibre of this kind (low–medium strength and high extensibility) is a long way from being suitable for use as a structural material. But it is easy to see that if the polyethylene had been irreversibly stretched by 33 per cent, then its tensile strength would have become $4/3 \times$ 23,500 lb./in.^{2}, its extensibility would have become $3/4 \times (40–33)$ or 5·2 per cent and its Young's modulus would have been

$$\frac{4/3 \times 23{,}500 \text{ lb./in.}^2}{0{\cdot}052} \text{ or } 602{,}000 \text{ lb./in.}^2$$

and its specific modulus would have been 650,000 lb./in.^{2}. This brings it nearer to being a material suitable for structural use, but it is still a long way off. And so we might expect to find that fibres that are highly stretched have high values of Young's modulus and are thereby more suitable for structural work than unstretched fibres. Values for some of our orthodox fibres are shown in the following table:

Fibre.	Tenacity (gm./den.).	Specific gravity.	Tensile strength (lb./in.^{2}).	Extension at break (%).	Young's modulus (lb./in.^{2}).	Specific modulus (lb./in.^{2} $\times 10^6$).
Viscose rayon	2·6	1·52	51,000	15	337,000	0·22
Acetate rayon	1·5	1·32	25,000	25	101,000	0·08
Cotton . .	3·2	1·50	61,000	9	682,000	0·46
Nylon (industrial type) .	8·8	1·14	128,000	14	917,000	0·80
Dynel (Type 180) . .	3·5	1·31	59,000	34	173,000	0·13
Terylene .	7·0	1·38	124,000	10	1,240,000	0·90
Fortisan .	7·0	1·51	136,000	8	1,700,000	1·13
Glass . .	6·3	2·54	205,000	2	10,300,000	4·06

The specific modulus for glass at $4{\cdot}06 \times 10^6$ lb./in.^{2} comes gratifyingly near the value of 4×10^6 lb./in.^{2} given by the R.A.E. for " all normal structural materials ". But the closeness of agreement is adventitious, being at the mercy of the figure of 2 per cent for extension at break. This is an approximation, and if a value of 1·8 or 2·2 per cent had been taken instead, as it might well have been, then the agreement of the calculated value with the R.A.E. figure would have been not nearly so good—about 10 per cent away. But it would still have been very good indeed, and one can only look for agreement of the order of " close to 4×10^6 lb./in.^{2} ". What is also clear is that the values of specific modulus for all the fibres except glass are a very long way short of 4×10^6 lb./in.^{2}. Three of the fibres, Fortisan and industrial high-tenacity nylon and polyester, are approximately 1×10^6 lb./in.^{2}. None of these presumably can be counted as a normal structural material.

It is interesting for the author to recall at this point work he carried out in the 'thirties under the leadership of Henri Dreyfus. Cellulose acetate yarn was stretched and gave fibres with (p. 264) a tenacity of 4·5 gm./den. and 6 per cent extension at break. The continuing aim of the research was to find a way of increasing this extensibility to 10 per cent, so that when experiments occasionally produced, as they did, fibres with such low extensions at break as 2 per cent, they were shunned like the devil. But now, in the light of the interest in structural fibres, it is apposite to observe that these rejected fibres must have had the following properties:

Tensile strength	75,000 lb./in.^{2}
Young's modulus	3,750,000 lb./in.^{2}
Specific modulus	$2{\cdot}8 \times 10^6$ lb./in.^{2}

They must have been quite close to the specifications for structural materials in some respects at least. Furthermore, the value of the specific modulus could have been increased by saponifying the cellulose acetate to cellulose under a state of tension to yield a kind of high-tenacity, low-extension Fortisan. Perhaps such fibres might be worth looking at again today; they themselves might have structural applications, and alternatively, they might be very suitable starting material for making very high-tenacity, high-modulus carbon fibres somewhat similar to those that have already been made by R.A.E. and by Rolls-Royce.

The quantities, the headings in the table on p. 589, are worth one more glance. We already have derived (p. 12) the expression:

$$\text{Tensile strength in lb./in.}^2 = 12{,}800 \text{ tenacity (in gm./den.)} \times d$$

where d is the density or specific gravity.

Now

$$\text{Young's modulus} = \frac{\text{tensile strength in lb./in.}^2}{\text{extension at break}}$$

$$= \frac{12{,}800 \text{ tenacity in gm./den.} \times d}{\text{extension at break}}$$

so that the specific modulus which by definition

$$= \frac{\text{Young's modulus}}{d}$$

$$= \frac{12{,}800 \text{ tenacity in gm./den.}}{\text{extension at break}}$$

In relating tenacity (gm./den.) to specific modulus, the d term has disappeared. To take one example from the table on p. 589: industrial nylon has a tenacity of 8·8 gm./den. and an extension at break of 14 per cent (*i.e.*, strain is 0·14). Hence its

$$\text{Specific modulus} = \frac{12{,}800 \times 8{\cdot}8}{0{\cdot}14} \text{ lb./in.}^2$$

$$= 804{,}600 \text{ lb./in.}^2 \text{ or } 0{\cdot}80 \times 10^6 \text{ lb./in.}^2$$

If the arithmetical constant is omitted, what is left is

$$\frac{\text{tenacity in gm./den.}}{\text{extension at break}} = \frac{8{\cdot}8}{0{\cdot}14} = 63 \text{ gm./den.}$$

and this quantity is used by fibre scientists to indicate the property which they call " average stiffness ". Values of average stiffness of different fibres are directly proportional to their specific modulus. By way of illustration, Dynel Type 150 has a breaking strength of 2·4 gm./den. and an extension at break of 60 per cent; its average stiffness is 2·4/0·6 or 4·0 gm./den. But Dynel Type 180 has a breaking strength of 3·5 gm./den. and an extension at break of 34 per cent; its average stiffness is 3·5/0·34 or 10·3 gm./den. The stronger yarn is the stiffer.

Initial Modulus. Solids which are truly elastic recover their shape when a stress is released. If a fibre is subjected to a stress which strains it so that its length is increased by 1 per cent and then the stress, the stretching force, is released the fibre recovers its original length; it is truly elastic. For small strains such as this, within the elastic limit, the stress is proportional to the strain, and the ratio stress : strain or Young's modulus is constant. For small stretches, stress and strain are proportional, or as Hooke's law puts it " the linear extension is proportional to the stretching force ". Eventually there comes a yield point when strain is no longer recoverable; this happens fairly early with most fibres, and if they are stretched more than say 5 per cent (the exact amount of stretch varies from one kind of fibre to another) they do not completely recover their original dimensions, but have become permanently deformed or stretched. The characteristic curves (load–elongation *i.e.*, stress–strain) of fibres are indicative of this. If the fibre were truly elastic the characteristic curve would be a straight line, whereas in fact it is usually a curve, with the strain becoming bigger than Hooke's law demands (*cf.* Figs. 7, 72 and the many other characteristic curves in this book). The Young's modulus is continually changing: at break it is very different from what it was initially. Often fibre scientists use the " initial modulus ", the ratio of stress : strain for very small stretches as a measure of fibre properties (*cf.* p. 309). Low values of this initial modulus are interpreted as low " stiffness ", and in practice are usually accompanied by a soft pleasant handle. But for structural reinforcement this is not what is wanted, rather is it a high " stiffness " that is desirable. In order to facilitate comparison with the R.A.E. values for their new high-strength fibres, the practice has been adopted in this chapter of considering not the initial modulus or Young's modulus for low stretches (small strain) but the Young's modulus at the other end of the stretch, the breaking point. For some structural materials, with low extensibilities, the Young's modulus may not be very different at break from what it is at the commencement of stretch, but for fibres it is definitely very different. For the purposes of comparison with the new strong carbon fibres we will consider the Young's modulus at, or just before, fibre breakage.

Mechanical Properties of Carbon Fibres

The fibres made by Rolls-Royce at 3,000° C. had average values of Young's moduli greater than 60×10^6 lb./in.^{2} with peak values of more than 100×10^6 lb./in.^{2}. Tensile strength of the fibres averaged about 250×10^3 lb./in.^{2}. These values are much lower than those that have been reported for graphite whiskers, so that there still is plenty of room for improvement, but even so, they are " sufficiently high to make the fibres outstanding ". The R.A.E. workers showed in 1967 that if fibres with a very high Young's modulus are required, then the final heat treatment should be at 2,500° C. or higher. If, on the other hand, fibres with high tensile strength and high breaking strains or elongation were wanted the optimum heat-treat temperature was only 1,500–1,600° C. The highest elongation at break that they found was 1·3 per cent, occurring as a result of heating to 1,500° C.; this is more than twice that (0·6) found for fibres of high Young's modulus which had been obtained by heating to 2,500° C. The strength of fibres heated to 1,500° C. was about 440×10^3 lb./in.^{2}. If testing was done on 1 cm. lengths instead of on 5 cm. lengths, results were about 10 per cent higher (they always are on textile yarns). The highest individual tensile strength found for any one fibre treated to 1,600° C. was 745×10^3 lb./in.^{2}; this was only about one-quarter of the maximum tensile strength $2{,}840 \times 10^3$ lb./in.^{2}, measured by Bacon on his graphite whiskers.

A few years ago (1966) Union Carbide Corporation announced the development of a strong and stiff carbon fibre available in continuous yarn and called Thornel graphite yarn; it has a tensile strength of 180×10^3 lb./in.^{2} and a modulus of elasticity of 25×10^6 lb./in.^{2}. For comparison the fibres developed by the R.A.E. (heated at 2,500° C.) had even higher strength and modulus, *viz.*, 300×10^3 lb./in.^{2} and 60×10^6 lb./in.^{2}, respectively. These fibres are polycrystalline, but oriented with the graphite planes (the planar ring molecules of graphite slide over each other easily, so that it feels greasy and is used as a lubricant) lined up parallel to the fibre axis.

In a similar context it may be noted that as well as carbon, boron fibres have been made: very strong with a tensile strength of 366×10^3 lb./in.^{2}; these were tested on 1-in.-long test pieces, so that 5-cm. test lengths would probably have given strengths of about 350×10^3 lb./in.^{2}.

The relevant properties of these various fibres are assembled in the following table; two ordinary fibres, viscose rayon and industrial nylon are included as markers and at the bottom high-tensile steel is shown as another marker.

Fibre.	Tenacity (gm./den.).	Specific gravity.	Tensile strength (lb./in.^{2} $\times 10^3$).	Extension at break (%).	Young's modulus (lb./in.^{2} $\times 10^6$).	Specific modulus (lb./in.^{2} $\times 10^6$).
Viscose rayon	2·6	1·52	51	15	0·34	0·22
Nylon (industrial)	8·8	1·14	128	14	0·92	0·80
Glass E Type	7·8	2·54	250	2·5	10	3·9
Glass HTS	11·7	2·54	380	3·2	12	4·7
Thornel graphite yarn	9·8	1·43	180	0·7	25	17
Carbon, Rolls-Royce (3,000° C.)	9·8	2·0	250	0·4	60	30
Carbon, R.A.E. high modulus (2,500° C.)	11·7	2·00	300	0·5	60	30
Carbon, R.A.E. high tenacity (1,500° C.)	19·1	1·74	430	1·3	33	19
Carbon whiskers (Bacon)	111 †	2·0 *	2,840	2	142	71†
H.T. steel	1·9	7·87	190	0·6	30	3·8

* Estimated figure but probably nearly accurate.

† Derived from the estimated figure marked by an asterisk; their only uncertainty is that which derives from the specific gravity estimate.

What is one to say of the table's contents?

1. It is a remarkable experience to hear of a fibre with a tenacity of more than 100 gm./den. This sets a new target for fibre synthesists.

2. The aim of the work, which was to make materials with a much higher specific modulus than that possessed by most structural materials, has been successful. Several carbon fibres with values of 3–8 times the specific modulus of such structural materials as steel, aluminium and glass have been made.

3. It is clear that orthodox fibres, even such as high-tenacity nylon, do not attain more than one-third of the stiffness of the usual structural materials.

4. There might be a possibility of attaining the required values of stiffness in orthodox fibres by stretching them so that their tenacity was brought to a maximum and their residual extension was very low indeed. This should be a promising field for research.

5. The relative weakness of steel on a " grams per denier " basis has long been known to fibre scientists. All that it proved to most of them was that " grams per denier " was not everything. But it does serve as a pointer to the possibility that new applications might develop for orthodox fibres with a very short break (low extensibility).

6. It should not be forgotten that a very high stiffness, although desirable for certain industrial applications, would be most unsuitable for apparel and most household uses. For our clothes we want something soft and kindly, something that bends and " gives " easily to conform to the wearer's body and movements.

Reinforced Plastics

The research at the Royal Aircraft Establishment was started in order to make lighter stiffer new materials. Because those materials commonly used for structural work had a specific modulus of around 4×10^6 lb./in.^{2} it was necessary to make, if the end was to be achieved, materials with a high Young's modulus and low specific gravity. As can be seen from the preceding paragraphs, and particularly from the table on p. 589, this aim has been achieved. The new carbon fibres are two or three times as strong as steel and twice as stiff (Young's modulus) but only one-quarter as dense.

According to W. Watt, L. N. Phillips and W. Johnson of the R.A.E., when plastics have been reinforced with these new high-strength, high-modulus carbon fibres, composite materials have resulted which are much stiffer than any reinforced plastics hitherto available and which have a stiffness : weight ratio surpassing that (4×10^6 lb./in.^{2}) of metals.

The new super-strong carbon fibres have been used with various resinous matrices. Thus such thermosetting polymers as polyesters, epoxy, phenolic and Friedel–Crafts resins have been used for embedding the fibres. The carbon fibres have a chemically inert surface, and they neither accelerate nor inhibit the resin cure; that is a good feature.

It is important in making such reinforced resins to achieve a high-volume fraction of fibre, *i.e.*, a high proportion of fibre to matrix. The carbon fibre is wetted with sufficient of a dilute resin solution to leave it coated thinly with resin after evaporation of the solvent. Heat and pressure are then applied to consolidate a linear assembly or bundle of coated fibres and to remove voids. In this way uni-directional composites are produced with a fibre content of 50 per cent by weight and 40 per cent by volume, *i.e.*, roughly as much fibre reinforcement as resin.

Typical test values on such a composite are:

Young's modulus	22×10^6 lb./in.^{2}
Tensile strength	105×10^3 lb./in.^{2}
Flexural strength	good
Density	1·54 gm./cm.^{3}

The unidirectional glass-filament resin mouldings which have hitherto been used had a specific gravity of 2·0 and a Young's modulus of 6×10^6 lb./in.^{2}. It can be seen that the specific modulus of the carbon fibre composite $14{\cdot}5 \times 10^6$ lb./in.^{2} (*i.e.*, $22 \times 10^6/1{\cdot}54$) is much higher than the corresponding figure of $3{\cdot}0 \times 10^6$ lb./in.^{2} (*i.e.*, $6 \times 10^6/2$) for glass-fibre-reinforced resin.

A comparison of the properties of an epoxy resin in which carbon fibres have been used for reinforcement with those of an aluminium alloy is also interesting.

	Density (lb./in.^{3}).	Tensile strength (lb./in.^{2}).	Young's modulus (lb./in.^{2}).
Aluminium	0·100	45,000	$10{\cdot}5 \times 10^6$
Carbon-fibre-reinforced epoxy resin	0·061	90,000	40×10^6

The fibres have also been used to reinforce polyimide resins, and this is one of their most useful applications. The new fibres open up possibilities of making strong plastics which are stiffer and lighter than it has hitherto been possible to make.

Other Properties

The damping properties of the composite resins are good because of hysteresis in the resinous matrix. How do they stand up to long use? Do they show fatigue? These are important questions to be asked of aero materials. Apparently they have good endurance, and more than 20 million cycles at a loading of 10,000–50,000 lb./in.^{2} have been sustained; this is quite high compared with a tensile strength for the composite of about 100,000 lb./in.^{2}. Part of the explanation of this long endurance is the good heat conductance of the carbon fibres; heat generated locally is rapidly dissipated. The resins that are reinforced with carbon fibres show less strain (deformation) than do the glass-fibre-reinforced resins hitherto used.

When the carbon-fibre-reinforced resins are heated between 20°

and 100° C. the unidirectional coefficient of expansion is slightly negative, $-0{\cdot}73 \times 10^{-6}$; the contraction is very small, but it reflects the dominant effect of the carbon fibres in the resin. The carbon fibres can be boiled in water without change of strength, and the resins that contain them have been given long immersion tests in water without loss of stiffness.

Structural uses are envisaged not only in the aerospace field but also more generally where a high stiffness : weight ratio is required. The resins may provide bearing materials and be suitable for marine use, where prolonged resistance to water is imperative. Another possibility is in chemical plant, where resistance to corrosion and swelling is important. But the outstanding feature of the new carbon fibres is their great strength combined with lightness, and this points unwaveringly to aircraft and space vehicles.

Thornel Yarns

Union Carbide Corporation introduced their first structural graphite yarn—Thornel 25—in 1966. Since then they have introduced two other types known respectively as Thornel 40 and Thornel 50. The terminal number is an indication of the modulus of elasticity of the filaments of which the yarn is composed, expressed in million lb./in.^{2}. A comparison of some of the properties of the filaments of which these three yarns are composed is shown below.

	Specific gravity.	Tensile strength (1,000 lb./in.^{2}).	Elongation at break (%).	Young's modulus of elasticity (million lb./in.^{2}).
Thornel 25	1·43	180	0·72	25
Thornel 40	1·56	250	0·63	40
Thornel 50	1·63	285	0·57	50

The strength tests have been made on 1-in. lengths of individual filaments. They correspond approximately to tenacity values of 10, 12·5 and 13·5 gm./den. And yet the tenacity of Thornel 40 *yarn* is given by the makers as only 4·7 gm./den. and of Thornel 50 *yarn* as the same. The explanation must be that there is considerable variation in the tensile strength of the individual filaments that make up the yarn; that the weakest break early, *i.e.*, at relatively low loads, and that when once rupture of the constituent filaments of the yarn has started, it goes progressively faster. Furthermore, the practice of testing filaments on test lengths as short as 1 in. gives high strength values; if the filaments had been tested on longer lengths, as the yarn doubtless is, then much lower tenacity figures would have resulted. This observation applies to all yarns, but if the yarns were very nearly uniform the difference between test results on long and short test lengths would not be great. If these test differences are great, as they apparently are in the graphite yarns, then that is a sure indication of irregularities and of the occurrence of weak places in the filaments. Only in the case of an ideal perfectly uniform yarn would the strength tests on long and short test pieces be the same. In fact, they never are; it is customary in large spinning factories to make routine strength tests on 50-cm. test lengths of the product; if additionally some tests are made on 5-cm. test lengths these will always be found to give higher-strength results. Thus, although the conditions chosen for testing Thornel filaments yield tenacity values of 10–14 gm./den., yet when the yarns are tested under more normal conditions they are found to have tenacities of only 5 gm./den. The strength of a chain is that of its weakest link.

It is a considerable achievement on the part of Union Carbide Corporation to have produced carbon fibres with the unique property of flexibility—yarns which can be handled and fabricated as easily as conventional glass fabrics and yarns. The carbon products are not fabricated as easily as conventional textile fabrics and yarns. The carbon products are not cheap; in 1968 Thornel 40 cost $325 per lb. and Thornel 50 cost $350 per lb. on standard 1-lb. packages, which are cylinders about 11 in. long by 3 in. inside diameter. The Thornel 25 variety is supplied in packages holding 0·1 lb. at a price of $50.

Thornel yarn is intended for use in composites, *e.g.*, as a reinforcement for epoxy resins. The composites are suitable for use in highly stressed structures where light weight matters. In missiles and rockets, for example, they are used for rocket-motor cases, nozzles, tanks and re-entry vehicle structures. In aircraft their uses are for wing skin and structures, fuselages, rotor blades and for turbine compressor blades, for naval use their outstanding potentiality is for deep-sea diving bells.

The Carbonising. Cellulose has the empirical formula $(C_6H_{10}O_5)_n$, so that the carbon content is 72/(72 + 10·08 + 80) or 44·4 per cent. In fact, the weight loss accompanying the carbonisation is more than 75 per cent, some of the residual carbon being lost. The carbonisation is thought to take place in the following four stages:

1. Desorption of physically adsorbed water.
2. Dehydration of the cellulose molecule through elimination of hydroxyl groups.

3. Thermal breakdown of bonds in the cellulose ring and of bonds connecting the rings, accompanied by loss of water, carbon monoxide and carbon dioxide. All this results in considerable weight loss.

4. The knitting together of the residual products from (3) into graphite-like rings, this being accompanied by the loss of methane and hydrogen.

Provided that the cellulose source, the original fibre, is one that will not melt, that it is, for example, viscose or Fortisan and not cellulose acetate, there will be some structural features that will persist right through all these changes, right through the fibre being heated to temperatures in excess of 2,000° C. If, for example, the original fibre was Fortisan, then features of its high orientation can be found by electron microscopy in the carbonised Fortisan, although this weighs only one-fifth of what the original fibre did. Apparently there is a microfibrillar structure which retains its shape and position, although inevitably it loses most of its substance and most of its weight in the carbonising. The process by which Thornel yarns are made is continuous and it is possible that throughout the carbonisation the yarn is maintained under tension, even though only a slight tension. This would help retention of structure and doubtless, too, retention of strength.

At present the fibres are of most interest for reinforcement of resins and plastics to make very strong and light components for space rockets. But it should not be overlooked that the fibres will be non-flammable and that they might ultimately be of interest for nursery nightwear. Black? Well, that does make one pause. Transformation to white and pastel shades might be possible by some spray-painting process. The really difficult step of making cellulose non-flammable has apparently been overcome by carbonising it. True, much of the cellulose has been lost, but it is a subject that might repay a little thought.

FURTHER READING

R. Bacon (Union Carbide Corp.), " Growth, Structure and Properties of Graphite Whiskers ", *J. appl. Physics*, **31**, (2), 283–290 (1960).

A. E. Standage and R. Prescott (Rolls-Royce), " High Elastic Modulus Carbon Fibre ", *Nature*, **211**, 169 (1966).

R. Moreton, W. Watt and W. Johnson (R.A.E.), " Carbon Fibres of High Strength and High Breaking Strain ", *Nature*, **213**, 690–691 (1967).

W. Watt, L. N. Phillips and W. Johnson (R.A.E.), " High-Strength High-Modulus Carbon Fibres ", *The Engineer*, **221**, 815–816 (1966).

Union Carbide Corp. *Technical Information Bulletin No. 465–202 GG:* " ' Thornel ' 40 Graphite Yarn ".

R. Bacon, A. A. Pallozzi and S. E. Slosarik (Union Carbide Corp.), " Carbon Filament Reinforced Composites ". Paper presented to the Reinforced Plastics Division of The Society of the Plastics Industry, Inc., at their 21st Annual Meeting at Chicago.

Graphite Fiber Composites, Symposium on. Presented at American Society of Mechanical Engineers, Winter Annual Meeting, Pittsburgh, Pa., November 1967.

CHAPTER 41

METALLIC YARNS

METALLIC threads were the first man-made fibres of all; they came thousands of years before nylon or rayon. Ancient and highly developed civilisations, expert in the art of hand-weaving, devised patterns of gold that have persisted until to-day. The Persians made fabulous carpets with gold thread, and the Indians ornamental saris with it. The metal threads were ribbon-like in section, and were often doubled with or twisted round some other thread, such as cotton. Except when made of gold, which is impossibly expensive for most purposes, they have been subject to the disadvantage that they tarnished or dulled on exposure to the air; another disadvantage to which all of them have been subject has been that they would cut each other and were correspondingly sharp or rough to the skin. Decorative appeal has always been the biggest attraction of metallic yarns; something beautiful and bright, cheerful and colourful is likely to enjoy consumer acceptance for a long time if people can afford it. Whereas the older yarns consisting of 24 carat gold wrapped around fine copper wire cost about £20 per lb., to-day, yarns with the same appeal and looking like gold, but actually containing none are to be bought for about £2–£3 per lb.

Chemical Nature

The modern and cheap metallic yarns consist of filaments of aluminium covered with plastics; two kinds of plastics are mainly used for the covering. The first of these and the commonest is cellulose acetate–butyrate and the second and the better is Mylar, du Pont's polyester film which is very similar chemically to Dacron and Terylene. The mixed ester of cellulose with acetic and butyric acids is more popular for use in plastics than is straight cellulose acetate, largely because it has a lower melting point (although none of the esters of cellulose can be melted without decomposition) and is more easily worked.

Manufacture

The raw material is a roll of aluminium foil of 0·00045 in. (nearly " half a thou ") thickness, and about 20 in. wide. To both sides of the sheet there is applied a thermoplastic adhesive to which has already been added the required colouring matters. The adhesive-coated foil is heated to about 90–95° C. and a sheet of cellulose acetate–butyrate transparent film is laminated to each side of the foil by passing through squeeze rollers at a pressure of 2,000 lb./in.^{2}. The laminated material is then slit into filaments of the required width, the most popular width being $\frac{1}{64}$ in., although other sizes from $\frac{1}{8}$ to $\frac{1}{120}$ in. are also made.

The nature of the adhesive that is used is important and is not ordinarily disclosed. Of the colours, gold is the most important, as it accounts for more than half the production; it is produced by the addition of an orange-yellow dyestuff to the adhesive. Silver, the next most popular colour, is simply the colour of the aluminium itself which does not tarnish. Other colours are available, such as bronze, gunmetal, emerald, peacock blue and dragon red, and to all of them a glitter is added by the underlayer of aluminium. Sometimes opaque porcelain colours are used instead of ordinary dyestuffs to give a less glittering and more subtle sheen. Multi-coloured effects, *e.g.*, red and green alternating irregularly along the length of the yarn have been obtained by pre-printing the plastic film and laminating in the usual way.

Properties

Brightness. Outstanding, of course, is the brilliance of appeal to the eye. Because the metal basis is aluminium there is no tarnishing and because it is sealed in, it cannot be affected by salt water, chlorinated swimming pools or climatic conditions. Because the covering film is plastic the yarns are flexible and have some extensibility.

Strength. Strength of the acetate–butyrate plastic covered yarn is not very high, but is sufficiently good to enable it to be used as warp or weft unsupported; if considerable strength is required in the finished fabric the metallic yarn can be supported with another fibre such as rayon or nylon round which it can be spirally twisted. The Mylar coated yarns are much stronger than the acetate–butyrate coated yarns, deriving most of their strength from the Mylar polyester; they can be used successfully for weaving and for knitting and tricot operations. Durable fabrics of 100 per cent Mylar metallic yarn have already been made.

Heat. The acetate–butyrate coated metallic yarns can be washed at temperatures as high as 70° C., but not higher, otherwise delamination occurs. The Mylar coated yarns can be washed at the boil, and are safe up to 145° C.

Colour Fastness. The colours used are very fast to light.

Dry Cleaning. Care must be exercised that the acetate–butyrate coated metallic yarns are cleaned with agents safe for cellulose acetate.

Sewing. Fabrics containing metallic yarn can be sewn satisfactorily and the yarns can be knotted.

Embossing. By virtue of the thermoplastic nature of their covering, the metallic yarns can be easily and permanently embossed.

Resistance. All the metallic yarns are mothproof; chemical and biological resistance depend on the nature of the coating; if this is Mylar the resistance of the yarns is excellent.

Elasticity. One kind of metallic yarn has special properties, that is an elastic yarn made by wrapping a metallic yarn round a rubber thread.

Whereas the older metallic yarns were expensive, subject to gradual tarnish, heavy, harsh to the handle and difficult to clean, the modern metallic yarns are free from all of these defects.

Dyeing

Best colour effects are available when yarn-dyed materials are used, and this is the method that should be adopted whenever practicable. If it is necessary to dye the metallic yarn in fabric form it can be done all right; if Mylar is the coating it can be dyed at the boil; if cellulose acetate–butyrate is the coating then the temperature of the dyebath should not exceed 70° C. Usually the purpose of the dyeing is to dye a fibre such as cotton or nylon with which the metallic yarn has been woven, and when this is so, all that is necessary is to choose dyestuffs which will not dye or stain the metallic yarn, *e.g.*, disperse dyes should be avoided. Where it is desired to dye the metallic yarn as well then dyes suitable for dyeing the plastic coating, as for cellulose acetate, or as for Terylene are used. In general, the use of dyed yarns is greatly to be preferred.

Uses

Metallic yarns are used wholly for decorative purposes. The leading American producers and the names of their products are:

Metlon Corporation:	Metlon and Metlon with Mylar.
Dobeckmun Co.:	Lurex, Lurex MF, Lurex MM.
Reynolds Metals Co.:	Reynolds aluminium yarns.
Standard Yarn Mills:	Lamé.

Annual production is in all about 3 million lb., worth about £3·5 million. The yarns are at present extremely popular, and one is bound to wonder if this is just another whim of fashion; the view in the trade is that this is not so, that the yarns have come to stay and that such beautiful and inexpensive materials will outlive their novelty. Very likely they will, but probably in modified forms different from those we know to-day. The original artificial silk was esteemed for its brilliance of lustre and its continuity of filament; nowadays most of it contains a delustrant and is cut up into short pieces an inch or two long. What is in store for metallic yarns is a matter for surmise, but perhaps cellulose acetate–butyrate will give way to hardier plastics such as Teflon and to cheaper ones from the vinyl field.

One of the biggest applications of metallic yarns to-day is for car upholstery; in America, Ford, Studebaker, Buick and Cadillac are using it—the first three are pre-eminent in cheap popular cars and the Cadillac serves a sophisticated class.

In women's dresses, blouses and skirts the glittering fibres give an other-worldly touch of swank and glamour. Upholstery, shoe-laces, bathing dresses, table linens, an excelsior-like packing material for such gifts as perfume and liqueurs—metallic yarns can crop up almost anywhere. They serve irreproachably the pomp of ornate ecclesiastical vestments. They have been chopped into very short pieces (already!) and used as confetti and under the apt name " Wink " as an add to vinylite floor-coverings to give bright sparkling flecks. Elastic metallic yarns have found great favour for women's sandals.

According to a recent fashion note, " At any cocktail party in Paris just now you will see evening shoes of white rayon, intricately embroidered with silver glitter thread, or white façonné embroidered with gold. In the daytime a plain black hopsack suit may be enlivened by sparkling jet-black Lurex, which gives the effect of minute raindrops resting on top of the fabric." The extravagance of lustre of these new yarns seems to have dictated an austerity of style and an absence of jewellery to offset it. The pure gold wire which was beaten into dress fabrics for the queens of Babylon cost a king's fortune; to-day their equivalents are accessible to the factory girl.

Lurex MM

Metlon and Standard Lurex, sometimes called Butyrate Lurex, are similar in that they consist of a sandwich of aluminium between two films of cellulose acetate–butyrate.

" Metlon with Mylar " and Lurex MF again are similar in that they consist of aluminium between two Mylar films.

Lurex MM is different and requires special consideration. The basis of it is Metallised Mylar produced by the vacuum deposition of aluminium on Mylar film; a layer of metallised Mylar is either

(*a*) bonded to one layer of clear Mylar or

(*b*) sandwiched between and bonded to two layers of clear Mylar.

Colour is introduced with the adhesive. The important difference is that the metallic layer in Lurex MM consists of a multitude of discrete particles and not, as in the other metallic yarns discussed, of a continuous ribbon. This construction gives Lurex MM particular softness and thinness, and it affects some of its other properties, too, as may be seen below. Strength is satisfactory.

Chemical Resistance. All of the metallic yarns, although protected top and bottom of their flat sides, are vulnerable at their cut sides; because the area exposed is small, tarnishing due to atmospheric exposure is negligible, and chemical attack is not serious unless the chemical is one that dissolves aluminium. Any one of the three Lurex yarns immersed in caustic soda loses metal due to the aluminium dissolving in the caustic soda on being attacked through the cut side of the yarn. But the attack may be a little less with Lurex MM than with Lurex MF, for example the former is unaffected by 2 per cent hydrochloric acid at 99° C. for 2 hours whereas the latter loses metal. Sometimes, though, the MF seems to have the edge over the MM in chemical resistance. Both are delaminated by trichlorethylene which is sometimes used in dry cleaning, but must not be used with Lurex MM or Lurex MF; both are safe in perchlorethylene. The regular Butyrate Lurex is safe in either.

Abrasion Resistance. Lurex MM has excellent abrasion resistance. Lurex MF has " good " and Butyrate Lurex " fair ".

Flex Life. Here as would be expected Lurex MM excels with four times the flex life of Lurex MF and 70 times that of Butyrate Lurex.

Electrical Conductivity. The conductivity of Lurex MM is much lower than of the other two, which because of their aluminium foil base are good conductors. Lurex yarns should not be used on looms with stop motions which are activated by contact with a conductor.

Tenacity. That of Butyrate Lurex is about 0·3 grams per denier and of Lurex MF about 0·7 and of Lurex MM about 1·25 grams per denier.

Extensibility. Whereas Butyrate Lurex has an extensibility at break of about 30 per cent, Lurex MF and MM both have a high extensibility of about 140 per cent. They are better too in their recovery from short stretches than is the Butyrate Lurex. Over-stretching is the commonest cause of spoiling the appearance of metallic yarns. But the main difference between Lurex MM (vacuum deposited metal) and the other is its greater softness and pliability.

Gauge

Metallic yarns are described by width, *e.g.*, $\frac{1}{64}$ in. or $\frac{1}{32}$ in. and by gauge. The gauge is the thickness in one hundred thousandths of an inch of the two layers that form the Lurex sandwich. The gauge figure does not indicate total yarn thickness because it takes no account of the thickness of pigmented adhesive nor of that of the aluminium. For example, 260 Butyrate Lurex consists of two layers of 0·00130 in. cellulose–acetate butyrate with a 0·00045 in. aluminium foil and adhesives between, but its total thickness is 0·0032 in., indicating that the two layers of adhesive must each be about 0·00008 in. A 260 gauge $\frac{1}{64}$ in. yarn yields about 10,500 yards per lb. corresponding roughly to 430 denier.

Prices

As the metallic yarns are improved in quality, their prices rise. Early in 1962 typical prices for $\frac{1}{64}$ in. yarns were:

Standard Lurex (butyrate and foil) . . .	36*s*. per lb.
Lurex MF (polyester and foil) . . .	60*s*. per lb.
Lurex MM (metallised polyester) . .	104*s*. per lb.

but they were still the same in late 1965.

CHAPTER 42

FIBRES OF THE FUTURE

CASTING one's mind back to the 'thirties, it is easy to recall the surprise with which nylon was received. Apart from those working on the polymers and the people associated with them, there was a general ignorance of its imminence. Even in most of the British and European textile laboratories there was no awareness that du Pont were on the brink of making a fully synthetic fibre. In retrospect it is clear that there was some evidence available, notably the classic papers of Carothers and Hill on fibre-forming polymers in the 1932 volume of the *Journal of the American Chemical Society*, and there were too some patent specifications. But most of us missed their significance.

When Terylene came in 1941 it was similarly unheralded. True there was a patent specification (B.P. 578,079) available, but there are so many patent specifications published, and probably not one in a hundred has any practical significance. So it aroused little attention. Carothers had looked at the polyesters and had not been much impressed; it was unlikely that anyone would do much better than he had done.

Similarly, when du Pont introduced Orlon, the homopolymer of acrylonitrile, there was little general knowledge of its coming until pilot-plant production had started in 1945; certainly there was no inspired anticipation of the fibre.

These three—nylon, Terylene and Orlon—formed the backbone of the synthetic-fibres industry fifteen or twenty years ago, and they still do so to-day. Yet the appearance of all three was in the nature of a surprise. There have been developments and modifications of them all, and these could more easily have been predicted. For example, the modification of a polyacrylonitrile fibre to include dye-receptive groups has attracted attention ever since Orlon appeared. But the basic acrylic fibres came unheralded.

So the prognosis for the prediction of fibres which are basically different from those that we have now is accordingly unfavourable. Having failed to anticipate the coming of nylon, Terylene and Orlon, it seems unlikely that the fibre industrialists will be luckier with what is perhaps just round the corner. There are three ways in which the problem can be approached:

1. By a consideration of the shortcomings of existing man-made fibres and of those fibre qualities that would be desirable but are not yet available in a fibre. What is needed here is the fibre scientist with sufficient intuition to be able to recognise those structural essentials that make the best natural fibres what they are, and the ability to incorporate them in synthetic fibres.

2. By reviewing the relevant published work of universities and technical colleges and industrial research establishments. To this can be added notes and observations from the patent literature, although here the ground is heavy.

3. From announcements that the manufacturers make of their intentions.

The Desirable Fibre

Silk, wool and cotton still remain as the best apparel fibres. The synthetic fibres have qualities which are not possessed by the natural fibres, especially in respect of strength, durability, dimensional stability, and resistance to chemical and biological, especially micro-biological, attack. Such aspects of superiority led us at one time to think that we had left the natural fibres far behind. And so we had, for industrial purposes. But for apparel, for appearance and kindliness, silk and wool are still supreme. We must make a partial exception of nylon which has proved to be ideal for stockings.

The properties of fibres can always be closely related to their structure. Wool, silk and nylon are all polyamides; the essential difference between them is that in the natural protein fibres the amide linkages are very close together, separated by only one carbon atom, whereas in nylon they are separated alternately by 6 and 4 carbon atoms in nylon 66 and by 5 carbon atoms in nylon 6.

Real silk is mainly a polymer of glycine and alanine, both amino-acids of general formula $NH_2CHRCOOH$, where R = H in glycine and R = CH_3 in alanine, so that real silk fibroin has the constitution

$$\left[-NH\overset{\overset{R}{|}}{C}HCONH\overset{\overset{R}{|}}{C}HCO- \right]_n$$

where R is either H or CH_3. It should be noted that the polar groups, the —CONH— groups, account for 86/128 or 67 per cent of the total weight. Real silk is easy to dye, it is receptive to dyes, it has plenty of polar groups for the dye molecules to cling to; it has a high moisture content, picks up and releases moisture easily and is compatible with living organisms and in particular with the human body. Its appearance, partly due to its fineness of filament, is wonderfully attractive, and perhaps the attractiveness gains a little from the slight dissimilarities between one filament and the next. These slight differences average out so that a silk fabric looks uniform, but they, especially those differences of cross-sectional shape and size, may give it the sheen and glitter which Antron nylon (p. 346) tries to copy. There are, in fact, some other amino-acids in real silk, notably tyrosine, but they are present in quantities so small that they can be disregarded in our present considerations. What the scroopy handle of real silk is due to, it is hard to say, but very likely it could not be imitated in a fibre which was very different from silk in its molecular structure.

Wool and hair consist of keratin which is a much more complex protein than silk fibroin. It still has the same general formula

$$\left[-NH\overset{\overset{\displaystyle R}{|}}{C}HCO- \right]_n,$$

but R exists in at least twenty varieties. It may have a value of H as in glycine, or of CH_3 as in alanine, or CH_2OH in serine, or $CH_2CH(CH_3)_2$ in leucine, or of $(CH_2)_4NH_2$ in lysine or of CH_2COOH in aspartic acid. There are, too, some cystine molecules in which two amino-acids are joined together by a disulphide linkage thus:

$$\begin{array}{ccccc} COOH & & & & NH_2 \\ | & & & & | \\ CH & -CH_2-S-S-CH_2- & & & CH \\ | & & & & | \\ NH_2 & & & & COOH \end{array}$$

It is easy to see that such a molecule as cystine can enter into the structure of two more or less parallel amino-acid polymer chains and cross-link them through the disulphide linkage, as in fact it does in wool.

$$\begin{array}{ccc} | & & | \\ CO & & NH \\ | & & | \\ CH & -CH_2-S-S-CH_2- & CH \\ | & & | \\ NH & & CO \\ | & & | \end{array}$$

These cross-linkages give to wool its power of recovery. Hang a crumpled worsted suit in the wardrobe for a week, and it comes out looking a lot more like itself. Do it with a man-made fibre suit, and if you expect the same result you will be disappointed. Durability, resilience, toughness and character largely depend in wool on these cross-links, and they would seem to be so obviously desirable that one wonders why they have not been put into our commercial synthetics. As regards the proportion of reactive polar groups in wool, the calculation is a shade more involved than in the case of fibroin, because some of the reactive groups are in the side-chains (the R groups), for example in lysine (free NH_2) and in aspartic acid (free COOH) and in some other amino-acids, and also because the side-chains are more complex, more various and heavier than they are in fibroin. Nevertheless, one must not paint the picture more complex than it really is; about half (by number not by weight) of the amino-acid residues (*i.e.*, the amino-acids with their original —COOH groups shortened to CO— and their original $—NH_2$ groups shortened to —NH—) are of glycine, alanine, serine, and glutamic acid. According to Speakman's classical work (*J. Text. Inst.*, **32** (7), 1–28, 1941), the average residue weight is 118 (the glycine residue $—NH{\cdot}CH_2CO—$ is 57 and the alanine residue $—NH{\cdot}CHCH_3{\cdot}CO—$ is 71), and each and every residue must contain 43 parts by weight of polar groups as a consequence of its —CONH— amide linkage. Additionally, serine and tyrosine contribute hydroxyl groups, tryptophane and histidine contribute imine groups, aspartic and glutamic acids carboxyl groups, and lysine and arginine amino groups.

Taking all these into account, the polar group content of keratin comes to 77 per cent of the total weight. There is not very much of the wool molecule that does not consist of reactive polar groups, rather less than one-quarter in fact.

If we consider nylon 66 similarly we can see that in its molecular repeat

$$—NH(CH_2)_6NHCO(CH_2)_4CO—$$

the reactive groups account for 86/226 or 38 per cent of the total weight. Similarly, in nylon 6, in the repeat

$$—NH(CH_2)_5CO—$$

the reactive groups account for 43/113 or 38 per cent of the total weight; this is the same proportion as in nylon 66.

Accordingly, we can draw up the table on p. 610.

It is easy to see why wool will pick up about 18 per cent of its weight of moisture, real silk 11 per cent, whereas nylon picks up only 4 per cent; it is easy to see why wool and silk are kindlier to the living organism. For comparison, cotton picks up about 7 per cent moisture and mercerised cotton about 12 per cent. All these figures are for standard conditions, *i.e.*, 70° F. and 65 per cent R.H. If the fibres are saturated the differences are even more marked; nylon soon feels and looks damp, wool does not feel damp, even if it holds one-third of its own weight of moisture. Wool, particularly, and real silk are much easier to dye than is nylon. The great amount of work that has been devoted to nylon dyeing has cleared away most of the difficulties, but it is still true to say that the affinity of nylon for dyes is much lower than is that of silk or wool.

Fibre.	Percentage of weight consisting of:	
	Polar groups.	Non-polar groups.
Real silk . . .	67	33
Wool	77	23
Nylon 66 . . .	38	62
Nylon 6 . . .	38	62

Those fibres that have just been discussed have all been polyamides. If the argument is extended to polyesters not much can be done with it, because there is only one polyester that has been made on a large scale, and there is no natural polyester fibre at all. In its approach to the natural protein fibres (wool and silk) nylon comes very much closer to them than does any polyester or acrylic, or olefine fibre; the simple reason is that nylon has, in common with wool and silk, a large number of amide groups.

If it is admitted, as a lot of people would admit, that closer approaches to real silk and wool constitute the most desirable avenue of fibre development, then it is reasonable to infer: (1) that desirable new fibres should be polyamides, (2) that they should have a very high content of polar groups, and (3) that for the wool approach at least, some cross-linking is desirable.

The Lower Nylons

This is straightforward enough and points an unwavering finger at the lower nylons: nylons 1, 2, 3, 4 and perhaps 5. Or perhaps at synthetic polypeptides made by condensing together a variety of amino-acids. Either way will give polyamides with a high content of polar groups. How this content varies may be seen from the table on p. 611.

For the manufacturer who wants to make something like real silk or wool, nylon 2 looks to be much the most promising, and it will be discussed in detail. But first let us consider nylon 4, nylon 3 and dimethyl nylon 3, which have all been the subject of commercial, if limited, development.

Fibre.	Percentage on total weight of weight of polar groups.
Rilsan (nylon 11)	24
Enant (nylon 7)	34
Nylon 6	38
Nylon 5	43
Nylon 4	51
Nylon 3	61
Dimethyl nylon 3	43
Nylon 2	75
Nylon 1	100

Nylon 4

Recently a nylon 4 has been announced by the General Aniline and Film Corporation; it still remains to be seen if it will be made on a large scale as a fibre, but its chemical constitution is of interest and suggests that it may have some advantages over nylon 6. It is made from 2-pyrrolidone, which in turn is made from acetylene, so that the economic position should be favourable; when it is polymerised with alkaline catalysts it gives polypyrrolidone or nylon 4. The process will be understood readily by comparison with the polymerisation of caprolactam to nylon 6:

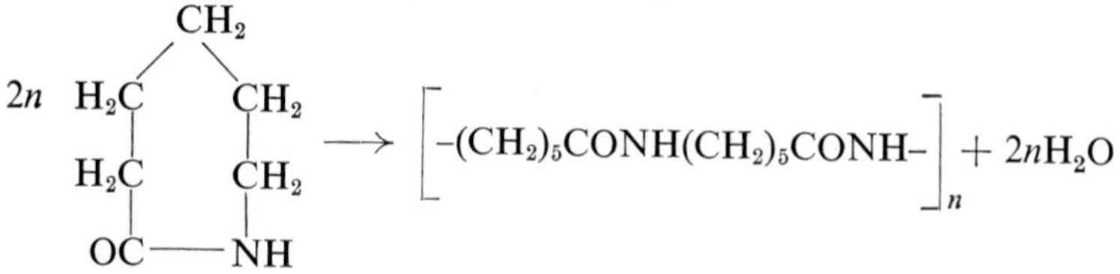

$$2n \ \text{(caprolactam ring: } CH_2, H_2C, CH_2, H_2C, CH_2, OC\text{—}NH) \longrightarrow \left[-(CH_2)_5CONH(CH_2)_5CONH- \right]_n + 2nH_2O$$

Caprolactam. Nylon 6 (Perlon).

$$2n \ \text{(2-pyrrolidone ring: } CH_2, H_2C, CH_2, OC\text{—}NH) \longrightarrow \left[-(CH_2)_3CONH(CH_2)_3CONH- \right]_n + 2nH_2O$$

2-Pyrrolidone. Nylon 4.

As its constitution would suggest it has a higher moisture regain than ordinary nylon, and accordingly absorbs moisture better and is relatively free from static. It has been reported that its moisture pick-up is similar to that of cotton; if it really is so, then nylon 4 should eventually be made on a big scale. Its development has been retarded by the proneness of the molten polymer to decompose, so that spinning from the melt is, so far, difficult and erratic.

Nylon 3

Nylon 3 can be made from acrylamide as described by A. S. Matlack in U.S. Patent 2672480 (1954); basically the reaction is a molecular rearrangement

$$n\ CH_2{:}CHCONH_2 \longrightarrow [-CH_2CH_2CONH-]_n$$

Acrylamide — Nylon 3 or Poly(β-alanine)

Fibres have not been made on a commercial basis from nylon 3 but they should have useful properties provided that they are stable to light.

Dimethyl Nylon 3

Fibres have been made by Farbwerke Hoechst A. G. Frankfurt/Main from dimethyl nylon 3 which is obtained by polymerising 4,4-dimethyl azetidin-2-one, thus:

$$n\ \begin{array}{l}(CH_3)_2C\text{---}CH_2 \\ \quad\ \ | \qquad\quad | \\ \quad HN\text{---}CO\end{array} \longrightarrow [-NHC(CH_3)_2CH_2CO-]_n$$

4,4-Dimethyl azetidin-2-one — Dimethyl nylon 3

The monomer is a lactam and the process is essentially the same as the preparation of nylon 6 from caprolactam. The polymer dimethyl nylon 3 melts only at 300° C. with decomposition so that melt spinning is impracticable. It has been successfully spun into water (coagulant) from a 12–15 per cent solution in methanolic calcium thiocyanate. A preliminary evaluation suggests a tenacity of about 3 grams per denier, an extension at break of about 40 per cent and a disappointingly low moisture regain at 4·5 per cent. The fibre is said to be resistant to oxidation.

Nylon 2

On structural grounds this is the most promising of all the nylons. With a structural repeat of $-NHCH_2CO-$ it is very close to real silk. It is polyglycine. There are two ways in which it has been made:

1. By heating glycinamide, the amide of glycine, in a closed vessel at 100° C. for 20 hr.

$$nNH_2CH_2CONH_2 \longrightarrow [-NHCH_2CO-]_n + nNH_3$$

Glycinamide. — Polyglycine.

2. By treating anhydrocarboxyglycine with water. The following formulae will indicate the relationship of this substance to glycine itself:

$$NH_2-CH_2-COOH \qquad \begin{array}{l}NH-CH_2-COOH \\ | \\ COOH\end{array} \qquad \begin{array}{l}NH-CH_2-CO \\ | \qquad\qquad\ \ | \\ CO\text{------}O\end{array}$$

Glycine. — Carboxyglycine (unknown). — Anhydrocarboxyglycine.

When it is treated with water anhydrocarboxyglycine yields nylon 2.

$$n \begin{array}{l} NH-CH_2-CO \\ \vert \qquad\qquad\quad \vert \\ CO\text{———}O \end{array} + H_2O \longrightarrow [-NH-CH_2-CO-]_n + nCO_2$$

Nylon 2.

Anhydrocarboxyglycine was first prepared by Leuchs in 1906. Speakman at Leeds used it as a means of providing a protein layer to apply to wool, to cover up its scales, and make it shrink-resistant. Speakman was very discerning in his choice. What could be better to apply to wool than a polyamino-acid with the same proportion of polar groups in it as wool itself had?

A related compound is thiazolid–2·5 dione which also reacts with water to yield nylon 2, and it also has been applied to wool.

$$n \begin{array}{l} NH-CH_2-CO \\ \vert \qquad\qquad\quad \vert \\ CO\text{———}S \end{array} + H_2O \longrightarrow [-NH-CH_2-CO-]_n + nCOS$$

The synthesis used by Leuchs consisted of the reaction of glycine with methyl chloroformate at 0° C. in aqueous alkaline solution:

$$CH_3OCOCl + NH_2-CH_2-COOH \longrightarrow CH_3-OCONHCH_2-COOH$$

hydrogen chloride being split off and taken up by the alkali. The ester was then precipitated by adding acid, and it was reacted with thionyl chloride at 60° C.

$$CH_3OCO-NH-CH_2-COOH \xrightarrow{SOCl_2} CH_3-OCONHCH_2-COCl$$

$$\longrightarrow \begin{array}{l} CH_2-CO\diagdown \\ \vert \qquad\qquad\quad O \\ NH-CO\diagup \end{array} + CH_3Cl$$

The second stage of this reaction takes place spontaneously.

Other reactions have been devised to make anhydrocarboxy-amino acids, notably by the reaction of phosgene with the amino-acid usually in organic solution, and such a reaction has been described by Farthing and Reynolds of I.C.I. It may be represented:

$$\begin{array}{l} CH-R-COOH \\ \vert \\ NH_2 \end{array} \xrightarrow{COCl_2} \begin{array}{l} CH-R-COOH \\ \vert \\ NH-CO-Cl \end{array} \longrightarrow \begin{array}{l} CHRCO\diagdown \\ \vert \qquad\qquad O \\ NH-CO\diagup \end{array}$$

Amino-acid. — Carboxyalkyl carbamyl chloride. — Anhydrocarboxy-amino acid.

They prepared in this way the anhydrocarboxy derivatives of:

glycine, where $R = H$
β-phenylalanine, where $R = CH_2—C_6H_5$
valine, where $R = CH(CH_3)_2$ (CH bearing two CH_3 groups)
and leucine, where $R = CH_2—CH(CH_3)_2$

There are several papers that have been published by Ballard and Bamford (Courtaulds) on the mechanism of the polymerisation of the anhydrocarboxy-amino acids, and these may have helped to form a base from which new fibres can and will be developed. Progress is slow; there are apparently no short cuts that inspired guesses have indicated. Structural complexity is increased by the presence of asymmetric carbon atoms in those α-amino-acids higher than glycine. The field of activity is a good one for the long-term development of polypeptide fibres. Possibly the proliferation of chemical compounds, the preparation of as many polypeptides as possible (not just half a dozen), might reveal useful guide lines. So far, the nylon 2 that has been made has been a low polymer with a D.P. of about 20, but it should be possible to increase this to, say, 100. Such an increase would be necessary to yield the high strength needed for a good fibre. Nylon 2 seems to have the best chance of all of yielding a good silk-like fibre or in staple form one that is wool-like.

But although nylon 2 itself has never got very far, there are two derivatives of it that have been made and tried as far as fibres. They are discussed below.

Methyl Nylon-2 or Polyalanine. If anhydrocarboxyalanine is used as the starting material instead of the corresponding glycine derivative, then the product after treatment with water will be α-methyl nylon 2, *i.e.*,

$$[—NHCH(CH_3)CO—]_n.$$

The application of this polymer to wool has been reviewed by the author in *Wool Shrinkage*, pp. 374–7, 1953, London, National Trade Press. Furthermore, the fibre possibilities of this material have been examined. The anhydrocarboxyalanine was polymerised in nitrobenzene solution using a primary base such as hexylamine as a reaction initiator. Methyl nylon 2 or polyalanine is formed and separates as a gel from the reaction solution, but very slowly, taking several days. If it is separated, washed in ether, dried and then dissolved in dichloroacetic acid the resulting solution can be used for spinning. This spinning solution is extruded through a spinneret into a coagulating bath of methanol. The so-obtained fibres can be stretched either cold or in steam. Filaments with 3 gm./den. tenacity and 20 per cent extension at break are reported to have been obtained.

Poly-γ-Methyl-L-Glutamate. This is another polymer of the substituted nylon 2 class, one which has actually been made into fibres for evaluation. The fabric from them is said to have been white, to have been thermally stable and not susceptible to yellowing. It was, however, completely soluble in strong aqueous alkalis.

The raw material for the fibre is L-glutamic acid, which is made in Japan and America for use as a taste potentiator; sodium glutamate when added to soups gives them a chicken flavour. The half ester of glutamic acid is made and is treated with phosgene to give the anhydro-carboxy compound.

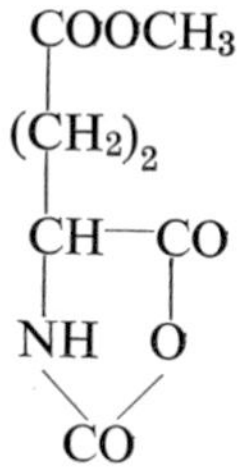

$$\begin{array}{l} COOCH_3 \\ | \\ (CH_2)_2 \\ | \\ CH—CO \\ |\quad\quad | \\ NH\quad O \\ \searrow\ \swarrow \\ \ \ CO \end{array}$$

Thereafter the treatment is similar to that given to methyl nylon 2. Polymerisation was carried out in 80/20 methylene chloride/dioxan with tributylamine as the catalyst. This gave a 7·5 per cent solution of poly-γ-methyl-L-glutamate. This was wet spun into a coagulating bath of acetone containing 5 per cent water. The yarn filaments were collected, washed, dried and stretched. The yarn was quite strong with a tenacity of 3·1 gm./den. (2·7 wet) and an extension of 11·8 per cent at break (15·0 wet). The thread was no longer soluble in methylene chloride; presumably the stretching had changed its structure. The filaments were circular in cross-section. The fibre made from the yarn had a silk-like appearance, was warm and had a scroopy handle, and was resistant to creasing. It could be ironed at a temperature that would scorch linen. Why has it not been made? Too costly? Susceptible to repeated laundering in weak alkali? Unstable to the ultra-violet component of light?

The formulae of glycine and of nylon 2 are:

$$NH_2—CH_2—COOH \qquad [—NH—CH_2—CO—]_n$$

Of any other substituted glycine, *i.e.*, any other α-amino-acid, the formulae of monomer and of substituted nylon 2 are:

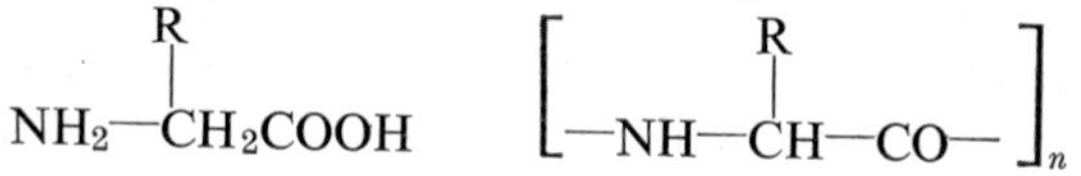

$$NH_2—\overset{\displaystyle R}{\overset{|}{C}}H_2COOH \qquad \left[—NH—\overset{\displaystyle R}{\overset{|}{C}}H—CO—\right]_n$$

When glutamic acid is the amino-acid that is used, R has the value $CH_2—CH_2—COOH$ so that the formulae of glutamic acid, its half ester and the substituted nylon 2 derived from it are:

$$NH_2—\underset{}{\overset{CH_2—CH_2—COOH}{\overset{|}{CH}}}—COOH \qquad NH_2—\overset{CH_2—CH_2—COOCH_3}{\overset{|}{CH}}—COOH$$

$$\left[—NH—\overset{CH_2—CH_2—COOCH_3}{\overset{|}{CH}}—CO—\right]_n$$

This is quite advanced. But a long time has elapsed since the polymer was first made, and not much has come of it. Those who are interested by this fascinating work should read B.P. 864,692 (4*s*. 6*d*. from The Patent Office, 25 Southampton Buildings, London, W.C.2).

Nylon 1

Perhaps here we are on more doubtful ground. Some mono*iso*cyanates have been polymerised at low temperatures in the range −20° to −100° C. in dimethyl formamide solution in the presence of an anionic catalyst, and they have yielded substituted nylon 1 of the type

$$—[NR—CO—]_n$$

When R is the butyl radicle the polymer has a satisfactory melting point of 209° C. and will dissolve in benzene, from which solution clear tough films can be cast and probably fibres could be spun. The unsubstituted nylon 1, which would be $(—NHCO—)_n$, cannot be made, at least not by this method. Perhaps nothing quite suitable for a fibre has yet been made in this way, but some of the polymers had molecular weights in excess of 100,000 and must have had the high D.P. which is the first essential for fibre formation.

Nylon 6-T

One cannot leave this subject without making reference to nylon 6-T, made from hexamethylene diamine and terephthalic acid. This has got further than any of the other fibres discussed in this chapter. One feels that it must be a success commercially in the future. It has been discussed in more detail on p. 380.

The Nylons

So much for the nylons of the future. By their close resemblance to the natural polypeptides they seem to be the most likely candidates for commercial fibre success, and of them all, nylon 2 and nylon 3, perhaps substituted, look the most promising. But research on polyamides has been going on for decades in the laboratories of the fibre giants. What they have made and what they have found is not usually disclosed until something is ready for marketing. One guesses that the road has been harder than might have been expected, and that the dividend on the huge financial outlay for research has been only small. Polyamides, polypeptides and proteins constitute a field of endeavour in which it is only too easy to become embedded. Clear thinking, and a determination to keep the work as simple as possible, are necessary if useful results are to emerge.

Some forthcoming fibres have been announced. Allied Chemical are reported to have a new one called " Source " in the offing, and they certainly carried out and disclosed a lot of work on nylon 4 some years ago. It transpires that Source is probably a nylon 6 in which polyester fibrils are embedded longitudinally.

Du Pont's Quiana is a polyamide and is claimed to be silk-like. We shall see. But in assessing what it is likely to be we must remember that du Pont have done more for synthetic fibres than most, more than anybody, in fact, with their nylon, Orlon and Teflon. Thirty or forty years ago they did for synthetic fibres what Samuel Courtauld had done for rayon thirty years earlier.

The Japanese firm Toyobo are reported to have a new fibre near to production—one called K-6, a graft co-polymer of protein and polyacrylonitrile. This fibre is reported to be silky. Again, one can only wait and see. But it is certain that all these fibres will be good fibres. No chemical giant is going to spend the vast sums necessary to get a new fibre going on anything that is of doubtful quality. Furthermore, a large number of different chemical polymers will have been made by each manufacturer, and the one polymer chosen for commercial development must at least have come top of this class. Whether the publicity will be lived up to, and whether the new fibres will be worth their price, is what we shall have to wait and see.

Polyimides

The structure of an amide may be represented $R{-}C{\overset{\displaystyle O}{\diagup\!\!\diagup}}\!\!\diagdown NH_2$.

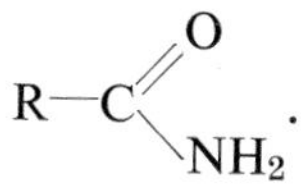

If it is a secondary or a tertiary amide it will be:

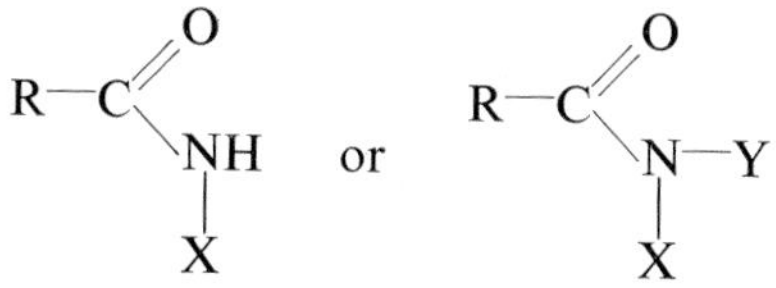

$$R{-}C({=}O){-}NH{-}X \qquad \text{or} \qquad R{-}C({=}O){-}N(X){-}Y$$

where X and Y are such substituent groups as alkyl (methyl, ethyl, etc.) or aryl (phenyl, tolyl, etc.) or aralkyl such as benzyl. In a polymeric amide the fundamental structure is:

$$-\overset{\overset{\displaystyle O}{\|}}{C}-NHRNH\overset{\overset{\displaystyle O}{\|}}{C}-R'-,$$

where R and R′ are divalent groups such as phenylene or methylene.

The fundamental imide structure is different. In it, two acid groups are united to one nitrogen atom. The most common is phthalimide:

$$C_6H_4\begin{matrix}\diagup C(=O) \diagdown \\ \diagdown C(=O) \diagup\end{matrix}NH$$

Only the *ortho* or common phthalic acid forms an imide (or an anhydride), the *meta* compound called *iso*phthalic acid, and the *para* compound terephthalic acid will form neither anhydride nor imide. At first glance phthalimide would seem to be a poor start for synthesising a fibre, but pyromellitic imide (see below), which is closely related to it, is much more promising.

Polyimides are made from dianhydrides and diamines by way of a soluble and fabricatable (capable of being made into a fibre, for example) polyamic acid. The polyamic acid is soluble, and it is spun into a fibre and is then treated to convert it into an insoluble and thermally stable imide fibre. The novel part of the process is that the fibre is made from the soluble polyamic acid and is then (after fibre formation) converted to the intractable polyimide.

Thus pyromellitic dianhydride and *bis*-(4-aminophenyl) ether will react to give a polyamic acid:

$$n\,O\begin{matrix}\diagup OC \diagdown \\ \diagdown OC \diagup\end{matrix}C_6H_2\begin{matrix}\diagup CO \diagdown \\ \diagdown CO \diagup\end{matrix}O + n\,NH_2-C_6H_4-O-C_6H_4-NH_2 \rightarrow$$

Pyromellitic dianhydride. *bis*-(4-aminophenyl) ether.

$$\left[\begin{matrix} HOOC \diagdown & & \diagup CO-NH-C_6H_4-O-C_6H_4- \\ & C_6H_2 & \\ -NHOC \diagup & & \diagdown COOH \end{matrix}\right]_n$$

Polyamic acid.

This acid or perhaps one of its salts is spun into fibres. These are converted to polyimides by thermal treatment at, say, 300° C., yielding from the polyamic acid just above:

$$\left[-N\begin{matrix} OC \\ OC \end{matrix} C_6H_2 \begin{matrix} CO \\ CO \end{matrix} N-C_6H_4-O-C_6H_4- \right]_n$$

Polyimide.

It is thought that some cross-linking between molecules occurs, doubtless because any one molecule of diamine is just as likely to combine with the anhydride groups on two different molecular chains as with two anhydride groups on the same molecular chain. The polymers have remarkable thermal stability, apparently greater than that of aromatic polyamides (Chapter 21), and they are very resilient. Doubtless the cross-linking gives toughness and resilience and elasticity to the fibres. To the author they look rather more like industrial than apparel or household fibres of the future.

Flameproof Cellulose

Perkin devised a method of flameproofing cotton, but it was not good enough to stand the test of time. One of the greatest needs of the textile industry in the 'twenties and 'thirties was considered to be a method of making cellulosic fibres, notably viscose rayon, flameproof. It is still one of our greatest needs. Many people have worked at the problem, and some partial successes have been achieved, but nothing good enough to last. The need is obvious, especially for kiddies' nightwear, and indeed for the apparel of old people; it is the very young and the very old who are most prone to accidents due to "clothes catching fire". Anyone who can find a good method of flameproofing cellulose will earn a halo which he will be able to wear not only in Britain but anywhere in the world where there are young children and fires. The basic requirements are two-fold: (1) the treatment must not make the cloth hard or boardy; the handle must remain soft; (2) the treatment must be fast to repeated washings and launderings. The second point is especially important, because if a fabric originally flameproof should lose its proof when washed, a false sense of security may be engendered, the consumer thinking that a garment is flameproof when it no longer is so. How many unsuccessful trials have been made to find a satisfactory way of flameproofing rayon and cotton! The author alone has made hundreds and has no doubt that others, especially industrial research establishments, must have made hundreds of thousands. If a good process ever does come, it will be of inestimable value. It is the sort of problem that might respond to a little new thinking, a fresh start; perhaps it might capitulate to some clever new graduate just starting to earn a living. There is no reason to think that the problem is insoluble. In this connection there is practically no fire risk with wool; fibres of the nylon 2 type will probably be similar in this respect and will be especially useful on that account. But it still remains that there is a crying need for a good method of flameproofing viscose rayon and cotton. Let us hope that flameproof viscose rayon will be one fibre of the future.

Polyhydrazides

Another way of making polymers with a high content of polar groups is by reacting a diester with hydrazine. For example, oxalic ester will react in this way with hydrazine.

$$n\mathrm{EtOOCCOOEt} + n\mathrm{H_2NNH_2} \longrightarrow -[\mathrm{CONHNHCO}-]_n + 2n\mathrm{EtOH}$$

The product is a nice white polymer and worth examination.

Similarly, dimethyl terephthalate will react with hydrazine to form a polymer:

$$n\mathrm{MeOOC}-\mathrm{C_6H_4}-\mathrm{COO\,Me} + n\mathrm{H_2NNH_2} \longrightarrow \left[-\mathrm{OC}-\mathrm{C_6H_4}-\mathrm{CONHNH}-\right]_n + 2n\,\mathrm{MeOH}$$

This again is probably worth following up. All that is necessary to make either polymer is to shake the ester with aqueous hydrazine. If the polymer is a solid it can helpfully first be dissolved in methylene chloride or some other inert solvent. The work should be done under qualified supervision, as hydrazine derivatives can be dangerous.

Polyureas

The Japanese polyurea fibre Urylon has been discontinued, and Russian work on this same subject seems to have stopped. The Japanese fibre has been dropped because of the difficulty of making nonamethylene diamine cheaply. The people who made the fibre were specialists in the production of urea, which is abundant and cheap. Its structure, $\mathrm{H_2NCONH_2}$, makes it look very promising material for linear polymers having a high content of polar groups. If it could be induced under pressure to combine with acetylene one might obtain such polymers as

$$[-\mathrm{HNCONHCH_2CH_2}-]_n$$

which would have the same polar group content as real silk.

One must not let the imagination roam too far. The possibilities are legion. Although some of the work indicated requires costly equipment to be found only in large establishments, other parts of it are less demanding and could be undertaken by technical college evening classes that were desirous of doing some fibre research. Their work might easily turn up a winner; it would be sure to turn up new knowledge.

Chameleon Camouflage

One last development in the fibres scene has been the attempt to make chameleon camouflage. The development of a dynamic camouflage system, one that would change colour with the individual soldier's surroundings, is a project that has been undertaken by the United States Army Laboratories' Clothing and Organic Materials Division. Ordinarily the dyer of fabrics wants his colour to be stable and unchangeable. Some colorants are photochromic; they change colour with the light. One state, *A*, is stable, but if material dyed with that colorant is exposed to light the colorant changes to an unstable form *B*. When the light is turned off it reverts to its original state. In some photochromic substances the two states *A* and *B* are different in hue. For example, the metal complexes of dithizone (short for diphenylthiocarbazone) showed large spectral differences between the stable *A* and the metastable *B* forms. The formula of dithizone is:

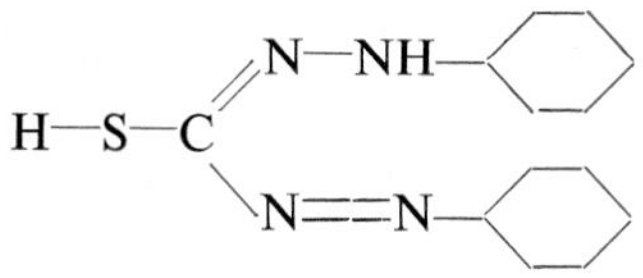

$$\mathrm{H{-}S{-}C}\begin{matrix} \nearrow\!\!\!\!\!= \mathrm{N{-}NH{-}C_6H_5} \\ \searrow \mathrm{N{=}N{-}C_6H_5} \end{matrix}$$

and the hydrogen atom at the extreme left can be replaced by a metal just as it can be in a mercaptan HSR. By way of example the mercury compound

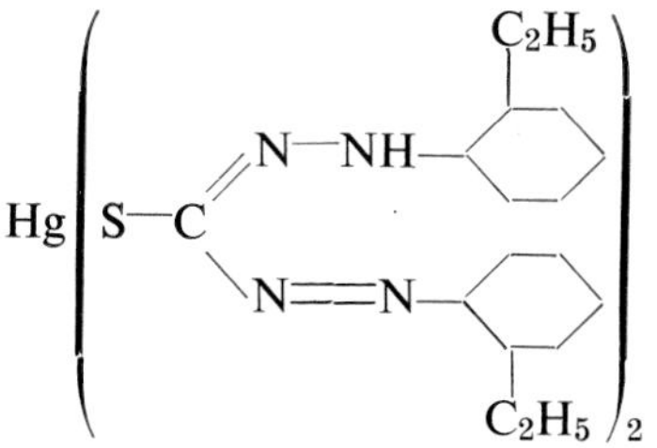

$$\mathrm{Hg}\left(\mathrm{S{-}C}\begin{matrix} \nearrow\!\!\!\!\!= \mathrm{N{-}NH{-}C_6H_4{-}C_2H_5} \\ \searrow \mathrm{N{=}N{-}C_6H_4{-}C_2H_5} \end{matrix}\right)_2$$

is yellow in the *A* form but blue in the *B* form. The palladium compound

$$\mathrm{Pd}\left(-\mathrm{S}-\mathrm{C}\begin{array}{l}\mathrm{CF_3} \\ /\!\!/\ \mathrm{N-NH-} \\ \backslash\ \mathrm{N{=}N-} \\ \mathrm{CF_3}\end{array}\right)_2$$

is green in the *A* form, yellow in the *B* form. Doubtless it will take some time to catch up on the chameleon, but a start has been made. One day we may have chameleon fibres, fibres pigmented with photochromic pigments. One obvious approach to this end is to examine the pigments in the skin of the chameleon. Are the chameleon's colour changes biological or purely physical? Do they take place after he is dead? If they do not, the problem is going to be difficult. It will be the soldier who will have to be dyed, not his uniform.

FURTHER READING

R. W. Moncrieff, " Fibres of the Future? (Nylons 1, 2, and 3) ", *The Dyer*, pp. 221–226 (22 February 1963).

C. H. Bamford, L. Brown, A. Elliott, W. E. Hanby and I. F. Trotter, " β-Forms of Fibrous Proteins and Synthetic Polypeptides ", *Nature*, **171**, 1149–1151 (1953).

A. C. Farthing and R. J. W. Reynolds, " Anhydro-N-Carboxy-DL-β-Phenylalanine (Preparation from amino-acid) ", *Nature*, **165**, 647 (1950).

D. G. H. Ballard, " Improvements in or relating to the Production of N-Carboxy Alpha Amino Acid Anhydrides ", B.P. 854,139 (1958).

D. G. H. Ballard and C. H. Bamford, " The Polymerization of N-Carboxy-α-Amino Acid Anhydrides—The Chain Effect ", *Proc. Roy. Soc. London, Ser. A.*, **223**, 455 (1954), and **236**, 384 (1956).

R. W. Moncrieff, " Chameleon Camouflage ", *The Dyer*, **136** (7) 518, 521 (1966).

Courtaulds Ltd. (inventor D. G. H. Ballard), " Improvements in and relating to the Production of Artificial Threads ", B.P. 864,692 (1961).

PART 4

PROCESSING

CHAPTER 43

THE CONTROL OF STATIC

WHEN yarns are being woven or knitted, the rubbing of fibre on machine will generate electrostatic charges if the fibre has reasonably good insulating properties. This " static " can be troublesome even with some of the natural fibres, and on a dry frosty morning, when the fibre and the surrounding air carry only a little moisture, the starting of wool weaving may be difficult. Cellulose acetate, with its reduced moisture regain (6·5 per cent compared with 11 per cent for viscose rayon), is even more troublesome, and as it was the first of the fibres to have a low regain it encountered all the initial difficulties, although these were overcome in the course of time. When nylon came along it had a still lower regain and was a good insulator and was even more prone to static; more recently some of the other synthetics with regains that are much lower still have presented new problems in static elimination.

For the most part, except in cold frosty weather, the natural fibres contain sufficient moisture for any electrostatic charges generated to leak away through them to earth, but the excellent insulating properties of the synthetics do not allow this, and unless suitable precautions are taken they will give a lot of trouble with static. There are four ways in which the trouble has been defeated.

1. *Atmospheric control.* By the maintenance of warm, humid conditions in weaving or other sheds. Static charges then leak away to earth through the moist fibre. This, of course, is an old and successful device and was used by the best weavers before the advent of the synthetic fibres.

2. *Anti-static finish.* By the application to filaments, usually by passing them over a wick, just prior to their take-up on the bobbins of the spinning machine, of a conducting finishing agent. A solution of triethanolamine oleate is one that has been largely used. If the finish is subsequently removed by any wet process, it is usually necessary to apply a similar finish again before the processing can be satisfactorily continued.

3. *High-voltage atmospheric ionisers.* These are small devices which are placed near the loom or warping mill. Very high electric potentials are applied to fine points from which they discharge into the air, locally ionising the air and allowing the charges on the fibre

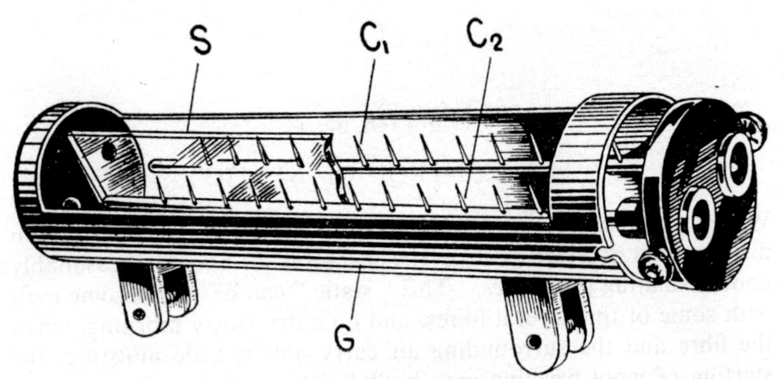

[*The British Cotton Industry Research Association*]

FIG. 198.—Diagram showing the construction of the Shirley Static Eliminator (manufactured by The Record Electrical Co. Ltd.).

C_1, C_2. Discharge points or electrodes positive and negative.
S. Separator (cut-away) of insulating material such as cellulose acetate.
G. Earthed metal shield, usually of aluminium.

The electrodes C_1, C_2 are fed by D.C. leads giving an output of 12,000 volts, positive and negative, respectively.

[*The British Cotton Industry Research Association*]

FIG. 199.—The Shirley Static Eliminator in use on a warping mill.

The point discharge from the electrodes ionises the surrounding air and so makes it conducting. Any static charge which develops on the fibre can leak away through the conducting air. The effective range is from 6–24 in. from the electrode.

to leak away through it. The Shirley Static Eliminator (Figs. 198, 199) has already found considerable use.

4. *Radioactive eliminators.* Some naturally occurring elements, notably radium, emit α-particles, which are really the nuclei of helium atoms, and the α-particles ionise the air through which they pass. Plates coated with a radium compound have been used to ionise the air near a loom and so allow static to dissipate, but α-particles are dangerous and can induce the growth of tumours in people exposed to them, so that it proved difficult to have sufficient radioactivity to prevent static without at the same time offering a health hazard. The difficulty has been overcome by the use of a radioisotope thallium 204 instead of radium; this is an artificial element made by exposing natural thallium, which is a mixture of isotopes 203 and 205, in an atomic pile where it is subjected to nuclear bombardment; the nuclei of the natural thallium 203 atoms absorb a neutron apiece and are converted into the rather unstable thallium 204 atoms. These unstable atoms continuously give off β-particles, which are high-speed electrons and consequently of low mass; they are very much less dangerous than α-particles, and in sufficiently small quantities β-particle emitters, of which thallium 204 is one, are safe to use. The Rase (Radio Active Static Eliminator) unit was developed by the British Cotton Industry Research Association and is made by the Record Electrical Co. Ltd.; it consists of a circular plate some 2 in. in diameter which carries 2 millicuries thallium 204. It is attached to a loom and is ordinarily kept in a cup with a lid during the working day, the cup absorbs the β-particles so that none reach the operative and, of course, at this time the unit is ineffective; before the operative goes home at the end of the day he takes the radioactive plate out of its cup and allows it to hang over the loom, and in the morning he replaces it in the cup. The consequence is that during the night, β-particles are being emitted into the air surrounding the loom and ionise the air, *i.e.*, they make it conductive so that it can conduct away to earth the electrostatic charges on the warp or fabric in the loom. If this is not done the charged fibre will attract to itself electrostatically, countless minute particles of dirt from the air, and these become so intimately embedded in the fibre that they cannot be removed by any practical process that may subsequently be carried out; the soiling that they produce is called "Fog Marking". Fig. 200 shows warp-knit fabric which has not been protected by a static eliminator and which has been badly soiled by fog marking, together with fabric which has been protected by the presence of the β-emitting thallium 204 and which is satisfactory. It has to be added that the thallium 204 atoms, breaking

down as they continuously are, have only a limited life; their "half-life" (time to fall in radioactivity to one-half of their original value) is 3½ years; it is anticipated that the Rase units will have an effective life of about five years before needing renewal. The use of strontium 90 instead of thallium 204 was adopted by British

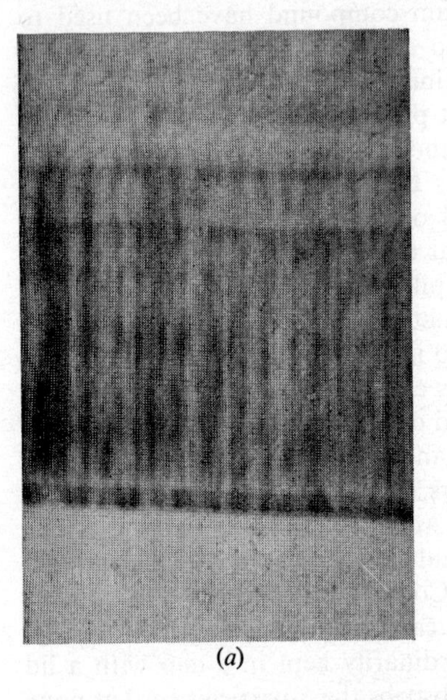

(*a*)

(*b*)

[*Record Electrical Co., Ltd.*]

FIG. 200.—Fog-marking on warp-knitted fabric: (*a*) without static eliminator; (*b*) with Shirley Radio Active Static Eliminator (RASE).

Celanese Ltd.; this has a half-life of 25 years so that eliminators containing it should have an effective working life of 40 years; furthermore, the emission of β-particles from strontium 90 is more energetic than from thallium 204, so that the units are more powerful and more effective. Because they are more powerful, greater care has to be taken to ensure that weavers are not exposed to a toxic dose of radiation.

FURTHER READING

Report on Textile Institute Conference, Zurich, Sept. 1956: "Static Electricity in Textiles". Manchester (1956).

CHAPTER 44

TEXTURED YARNS

FANCY yarns which provide bouclé, corkscrew, slub and similar effects have for generations been part of the stock-in-trade of the cotton and worsted spinner; they are made by various combinations of twist and sometimes by the union of two fibres, *e.g.*, a cellulose acetate slub may be caught at intervals in a cotton yarn and contributes slubs with differential dyeing properties which may give a pleasing effect in a furnishing fabric. Such fancy yarns have had only a limited application, because novelties cease to become novelties when they become ubiquitous and most of the appeal of fancy yarns has come from novelty rather than from intrinsic beauty and performance. The man-made fibres can be used for such yarns equally as well as natural, but apart from the use of crêpe and voile yarns have not commanded much attention; the man-made fibres have relied on their intrinsic characteristics for their wide application. There have, however, been partial exceptions to this rule, notably in the production of crêpe and of slub yarns. These, together with those fancy or textured yarns, whose qualities depend on some novel property such as thermoplasticity, of the man-made fibres will be considered briefly.

CRÊPE YARNS

The first crêpe yarns were made from real silk. A very high twist is inserted, and the finer the yarn the higher the twist must be; something of the order of 70 t.p.i. will be suitable for 75 denier yarn. Crêpe yarns are usually woven as weft in a plain warp, and when the fabric is wetted the filaments swell, and then, because of their high twist, they snarl and distort, so that they give " pebble " or " figure " to the fabric. For most marocains and crêpe-de-chine fabrics the crêpe weft is woven alternately two picks S twist and two picks Z twist. In some looms alternate picks will be S and Z. When this technique was borrowed for the man-made fibres from the silk industry, it was found that viscose rayon was eminently suitable for crêpes and, indeed, has been used in tremendous quantities. Cellulose acetate when twisted to crêpe twist (approximately 60–70 t.p.i. in 100 denier) did not give a good crêpe figure to the fabric it was woven in; the fabric did not shrink when finished, as it should have done; eventually it was found that if the cellulose acetate yarn was twisted in an atmosphere of steam, preferably under a little pressure, the crêpe yarn that resulted would cause the fabric in which it was woven to shrink and to exhibit a good crêpe figure when finished. The refusal of the cellulose acetate yarn, twisted normally (in the absence of steam) to crêpe, was partly due to its low swelling capacity in water and partly due to its plastic and easily deformable nature; the steam-twisting made no difference to the low water imbibition, but it did orientate the yarn and give it some rigidity, so that the cellulose acetate crêpe yarn behaved more like a highly twisted yarn and less like a piece of mangled cellulose acetate. It was in fact the thermoplastic nature of cellulose acetate which permitted its conversion into a good crêpe yarn, and it was this same property which permitted its conversion to the highly stretched form described on p. 262.

Voile Yarns

Yarns that have been twisted to about one-third or one-half of the twist of crêpe yarns, say about 20 t.p.i. in 100 denier or 25 t.p.i. in 75 denier, are known as voile yarns. They are ordinarily used only in one way, say S, of twist, and simply make the yarn more compact. They are used as the name suggests in voiles, and in light mesh-like fabrics such as curtainings. They are commonly made only from filament, not from staple, yarn.

Slub Yarns

Rayons have been spun in continuous filament with their filaments changing in denier. Ordinarily in spinning rayon it is important to maintain the denier constant within limits of about 3 per cent so that the fabric made from yarn will be even and regular. The slub yarns that were spun had deniers which varied by several hundred per cent; they consisted of yarns with thick and thin places which alternated every few inches to simulate a slub yarn. Meubalese is such a yarn made from cellulose acetate and it has found a certain popularity for furnishings.

Heat-set Stretch Yarns

Some of the synthetic fibres have thermoplastic properties that are much more pronounced than those of acetate, notably nylon and Terylene, and their thermoplastic nature is commonly made use of to heat-set fabrics made from them and to give them dimensional stability. The same property, that of thermoplasticity, has also been made use of to convert these fibres, so far mostly nylon, but Terylene is following, into yarns which have new properties, and these yarns with new properties are important in that they appear not so much to come into the category of fly-by-night novelties but into that of devices for making bread-and-butter articles such as men's socks and ladies' stockings with novel and desirable qualities. Most men at some time or other will have worn knitted woollen socks, and those who have also tried socks made in the conventional way out of nylon and Terylene yarns cannot have failed to notice how they have lacked the give and spring of wool. But some of the new yarns known collectively as " stretch " yarns, made from nylon, have more give and springiness than any wool sock that was ever made. This is not equivalent to saying that in all respects, or even on balance, they are superior to woollen socks, but certainly in respect of ease of putting on and in snugness of fit (and of course in length of wear) they are superior.

There are in principle three main methods of making stretch yarns. They are:

1. *Twisting* (either false twisting or conventional). Usually twist is inserted, the yarn in the twisted state is heat-set, and then it is untwisted. It kinks because of the " set " distortion. The fibre is crimped sinusoidally. Typical examples are Helanca, Fluflon, Superloft, Saaba and FT(ARCT) yarns.

2. *Drawing over Hot Knife Edge* (or over a hot plate and then a knife edge). This imparts a spiral-like curl to the fibre, because one side of each fibre (that touching the hot knife) is treated differently from the other side. In one sense this copies the bicomponent structure of wool (p. 676). A typical example is Agilon.

3. *Stuffing Box Method.* Filaments are forced into a stuffing-box or similar chamber, are crimped and the crimp heat-set in them. This imparts a saw-tooth crimp and gives the greatest bulk of all. Typical examples are Ban-Lon (Textralized) and Spunize. In the Spunize version the yarn is fed into the box in warp form so that, say, 400 ends are treated at once; this requires a final process of end separation but enormously increases the rate of production.

Some of the best known of the stretch yarns are described below. All told their production may amount to 200 million lb. per year.

Helanca (Twist Method)

Helanca is a product of Heberlein & Co. A.G., Switzerland, and is made under licence in other countries too. Usually it is made of nylon, but other synthetic yarns can be used, provided that they are thermoplastic. It possesses amazing stretchability; it does not sag and it is soft to the handle. Helanca yarns are made by twisting continuous filament yarns to high twist levels, heat-setting, and then removing the twist, probably in several steps. The yarn is usually two-ply with the plies having been twisted in opposite directions to balance the yarn construction and to prevent excessive liveliness. The filaments of Helanca are crimped sinusoidally, and when they are straightened by stretching the yarn they increase in length by some 400 per cent; on release of the tension they revert to their original length. Fig. 201 shows (*a*) an ordinary filament nylon yarn,

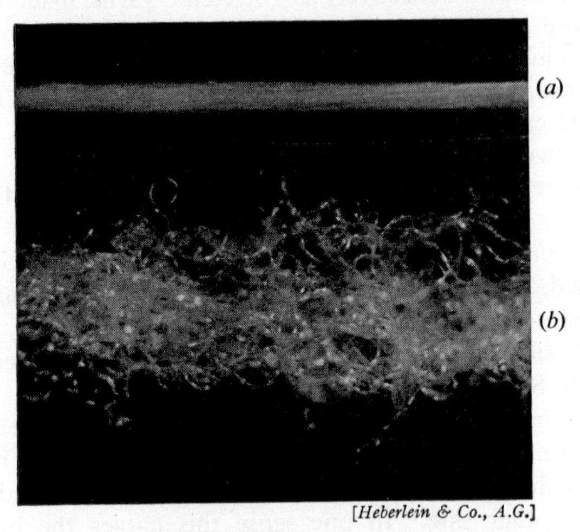

[*Heberlein & Co., A.G.*]

FIG. 201.—(*a*) Multi-filament nylon yarn; (*b*) the same yarn after conversion to Helanca.

(*b*) the same yarn after conversion into Helanca. Helanca was introduced in 1947, and by 1952 had achieved international success (it made stretch socks famous). It overcomes the " lack of stretch " defect of the synthetic fibres; it is as easy to put on a Helanca nylon sock as a woollen sock. So great is the stretch that only two sizes, large and small, need to be made, a matter of convenience and saving in financial outlay to manufacturer, wholesaler and retailer, and something that eliminates the gift-buyer's problem of size. Moisture absorption is high, due to mechanical entrapment, and not to any increase in the regain of the nylon (it would be impossible to bring this about by mechanical means). For socks, ladies' stockings, intimate underwear and girdles, Helanca has proved to be a great success; in addition to its own special virtues it has all those of the nylon from which it is made—strength, durability and quick-drying. A pair of Helanca nylon socks, equivalent to an ordinary size 11, weighs only 21 grams. The production process is slow and the approximate production from 1 spindle on 70 denier nylon for a month (three shifts, no week-ends) is only 0·7 lb.

Helanca is a great achievement, but it still lacks, in the author's view at least, the warmth and comfort of wool. Helanca, too, is the textured yarn that showed the way; it was the success achieved by Helanca that led to the production of the others. Helanca is used mainly for ladies' underwear, stockings, men's socks, men's underwear, foundation garments, swim-wear, surgical bandages and ski-wear.

Fluflon (Twist Method)

Fluflon is a stretch bulk yarn of good uniformity. The process for making it is licensed by Marionette Mills Inc. It is made by a continuous process on the false-twist principle. Probably two yarns are fed to a twister which inserts extra twist to double them, heat-sets them continuously in a hot chamber and then untwists the two yarns and winds them up on two separate packages. Production is high and 1 spindle will yield 67·5 lb. per month (three shifts, no week-ends). Very light-weight garments can be made from it; a pair of socks may weigh only 21 grams. Fluflon can be stretched 400 per cent and will return to its original length when released. As a rule Fluflon is two-ply, with the plies having been twisted in opposite directions to balance the yarn construction and to prevent excessive liveliness; the doubling is carried out as a separate operation. Yarn morphology and properties of Fluflon are very similar to those of Helanca. The high production per spindle of Fluflon is greatly to its advantage. Fluflene is a similar yarn made from Terylene.

Agilon (Knife Edge Method)

Agilon is a stretch bulk yarn made by a process controlled by the Deering Milliken Research Corporation of Pendleton, South Carolina. It is a non-torque yarn, made not by twisting, as are Helanca and Fluflon, but by drawing nylon yarn, *e.g.*, 15 denier monofil, over a hot sharp knife-blade. Agilon yarns will stretch 400 per cent and return to their original length when tension is released; they are similar in appearance to Helanca and Fluflon, but are not inherently twist lively. So far, Agilon stretch yarns have been used mostly for stockings, but socks, circular knit fabrics, women's sweaters (resistant to pilling), upholstery fabrics and carpets have also included it. The carpets with a cut curly pile and with a rubber backing to hold the spiralling fibres firmly are particularly effective. Fewer sizes—usually two—of stockings and socks are sufficient.

Whereas torque yarns usually have to be two-ply, or at any rate an even numbered ply, non-torque yarns can be used satisfactorily as singles and as an odd-numbered ply if required.

Ban-Lon (Stuffing Box Method)

The process for making Ban-Lon yarns is owned and controlled by J. Bancroft & Sons Co. of Wilmington, Delaware. Essentially it is one of inserting crimp into a thermoplastic yarn and heat-setting it at the same time. Nylon, which is the fibre to which the Ban-Lon process has mostly been applied, is crammed into a stuffing box, where the filaments adopt an accordion-like crimp and this crimp is heat-set. The nature of the crimp can be seen in Fig. 202, which

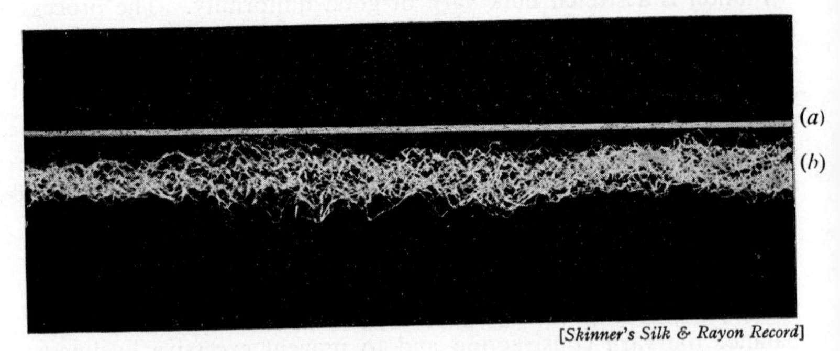

[*Skinner's Silk & Rayon Record*]

FIG. 202.—(*a*) Multi-filament nylon yarn; (*b*) the same yarn after conversion to Ban-Lon.

shows the Ban-Lon yarn and above it the ordinary filament nylon yarn from which it was made. The fibre has an extensibility of over 200 per cent due to straightening out the crimp, and it recovers this stretch when the tension is released so that a Ban-Lon yarn can be stretched and released almost like a strand of rubber. The degree of stretch is " moderate "; it is less than that obtained with Helanca and Fluflon, but is still very considerable, as can be seen from Fig. 203.

Apart from its elasticity, the yarn has, and so, of course, have garments that have been made from it, a moisture absorption of up to thirteen times their own weight. Remembering that the regain of nylon is only 4 per cent, this figure of 1,300 per cent at first seems incredible. The water absorption is not due to any increase in the regain of the nylon but is due to the moisture being held mechanically in the interstices between the crimped fibres. In just the same way a cotton terry towelling may be capable of absorbing some five times its own weight of moisture even although the regain of the cotton fibre is only about 9 per cent, and indeed a hank of ordinary nylon yarn immersed in water and removed therefrom may carry with it more than its own weight of water—there is a vast difference between regain and mechanical attraction and holding due to capillarity. It is claimed that the ease with which Ban-Lon textured yarns pick up moisture increases the comfort of garments made from them and side-steps one of the main defects of synthetic fibres, namely their

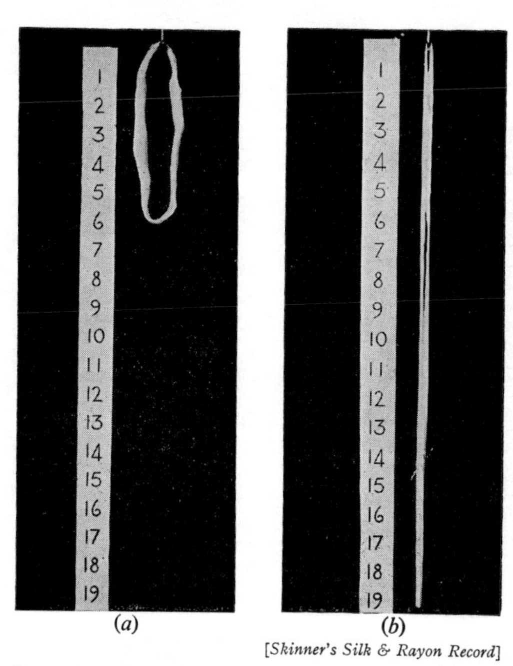

(*a*) (*b*)

[*Skinner's Silk & Rayon Record*]

FIG. 203.—(*a*) Ban-Lon yarn without tension; (*b*) the same yarn under tension (extension is above 200 per cent).

hydrophobic quality. Fig. 204 illustrates how a Ban-Lon nylon absorbs water. In (*a*) 10-gram hanks of 70 denier Ban-Lon, 2/40 s worsted and 70 denier continuous nylon yarn are shown; each is immersed in 90 ml. water, and in (*b*) it can be seen that the Ban-Lon fills the cylinder and apparently absorbs all the water; in (*c*) when the hanks are removed the Ban-Lon carries with it seven times, the worsted four times, and the nylon one and a half times its own weight of water. In a similar way air is retained in the fibre interstices, and this gives thermal insulation and warmth to the garment. Resilience is an intrinsic property of Ban-Lon yarns and resistance to abrasion

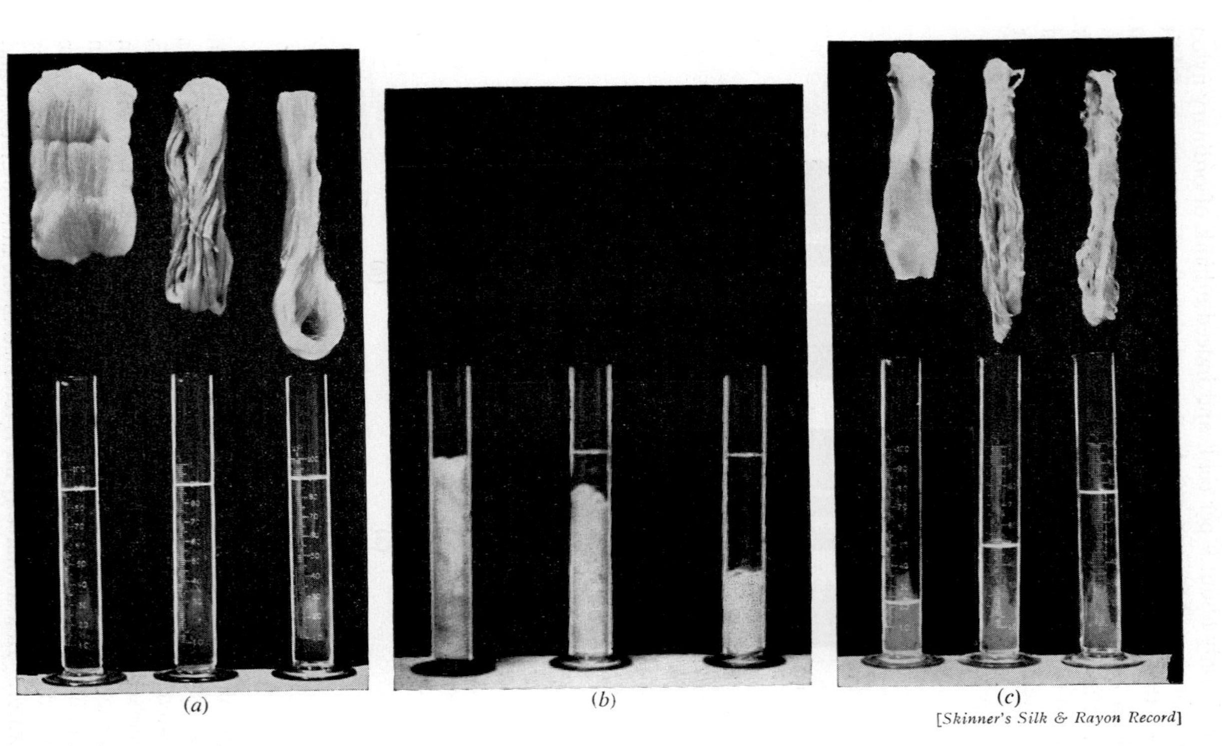

(a) (b) (c)

[*Skinner's Silk & Rayon Record*]

FIG. 204.—Water absorption by Ban-Lon nylon, wool, and ordinary nylon.

is good. Because they are continuous filament, pilling does not occur in garments made from them.

They are used for sweaters, where their elasticity, bulk, warmth and freedom from pilling are advantageous, and in tricots, socks, gloves, sports shirts and underwear. Gloves seem to be an ideal application for yarn with the elasticity and snugness of fit of stretch yarns.

Ban-Lon yarns have been made from Dacron and Orlon as well as from nylon.

Garments made from Ban-Lon yarns should be set at about 130° C. for 20 minutes to give the sweater, for example, complete stability against stretching and shrinking and to eliminate dye wrinkles and to improve the texture of the fabric. Dyeing is most easily done with disperse dyes at about 85° C. Acid dyes give better fastness but less level results.

Crimplene

Crimplene is the registered name given by I.C.I. Fibres Ltd. to bulked Terylene. The method of texturing is one that was devised by Cheslene and Crepes Ltd., from whom I.C.I. bought the patent rights. The main features of Crimplene are: (1) it is a stable non-torque yarn; (2) when its bulk is developed by wet processing (dyeing) after knitting, excessive shrinkage does not occur; (3) it has a fair bulk but is not too stretchy; (4) it is suitable for package dyeing. The yarn is used very extensively by the knitting industry for skirts, blouses and jersey fabrics. A glance round one of Marks and Spencer's stores will indicate how popular Crimplene has become.

Because it is so stable (it has been thermally set), fabric made from Crimplene can be dyed, *e.g.*, on high-temperature pressure winches without the need of having been pre-set. Some similar textured polyester yarns that are made on the Continent are less stable than Crimplene, and they do require pre-setting. Plain Crimplene fabrics have been made in such enormous quantities that they have become monotonous, and this has led to the manufacture of patterned knitted Crimplene fabrics. These tend to be flattened and to lose their pattern if dyed on the beam because of the pressure they are subjected to. This difficulty has accelerated the coming of high-temperature winches on which such fabrics can be dyed satisfactorily. Because Crimplene is a polyester it needs either high-temperature or carrier dyeing.

Taslan Textured Yarns

The process of making Taslan yarns is due to du Pont, who license their customers to practise it. It is in principle as follows:

Continuous filament yarn, usually nylon although Terylene also is coming into use for the process, is fed through an air jet to take-up rollers which draw off the yarn at a speed lower than the speed at which it is fed to the jet. Because the take-up speed is slower than the feed, the air-jet forms numerous randomly spaced loops (to take up the slack) in the filaments, and so brings about an increase in *yarn* denier, but not of course in filament denier; the increase in yarn denier is due solely to the looping of the component filaments. Although first used on nylon, the process does not depend on the thermoplasticity of the fibre and can be applied to any continuous multi-filament yarns, *e.g.*, viscose rayon. In fact the Taslan process has found its greatest application in the texturising of glass yarns for upholstery and drapery fabrics. Taslan yarns are not stretch yarns; they are called " bulked " because the looping gives them an improved cover. Only looped (as opposed to crimped and curled texturising) yarns can be successfully made from non-thermoplastic filament yarns. Operating conditions such as feed speed, jet design and adjustment, air pressure, take-up speed and tension can be varied to produce a range of loop sizes and frequencies from a single starting yarn, so that the process is versatile. The notation, perhaps a little clumsy, used to designate Taslan textured yarns is illustrated by the following two examples:

96T/80–68–15½ Z textured Dacron yarn =

One end of 80 denier 68 fils Dacron twisted 15½ t.p.i. Z twist and then Taslan textured to a 20 per cent increase in denier (80 increases to 96).

96T/2 (40–34–½ Z) 12 Z textured nylon yarn =

Two ends of 40 denier 34 fils, ½ t.p.i. Z twist nylon plied (doubled) to 12 Z t.p.i. and then textured to a 20 per cent increase in denier.

Most Taslan yarns have been textured to give a 10 or 20 per cent increase in denier, but their increase in voluminosity may be 50 to 200 per cent, because the loops on the filaments greatly increase the diameter of the yarn by separating the filaments. It is this increased bulk that gives to Taslan yarns their attraction. Fig. 205 shows a photomicrograph of a fabric made from Taslan textured yarn, and the bulkiness of the yarn can be seen; although the filaments are still continuous, the fabric has much of the appearance of one

[*E. I. du Pont de Nemours & Co., Inc.*]

FIG. 205.—Fabric made from Taslan textured yarn (half-way to staple fibre).

woven from staple fibre. As can be seen from Fig. 206, the characteristic curve for a Taslan-textured yarn is intermediate between those for filament and staple yarns. This is in sharp contrast to the behaviour of yarns like Helanca, which will stretch several hundred per cent under a load of 0·1 gram per denier.

Fabrics made from Taslan-textured yarn should be dyed and finished in the same way as continuous filament yarn; the loops on the filaments and the bulkiness of the yarn tend to cover up irregularities and to give a dyeing that is apparently more level than when a filament yarn is used. Handle is drier and crisper than that of staple fibre fabric; the slight tendency to harshness may be disadvantageous.

In the following table the bulk densities of some woven fabrics of similar construction made from filament, staple and Taslan yarns are compared.

Fabric.	Density (gm./c.c.) of fabric made from:		
	Filament yarn.	Staple yarn.	Taslan yarn.
Nylon shirting . . .	0·58	0·45	0·39
Orlon shirting . . .	0·46	0·34	0·28
Acetate suiting . . .	0·61	0·51	0·46
Viscose rayon suiting . .	0·54	0·46	0·46

The density of the fabric made from Taslan yarns is always the lowest. The light transmission of Taslan yarns is lower than of

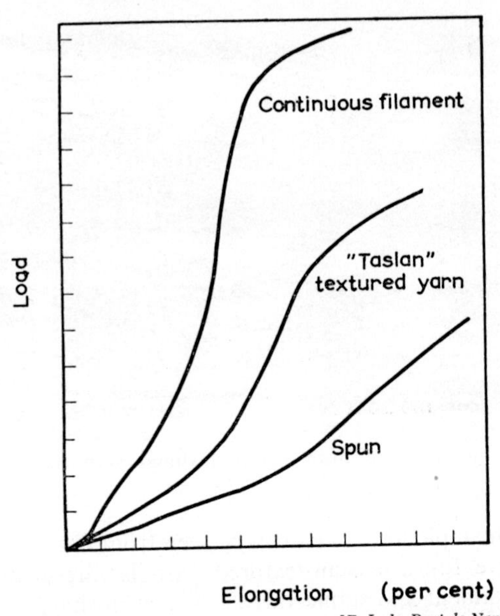

[*E. I. du Pont de Nemours & Co., Inc.*]

FIG. 206.—Stress–strain curve for Taslan textured yarn (half-way to staple fibre).

filament yarns. Of two nylon tricot (two-bar jersey stitch) fabrics of similar weight per square yard, the one made from Taslan yarn had a thickness of 0·014 in., whereas that made from filament yarn was only 0·010 in. thick, *i.e.*, the Taslan texturing has given a 40 per cent increase in fabric thickness.

CLASSIFICATION OF TEXTURED YARNS

The following tentative definitions were published in the 1956 *Journal of the Textile Institute* (Vol. 47, P. 280).

" Bulked Yarns

Yarns which have been treated physically or chemically so as to have a notably greater ' apparent ' volume or bulk sufficiently stable to withstand yarn processing tensions and the normal forces exerted on garments during wear."

This group would appear to include Taslan specifically and really all the stretch yarns are bulked as well. Taslan is mentioned separately because it is not a stretch yarn.

" Bulky Yarns

1. Yarn in which the apparent density of the filaments is much lower than the real density.

2. Spun yarns made from staple fibres having a high degree of resiliency."

Clearly (1) would include hollow filament yarns such as Bubblfil and Celta, whilst (2) might include the Hi-Bulk Acrilan and the corresponding Orlon yarns.

" Stretch Yarns

Yarns made from thermoplastic fibre, usually in the form of continuous filament, which are capable of a pronounced degree of stretch and rapid recovery. This property is conferred on yarn (having one or more filaments) which has been subjected to an appropriate combination of deforming, heat-setting and developing treatments."

These stretch yarns are also bulked yarns. There are two main classes of stretch yarns:

1. Twist yarn (torque), *e.g.*, Helanca, Fluflon.
2. Crimped yarn (non-torque), for example Agilon, Ban-Lon, and crinkled yarn (unravelled from heat-set fabric).

But although Agilon is made not by twisting, but by drawing over a blade, it resembles Helanca and Fluflon morphologically; in all three the individual yarn filaments are curled, looped and twisted. Ban-Lon, the true crimped yarn, is characterised by the saw-tooth crimp and is free from loops and spirals.

Set Thermoplastic Yarns

All of these depend on the straight filament yarn being converted into a not-straight form by being crimped or twisted or otherwise mechanically distorted and being heat-set in that distorted condition;

then, when the yarn is wound straight on to a cone or other package, it possesses built into it the required potential shrinkage. An early method of making such yarns was to knit nylon yarn on a small-diameter-rib knitting machine, then to set the fabric by heat and afterwards to unravel it and wind the yarn on cones. When this yarn was subsequently knitted, the *set* stitch gave the fabric a crêpe-like appearance. The method is obviously clumsy and has been replaced by others more practicable.

Stretch yarns when under tension look much like any other yarn, and fabric knitted from them while held under tension on the machine looks normal, but when it comes off the machine the yarn crimps or crinkles and so shortens its effective length, and the fabric shrinks, and at the same time the interstices in the fabric are partly filled in to give a much fuller fabric with better cover, and one that is much less transparent to light. The nature of the distortion to which the filaments in the various bulk yarns have been subjected depends on the way in which they have been treated. To recapitulate:

Agilon is characterised by its filaments having a crimp that is spiral in shape and of high frequency. Plying is not necessary, as it is desirable with Helanca, because the crimp is non-torque derived.

Ban-Lon has a saw-tooth shaped crimp, without spiralling; this derives from its compression shrinking in a stuffing box.

Fluflon, which is made on the false-twist principle, has a sinusoidal crimp of very high amplitude and frequency, and because of the torque some of the filaments have become looped and spiralled; it is a high stretch yarn and will stretch 400 per cent and return to its original length.

Helanca has a sinusoidal crimp of very high frequency and amplitude, and some of its filaments spiral and loop, a feature deriving from its twist method of manufacture; it can be stretched 400 per cent and will return to its normal length.

Taslan is not a stretch yarn; its filaments are looped irregularly, and as the loops cause the filaments to stand separate, it has a high cover.

High Bulk Acrylics

High bulk acrylic fibres have become very popular. They are made by combining highly stretched fibres with unstretched; when wetted the highly stretched shrink more and cause the unstretched to curl and cockle, so giving voluminosity. High Bulk Orlon is described on p. 514 and Hi-Bulk Acrilan on p. 528. Fig. 207 shows the Hi-Bulk Acrilan before and after wetting.

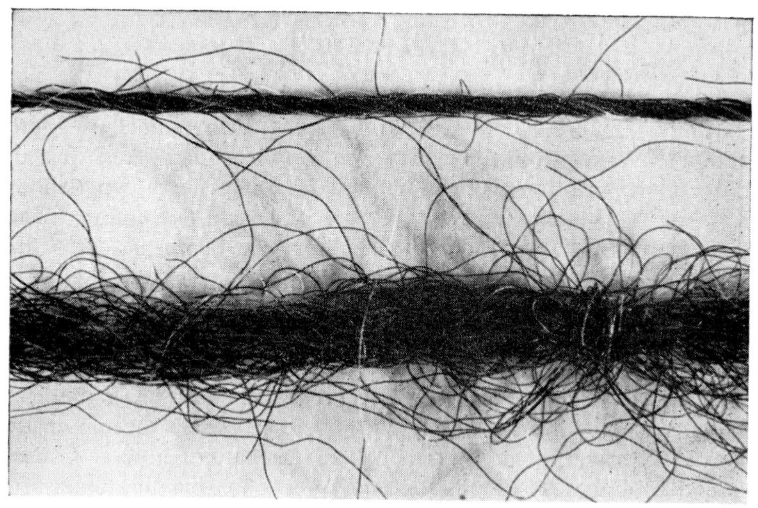

[*Chemstrand Corporation*

FIG. 207.—Hi-Bulk Acrilan yarn before (*above*) and after (*below*) relaxation.

FURTHER READING

J. T. Rivers, " ' Taslan ' Textured Yarn ". Paper given to Fourth Canadian Textile Seminar, The Textile Technical Federation of Canada.
L. G. Ray, " Textured Filament Yarns ", 41st Annual Conference of Textile Institute, pp. 33–44 (1956).
J. H. Kennedy, *Text. World*, **104**, No. 11, 140 (1954).
J. C. Field, *Text. World*, **104**, No. 7, 75 (1954).
J. H. Blore, *Text. World*, **105**, No. 6, 94 (1955).
Text. Industr., **119**, No. 10, 104 (1955).
Skinners' Silk & Rayon Rec., **29**, No. 3, 298 (1955).

CHAPTER 45

STAPLE FIBRE

AMONG the natural fibres, only real silk is continuous in filament; all the other fibres—cotton, wool, flax, jute, hemp, ramie, *etc.*—are composed of relatively short fibres, some longer than others, but all fairly short. Cotton averages about 1 in. staple length, wool about 3–4 in., the linen strand may be 12–20 in. or more. The very manner of producing artificial fibres or rayons leads to the formation of continuous filaments. A thick, viscous liquid is metered by a pump through tiny orifices, beyond which it sets solid in the form of filaments. The process is continuous; the filaments can be, and frequently are, miles long. At first, one imagines, this was considered a great advantage, for a close resemblance, in physical shape at least, was immediately made possible between the man-made fibre and real silk, most prized of all the natural fibres. Consequently, Man was content for many years to spin his rayon—or artificial silk, as he then called it—in continuous filaments. It is true that a Frenchman, by name A. Pellerin, did at a very early stage (1907) patent the manufacture of discontinuous filaments of cellulose, but as it was then very difficult to make satisfactory rayon at all, and as the emphasis in rayon manufacture was still on *artificial silk*, nothing came of the idea. During the 1914–18 War the Germans are reported to have cut viscose into short staple lengths and spun it, but apparently the results were unsuccessful, as the process was not persevered with.

Development of Staple Fibre Viscose Rayon. It was not until 1920 that the matter was seriously raised again. About that time large quantities of continuous filament viscose were being used, and consequently a considerable amount of waste viscose rayon was being made in the weaving-mills. What was to be done with this waste? It was torn and shredded so far as could be into fibres, and was spun combined with wool. But, because the viscose waste was a mixture of different filament deniers and staple lengths, it spun very badly and gave only a poor, uneven yarn. Nevertheless, these attempts to work up waste viscose rayon drew new attention to the possibility of spinning viscose on the cotton or worsted system. The rayon manufacturers prepared staple viscose rayon uniform in filament denier and in staple length by spinning continuous filament viscose, and cutting it into short lengths; they then spun it on the cotton-spinning system themselves. Courtaulds, Ltd. started to spin staple-fibre viscose rayon at their Wolverhampton and Flint factories in 1925. Later it was spun at their new Greenfield works in North Wales and they acquired a Lancashire cotton mill in which to spin Fibro into yarn. It was soon shown that a good staple yarn could be made, and by 1925 Lancashire mills were beginning to use staple fibre viscose. In 1929, the world production of staple fibre viscose was about 3,000 tons. When the Excise Duty on rayon was removed in 1935 an enormous expansion of the use of staple fibre came about, and in 1941 world production had risen to about 700,000 tons; by this time it was one of the world's major fibres. It is interesting to note that this enormous increase in production was brought about by attempts to find a use for waste continuous-filament viscose rayon. The way in which the use of staple fibre has caught up with and overtaken that of continuous filament viscose rayon is shown by the following figures:

Year.	Thousands of tons of viscose rayon made.	
	Staple fibre.	Continuous filament.
1934	24	350
1935	60	430
1936	130	450
1940	620	520
1960	1,380	920
1963	1,740	960
1967	1,830	1,000

It is interesting to note that in 1939, and probably at least as late as 1942, Germany was producing 40 per cent of the world's total rayon staple. In 1947 her contribution was only about 11 per cent of a world total which had fallen to 330,000 tons. (U.S.A. produced 32 per cent and the United Kingdom 12 per cent.) The figures given on p. 840 show that world production is rising.

Chemical Nature

Chemically, staple fibre is no different from continuous filament. Staple fibre viscose rayon is the same chemically as continuous viscose rayon; staple cellulose acetate is the same chemically as continuous cellulose acetate. The only point that may be noted in this connection is that staple fibre viscose is frequently impregnated with a U.F. or similar resin condensate in order to make it anti-crease and to increase its fullness of handle. This application is normally made to the fibre in fabric form, and will be considered later. It should be noted, though, that there are sometimes differences in physical properties between staple and continuous synthetic filaments, as there are for example in Orlon and Dacron.

Probably these are mainly due to differences in the degree of stretch that has been imparted after the spinning operation. Nylon for tyres has very different characteristics from apparel nylon.

Manufacture

Until the viscose is spun, the method of manufacture is the same as that for continuous filament yarns. Then there is a difference, for whereas continuous filament viscose is spun from a fairly small number of orifices, say 50, staple fibre is spun from spinnerets with many thousands of orifices. Instead of collecting together 50 filaments to make a yarn, tens or perhaps hundreds of thousands of filaments are drawn together to make the tow which is destined for staple fibre. There are three general methods by which tow is converted to staple:

1. Wet cutting.
2. Dry cutting.
3. Continuous processes.

The first two result in bundles of fibres, pointing in all directions, which are baled; it is the special virtue of the third general method that the parallel arrangement of filaments is retained.

[*Oscar Kohorn & Co., Ltd.*]

FIG. 208.—Staple fibre cutter as installed by Oscar Kohorn & Co. Ltd., New York.

Wet-Cutting. The thousands of filaments are drawn together in rope form, and, a few moments after coagulation, the rope is cut automatically and precisely into uniform lengths. One make (Kohorn) of cutter is designed to give staple lengths from a fraction of an inch up to 8-in. lengths. The cut staple is washed, desulphurised, bleached and oiled. It is dried in an open condition, and then conditioned until it has a 10 per cent moisture content [11·1 per cent regain, as $90 + \left(\frac{11\cdot1}{100} \times 90\right) = 100$]. In some machines the fibre is again " opened " after it has been dried. It is then baled for despatch to textile spinners. Note that during the wet processes the cut staple has the opportunity to shrink a little and develop a little crimp—a quality that is appreciated, particularly by worsted spinners.

The Gru-gru cutter has been very largely used, especially in Germany, for cutting rayon into staple fibre. The rayon, which has already undergone some stretch between the godets, is stretched again by porcelain rollers and then goes into the Gru-gru cutter. This consists of two wheels with slotted rims and of intersecting knives which pass through the slots as the wheels turn and sever the cable of filaments which is between the wheel rims. The knives are made of hard stainless steel, but have to be re-sharpened every few hours.

Dry-Cutting. In this method the viscose is carried through all the wet processes as a rope, and when this has been purified and dried it is cut. This method has the disadvantage that there is no opportunity for crimp to develop.

Those processes just discussed for the preparation of staple fibre consist essentially of:

1. Spinning (extrusion spinning) continuous filaments.
2. Cutting them up into short lengths and separating and mixing them.
3. Spinning (textile spinning) the short lengths into a continuous yarn by laying them parallel, drawing them out and twisting them.

The possibility of cutting out one of these processes by maintaining the parallel and linear arrangements of the filaments from extrusion to finished yarn has always been attractive. It certainly seems an odd business to have a parallel arrangement, to randomise it and then to restore it. Nevertheless, to-day, with existing machinery, this roundabout method affords the cheapest way of making staple fibre yarns in some countries. Eventually, one would think, it is bound to be displaced by one or other of the continuous processes in which parallel arrangement of filaments is retained throughout. With these newer processes, the filaments are cut or broken in such a way that the bunch (as extruded) of filaments is never severed; some filaments are severed at one point, others half an inch farther along the bunch of filaments, others again an inch farther along and so on. Every filament is severed at frequent intervals, but the points of cut or break of all the filaments never coincide, so that the bunch of filaments never falls apart, although it can be drawn out by the gentle tensions imposed by the textile spinning processes.

Three methods of severing the individual filaments have been used: abrading, cutting and breaking. Of these the last two are used on tow-to-top processes.

Abrading. Continuous filament cellulose acetate yarns have been passed through tubes lined with emery paper to roughen the yarns and break the filaments irregularly; the same effect has been obtained by passing the yarns through finely fluted rollers. Filament rupture does not occur regularly; sometimes a filament will not be cut or broken for 20 or 30 in. and yet the filaments next to it may be only a fraction of an inch long. The abrasive tube can be attached to the spinning machine so that a roughened " staple fibre " can be spun directly. Because of the irregularity of rupture, the strength of the yarn is variable and the yarn is usually weak. A good feature is that the denier of the staple yarn is as uniform as that of the continuous filament yarn from which it has been made. The process is usually carried out on 150 denier (35 s cotton counts).

Tow-to-Top

Cutting. There are two well-known methods of making top from tow by cutting the filaments; in either case the filaments are cut in the form of a tow which may contain one or two hundreds of thousands of filaments, spun and collected together as a rope or tow. The tow is cut with a revolving knife so that the cuts are made diagonally with the length of the tow; in this way the continuity of the rope or tow is not disturbed, although all of the individual filaments are cut. The two best known methods are the Greenfield and the Pacific Converter. Tow-to-top methods eliminate the carding operation.

The Greenfield Top. The Greenfield Top consists of a sliver of rayon staple, first made in Courtaulds' Greenfield factory in 1939 and resembling a worsted top. The initial material is a tow of perhaps some hundreds of thousands of continuous viscose filaments. It is fed from cans or balls under a tensioning bar, *A* (Fig. 209), round feed nip rollers, *B*, which feed it uniformly round a

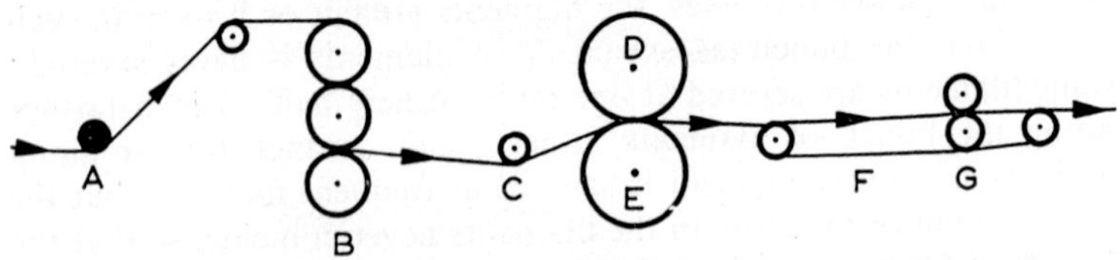

Fig. 209.—Diagrammatic view of cutting stage in Greenfield Tow-to-Top Process.

tensioning roller, *C*, to the helical cutter, *D*, which rotates against a rotating steel boss *E*. Here the filaments are cut within a chosen angle of 5° to 15° to the tow length. The cutting rollers carry the sheet of cut fibres forward and deposit it on an endless feed apron, *F*.

The apron feeds the sheet of fibres to an orthodox gill-box (back rollers of the gill-box are shown at *G*) of the kind used in worsted spinning, and this is set so as to give sufficient draft and pin action to convert the sheet of fibres into sliver or top form; this top is delivered in the usual way into a can. The top can then be used in just the same way as a worsted top; it is similar to it in that all the fibres are parallel, are untwisted and are, in fact, in the form of a light, open, untwisted rope. Subsequent normal spinning operations successively thin this rope down and insert twist until finally after several such stages a yarn results. It is a good feature of the Greenfield top that, having been made by cutting and not by breaking the filaments, their staple length is uniform; furthermore, they have not had all their extensibility pulled out of them.

The Pacific Converter. This machine, which was invented in 1950 by the Pacific Mills Corporation of Lawrence, Massachusetts, has been developed and built by the Warner and Swasey Company of Cleveland, Ohio. The flowsheet, Fig. 210, and the schematic diagram, Fig. 211, show the principle of operation, and Fig. 212 shows the machine itself. It affords the second example of a *cutting* continuous process. It converts a tow of continuous filaments into a top of staple without the use of a card (because the filaments are already parallel) by means of the following processes:

1. Viscose tow is usually about 190,000 denier, acetate tow about 120,000 denier. Dynel three-filament denier-tow is supplied in 180,000 total denier tow; a sufficient number of these tows is run together to feed the Pacific Converter with 1,800,000 denier. The tow is fed from very large spools or ball warps on which it has been collected, through a guide-frame to feed rolls, from which it emerges as a flat web still continuous, *i.e.*, the rope is converted to a flat web.

2. The web can be stretched uniformly to any desired degree between two pairs of electrically heated rollers, the second pair travelling faster than the first. This stretching is necessary only for synthetic fibre tow.

3. The web is cut into oblique strips extending at an angle of about 10° to the line of feed. This cutting is done by a helical steel thread on the periphery of a roller, this pressing strongly on the web, which is supported by a revolving anvil. Different kinds of cutter rolls are supplied so that the fibre can be cut into 3-in., $4\frac{1}{2}$-in. or 6-in. lengths; also fibre can be cut if required with variable lengths, *e.g.*, $1\frac{1}{2}$- to $4\frac{1}{2}$-in. variable or 3-in. to 6-in. variable.

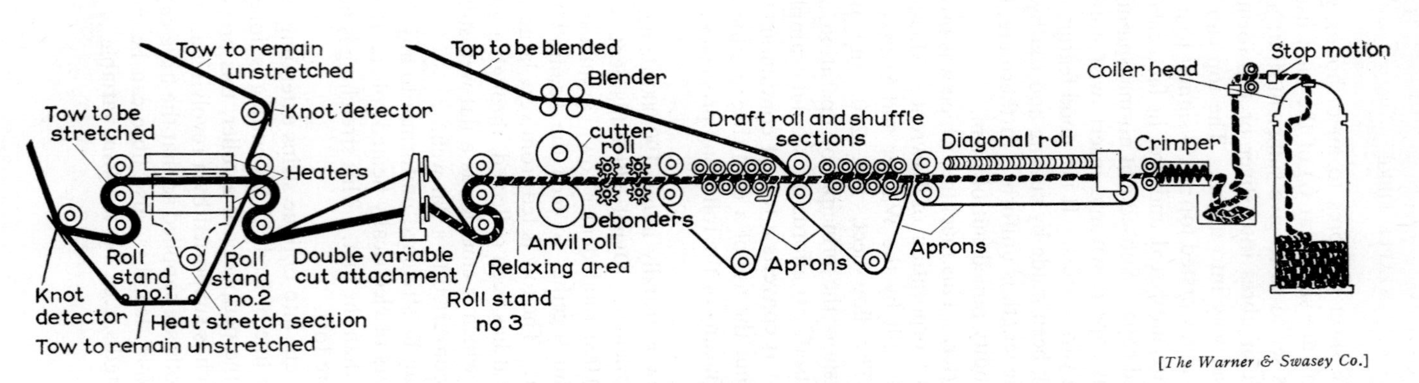

[*The Warner & Swasey Co.*]

FIG. 210.—Flowsheet through Pacific Converter.
Tow is fed in at the extreme left and Top is collected at the extreme right.

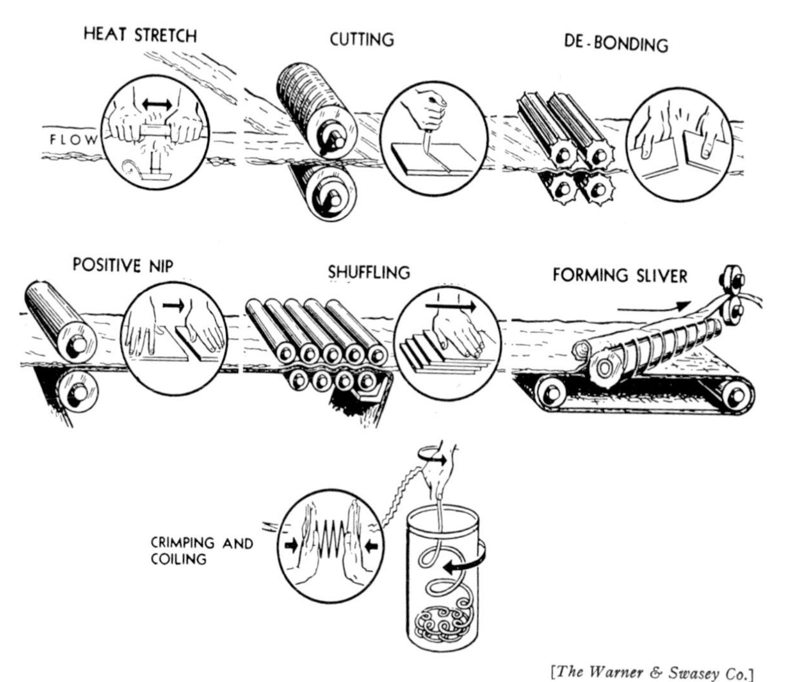

[*The Warner & Swasey Co.*]

FIG. 211.—Schematic view of sequence of operations in the Pacific Converter.

[*The Warner & Swasey Co.*

FIG. 212.—The Pacific Converter.
Tow is fed in at the left, Top collected at the right.

4. The rollers feed the cut web forward; they can do this because the space between the threads of the helical cutting roller is filled with rubber.

5. The staple fibres are now "debonded" or separated especially at their cut ends where they have been crushed together so that they adhere. This is done by passing the web through fluted rollers which work and bend the staple fibres up and down.

6. The fibres are passed between fluted rollers and a leather apron so that the top of the web is advanced relatively to the bottom; this results in a shuffling action (rather like shuffling a pack of cards) and further separates the fibres and makes them non-conterminous.

7. The web is drafted by the greater surface speed of the rolls or aprons with which it is successively engaged.

8. The web is rolled into a sliver by the diagonal action of the tapes.

9. Rolls at the delivery end of the apron crimp the fibre and a coiler head lays the sliver into the can ready for further processing.

The essential features of the process are:

(*a*) heat stretching of synthetics (if needed);

(*b*) formation of a flat web;

(*c*) blending with wool or a natural fibre (if needed);

(*d*) cutting the web by a driven helix blade cutter;

(*e*) moving the cut fibres relative to each other so that they are non-conterminous;

(*f*) rolling the web diagonally (Fig. 211) into a continuous sliver of staple fibre.

As can be seen from Fig. 210, there are three places in the Converter at which fibre can be fed in:

1. At the extreme left where tow that requires stretching is fed.

2. Immediately after the heater section where tow that does not need to be stretched (*e.g.*, rayon) can be fed in.

3. After the cutter roll and debonding unit where top (*e.g.*, wool) that has to be blended with tow is fed in.

Output from a Converter is about 100–130 lb. per hour or up to 180 lb. if top is blended after cutting the tow. One operative can look after two Converters. The top that comes from the Converter can be used for the Bradford system of manufacture, or it can be treated in a Warner and Swasey Pin Drafter; three or four passes through this machine make it into a form suitable for roving and then spinning.

Breaking—The Perlok System. This also is a method for processing tows of continuous filaments directly into slivers of staple fibre. The Perlok system, and the American Turbo-Stapler which is a development of it, afford the outstanding example of the breaking of filaments, as distinct from cutting. In its original form it consisted of drafting rollers (front rollers going faster than the back ones) which *broke* the filaments by a straight longitudinal pull, the filaments breaking into irregular staple lengths. Each breaks at its weakest point; all do not break at the same place, so that although all the individual filaments are broken, the continuity of the tow is not destroyed. Defects of this method were:

1. There was a considerable loss of extensibility in the fibre.
2. The staple length was not uniform.
3. The heavy tows now made of some 200,000 denier are difficult to break solely by drafting.

The process has accordingly been modified so that tow breaker wheels are interposed between the drafting rollers; these wheels are like cog wheels which just do not touch or " bottom " and they have sharp protruding edges. The tow passes between them under considerable tension (due to the difference in speed of the drafting rollers) and the filaments break on the sharp edges of the breaker wheels; the length of staple is equal to the distance from the edges of the breaker wheels to the nip point of the front rollers. The principle is similar to that used in breaking a piece of string over a sharp edge. The tow should be fed into the Perlok machine in the form of a broad *flat* ribbon, through two sets of feed nip rollers, *A* and *B* (Fig. 213), which ensure uniformity of feed, then round the

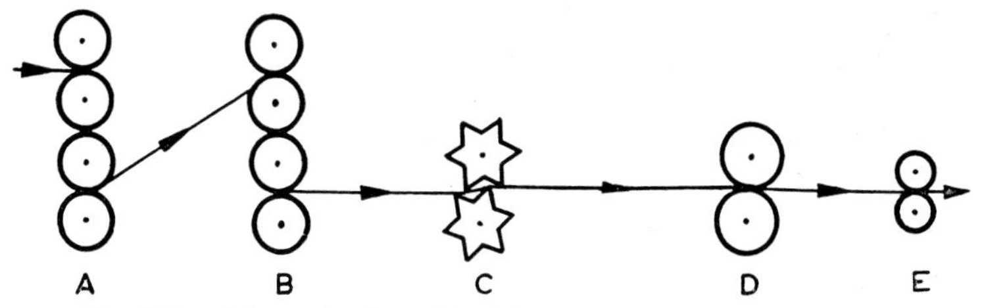

FIG. 213.—Schematic view of Perlok Tow-to-Top Stretch Breaking.

deeply fluted deflecting rollers, *C*, where the filaments break under the tension exerted by the output rollers, *D*, which rotate at a speed to give twice the linear speed of the feed rollers, *B*. Thence the broken filaments, still parallel and in sheet form, pass to the crimping rollers, *E*.

Stretch breaking was never so attractive as cutting for the regenerated cellulose fibres, mainly viscose rayon, because it involved the expensive procedures of wetting and drying to relax the fibre strains and to recover, before spinning, the residual fibre shrinkage. The system has, however, been adapted to nylon, Terylene, and Orlon with great success; these fibres are stretch-broken, the sliver is mechanically crimped and the crimp heat-set and the fibre relaxed by a comparatively simple steam heating process, thus recovering residual shrinkage and resilience before (or, if preferred, after) spinning. It was this process used in the Turbo stapler which led to the discovery that the bulking properties of the acrylics could be greatly improved by blending in one yarn fibres (*a*) stretch-broken, crimped and relaxed; with (*b*) stretch-broken and crimped but not relaxed. With a blend of 60 (*a*)/40 (*b*) sweaters can be knitted and then boiled off to recover up to 20 per cent residual shrinkage in the unrelaxed fibre component of the yarn in the fabric. This gives the high bulk which is liked in sweaters. (J. L. Lohrke, " Progress in the Perlok System ", *Rayon Textile Monthly*, **25**, 437 (1944).)

Advantages of Tow-to-Top System. Advantages of the " tow-to-top " process over the cut staple process are:

1. The yarns are 20 per cent stronger.
2. They are more uniform.
3. They can be spun to finer counts, even down to 80 s worsted.

The reason for the advantage of strength is that in the tow-to-top process the fibres are subjected to less wear and tear than they are in the carding and combing operations, which are normally subsequent to the staple cutting. In addition, there is less waste in the tow-to-top process, and it is cheaper. It is easier to dye top than it is loose staple fibres. It has to be added, however, that there is no opportunity for blending, to secure uniformity, as there is in the orthodox staple fibre which is blended by mixing the contents of different bales as the fibre is fed to the card.

Glass Staple. For the methods of manufacture see Chapter 39.

Properties

Rayon staple is usually cut of uniform length, which may be anything from 1 to 8 in. For material that is to be spun on cotton textile machinery a staple length of $1\frac{7}{16}$ in. is most usual. Courtaulds' staple fibre viscose is known as " Fibro ". Three filament denier Fibro of $1\frac{7}{16}$ in. staple can be spun to 20 s cotton counts, and the same filament denier in $2\frac{1}{2}$-in. staple to 30 s. Generally, fine filament staple gives a yarn stronger than does coarse filament. There appears to be an optimum staple length for strength of the spun yarn; in the case of $1\frac{1}{2}$ filament denier Fibro it is $2\frac{1}{2}$ in.

Staple fibre rayon has been made of *varied* staple length (some fibres longer than others), the idea being that it would mix better with wool, which is itself variable in staple length, but nearly all to-day's production is cut to uniform length. A strong Fibro usually of 1·25 filament denier and $1\frac{7}{8}$-in. staple, and with a higher tenacity than usual, is also made. The extra strength enables it to be spun to finer counts.

Fibroceta. Fibroceta is a Courtaulds' product, a staple fibre of cellulose acetate, made in $2\frac{1}{2}$ and $4\frac{1}{2}$ filament deniers. It can be blended with viscose staple to give marl or speckled effects. Its regain is 6 per cent.

Spinning of Staple

The staple fibre, when once it has been made, is ready for "spinning" by traditional textile processes. This kind of "spinning" consists of running a large number of fibres together, gradually drawing them out finer and finer, and at the same time twisting them together. There are often many operations between the first sliver and the final yarn, for the drawing out has to be done bit by bit; in worsted spinning it is, for example, not uncommon that there should be eight or ten separate operations before the final yarn is made.

Natural fibres vary very considerably; those of wool are longer than those of cotton, whilst those of flax (as spun) are longer than either. It is, however, a virtue of staple fibre rayon, and one that should be emphasised, that it can be spun on any of the systems in general use, *e.g.*, cotton, worsted, wool, jute, flax, *etc.*, and this virtue has very greatly helped the development of staple fibre rayon and staple synthetic fibres. Naturally, the staple length suitable to the spinning system is chosen; whereas $1\frac{1}{4}$ in. is very suitable for cotton spinning, 3-in. staple is more suitable for use on the worsted system.

A point to which attention should be directed is that most staple fibre is supplied as loose fibre and has, accordingly, to be carded in order to parallelise the fibre and give a sliver which can then be spun traditionally. If, however, the tow-to-top method is used, then the fibres are already parallelised, and consequently they do not need carding, but can go straight to the later operations of drawing and roving. It is abundantly clear that as when the fibre is first spun (*i.e.*, spun into a fibre from viscose, *etc.*), the filaments emerge parallel from the spinneret, it is a great pity subsequently to bundle them up in any sort of order and then have the work of carding to sort them out parallel again. There can, in fact, be little doubt that those processes whereby staple fibre is produced without loss of parallel arrangement (*e.g.*, tow-to-top) will eventually oust the others.

Direct Spinning. In France, there has been developed a method of direct spinning of continuous tows of 1,000–8,000 denier to give yarns of fine counts from which fabrics with good handle can be made.

Dyeing

Staple fibre yarns and fabrics nearly always are dyed in exactly the same way as the corresponding continuous filament products, but some staple fibre yarn is now made already dyed. The dyestuff is added to the viscose solution before spinning, so that the dyestuff is internally held by the fibre. Such dyeings have very good fastness to light and washing. A range of colours which includes three depths of blue, an apple-green, malachite-green, sulphur, Indian-yellow, pink, red and black is available. Casein staple fibre, Merinova, is also supplied in a wide range of fast colours and so indeed are most fibres excepting the polyamides and polyesters.

Crease-resistant Finish

When staple fibre viscose was first made it had great difficulty in competing with wool and cotton. This was partly because it was subject to an Excise Duty, which has since been done away with, but partly also because it creased badly.

The urea-formaldehyde resin finish invented by the chemists of Tootal Broadhurst Lee Co., and known under the registered name " Tebilized ", was originally applied to cotton, but has been found extremely suitable for application to staple-fibre viscose. It gives fabrics made from staple viscose a fuller and richer handle and prevents creasing, which is a serious defect in a textile fibre. Melamine-formaldehyde resins have also been widely used for a similar purpose.

A resin anti-crease finish must not make a fabric boardy, the treated fabric must drape well, and the finish must be fast to washing. In order to meet this specification, the resin must be inside the fibre, not simply applied to its surface. The urea-formaldehyde finish is satisfactory in these respects.

Urea is first condensed with formaldehyde. If one molecular equivalent of formaldehyde is taken for one molecular equivalent of urea, then a monomethylol urea results:

$$\begin{array}{c} NH_2 \\ | \\ CO \\ | \\ NH_2 \end{array} + HCHO \longrightarrow \begin{array}{c} NHCH_2OH \\ | \\ CO \\ | \\ NH_2 \end{array}$$

If two equivalents of formaldehyde are used for one equivalent of urea, dimethylol urea results:

$$\begin{matrix} NH_2 \\ | \\ CO \\ | \\ NH_2 \end{matrix} + 2HCHO \longrightarrow \begin{matrix} NHCH_2OH \\ | \\ CO \\ | \\ NHCH_2OH \end{matrix}$$

The relative quantities found to give the best results are 1·6 molecules of formaldehyde to one of urea, which will give a mixture of mono- and di-methylol ureas. If more formaldehyde is used, the product is springier but harder.

Application. The formaldehyde and urea are pre-condensed, being allowed to react at room temperature for about five hours until the viscosity is 6 centipoises at 20° C. The resin is diluted to a solids content of about 25 per cent, and an acid catalyst, such as ammonium dihydrogen phosphate $NH_4H_2PO_4$, is added. The fabric is impregnated in this resin solution, mangled, dried and then heated to 120° C. for about three minutes. This is called " curing ", and in this process the resin polymerises rapidly and changes from a low-molecular-weight water-soluble resin to a high-molecular-weight water-insoluble resin. The resin in its water-soluble form penetrated the fibres during their impregnation, and by curing, it has been made insoluble, so that locked in the interior of each fibre is a water-insoluble resin. The drying is best carried out on a pin stenter.

Note that during the heating ammonia is driven off from the ammonium dihydrogen phosphate, and free phosphoric acid is formed. The presence of some such acid in sufficient quantity to make the resin *p*H below 4·5 is essential. The treatment makes the fabric crease-resistant and, because the resin is insoluble, this property of crease-resistance is retained even after repeated washing. After curing it is advisable to give a rinse to remove resin from the *surface* of the fibres, as if left on it is inclined to give a boardy handle.

Location of Resin Finish. Doubt was sometimes expressed as to how much of the resin was really inside the fibre, but in 1960 Park showed that the resin really was uniformly distributed through the filaments. His method was to use formaldehyde in which some of the hydrogen was hydrogen of atomic weight 3; this isotope is called tritium and it gives off β-particles as it decays which it does all the time. These β-particles will expose a photographic film. The " labelled " formaldehyde (*i.e.*, labelled with tritium) was reacted with urea, the resulting dimethylol urea being applied to viscose rayon fabric and cured in the fabric. Then sections of the filaments were covered with photographic film, kept in the dark for two days and developed; blackening of the film indicated the presence of tritium. The results showed that the tritium (and therefore the U.F. resin that contained it) was uniformly distributed through the body of the filaments.

Beetle Resins. These afford good examples of the resins commercially available. BT 321 (B.I.P. Chemicals Ltd.) gives rather better results than a melamine-formaldehyde resin BT 309 from the standpoint of crease resistance, although if a fabric is expected to be subjected to very severe washing, the melamine preparation may be preferred.

BT 321 (90 per cent resin solids) is a modified U.F. resin in solid form; the method of application of this resin to spun viscose fabrics is as follows:

1. Scour the fabric, making sure that all the size is removed;
2. Pad in a bath containing

BT 321 U.F. resin	10 parts
Accelerator (acid catalyst)	0·5 parts
Water	to 100 parts

arranging that the expression is about 80–100 per cent so that the fabric picks up 7·2 to 9 parts resin solids. The bath is not stable, the U.F. resin tending to polymerise, and it should be freshly made up several times a day.

3. The impregnated fabric is dried on a stenter without undue tension, otherwise the fabric will finish " boardy ".

4. The fabric is baked to condense the resin, and thereby to develop crease-resistance and to give durability to washing and dry-cleaning. The cure may be for 3 minutes at 150° C. or for 5 minutes at 140° C., but these times and temperatures depend on the quantity of accelerator that has been used.

5. The fabric is washed in 2 gm./litre soap and 1 gm./litre soda ash at 40–50° C.; this wash neutralises and dissolves any acidic by-products of the cure, it removes *surface* resin to give a soft and pliable hand.

6. The fabric is rinsed.

This processing should give a full, soft handle, good resistance to creasing, a measure of dimensional stability and added weight to spun-viscose rayon fabrics.

Chemical Effect of Curing. The change undergone by the methylol ureas during curing is known as polymerisation. A large number of small molecules unite together, water being split off, to give a large molecule. The insolubility of the cured resin is due to the great increase in size of molecule. Whereas before curing the molecules of methylol urea were small and could penetrate into the fibre, after curing the molecules are so large that they cannot find their way out, even if the fabric is washed. This change in molecular state enables the intermediates for the resin to be applied so that they penetrate into the fibre, and, when there, to be treated so that they are fixed in. If an attempt was made to cure or polymerise the resin and then apply it to the fabric, the large molecules of resin would never be able to find their way into the fibres, with the result that the resin would be applied only externally to the surface of the fibres, and the fabric would be hard and boardy. The chemical changes that take place may be expressed as follows: First one of the dimethylol urea molecules will combine with one of the monomethylol ureas:

$$\begin{array}{lcl}
\mathrm{NHCH_2\,OH} & & \mathrm{H\,NCH_2OH} \\
| & & \quad | \\
\mathrm{CO} & + & \quad \mathrm{CO} \\
| & & \quad | \\
\mathrm{NHCH_2\,OH} & & \mathrm{H\,NH}
\end{array}
\longrightarrow
\begin{array}{lcl}
\mathrm{NHCH_2} & - & \mathrm{NCH_2OH} \\
| & & | \\
\mathrm{CO} & & \mathrm{CO} \\
| & & | \\
\mathrm{NHCH_2} & - & \mathrm{NH}
\end{array}
+ 2H_2O$$

Then two of these will unite,

$$\begin{array}{lcl}
\mathrm{NHCH_2} & - & \mathrm{NCH_2OH} \\
| & & | \\
\mathrm{CO} & & \mathrm{CO} \\
| & & | \\
\mathrm{NH} & - & \mathrm{CH_2NH}
\end{array}
+
\begin{array}{lcl}
\mathrm{NH} & -\mathrm{CO}- & \mathrm{NCH_2OH} \\
| & & | \\
\mathrm{CH_2} & & \mathrm{CH_2} \\
| & & | \\
\mathrm{NH} & -\mathrm{CO}- & \mathrm{NH}
\end{array}
\longrightarrow$$

$$\begin{array}{lllll}
\mathrm{NHCH_2} & \mathrm{NCH_2} & \mathrm{N} & -\mathrm{CO}- & \mathrm{NCH_2OH} \\
| & | & | & & | \\
\mathrm{CO} & \mathrm{CO} & \mathrm{CH_2} & & \mathrm{CH_2} \\
| & | & | & & | \\
\mathrm{NHCH_2} & \mathrm{NH} & \mathrm{NH} & -\mathrm{CO}- & \mathrm{NH}
\end{array}
+ H_2O$$

Probably, too, ether linkages are formed by the elimination of water from the CH_2OH group in this way

$$-CH_2OH + HOCH_2- \longrightarrow -CH_2OCH_2- + H_2O$$

Many mechanisms will operate, but all will conduce towards building up very large molecules. Doubtless, too, some of the hydroxyl groups of the viscose rayon staple take part in the reaction, so that the resin cross-links the cellulose chains together. This is one of the main reasons why the treatment makes the viscose rayon uncreasable; it introduces cross-linkages between the linear molecules. It is well known that wool and hair fibres are the most naturally crease-resistant fibres that we have, and it is believed that this is due to their unique possession of cross-linkages. In their case cystine molecules cross-link the polypeptide chains together in this way:

$$\overset{|}{\underset{|}{C}}HCH_2SSCH_2\overset{|}{\underset{|}{C}}H$$

Consequently, when wool fibres are subjected to deformation the cross-links are under strain, and as soon as the tension is released they regain their original positions. This power of recovery reduces the chance of a *permanent* deformation—*i.e.*, of creasing. When the urea-formaldehyde resin is polymerised in the viscose rayon staple, cross-linkages are formed between the cellulose molecules as shown below:

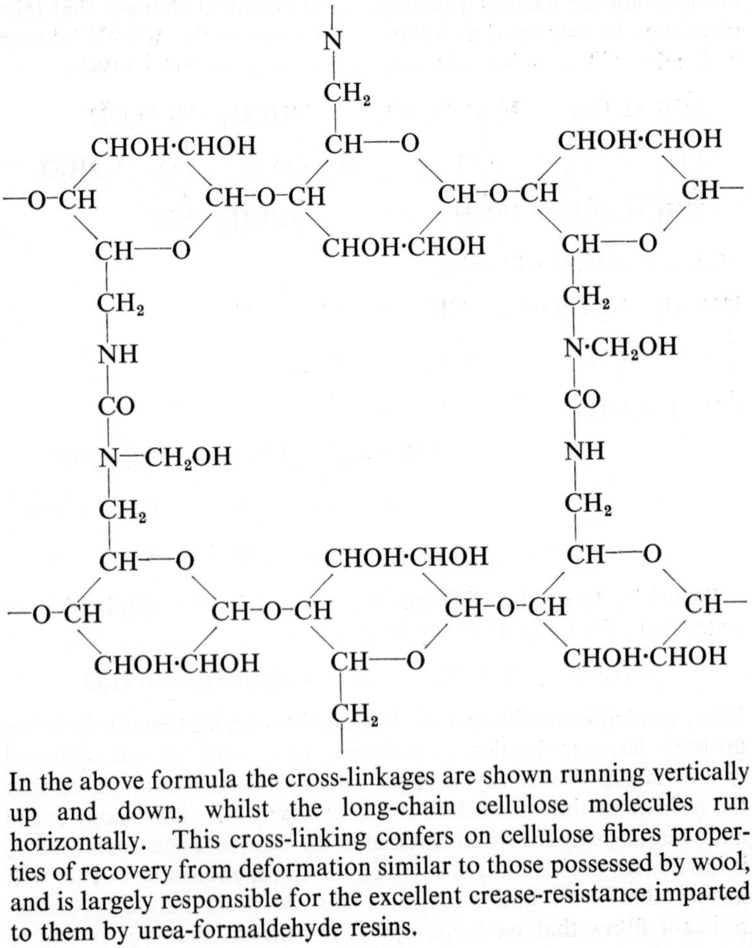

```
                            |
                            N
                            |
                           CH2
                            |
       CHOH·CHOH           CH——O                 CHOH·CHOH
      /         \         /     \               /         \
—O-CH            CH-O-CH         CH-O-CH                   CH—
      \         /         \     /       \               /
       CH——O               CHOH·CHOH     CH——O
       |                                 |
       CH2                               CH2
       |                                 |
       NH                                N·CH2OH
       |                                 |
       CO                                CO
       |                                 |
       N—CH2OH                           NH
       |                                 |
       CH2                               CH2
       |                                 |
       CH——O               CHOH·CHOH     CH——O
      /     \             /         \   /     \
—O-CH        CH-O-CH                 CH-O-CH        CH—
      \     /       \               /       \      /
       CHOH·CHOH     CH——O                   CHOH·CHOH
                     |
                     CH2
                     |
```

In the above formula the cross-linkages are shown running vertically up and down, whilst the long-chain cellulose molecules run horizontally. This cross-linking confers on cellulose fibres properties of recovery from deformation similar to those possessed by wool, and is largely responsible for the excellent crease-resistance imparted to them by urea-formaldehyde resins.

Melamine Resins. In addition to urea-formaldehyde, other resins, such as melamine-formaldehyde and phenol-formaldehyde, have been tried to make cellulosic fibres crease-resistant. The phenolic resins are usually coloured, and this constitutes a great deterrent to their use for textile purposes.

Melamine-formaldehyde resins have been used more for application to wool to make it unshrinkable than to viscose rayon, but they are excellent. They are, however, relatively expensive, and their only advantage to offset the increase in price so far as application to viscose rayon is concerned is a slight improvement in dimensional stability to severe washing.

The formula of melamine is

```
              NH2
               |
               C
            //   \
           N      N
           |      ||
    NH2—C         C—NH2
            \\   /
               N
```

It combines with formaldehyde to give various methylol ureas of which trimethylol melamine

```
              NH2
               |
               C
            //   \
           N      N
           |      ||    /CH2OH
   HOCH2NHC       CN<
            \\   /      \CH2OH
               N
```

may be taken as typical. On being heated or cured, these methylol melamines polymerise in a way very similar to the urea-formaldehyde resins, and giant three-dimensional polymeric molecules, which are infusible and insoluble, are formed in the fibre, to which the methylol melamine has been applied. Sometimes methylated methylol melamines, which have the advantage (before curing) of greater solubility, are used. In this case some of the $—CH_2OH$ groups in the formula given above are converted into $—CH_2OCH_3$.

Ethylene Urea-Formaldehyde Resins. A disadvantage of U.F. and melamine-formaldehyde resin finishes is that they are chlorine retentive; if resin finished fabric is treated with a chlorine bleach, it yellows; the chlorine reacts with the active hydrogen atoms attached to the nitrogen atom of the urea or melamine. If, instead ofurea, ethylene urea is used, then the dimethylol ethylene urea (the resin pre-condensate) contains no such reactive hydrogen, and the resulting resin is much more resistant to chlorine. The formulae for the two pre-condensates illustrate the difference:

$$O{=}C\begin{cases}NHCH_2OH\\NHCH_2OH\end{cases}$$

Dimethylol urea.
(Reactive hydrogen atoms are shown bold.)

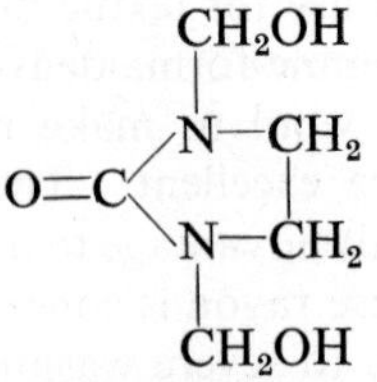

$$O{=}C\begin{cases}N(CH_2OH){-}CH_2\\ \quad\quad\quad\quad\quad\ \ |\\N(CH_2OH){-}CH_2\end{cases}$$

Dimethylol ethylene urea.
(No reactive hydrogen atoms.)

Melamine-formaldehyde resins are not much used on shirting fabrics nowadays, the ethylene urea-formaldehyde resins are preferred for the reason given above.

Advantages of Crease-Resistant Fabrics. Other advantages of the crease-resistant Fibro fabrics over untreated fabrics are:

1. They are dimensionally stable—*i.e.*, they do not shrink or stretch when washed.
2. They have gained in weight, due to the addition to the fabric of the resin.
3. They are more resistant to degradation by light.
4. In the case of thin fabrics, there is a greatly reduced tendency to " slip "—*i.e.*, for the warp threads to slip over the weft and leave an ugly thin place.
5. Their strength is a little higher.
6. If a permanent crease is required and a crease is inserted before the resin is applied, then after the resin has been applied the crease will be permanent.

The process is extremely good, and has a most beneficial effect on Fibro fabrics.

Blending

Mixtures of different fibres have been used in fabrics made from continuous filament yarns, but when this has been done, almost always one fibre has constituted the warp of a fabric, the other fibre has been the weft. For example, viscose rayon warps and cellulose acetate wefts were very popular at one time for the production of shot effects; cellulose acetate warps and crêpe viscose rayon wefts were widely used for marocains and crêpe-de-chines—the cellulose acetate provided the handle and drape, the viscose rayon provided the crêping power. But mixtures of two fibres in either the warp or the weft were unusual.

Staple fibre opens the door to many more blends of different fibres, because it is much easier to mix fibres to make a yarn than it is to mix yarns to make a fabric.

It is a general rule when blending fibres to blend as early in the process as possible, thereby obtaining the most intimate mixing; this applies especially if the two fibres that are being used have different dyeing properties. Usually this means blending during carding; the two kinds of fibre, say viscose rayon staple and nylon staple, will be fed simultaneously into the hopper of the card, and the resulting sliver will be a fair mixture of the two. This method has the additional advantage that it is much easier to card a mixture of viscose rayon and nylon staple than nylon staple alone.

Sometimes rayon staple of different filament deniers will be blended. Generally, coarse filament deniers can be spun only to coarse (not to fine) counts, whilst fine filament deniers give yarns with a soft limp handle, but by blending the two, yarns which possess intermediate characteristics can be spun.

If a strong fibre is blended with a weak fibre it is preferable to have the weak fibre present in a heavier filament denier, so that the breaks on the card will be equal between the two fibres. For example, a cellulose acetate fibre such as Fibroceta is weaker than a viscose rayon staple, but good results can be obtained by blending $4\frac{1}{2}$-denier $2\frac{1}{2}$-in. acetate staple with 3-denier $2\frac{1}{2}$-in. viscose staple. The acetate contributes its characteristic handle and draping quality to the blend.

Effect of Blending on Tenacity and Extension. The two physical characteristics of a yarn that are most frequently measured are the tenacity and the extension at break. When both of these values are high the yarn will be a good yarn, but if one is high and not the other the yarn may well be useless. Both qualities are necessary for ease of spinning and for durability of the finished product; probably if a separation is to be made at all, it is safe to say that ease of spinning is in the main determined by the possession of a high extension at break, whilst durability (wear resistance) is largely determined by tenacity. One of the two outstanding advantages that might be expected to accrue from the admixture of a synthetic to a natural fibre is an increase in tenacity; in many mixtures this increase is to be found, but the increase is often not what might be expected from first principles, but varies from mixture to mixture.

Very often, mixtures of two fibres have a lower breaking strength than either fibre has individually; this is due to one of the components taking all the initial strain. The fibre with the higher Young's modulus carries more than its share of the load and breaks before the other fibre, which may be the stronger, is taking its fair share of the load. Thus a blended yarn of 75 per cent viscose rayon and 25 per cent Dacron has a lower breaking load than has viscose rayon, because the relatively high Young's modulus of the rayon ply up to its breaking point causes it to carry more than its share of the load and to break at a lower total load than would a 100 per cent viscose rayon yarn of the same counts as the blended yarn. As the proportion of the stronger fibre Dacron is increased in blends, the strength rises, and a 50 viscose rayon/50 Dacron blend is as strong as 100 per cent viscose rayon, whilst a 25 viscose rayon/75 Dacron blend has a much higher strength.

Nylon also has a lower Young's modulus than viscose rayon, and blends of viscose rayon/nylon do not show any strength advantage until the percentage of nylon exceeds 50.

Cellulose acetate, however, has a much lower Young's modulus than either viscose rayon or Dacron; accordingly when Dacron is blended with cellulose acetate staple, there is from the beginning an increase in strength, although the increase is not quite so great as would be expected from the arithmetic mean. Similarly, if nylon is blended with acetate, the addition of the nylon gives greater strength.

Wool has a low Young's modulus, and consequently the blending of either Dacron or nylon with it increases the strength, a property which has been put to a considerable advantage.

In brief, even small proportions of the strong fibres, nylon and Dacron will increase the tenacity of acetate yarns or of wool yarns; such small additions will, however, decrease the strength of viscose rayon, and in fact large proportions of the strong fibre have to be used in order to make viscose rayon stronger. So they do for cotton, too, and spun blends of cotton and nylon up to about 80 per cent nylon are weaker than all-cotton and mixtures within the range 40–60 per cent nylon are much weaker; extensibility of the blend is never lower than that of the cotton (8 per cent), increases a little to 9 per cent when 40 per cent nylon is included, and thereafter rises fairly fast to about 17 per cent at 80 per cent nylon and to 20 per cent for all-nylon.

An odd point of interest concerns the blend of two parts cotton with one part viscose staple. Cotton has a higher strength wet than dry, whilst viscose rayon has a much lower strength wet than dry, but a two-thirds cotton one-third viscose staple blend has the same strength wet as dry; the dry strength of this blend is, at the same time, lower than the dry strength of either cotton or viscose rayon staple.

Extension at break of a blended yarn is always intermediate between the extensions at break of the component yarns. It is not, however, the arithmetic mean of the two, but almost always considerably lower than this. It is not easy to predict the extension of a blended yarn from a knowledge of the extensions of the component fibres.

Effect of Blending on Durability. The synthetic fibres, nylon and Dacron, have excellent durability, and a measure of this durability can be obtained by noting the resistance to abrasion. Wool and rayon have relatively poor resistance to abrasion, especially when wet, but if a little nylon or Dacron is blended with the wool or rayon, the abrasion resistance is enormously increased. Nylon has the greatest effect; Dacron has a less but still outstanding effect; Orlon gives only a slight advantage. Some comparative figures obtained by the Stoll wet flex test by Dennison and Leach are given in the following table:

Fabric composition.	Resistance to abrasion (number of flexes).
All wool	90
All rayon	94
All Dacron	1,570
All nylon	2,520
All Orlon	330
50 Dacron/50 rayon	996
50 nylon/50 rayon	2,360
50 Dacron/50 wool	800
50 Orlon/50 wool	290

The enormous advantage that derives from the admixture of nylon or Dacron to rayon or wool is very evident. The improvement given by nylon is so outstanding that it is sometimes included in commercial worsted suitings to improve their wearing properties.

Points to note about the addition of the strong synthetic fibres to wool or rayon are:

1. The resistance to abrasion increases at once even with small additions; there is no initial fall as there is in the tenacity of blends;
2. Blending should be done at the earliest stage to obtain the maximum benefit. A fabric made from Dacron and rayon which have been stock-blended, *i.e.*, blended as loose fibre, will have a noticeably greater resistance to abrasion than a similar fabric made from plied yarns, one ply being Dacron and the other rayon. An intimate intermingling of fibres gives the best results.

Uses

The conversion of rayon into staple fibre has enormously extended the field of use of rayon. Particularly is this so in the heavier fabrics, such as suitings, which traditionally have been made from wool.

Viscose Rayon Staple. Viscose staple fibre itself can be used alone, or blended with wool, mohair, cotton or linen. It can be used for dress-goods, plain dyed or printed, and for shirtings. Mixed with cotton it is very suitable for sports shirts. When given an anti-crease finish it is suitable for light suitings. When made, as it can be, with slubs it gives an imitation linen. It is knitted for underwear and for jumpers and jerseys. It is used for furnishings, handkerchiefs, sheetings and tablecloths. Finally it can be crêpe-twisted and woven into marocains and crêpe-de-chines. It has, in fact, successfully invaded most textile fields.

Medical. Special mention may be made of some of the medical uses to which viscose rayon is put. The properties which make these possible are that it is highly absorbent and will imbibe 90–95 per cent of its own weight of water, and that it is a clean and hygienic material containing no extraneous waxes, seed-pods or sand as cotton is liable to. Furthermore, because it is so hydrophilic it is not liable to generate static electricity, and can be used safely in operating theatres, where the presence of ether fumes may make the hydrophobic synthetic fibres, which are liable to generate static and liable, even if only remotely, to start a fire, dangerous. Rayon lint is composed wholly of viscose staple and is similar in appearance and handle to cotton, but is superior to it in being softer and kinder to the wound and more absorbent; it reacts favourably to antiseptics and can be sterilised without losing its softness. Absorbent gauze made from viscose rayon staple does not adhere to the surface of wounds to the same degree as cotton gauze and is more easily removed; it is also suitable for swabs and the use of rayon for tubular gauze is developing. Rayon tulle net used for paraffin gauze dressings absorbs more paraffin than cotton and is more easily removed from a wound. Other outlets for viscose staple are for elastic and crêpe support bandages, for sanitary towels and cotton "wool". Blends of viscose rayon and cotton have been used for hospital clothing, nurses' dresses, aprons and caps. The appearance of uniforms can be enhanced by the use of a fusible Courlene interlining in semi-stiff collars, cuffs and belts. Nurses' uniforms made from 50/50 Fibro/cotton have given better wear than the traditional all-cotton fabrics, especially in resistance to abrasion such as takes place where the stiff collar rubs on the shoulder of the garment.

Industrial. Overalls and denims made from 100 per cent Fibro

Duracol (spun dyed Fibro) and from 50/50 mixtures of it with cotton have been adopted by government departments for use by postmen, coastguards, policemen and firemen. Duracol gives a fastness equal to vat dyeing at a considerably lower cost for heavy shades such as navy and black.

Man-Made Flock. Viscose rayon flock, when cut into very short lengths, *e.g.*, from 0·5 to 3 mm., has been used to give suede, pile and nap effects. If a plain dyed marocain fabric is printed in a pattern with gum, and Fibro flock in a bright shade is blown on or attracted electrostatically to it, it adheres only to the gum-printed areas; to these it gives the appearance of velvet, and the whole piece looks as if it is a cut-away pile fabric. Carpets are even now being made with viscose flock of 3–7 mm. length, produced by Vereinigte Glanzstoffabriken A.G., on the Eloflock machine: the fibre flock is given a positive charge and is projected on to an adhesive covered fabric resting on a negatively charged (high voltage) metal plate. Perlon and nylon flock has been used as well as viscose. The trick is to keep the flock vertical in its passage to the adhesive coated fabric so that it becomes anchored only at one end and stays erect. About 300 grams of Perlon flock is applied per square metre for carpet. The production from such machines can be enormous; if the machines do succeed in reaching the stage of producing good quality carpeting regularly they might threaten traditional carpet weaving. Pile fabrics, imitation furs and teddy-bear fabrics can be made similarly.

Acetate Staple. Cellulose acetate staple fibre is mostly used as a blend with cotton, and to a smaller extent is blended with wool. The outstanding characteristics of cellulose acetate staple are good handle and draping qualities with fairly good recovery from creasing. When cellulose acetate staple/cotton blends are dyed, the cellulose acetate is often reserved and the mixture yarn is used in suitings.

When acetate staple is blended with viscose staple the tenacity falls and the extensibility of the yarn increases as the proportion of acetate increases; the figures in the following table exemplify this trend:

Parts of $1\frac{7}{16}$ in. $2\frac{1}{2}$ denier dull acetate staple.	Parts of $1\frac{7}{16}$ in. 3 denier matt viscose staple.	Tenacity of blended yarn (gm./denier).	Extension at break of blended yarn (per cent).
30	70	1·29	11·7
50	50	0·87	14·3

All cellulose acetate staple is spun from a pigmented (titanium dioxide) spinning solution. Acetate/viscose staple blends have been used in tropical suitings, and it is claimed that fabrics made from such blended yarns have good dimensional stability.

When cellulose acetate staple is blended with wool the staple length is either 2½ in., 4 in. or 5 in., and 4½ is the usual filament denier. The addition of acetate staple to wool is said to decrease the felting power of wool; a blanket made to the following specification

Warp . .	80 acetate staple 5-in. 5·4-denier/20 wool blend
Weft . .	All wool

shrunk only about 5 per cent in area after ten washes, whereas an all-wool blanket of otherwise similar construction shrunk 10–11 per cent after the same number of washes.

Acetate/nylon blends have been spun on the cotton system, but the tenacity of the resulting yarn is rather low, as shown in the table below.

Parts of $1\frac{7}{16}$ in. 2½ denier dull acetate staple.	Parts of 1½ in. 3 denier nylon staple.	Tenacity of blended yarn (gm./denier).	Extension at break of blended yarn (per cent).
60	40	1·17	16·0
70	30	0·95	13·1
85	15	0·76	11·2

If longer staple is used and spun on the worsted system there is a significant increase in tenacity; using 70 parts cellulose acetate 4 in. 4½ denier staple and 30 parts nylon 4 in. 3 denier, the blended yarn has a tenacity of 1·15 grams per denier (*cf.* with 0·95 in above table) and an extension of 13·6 per cent.

Dress, blouse and shirting fabrics and tropical suitings are the main end-products of acetate blends. Always there is the possibility of obtaining " mixture " dyed effects by reserving the acetate.

Nylon Staple. As yet, nylon has been blended mostly with wool, perhaps partly because both fibres are relatively expensive and also because of the related chemical constitution of the two fibres (both consist essentially of long polyamide chains) and of their similarity in extensibility and elasticity. To such mixtures nylon contributes a high breaking load and better resistance to abrasion. Usually a 4 in. 3 denier staple nylon is mixed with botany wool, and a 6 in. 6 denier staple nylon with crossbred wools.

Textile processing is improved in efficiency by the addition of staple nylon to wool. It has been reported that the addition of

5 per cent nylon improves the spinning efficiency of wool, and that the weaving efficiency on automatic looms rises from 75 to 90 per cent when 10 per cent of nylon staple is mixed with wool. When nylon staple is blended with cashmere, the knitting efficiency of the blend is higher than that of the cashmere alone. In the woollen (as opposed to the worsted) section, nylon/wool blends enable finer counts to be spun; the inclusion of 25 per cent nylon will enable a wool which could only be spun to 32 s woollen counts to be spun to 45 s woollen counts.

The biggest advantage that derives from blending nylon staple with wool is the increase in life of the garment; military socks made from a blended yarn of 75 wool/25 nylon staple have been reported to wear five times as long as all wool. "Nylox" is the name of one blend of wool and nylon staple which has been marketed. There are four defects associated with the use of blended nylon and wool; these are:

1. If made into socks, the socks are much less comfortable, in the author's view at least, to wear than all wool; they are inclined to feel cottony and rather harsh.
2. Socks so made are not so easy to put on, the socks do not stretch laterally so easily or so much as all-wool socks.
3. There is a migration of the nylon fibres out of the sock, particularly at the heel where abrasion is greatest; if the sock yarn was a blend of natural nylon staple and black wool, it is easy to observe the natural nylon fibres on the heel of a worn sock.
4. If made into jumpers, cardigans, *etc.*, pilling is much more evident than it is with all wool; this again as in 3. is due to migration of the nylon fibres out of the yarn and of the garment.

In the author's view so far as nylon blends in hand-knitted and next-to-the-skin garments are concerned, a heavy price—that of loss of softness of handle and of increased pilling—has to be paid for the greater durability that nylon contributes.

For other garments, such as riding twills and for suitings and shirts, these disadvantages are less noticeable. Riding twills can be reduced in weight by the blending of nylon, the British Services have used fabrics of 75 wool/25 nylon staple for suitings and shirts, the U.S. authorities have used an 18 oz. worsted serge made of 85 wool/15 nylon. A staple nylon/staple rayon/mohair blend has been used for tropical wear. Flags and buntings made of 25 wool/75 staple nylon last ten times as long as the same articles made entirely from wool.

A tendency has been evident to use lower-quality wools, and even shoddy in blends, reliance being placed on the nylon staple for strength. The facility with which crimp can be imparted to nylon by steam- or heat-setting increases its suitability for blending with wool. The admixture of nylon with wool will not prevent the wool shrinking or felting, nor will it protect it from the attack of moth larvae. It will, however, *reduce* the shrinking and felting.

Nylon staple has been blended also with cotton, but this blend is not so attractive as the wool blend, for two reasons:

1. Fabrics containing a high proportion of cotton usually group themselves in a lower price range than those containing a high proportion of wool.
2. There is no considerable gain in dry strength until the nylon content exceeds 50 per cent of the blend.

Nylon staple has also been blended with viscose rayon staple on cotton-spinning machinery with happier results. There are two factors advantageous to the production of such rayon/nylon staple blends, *viz.*:

1. There is a great deal of cotton-spinning machinery available, and if it can be used for the production of a new product it is more likely to find continuous employment than if it continues to rely solely on the manufacture of cotton yarns, which can now be made in most parts of the world.
2. The resistance to abrasion of the rayon staple is considerably improved by blending with nylon staple.

If, for example, 3 denier $1\frac{7}{16}$ in. matt Fibro is blended with 3 denier $1\frac{1}{2}$ in. nylon staple, the resistance to abrasion (tested on a Ring Wear Tester) of fabrics made from the resulting yarns increases as shown in the following table:

Nylon staple (per cent).	Fibro (per cent).	Resistance to abrasion (arbitrary units).
0	100	1
10	90	1·4
20	80	1·9
40	60	2·7
60	40	3·7
80	20	7·6
100	0	8·5

Terylene. Terylene staple fibres in current production are made for various systems, for example:

3 denier $4\frac{1}{2}$ inch for the worsted system.
4 denier $4\frac{1}{2}$ inch for the worsted system.
3 denier $2\frac{1}{2}$ inch for the woollen system.

All of them have a heat-set crimp, and the crimp gives a loftiness which is advantageous; it also gives a closer approach to natural wool fibres which always contain crimp. Fine wool fibres usually have much more crimp than coarse ones, and as it is usually with the finer fibres of say 60 s to 80 s quality that Terylene will be blended, a pronounced crimp is desirable to improve the uniformity of the blended yarn. Furthermore, the crimp assists the carding and subsequent operations. All Terylene staple fibre is pigment delustred.

The physical characteristics of Terylene staple fibre are different (*cf*. p. 406) from those of filament yarn, the tenacity being relatively low at 3·5 to 4·0 grams per denier and the elongation correspondingly high at 40–25 per cent; in other respects, such as resistance to heat, light, chemical and microbiological attack, the staple fibre resembles the filament yarn. The purpose of the high extension is to match that of wool, which averages about 38 per cent; the tenacity of wool is, of course, much lower at about 1·4 grams per denier dry and about 1·1 grams per denier wet. So far as moisture regain is concerned, wool and Terylene have nothing in common; wool has a very high regain of 16 per cent and Terylene a very low one of 0·4, both at standard conditions 65 per cent R.H. and 25° C. Staple Terylene is stiffer than wool; although in 4 denier it has about the same diameter as 70 s wool, it is more like 64 s wool in handle. Resistance of the staple fibre to bacteria, fungi and moth larvae is excellent, just as it is with filament Terylene.

Terylene staple fibre, as supplied, carries an anti-static finish, and if this is removed in one of the spinning preparatory processes, *e.g.*, during backwashing or slubbing dyeing, it must be replaced; this can be done suitably by applying 1 per cent of a combing oil containing 5 per cent (on the weight of the oil) of the anti-static material Lissapol NX. Terylene staple fibre is spun on the worsted and woollen systems.

Worsted System. When the bale of Terylene staple is released from the restraining bands it flies apart. It is first carded; the relative humidity should not be lower than 55–65 per cent during carding. Backwashing is unnecessary. The card sliver is gilled several times, then combed in a Noble comb and the comb sliver is finisher-gilled and converted into 7-lb. balls. Drawing is done normally, but because of the greater force required to cause the fibres to draft past each other, less twist is necessary at all stages than with wool. After drawing, the fibre can be spun on cap, ring and flyer frames with a lower twist than is usual for the corresponding worsted counts. Terylene 4 denier staple has been spun commercially to 60 s worsted counts and 3 denier to 80 s. Whilst a lower twist gives a fuller handle it is accompanied by a tendency in Terylene or Terylene/wool blends to " pill ", and this tendency can be reduced by inserting a higher twist during spinning, but high or even normal twists should be avoided in the preparatory stages because they cause tight ends and uneven drafting. If, as for a tropical suiting, a firm crisp handle is required, then higher twists than usual can be inserted during spinning.

Wool can be blended with the Terylene during carding, at the can gill box or during combing.

Woollen Systems. If the Terylene staple fibre has been stock dyed, the anti-static agent must be replaced by the application of combing oil containing Lissapol NX, or if it is blended with wool prior to carding, the two fibres can be stacked in sandwich form and about 10 per cent of oil containing some Lissapol NX applied to the wool component only. After stacking, the blend should be passed through an opener two or three times before feeding into the hopper of a scribbling and carding set. The card web is condensed by tape or ring methods and then spun on mule frames. In order to minimise pilling in the finished fabric, a higher twist than usual for woollen spun yarns should be inserted into yarns containing Terylene.

Properties. Staple fibre Terylene yarn spun on the worsted system shrinks about 6 per cent in boiling water; this shrinkage is nearly all due to relaxation of mechanical strains imposed during the preparatory and spinning operations because the fibre before treatment will shrink less than 1 per cent. The tenacity of a 100 per cent Terylene staple yarn will vary according to its construction from 2·0 to 2·8 grams per denier (worsted 0·6–0·95) and the extension from 30 to 20 per cent (worsted 25–15); the Terylene yarn is as strong wet as dry, whereas worsted loses about 15 per cent of its strength on wetting. Terylene yarn spun on the woollen system will have a tenacity of about 2·2 grams per denier (wool 0·45 grams per denier) and an extension of about 33 per cent (wool 19 per cent).

Dacron. This fibre is produced in several deniers and staple lengths, all semi-dull.

Overriding all other considerations in respect of its use as staple fibre is the warmth of handle of garments made from it. Socks, jumpers, and dresses made from Dacron might well be mistaken for wool in this respect; in other respects, notably softness, there is a very considerable difference. So there is in moisture absorption, heat of wetting, wicking and pilling.

Like nylon, Dacron staple when blended with wool greatly increases its strength; the increase in resistance to abrasion which Dacron confers on wool (p. 665) is significant, but much less than that conferred by nylon. Because of the extremely low regain of Dacron (the physical properties of Dacron when thoroughly wet are about the same as when dry) its blends with wool have appreciably more dimensional stability to changes of relative humidity. Dacron/wool blends also press, and retain their crease, very well. Dacron is very subject to pilling, more so even than wool, and this constitutes a considerable defect.

When blended with viscose rayon, Dacron gives some improvement in resistance to abrasion, but its main contribution is to enable the rayon blend to recover well from wrinkling and to retain a pressed crease well.

All Dacron fabrics are very subject to hole burning, *e.g.*, with a cigarette end, but blends of Dacron with viscose rayon staple, cellulose acetate staple or with wool are much more satisfactory in this respect.

A detailed study of the behaviour of summer-weight tropical suitings made from Dacron and Dacron/wool blends has been made by the U.S. Air Force. The fabrics were found to be acceptable from the standpoint of comfort as compared with an all-wool fabric, although two complaints were made of an all-Dacron fabric clinging due to static; wrinkle or crease-resistance was good, but there was no saving in the frequency of cleaning and pressing. Other adverse comments related to the ease with which cigarette ashes would burn a small hole in the all-Dacron fabric and also to the proneness of the fabric to pick up dirt quickly from a chair along the folds and creases. On the whole, the results were good, and there is no doubt that Dacron will find extensive use in suitings. Dacron staple has also been used as a surface yarn for carpets; it should give very long wear.

Pilling

Not only Dacron, but several of the newer fibres are subject to pilling. Everybody must have noticed little fluffy balls or pills which work up on the surface of a woollen sweater or jumper. The author once examined some of these and found that mostly they consisted of the short fibres—all wool contains some short fibres—and the probability is that, being short, they are tied down and held in position less securely than the other fibres, that they migrate during wear and tangle up with each other into little balls. Pilling is far too serious a defect in wool for anyone to want to run the risk of making it worse, and as a rule much can be done by suitable structure of yarn and fabric to minimise it.

An interesting method of examining the structure of pills has been devised by the makers of Dynel. A 65/35 wool/Dynel fabric was found to form pills on an abrasion test; two pieces of the pilled fabric were taken, and the Dynel was dissolved out of one of them by treatment with acetone, the residual wool fabric still showed the pills. The wool was dissolved out of the other piece by treatment with alkali; the residual Dynel fabric showed no pills and the conclusion was drawn that the blend would not be expected to pill any more than all-wool. Nevertheless, the pills can usually be removed easily from an all-wool fabric; often, indeed, they fall off in wear, whereas they may be held firmly to the body of the fabric in a blend of wool and a strong fibre. For example, the pills that formed on an 85/15 wool/nylon fabric were very difficult to remove, but when the nylon component was dissolved out it left a wool fabric from which the pills fell away easily. Reduction in the number of short fibres in a yarn is desirable to minimise pilling, but it is impossible in practice to make a yarn without some short fibres; if, for example, the combing process is continued to remove all short fibres, the same process breaks some fibres, so introducing new ones. There seems to be no complete remedy for the trouble at present; in practice, manufacturers find that some blends are more liable to pilling than others, and these are naturally avoided.

[A corollary of the process suggested by the Dynel manufacturers is that by using suitable solvents, pieces of a fabric made of two different blended fibres can be converted into fabrics of each fibre only, and if pieces of the same dimensions are used it can be shown that the two separate fabrics made of (*a*) Dynel and (*b*) wool, together weigh just as much as the original fabric of the blend of the two fibres.]

A study of pilling of fabrics made from spun nylon and nylon blends has led to the following conclusions:

1. Pilling even on the same cloth and type of garment varies with the wearer, some people show more pilling than others, usually those that are more active. Boys' trousers and girls' skirts were made from the same blend and much more pilling developed on the trousers than on the skirts.
2. Pilling is caused by protruding fibres which become

entangled when the fabric is rubbed. If these protruding fibres are removed by singeing or by cropping, pilling is reduced.

3. Heavy filament deniers give less pilling than light ones.
4. The spinning system affects the amount of pilling; best is silk, then the worsted system, intermediate is the cotton system and worst of all the woollen system.
5. Increase of yarn twist reduces pilling.
6. Increasing the yarn intersections in a fabric by changing from twill to plain weave or by putting in more ends and picks reduces pilling.

The problem is more important commercially than one might think, and it has been aggravated by the use of strong fibres such as nylon and Terylene, because pills which contain these very strong fibres are less likely to drop off than those containing only wool fibres.

Nothing Like Wool

First attempts to make rayon concentrated on the *continuity* of filament; the reason was evident—real silk had continuous filaments, real silk was the most highly esteemed fibre; it excelled in strength, fineness, and lustre, and the first rayons were called " artificial silk " without any hedging over nomenclature. In our aim to make a synthetic silk we have succeeded; nylon and Terylene are, on the whole, better fibres than real silk—they are more useful in that they are stronger and more durable and are more attractive in that they can be spun finer or coarser, are more uniform and can be produced with whatever lustre is required. Even so, the victory is no walk-over and there are still plenty of women of discernment who, for their own wear, prefer natural silk to any synthetic fibre.

But there is another natural fibre which is much more useful and really more beautiful even than real silk and that, of course, is wool. Wool, however, was for a long time cheap and plentiful, and there was no special incentive for us to try and make an artificial wool. In other years wool has been dear and the supply of it has been unequal to the demand. Logically, therefore, we have tried to make an artificial or synthetic wool, and the first step towards this has been to cut up the continuous filaments that earlier we were at such pains to make, into short filaments of about the same length as wool fibres. Oddly enough, we now use more of our artificial fibres after they have been cut into short fibres, *i.e.*, as staple, than we do in the form of continuous filament.

Crimp. Furthermore, in order to make the cut fibres as similar as may be to wool, manufacturers now crimp the fibres, *i.e.*, they run them through suitable rollers, so that instead of being straight

the fibres are crimped

Then the fibres are spun on the usual worsted spinning machinery, and a yarn results which superficially resembles, say, a hand-knitting wool. Nevertheless the resemblance is only superficial, and it will probably be a very long time indeed before an artificial fibre which is a good match for wool can be made.

The poorness of the resemblance of mechanical crimp in man-made fibres to the natural coil-like crimp in wool has led to the

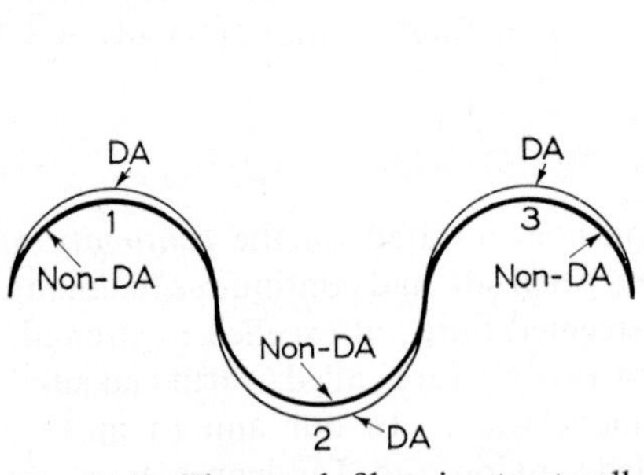

FIG. 214.—The wool fibre is structurally asymmetrical. The outside of the curvature of the crimp is DA (dye accessible) to Janus Green but the inside of the curvature is not (non-DA). The drawing represents an enlargement of about 50 times of the wool fibre.

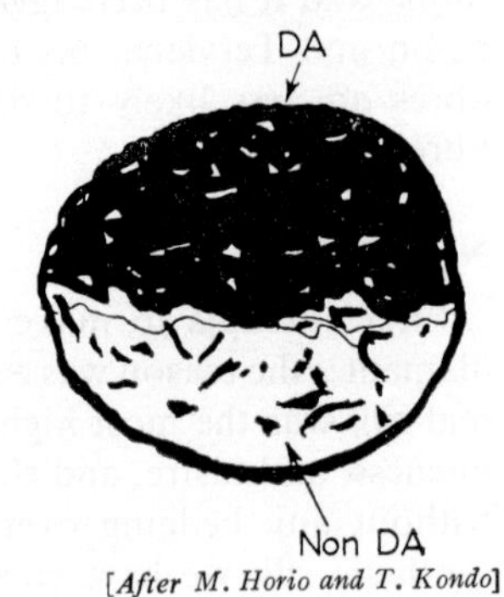

[*After M. Horio and T. Kondo*]

FIG. 215.—Cross-section of a wool fibre, stained with Janus Green to show its asymmetry in respect of dye accessibility. Magnification about × 1,000.

introduction in viscose rayon of chemical crimp (p. 181) which is a better match for wool crimp but still not altogether satisfactory. Knowledge of the subject made progress when a structural asymmetry of wool was demonstrated by Horio and Kondo of Kyoto University. They stained wool fibres with Janus Green under precise *p*H control, and showed that the outside of the curvature of the crimp was always stained a deeper shade than the inside; they termed the outside Dye Accessible (DA) and the inside non-DA. If the wool fibre is washed with dilute acid just after staining, the colour difference is increased. What is meant by this differential staining can be seen from Fig. 214 which is a sketch of a single wool fibre and Fig. 215 a representation of the microscopic appearance of a wool fibre which has been stained and which very clearly shows the asymmetry of the fibre.

A half turn of twist (180°) occurs between points 1 and 2 (Fig. 214) and another half turn in the same direction between points 2 and 3.

If instead of Janus Green, an acid dyestuff, *e.g.*, Ponceau 2R, is used the outer curvature becomes non-DA and the inner becomes DA, *i.e.*, a complete reversal.

The asymmetry of structure is accompanied by asymmetry of swelling on wetting, and such single filaments will coil when wetted. But if there are many fibres, each restrained by others, as in a lock of wool, the movement is one of crimping rather than coiling.

The asymmetry of chemical crimp viscose rayon (p. 181) is one way of copying the asymmetry of the wool fibre; the method of conjugate filaments (p. 183) is another and perhaps this is at its best in the production of Orlon Sayelle (p. 516). At the National Bureau of Standards it has been shown (*Chem. Engng News*, **39**, No. 43, p. 50 (1961)) that if cross-linkages of the type $-CH_2-S-CH_2-S-CH_2-$ (joined to N atoms in polyamide chains) are inserted in nylon 6, the fibre will subsequently coil and crimp in a swelling agent. In relating this behaviour to that of wool we remember that wool contains disulphide cross-linkages and that water is a swelling agent for wool. Is it possible that the asymmetry of wool demonstrated by Horio and Kondo depends on differential pulls on the two ends of the cystine linkages?

Wool Properties. The five outstanding characteristics of wool are:

1. Its warmth.
2. Its softness to the hand.
3. Its high extensibility (especially when wet) and good recovery therefrom; these properties are mainly responsible for the way in which wool garments will stand up to hard wear, not tearing when snagged, and will lose their creases and regain their shape when hung up after wear.
4. Its ability to felt; although this last can be a great nuisance when light woolly garments are washed, it is the process which makes possible the production of "fulled" overcoatings and other heavy cloths, as well as felt hats and many other felts such as are used for floor-coverings and typewriter pads.
5. Its ability to take up a third of its own weight of moisture without feeling damp.

These five outstanding properties of wool cannot be met by any single man-made fibre at present, nor indeed can any three of them.

The ability to felt is a unique characteristic of animal hair fibres of which wool is the commonest; it depends on the system of overlapping scales which every hair possesses; these scales can be seen in Fig. 216, which is a photomicrograph of a wool fibre, and in Fig. 217 which is a simplified or idealised view of the scale system. When a fabric made of wool is washed all the little scale edges of one fibre grapple with those of other fibres, so that all the little fibres are drawn up into loops and entanglements; consequently, the fabric draws together or shrinks, becomes thicker and felts. The felting process is entirely due to the scale system and it is very unlikely

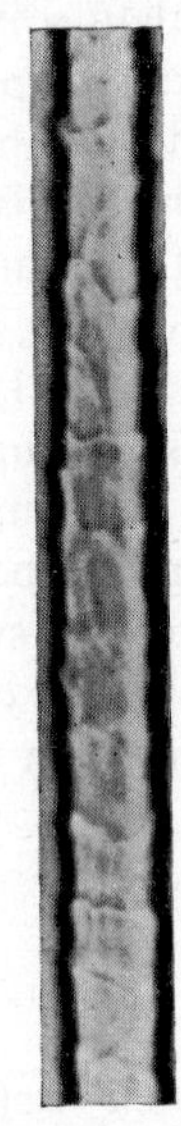

FIG. 216.—The scales of a natural wool fibre (× 400).

FIG. 217.—Simplified or idealised view of wool fibre scales.

that any fibre synthesist will be able to spin an artificial fibre whose surface is characterised by a series of overlapping scales.

The warmth and handle of wool are largely due to its chemical structure, its protein character, and those fibres such as Lanital, Ardil, Merinova, and Vicara, which have been derived from natural proteins, have a somewhat similar warm and soft handle; even so, their handle is different from that of wool because their frictional properties are different. Wool has a lower coefficient of friction if rubbed from root to tip, *i.e.*, in the with-scale direction, than if rubbed from tip to root, *i.e.*, in the anti-scale direction; it will be difficult, although probably not ultimately impossible, for the chemist to make an artificial fibre which possesses this directional frictional effect. Nevertheless the handle of the protein artificial fibres is not *very* different from that of wool, so that when these fibres are blended with wool, the handle of the wool may not be noticeably impaired; it is, however, still quite impossible to make a 100 per cent artificial fibre which has a handle like that of wool. The regenerated protein fibres, too, lack the cystine cross-linkages which join the long molecular chains of wool together like the rungs in a ladder, Fig. 218, and which when the fibre is stretched are dislocated as in Fig. 219; on release of the stretching tension the cross-linkages pull back the long molecular chains to which they are attached to their original positions, so that the wool fibre regains its original dimensions, and so that the garment which is made of the wool fibres loses its creases and regains its original shape. So

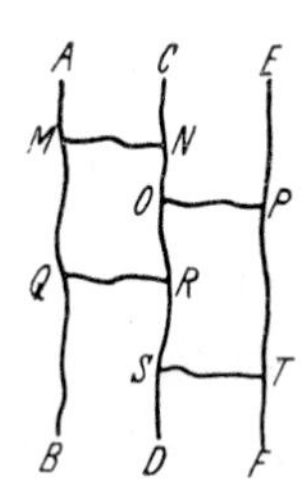

FIG. 218.—Long polypeptide chains AB, CD, EF are held together by cross-linkages MN, OP, QR, ST.

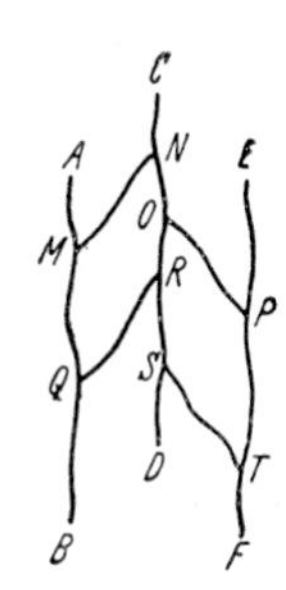

FIG. 219.—Some dislocation occurs under tension.

important are these cross-linkages that many attempts have been made to provide them in artificial fibres; no man-made fibre embodies the cystine linkages that characterise wool, but most of the protein regenerated fibres have had other kinds of cross-linkages provided by treatment with formaldehyde—these, however, are less effective than are cystine cross-linkages.

One other feature of the wool molecule is that it is coiled, so that when stretched in water or under moist conditions, the molecules first uncoil, and this uncoiling results in wool having an extensibility in the wet state of 70–100 per cent. This high extensibility is valuable, as it enables wool fibres to take up sudden strains without breaking. Whilst this property is shared by the regenerated protein fibres, it is quite unknown amongst the true synthetic fibres.

In brief, the regenerated protein fibres have a handle and warmth that are sufficiently near to those of wool to allow them to be used very satisfactorily in 50/50 blends with wool, but they lack the chemical stability (due to the cross-linkages) of wool, they lack its good recovery, its strength (particularly its wet strength), and they lack the peculiar properties of wool that derive from its scale structure, notably its unique handle and its felting propensities.

Of the synthetic fibres, Terylene or Dacron is outstanding in its wool-like warm handle, and it is this property more than any other which will contribute to the popularity of this fibre. It has many

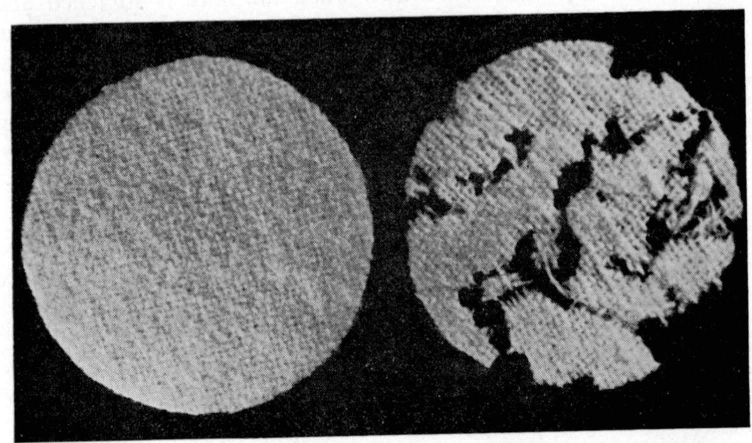

FIG. 220.—Wool is subject to attack by carpet beetles and moths as these two pictures of a wool fabric before and after attack show. Synthetic fibres are resistant to insect attack, and this is one of their valuable features.

advantages over wool in respect of strength, resistance to abrasion, and chemical stability, but it lacks the high wet extensibility of wool and, of course, it lacks completely the scale structure, which confers a unique handle and felting properties; furthermore, it lacks the high moisture pick-up of wool, which derives from the multiplicity of hydrophilic amine, hydroxyl and carboxyl groups in the wool molecule, and which enables wool to absorb 30 per cent of its own weight of water without feeling damp—a property of inestimable value for underclothing.

The complexity of Nature is often purposeful; the wool molecule is a co-polymer of eighteen different amino-acids in more or less constant proportions, and probably in fairly definite sequence of arrangement; most of our synthetic fibres are straight polymers of one compound or co-polymers of two. Morphologically, too, there are tremendous differences; the wool fibre consists of epi-cuticle, scales, sub-cuticle, cortex, and sometimes of medulla, whereas the synthetic fibre is a simple cylindrical (or collapsed cylindrical) rod.

The fibre synthethists have done marvellously well, but they have still a very long way to go before they can make a fibre that will feel, look, and behave like wool. It is far more difficult to make an artificial wool than an artificial silk. Most likely in fifty years' time, it will still be true to say there is " nothing like wool ". In respect of freedom from insect attack the synthetic fibres have a big advantage. Fig. 220 shows what can happen to wool.

FURTHER READING

B.I.O.S. Final Report No. 1096. Item Nos. 22 and 31. H.M.S.O. (description of German tow-to-top processes).
" British Rayon Manual ", 76–91, Manchester (1947).
" Rayon Staple in the Cotton Industry ", Courtaulds, Ltd. (1948).
C. M. Whittaker, " The Fibro Manual ". London (1949).
" Talks on Rayon ", 57–76 (The Cotton and Rayon Merchants Association).
" Nylon Staple, its Manufacture, Handling and Uses ", *Rayon Text. Monthly*, **28**, 68 (1947).
R. A. Farlane and C. C. Wilcock, *J. Soc. Dyers & Col.*, **65**, 145 (1949) (dyeing of Fibro).
" Processing of Rayon Cut Staple ", *Textile Mercury and Argus*, **120**, 959 (1949).
American Viscose Corporation, " Rayon Technology ", 2–84. New York (1948).
R. W. Dennison and L. L. Leach, " Blends Containing the New Man-made Fibres ", *J. Text. Inst.*, **43**, P 473 (1953).
S. A. G. Caldwell, " Rayon Staple Fibre Spinning ", Emmott (1953).
Textile Institute Annual Conference on Fibre Blends, Edinburgh, 1952. *J. Text. Inst.*, **43**, No. 8 (1952).
J. B. Quig and R. W. Dennison, " Functional Properties of Synthetics ", *Ind. Eng. Chem.*, **44**, 2176 (1952).
H. Ashton, " Greenfield Top—Rayon Tow to Top ", *Text. Manfr.*, **69**, 259 (1943).
" Tow to Yarn Spinning Frame ", *Text. Recorder*, **65**, 43 (Dec. 1943).
Brit. Pats. 511,587, 518,995, 523,579, 535,793, 537,742 (these relate to Greenfield Top).
" Perlok System ", *Rayon Text. Monthly*, **25**, 437 (1944).
J. W. Fairbairn and T. D. Whittet, " Rayon Dressings, uses and identification ", *Pharmaceutical J.*, 161–163 (1951).
" Rayon Lint " Prescriber's Notes, Min. of Health (Dec. 1952).
" Electrical Safety in Hospitals ", Min. of Health, interim note (19 June 1953).
M. E. Baird, P. Hatfield and G. J. Morris, " Pilling of Fabrics ", *J. Text. Inst.*, **47**, T 181–T 201 (1956).
M. Horio and T. Kondo, " Crimping of Wool Fibres ", *Text. Res. J.*, **23**, 373 (1953).
E. H. Mercer, " The Heterogeneity of the Keratin Fibres ", *Text. Res. J.*, **23**, 388 (1953).
G. S. Park, " An Autoradiographic Method Based on Tritium for Locating Resin Finish in Textiles ", *J. Soc. Dyers & Col.*, **76**, 624 (1960).
" Electrostatic Flocking ", *Man-Made Textiles*, p. 49 (April, 1961).

CHAPTER 46

NON-WOVEN FABRICS, FELTS AND PAPERS

PREHISTORIC man is reputed to have made clothes from the inner bark of the mulberry tree. Probably he rated them as rather unsatisfactory and as an indifferent substitute for the furs and skins that he preferred and which were as good as anything we have to-day. Furs were possibly often difficult to acquire, bark was probably easy, but neither needed to be made, and it was only with the coming of civilisation that work was organised and that the laborious processes of spinning and weaving and knitting were discovered.

The labour of weaving or knitting is immense; there will be about six or seven *million* yarn intersections in a square yard of an ordinary woven jappe, and whilst the loom makes these fairly efficiently, their number sets a limit to the speed with which fabric can be produced; somewhat similar considerations apply to knitting. There has, therefore, been a considerable incentive to make fabric by methods which avoided the onerous processes of weaving and knitting fabrics either

(*a*) from film which can be spun out of polyvinyl chloride or polyethylene much faster than mulberry bark can be grown;

(*b*) from felts which normally have their own special uses which they fit ideally but which are unsuitable for most of the uses to which we put fabric;

(*c*) from fibres which are bonded or stuck together with some dry setting adhesive to form a felt-like sheet; or by

(*d*) the preparation of similar felt-like sheets on a rubber or plastic backing, usually for carpets.

Film has not proved to be ideal for clothing; it tears too easily and does not ventilate the body, although plastic " macs " which will fold up and go into the pocket have proved popular and useful. Felts have been used for a very long time; their most recent and unusual application has been for skirts. The bonded fabrics have been made with far greater difficulty so far than woven fabrics, and they have not been nearly so good, but a good deal of effort and persistence is being applied to their production and ultimately it is likely that something useful and permanent will emerge. Even so, spinning, weaving and knitting, laborious though they are, give useful and beautiful products with unique characteristics, and it would seem wise before spending so much energy in trying to cut them out, to see if they could not be simplified. Much, of course, has been done, the tow-to-top and direct spinning methods cut out laborious carding and combing operations; even better, the continuous filament spinning cuts out textile staple spinning altogether; automatic looms that are precision machines are available and warp knitting machines that go many times as fast as they quite recently did are in use. Perhaps the development that seems most fundamental in the weaving process and one that seems to be neglected is progress with the circular loom. There is something to be said for the view that research and development on the circular loom should ultimately prove more profitable than work on non-woven fabric. But there are others who think that spinning and weaving are intolerably onerous and that whatever the cost, it would be good to be delivered from them.

Bonded Fabrics

A non-woven fabric is composed of two parts, the web of fibres and the bond. Preparation of the web requires great care; bonding is relatively easy. One pound (weight) of fibres will make perhaps five or six yards of an average dress fabric, and it contains about a hundred million fibres. The problem in making bonded fabric is first to separate or open these fibres and then to lay them flat and (sometimes) in random direction. Two kinds of web, *i.e.*, flat arrangements of fibres are used, (*a*) oriented and (*b*) random.

Oriented Webs (Parallel Laid)

These are produced by conventional textile methods: opening, picking and carding. Opening consists of the mechanical release of the fibres if baled, as are most natural fibres and most staple rayon; the opened fibres may occupy fifteen times the volume that they did in the bale. The picker completes the opening or separation of the fibres, and arranges the fibres in lap form in which they can conveniently be fed to the carding machine. The card separates the fibres that are still in bundles or tufts down to individuals, and converts the fairly heavy lap into a lighter web; it is the virtue of the card from the standpoint of the cotton and worsted spinner that it arranges all the fibres parallel or orients them. But, because of this, it can be used only for oriented webs. The fundamental defect of oriented webs is that they are weak across their width, the fibres all running parallel with the length of the web are easily pulled apart sideways. Sometimes, oriented webs are used in that condition as they will satisfactorily fulfil certain requirements, other times they are cross-laid.

Cross-Laid Webs. One web is laid on another, ideally at right angles, giving if this process is continued for a few layers a laminated structure which after having been bonded has equal strength transversely and lengthways.

In practice the alternate layers are not laid quite at right angles although they are angled. One of the basic advantages of bonded fibre fabrics is that they can be made at high speeds, at yards per minute, at speeds associated with paper-making rather than with weaving or knitting. With the parallel-laid oriented webs, high speeds present no difficulty, but it has been impossible to devise a continuous process in which one web can be laid at right angles on another. A compromise is made; one lap of parallel-laid fibres is covered continuously with another in S form as shown in Fig. 221,

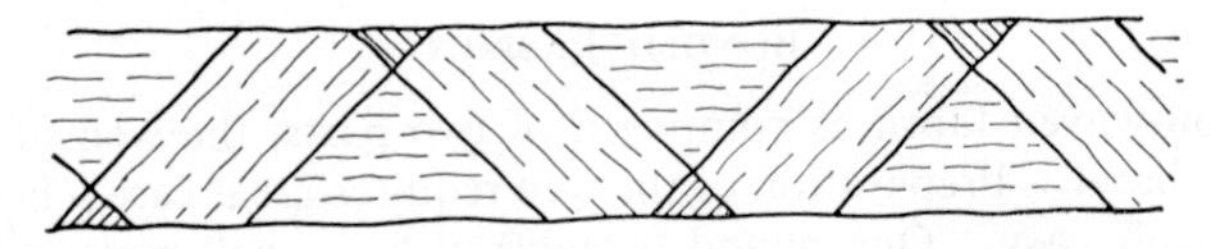

FIG. 221.—Cross-laid web. The shaded portions represent double thickness and an attempt is made to spread these out evenly as the successive layers build up to form the fabric.

then another parallel-laid web is applied, then another in S form and so on. In this way, the continuous character of the process is retained. Endeavour is made to spread out the double thickness areas (shaded) evenly in successive layers, so that the final fabric is uniform in thickness. Transverse strength is much better than in a parallel-laid bonded fabric, but not so good as in one made from a random web.

Random Webs

These are made on special equipment and do not need the help of conventional textile machinery. A commercial machine made by the Curlator Corporation (New York) consists of two parts, a feeder and a webber, called by their maker Rando-Feeder and Rando-Webber (Fig. 222). The feeder is somewhat similar to a picker, it eliminates the trash from the fibre, and feeds fibre in the form of a mat to the webber, in which the fibre is picked up by an air stream which separates clumps, caught by another air stream which mixes the fibre and then by a third which throws the fibre against a collecting screen. This, shown as *A* in Fig. 222, is where

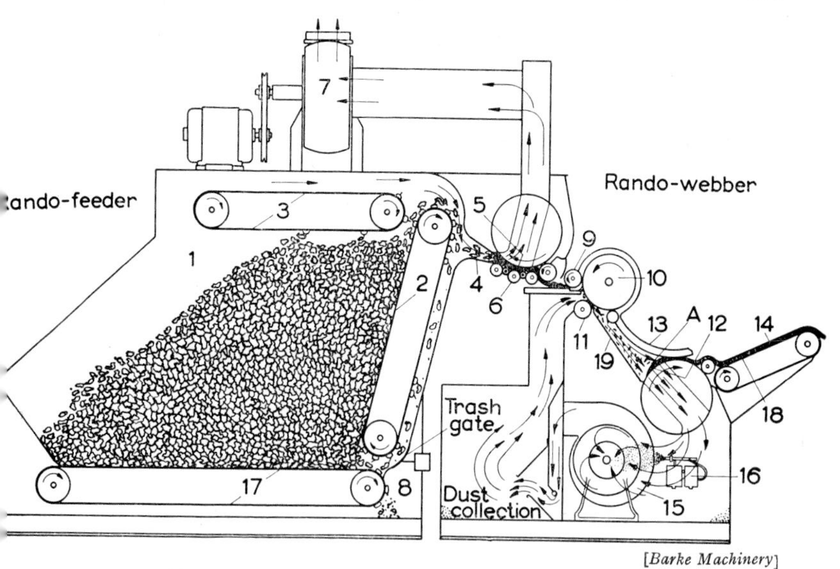

[*Barke Machinery*]

FIG. 222.—The Rando-Feeder and Rando-Webber.

1. Opener and tuft former.
2. Elevating apron.
3. Stripper apron.
4. Air bridge.
5. Feedmat condenser screen.
6. Roller conveyor.
7. Air bridge fan.
8. Trash chamber.
9. Feed roll.
10. Lickerin.
11. Saber tube.
12. Condenser for forming rando-web.
13. Adjustable duct cover.
14. Rando-web.
15. Webber fan.
16. Humidifier.
17. Creeping delivery apron.
18. Delivery conveyor.
19. Venturi.

the airborne fibre is trapped by the screen on which it forms a random web. The collected web is calendered and is then ready for bonding. Because the web was formed aerodynamically it is truly random, and, furthermore, the fibres have undergone very little textile manipulation so that they have not suffered damage. Short fibres give better results than long ones. Production is fairly high, the machine having a capacity of about 2½ cwt. per 8 hour day on a 40 in. width basis. The web can vary from " paper thin " (not so easy) to 2 in. thick. Advantages of this process are:

1. Equal strength in all directions without cross laminating.
2. Thickness of web easy to adjust.
3. The operation is continuous.

Bonding

When the fibres have been arranged conveniently in the form of a web they still have to be bonded together, as the unbonded web has only a negligible strength and would pull apart like a lump of cotton wool. The binder may be of two kinds:

1. an adhesive applied from solution or dispersion, one which dries and hardens when heated, or
2. it may be a thermoplastic fibre which constitutes part of the web.

If a solution or dispersion is used it may be applied by running the web through it, or sometimes it is applied by a printing technique (preferably in a diamond or dot pattern) to perhaps only a half of the total area of the web; this latter process is called "discontinuous bonding" and because a large number of small areas of the fabric are unbonded, the fabric has a softer handle, but it is not so strong as when continuously bonded.

Bonding Agents

Early ones used included starch, glue and casein. Later ones have been polyvinyl acetate, viscose, prevulcanised rubber latex, urea-formaldehyde and urea-melamine resins, various dispersions of polyvinyl chloride, polystyrene, and polyacrylates, and also such water-soluble polymers as carboxymethylcellulose, methyl cellulose, polyvinyl alcohol, and so on. The web is impregnated with a solution or dispersion of the binder and is dried. If a resin monomer (such as U.F.) is used, a catalyst necessarily has to be included with it to ensure that polymerisation will take place during the heat cure. If viscose is used as the binder a suitable coagulant must be applied to regenerate the cellulose. Cellulose bonded non-wovens have high wet strength and the bonding agent, the cellulose, is resistant to most solvents. In the U.K. a large variety of bonded fibre fabrics is made by Bonded Fibre Fabric Ltd. In America, the Visking Corporation at North Little Rock were one of the first makers.

Thermoplastic Fibre Bonding

The thermoplastic fibres are blended in at the feeder with the other fibres. The web is passed (without binding agent in the ordinary sense) between heated rollers to soften the thermoplastic resin and so bind the web together. It is possible, but not usual, to make the bond with solvents instead of heat. Two fibres that are widely used for bonding are Vinyon and cellulose acetate, which have softening points of 75° C. and 175° C. respectively.

Dyeing

Non-wovens are usually dyed with direct, acid or basic colours, not with vats or sulphur colours. Prints have also appeared; in

America a fairly heavy bonded fabric is printed and made up into standard size curtains; the idea is that the householder puts them up, keeps them up for six months or a year until they are dirty, then throws them away and buys a fresh set. They really look too good to be thrown away. Crease-resist finishes have also been applied, but they have to work very hard.

Uses

Major uses for bonded fabrics are disposable diapers, dish-cloths, vacuum-cleaner bags, interlinings, filter-cloths, cheese bandages, *etc.* In the clothing trade the big use for bonded fabrics is as interlinings; some reluctance to discard old and tried woven materials is naturally evident, and the bonded fabrics will have to prove themselves; at the end of 1961, there were still one or two "ifs" and "buts" although the manufacturers seem to be confident of their ability to overcome them quickly. It remains to be seen; often the most difficult part of invention and development is the last bit, from the 90 per cent satisfactory to the completely satisfactory stage. In the shoe trade, too, bonded fabrics have made progress as linings and insoles. Bonded fabrics are light in weight and cut edges will not fray; they do not need to be hemmed. Their porosity to air is considered hygienic and has led to one special use for burn dressings: bonded fabric is coated with a very thin film of aluminium to prevent the fibres swelling when wet and backed with wadding to absorb the liquid that always oozes from burns; the aluminised fabric is applied directly to the wound which it protects and allows the liquid to pass through to the wadding; the result is that the dressings can be changed painlessly because they do not stick. But one must keep a sense of proportion; dressings anyway can often be changed without inflicting much pain. Filtration is a good use and a major one for bonded fabrics; they present a large number of fine fibres to the passage of the liquid, not a small number of relatively much coarser yarns; bonded fabrics are used to strain the tea from the tea leaves in the cup-of-tea slot machines that are becoming popular.

Some of the laminated bonded fabrics can be pulled apart very easily, *i.e.*, peeled off layer by layer. A needle-punching device, which takes some of the fibres through from one side of the fabric to another, helps to avoid this defect and is sometimes used. Most bonded fabrics that are made still look pretty poor; occasionally they are given a lace-like appearance by piercing the web with jets of water before bonding.

There can be no doubt that the days of laborious and slow flat weaving must come to an end sometime. There are various possibilities evident as to how cloth will be made in the future. It seems extremely unlikely that bonded fabrics can ever be really widely developed except for inferior uses; adhesion of filaments or fibres is fundamentally bad in the textile industry; it ruins the handle.

Non-woven fabrics were hardly made (apart from felts) 15 years ago. Production in 1956 was about 60 million lb. in America alone and was expected to be about 600 million lb. in 10 years' time. Development has not moved so fast as expected. According to the U.S. Census Bureau the production of bonded fibre fabrics in 1965 was 128·9 million lb. and in 1966 was 146·9 million lb. Of this last total about half was composed mainly of cotton and rayon, and half of other fibres. Additionally there was a U.S. production of punched felts of 63·7 million lb. in 1968, but of this 53·7 million lb. was of hair and jute and the remaining 10 million lb. was divided between wool and man-made fibres. These non-woven fabrics have attracted more attention than they deserve. A structural analogy between bonded non-woven fabrics and polymer macromolecules has been drawn by Russian scientists.

Felts

Traditionally, felts have been made from wool and from other animal hair fibres, all of which have an imbricated scale structure so that the frictional coefficient root-to-tip is much lower than that in the tip-to-root direction, the difference between the two being described as the directional frictional effect.

When a fabric made from wool fibres is wet and is subjected to mechanical treatment of some sort, such as rubbing, pressing, squeezing, twisting or pounding, the fibres move under such mechanical stimulus but because of their scale structure they move in one direction only. Fig. 223 illustrates the mechanism of unidirectional movement.

The unidirectional movement is the primary cause of wool felting. If a fabric is made up of millions of fibres which have the property of moving in only one direction, and if it is subject whilst in a readily deformable condition (that is, wet) to forces from all directions caused by squeezing and pounding or beating, thousands of fibres will start moving. They will pull with them thousands of other fibres which can, by moving, bend and so shorten the distance between tip and root, not by contraction (for the wool *fibre* does not shrink) but by bending. If a large proportion of the fibres that constitute a fabric are effectively shortened, what happens to the fabric? It shortens too; in other words it shrinks or felts.

No rayon or synthetic fibre is covered with scales as is every hair and wool fibre, and consequently man-made fibres will not felt in the same way as wool does. Usually this is considered an advantage and the dimensional stability that characterises fabrics made from 55/45 Orlon/wool or 55/45 Terylene/wool is one of their most highly prized features. But when felts are wanted from man-made fibres some new mechanism of felting must be invoked, for there is no man-made fibre that will migrate unidirectionally. Fortunately

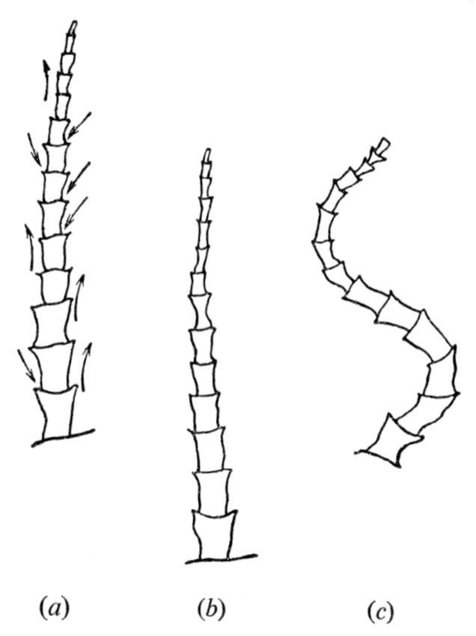

FIG. 223.—Illustrating how the scale structure causes unidirectional movement of the fibre.

(*a*) The fibre in its initial position. It is subject to the action of both upward and downward forces. The upward forces slide over the scale edges and are inoperative; the downward forces act on the ratchet-like scale edges and push the fibre towards its root-end. (*b*) The fibre being free to move has been pushed root-end first to its new position; it has migrated unidirectionally. (*c*) If in position as in (*a*) the root-end of the fibre was held fast by another fibre so that it could not move, the fibre bends under the action of the downward forces. This reduces the distance from tip to root and when this happens to the fibres in a fabric, the fabric shrinks.

for the would-be felter some of the man-made fibres will shrink, and if a bundle of such fibres are caused to shrink the bundle shrinks too; the essential difference from wool felting is that the man-made fibre shrinks and its denier increases, whereas the wool fibre bends without increase in denier. Any man-made fibre that will shrink appreciably, usually by treatment in hot water, can be made to give a felt. The use of P.V.C. fibre for this purpose has been described on p. 478, another approach to the practice of making man-made fibres felt is described hereunder:

Felts from Dacron were required because of the durability, dimensional stability, strength and chemical and biological resistance of this fibre; but Dacron does not shrink in hot water. It was therefore specially manufactured so that it possessed a high inherent shrinkage of some 50–75 per cent at temperatures approaching 100° C.—a fibre, it will be observed, very different from standard

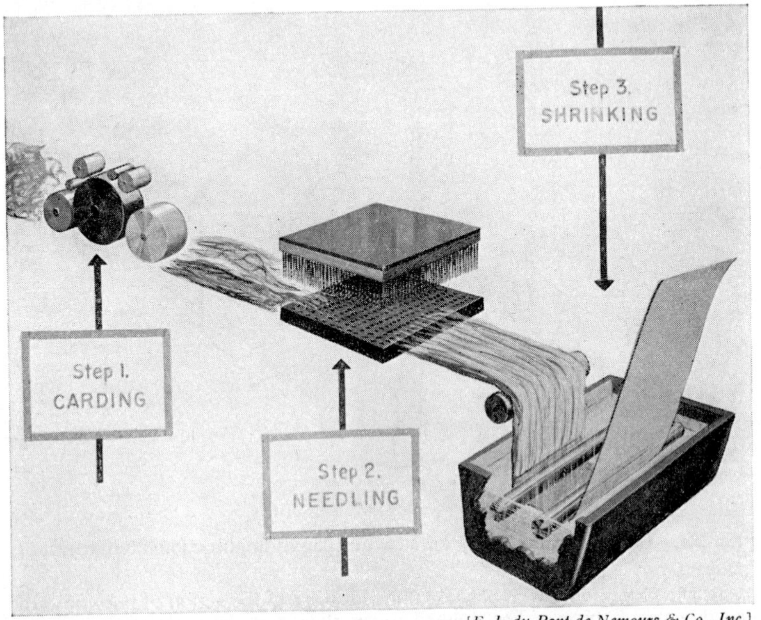

[*E. I. du Pont de Nemours & Co., Inc.*]

FIG. 224.—Preparation of a Dacron felt.

Dacron or Terylene. This fibre is then subjected to the processes shown in Fig. 224:

(*a*) It is carded to bring the fibres into a uniform pad.

(*b*) The pad is needle punched. A needle loom is commonly used to secure loose fibres to a woven base by punching with multiple needles, and is also used to make " felts " for carpet underpads from inexpensive fibres in a similar way. In the preparation of the Dacron felt, no base fabric was used; the needle loom contains many *barbed* needles and these reciprocate up and down so that some of the fibres are pushed through the batt, and when the needles are withdrawn the fibres remain laterally extended through the batt thickness and promote entanglement of the otherwise parallel fibres. The needling operation is a means of securing such an entanglement of fibres as occurs naturally when wool fibres are washed and rubbed.

(*c*) The entangled batt is passed through a hot water bath which causes the fibres to shrink and which shrinks and consolidates the batt into a felt.

The process seems at first sight to be rather involved compared with the elegant simplicity of wool felting, but the felt that it produces has special properties, and these are so useful that the process which was developed by du Pont is now being used by felt manufacturers in America.

The main advantages of the Dacron felt are: durability, high strength, high splitting strength, resistance to further shrinkage (wool felts will continue to felt and shrink in use under moist conditions) and resistance to microbiological and chemical attack. Felts can be made to required thicknesses and degrees of compactness. Suggested uses for Dacron felts have been for filtration (chemical and heat resistance and ease of cleaning), wicking (constant wicking rate), polishing (abrasion resistance and freedom from grit), and medical (can be boiled, non-swelling). So far as concerns polishing, heavy denier Dacron felts are superior to light deniers for polishing plate glass. Filtration of pollutants from air has been a major use.

Other fibres will doubtless be used for making felts. It is important to note that only those fibres such as P.V.C. and polyethylene which shrink very considerably in hot water are suitable, as ordinarily made, for felting, and that other fibres such as Dacron must be manufactured by a process so modified as to result in the fibres having a high potential shrinkage in hot water; the felting process could not be used successfully on standard Dacron or nylon.

Most of the felt skirts that were so popular a few years ago were real wool felts, but although they seem to have worn well enough, their dimensional stability is suspect and some cleaners have refused to accept them at all.

Superfine Fibres for Aerosol Filters

For defence purposes the filtration of aerosols may be important; this consists of the removal of extremely small droplets, probably of some toxic material, from air. It is known that extremely small diameter fibres, suitably arranged, possess aerosol filtration efficiencies far in excess of ordinarily used materials, and research work was undertaken at the Naval Research Laboratory, Washington, to see what degree of fineness could be obtained. As a standard we can take (*cf.* p. 22) 3 denier nylon which has a diameter of about 19–20 μ. Using the method developed at Washington, fibres have been produced in the form of a mat or a felt with diameters as low as 0·1 μ, which is of the same magnitude as the wave-length of ultra-violet light. The apparatus used is shown diagrammatically in Fig. 225; granules of a fibre-forming material such as nylon are

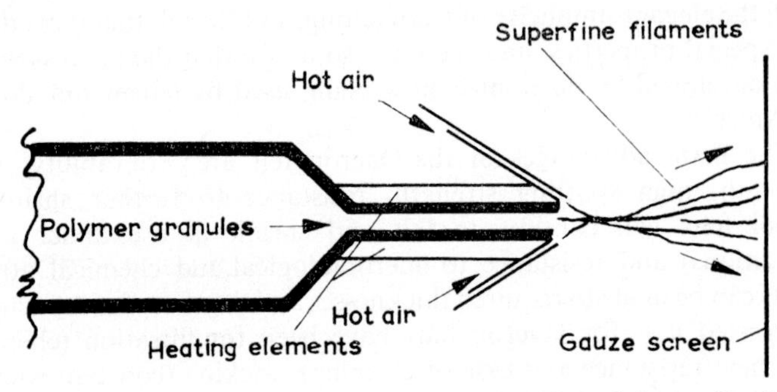

[*After Wenter, Ind. engng. Chem.*]

FIG. 225.—Preparation of superfine fibres for aerosol filters.

squeezed by a ram into the fine heating element, where they encounter a temperature of about 320° C. which not only liquefies them (normal spinning temperature of nylon is about 285° C.) but reduces the viscosity of the melt; within a matter of seconds the thin melt is squeezed through the tiny orifices of 0·014 in. diameter. Here the extruded melt is swept up by the two streams of hot (320° C.) compressed air (50 lb./in.^{2}) which attenuate the filaments and carry them on to a moving gauze screen, from which they can be removed as a random network of fibres. The greatest randomness of fibre distribution in the net occurs when the two streams of hot air at the jet meet each other at an angle of 90°; the greatest degree of attentuation of the fibres occurs when the two air streams meet at 30° but this results in a good deal of undesirable parallelisation of the filaments; an angle of 60° between the air jets gives useful results. The filaments are much too fine to be photographed optically but electron micrographs of a nylon pad show that the diameter of the nylon ranges from 0·11 to 0·22 μ. Such very fine fibres have been made from nylon, nylon 610, nylon 6, polyethylene, Terylene, Perspex (polymethyl methacrylate), polystyrene and polytrifluorochloroethylene with nozzle temperatures varying from

320–380° C. and rather higher air temperatures. Polyvinyl chloride and polyacrylonitrile did not yield fibres by this process because they decomposed at temperatures well below those necessary to give adequate fluidity. The other polymers from which the superfine fibres were successfully made would not have withstood the high spinning temperatures of over 300° C. for long, but they did withstand it for the very short time necessary in the melting section.

The fibre mats formed by the projection of the filament as spun against a gauze had excellent filtering properties for fine aerosols. Polystyrene, poly(methyl methacrylate) and polytrifluorochloroethylene are not ordinarily considered to be fibre-forming materials but they give good results by this process; polytrifluorochloroethylene might be particularly useful because of its good thermal and chemical resistance.

The essential features of this process are:

1. Heating the polymer well above its melting point until its fluidity is high.
2. Catching the just extruded filaments in high-speed hot air blasts to attenuate them. This last feature is reminiscent of the glass process (p. 576).

Papers from Synthetic Fibres

Paper is ordinarily made from cellulosic material, and the processes used depend on two of the properties of cellulosic fibrous material:

1. The ease with which it can be converted to a slurry with water, and this in turn depends on the hydrophilic character of the cellulose.
2. The fibrillation which cellulose fibres undergo when beaten; the fibrils that are formed intertwine, and in the subsequent drying they interlock one fibre with another.

Synthetic fibres are not hydrophilic but hydrophobic; nylon with a moisture regain of 4 per cent might be said to occupy an intermediate position, but Orlon (1 per cent) and Terylene or Dacron (0·4 per cent) are both definitely hydrophobic. Accordingly, the preparation of slurries in which these fibres, cut to a suitable length, are dispersed, presents some difficulties. Secondly, when synthetic fibres are beaten they do not generally fibrillate, but rather shorten in length so that no mechanism is provided for subsequent bonding—there are no fibrils which can interlace and on being dried out can form a bond. It is the bonding of cellulose fibres which gives strength to paper and there is no strength without the bonding. Consequently, bonding of the synthetic fibres must be brought about to get a paper and since it does not come via fibrillation or beating, other methods of bonding have had to be devised. Fig. 226 shows an unbonded Orlon waterleaf; the smooth, unfibrillated fibres are simply interlaced with very little mechanical and no chemical bonding of the structure.

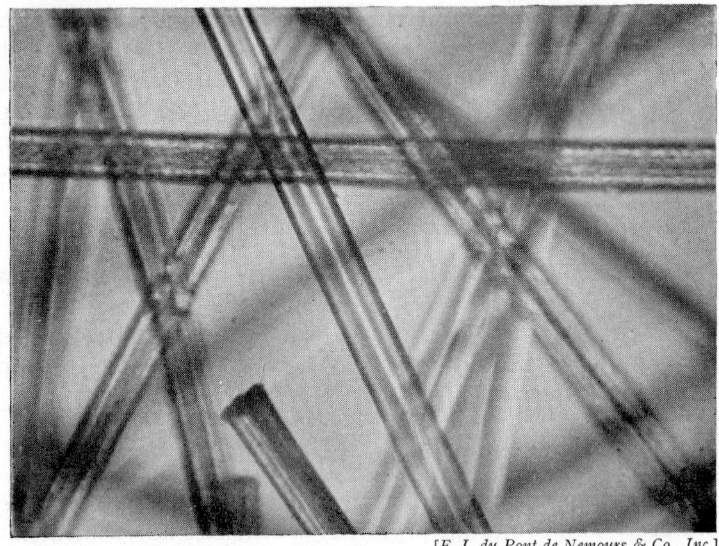

[*E. I. du Pont de Nemours & Co., Inc.*]

FIG. 226.—Unbonded Orlon waterleaf (× 250).

Salt Bonding

This is the first of the two methods of bonding; it depends on the swelling and solvent (or near-solvent) action that some concentrated salt solutions have on the synthetic fibres, and on their loss of these actions when diluted. For nylon and Orlon, suitable salts are calcium and lithium bromides and calcium and magnesium thiocyanates; for Dacron, too, calcium and magnesium thiocyanates are suitable; usually 5–10 per cent solutions are used. The salt solution is applied to the waterleaf (Fig. 226) and the water is evaporated so that the salt solution concentrates, and tends to localise preferentially at the crossover points (just where it is wanted) of the fibres, ultimately becoming sufficiently concentrated to gelatinise small portions of the fibre surfaces at the crossover points: on complete evaporation of the water, bonding becomes complete, as can be seen in Fig. 227, especially at the points indicated by the small arrows. With salt binding, the binding agent is the fibre itself, the part that was gelatinised, so that homogeneous paper results, fibre and binder are one and the same substance.

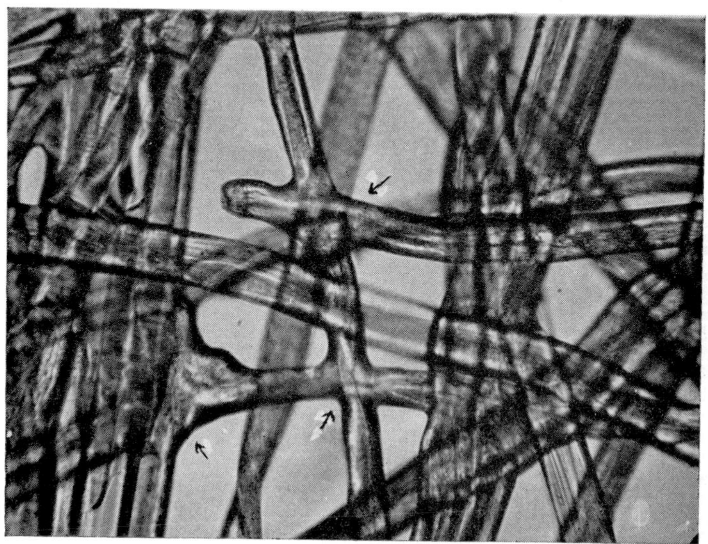

[*E. I. du Pont de Nemours & Co., Inc.*]

FIG. 227.—Salt-bonded Orlon waterleaf (× 250).

Synthetic Polymer Bonding

The waterleaf to be bonded is impregnated with a dispersion or solution of the binding polymer. On evaporation of the liquid, in which the polymer was dissolved or dispersed, the polymer concentrates at the crossover points of the filaments by capillary action; when all the liquid has been evaporated, the polymer has bound the fibres at their points of intersection. Various polymeric materials can be used as binders; if they resemble the fibre chemically so much the better, *e.g.*, a polyamide binder is good with nylon, and a polyester binder with Dacron, and various synthetic latexes such as Goodyear's " Chemigum " with Orlon.

Preparation

Synthetic fibre papers have so far been prepared mainly on an experimental scale. The method is to cut the fibres to the desired length and to make a 0·1 per cent slurry of them in water, adding a little shrinking agent if necessary. The slurry is deposited on a screen and the liquor drained away, leaving a waterleaf which is impregnated with either a salt or a polymer bonding agent and finally hot pressed at about 160° C. and 200 lb./in.^{2} pressure for 30 seconds, to form a sheet of paper.

Properties

The properties of the papers made from synthetic fibres are unusual, as the following table illustrates:

Paper.	Tensile strength (lb./in.).	Elongation (per cent).	Tear strength (grams).	No. of folds to rupture.
Kraft (cellulosic), 0·16 mm. thick	12	3	280	1,200
Nylon 3 den. fibre, $\frac{5}{8}$ in. long (salt-bound)	32	15	1,056	52,000
Nylon 3 den. fibre, $\frac{5}{8}$ in. long (polymer bound)	33	33	1,228	850,000
Orlon Type 81, 2·5 den. fibre, $\frac{5}{8}$ in. long (salt-bound)	31	6	298	51,000
Orlon Type 81, 2·5 den. fibre, $\frac{5}{8}$ in. long (polymer bound)	29	15	246	65,000
Dacron 2 den. fibre, $\frac{1}{2}$ in. long (salt-bound)	22	8	384	68,000
Dacron 2 den. fibre, $\frac{1}{2}$ in. long (polymer bound)	32	53	928	53,000

The dimensional stability on repeated wetting and drying is better in the synthetic fibre papers than in cellulosic paper.

Fibrids

Fibrids are being offered experimentally (in wet form, 20 per cent solids) by du Pont to be used with synthetic fibres for making paper which is very strong; papers so made are called " textryls ". The fibrids are filmy or fibrous binder particles about 1 mm. long and their twig-like morphology gives them some of the self-bonding properties of wood pulp; they are, in fact, imitation wood pulp made in nylon, Dacron or Orlon. A paper or textryl can be made from 30 per cent 201 fibrids (Dacron type) and 70 per cent Dacron fibres of about $\frac{1}{4}$-in. length, and then subjected to calendering at about 210° C.; this fuses the fibrids so that they seal the fibres together. A photomicrograph of a Dacron textryl is shown in Fig. 228. Nylon fibres can be similarly bound with 101 (nylon-like) fibrids, and Orlon fibres with 302 fibrids. A Dacron textryl is rather like parchment to look at and to touch, and rather like blotting paper if it is written on. Whereas it takes (table on p. 696) only about 60,000 folds to rupture a Dacron paper, salt or polymer bound, it takes 400,000 foldings to rupture a Dacron textryl.

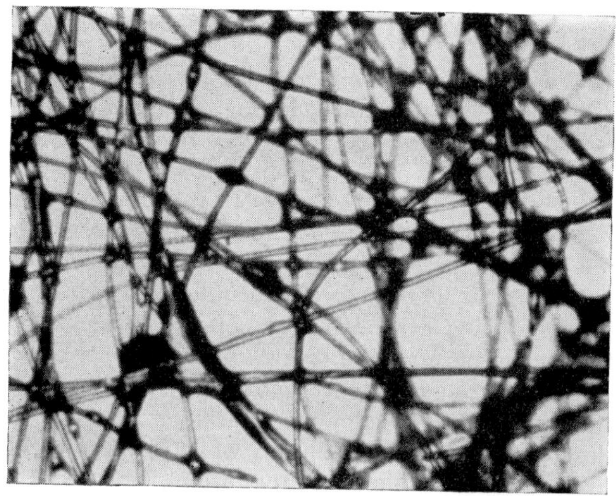

[*E. I. du Pont de Nemours & Co.*]

FIG. 228.—Photomicrograph showing structure of a textryl made from 70 parts Dacron polyester fibre and 30 parts 201 fibrids (Dacron type) bonded together. Magnification × 50.

Uses

Uses originally envisaged for synthetic paper were for currency (pounds and dollar notes) and for maps, as well as for filters for corrosive chemicals. Textryls have also been used as interlinings and as electrical insulation materials.

FURTHER READING

H. G. Lauterbach, " Felt from Man-made Fibres ", *Text. Research J.*, **25**, 143–149 (1955).

H. L. Leventhal, " Production and Utilization of Non-woven Fabrics ", *Amer. Dyest. Rep.*, **44**, 464–466 (1955).

R. W. Moncrieff, " Wool Shrinkage and its Prevention ". London: National Trade Press (1954).

V. A. Wente, " Superfine Thermoplastic Fibers ", *Industr. Engng Chem.*, **48**, 1342–1346 (1956).

R. A. A. Hentschel, " Some New Developments in Synthetic Fiber Fabrics ", *Tappi*, **41**, No. 1, 22–26 (1961).

Anon., " Textryls ", *Chemical Processing* (August, 1961).

R. W. Moncrieff, "An Analogy between Bonded Non-woven Fabrics and Polymer Macromolecules ", *Text. Manufacturer*, 385–7, September (1965).

E. T. Ustinova and S. S. Voyutskii, *Izh. Vysshikh Uchebn. Zavedini, Teknol. Tekstil'n. Prom.* (1), 104–10 (1965) (Latex bonding).

CHAPTER 47

DYEING AND FINISHING

THE dyeing of each kind of fibre has been considered in the relevant chapter, but there are some considerations common to the dyeing of all fibres that may be discussed here.

Identification

Before any attempt is made to dye an artificial fibre, whether loose, in yarn or in fabric, it is essential to identify it, if its identity is not already known. Suggestions for the identification of fibres are given in Chapter 48. It will probably be possible to identify the fibre beyond doubt by burning and solubility tests. The most frequent problem that comes up in practice is to determine if a fabric is viscose or cellulose acetate. The solubility in acetone is a good indication of cellulose acetate, and if the fibre is white or light in colour, the stains given with Shirlastain A are very useful, *viz.*:

Viscose—Pink
Cuprammonium—Blue
Cellulose Acetate—Greenish Yellow

If in doubt, reference should be made to further tests. It has to be remembered, though, that the kind of dyestuff for which a fibre shows affinity is itself a good indication of the nature of the fibre, so that if a small piece of the fibrous material is available it may be tried with direct dyestuffs, acid dyes or dispersed acetate colours.

Dull Lustre

When rayon was first produced, its high bright lustre was admired; later, fashion changed and brightness was disliked. Various ways of producing a dull (matt) or a semi-dull (pearl) appearance were devised; *e.g.*, an early viscose rayon called Dulesco was dulled by the inclusion of tiny oil droplets dispersed in it, and cellulose acetate could be dulled by boiling water especially with soap or with a little phenol. But there were disadvantages attached to these methods, and for many years now, yarn has been spun dull, by the inclusion in it of a little titanium dioxide; this gives a permanent dullness. The advantages of using titanium dioxide as a delustrant are:

1. It is chemically inert, and unaffected by wet processes to which yarns or fabrics are likely to be subjected.

2. It has an excellent covering power, its refractive index is 2·7, very different from those of the common fibres.
3. It is readily obtainable in particles of uniformly small size, about 0·8 μ, which will not choke the spinneret. An orifice may be 0·05 mm., which is 50 μ, so that the particle is much smaller than the spinning hole. The pigment particles in a dull Dacron yarn can be easily seen in Fig. 148.

The disadvantages attached to the use of titanium dioxide as a pigment are:

1. Its incorporation (usually about 1–2 per cent on the weight of the fibre) reduces the strength of the fibre by about 5 per cent.
2. The matt fibre may, owing to the solid particles it contains, exert an abrasive action on the parts of machinery in which it is worked, such as the healds and reed of a loom, and wear them out rapidly.
3. It reduces the tinctorial quality of the dyed fibre. When fibres containing particles of titanium dioxide are dyed, the titanium dioxide remains white, and so reflects considerable white light from the surface of the fibre, diluting the coloured light reflected from the surface of the dyed fibre. Accordingly, it is necessary to use more dyestuff to obtain the same depth of shade on a pigmented fibre than it is on a normal fibre.
4. It increases the rate of photo-tendering of cellulosic fibres.

Fine and Coarse Filaments

Because of the much greater surface offered by fine filament yarns, they invariably require more dyestuff than do coarse filaments to reach the same shade.

Dyeing Affinity

Extremes of dyeing affinity are found amongst the artificial fibres. Those which are synthetic and highly oriented have an extremely low affinity, partly because the fibre molecules are packed together too closely to allow ingress of the dye-liquor. In addition, such fibres as Vinyon and Saran have no reactive groups such as hydroxyl, amino or carboxyl, which are normally suitable sites for dyestuffs. Contrarily, the regenerated proteins swell readily in water and have an affinity for acid dyestuffs greater than that of wool itself, and often greater than that of chlorinated wool. The regenerated protein fibres have an abundance of amino and carboxyl groups which can act as acceptors for dyestuff molecules.

Viscose has the same number of hydroxyl groups as cotton, but has a much higher affinity for direct dyestuffs than has that fibre, and this may be attributed in part to the degradation of the cellulose molecules which has taken place in the conversion of wood pulp to viscose, but mainly to the change from Cellulose I to Cellulose II (see also p. 127). Cuprammonium, because of its usually finer filaments, although it has a very high affinity for direct dyestuffs, requires more dyeware to reach the same shade than does viscose.

Cellulose acetate, because of its lack of reactive groups, when first made could hardly be dyed at all, because most of the well-known dyestuffs had little or no affinity for it. A new method of dyeing had to be introduced in the " dispersed insoluble dyestuff " process. This consists in the dispersion with the aid of Turkey-red oil, or some other wetting agent, of the finely dispersed solid dyestuff in water, the impregnation of the fabric (or yarn) in this dispersion, and the solution of the dyestuff in the fabric. The process has fortunately lent itself very well to use with the synthetic fibres such as nylon and Vinyon which, because they also suffer from paucity of reactive groups, cannot be dyed by chemical methods—*i.e.*, methods where the dyestuff combines with hydroxyl or amino-groups on the fibre. When cellulose acetate rayon was first produced there existed no satisfactory way of dyeing it, and its commercial development waited for the discovery of the dispersed colour method of dyeing. Nylon and Vinyon were more fortunate in their advent, as this process, already well-established, proved suitable for these fibres.

The dyeing affinity of synthetic fibres has proved to be capable of modification. Whereas the original Orlon Type 81 was pure polyacrylonitrile and was practically undyeable, " modified " acrylic fibres with better dyeing properties have been produced, notably Orlon Type 42, Acrilan and Zefran. Usually, dyeability is obtained at the cost of some other property, often chemical resistance; it is too much to hope that a fibre can be resistant to all chemicals except dyestuffs. Teflon is resistant to practically everything including dyestuffs, and those fibres that have the best dyeability, *e.g.*, reconstituted protein fibres and viscose and cuprammonium rayons, do not have unusually good resistance to other chemicals.

Internal Dyeing

Some fibres are dyed " internally "—*i.e.*, dyestuff or pigment is mixed with the dope or the solution before it is spun. This method is especially valuable in the case of the highly oriented and non-polar fibres which are difficult to dye in the ordinary way. The difficulty, which is often experienced with these fibres, of obtaining satisfactory penetration of the fibre by the dye-liquor is thus avoided.

Also, because of the excellent distribution of dyestuff throughout the fibre, the dyeing is unusually fast.

Application to Viscose Rayon. This method of internal dyeing, largely because of the advantages of uniformity from one bobbin to another, and of improved fastness, has also been applied to viscose rayon. The Coloray fibres (Courtaulds, Ltd.) are an example of internally dyed viscose rayon. In their manufacture, highly coloured pigments in a very fine state of division are dispersed in the viscose spinning solution. The yarn spun from this solution contains the pigment dispersed evenly right through the filament, thus giving a solid shade, permanent and fast.

Pigments Used. The pigments used for these yarns must be fast to strong alkali and strong acid, and also to sulphur compounds, in order to resist attack in the viscose solution and in the spinning bath. They must also be fast to subsequent chemical treatments such as desulphurising, bleaching, soaping, etc. Clearly pigments that are chosen to meet these exacting requirements will be fast to all conditions normally encountered in use. Tests undergone by Neochrome yarns, an early type, included:

1. No trace of staining on to white yarn, and no reduction of shade after thirty minutes' immersion in $\frac{1}{4}$ per cent soap solution at the boil.
2. No change after immersion in $\frac{3}{4}$° Tw. sodium hypochlorite solution at room temperature for thirty minutes.

The name Neochrome is now used for spun dyed colours on the acrylic fibre Courtelle.

Amines. In 1964 in *Khim. Volokna* **2**, 41–3, A. N. Bykov *et al.* described the colours that had been used for dope-dyeing Kapron (nylon 6) at Ivanov. In principle the method consisted of the inclusion of a coloured amine of the anthraquinone class in the caprolactam which was being polymerised. This, of course, is different from the usual method of mass coloration of the molten polymer with insoluble pigments. These last give good fast colours which are uniform in shade, but are costly because the pigments are dear and their use involves an extra process for the fibre manufacturer. However, that is the method that is used throughout the world. The new Russian method is said to be much less expensive. It depends on a chemical reaction between the terminal carboxyl groups of the polycaprolactam and the amine groups on the anthraquinone derivative. So long as there are any free carboxyl groups on the polymer, the amine will react with them; when they have all been used up, no further polymerisation is possible and the polymer is stabilised. Because amines of this kind are deeply coloured, the Kapron is coloured by them. Two amines specifically used for the simultaneous stabilisation and coloration of the polyamide are:

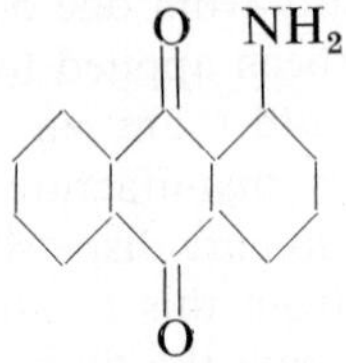

[Structural formula: O, NH_2 — anthraquinone ring system with O]

α–aminoanthraquinone and

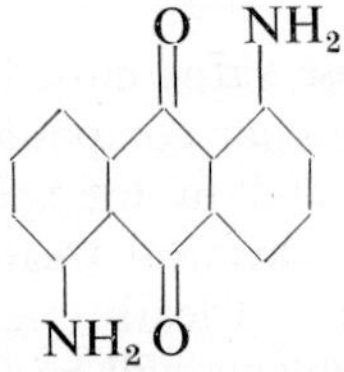

[Structural formula: O, NH_2 — anthraquinone ring system with NH_2, O]

1·5–diaminoanthraquinone

Both of these are red, although they are not ordinarily considered to be dyes. They colour the fibre satisfactorily, it is reported, towards the action of light, laundering, perspiration and organic solvents. The addition of 0·2 per cent of one of these amines will stabilise the polycaprolactam at an average molecular weight of 22,000, whereas 1 per cent of amine stabilises the polymer at 16,000 mol. wt. The effect on physical properties is insignificant with less than 1 per cent amine, thereafter it causes a slight loss of strength. (See also *Textile Mercury*, 19 February 1965, pp. 33–34.) It is a new development that will be watched with interest.

Cost

If the spinner produces such fast dyed yarns he helps the weavers and knitters, but he takes on a lot of extra work himself owing to the necessity of segregating pipe-lines used for one colour of spinning solution from another. The prices that follow are for late 1968.

Examples of viscose continuous filament yarns that are solution dyed are: Courtaulds' Spun Dyed, American Enka's Jetspun, American Viscose Corporation's Spun Dyed; and in staple fibre Courtaulds' Fibro-Duracol, American Viscose Corporation's Spun Dyed, Courtaulds (Alabama) Coloray. Premium price for colours in continuous filament is about 25–35 cents per lb. over natural in America and about 18–30*d.* in U.K. depending on the package. In staple (natural is 28 cents in U.S.A., 23*d.* in U.K.); Coloray is 42 cents for Tan, 61 cents for Red; A.V.C. charge 43 cents for Rosewood, 59 cents for Apple Red; in Britain the price for Fibro-

Duracol varies from 31*d.* to 42*d.* according to colour. Jetspun (American Enka) 100 den. 40 fil. 3 t.p.i. is 142 cents on cones compared with 104 cents for the natural continuous filament. Celanese Fibers charge an extra 35 cents for black on their Fortisan yarn; the price varies with denier, but is $2.70 for 90 denier black.

Cuprammonium. Beaunit Fibers do their Bemberg cuprammonium yarn in black at a premium of 15 cents and in some colours at 25–35 cents (Cupracolor).

Acetate. In U.S.A. Celaperm the filament acetate of Celanese Fibers Co., and Eastman Chemicals Chromspun are all available. Typical prices per lb. (late 1968) are 85 cents natural, 97 cents black, $1.22 colours in 100 denier. In Britain, Celafibre spun black is 53*d.* against 41*d.* for natural.

Triacetate. Triacetate staple is available in U.K.: black at 7*d.* premium, other shades at 10*d.*–16*d.*

Acrylics. In U.K. Acrilan is available in some colours. Typical prices for staple are 80*d.* natural, 84*d.* black, and 90*d.*–100*d.* for different colours. Courtelle-Duracol black is 84*d.* against 80*d.* natural staple. In America, Monsanto offer a wonderful range of solution-dyed Acrilan staple; there is an 11 cents premium for light colours such as Antelope and Cadet-Grey and a 21 cents premium for dark colours like Midnight Blue and Erin Green; the dispersion of pigment is uniform throughout the filaments, as can be seen from photomicrographs of cross-sections. Fig. 229 shows fabrics made of solution-dyed Acrilan undergoing weathering tests. Dow Badische make an acrylic fibre at 82 cents in staple and a similar coloured fibre at 98 cents. In Germany Dolan is spun-dyed in 48 shades. Orlon, first of the acrylics, appears not to be offered spun-dyed; of the modacrylic fibres Dynel coloured staple is 95 cents whereas natural is 75 cents. In late 1969 Courtelle natural is 63½*d.*

Polyolefines. Both polyethylene and polypropylene are available spun-dyed from several manufacturers. These fibres can be melt-spun at much lower temperatures than nylon or Terylene, so that not such excessive demands are made on the heat stability of the pigment used. In Britain Courlene X3 is available coloured; in U.S.A. Dawbarn Bros. Inc. do both polyethylene and polypropylene in a range of colours. So do other manufacturers such as Hercules Power Co. (15 cents premium for colours) and American Thermoplastic Products, who do a polypropylene cordage yarn at only 2 cents (on 60 cents) extra for black, white, yellow, blue and green, and at only 5 cents extra for red and orange. Phillips Fiber Corporation charge about 15 cents extra (on 85 cents) for colour in special polypropylene yarns. Dawbarn are especially interesting because they are a small-to-medium sized company who make a success of man-made fibre spinning. Elsewhere it is almost a prerogative of the giants.

Saran. Many manufacturers spin dope-dyed Saran. Particularly interesting are National Plastic Products Co. of Odenton, Md., who spin a 750 denier for industrial use only at a premium for colours of only 5 cents. Enjay Chemical Co. spin Saran monofils at about 80 cents, with 6 cents extra for red and 3 cents only for other colours.

[*Chemstrand Corp.*]

FIG. 229.—Weathering tests being carried out on awning fabrics of solution-dyed Acrilan. Here is shown one of the test stands.

They have a similar listing for the polyethylene that they make. Sometimes a minimum quantity is required for a special shade; this may be as much as 8,000–25,000 lb.

Nylon. The spinning temperature of 290° C. for nylon 66 has been too much for the pigments available. Only black is offered; du Pont sell a Color Sealed black filament nylon at a premium of about 35 cents per lb. Mass pigmented Kapron, which is nylon 6, has been described by Borik and some of his colleagues at the Klinskii Institute. Quantities of pigment used in the yarn are of the order of 0·6 per cent for blue, 0·8 per cent for scarlet, 0·5 per cent for brown and 2 per cent for black.

Terylene. In Britain, Terylene black staple is available at a premium of 28*d.* Colour-sealed black Dacron is also available in America.

What this comes to is that cellulosic and acrylic fibres are available in a wide range of shades, but one often has to pay two to three shillings a pound extra for the colour. Reckoning 5 yards of fabric to the pound, the use of spun-dyed yarn is going to cost 6*d.*–9*d.* per yard extra; a figure comparable with the cost of piece-dyeing natural fabric. The gain is fastness, the loss is flexibility. In the polyamides and polyesters, not much spun colour is available: black is to be had, but practically nothing else. A start has been made abroad with spun-dyed nylon 6 and one would think it cannot be long in becoming generally available. If it did, this more than anything else might advance nylon 6 at the expense of nylon 66. An exception is that shades in staple rayon are less costly, about 4*d.*–15*d.* per lb.

Bleaching

All the man-made fibres are bleached during manufacture, so that as a rule bleaching in the yarn or fabric is unnecessary. Some of the fibres, *e.g.*, Saran, Dynel and Orlon, were yellow or fawn in colour when first produced, and some others were quite a long way off white, but most fibres are now available in a good white.

Viscose rayon may be bleached with a dilute solution of hydrogen peroxide (about 0·5 vol.—*i.e.*, about 0·15 per cent H_2O_2). Cellulose acetate may also be bleached with dilute peroxide.

Cuprammonium is liable to be damaged by hydrogen peroxide, and is best bleached with dilute solutions of hypochlorite. Peracetic acid, $CH_3COO{\cdot}OH$ is a useful bleach for nylon.

In general, however, whatever bleaching is possible on the fibres has already been carried out by the fibre manufacturer, and the user is not often called on to bleach man-made fibres further, except in the case of mixtures. Viscose rayon, for example, is often spun mixed with wool, and the wool has to be bleached. No harm comes to the viscose rayon in a dilute solution of hydrogen peroxide in which the wool can be satisfactorily bleached.

Dyeing Machines

Loose Fibre

Loose fibre may be dyed in open becks in which continuous movement of the fibre mass in the liquor is ensured by bubbling air through the liquor.

Circulating machines are perhaps preferable. The Obermaier machine is suitable. Loose fibre or slubbing, held together in net bags, is packed in a perforated metal basket round a central perforated

[*Longclose Engineering Co., Ltd., Leeds*]

FIG. 230.—The conical pan of a dyeing machine being filled with loose stock, with the aid of a Stamper/Loader unit.

cylinder through which the dye-liquor is pumped. The liquor is steam-heated and, as it passes through the fibre or slubbing, it dyes it. The Longclose pressure dyeing machine also is widely used, and is often made in such a size that it will dye about 5 or 6 cwt. at a time. The fibre is packed in a drum which can be lifted in and out of the machine and, as spare drums are usually kept available, they can be packed and kept ready for loading, as soon as the preceding drum is removed from the machine. Provision is made for reversal of flow so as to avoid " channelling " and ensure uniformity of distribution of dyestuff through the fibre mass. Package dyeing machines have the advantage of using a short liquor: goods ratio, often of only 5:1 compared with a 20:1 or 30:1 ratio which is more usual in other types of machines and thereby of saving dyestuff. Figure 230 shows one of the conical pans of a Longclose machine being filled with loose stock, *i.e.*, staple fibre. The fibre is stamped down, after having been wetted out, by stainless-steel feet, and the stamping unit can be moved if required on a conveyor system. One of these pans will take 750 lb. fibre; it is particularly useful for hard-twisted carpet yarn. In the illustration the wet stock is on the left-hand side of the operation; when the pan has been filled it will be transferred to a dyeing machine (at either atmospheric pressure or a pressure machine to run at temperatures over the boil).

Yarns

Most yarn nowadays is dyed in package form on cakes, cheeses, muffs or bottle bobbins, and there are in the world some seventy designers and manufacturers of dyeing machinery who have made ingenious and widely used machines to dye yarn in package form. The difficulty with package dyeing is to get perfectly even dyeing, and the opinion is widely expressed that processing package-dyed yarns into self-shade goods will always be rather risky. That is the sternest test of all, to make a plain fabric from dyed yarn and get it level without warp streaks or weft bars. Knitted fabric, or woven fancies such as brocade or striped or check shirtings, are less demanding. Carpet yarns and acrylic yarns are still preferably dyed in the hank, since this method gives the best levelness of all and, furthermore, allows the yarns to shrink freely. It is really evident enough that yarn spread out in hanks is far more accessible to the dye-liquor than yarn that is wound on packages. It is a simple obvious truth that in these days of mechanisation and automation is often overlooked, that the oldest and simplest method of yarn dyeing-in-hank form is still the best.

As regards its practice the earliest and simplest method was to thread hanks of yarn over poles and suspend these across the top of an open dye-bath, working the poles by hand so that the hanks of yarn are caused to rotate round them. The method is still used, and vat dyestuffs in particular are still dyed by this method. Two men are required to work the sticks or poles along the bath. One vat is kept for reds, one for blues and so on. This traditional method has largely given place to stainless steel instead of wood vats and to mechanical (Fig. 231) instead of manual turning of the hanks. Nowadays, the manual method is mostly found in old family businesses which have been impoverished by death duties and have little or no capital available for new machinery. But such old-fashioned businesses often turn out very good work; the operatives are craftsmen and are honest servants who do their work well. The manufacturer of dyeing machinery with the help of his electronic friends tries to build machines which need only the touch of a switch to start them and which will then carry the dyeing to completion without further attention. It seems to be an uncertain avenue along which to make progress: perhaps it might be better to let men do the best work they can; there is more to a man than a collection of resistances, transistors, clocks and thermistors.

[*Courtaulds, Ltd.*]

FIG. 231.—A Gerber Hank Dyeing Machine at Courtaulds' Droylsden works, dressed ready for dyeing.

The Klauder-Weldon machine is one in which a very large wheel rotates in the dye-bath. Poles on which the hanks of yarn are suspended are attached to the periphery of the wheel and carried through the liquor. This machine gave very good results, and it is perhaps to be regretted that its use has gone out of fashion. It should be noted that in it the yarn is moved through the liquor.

A modification of this early method is to be found in a machine extensively used for dyeing rayon in hank form. It consists of a long, shallow trough of dye-liquor over which project numerous porcelain rods that carry the hanks of yarn. Porcelain is used so that the rayon will not be damaged. Each rod will hold about 2–3 lb. of rayon. In operation the rods rotate, and also may move in an elliptical orbit, so that the hanks are subject to continual turning. The rods also reverse automatically after so many turns. The lower half of each hank dips in the dye-liquor. Although the hanks are not nearly fully immersed, as they are in hand-dyeing methods, the constant rotation ensures uniformity of treatment. Fig. 231 shows hanks on a Gerber machine ready for dyeing; Fig. 232 shows them actually being dyed.

[*Courtaulds, Ltd.*]

FIG. 232.—Yarn being turned whilst the rollers revolve on a Gerber Hank Dyeing Machine at Courtaulds' Droylsden works. Note the elliptical and ridged section of the rods.

In the Hussong machine the yarn is suspended on poles in a bath, and the liquor is circulated by a propeller situated in a chamber adjacent to that in which the yarn is hung. Provision is made for reversal of the direction of circulation of the liquor. An example which has some unusual features of this type of machine is shown in Fig. 233. This Pegg Pulsator machine is so called because the flow of dye-liquor is one way only and is intermittent; the pulsing gives penetration; this unidirectional flow avoids tangling of the yarn. Nowadays, the Pulsator is used mainly to dye cotton and wool hanks of yarn. It can also be used for dyeing loose stock, *i.e.*, cotton or wool fibre, before it has been spun. Such material is placed on a false bottom specially supplied for this work. Sometimes the vats are sunk partly below floor level, so making access easier.

Machine for High-bulk Acrylic Yarns. Pegg's also make a hank dyeing machine (Model GSH) which is particularly suitable for high-bulk acrylic yarns and fine yarns generally. Liquor circulation is one-way, but is continuous, and the liquor is circulated through the hank poles which support the yarn; these poles are slotted, and the liquor flowing through them penetrates the hank uniformly. The flow through the hank poles, which are integral with the machine and cannot be lifted off, is shown in Fig. 234. The hank of yarn

[*Samuel Pegg & Son*]

FIG. 233.—Pegg Pulsator Hank Dyeing Machine, shown with the yarn raised above the bath, ready to be dyed.

This particular machine is recommended to be used with single-direction (downwards through the yarn) intermittent flow. Liquor : goods ratio about 10 : 1. Made in capacities from 10 lb. to 1,200 lb.

illustrated is supported by the pole. The reason for this device is that, ordinarily, with solid poles that part of the yarn which is touching the pole and supporting the weight of the hank may be underdyed. With the liquor circulating through the poles the yarn rests not so much on the poles as on the liquor, which, thrusting upwards, supports the hanks. But it has to be remembered that the weight of hanks in a bath full of dye-liquor is only small because their specific gravity is not much greater than that of water (unity). Figure 235 shows the circulation system of the GSH machine: the slotted hank pole is marked *A*, the hank frame adjustment for various lengths of hank is *B* and arrow *C* indicates the direction of

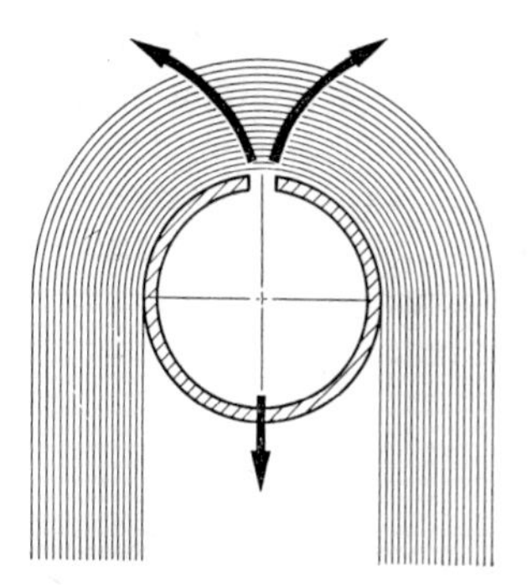

[*Samuel Pegg & Son, Ltd., Leicester*]

FIG. 234.—The dye-liquor flows *through* slots in the poles that support the hanks of yarn. This ensures more uniform penetration.

one-way circulation. Reversal of liquor flow is provided, but is intended by the makers only for mixing the dye-liquor before the hanks of yarn are entered in the bath. Nevertheless, some of Pegg's customers do reverse the flow during the actual dyeing. The

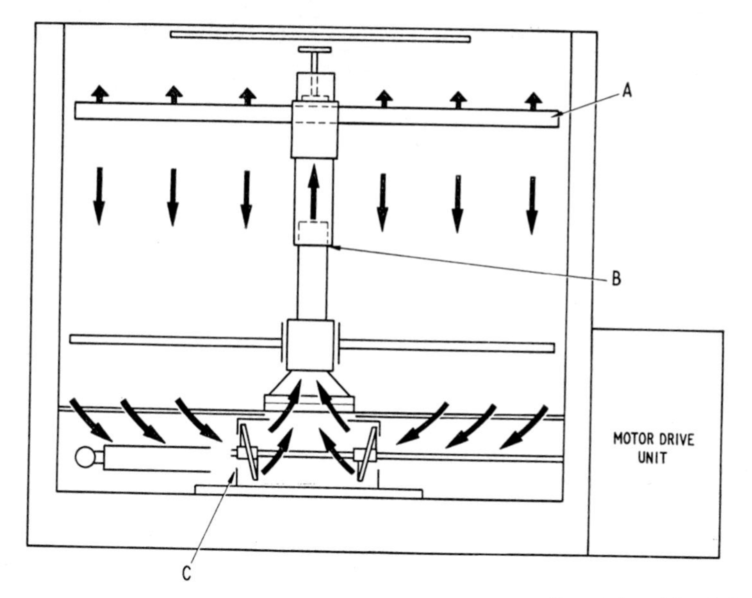

[*Samuel Pegg & Son, Ltd.*]

FIG. 235.—The circulation system of the slotted-pole hank dyeing machine.

advantage of reversal would presumably be better uniformity; the disadvantage would be one of tangling of the hanks, making their subsequent unwinding or reeling more difficult. The whole aim of the design of the machine is to ensure perfect penetration of the dye-liquor through the hanks and to avoid disturbing and tangling the hanks. Features that contribute to these ends are: (1) yarn cannot be trapped because the poles are integral with the machine; (2) the dye vat is completely filled with yarn to keep the liquor-to-goods ratio to a minimum; (3) there is a screw adjustment for long and

[*Longclose Engineering Co., Ltd.*]

FIG. 236.—Coupled pair of carpet yarn Hank Dyeing Machines (total batch weight 4,400 lb.).

short reel hanks so that the short hanks can be dyed in a smaller volume of liquor; (4) the liquor is heated indirectly by a closed steam coil. Whereas in the Pulsator machine the bottoms of the hanks are free and tend to bulge out as shown in Fig. 233, in the GSH machine there are bottom pegs to hold the bottoms of the hanks in place.

Carpet Hank Dyeing. Immense quantities of carpet yarn are hank dyed. Some of it is acrylic fibre, but it is relatively rarely that a straight acrylic is used—more often it is present in a blend. Figure 236 shows a coupled pair of carpet yarn hank dyeing machines (Longclose) with a combined capacity of 4,400 lb. On the right of the picture can be seen hanging some hanks apparently from a just-previous dyeing. Machines such as those shown in the illustration represent a lifetime's experience in design in their many special features. Two that are outstanding are:

1. A staggered stick (the sticks that support the yarn) arrangement as shown at (*a*) in Fig. 237 gives a machine loading of 25 per cent more yarn than the conventional loading shown at (*b*). This staggered arrangement also allows better liquor flow.

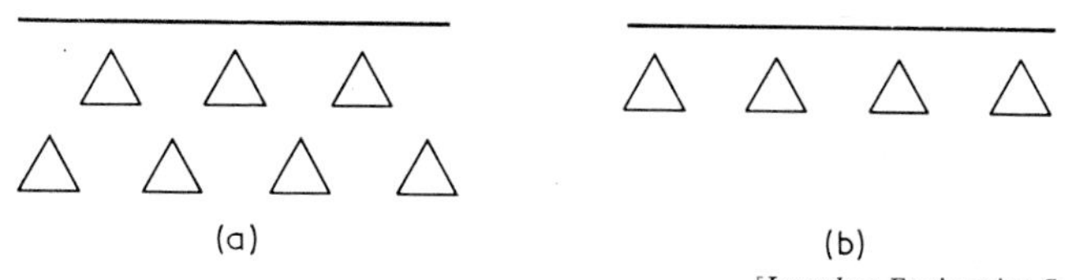

[*Longclose Engineering Co., Ltd.*]

FIG. 237.—Staggering of the poles that support the yarn increases the load of a dyeing machine: (*a*) staggered; (*b*) usual.

2. The propellers that are used to circulate the liquor are large and double scroll, so that even when running at slow speeds they give optimum high rates of liquor flow with minimum turbulence.

Pressure Dyeing

The Longclose pressure-dyeing machines are used extensively for dyeing yarn in various forms of package. They are made of stainless steel so as to eliminate metal effects, are fitted with combinations of time/temperature thermostatic control, and with automatic control of reversal of flow, *i.e.*, from outside-to-inside or inside-to-outside of package. In all of them the dye-liquor is pumped through the yarn package, which is supported on a perforated stainless steel spindle, the liquor flowing through the perforations and thence through the yarn package or *vice versa*, and dyeing the yarn as it passes through it. Clearly the dyestuff must be well dispersed in the liquid, otherwise the yarn will filter out particles of dyestuff; but early difficulties connected with the fine dispersion of acetate dyestuffs have been overcome and nowadays leading commission dyers dye acetate on cheeses. The three outstanding advantages of ordinary pressure dyeing of yarn are:

1. There is very little handling of the yarn—labour costs are low.

2. The liquor: goods ratio is low—often only 5:1 or 10:1 instead of say 30:1—representing a considerable saving on dyestuff.

3. The yarn is subject to less mechanical movement and friction than in hank-dyeing so that it is preserved in better condition, there are fewer broken filaments and less waste.

[*Longclose Engineering Co., Ltd.*]

FIG. 238.—Totally Enclosed Pressure Dyeing Machine. The frame above is loaded with Crimplene packages.

Figure 238 shows a Longclose pressure dyeing machine with a frame loaded up with textured polyester (Crimplene) yarn and ready to be lowered into the pressure container holding the dye-liquor. The packages on the perforated rods bed into each other so well that the dividing lines between them can hardly be seen. For the purpose of illustration these dividing lines have been inked in, in some of the columns. The difficulty that had to be overcome in the design of pressure-dyeing equipment was that of obtaining level results, and the achievement of this end has been dependent on three factors all of which are still important; they are:

1. The package, whether cheese, cake or bobbin, must be built up with a rapid cross-wind, as a parallel wound package is extremely dense and difficult to penetrate.

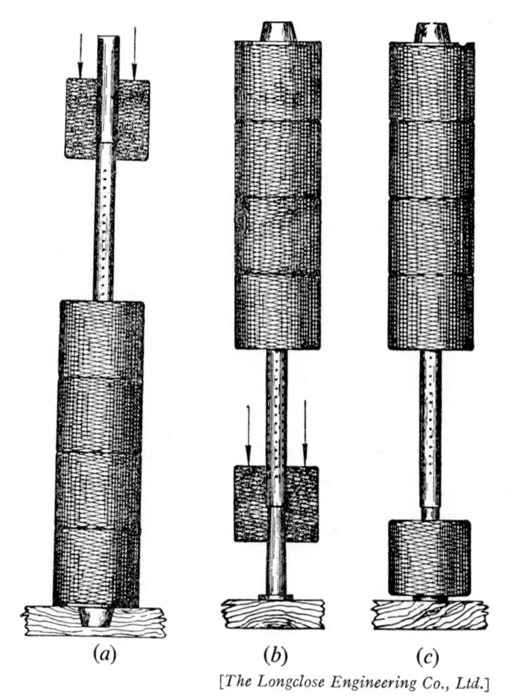

[*The Longclose Engineering Co., Ltd.*]

FIG. 239.—Transfer of rayon cheeses from wooden bobbins to stainless-perforated tube for dyeing, and after having been dyed from the tube to single-flange wooden bobbins.

2. The package must not have been wound too hard, *i.e.*, under too high a tension.
3. The pump must give a good flow of dye-liquor, not just a trickle, through the package.

Channelling of the liquor through the package must be avoided, it is usually the result of bad winding; the proper preparation of the package is nine-tenths of the battle for good package dyeing.

Some of the ways in which man-made fibres are pressure-dyed are as follows:

Cheeses. Used particularly for spun viscose rayon. The cheeses are wound on wooden bobbins, slipped from these to perforated stainless-steel spindles (Fig. 239 *a*) for dyeing, and subsequently slipped to smaller bobbins with a flange at one end after having been dyed (Fig. 239, *b* and *c*). The initial wooden bobbin can be seen at the top of Fig. 239 *a*, the final wooden bobbin at the bottom of Fig. 239 *b*. The cheese frame ready loaded is shown at the top of Fig. 238.

[*The Longclose Engineering Co., Ltd.*]

FIG. 240.—Rayon cake-dyeing frame showing good condition of cakes after dyeing.

Cakes. Particularly important for viscose filament cakes straight from the Topham Box; dyeing the primary spinning packages saves a great deal of labour and of damage (during the old processes of hank-winding and back winding) to the yarn. The cakes are supported by plastic or stainless steel carriers (Fig. 240). Fig. 241 shows the operation in progress. It is essential in cake-dyeing to maintain a fast flow of dye-liquor through the cake and to keep the dye-liquor free from suspended matter; usually the addition of 2 oz. Teepol or 4 oz. Calgon T per 100 gallons of liquor will avoid the appearance of suspended matter. Many direct dyes can be used successfully; those that are sensitive to alkali should be avoided. Sometimes unlevel results are obtained when viscose rayon cakes are dyed in circulating liquor machines, and a main cause of this can be unsuitable selection of dyestuffs. Fine filament yarn is more troublesome to get level than coarse filament and is particularly prone to unlevelness when it is cake-dyed. An investigation has been made by I.C.I. into the suitability of individual direct dyestuffs

[*The Longclose Engineering Co., Ltd.*]

FIG. 241.—Rayon cake-dyeing in progress.

for the dyeing of viscose rayon cakes and the results of their tests have been incorporated in their Technical Circular No. 581 (September, 1960). This divides the dyestuffs that were tested into four groups:

1. Dyestuffs which gave perfect or nearly perfect levelness (21 dyestuffs, *e.g.*, Chrysophenine G).
2. Dyestuffs which showed a slight degree of unlevelness (28 dyestuffs, *e.g.*, Durazol Sky Blue G).
3. Dyestuffs which showed a marked degree of unlevelness (39 dyestuffs, *e.g.*, Primuline A).
4. Dyestuffs which showed very poor penetration of the cake, the less accessible portions being practically undyed (30 dyestuffs, *e.g.*, Chlorazol Azurine G).

The distribution of black dyestuffs is interesting; there are none in Group 1, two in Group 2, four in Group 3, and seven in Group 4. The test conditions were purposely made severe; under easier conditions better results can be obtained with some of the dyestuffs. Nevertheless it behoves the dyer to make his selection of colours from the first two groups whenever he can. Cakes must be carefully selected; the aim is to get a high liquor flow through them at a low

[*The Longclose Engineering Co., Ltd.*]

FIG. 242.—Nylon muffs before (*left*) and after (*right*) dyeing.

pressure; high pressure pumping may distort the cakes and lead to channelling.

Vat dyes are troublesome because the viscose rayon has such a high affinity for the *leuco* compound which is inclined to over-dye on the outside of the package. The trouble has been overcome by some dyers in the following way: The cake is pigment-padded with controlled exhaustion, by the addition of electrolyte to the dye-bath; this enables the unreduced colour to be dispersed fairly evenly throughout the package prior to reduction; the dyestuff is reduced *in situ* and the *leuco* compound formed is taken up by the fibre very rapidly; levelness is obtained by running the liquor through the cake for about 2 hours, and finally the dyestuff is oxidised to the pigment form. The vat dyed cake is in good condition. The rayon cake-dyeing business has declined, because so much of today's dyed yarn is synthetic, *e.g.*, nylon. Much less rayon is cake-dyed than formerly, but what is done is still done on machines of the type shown in Figs. 240, 241.

Muffs. Another development has been in dyeing the latest forms of crimped high tenacity nylon. Some difficulties were experienced at first with this type of yarn but it is now successfully dyed in small hanks wound on a wrap reel, removal from which results in the formation of a small " muff ". These muffs (Fig. 242) are usually packed in a perforated wall cage, round a central perforated cylinder through which the dye-liquor is pumped—rather in the same way as loose fibre is pressure dyed. Muffs of about 1 lb. weight have been made of stretch nylon and also of High Bulk Orlon; they permit plenty of yarn shrinkage during dyeing.

In 1969 muffs are still used for dyeing, although they are less popular than six or seven years ago. Spinners who make textured polyester yarn, *e.g.*, Crimplene, have found that it does better on cheeses than on muffs. The problem with this type of yarn on muffs is that the loading is low, and one can put more of the yarn on to a cheese. For this type of work cheeses are used up to 9 in. in diameter and carrying $1\frac{1}{2}$ lb. of yarn.

It should be noted that one and the same Pressure Dyeing Machine can be used with various types of frames, notably:

1. A cheese or cone dyeing frame on which either cheeses or cones can be dyed as required (for cones removable adaptors are used).
2. Cakes and "mock-cakes" (see p. 722).
3. Muffs.

Springs. Spring centres of the kind shown in Fig. 243 were originally used for holding polyester yarn, which was then dyed. The method has gained in popularity and importance and is now used not only for polyester but also for nylon and acrylic yarns—although not for high-bulk yarns. These centres ease the handling of packages; they slip over the perforated tubes of the dye machine and allow the cakes to bed into each other as shown in Fig. 238 (very important), and so even out slightly differing densities between one package and another. The interlaced type of spring shown in Fig. 243 prevents the fibre from falling between the turns of the main helix.

Tops. It has been reported that in America 7 lb. tops of the synthetic fibres Orlon, Dynel, Dacron and Vicara are being carrier dyed with a pressure of 25 lb. and that two runs are obtained in an 8-hour day. In Britain Bradford-type ball tops of 6–9 lb. are dyed on the Longclose system over a perforated spindle with the outer layer of the sliver of fibres protected by a perforated cage. There are also large coiled tops weighing about 40 lb. each, four of which can be dyed together; these may be of either synthetic fibre or wool,

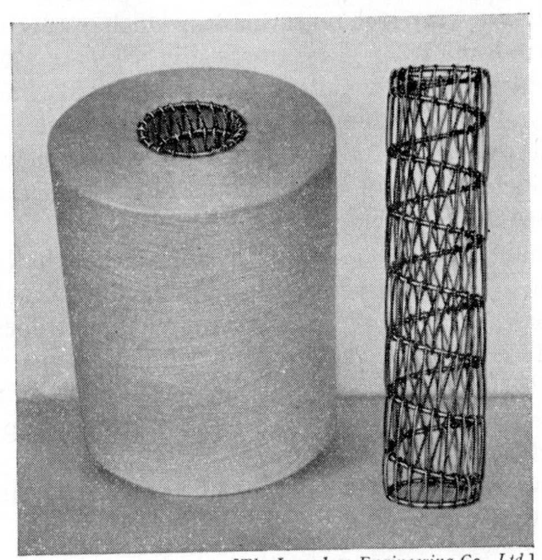

[*The Longclose Engineering Co., Ltd.*]

FIG. 243.—Cheese wound on spring centre.

and are used mainly in the worsted industry. The pressure system of dyeing is applicable to almost all forms in which fibres are assembled.

Automated High Temperature Pressure Dyeing. Figure 244 shows an automated high temperature pressure dyeing machine using a Vaco-Pilot electronic programme controller. It can be seen that there is a modern clean layout. Automation has become fashionable in the textile industry, and dyeing provides it with a wide field for use and expansion. Dyeings can be made with lots up to 1,100 lb. weight; matching samples of yarn can be removed from these above-the-boil machines without releasing the pressure or dropping the temperature, a small sample holder which can be temporarily isolated by taps from the main machine being built in. Pressure machines, permitting the use of temperatures well above the boil, up to 140° C. but usually used at 130° C., are of tremendous advantage in dyeing Terylene. This fibre cannot be satisfactorily dyed to medium-heavy shades at 95–100° C. except in the presence of carriers, and it is of considerable use with other fibres. It is, of course, the introduction of the synthetic fibres such as Terylene and Orlon which are very difficult to dye under ordinary conditions, but which can, owing to the greater mobility of their molecules at 120° C., be satisfactorily dyed at that temperature, which has led to rapid

[*Longclose Engineering Co., Ltd.*]

FIG. 244.—High Temperature Pressure Dyeing Machine using an electronic programme controller.

development of above-the-boil dyeing. Not only has it been found almost indispensable for Terylene and Orlon, but it has also been used to dye some fibres which can be satisfactorily dyed at atmospheric pressure (*i.e.*, at dye-bath temperatures not exceeding 100° C.) even better. It may be observed that of the two possible ways of getting medium and heavy shades on Terylene, *viz*. use of carriers or of temperatures of 110–120° C., it is the latter method—the high-temperature pressure method—which is being most widely adopted for the bulk of the production. This is not surprising, because carriers are a constant potential source of trouble with inequalities, toxicity, residual odour and so on, whereas, when once the apparatus for high pressure dyeing has been installed, when once the capital outlay had been made, the process is easy to operate and gives good results and good reproducibility. Specific examples of the use of " high temperature " (*i.e.*, above-the-boil) package dyeing on man-made fibres are as follows:

Terylene. Terylene is now being dyed as spun yarn, continuous filament yarn, as loose fibre, as combed slubbing, and as thrown braids and cords on high-temperature pressure dyeing machines. The yarn can be dyed in cheese form, preferably wound on stainless steel interlaced spring centres (Fig. 243). Such springs are used for many kinds of fibres, and with the higher qualities a coarse stockinette sleeve is placed over the wire centre before winding. When yarns such as nylon and Terylene are cheese dyed, they shrink considerably at the high temperatures used; consequently, these yarns must be shrunk and pre-set and then re-wound to give a final dyeing package; the yarn is scoured shrunk and set on the cheese, then re-wound and dyed, all on the same machine.

Terylene and nylon can also be high-temperature dyed on a " mock-cake " which has a large internal bore which can be adjusted to be a comfortable fit over the barrel of the former after the yarn has been fully shrunk; this obviates the need for re-winding.

It was at one time thought that high-temperature dyeing of Terylene under pressure might cause trouble with fusion of filaments one to the other; in fact, such fears have proved to be groundless, and high-temperature dyeing of Terylene is unquestionably the best means of getting full shades in a reasonable time. Whereas the use of carriers and normal temperature (not above-the-boil) equipment means a very protracted process usually of the order of six hours, two batches can be processed on high-temperature equipment in this time.

As a result of a detailed study made by I.C.I. of the high-temperature dyeing of Terylene with disperse dyestuffs, the following conclusions have been drawn:

1. The fibre is undamaged in dye-baths between *p*H 2·8 and 7·0 for 2 hours at temperatures as high as 160° C., but if the *p*H exceeds 7 (if alkalinity supervenes) damage takes place at any temperature above 100° C.
2. Dyestuffs shade and fastness are not as a rule affected by 2 hours at 130° C.
3. Terylene can be dyed at 120° C. as fast as acetate at 85° C.; no advantage is gained by increasing the temperature from 120° to 130° C.
4. Better depth of colour can be obtained at 120° C. without a carrier, than at 100° C. with a carrier.

5. With some exceptions it is undesirable to use a carrier at 120° C.

6. Some of the dyestuffs are unstable in alkaline baths, but in every case the bath should be kept neutral, or very slightly acid for the sake of the fibre—to prevent its degradation.

Viscose rayon. Cake-dyeing is the most economical method of dyeing viscose rayon and if the process is carried out at above-the-boil temperatures there is a very considerable saving in time; whereas 4–6 hours is usual for direct dyestuffs at 95–100° C., 1–1½ hours is sufficient at 110–120° C. Vat dyestuffs, applied by a pre-pigmentation technique followed by reduction on the fibre, levelling on the fibre and oxidation by blowing air through the package can also be applied to viscose rayon at high temperatures, with advantage derived from the shorter dyeing time necessary and from the better levelling obtained. Staple fibre viscose packages are always easier to package dye than filament viscose rayon, because of the greater quantity of air entrapped in winding, and the resulting greater porosity of the package.

Nylon and Perlon. These fibres can be dyed well enough at the boil, and the use of high-temperature techniques is dictated not by the dyeability of the fibre, but simply by the desire to save time and gain increased levelness.

Orlon. Less experience has so far been gained with Orlon than with Terylene but the considerations that apply to Terylene high-temperature dyeing will apply equally to Orlon Type 81 which is even less " dyeable "; Orlon Type 42 has, as described elsewhere, been satisfactorily dyed in fabric form at 107° C.

Advantages and disadvantages of high-temperature dyeing. The advantages (additional to those of pressure dyeing at temperatures not higher than 100° C.) are mainly:

1. Shorter dyeing time.
2. Makes full range of colours possible on hydrophobic synthetic fibres without the use of carriers.
3. Better levelling properties, *e.g.*, with vat dyes on viscose rayon cakes.

Disadvantages associated with the process are:

1. The higher cost of the installation.
2. The greater care needed in the preparation of the package.
3. The need to select dyestuffs—not all are stable at high temperatures.

The technique should not be regarded as a panacea for all the ills of dyeing, but in the author's view its advantages greatly outweigh its disadvantages and it is likely to be used more and more. It may be noted here that the Thermosol method of dyeing (p. 417) is itself a variety of high-temperature dyeing.

Ribbons. Ribbons of viscose rayon can also be dyed in package form.

Drying the Dyed Yarn or Fibre

Package dyeing machines may include provision for dropping the dye-liquor at the end of the dyeing period and giving a wash to the yarn, usually with hot and then with cold water. There is nothing difficult about this, it is simply a question of having pipeline supplies of hot and cold water, and drains and valves or taps which can be turned on and off. With the advent of automation, the valves are often operated by an electronic controller; the whole sequence of operations—raising the temperature of the dye-liquor, dyeing, turning off the heat, stopping the circulation, draining, letting in hot water, circulating it, running it off, letting in cold water, circulating it and running it off—can be, and very often is, looked after automatically. The operatives who are still employed spend their time preparing the next batch to go into the dye machine as soon as one comes out. The machines are costly and must be kept producing to pay for themselves.

The load of yarn, dyed and washed, that comes out of the package dyeing machine has to be dried. This operation is usually carried out in a separate machine. The cradle which carries the packages is lifted out of the dyeing machine by a crane and transferred into a drying machine. Figure 245 shows a high-pressure rapid dryer in which the surplus water is initially extracted and then hot air is recirculated. The complete cycle of drying operations, including unloading again by crane, may take only 40–45 min.; consequently one drying machine may serve several pressure dyeing machines. The machine gives even drying from package to package and from inside to out. Contamination from dirty atmospheric air is eliminated.

The system of package dyeing and drying which has been described possesses several advantages, but one of the greatest of these is that handling of the yarn is reduced to a minimum. The yarn is handled when the packages are put on to the perforated dye-liquor pipes and not again until the packages have been dyed, washed and dried, when they are unloaded. Labour is more costly than it ever was and has to be used as economically as possible. Another point, too, is that

[*Longclose Engineering Co., Ltd.*]

FIG. 245.—High Pressure Recirculatory Rapid Hot Air Dryer for packages of dyed yarn.

the less handling yarn receives, the better; the risks of fraying and breaking filaments are reduced. Thus the pressure system of dyeing, whereby the yarn is loaded on to the perforated tubes before dyeing commences and only taken off them after dyeing and drying are complete, is very good indeed.

Fabrics

The jig (Fig. 246) is suitable for most plain fabrics; it has the advantage that the fabric is kept in the flat state all the time it

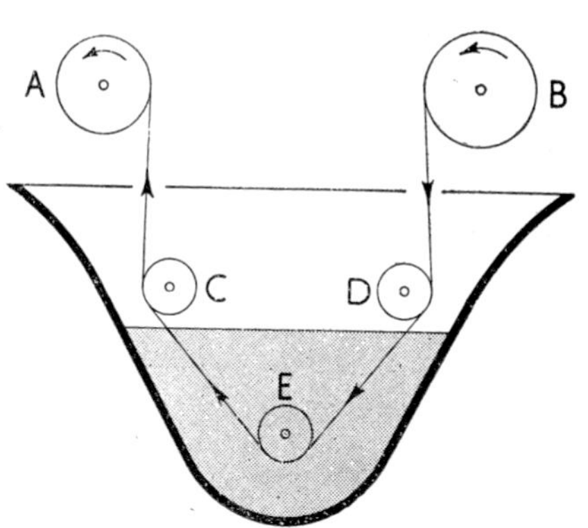

FIG. 246.—Diagram showing principle of dyeing jig.

Fabric from roller B passes round the guide rollers D, E, C, through the dye-liquor, and is taken up on roller A. The dye-liquor is steam-heated. The number of times the fabric is wound from one roller to another is the number of " ends."

is being dyed, so that there is no opportunity for rope-creases to develop. In the jig (or jigger) the fabric is passed from one roller to be wound on another, and as it passes from one to the other roller it goes through the dye-liquor, which is steam-heated. It is passed backwards and forwards until dyeing is complete, each pass being called an " end ". Disadvantages are that the fabric is out of the liquor for most of the dyeing time and that the fabric, during dyeing, is under tension. This tension may damage weak fabrics. Sometimes, too, the selvedges of fabric may cool down faster than the

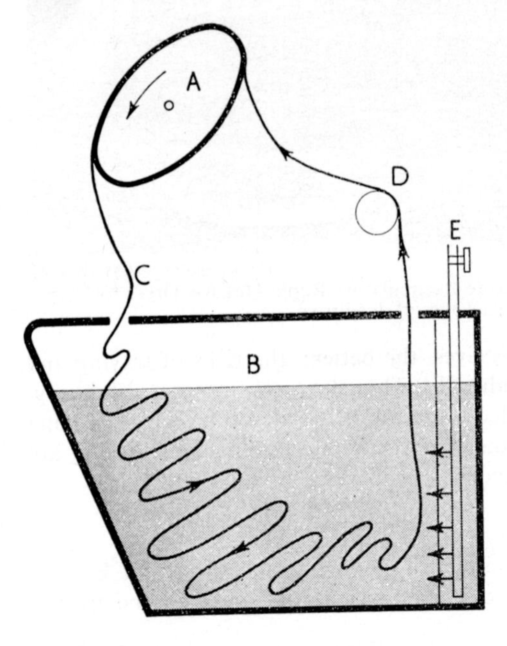

FIG. 247.—Diagram illustrating the principle on which a winch works.

A. Elliptical roller.
B. Dye-liquor trough.
C. Fabric.
D. Guide roller.
E. Steam heating pipe.

Note that the fabric plaited by the elliptical roller spends most of the time in the liquor—a big advantage over the jig. A winch is suitable for plain and crêpe fabrics. In some cases it may give " rope-creases " due to the fabric being bunched in the form of a long rope.

bulk of the fabric on the rollers and so come out weak in colour. Similar heat losses may occur from the roller itself absorbing heat from the fabric. Devices, designed to overcome such dyeing defects, have been introduced into modern jigs; such are: (1) internal heating of the draw rollers; (2) spraying the batch edges (selvedges) with hot dye-liquor. Nowadays, too, many jigs are enclosed, *i.e.*, they have covers, to prevent heat losses. This has had the additional advantage of keeping the dye-house atmosphere clearer; at one time the steam rising from open jiggers lent a permanent misty fog to some dyehouses.

The Winch. The winch (Fig. 247) is used for better-quality fabrics, especially those which must not be subjected to tension during dyeing. It consists of a trough containing the heated dye-liquor, across and over which is fitted a roller, sometimes circular in section, but better elliptical. The fabric, sewn in the form of an endless band, is passed over the roller, and as this rotates it is moved round. The fabric spends most of its time during dyeing in the liquor. The elliptical winch spreads the fabric better at the bottom of the trough than does the circular winch, because it has a plaiting action. Nevertheless, a readily extensible fabric, and knitted fabrics are often of this kind, is preferably dyed on a circular winch, whereas more stable fabrics are best suited by elliptical reels. In order to cope with both kinds, Leemetals Ltd. manufacture a " variable geometry " winch (Fig. 248) which can be adjusted from circular to

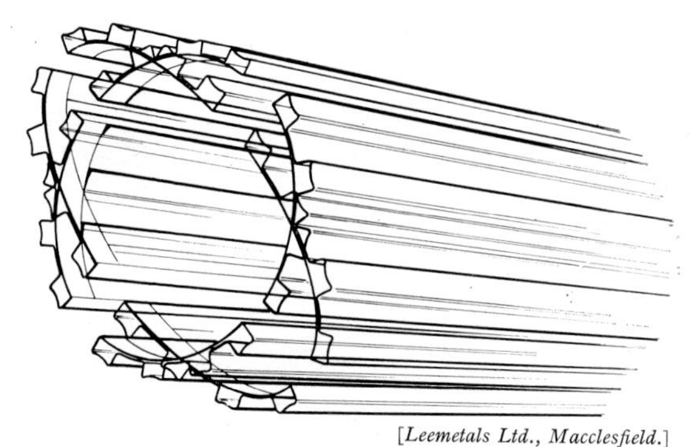

[*Leemetals Ltd., Macclesfield.*]

FIG. 248.—Variable geometry winch which can be adjusted from a circular to an elliptical shape.

elliptical without changing the circumference, so that fabric speed is unaltered. This winch, in common with most of the machines described in this chapter, is protected by British and foreign patents. Several pieces—perhaps six or so—of fabric will be threaded side by side on the winch and dyed together. Rope-creases are liable to develop in fabrics treated on the winch, particularly if the fibres from which they are made are thermoplastic, as, for example, are cellulose acetate and nylon. This difficulty can be avoided by pre-setting the fabric in the flat state in steam before dyeing. If this has been done, any creases that develop temporarily on the winch will come out when the fabric is removed from the winch.

A good example of a modern winch is the shallow draught (centre heat) machine made by Leemetals Ltd. It is an enclosed (this saves steam) but not a pressure machine, and is illustrated in Fig. 249. The fabric, either knitted or woven, has a long pass through the liquor, and owing to the shallow draught the volume of liquor is still small. The heating unit is in the centre of the machine, and this makes possible a high liquor temperature. If circular knit fabric is being dyed the tubular fabric balloons out as it passes over the heater; the fabric, when the balloon collapses, re-forms in a different fold and thus reduces the chance of rope-marking or creasing and also improves the levelness of coloration.

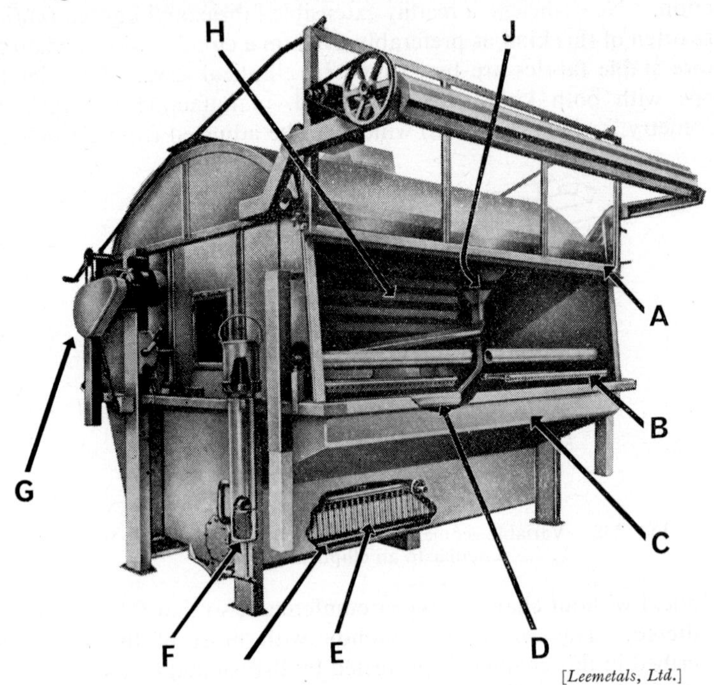

[*Leemetals, Ltd.*]

FIG. 249.—The Leemetals Shallow Draught Winch. A specially advantageous feature is that circular fabric often balloons in it.

A. Full view, full access doors.
B. Auto knock-off peg rail.
C. Shallow front shelf.
D. Adjustable midfeather.
E. Heat exchanger unit.
F. Colour introduction and filter.
G. Variable winch speed.
H. " Advance " type winch.
J. Continuous rope attachment.

Carpets. Carpets are a special kind of fabric. They are extra wide, up to 15–16 ft., and they are very heavy. Longclose make a winch (Fig. 250) of extra heavy construction and of extra width for tufted carpet dyeing. It will dye up to 150 yd. in a batch. Clearly, in such a wide machine special measures will have to be taken to ensure uniformity of dye content of the liquor. In practice, this is achieved by having cross-circulation of liquor along both diagonals of the vessel. Similarly, the heating system has to be fed from both ends of the machine to ensure uniformity of temperature.

[*Longclose Engineering Co., Ltd.*]

FIG. 250.—Sixteen-foot-wide Tufted Carpet Piece Winch Dyeing Machine.

The Padding Mangle. The padding mangle (Fig. 251) is suitable for tinting or very light shades. It consists of a trough containing dye-liquor and a mangle. The fabric is passed through the trough, then through the mangle and immediately to a dryer. The " padding " method, as it is called, has only a very restricted use for the coloration of rayon.

The Étoile. The Étoile, or star-machine, is used for some crêpe fabrics. The fabric is spirally hooked on a round frame and the frame suspended in a dye-bath. This method is thought to offer least opportunity of mechanical damage to the fabric and is used for velvets often made today from synthetic fibres. It is a machine for the craftsman and the connoisseur; it takes a lot of labour and its output is not very big. It has traditionally been used for velvets.

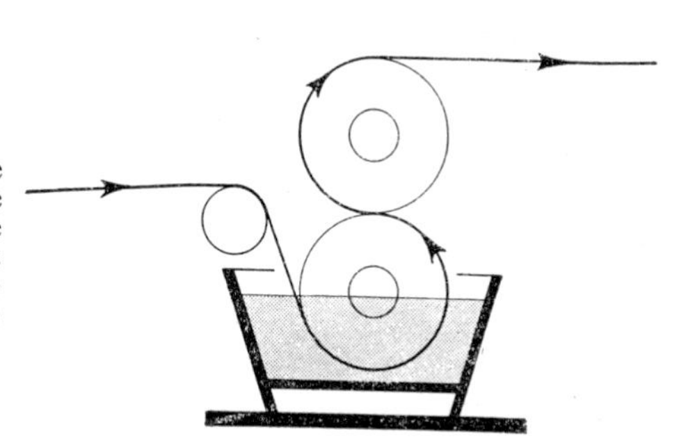

FIG. 251.—Padding mangle.

The two main rollers are under pressure so that the fabric, after its dip in the liquor, is squeezed. It is essential for the pressure to be uniform along the length of the rollers.

The Beam Dye Machine. Beam dyeing machines are best known for pressure-dyeing (below), but they are also made, *e.g.*, the Longclose machine for atmospheric pressure-dyeing, at temperatures up to 100° C. Naturally such machines are very much cheaper than pressure machines, as the construction can be much lighter. Those advantages that are associated with the pressure beam machines are common to the atmospheric beam dyer, except of course that the dyeing temperature cannot exceed 100° C., and this cuts down the range of fibres that can be dyed, notably the polyesters.

Pressure-Dyeing

Pressure-dyeing methods for fabrics, in which the fabric is wound on a perforated beam, were slow to develop despite the obvious attractions of the method in respect of economy of labour and probably of plant. The persistent difficulty was to secure uniformity throughout a piece of fabric, a difficulty partly caused by the fabric filtering out dyestuff. Development of the method received a considerable impetus from the growth of the practice of dyeing synthetic fibres above the boil; this was much the best way of getting medium and deep shades, and pressure was essential. Nowadays there are machines which will give good results on most fabrics.

Until the late 'fifties, machines that were open to the atmosphere were used, the fabric was rolled on a perforated beam and the dye-liquor was pumped at a temperature of about 90° C. through the perforations and thus through the fabric—the only pressure was that used to force the dye-liquor through the fabric. Since then totally enclosed machines have been developed, and in these the temperature of the dye-liquor may be held at say 130° C., well above the boiling point, so that the whole system is under pressure; as before, extra pressure is provided to pump the dye-liquor through the fabric. Not every kind of fabric can be dyed successfully on the beam but with fabrics of a fairly open structure the method has been very successful. The outstanding advantage of dyeing fabric rolled flat on the beam is that there will be no rope creases as there may be in winch dyeing. But first of all the fabric must be wound carefully and uniformly on to the beam; if creases are allowed to form at this stage they will be made permanent by the pressure dyeing. Machines that are available for " package " fabric dyeing are:

Burlington's Hy-Press Dye Machine. This is illustrated in Figs. 252, 253; it consists essentially of a perforated beam in a pressure vessel; in the illustration (Fig. 252) the head of the vessel and the beam are shown withdrawn from the autoclave; after the beam has been loaded with fabric they are run into the autoclave and the whole sealed (Fig. 253). This machine will work at tempera-

[*Burlington Engineering Co., Inc.*

FIG. 252.—The Burlington Hy-Press Dye Machine, unloaded, with beam withdrawn.

tures up to 150° C. corresponding to a working pressure of about 50 lb./in.^{2} and it will take loads of 2,000 to 4,000 yards of woven cloth, depending on the weight of the fabric, or of 1,000 to 1,500 lb. flat knitted fabric; it is particularly useful for tricot, which, whether

[*Burlington Engineering Co., Inc.*

FIG. 253.—The Burlington Hy-Press Dye Machine, loaded and ready for dyeing.

made from nylon, Dacron, Orlon or acetate, can be processed satisfactorily. The machine is equipped with two-way flow, either in-to-out or out-to-in and with instruments to control the direction of flow, temperature and rate of heating. The successful use of this machine, or any like it, depends on the correct preparation or batching of the fabric. A typical process being run by a manufacturer and dyer of nylon and acetate tricots is as follows:

1. The perforated beam of the Hy-Press machine is first wrapped with about twenty-four layers of cotton tricot fabric to prevent the appearance of perforation marks on the dyed acetate or nylon.

2. The tricot fabric is run from rolls through a wetting-out tank, spreader and pin-stenter on to the beam. The aim is to get the fabric spread flat and evenly on to the beam, but it should be added that some users do not wet out the cloth during batching; it is preferable to do so.

3. After the beam has been wound, a few turns of muslin or "tobacco cloth" are wrapped round to keep the fabric clean, and the muslin is secured with metal bands at each side of the beam. These bands also serve to prevent the tricot fabric slipping sideways. Beams are made with the perforated area of different widths, *e.g.*, 84 in. or 110 in. for tricot, and the perforations are usually $\frac{7}{16}$ in. diameter. Fabrics are wound with the selvedges about 2 in. beyond the perforations; if the fabric is wound too far beyond the perforations, its edges will not be dyed; if wound too close to the end of the perforations, the dye will be forced out of the ends of the roll without going through the fabric.

4. Acetate tricot is scoured in the dyeing machine with a solution of a wetting agent and 0·5 per cent soda ash at 70° C.; this is not necessary with nylon tricot which carries no water-insoluble material.

5. The goods (acetate or nylon tricot) are entered in cold water, a levelling or dispersing agent is added, then the dyestuff; the temperature is raised to 80° C. for acetate or to 110° C. for nylon, and the dye-liquor is circulated from inside to out until the shade is right. (In pressure dyeing at 110° C. a sample of fabric which can be isolated from the pressure system, for occasional inspection, is used as a control.)

6. The fabric is rinsed, dried and heat-set. Nylon is dried and heat-set in one operation on a pin-stenter, drying with hot air at about 160° C. and then being given a quick set

by a radiant heater with the cloth temperature at 210° C. Acetate is dried on a separate frame at a temperature of about 160° C.

7. The fabric is cooled, selvedges are cut off with a knife and the cloth is batched.

When woven fabrics are dyed, consideration must be given to the fabric construction; some measure of porosity is desirable. Even so, suitings of all-Dacron and 50/50 Dacron/wool blends have been successfully dyed. Even tufted carpets made of nylon have been dyed on similar machines, using a load of 3,000 lb. of carpet. Orlon Type 42 dyes well at 110° C. either with the Sevron and other cationic dyes or with disperse or acid dyes. Triacetate tricot fibres have been satisfactorily dyed, heat treated and set on the Hy-Press machine.

The Hy-Press machine, made by the Burlington Engineering Co. Inc. of Graham, North Carolina, seems already to have established itself for large production of acetate and nylon fabrics, and to have demonstrated its capabilities for dyeing Terylene and its blends and Orlon too. The following advantages and disadvantages of the machine may be cited:

Advantages

1. Easy to load (no stitching together of pieces).

2. Takes a large load, say 4,000 yards, of a light fabric —twice as much as would be treated in a batch by traditional methods.

3. Shorter dyeing cycles—the high temperature of 120° C. may enable the actual dyeing operation to be done in 30–60 minutes instead of perhaps 4–6 hours. (The total time spent at the autoclave, including loading, scouring, rinsing, setting and heating the dye-bath, dyeing, sampling, rinsing and removing, may be only about 3½ hours.)

4. Saving on capital cost.

5. Versatile machine on which scouring and heat-setting can be done.

6. Low liquor volume so that only half the usual quantity of dye-stuff may be needed for a given weight of fabric.

7. Perhaps most important of all is that the high temperatures available up to 150° C. enable dyeings to be made, *e.g.*, medium and heavy shades on Terylene and Orlon which it would not be practi-cable otherwise to make on fabric. Also used for heavy shades on nylon with acid dyes.

8. Fabric is maintained full width during dyeing so that rope creases do not develop.

Disadvantages

1. Limitation to porous loosely constructed fabrics—unsuitable for tightly woven fabrics. Nylon fabrics weighing 3 oz./sq. yd., are easy to dye, but those weighing 6 oz. are not, because circulation of the liquor is impeded. This effect can be overcome in laboratory experiments by co-batching a layer of porous cloth such as an Orlon Type 81 marquisette along with the nylon fabric, *i.e.*, running the two cloths together (like a double cloth) on to the beam at once.
2. Critical control of fabric width is necessary on the beam.
3. Some difficulty may be experienced with the re-dyeing of stripy lots.
4. Moiré effects can develop if much pressure due to shrinkage occurs; this can be avoided by pre-wrapping the beam with a 1 in. pad of knitted fabric; this acts as a cushion and allows the tricot fabric to shrink with less pressure being exerted by one layer of fabric on the next.

The Longclose Beam Machine. The Longclose high-temperature horizontal beam machine (Mark 3) is shown in Fig. 254 with the beam withdrawn from the casing. When beam machines were originally developed for dyeing fabric they were used for woven-filament yarn cloths and for warp-knit nylon. Now they are used for all types of cloths and often show big advantages, mainly in reduction of labour costs, over other types of dyeing machine. Then, too, the fabric which is being dyed on the beam is always in open width, so it is not subject to the rope-marks and creases which can be a trouble in winch dyeing. Very large quantities of fabric can be dyed on a beam. The quantity depends on the construction of the fabric and its porosity, but often the only limitation is the diameter of the tank; a loading of 3,000 yd. of knitted fabric is not uncommon. Dyeing time is often fairly short because of the high temperature of around 130° C. that is used. All types of fibres, including nylon and polyester, can be happily dyed level on the beam.

Beams are good for production dyeing; they help large outputs. They keep the fabric flat, so that provided it has been carefully wound round the beam, no creases will be introduced by the dyeing process. Perhaps, however, it compresses the fabric, so that handle may be a shade boardy, and definitely it flattens it, so that fabric carrying a woven or knitted pattern or design is unsuitable for beam dyeing.

One development of beam machines has been to use beams of greater and greater diameter; this has the advantage of affording a bigger area of perforated wall inside the beam, through which the dye-liquor can be fed to the fabric. There is, too, the additional advantage that a given length of fabric requires fewer laps on a large

[*Longclose Engineering Co., Ltd.*]

FIG. 254.—A High Temperature (pressure) Horizontal Beam Machine (Longclose Mark 3).

diameter beam than on one of small diameter, so making penetration of fabric by the liquor easier. A good result of this development is that larger lots and heavier loads can be dyed more level. When processing acrylic double knit (jersey) fabric a beam diameter of at least 18 in. is desirable for large lots. Temperatures of 102–110° C. are suitable for dyeing this class of fabric.

Bibby Beam Dyeing Machine. Bibby's of Halifax have produced a horizontal beam dyeing machine " for a crease-free finish " on nylon and Terylene fabrics. A range of beams for dyeing material of various widths is available. The perforated beams themselves are made rigid by cast stainless-steel end plates and are perforated according to the width of fabric that is being dyed. Clearly, the perforations must not extend beyond the fabric, or all the liquor that was pumped would take the easy way out through the exposed holes, rather than the hard way through the fabric. Ordinary nylon and Terylene fabrics weighing about 3–4 oz./yd.^{2} can be loaded on to the Bibby beam with an average length of 2,000–3,000 yd., and any width of fabric from 49 to 120 in. can be dyed.

The Barotor. The Barotor was developed by du Pont and manufactured by their licensees, *e.g.*, in Belgium by Toleries Gantoise,

Ghent. It was intended for dyeing fabric at temperatures up to 120° C. But it was intricate and complex in design and has been given up. The work which it could do has largely been taken over by beam pressure machines and by the new pressure winch machines.

High-pressure Winches

The winch (p. 726) is a very nice machine indeed for dyeing fabric, and enormous quantities of fabric are dyed on it. It is easy to operate, to load and unload, the minder can see what is going on, large batches of fabric can be dyed at once, and on the whole dyeing is good and level. The disadvantage is that rope-creases sometimes form; the fabric as it goes round and round the machine is gathered up like a rope, and there is often little or no change in the position of the creases, and in thermoplastic fabrics such as cellulose acetate the creases may show overdyed in the finished fabric. Then the creases are, in fact, permanent. Naturally, dyers find out how to avoid the creases; indeed, if they are to stay in the dyeing business they have to. One good method is to pre-set the fabric before it is dyed, then any creases that are formed will come out and not be permanent, for the fabric will always endeavour to regain that perfectly flat state in which it was set.

With the coming of the synthetic fibres and particularly of polyesters, the ordinary winch is inadequate. Carriers can be used, but they have their disadvantages: they are costly, sometimes odorous and difficult to remove completely. The challenge to the dyeing machinery makers was obvious. Could they make an enclosed winch in which fabric could be dyed at temperatures up to 130° C. or 140° C? It was, of course, a much more difficult task to enclose a rotating winch, with long lengths of fabric moving round and round it, than it had been to make pressure vessels of the Obermaier type, in which superheated liquid was pumped through stationary packages of yarn, but it was perhaps not very much more difficult than had been the problem, so nicely solved by Burlington, of making an enclosed beam dyeing machine. However, the dyeing machinery makers solved their problem, and at least three good high-pressure winches have been made in the United Kingdom. They will be described briefly.

Pegg Spray Jet High Temperature Winch. This is made in stainless steel and can be used at dyeing temperatures up to 140° C. There is an internal downwards spray system and liquor circulation so arranged as to give crease-free fabric. The fabric is not wholly immersed in the sense that the machine is full of liquor. The special feature of this machine is that the whole of the working parts will slide out of the shell (Fig. 255), so giving ease of loading and unloading; one can load up outside. It is suitable particularly for dyeing textured polyester fabric. It can be monitored by a punched-

[*Samuel Pegg & Son, Ltd.*]

FIG. 255.—A High Temperature Winch Machine made by Pegg's. A useful feature is that the winch will slide out of the shell so facilitating loading.

card control if required; in that event, the initial cost is much greater, but in operation satisfactory working is much less dependent on careful operative control.

Leemetals Shallow Draught High Temperature Winch. This machine is a development of the enclosed but not pressurised

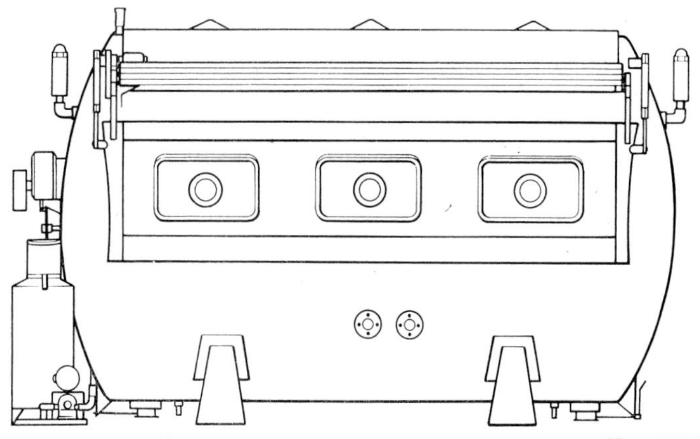

[*Leemetals, Ltd.*]

FIG. 256.—The Leemetals Shallow Draught High Temperature Winch—general appearance.

machine shown on p. 728. The features of shallow draught and long pass of fabric through the liquor are retained. The heating again is in the centre of the vessel. This machine is particularly suitable for texturised polyester fabric of the Crimplene type; this fabric is not ordinarily set dimensionally (pre-set) before dyeing, so that the avoidance of rope-marks and creases is an ever-present problem. If the fabric is tubular and will balloon, as it does so easily in the atmospheric pressure winch, such conditions are ideal. Fabrics can also be made to balloon, although it is more difficult to effect in the pressure winch. After they have been dyed the tubular fabrics are normally cut open and heat-set on a stenter. The exterior appearance of the winch machine is shown in Fig. 256. Fabric speeds inside the machine can be varied between 40 and 110 yd./min. Access for loading and unloading is through the doors shown in the illustration.

Longclose High Temperature Pressure Winch Machine. This machine, together with a control panel (on the right), is shown in Fig. 257. The fabric is loaded into the machine through the two large manholes, is thrown over the top of the winch (the revolving part inside) and fished round. In practice, this does not present any difficulty. It can be used at temperatures up to 140° C.; the outer vessel is of large diameter, with a flat shallow dyeing compartment through which there is a positive liquor circulation. It is made in various sizes; one, the largest, has a useful length of dyeing compartment of 12 ft. with a liquor capacity of 1,000 gal. and intended to have 12 ropes of fabric arranged as is usual in parallel. Various practical advantages are built in: steam coils and shaped perforated false bottoms for unobstructed even heating, variability of speed of winch, overflow wash-off facility and time/temperature control and recording with pneumatically operated valve controls (instead of turning a wheel by hand) arranged on a control panel. A fabric sample for control shade examination is included. It is a good machine and is intended for knitted and woven man-made materials. One kind of fabric that it is often used to dye is that for woven rainwear.

Most of the good features are common to all three machines—the accessibility of the Pegg, the ballooning of fabric on the Leemetals and the advanced control system of the Longclose are to be specially noted.

One point which might be raised is why should one build and use an expensive machine like a pressure winch when beam dyeing under pressure is available and is cheaper in that much larger quantities of fabric can be dyed at a time. Beam machines will take up to

3,000 yd. of knitted fabric, and for plain fabrics they are very good. But for fabric with a knitted design or with an embossed design the pressure of the fabric, due to its weight, flattens it and may make it boardy and harsh and leaner. Similarly, for bulked polyester yarns that have been knitted or woven into fabric the winch is more suitable because it permits fabric shrinkage. These, then, are the points in favour of the pressure winch: it permits shrinkage, does not spoil a knitted or embossed design, gives a full and soft handle. Its

[*Longclose Engineering Co., Ltd.*]

FIG. 257.—High Temperature Pressure Winch Machine and control panel, inclusive of injection tank.

disadvantages are of capital cost, of smaller load and consequently of increased labour cost. High-temperature pressure winches are selling well, report the makers, and that in itself is convincing. One trend in fabric design from plain to fancy Crimplene fabrics has stimulated the demand for such machines.

Hisaka Circular Dyeing Machine

It follows, normally, that those countries which engage in fibre manufacture soon become engaged also in the manufacture of machinery in which to dye their fibres or fabrics made from them. Japan has a very good record of man-made fibre manufacture, and now that country is beginning to make its mark in dyeing machinery. C. Itoh & Co. of Osaka (and Camomile Street, London) have made a " Circular " dyeing machine which has some interesting features. A cross-section of the machine is shown in Fig. 258 and the path of fabric in it is also indicated; the machine consists of a cylinder *A* in which there is an eccentrically driven roller *B* around which the fabric passes. The fabric then passes through the U-tube *C* back to the eccentric roller. The machine is made of stainless steel and is pressurised for high-temperature, up to 140° C., work. The U-tube, and there are three side by side on a machine, takes the place of the large chamber of a winch, and thereby cuts down enormously the total liquor volume and the liquor-to-goods ratio. The fabric is

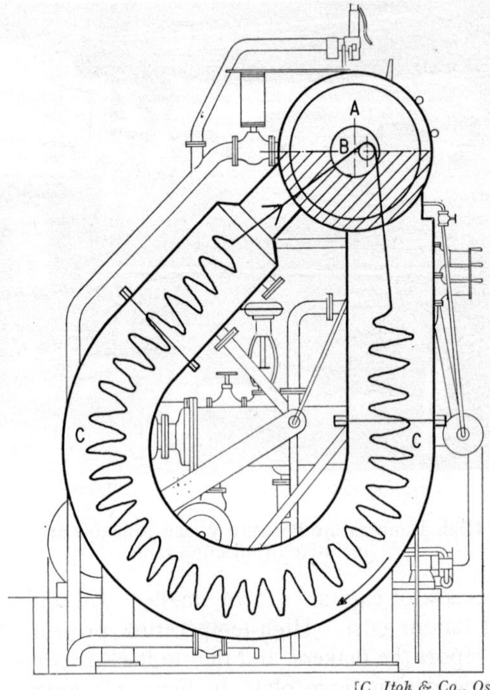

[*C. Itoh & Co., Osaka and London*]

FIG. 258.—Hisaka Circular Dyeing Machine. It resembles a pressure winch but embodies some valuable new features.

circulated in liquid most of the time (in a winch it is in liquid for only a small part of the time of revolution), and this considerably reduces the tension on the fabric. Such near-absence of tension is extremely valuable; it enables the softness of fabric handle to be retained and also the roundness of fibre compared with the jig or beam machines. Creasing of fabric is avoided by the spreading action of the driving roller; there is no opportunity for moiré effects or differential selvedge dyeing to occur. The machine has several advantages, but the greatest of these is that the fabric is immersed in liquor for most of the time, and is consequently nearly weightless so that no stretching takes place. A soft full handle in the dyed fabric should be the result. As the makers say: " Super quality of its dyed fabrics as well as its economical feature, will surely make you satisfy at all." But, however it is expressed, the reduction of tension on the fabric is bound to be good.

Garments

Some rayon materials are dyed in garment form. Garments, particularly ladies' hose, are very susceptible to damage, and must be treated very carefully. Small machines in which the garments and liquor are moved by a paddle are sometimes used. At other times air circulation through the liquor is used to obtain the agitation necessary to secure uniform treatment and yet avoid any agitation that would cause mechanical damage to the goods, such as fractured filaments and resultant hairiness.

Hose are often dyed in a rotary machine. They are first turned inside out to afford some protection to the surface, and then bundled in mesh bags, several dozen in a bag. The bags are packed in a container which rotates first in one way, then in the other, in an outer bath of dye-liquor.

The Pegg Toroid is one of these garment dye machines. The liquor circulation causes the garments to be tumbled continuously over and over towards the distributor head, and at the same time they are circulated gently round and round the dye vessel. When used with a pressure lid the machine can be used up to temperatures of 140° C. The Toroid consists essentially of a stainless-steel pressure vessel with speedilock arrangement for sealing the kier cover and a pump to give a liquor flow infinitely variable between zero and maximum. The operation of locking, raising and lifting the kier cover are done through pneumatic control. Temperature-control equipment is provided. A dye addition tank is fitted to allow addition of dyestuffs and chemicals without releasing the working pressure, and there is a sample kier for shade matching during the dyeing process. The machine has been produced in response to the popularity of textured polyester knitwear in Europe and the increasing interest of such knitted materials in the United Kingdom.

The " oval paddle " machine made by Bibby's of Halifax is intended for dyeing Ban Lon and Orlon knitwear and nylon hose. The largest of several sizes has a capacity of 900 gal., and the liquor can be brought to the boil in 30 min. if there is steam at a pressure of 1,000 lb./in.[2] available. It will take 1,000 doz. nylon hose at a time or 45 doz. knitted garments with a total weight of 350 lb.

Continuous Dyeing

Continuous methods of dyeing have much to be said for them in respect of output and rapidity. One of the earliest good continuous methods was the Standfast Molten Metal Method (pp. 199–201), which has been used very successfully for vat dyeing on viscose rayon. The lack of affinity of such fibres as acetate, nylon and Terylene for vat dyes has restricted the use of the Standfast method on such fabrics. The American Pad-Steam Continuous Dyeing Process (p. 199) is another continuous process for putting vat colours on to viscose.

The Thermosol process (p. 417) was introduced to dye polyesters; it consisted essentially of padding the fabric in a dispersion of an " acetate " dye, drying the fabric at about 70° C. and then passing it through an oven at 175–200° C., so that it was exposed to this temperature for 1 min. Essentially the Thermosol method was pad, dry, bake.

Pad, Dry, Bake. The method has been widely used for triacetate, Terylene and nylon; for the first two the bake is usually for 1 min. at 180° C. and for nylon it is for 1 min. at 190° C.

A suitable pad mixture is, for example, for triacetate:

Dispersol Fast Orange B	90 gm./l.
Urea	200 gm./l.
Migration inhibitor	10 gm./l.
Non-ionic surfactant	2 gm./l.

The fabric is padded in this mixture and mangled to retain 50 per cent of liquor on its own weight. The fabric is dried and then baked for 1 min. at 180° C. to give a good golden-orange coloration.

Duranol Blue G liquid used at 80 gm./l. gives a very fine Royal Blue on Terylene, and the Procinyl colours are suitable for application to nylon. The process is excellent.

Pad, Dry, Steam. A modification of the Pad, Dry, Bake method is I.C.I.'s Pad, Dry, Steam at high temperature. It is essentially the same as the Pad, Dry, Bake, except that instead of dry baking for 1 min. at 180° C. or 190° C., the padded and dried fabric is exposed to high-temperature steam for 1 min. at 180° C. or 190° C. The sequence of operations in the two processes is shown in Fig. 259, The colours obtained by Pad, Dry, Steam are a little deeper in shade than that given by the Pad, Dry, Bake method. The two methods just described can usefully be compared with two very similar continuous methods used for the application of Cibacrons to cellulosic fabrics and described on p. 780.

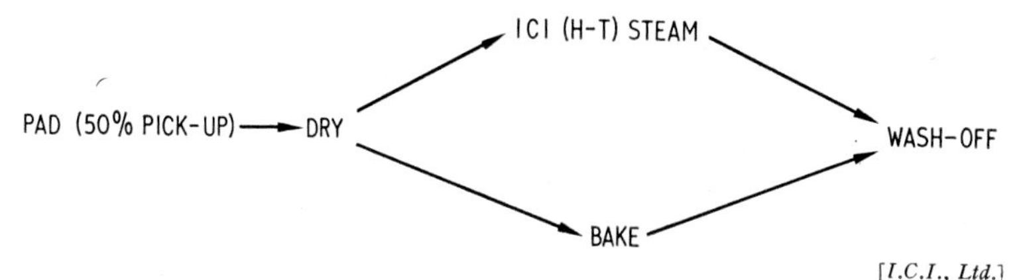

[*I.C.I., Ltd.*]

FIG. 259.—The sequence of operations in the Pad, Dry, Bake and Pad, Dry, Steam dyeing methods compared.

Materials

Old dye-vats were made of wood. They were difficult to clean, and as they absorbed dyestuff there was the chance of contamination of one colour by that previously used. They are not recommended for rayons because of the chance of staining and of mechanical damage if the rayon contacts the bath. It must not, of course, be forgotten that for tens of years enormous quantities of material were dyed satisfactorily in such baths. For rayons, stainless steel is recommended, and is used exclusively by some makers of dyeing machines. The type known as FMB contains about 3–4 per cent of molybdenum, which gives it a greatly increased resistance to chemical attack. It is a little more expensive than the more usual FDP, but is, in the writer's view, well worth the slight extra cost.

Automatic Control

Various devices are available for fitting to dye-baths. As a rule, these are designed to open and close steam valves automatically on the steam heating supply to the vat. One of these devices will, for example, raise the temperature of the bath from, say, 20° C. to 70° C. uniformly over a period of twenty minutes, maintain it at 70° C. for twenty minutes, raise it to 80° C. over a period of ten minutes, keep it at 80° C. for thirty minutes, and then shut the steam off. This only by way of illustration, for any other similar sequence of time-temperature changes could be carried out. At the same time as it controls the temperature, the instrument records it with a pen on a circular chart. These charts are useful to keep, so that in the event of a complaint of any dyed lot being later received, the chart for that lot can be examined to see if there were any irregularities—an " inquest " can be held. The sequence of events in the time-temperature changes can be pre-determined by cutting a cam in very thin

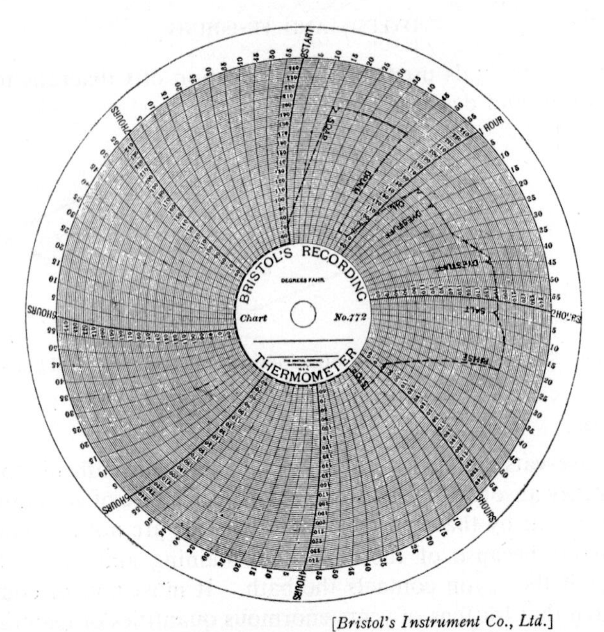

[*Bristol's Instrument Co., Ltd.*]

FIG. 260.—Typical time-temperature chart for a dye-vat automatic controller.

[*Foxboro-Yoxall, Ltd.*]

FIG. 261.—A large penthouse automatic control room in an American fabric dye-house.

sheet metal or plastic; the instrument is so designed that this cam will form a pattern for the recorder. By altering the shape of the cam, any desired sequence of time-temperature changes can be obtained. The use of these automatic machines, which are made by Foxboro-Yoxall and Bristol's Instrument Co. Ltd., amongst others, is worthwhile because they cut out human fallibility—*e.g.*, the dye-house operative may leave his machine temporarily unattended, and during the interval the machine if not automatically controlled may have boiled over, or perhaps cooled. A faulty dyeing results, but there is no record of how or why it happened. In addition, the use of automatic control reduces the work of the dye-house operative and opens up the possibility of his being given more machines to look after, so reducing labour costs. Figure 260 shows a typical chart, which follows exactly the shape to which the cam has been cut. Figure 261 shows the control room for a large piece goods dye-house. Each of the many dye-baths is controlled automatically.

Automation has made great strides since the previous paragraphs were written. They are still true, controllers of the kind in which a sequence of operations is determined by cutting a cam appropriately are still widely used as they have been for twenty years and more, and they still provide a simple and effective means of controlling some of the factors involved in the dye-house. But automation can be more ambitious today; its functions and development have been summarised and knowledgeably described by E. A. Syson (see Further Reading).

Nowadays, the automatic controllers will fill and empty tanks; they must have full-capacity reserve tanks to supply the liquid immediately it is needed, as well as air-blowing and vacuum-extraction to operate pneumatic valves. No longer need the operative turn wheel valves by hand. Such instructions as temperature control, duration of dyeing and flow control, including periodic reversal of the direction of flow, may be built into a controller and can be changed if a new dyeing operation is to be used. The controlling instruments are very versatile. It is said that the information relating to the control of dyeing cycles that are used to-day is too complicated to be handled satisfactorily by an operative. Doubtless much depends on the operative. It does seem to be the experience of dyers who use automation that there is less time wasted between cycles, and that they may even get an extra turn out of their dyeing machinery in a day. If this is done, then it is a big feather in their cap, because modern dyeing machinery is costly and must be kept going to pay for itself.

Automation has found its way into many aspects of textile and fibre processing (*cf.* p. 190), not only into the dyeing operation, although it is this last that lends itself best to, or calls out most urgently for, automation.

Some of the advantages that have been claimed for the use of automation in the dyeing and finishing processes are listed below:

1. The dyeing process time is reduced to a minimum (no waiting for operatives who may be busy on other jobs).
2. As a result of (1) the equipment has a higher productivity.
3. The labour needed for operating the dye machines is reduced (labour is still needed for loading and unloading and examination of shade, but not for filling and emptying tanks, controlling temperatures, opening and closing valves).
4. The reproducibility of shade from one batch to another is better, because there is less opportunity for operator errors.
5. Temperature changes can be controlled exactly, *e.g.*, a rate of temperature rise of 1° C. or 2° C./min. can be faithfully followed automatically, but not so easily manually.
6. Dyeings can be done with pre-set clocks before or after normal working hours; the operator can go home and leave the automated machine to finish a dyeing.
7. All operations from the first wetting-out to the final rinse can be controlled. Included in the programme there may be pre-heating, transferring measured quantities of water to the machine, injecting chemical and dye solutions in a correct sequence.
8. There are push-buttons for remote control, with motor starters, pressure gauges and thermometers all mounted for convenience and simplicity of operation on a panel which may be some distance from the dye-bath.
9. The state of the process—the stage that has been reached in a dye cycle is continuously indicated by the automatic controller.
10. The advanced controllers mostly work through transistor circuits; these do not give much trouble. Automatic electronic devices are so wired as to make servicing, and if necessary trouble location, easy by ordinary electricians.
11. Machines can be supplied by the dye-house machinery makers, complete with automatic controllers. The best of them are not cheap and may double the cost of what is ordinarily a £5,000 dyeing machine.

If something goes wrong, the automatic machine buzzes and flashes for attention. Probably most of these distress calls are due to a failure of services; if the steam supply drops, the automaton cannot keep to the programmed temperature rise and draws attention to the unsatisfactory working conditions as quickly as any shop steward could.

[*Taylor Ramstetter.*]

FIG. 262. The Taylor Ramstetter control panel. It controls and brings into operation at the appropriate times, a series of process control instruments.

Looking through the eleven advantages listed above, it will be seen that there is some overlap amongst them; probably something should also be discounted for enthusiasm, and the automatic instruments seem to be very costly indeed. But those dyers and finishers who have spent their money on such costly equipment seem to be pleased that they have done so, and that would seem to be the answer to any adverse criticism. As the author sees it, automatic control of dyeing and finishing has proved to be successful, much more successful than one might have thought likely. All that remains now, or the main thing, is for the instrument makers to bring down their prices. It is not really all; without doubt attempts will be made to provide exact colour matching automatically, with the inclusion of colorimeters and perhaps with connection to spectrometers. So much has already been done that confidence has been engendered for an assault on these more difficult problems. Nevertheless, it will be a colossal task to match instrumentally the discriminatory powers of the eye.

A central control panel made by Taylor Ramstetter is shown in Fig. 262. Courtaulds Engineering make a Celcon automatic dye-cycle controller which is actuated by a punched card; each punched hole in the card automatically initiates and regulates one or several steps of the dyeing cycle. Both of these controllers are versatile; they will cope with almost any requirements.

It should be added that in addition to control of dyeing cycles, automatic controllers are used in many other departments of the textile industry. Foxboro-Yoxall have controllers that are used for the control of conductivity, Redox potential, *p*H value, moisture content, humidity, density, viscosity and turbidity, all in the textiles industry. Advance is rapid, and it cannot be long before all those qualities that are easy to measure, such as *p*H, density, temperature and time and a dozen others will be controlled automatically. When it comes to controlling qualities which are not easy to measure, such as the quality of a merino wool, the brightness of a baby pink dyeing, the handle of a spun fabric, the scroop of a silk or the aesthetic appeal of a fabric and a hundred others, then their control will be more difficult and longer in coming.

Water Supply

Soft water is desirable in a dye-house, and is essential for scouring when soap is used. Water of commercial zero (actually $\frac{1}{2}°$) hardness is obtained by the base-exchange process. Suitable equipment is made by the Permutit Co., Ltd., amongst others. Water entering a dye-house should be neutral or very slightly alkaline; *p*H 7–8 is best. Automatic controllers which inject sodium carbonate into the water supply to keep the *p*H steady are available. The Multelec *p*H recorder and controller made by Geo. Kent, Ltd., is suitable for this type of work.

Water should be iron-free for the finest class of work. Various methods are available for making it so, most of which depend on aeration and on treatment with lime and alum. Perhaps the simplest method of obtaining water iron-free is to subject it to filtration through manganese zeolite. This removes the iron; the zeolite has to be regenerated at intervals by treatment with a solution of potassium permanganate. It may be noted that water used in the manufacture of rayons also must be free from iron, and that this method of removing iron by filtration through manganese zeolite is used by some rayon manufacturers. The material and plant in which it can be operated are made by the Permutit Co., Ltd.

One knows that there is no water to speak of in Kuwait and very little in the centre of Australia, but it is surprising to learn that a chronic shortage of it is threatened, and indeed already exists, in the south-east of this country. The consumption of water in England and Wales is about 4,500 million gal./day or $1{\cdot}65 \times 10^{12}$ gal. in a year. This is equivalent to a precipitation of 2 in. over the whole country, and the actual rainfall averages 36 in. in a year—less in the south, more in the north and west. To collect 2 in. out of 36 looks easy enough, but in fact more than half of what falls is lost by evaporation from grass, leaves and other vegetation, and there are, too, other serious sources of loss. It may be possible in the fairly near future to double the procurement of water, but this commodity will inevitably become dearer as more difficult sources are tapped.

Water for the dye-house must still be soft, which at first sight seems odd now that soap has been almost completely replaced by surface-active detergents which are not precipitated by calcium and magnesium ions. Unfortunately, these ions are powerful " salting out " agents, and their presence tends to drive any dyestuff out of the solution and on to the fibre; these ions are much more powerful than sodium ions, and their presence in water is undesirable if repetition dyeing is to be carried out. Furthermore, there are some individual dyes that are precipitated by calcium and magnesium. Ion exchange is usual for softening dye-house water, and it costs about 5–15*d.*/1,000 gal. Demineralisation is excellent, but too costly at 50–150*d.*/1,000 gal. except for very special purposes, *e.g.*, processing in the production of electrical insulation fabrics to a rigorous specification for ionic impurities.

Ion Exchange. Either a natural or a synthetic zeolite or a synthetic anion-exchange resin is used. The softening reaction may be represented

$$[E]Na_{2n} + n\ Ca(HCO_3)_2 \longrightarrow [E]Ca_n + 2n\ NaHCO_3$$

where $[E]Na_{2n}$ represents the zeolite. The calcium is thus removed from the water. A similar equation may be written with magnesium instead of calcium, and for sulphate instead of bicarbonate, thus:

$$[E]Na_{2n} + n\ MgSO_4 \longrightarrow [E]Mg_n + n\ Na_2SO_4$$

The water so treated has all its calcium and magnesium replaced by sodium; it will contain some sodium chloride, sulphate and bicarbonate, and perhaps some CO_2. The zeolite is regenerated to be used again, this repeatedly, by treatment with common salt:

$$[E]Ca_n + 2n\,NaCl \longrightarrow [E]Na_{2n} + n\,CaCl_2$$

It will be seen that the water that has been softened by ion exchange is likely to contain sodium bicarbonate. If such water is boiled, as the baths in jigs and winches frequently are, it loses CO_2 and becomes alkaline:

$$2\,NaHCO_3 \longrightarrow Na_2CO_3 + CO_2 + H_2O$$

The danger here is that some dyes are reduced by cellulose (viscose rayon) in alkaline baths. Thus the dyestuff Chlorantine Fast Green BLL contains three diazo groups (the diazo group is the vulnerable part, susceptible to reduction) in its molecule. After being boiled for two hours in slightly alkaline liquor one of these three groups is reduced and the green dye correspondingly breaks down to two smaller-molecule dyes, one violet and one yellow; on still longer boiling these smaller-molecule dyes also break down to colourless materials. It has been found that the trouble can be avoided by adding ammonium sulphate to the bath; then when the bath is boiled, ammonia and carbon dioxide boil off and leave behind neutral sodium sulphate in the bath.

$$2\,NaHCO_3 + (NH_4)_2SO_4 \longrightarrow Na_2SO_4 + 2NH_3 + 2CO_2 + 2H_2O$$

This danger of reduction of dyes is relatively new. Formerly, when open jigs and winches were used plenty of air circulated over them, and the oxygen in it made reduction of the dye very unlikely. But nowadays covered jigs and winches are used to save steam, so that less air has access to the dye-liquor.

The dyer must never overlook his water supply. It can make or mar his work. In foggy weather it may become covered with a film of smoke which will soil his pastel shades.

Demineralisation. This process consists essentially of two stages of ion exchange; in the first the water is passed through an acid-regenerated exchanger, where the cations (calcium and magnesium) are absorbed thus:

$$[E_1]H_{2n} + n\,CaCl_2 \longrightarrow [E_1]Ca_n + 2n\,HCl$$

regeneration of the exchange resin $[E_1]H_{2n}$ being effected with hydrochloric acid thus:

$$[E_1]Ca_n + 2n\ HCl \longrightarrow [E_1]H_{2n} + n\ CaCl_2$$

In the second stage the acid anions are absorbed on a different zeolite

$$[E_2](OH)_{2n} + 2n\ HCl \longrightarrow [E_2]Cl_{2n} + 2n\ H_2O$$

and the regeneration of this zeolite is effected by treatment with caustic soda.

$$[E_2]Cl_{2n} + 2n\ NaOH \longrightarrow [E_2](OH)_{2n} + 2n\ NaCl$$

Rinsing. With the tightening of water supplies, there must come big changes in the rinsing of textiles; whereas 500 l. water might be used in practice and wastefully to rinse 1 kg. of a synthetic fibre, it should be possible theoretically to use only 1 l. Experiments on rinsing dyed fabrics in jigs have shown that big economies in the volume of water that is used can be made without much detriment to the efficiency of washing. To-day's dye-house methods are often wasteful of water, and they had their origin in times when it was abundant and cost next to nothing.

Surfactants. Surface-active agents are used almost everywhere in the dye-house, and they find their way into the effluents too. Cationic surfactants are costly and only little used; non-ionics are now used in enormous quantities and so are anionics. The older type of anionics, such as PT–ABS, propylene tetramer–alkyl benzene sulphonate, have proved to be resistant to most sewage treatments and have given a lot of trouble in sewage works, and their use has even led to foaming on some rivers, such as the Aire and Calder. Such branched-chain detergents have now been largely replaced by the LAS, linear alkyl benzene sulphonates, which are biodegradable and will succumb to activated sludge treatment in the sewage works. Non-ionic detergents usually embrace a poly(ethylene oxide) chain in their composition, and the longer this chain is, the more resistant they are to degradation.

Those who have had hard practical experience do not consider that textile effluents should present much difficulty to sewage works, and they have often pressed this point home in discussions with the managers of such works. Sometimes, however, textile effluents can be deficient in nitrogen, so that the bacteria which are expected to work on them are starved; the remedy is simple, just to add some ammonium sulphate to the effluent.

Typical structural formulae for the different types of surfactants are shown below.

Cationic (not much used in the dye-house—too costly). These have various structures, but all have the negatively charged Cl^- or other halogen ion. This, being negatively charged, is attracted to the anode, which is positive. The student is recommended to remember that the *anode is positive*; knowing this for certain, it is easy at any time to thread one's way between anions and cations. The halogen ion is negative or anionic and the main part of the surfactant is positive or cationic; it is drawn (unlikes attract) to the negative cathode. A typical formula is:

$$\left[R_1 - \overset{\displaystyle R_2}{\underset{\displaystyle R_4}{N}} - R_3 \right]^+ Cl^- \text{ or more specifically } \left[CH_3 - \overset{\displaystyle C_{16}H_{33}}{\underset{\displaystyle CH_3}{N}} - CH_3 \right]^+ Br^-$$

Cetyl trimethyl-ammonium bromide.

Unexpectedly, the cationic surfactants have not shown very good detergent properties, but they have been found to give a soft handle to fabrics that are given a final rinse in their solutions, and they have also been used for " fixing " dyes on fabrics. These properties are shown by more complicated members, but still with the same basic structure; some of the Sapamines and Fixanol are of this type.

Anionic (used on a large scale mainly as detergents). The cation is usually the sodium ion; the rest of the molecule, the long-chain part, is the anion. Ordinary soap is an anionic detergent, with sodium as the cation and stearic or oleic ions as the anions. Of the synthetic anionic detergents those with branched chains, such as PT–ABS, short for the propylene tetramer–alkyl benzene sulphonate,

$$CH_3CH_2 - \overset{\displaystyle CH_3}{CH} - \overset{\displaystyle CH_3}{\underset{\displaystyle C_6H_4SO_3Na}{C}} - \overset{\displaystyle CH_3}{\underset{\displaystyle CH_3}{C}} - \overset{\displaystyle CH_3}{CH}CH_3$$

were formerly used on a very large scale. It should be noted that sulphonates are characterised by the $-SO_3Na$ group and the sulphates by the $-OSO_3Na$ group, both being connected to carbon atoms. The branched-chain sodium alkyl sulphonates were not biodegradable, and when they came to be used on a really large scale around 1962–63 they gave a lot of trouble. It has been found that if the alkyl group has a straight chain, then the resulting detergents, linear alkyl sulphates and sulphonates (LAS), are easily broken down by bacteria at sewage farms, and they are accordingly known as biodegradable. Only biodegradable surfactants should be used on a large scale. A typical LAS detergent has the structure

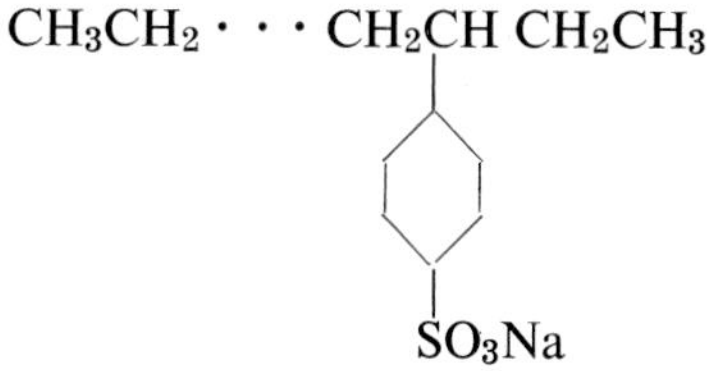

$$CH_3CH_2 \cdots CH_2\underset{\underset{\displaystyle SO_3Na}{|}}{\underset{|}{\bigcirc}}\!\!\!\!\!\!\!\!\!\!\!\!CH\ CH_2CH_3$$

The linear LAS compounds include alkyl groups which are straight chain, whereas the earlier detergents had branched alkyl chains. The linear compounds are biodegradable; the branched-chain compounds are not.

Non-Ionic. The non-ionic detergents are used on a very big scale. Essentially they are synthesised from a fatty compound and a polymer of ethylene oxide, or propylene oxide. One of the first, and still one of the best, was Lissapol N; it was said to be an alkyl phenol polyethanoxy ethanol and to have a structure similar to:

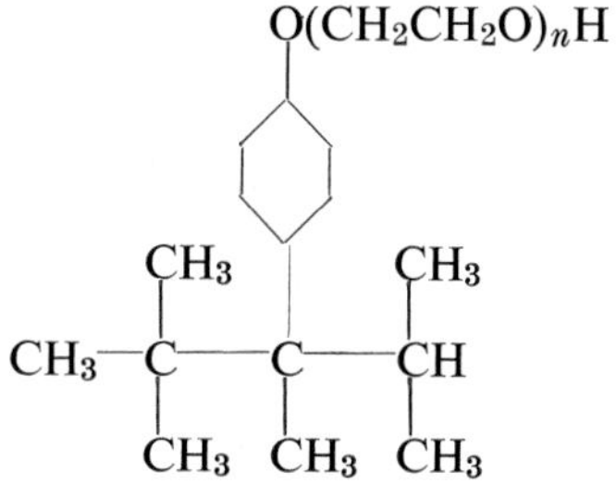

$$\begin{array}{ccccc} & & O(CH_2CH_2O)_nH & & \\ & & | & & \\ & & \bigcirc & & \\ & CH_3 & | & CH_3 & \\ & | & | & | & \\ CH_3- & C & -\ C\ - & CH & \\ & | & | & | & \\ & CH_3 & CH_3 & CH_3 & \end{array}$$

The formula looks a little complex, but if the alkyl group is written as R we have

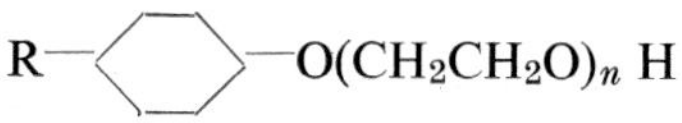

$$R-\bigcirc-O(CH_2CH_2O)_n\ H$$

It is the length of the poly(oxyethylene) chain which determines the properties of the detergent. The value of n which determines the length of the chain may be between 4 and 40, but is probably most often about 8–12. Non-ionic detergents are not easily biodegradable; the higher the value of n, the more difficult they are to break down. Non-ionic detergents are very prone to cause foaming, and in the presence of the smallest of quantities of anionic material the tendency of the non-ionics to foam is greatly increased, and is observable at concentrations as low as 0·4 p.p.m., a notable example of synergism.

Choice. In the U.K. there are some 400 different surfactants; in the U.S.A. there are some 3,000. They include detergents, emulsifiers, softening agents, bactericides and water-repellent finishes. Which one to pick? Factors to be taken into consideration are:

1. The type of fibre.
2. The nature of the soil, *e.g.*, oil, if this has to be removed.
3. Foaming is usually disadvantageous; it may cause baths to overflow, with loss of chemicals.
4. The hardness of the water.
5. Whether stability in the presence of acids, alkalis or metal salts is necessary.
6. Compatibility with the dyestuffs to be used.
7. Compatibility with the finish, *e.g.*, an anionic detergent must not be used with a cationic finish.
8. The absorption characteristics with special reference to the desired handle of the finished fabric.
9. Cost.
10. Physical form—liquid or powder.
11. Activity at very low concentrations.
12. Behaviour in effluents with special reference to pollution.

It is a formidable list. The dyer who feels a little at a loss can always get reliable technical help from dyemakers and indeed from the surfactant manufacturers.

Classification of Dyestuffs

Dyestuffs are grouped in several large classes. The classification is not hard and fast, and a dyestuff may well belong to two classes. The main groups are:

Direct Dyestuffs. These are applied to cellulosic fibres, cotton, viscose, viscose staple and cuprammonium as well as the strong viscoses. They have little or no affinity for cellulose acetate or the synthetic fibres (nylon, *etc.*) and usually only stain wool and regenerated protein fibres. The method of application is simply to dissolve the dyestuff in water, add sodium chloride—perhaps 15 per cent on the weight of the goods—enter the goods cold, warm up slowly to about 90° C. and maintain there for about forty-five minutes. The dyestuff has quite a high affinity for water as well as for the fibre, so that steps have to be taken to effect exhaustion. The addition of salt to the liquor is inclined to make the aqueous liquor a less attractive solvent for the dyestuff, and thus urge it on to the fibre. The use of a small liquor: goods ratio is also advisable to improve exhaustion.

Acid Dyestuffs. These are applied to wool and to regenerated protein fibres—*e.g.*, Merinova, Fibrolane BX and Lanital. They could also be applied to Rayolanda, which, although essentially a viscose rayon, had been resin-modified to give it an affinity for acid dyestuffs. Acid dyestuffs are mostly sulphonic acids, sometimes carboxylic acids, often sold by the manufacturer in the form of sodium salts. They are dissolved in the dye-bath, sulphuric or formic acid is added, and Glauber's salts. Approximate quantities used (all on the weight of the fibre) are:

Dyestuff.	For pastel shades, 0·1 per cent.
	For medium shades, 0·5–2 per cent.
	For heavy shades, 2–8 per cent.
Acid.	From 1 to 4 per cent, according to depth of shade.
Glauber's Salts.	From 2 to 6 per cent also according to depth of shade.

The more dyestuff is used, the more, correspondingly, of acid and Glauber's salts is used. Note that Glauber's salts is hydrated sodium sulphate, $Na_2SO_4,10H_2O$. Some dyers now use, instead, anhydrous sodium sulphate, Na_2SO_4, which is a dry powder and cleaner to handle; it is often described as A.S.S.

Acid dyestuffs have a very high affinity for protein fibres, as they combine chemically with the free amino-groups. If dyeing is allowed to take place too quickly, results are patchy or streaky. It is therefore slowed down by

1. entering the goods cold and raising the temperature only slowly;
2. using a high liquor: goods ratio of about 40:1;
3. adding sodium sulphate to the bath, as this has a retarding effect.

The fact that a salt retards the application of acid dyestuffs, but accelerates the application of direct dyestuffs, illustrates the essential difference between the dyeing of wool with acid dyes, which is mainly a process of chemical combination, and the dyeing of cotton with direct dyes, which is more a matter of adsorption.

Dyeing with acid dyes is usually complete after thirty minutes at the boil (or near-boil). Some regenerated proteins, owing to their low wet strength, are better dyed at lower temperatures.

Mordant Dyestuffs. These depend on the presence of a metallic group to unite the dyestuff to the fibre. In the case of the regenerated proteins, chromium is generally used for the application of " chrome " dyestuffs. The " chrome " dyestuffs are usually similar in type to acid colours. Perhaps six molecules of an acid colour will combine with the chromium atom of the mordant, and in this way dyestuffs with relatively small molecules are converted into very large molecule dyestuffs. Large dyestuff molecules cannot enter the pores of the fibres so readily as can small dyestuff molecules, so they have little chance of migrating from one pore to another—a movement which constitutes the levelling process for acid colours. However, once the large dyestuff molecules are in the pores which, like the fibre itself, expand in the hot dye-liquors, they are firmly trapped when the fibres are removed from the liquor and the pores contract. Consequently, chrome dyestuffs do not dye " level ", as do acid colours, but they have a very much superior wet fastness. On account of the difficulty of levelling, it is usual to apply chrome colours, not to the finished yarn, but to slubbing or top, so that any inequalities of shade are levelled out mechanically in the subsequent drawing, roving and spinning operations.

A chrome mordant can be applied in at least three ways:

1. The mordant is precipitated on the fibre before the dyeing process. This is known as the " chrome " or " bottom-chrome method ". When the fibre is subsequently dyed, an insoluble colour lake (or compound) is formed between the mordant and dyestuff. The mordant usually used is sodium bichromate. If the dyestuff is liable to oxidation, a reducing agent such as oxalic acid is added to the bath, and the mordant is known as " reduced chrome".

2. The mordant is applied at the same time as the dyestuff. This is known as the Metachrome process, and the mordant used is sodium chromate (*not* bichromate), in the presence of ammonium sulphate. This process is quicker and labour-saving, but does not give the same fastness as the other two.

3. The dyestuff is applied, and then the sodium bichromate which is used as a mordant. This process is largely restricted to navies and blacks. It is known as the Afterchrome process.

The " chrome " dyestuffs are used on the regenerated protein fibres, as well, of course, as on wool (pp. 303, 338).

Pre-Metallised Dyestuffs. Mordant dyestuffs in which a metal salt is applied, as well as a dye, to a fibre have been known for generations; after-chroming (see above) in particular has been a traditional process and a very useful one too, giving good fast dyeings. A development which was likely to come and was quite obvious to many dye chemists was to include the metal in the dye. There was, of course, no certainty that this would give such good results, and as events proved, some ways of including the metal were better than others, and some were unsatisfactory. The best way proved to be the 1 : 2 method, fostered by the late F. L. Goodall for the Geigy Company, who broke new ground with their Irgalan dyes about or just before 1950. Such dyes, then known as " pre-metallised " and now generally just as " metallised ", were originally introduced for dyeing wool, but they have since been widely used for dyeing synthetic fibres, particularly nylon and Perlon, in level shades of good all-round fastness. And also, as will be described later, they can be used on modacrylic fibres.

Metallised dyes contain one or more metal atoms built into the dye molecule. The preferred ranges are of the 1 : 2 kind, in which one atom of the metal, *e.g.*, chromium or cobalt, is associated with two molecules of an unsulphonated azo or azomethine dye. Azo dyes are those that contain the azo group $-N{=}N-$; azomethine dyes contain the carbon–nitrogen double bond $-N{=}C{<}$, *i.e.*, the azomethine bond. The association of the metal with these dyes is such that the azo or azomethine bond, $-N{=}N-$ or $-N{=}C{<}$, must unite two carbon atoms each of which is directly attached to a carbon atom carrying a hydroxyl group; thus:

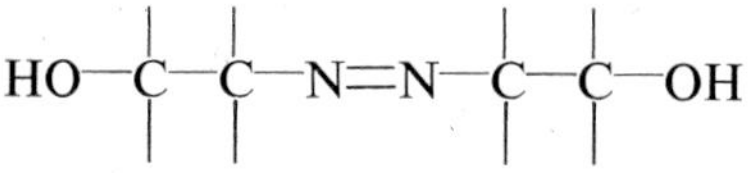

$$HO-\overset{|}{\underset{|}{C}}-\overset{|}{\underset{|}{C}}-N{=}N-\overset{|}{\underset{|}{C}}-\overset{|}{\underset{|}{C}}-OH$$

or

$$HO-\overset{|}{\underset{|}{C}}-\overset{|}{\underset{|}{C}}-N{=}\overset{|}{\underset{|}{C}}-\overset{|}{\underset{|}{C}}-\overset{|}{\underset{|}{C}}-OH$$

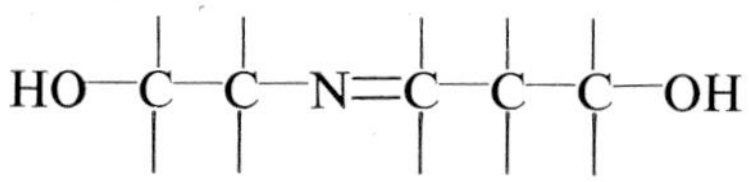

These arrangements look a little unfamiliar, but if they are transferred to an aromatic background substance, they are more easily recognisable. Thus, 2,2′-dihydroxyazobenzene contains the azo bond in the situation just defined. It has the structure:

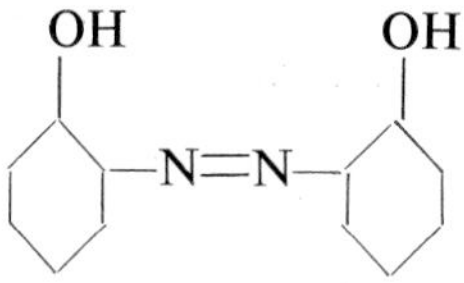

OH OH

$C_6H_4(OH)-N{=}N-C_6H_4(OH)$

If two molecules of this azo body are associated with one chromium atom, then we should have:

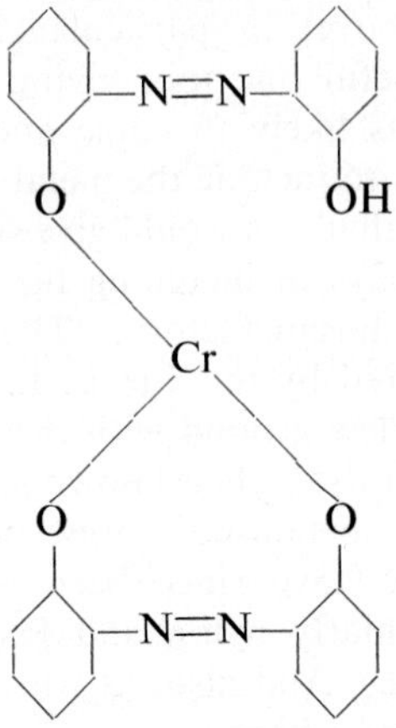

—N=N—

O OH

Cr

O O

—N=N—

This 1 : 2 chromium complex rearranges itself to adopt the structure:

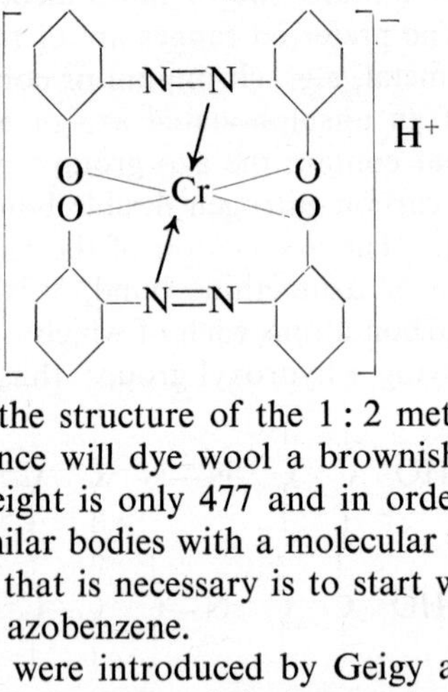

—N=N—

O Cr O

O O

—N=N—

H^+

This is, in principle, the structure of the 1 : 2 metallised dyes, and this particular substance will dye wool a brownish-red colour. In fact, its molecular weight is only 477 and in order to get the best dyeing properties, similar bodies with a molecular weight of at least 600 are needed. All that is necessary is to start with a substituted and therefore heavier azobenzene.

Such dyes as these were introduced by Geigy as Irgalans; they were easy to apply, they gave good dyeings and commercially they did very well. Other ranges of somewhat similar dyes were produced by other dyemakers; for example, du Pont's Capracyls, Badische's Vialons, Ciba's Cibalans, Sandoz's Lanasyns, Baeyer's Isolans and Badische's Ortolans, as well as Geigy's own Supralans. All these dye ranges consist of chromium or cobalt-containing azo dyes, containing no ionising lyophilic groups, and in which two dye molecules are combined with one metal atom, *i.e.*, they are 1 : 2 complexes.

An example of such a metallised dye is Vialon Fast Copper R (C.I. Acid Red 227), and a specific way in which it may be used on a modacrylic fibre has been described by Courtaulds in B.P. 940,372 (1963). According to this, the modacrylic fabric is immersed in 70 times its weight of water containing 2 gm./l. ammonium sulphate and 1 ml. of 0·88 (sp. gr.) ammonia per litre, a little of a dyeing assistant and 1 per cent on the weight of the fabric of Vialon Fast Copper R (BASF). Dyeing is carried out in a closed pressure apparatus raising the temperature to 110° C. in 30 min. and then continuing the dyeing at that temperature for 1 hr. After cooling, the fabric is removed, soaped for 5 min. in a 5 gm./l. soap solution at 40° C., rinsed in water and dried. A full brown shade is so obtained. A similar dyeing carried out at atmospheric pressure with a maximum temperature of 98° C. gives a similar but much paler shade; the high temperature (above the boil) is necessary to swell the fibre and thereby give entry to more molecules of dyestuff; on cooling the fibre the molecules of dye are trapped in it and they confer on it a good fast colour. It is of interest that this particular dye gives yellowish-red (not brown) dyeings on nylon and Perlon; quite often one and the same dye will give different shades on different fibres. It will be appreciated that the application of these dyes is a very simple process. At temperatures of around 90° C. and higher they (as their ammonium salts) form true solutions in water; at room temperature some molecular aggregation takes place and incipient turbidity may develop.

Dispersed Dyestuffs. The dispersed dyestuffs, as their name implies, disperse in water, provided a wetting agent such as Turkey-red oil is present. They dye the fibres by dissolving in them. The dyestuffs are organic in nature and dissolve in organic solvents. The fibres must be considered as solvents. Particularly are they useful for cellulose acetate, nylon, Vinyon and similar fibres. These fibres, which have few or no reactive groups, are readily dyed by this process of solid solution. The student must think of the particles of dispersed dyestuff being in a very unsuitable environment in the water, for which they have no liking, in which they are dispersed. When they have an opportunity of escaping from it into the fibres which provide a more suitable *milieu*, they move into the fibre and dye it. The process is described in some detail in Chapter 11.

For Polyesters. There is a particularly good range of dispersed dyes known as the Polycrons (L. B. Holliday) which are mainly used on polyesters but have secondary uses on triacetate and acetate. Dyed on to polyesters they have excellent fastness to light and washing. They are applied at any desired stage of textile processing by either High-temperature or Carrier methods.

For high-temperature dyeing the dye-bath is set with 1 gm./l. levelling agent and with the well-dispersed dye, at 50° C. and *p*H 6. The polyester material is entered and the temperature raised to 130° C. over 40 min., this being done, as it must be, in a pressure machine. The temperature is kept at 130° C. for 90 min. for dark shades and for only 30 min. for pale shades. The liquor is then run off and the polyester material is well rinsed and soaped. If it has been dyed in a heavy shade it is given a reduction clear treatment with 1–2 gm./l. hydrosulphite, and 2–3 gm./l. caustic soda (70° Tw.) for 20 min. at 70° C.

Instead of high temperature, a carrier may be used. In that case the dye-bath is set with a carrier of the *o*-phenylphenol type and 1 gm./l. of a levelling agent, as well as the dye. As regards the quantity of dye to be used on a bulked polyester fabric, 0·2 per cent of the dye Polycron Dark Blue T (on weight of fabric) gives a light saxe shade; 1·0 per cent gives a good cornflower blue and 6 per cent gives a navy. In carrier dyeing the polyester material is entered at 40° C. and the temperature raised over 50 min. to 100° C., and this temperature is maintained for 1 hr. (The carrier makes what otherwise requires 130° C. possible at 100° C.) The goods are rinsed and then, in order to remove any surplus carrier, they are heated to 190° C. for 1 min. Alternatively, they may be given an alkaline scour at 70–80° C. with a sulphated fatty alcohol which removes most of the carrier. Residual traces of carrier can cause the fabric to develop an odour. All in all, high temperature is preferable to carrier dyeing. The Polycrons can, however, be used with either method to dye polyester fibre or fabric, and dye it well.

Basic Dyestuffs. These dyestuffs give brilliant colours and are useful for ribbons, party frocks, *etc.*, but they are characterised by a poor fastness to light and to washing. Perkin's Mauveine—first of the coal-tar dyestuffs—belonged to this class. They combine with fibres carrying acid groups—*viz.* regenerated proteins, wool and silk. They will not dye cellulosic fibres unless mordanted. The mordanting process is described in Chapter 9, on Viscose. They are not used to any considerable extent, although some are unsurpassed for brilliance. The Rhodamines will, for example, give on wool a pink brighter than is obtainable with other colours. Their extreme brightness of hue is probably associated with their fluorescent properties. They absorb light energy and fluoresce, due to intramolecular vibration, but the molecules of dyestuff are themselves broken down. The reader will understand why brilliance of hue and fugitiveness to light often go hand-in-hand. Their use on various rayons is described in the appropriate chapters.

Basic Dyes on Acrylic Fibres. With the arrival of the acrylic fibres, basic (cationic) dyes have come into their own again. They do very well indeed on these fibres. It is clear that those particular dyes which sell best are the most likely to give the best results. The Panacryl range of modified basic dyestuffs is made by L. B. Holliday and Co. Ltd., of Huddersfield, and is finding a very large commercial uptake for dyeing acrylic fibres. These dyestuffs constitute a range of nice bright colours with very good fastness to light and excellent fastness to washing. A typical dyeing procedure to give a full shade of black on Courtelle, Acrilan 16 or Orlon is as follows:

The acrylic goods are scoured with a dilute solution of a non-ionic detergent and then entered into a bath at 55° C. which has been made up with:

Glacial acetic acid	1·5 per cent
Anhyd. sodium acetate	3·0 ,, ,,
Anhyd. sodium sulphate	5·0 ,, ,,
Levelling agent	2 gm./l.

The percentages are on the weight of the material being dyed.

After running the goods for a few minutes the previously dissolved dyestuff, say 3 per cent of Panacryl Black E, is added through a filter to the bath. The temperature of the bath is raised quickly to 75° C., then during 15 min. to 85° C. and in the next 15 min. to 100° C. Boiling is continued for 2½–3 hr. If the dye-bath is exhausted, *i.e.*, if all the dye has been taken up by the fibre, a clearing treatment should not be necessary; otherwise a clearing treatment to remove any surface colour from the fibre is desirable. The recommended clearing treatment is with 2 gm./l. Hydrosol for 20 min. at 70° C. Finally, a softening finish can be applied. Fastness is good, all-round 4, 4–5 or 5 points. Acrylic fibres have only a limited number of dyeing sites available for basic dyes; the " saturation value ", *i.e.*, the percentage of colour which will exhaust on to the fibre in the absence of other dyes and any retarding agent is as follows:

Panacryl Black E on Courtelle	4·5 per cent
Panacryl Black E on Acrilan 16	3·0 ,, ,,
Panacryl Black E on Orlon	4·0 ,, ,,
Panacryl Blue 5G on Courtelle (C.I. Basic Blue 3)	8·5 ,, ,,
Panacryl Blue 5G on Acrilan 16	6·0 ,, ,,
Panacryl Blue 5G on Orlon	7·5 ,, ,,

The percentages are on the weight of the acrylic fibre. The 2½–3 hr. boil referred to in the recipe can be reduced to 1 hr. for pale shades.

It is noteworthy that Acrilan 16 appears to have fewer dyeing sites available than either Courtelle or Orlon. Sometimes, *e.g.*, in the case of Panacryl Brilliant Red 4G (C.I. Basic Red 14), 1 per cent of colour on Courtelle gives as deep a shade as does 2 per cent on Acrilan 16. All three of these fibres—Courtelle, Acrilan 16 and Orlon—are believed to be at least 95 per cent polyacrylonitrile. A great deal of information on this and similar practical considerations is available from the dyemaker.

Sulphur Dyestuffs. These are made by melting organic materials with sulphur and alkaline sulphides. As they have to be applied from alkaline baths, they are unsuitable for use with cellulose acetate, regenerated protein fibres, alginate or other fibres that are sensitive to alkali. They are, however, used to a considerable extent on viscose, a use restricted to viscose slubbing, owing to their unevenness. As a rule they give fast, but not level, colours. Because they are usually insoluble in water they are applied from solutions of sodium sulphide; this substance not only dissolves them, but also reduces them to a more soluble form, which is subsequently oxidised to the insoluble form when the dyed goods are exposed to the atmosphere. These colours have also been used for nylon. Probably their greatest use is for blacks. Dinitro-phenol black is cheap, fast to light, washing and cross-dyeing. This last feature is of importance in the case of rayon. It has been found that sulphur colours often cause tendering of yarn and fabric, a defect which develops some time after dyeing, but as it has also been found that a very good wash-off greatly reduces this, the defect is probably not due to the dyestuff itself.

Temperature of dyeing with sulphur colours on viscose varies from 50° to 90° C., and dyeing takes about forty-five minutes. The lower temperature is necessary in some cases, such as greens, for bright results. In practice, the sulphur dyestuff is dissolved in a solution of sodium sulphide, so that for every pound of dyestuff 2 lb. of sodium sulphide crystals are used. Blacks require about 10 per cent dyestuff and no salt. Colours require less dyestuff, depending of course on the depth of shade, and usually about 5–15 per cent of salt, or of Glauber's salts. A wetting agent is usually added to the dye-bath.

Azoic Dyestuffs. The method of dyeing with these colours is to impregnate the fibre, yarn or fabric with a naphthol, and then to treat the impregnated material with a solution of an amine that has been diazotised with nitrous acid. The naphthol couples with the amine, and an insoluble azo colour is formed on the fibre. As any naphthol or phenol can be coupled with any base, a great number of combinations are possible and a wide variety of shades of good fastness can be obtained. After the azo-colour has been formed on the fibre it is essential to give a good soaping off to remove surface colour, which otherwise would give rise to rubbing and poor fastness to washing. The application on viscose of the Brenthol colours, which are azoics, is described in Chapter 9 on page 196.

The chemical reactions involved in the application of azoic colours are as follows:

The amine is diazotised with nitrous acid,

$$RNH_2 + HNO_2 \longrightarrow R{-}N{=}NOH \longrightarrow R{\cdot}N{=}NCl$$

The diazonium chloride can then be coupled with either

(1) a phenol, in which case the reaction is

$$R{\cdot}N{=}NCl + C_6H_5{-}OH \longrightarrow R{\cdot}N{=}N{-}C_6H_4{-}OH + HCl$$

$$\text{and } R{\cdot}N{=}NO{-}C_6H_5 + HCl$$

or (2) an amine, when the reaction is

$$RN{=}NCl + C_6H_5{-}NH_2 \longrightarrow R{\cdot}N{=}N{\cdot}NH{-}C_6H_5 + HCl$$

$$\text{and } RN{=}N{-}C_6H_4{-}NH_2 + HCl$$

Modacrylics. A specific example of a dyeing of a modacrylic fibre with an enolic azo coupling component and a diazotisable amine has been taken from B.P. 940,372. The dye-bath was made up of:

Brentamine Fast Red B Base (C.I. Azoic Diazo Component 5)	2 parts
Brenthol FR (*o*-anisidide of β-oxynaphthoic acid)	2 parts
Caustic soda	0·66 parts
Dyeing assistant (Lissapol C)	4 parts
Water	4,000 parts

This bath was adjusted to *p*H 6–7 (the caustic soda is used to get the Brenthol into solution to begin with). Fabric equal to 100 parts was introduced into the dye-bath at 65° C. in a closed vessel, and the temperature was raised to 110° C. in 16 min. and maintained there for 80 min. After cooling to 60° C. in the course of an hour the fabric was removed, rinsed, and diazotised (*i.e.*, the Brentamine base on it was diazotised) for 30 min. at 65° C. in a bath made up of:

Sodium nitrite	16 parts
Acetic acid (80%)	40 parts
Water	5,000 parts

The fabric was treated for 15 min. in a boiling 2 gm./l. solution of Lissapol C and dried. It had a red shade, which was much fuller than when the dyeing was done at 98° C. The high temperature gives depth of shade.

Vat Dyestuffs. Of all dyestuffs that have yet been made, the vat colours are the fastest. Their washing fastness is usually rated as 5 (out of a maximum of 5) and their light fastness varies from 6 to 8 (out of a maximum of 8). Only the reactive colours approach them for fastness. For window hangings, which are subject to severe exposure to light, and for furnishings and upholstery which have to last for a very long time, they are essential for satisfactory results, and are widely used.

Vat colours are quite insoluble in water and in order to apply them to textiles they are reduced with sodium hydrosulphite (hydros), $Na_2S_2O_4$, to their leuco compounds and dissolved in quite strongly alkaline solutions. The fibre is impregnated in these reduced solutions and then acidified and either treated with an oxidising agent or exposed to the air. In either case the reduced dyestuff, now in the fibre, is oxidised back to its insoluble form. Then if a good soaping is given to remove surface colour the dyeing will be very fast. Very often the reduced compound—the leuco compound—is colourless or only light in colour. A fabric which comes out of the dye-bath a nondescript beige colour may well, on oxidation, become red or blue. The colour of the leuco-base bears no relation to that of the oxidised dyestuff. Provided that a good soaping off to remove surface colour is given, very fast dyeings are obtained. Not infrequently, this final wash is scamped, with the result that complaints ensue about the lack of fastness.

Viscose is dyed in considerable quantity with vats both in yarn form and in fabric. Yarn is often dyed in the open beck or in baths when it is turned mechanically. Staple fibre viscose is dyed as bundles in circulating machines. The use of vat dyestuffs on viscose rayon is described on pp. 127–203.

There are different kinds of vat dyes; the best are the anthraquinonoid type, which are dyed from a strongly alkaline vat. Others of the indigoid type are dyed from weaker solutions of alkali, but have not the same superlative fastness. One of the fastest and most beautiful of all vat colours is Caledon Jade Green.

Acid Reduction. For some purposes, it is preferred to reduce vats in an acid instead of an alkaline bath and when this is so a reducing agent different from sodium hydrosulphite has to be used. One that has found wide application is thiourea dioxide—" Manofast " of Hardman and Holden Ltd.—which is produced by oxidising thiourea with hydrogen peroxide. It seems a little odd that such a highly oxidised substance should act as a reducing agent, and it may be that it does so because it undergoes a molecular rearrangement to a sulphinic acid which absorbs oxygen to change to a sulphonic acid.

$$SC(NH_2)_2 + H_2O_2 \longrightarrow (O)_2SC(NH_2)_2 \longrightarrow$$

Thiourea. Thiourea dioxide.

$$HO{-}S(=O){-}C(=NH)NH_2 \xrightarrow{O_2} HO{-}S(=O)_2{-}C(=NH)NH_2$$

(Formamidin) sulphinic acid. Sulphonic acid.

The use of acid leuco vats is particularly suitable for vat dyeing cellulose acetate which will not stand up to the strongly alkaline solutions used in the alkaline leuco processes. In particular " Manofast " as a reducing agent is very suitable for printing cellulose triacetate with vat colours. Moreover, Manofast is an excellent reducing agent for obtaining clear whites in discharge styles on cellulose (di)acetate dyed with vat colours.

Vat colours have not a great affinity for nylon, but it is reasonable to expect that this difficulty will be overcome. Curiously, many vats of light fastness 7–8 on viscose, have only a fastness of 3–4 on nylon. The only reason one uses vats is to obtain the superlative fastness to light and washing, and if vats on nylon give only indifferent fastness to light, their use on that fibre will not be a very attractive proposition.

Vat colours printed by the leuco acid method give sharper and less mottled prints on polyamide fibres than when the alkaline method is used; thiourea dioxide is a suitable acidic reducing agent. Nor does Terylene demand vat dyestuffs for the reason that disperse dyestuffs when once they have penetrated Terylene give good fastness, and vats will be warranted only in very special cases.

Mechanism of Alkaline Reduction. The group typical of vat dyestuffs is $>C{=}O$. This is converted by the sodium hydrosulphite to the leuco compound $>C{-}OH$, and by the alkali that is used to the soluble sodium leuco compound $\gg$C—ONa. The fibre, yarn or fabric is dyed with this soluble compound (because it is soluble, it penetrates the fibres), and then after the impregnation the fabric is acidified, which converts the $\gg$C—ONa group back to $\gg$C—OH, and then oxidised to give the original dyestuff $>$C=O which is now inside the fibre.

In the solubilised vats (Indigosols), which are more suitable for application to alkali-sensitive fibres, the $>$C=O group has been converted to $\gg$C—OSO_3Na; the sulphated group gives greatly increased solubility; these colours are ordinarily applied from an acid bath.

Liquid Dyes

Dyestuffs are usually supplied in powder form, and to make a uniform solution or suspension from such a powder, which is often difficult to wet-out, is no easy task. Usually the powder has to be made into a paste first with a small quantity of hot water and a surfactant, and then the paste can be more easily dissolved in a large volume of water. It is much the same method as the housewife used to, and sometimes still does, employ to dissolve starch. Other times, the dyestuff makers help the dyer by supplying their dyes in paste form, and these are usually easier to get into solution in the dye-bath or the printing paste. They have always been largely used for printing.

With the coming of more efficient surface-active materials and mills that grind finer than ever before, it has become possible for the dyemaker to supply some of his dyes in liquid form, and these are designed to eliminate any trouble that the dyer might otherwise find in preparing his dye-bath. They are excellent and they are popular, but it should not be forgotten that it is easier to weigh accurately a solid powder than it is to measure a volume (or a weight) of liquid. Furthermore, the liquid dyes are a little more fussy about the kind of water that must be used to make up the dye-bath. The dyemaker has selected his conditions for the preparation of the liquid dyes very carefully, and if these are departed from too far, then the liquid may become unstable and lose its homogeneity. Three examples will illustrate what is available.

B.A.S.F. (Badische) make Indanthren " Colloisal " Liquid dyes which are free-flowing and which avoid the need for pasting. They have only to be diluted with water and the dyestuff dispersion is ready for use. The Indanthrens have very good fastness and excellent stability in the liquid form. Badische say that these liquid dyes have a relatively low content of dispersing agent, and are therefore particularly suitable for continuous dyeing processes. Truly, the dyemakers do all that they can to help the dyer, but it still remains that it is the quality of the head dyer which determines whether good or bad dyeings will come out of the dye-house.

The same B.A.S.F. organisation also make Vialon Fast Dyes in liquid form. These are 1 : 2 metal complex dyes and are not dispersions but are true solutions; they contain dye solvents. When they are poured into warm soft water the dye comes out of solution, but in the form of a very fine dispersion which will re-dissolve when the water is heated. These liquid dyes are more susceptible to hard water and salts in the bath than powder dyes, so they should always be applied in soft water, and salt should not be added until most of the dye has gone on to the fibre. They find many fields of application, notably for nylon fabrics on beam dyeing machines and for nylon loose fibre stock intended for carpets. They can also be used in the one-bath, two-stage dyeing of polyamide/cellulosic fibre blends, the Vialon Fast dye dyeing the nylon, and direct dyes, which must be selected, dyeing the cellulose.

Also, the Basacryl range of dyes, again B.A.S.F., which are modified colours intended for dyeing and printing acrylic fibres, are in the majority of individual dyes available in liquid as well as in powder form. If the dyestuff strength in the liquid form is equal to that of the powder brand it obviates many of the difficulties of handling, such as dusting, weighing out, mixing and dissolving. Liquid dyes are a modern development, and they are particularly suitable for large dyeing units as well as for continuous dyeing systems.

Liquid Vats. Vat dyes in liquid form are now marketed by the following makers:

B.A.S.F. and Bayer	Colloisal liquid
Cassella	Stabilisol liquid
Ciba	Micro-dispersed liquid
Geigy	Liquid " M " dispersed
Sandoz	Ultra-dispersed liquid

Additional to those advantages already noted, it may be added that, on being stored, these liquid dyes do not tend to dry out as pastes do.

Liquid Disperse Dyes. Powder disperse dyes are largely used for dyeing polyesters. They are insoluble in water and require pasting up in lukewarm water to which a dispersing agent has been added, although nowadays often this has been included by the dyemaker in the powder. Disperse dyes in liquid form are a modern development; they will often give deeper shades than powders. Disperse dyes in liquid form suitable for polyester fibres are commercially available as follows:

B.A.S.F.	Palanil dyes liquid
Ciba	Terasil dyes micro-dispersed liquid 50%
Francolor	Esterophyl dyes liquid paste
Geigy	Setacyl dyes liquid
I.C.I.	Dispersol and Duranol dyes liquid
Sandoz	Foron dyes liquid
Yorkshire D.C.	Serisol dyes Fluisol 50.

The old-style colour shop in the dye-house is being superseded by systems in which liquid dyes are piped in, to be metered. Lastly, liquid dyes leave their empty containers clean, while pasted dyes do not.

Reactive Dyestuffs. There are really only three fundamental methods of colouring fibres and fabrics.

1. Inclusion of colour in the fibre or fabric in such a way that it is mechanically retained. The outstanding example is the inclusion of fine particles of insoluble pigment in the dope or solution before the fibre is spun, so giving spun-dyed viscose rayon, or Acrilan, or whatever other fibre is being spun. The vat dyes afford another example because the insoluble dyestuff is formed in the fibre; the same applies to azoic dyestuffs. Some high temperature pressure dyeings of Terylene (Figs. 149, 150) do much the same.

2. Sorption of the dyestuff on the surface of the fibre. When direct dyestuffs are applied to cotton and to viscose rayon the process is one of physical adsorption and the dyestuff is encouraged to leave the liquor and go on to the fibre by having plenty of salt in the liquor. Acid dyestuffs on wool behave similarly but there is, additionally to the physical adsorption, extra retentive power given by the formation of salt-links between the acid dyestuff and the basic groups on the wool keratin. There is, however, no covalent union between dye and fibre.

3. Reaction of the dyestuff with the fibre. Traditional processes provide only one example; when metal-complex dyes are used on protein fibres, the metal atom chelates both fibre and dye. The ideal of finding some dyestuffs that would react with cotton and viscose rayon so becoming part of the fibre is one that chemists have had in mind for long enough, but only recently has it been developed practically.

Historical. Farbwerke Hoechst and Ciba both possessed before 1950 dyestuffs that reacted with protein fibres and which would under alkaline conditions react with cellulose, but it is not clear that their potentiality of reacting with cellulose was appreciated. Both firms may have had something as good as the Procions in their hands for ten years without realising it. It can be a mistake to rely too much on logic in the application of science. We must be thankful that even to-day the unexpected often happens. It must have been a slightly rueful Dr. F. Osterlow who (*Melliand Textilberichte* **41**, 1533, 1960) told the VTCC-Hauptversammlung at Baden-Baden on 7 May 1960: " Im Jahre 1952 erschienen mit der Entwicklung der Remalan-Farbstoffe die ersten Reaktions-Farbstoffe, die mit der Faser eine chemische Bindung eingehen und so zu einem Teil des Fasermoleküls werden. Damit beginnt für die gesamte Färberei ein neuer und entscheidender Abschnitt."

There is no doubt that Hoechst's Remalan dyes did react covalently with wool and real silk before, long before, the introduction of the Procions, and the same applies to Ciba's Cibacrons. In fact, Ciba's British Patent 780,591 (applied for 1953, published 1957) related especially to monoazo dyestuffs of the formula:

NH_2
N
—N=N— —NH—C C—A
—OH Y N N
C
SO_3H
Cl

where Y was an alkyl, aralkyl or aryl sulphone group and A an alkyl or similar group bound to the triazine ring through an —NH— group. Here is the triazine ring with its reactive chlorine groups, which subsequently formed the basis of the Procion reactive dyes. In fact, cyanuric chloride is one of the starting materials for Ciba's dyes described in B.P. 780,591, and one of their claims was for " a dyeing or printing process in which a dyestuff claimed in any one of claims 1–4 is used." But in the text of the patent attention was directed to nitrogenous, not to cellulosic, fibres, and there is no mention of making the dye-liquor alkaline to bring about a reaction between dye and fibre. One cannot help feeling that I.C.I.'s contribution to the development of reactive dyes, decisive though it was, was no more than an embellishment of Ciba's earlier discoveries.

Properties of Reactive Dyestuffs. Before discussing in detail individual reactive dyestuffs or indeed individual ranges, it must be pointed out that a reactive is in the main an ordinary dye, one to which there has been added a special chemical group, one which will react covalently with a fibre. One may, for example, have a direct dye, DH, which is ordinarily applied to rayon cellulose and is adsorbed on to the rayon fibre (D stands for the main part of a dye molecule):

DH + Fibre	$\longrightarrow$	DH . . . Fibre
(Direct dye.)		(Direct dyed fibre.)

or, if the fibre is a cellulosic fibre, we can write

DH	+ HO — Cell.	$\longrightarrow$	DH . . . HO — Cell.
(Unreactive dye.)	(Cellulose.)		Dyed (not reactively) cellulose.

where HO — Cell. represents cellulose. There is no chemical reaction and no covalent union; there is, as indicated by the dotted bond, hydrogen bonding and attachment through Van der Waals forces. The dyeing is reasonably fast, and the cotton fabrics that have been dyed have for centuries or thousands of years been quite well dyed, especially so in the last hundred years since synthetic "aniline" dyes (*cf*. Perkin's mauveine, the first of them) came along. The nature of the union between dye and fibre, commonly referred to as adsorption, may be difficult to analyse precisely; it may have been mechanical, physical or chemical sorption, hydrogen bonding and so on, but it was certainly not covalent union.

One point that it is essential to grasp is that the covalent union which reactive dyes offer is not essential to get good dyeings. Very good dyeings were obtainable, *e.g.*, with vat colours, long before reactive dyes appeared.

If, however, the dye DH is made to combine with a substance which is very reactive, but in such a way that it does not absorb all the reactive groups of the new substance, a reactive dye results:

DH	+ ClRCl	$\longrightarrow$	DRCl	+ HCl
(Ordinary dye.)	(Reactive substance.)		(Reactive dye.)	(Hydrogen chloride split off.)

This is by way of example. Other kinds of reactive substances than ClRCl may be used, but this is the commonest. The reactive dye DRCl that is so formed will react with cellulose covalently, with the splitting off (in this case) of hydrogen chloride.

DRCl	+ HO—Cell.	$\longrightarrow$	DR—OCell. + HCl
(Reactive dye.)	(Cellulose.)		(Reactively (covalently) dyed cellulose.)

Now, we might reasonably ask what have we gained? With the old direct dye DH we obtained dyed rayon DH . . . HO—Cell. With the reactive dye DRCl we have obtained reactively dyed rayon DR—OCell. All that we have gained is covalent union (*i.e.*, chemical combination) between dye and fibre. As a result of this, fastness to washing and wet processing generally should be much better, and so it is. In reactively dyed cellulose DR—OCell. the dye is part of the fibre; it is impossible to wash it out or to extract it even with such powerful agents as aqueous pyridine. But towards light there will be no such improvement; the light fastness of the original dye DH will not have been much changed, and the reactively dyed fibre DR—OCell. will not be much better in its resistance to light than was the cellulose DH . . . HO—Cell. that had been dyed with the original direct dye. The wet fastness of reactive dyeings is much better than of ordinary dyeings, but in other respects the reactive dyeings may be no better than those given by the original dye DH from which the reactive dye DRCl was made.

The Starting Dye. A corollary of this is that if good all-round fastness is required in a reactive dye it must have been made from an ordinary dye which is itself fast to light, heat, rubbing and so on. A good ordinary dye when it has been converted will give a good reactive dye; a bad ordinary dye can yield only a bad reactive dye. Perhaps a reservation should be made that if a dye is bad only in respect of wet fastness but is good to light, heat, and rubbing, then it might still be converted into a good reactive dye. But more generally the principle holds that the properties of the reactive dye will be very much like those of the dye from which it was made, plus the very great advantage of a big improvement in wet fastness. If, for example, the starting dye is a copper phthalocyanine derivative which itself is a good fast bright green or blue, then a good bright and still faster green or blue reactive dye will result. Another point about reactive dyes which should not be overlooked is that they are easy to apply. Their two outstanding advantages are: (1) improved wet fastness, and (2) greater ease of application. Sometimes the second advantage is more to be desired than the first.

Procions. Reactive dyes became commercially important when I.C.I. introduced the Procion colours in 1956. Other companies have been quick to follow with Cibacrons (Ciba), Remazols (Farbwerke Hoechst), Reactones (Geigy) and Drimarenes (Sandoz). The Procion colours are built around cyanuric chloride; when one chlorine in this compound has been replaced by an amine group, the compound which is obtained has another chlorine atom still sufficiently mobile to react with a cellulose hydroxyl group, in an alkaline medium at *p*H 10·5–11.

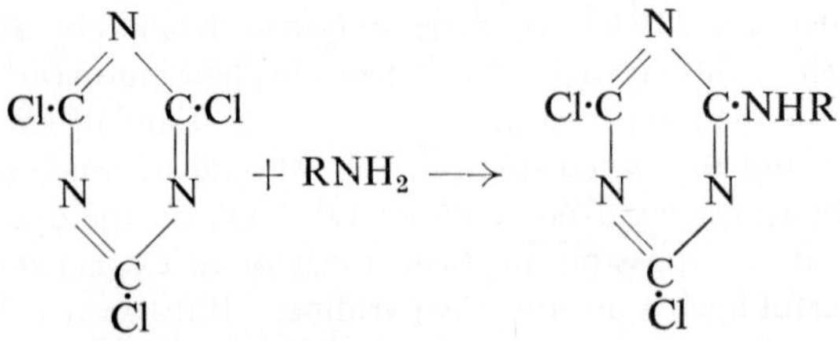

Cyanuric chloride + RNH_2 (Coloured amine) ⟶ Procion dyestuff.

(Cyanuric chloride: a triazine ring with Cl·C at three positions; Procion dyestuff: Cl·C, C·NHR, C·Cl.)

The Procion dyestuffs are readily soluble in water and diffuse readily through fabrics; they do not react with cellulose except in the presence of alkali, but at *p*H 10·5–11 reaction proceeds rapidly.

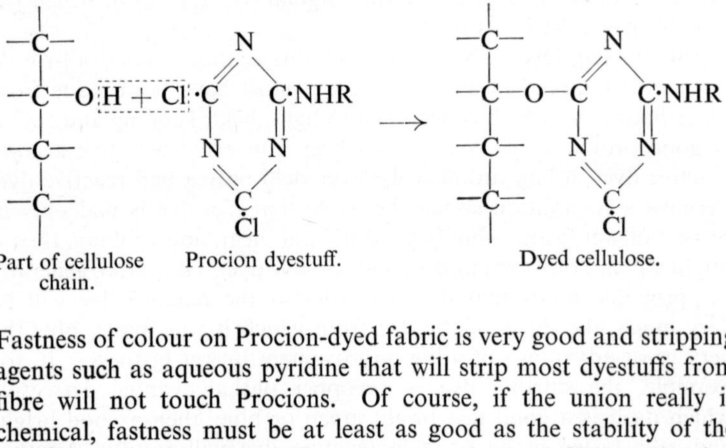

Part of cellulose chain (—C—OH) + Procion dyestuff (Cl·C …) ⟶ Dyed cellulose (—C—O—C …).

Fastness of colour on Procion-dyed fabric is very good and stripping agents such as aqueous pyridine that will strip most dyestuffs from fibre will not touch Procions. Of course, if the union really is chemical, fastness must be at least as good as the stability of the dyestuff or of the ether linkage between dye and fibre molecules. There is plenty of supporting evidence that covalent combination does take place between dye and fibre. The most elegant proof is that if a Procion dye which contains an azo link (*e.g.*, Procion Yellow R) is applied to viscose and then the dyed fibre is reduced, scission occurs at the azo link leaving a diazotisable amine group attached to the fibre, and this can be coupled in the ordinary way with an amine or a phenol to re-dye the fibre. When the Procion dyes were introduced there was a red, a blue and a yellow; since then the number has grown considerably, the method of their preparation is such that a large number of already available dye intermediates can be converted to reactive dyes and tried in the Procion range.

H (for Heat) Procions. The three chlorine atoms in cyanuric chloride are very reactive. When one of them has been reacted, *e.g.*, with an amine to form a Procion dyestuff, the two remaining chlorine atoms are not so reactive as when there were three, but still reactive enough to react in the presence of aqueous alkali with cellulosic hydroxyl groups at room temperature. Such Procion dyestuffs (already described) are called Dichloro and their characteristic is that they will combine with alkaline cellulose at room temperature. If, now, a second chlorine atom in a Procion dyestuff is reacted with phenol or aniline, to block it, then only one reactive chlorine atom of the original three remains; this, the last of the three chlorines, is much less reactive and will not react with alkaline cellulose at room temperature but will do so if heated, *e.g.*, to 70° C. Such Monochloro Procions are referred to as H-type Procions. They have advantages in practice; whereas ordinary Procion dyestuffs have only a very short life in printing pastes, the H-type Procions are much more stable and are very suitable for printing.

Importance is attached to printing paste stability in such newcömers to the range as Procion Brilliant Purple H3R and Procionrubin HBS.

Reaction with Water. The really surprising feature of the Procions is that they combine readily enough with the hydroxyl group on the cellulose chain but only very much less rapidly with the hydroxyl groups in the water medium from which they are applied. Cellulosic hydroxyl groups are not noted for their reactivity; cellulose is a stable compound and some quite severe steps have usually to be taken to make it react, for example to acetylate it. Water is so very reactive with most compounds containing reactive chlorine, that its use has to be avoided. Yet when Procion reactive dyes (containing reactive chlorine) are dissolved in water and offered to cellulose in alkaline conditions, the fibre reacts with at least twice, and probably four or five times as much dyestuff, as does the water. Why? It has been suggested that the dyestuff already adsorbed on the fibre is in a very favourable position to react with it, and does so preferentially to reacting with the water. More information will probably be forthcoming to resolve the seeming improbability.

Outstanding Features of Procions. Perhaps these may fairly be listed as:

1. They combine covalently with cellulose, this leading to excellent wash-fastness.
2. They do not react with water nearly so readily as with cellulosic hydroxyl in alkaline conditions, so that they can be applied from aqueous solution.

3. Reactivity of the dyestuffs can be reduced when desirable by blocking one of the reactive chlorine atoms giving H-type Procions.

4. Procions are dyes with small molecules; their molecules do not have to be very long as those of direct dyes do to match the distance between adsorption sites on the fibre. Short molecules bring two advantages: (*a*) clarity and brightness of hue, and (*b*) easy penetration and therefore good levelling.

5. Because there is some, even although not very much, reaction between Procion dyestuffs and water, it is very important to wash the dyed fibre thoroughly clean and free from the reaction product with water.

6. Procion dyestuffs can be applied to the fibre by various methods (see p. 203).

Procinyl Reactive Dyestuffs. In 1959 the Procinyl range was introduced, a range of reactive dyes for nylon; they are disperse reactive dyes made by reacting a disperse dye with cyanuric chloride, or some similar compound. In the dyeing process, the Procinyl dyes are dyed on to nylon in a neutral or acid, *p*H 4, bath, behaving just like other disperse dyes; then the system is rendered mildly alkaline, *p*H 10; this splits off HCl from the reactive chlorine on the dyestuff and hydrogen on the nylon. The first reaction of the reactive dyes with nylon is with the terminal amino groups; then when these are exhausted, hydrogen is abstracted from the amide groups along the chain. The method of application of Procinyl dyes to nylon is discussed on p. 338. Here, too, mention should be made of Procilan dyes introduced by I.C.I. for dyeing wool. Essentially they are 1 : 2 metal–dye complexes containing a weakly fibre-reactive group. Some of the reactive dyes for nylon do not depend on the possession of a mono- or di-chlorinated 1 : 3 : 5-triazine nucleus but rather on a vinyl sulphone group which owes its reactivity to its unsaturation. There are other groups, too, that are being incorporated in dyestuffs to gain reactivity; it is important to appreciate that reactivity in dyestuffs does not depend exclusively on their descent from cyanuric chloride.

Hoechst's Remalan Dyes. Farbwerke Hoechst had worked on the possibility of combining substances which were dyes or potential dyes from as far back as 1940. Their two chemists who were in charge of this work were Johannes Heyna and Willy Schumacher. Their names are not well known in this country at least, but Hoechst gives great credit to Heyna, who was apparently the leader and who probably made reactive dyes before anyone else did. One cannot be sure; analogous work was going on at Ciba in the 'forties, but to both schools great credit is due.

Hoechst's first reactive dyes were called Remalans; they were marketed by 1952, and had probably been used for development work at least since 1949. They were intended for dyeing wool and real silk in shades that would be fast to washing. They depended, in the eyes of their discoverers, on the possession by wool and real silk of amino groups with which the dyes could react. It is a pity that Heyna and his contemporaries did not realise that cellulosic fibres with their hydroxyl groups could offer a similar opportunity of reaction.

The Remalan dyestuffs were sulphonic acid esters of β-hydroxy-ethyl sulphones:

$$HO_3S—D—SO_2—CH_2—CH_2—OSO_3H$$

where $HO_3S—D$ represents an acid dyestuff residue, the main bulk of an acid dye molecule. This molecule will, in the presence of a basic group, lose sulphuric acid and yield an unsaturated and highly active vinyl sulphone:

$$HO_3S—D—SO_2—CH{=}CH_2$$

If the wool molecule is written $W—NH_2$, just to indicate separately one of its amino groups with which the dye may react, then we can write the reaction between Remalan dye and wool as:

$$2HO_3S—D—SO_2—CH{=}CH_2 + H_2O + W—NH_2 \longrightarrow HO_3S—D—SO_2—CH_2—CH_2—NH—W + HO_3S—D—SO_2—CH_2—CH_2{\cdot}OH$$

These reactions take place in neutral solution; they are wasteful in that only half of the dye combines with the fibre. But in acid solution only the first of the two, the one leading to the formation of the covalent compound $HO_3S—D—SO_2—CH_2—CH_2—NH—W$ of dye and wool, takes place. Two other reactions which may take place in the acid medium lead to electrovalent not covalent union between the Remalan dye and wool. They are

$$CH_2{=}CH—D—SO_3H + H_2NW \longrightarrow CH_2{=}CH—SO_2—D—\overset{-}{SO_3}—\overset{+}{NH_3}—W$$

and

$$HO_3S—CH_2—CH_2—SO_2—D—SO_3H + H_2NW \longrightarrow HO_3S—CH_2—CH_2—D—\overset{-}{SO_3}—\overset{+}{NH_3}—W$$

The Remalan dyes were ordinary acid dyes to which had been

attached a reactive group. This reactive group was a sulphonic acid ester of a β-hydroxyl ethyl sulphone

$$—SO_2—CH_2—CH_2—OSO_3H$$

and in the dye-bath it changed to a vinyl sulphone group:

$$—SO_2—CH=CH_2$$

This vinyl sulphone group and its ability to combine with amino groups was the basis of Hoechst's reactive dyes. The rest of the molecule was just an ordinary acid dye. The Remalan dyes were exclusively so known until after I.C.I. introduced their Procions in 1956. Then it was realised by Hoechst that their Remalans would also dye cellulose reactively in alkaline media, and some of their dyestuffs were marketed under the name Remazols for this purpose. Since then they have also been found useful for dyeing nylon.

But some dyes, for reactive dyeing of wool and silk, are still sold under the name Remalans. Part of the dyeing may be electrovalent, depending on the *p*H value of the dye-bath (F. Osterlow, *Melliand Textilberichte*, **41**, 1533, 1960). Generally, the *p*H value is adjusted to favour the covalent type of union between dye and wool or silk fibre.

To-day there are, too, Remalan Fast Dyes, a development of Farbwerke Hoechst's earliest Remalans. These fast dyes include the characteristic vinyl sulphone group to give reactivity, whilst the dyestuff residue is a 1 : 2 chromium or 1 : 2 cobalt dye complex. They are, in fact, reactive metalliferous dyes.

Remazol Reactive Dyes. The Remazol dyes of Farbwerke Hoechst in the sulphonic ester state lose the elements of sulphuric acid in the presence of trisodium phosphate (a quite strong alkali) to adopt the vinyl sulphone form. This group will react with the terminal amino group of the nylon which we may write $H_2N—E$, with E representing the residue of the nylon fibre molecule; thus:

$$\underset{\text{Remazol.}}{HO_3S—D—SO_2—CH=CH_2} + \underset{\text{Nylon.}}{NH_2E} \longrightarrow$$

$$\underset{\text{Dyed nylon.}}{HO_3S—D—SO_2—CH_2CH_2NHE}$$

More simply the Remazol dyes can be written

$$R—SO_2—CH_2—CH_2—OSO_3Na$$

where R stands for the sulphonic acid dye fragment $HO_3S—D$ and these by the loss of H_2SO_4 change to $R—SO_2{\cdot}CH=CH_2$ which is the vinyl sulphone arrangement. The first is the stable inactivated form of the dye as it is sold; the second is the active form in which it is used. The reaction of the Remazol dye in its active form with cellulose is an addition reaction, whereas the Procions, Cibacrons and Reactones undergo condensation reactions in which the elements of hydrogen chloride are split off (Cl from the dye, H from the cellulose). The reaction of a Remazol dye with cellulose may be written:

$$RSO_2CH{=}CH_2 + HOCell. \longrightarrow RSO_2CH_2CH_2OCell.$$

resulting by simple addition in the formation of a cellulose ether. The union is fast to acid hydrolysis, and Remazol dyeings are probably the fastest of all the reactive dyeings to acid agents.

Whilst Remazol colours are intended primarily for use on cellulose they can also be used on Creslan, Zefran and Acrilan. Whereas the conditions have to be alkaline to dye cellulose, they have to be acid for Creslan, Zefran and Acrilan. A cotton/Acrilan fabric dyed with acid Remazol colours dyes the Acrilan and leaves the cotton white. But if the bath is made alkaline instead, then the cotton is dyed and the Acrilan is left white. Acrilan 16, a special kind of Acrilan, dyes like cotton, so that fabrics which are made from Acrilan and Acrilan 16 will give interesting contrast effects.

It is possible to get solid shades on fabrics containing, for example, Creslan (non-cellulosic) and rayon (cellulosic) with Remazol dyes. An alkaline bath of Remazol dye, salt and trisodium phosphate is used, and this dyes the rayon but not the Creslan; then acid is added, the bath is heated and that part of the dye which was unused in the first part of the process now dyes on to the Creslan. Hence, solid colours are obtained in Creslan/rayon mixtures. Dyeing of Remazol dyes on cellulose is carried out at about 55–60° C., *i.e.*, in water that is just uncomfortably hot.

It is undoubted that the union between Remazol dye and cellulose is covalent, but is it so between dye and an acrylic fibre or between the dye and nylon? The evidence that it is so depends on the better fastness that is obtained when Remazols are used, better than when acid or metallised dyes are used on the nylon or acrylic fibre.

Cibacrons. The Cibacrons are the reactive dyes that Ciba make; they consist essentially of acid dyes to which a reactive group has been affixed. They are a development of the Cibalan Brilliant colours which were made in the early 'fifties for reaction with wool and silk. These dyes contained either a monochlorotriazine nucleus, like the Procions, or alternatively, a chloracetyl radicle. Some of Ciba's early dyes of this kind are described in British Patents 768,241, 775,308, 779,818 and 780,591. The suitability of such dyes for dyeing cellulose was recognised only after I.C.I. had introduced the Procions. The early Cibalans gave fast reactive dyeings on wool, but there was nothing in the patents about making the bath alkaline (which would not have done wool any good) for dyeing cellulose, whether, cotton or rayon.

Cibacrons, like other reactive dyes, are easy to apply, and they give bright colours with extremely good wet fastness. If a dyed fibre or fabric is boiled, *i.e.*, a small piece of it is boiled in a test-tube in a few ml. dimethyl formamide, and none of the colour comes out of the fabric to colour the solvent, it is almost certainly reactive dyed. There are a few vat and sulphur colours which will also withstand this test; they can be distinguished by their behaviour on reduction to the leuco compound and re-oxidation. Most of the fastness properties, other than wet fastness, of a Cibacron dye are derived from the acid dye, which constitutes most of the reactive dye molecule; the other part, the reactive part, usually a chlorotriazine nucleus, determines the reactivity of the dye and the wet fastness of dyed fibre or fabric. Some members of the Cibacron range carry the nucleus of a copper phthalocyanine instead of an acid dye in their molecule. The Cibacron dyes are stable, *i.e.*, they have a long shelf life; they are soluble in water, although their solubility is reduced by Glauber's salt and greatly reduced by common salt. These salts are therefore added to the dye-bath to increase the substantivity (affinity), which is otherwise rather low, of the dye and fibre; the presence of salt forces the dye out of the water phase on to the fibre. Dyeing is continued at about 70° C. for 45 min. and then the bath is made alkaline with soda ash to bring about the reaction between dye and fibre. Application of the Cibacrons is very simple.

An important feature of reactive dyes is that when once the reaction has taken place, *i.e.*, in the case of Cibacrons when the bath has been made alkaline, the dye molecules are firmly anchored to the fibre and they can no longer migrate. Levelling of dyeings of all kinds depends, as a rule, on a continuance of any dyeing process to allow the dye molecules to migrate from one spot of fibre to another, and if this migration is stopped, as it is by the fixation of Cibacron dyes, then no more levelling can take place. It is therefore important to allow plenty of time in the initial stage, before adding alkali, when the dye behaves much as any other acid dye, for migration and levelling to occur. Precise time and temperature conditions of dyeing depend on several factors, including which individual dyestuff is being used, and the makers' recommendations must be followed. The time taken for fixation is quite considerable: 60 min. for most of the dyes and 90 min. for combinations containing Cibacron Turquoise Blue brands, which consist of a copper phthalocyanine (instead of an acid dye) attached to a reactive group.

Cibacron dyes are suitable for viscose and cuprammonium rayon in both filament and staple form. They can be used on loose fibre, tow, hanks, yarns on cheeses or beams, and cakes. They can be used on any of the usual kinds of dyeing machines.

To summarise, the dyeing process is:

1. The Cibacron dye is pasted with cold water, is diluted with hot water and then boiled up with live steam, in order to obtain a dye solution.
2. The dye-bath is made up with soft water, and about 60 gm./l. of common salt or Glauber's salts and from 1 to 3 per cent (for medium shades) of dye on the weight of the fibre.
3. Dyeing is carried out for about 45 min. at 70° C. to level up.
4. The bath is made alkaline with sodium carbonate (15–20 gm./l. soda ash) and the dyeing is fixed in this bath for 1 hr. (90 min. with Turquoise) at 70° C.
5. Any unfixed dye is washed off (this if left on would not be really fast to washing, so that the fastness expected of the reactive dyeing would not be apparent).

Cibacron Catalyst CCB. Ciba is the first to use catalysts in the application of reactive dyes. Cibacron Catalyst CCB quaternises the dyestuff, and the addition compound that is thereby formed between the catalyst and the reactive dye is considerably more reactive than was the reactive dye itself. Accordingly, if the catalyst is used the temperature of both dyeing and fixation can be much lower: 35° C. instead of 70° C. Other advantages claimed for the use of the catalyst are that it eliminates differences in the degree of fixation caused by temperature fluctuation and that it enables the unfixed portion of the dye (which could lead to apparent lack of wet fastness) to be removed more easily.

Fabric. Pad-Steam and Pad-Dry fixation are two preferred methods that are used for dyeing Cibacrons on fabric, but there are many variations. More elaborate and expensive methods of dyeing can be used, but there is no point in using them when the simple and cheap pad methods give such good results. It should be remembered that ease of application is one of the main advantages of reactive dyes.

In the Pad-Steam method, which is very suitable for large yardages of one shade, the fabric is padded in a two-bowl mangle, the dye liquor being at 60–80° C. in the first bowl and a solution of Glauber's or common salt and caustic soda and a surfactant in the second bowl at room temperature. Mangling is done to give a 60–80 per cent retention of liquor on the weight of the fabric. It is very advantageous but not absolutely essential to dry the fabric between the two bowls. Steaming is then done at 100–102° C. for 1 min. The lay-out for the Pad-Steam method is shown diagrammatically in Fig. 263. In the Pad-Dry fixation method the fabric is led through a

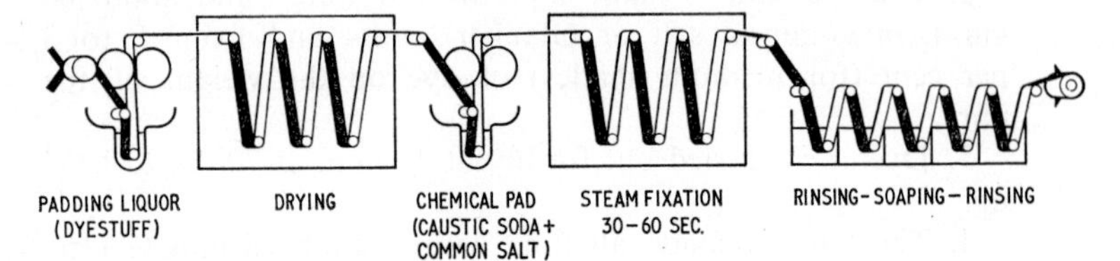

[*Ciba, Ltd.*]

FIG. 263.—The sequence of operations in the Pad-Steam method of dyeing.

pad mangle which contains dye and alkali in the one bowl, then dried and then thermo-fixed by exposure to dry heat at 160° C. for 5 min., and is finally rinsed, soaped and rinsed. The process is shown diagrammatically in Fig. 264. Nothing could be much

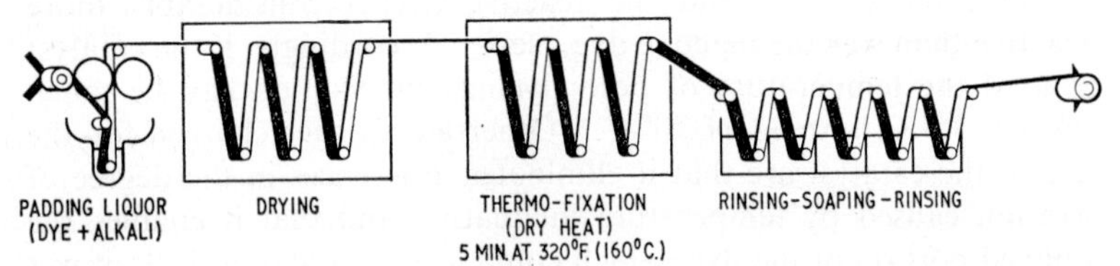

[*Ciba, Ltd.*]

FIG. 264.—The sequence of operations in the Pad-Dry fixation method of dyeing.

simpler than these two methods. They should be compared with the Pad, Dry, Bake and Pad, Dry, H.T. Steam methods for triacetate, polyester and nylon described on p. 743.

Reactone Dyes. Geigy's Reactone dyes are chlorpyrimidine derivatives. The relation between cyanuric chloride and tetrachlorpyrimidine is very close:

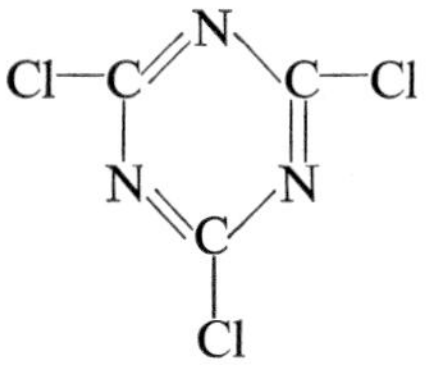

N
Cl—C C—Cl
N N
C
Cl

Cyanuric chloride.

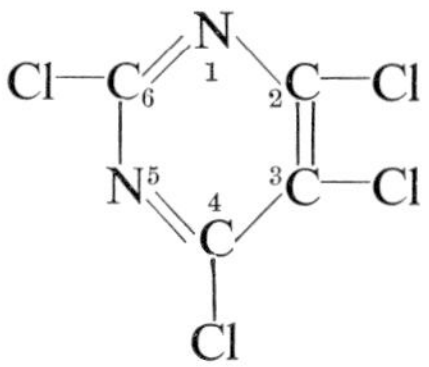

N
Cl—C 6 1 2 C—Cl
N 5 4 3 C—Cl
C
Cl

Tetrachlorpyrimidine.

The conventional numbering system of the pyrimidine nucleus is also indicated. It is the Cl in the 4 position which is the most active because it is reinforced by the Cl atoms in positions 2 and 6 and is opposed only by the *ortho* Cl in position 3, *ortho* opposition being weaker than *para*. If the chlorpyrimidine is reacted with a dye the chlorine in the 4 position splits off and a reactive dye of the Reactone type results:

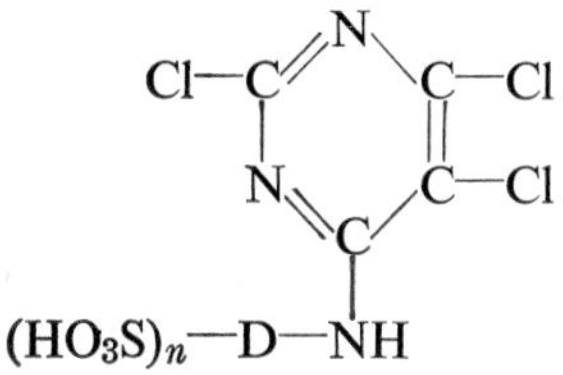

N
Cl—C C—Cl
N C—Cl
C
$(HO_3S)_n$—D—NH

As before, D represents the main part of a dye molecule. Such a dyestuff will react with and thereby dye cellulose through its 2 or 6 chlorine atom thus:

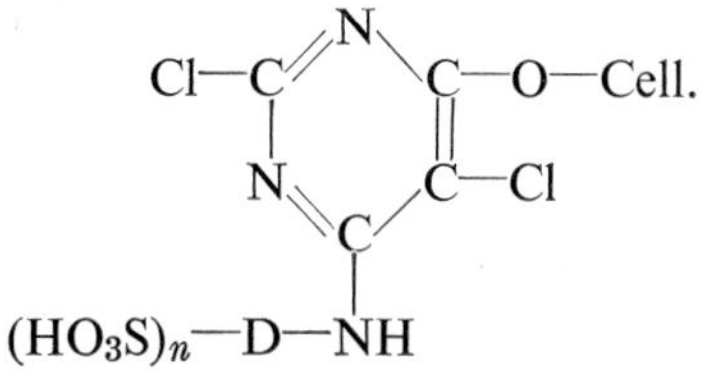

N
Cl—C C—O—Cell.
N C—Cl
C
$(HO_3S)_n$—D—NH

In relation to this formula HO—Cell. as usual represented cellulose. The hydrogen of the hydroxyl group of cellulose has combined with the chlorine atom in position 2 of the pyrimidine nucleus and has split off hydrogen chloride to leave the Reactone-dyed cellulose.

Reactone " S " dyes are a newer modification; they have good exhaustion properties in the presence of an electrolyte and can be dyed by substantive techniques. In the absence of electrolytes their substantivity is lower, so that unfixed dye is easily removed in the wash-off.

Levafix and Levafix E Dyes. The original Levafix reactive dyes were introduced in 1958. Their reactive group was a sulphuric acid semi-ester of β-hydroxyethyl sulphonamide. Thus they had the structure:

$$D-SO_2-NH-CH_2-CH_2-O-SO_3H$$

(As usual D represents the molecule of an ordinary or non-reactive dye.) The reaction of a dye of this composition with cellulose involved first the splitting off of sulphuric acid to yield the ethylene imine derivative

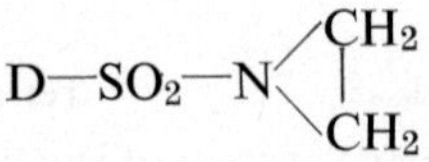

$$D-SO_2-N\begin{cases}CH_2\\ |\\ CH_2\end{cases}$$

and this reacted additively with cellulose, *e.g.*, viscose or cupra-ammonium rayon HO—Cell., to yield:

$$D-SO_2-NH-CH_2-CH_2-O\ Cell.$$

which is in effect the Levafix-reactively dyed cellulose. These Levafix dyes were specifically intended for textile printing on account of their low substantivity and their retarded reactivity, which only became fully active at high temperatures. Their manufacturers, Farbenfabriken Bayer, recognised the need for a range of highly reactive dyes for dyeing by the cold pad bath method, and in 1961 they introduced the Levafix E dyes, which are intensely reactive. Levafix E dyes are much more reactive than Levafix dyes.

Levafix E. These dyestuffs are based, in the sense that the Procions are based on a chlorinated triazine ring, on a chlorinated quinoxaline ring:

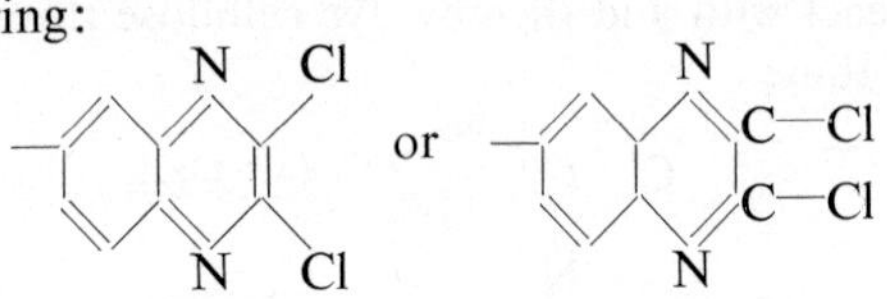

[Structural formulae: 2,3-dichloroquinoxaline ring (N, Cl, N, Cl) or the same ring drawn with N, C—Cl, C—Cl, N]

They are 2,3-dichloroquinoxaline-6-carboxyamides and they have the constitution

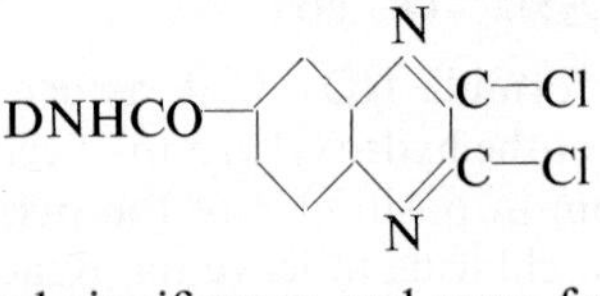

[Structural formula: DNHCO— attached to quinoxaline ring bearing N, C—Cl, C—Cl, N]

where D has its usual significance and may, for example, represent the nucleus of an azo dye. Thus, a typical member of the Levafix E range has the constitution:

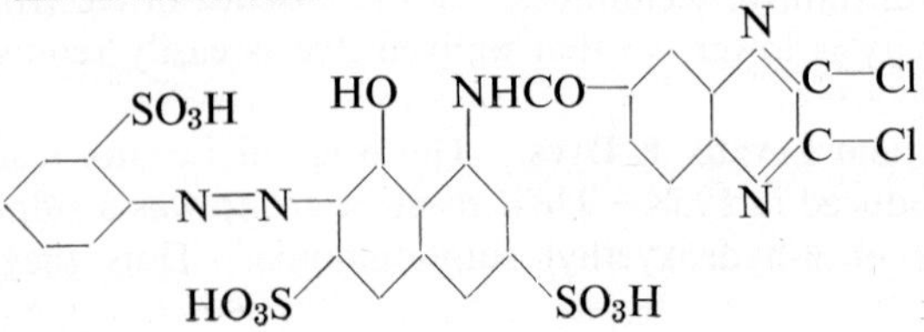

[Structural formula: benzene ring with SO_3H and —N=N— linked to naphthalene ring bearing HO, NHCO—(2,3-dichloroquinoxaline: N, C—Cl, C—Cl, N), HO_3S and SO_3H]

Such dyes are synthesised by the action of 2,3-dichloroquinoxaline-6-carbonyl chloride on a dye which contains a primary or secondary amine group. Thus the dye with the constitution shown just above would be made from these intermediates:

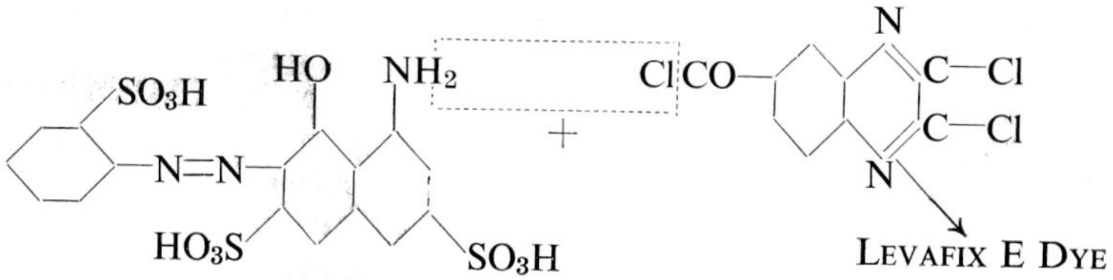

SO_3H HO NH_2 + Cl CO— N C—Cl C—Cl N

—N=N—

HO_3S SO_3H LEVAFIX E DYE

The Levafix E dyes have a very high reactivity and can be applied cold, perhaps most suitably at 35° C. with alkali to bring about the reaction. With cellulose this reaction may be represented:

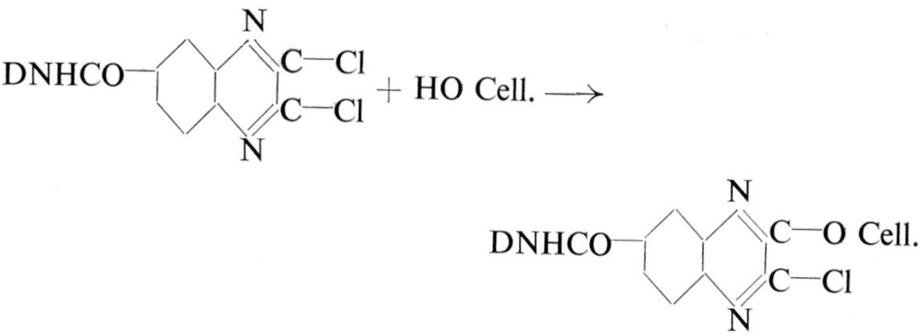

DNHCO— N C—Cl C—Cl N + HO Cell. ⟶

DNHCO— N C—O Cell. C—Cl N

There is inevitably some dyestuff lost, *i.e.*, not made good use of, by reacting with the water of the dye-bath; the constitution of the hydrolysis product so formed is:

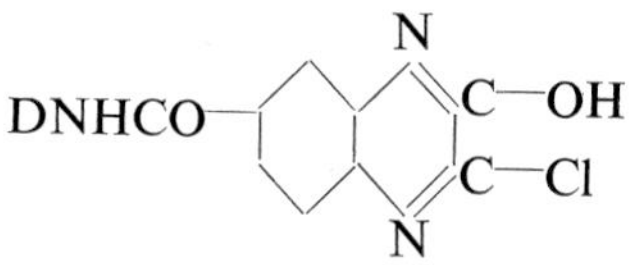

DNHCO— N C—OH C—Cl N

The Procions and Cibacrons depend on the cyanuric chloride or triazine ring, the Reactones depend on the tetrachloropyrimidine ring and the Levafix E dyes on the dichloroquinoxaline ring. There is a very strong family resemblance between them all. Fundamentally, all depend on the proximity of nitrogen ring atoms, suitably oriented, to make a chlorine atom extraordinarily reactive.

Reacna Dyestuffs. The Reacna C dyes are a range of hot-dyeing reactive dyes particularly suitable for dyeing and printing cellulosic rayon. They are nice bright colours, and the Reacna Turquoise C-3GL and the Reacna Scarlet C-GS particularly take the eye. They are made by A.C.N.A. (Aziende Colori Nazionali Affini), who are the dyestuffs division of Montecatini Edison.

Method. The method of dyeing yarn is as follows: the dye is dissolved in hot water and the solution is diluted with cold water until the temperature is about 40° C. The dye-bath is prepared with about 80 gm./l. anhydrous sodium sulphate and 20 gm./l. sodium carbonate and is made up to the desired volume with cold water, and the dye solution which has been previously prepared is added in the desired quantity. The yarn is entered into the dye-bath at 25–30° C.; dyeing starts at this temperature, and the temperature is raised slowly to 85–95° C. (depending on the particular dye) over 45 min. and is maintained at that temperature for 1 hr. The yarn is rinsed in hot, then cold, water and is soaped at the boil for 20–30 min. in the presence of 2 gm./l. surfactant. Dyeing of fabric on the jig is carried out very similarly. Pale shades require 0·5–1 per cent (on the weight of yarn or fabric) of dye, medium shades 1–2 per cent and dark shades 2–4 per cent. The dyeing process as outlined above is delightfully easy, and this is one of the big advantages of reactive dyes.

Constitution. Manufacturers seldom say just what the constitution of their dyes is: for older dyes much can be gleaned from the Colour Index; for new dyes one must look at the patent specifications. It seems that the Reacna dyes are derivatives of cyanuric chloride. In B.P. 1,035,444 A.C.N.A. point out that the general formula for dyes derived from cyanuric chloride is:

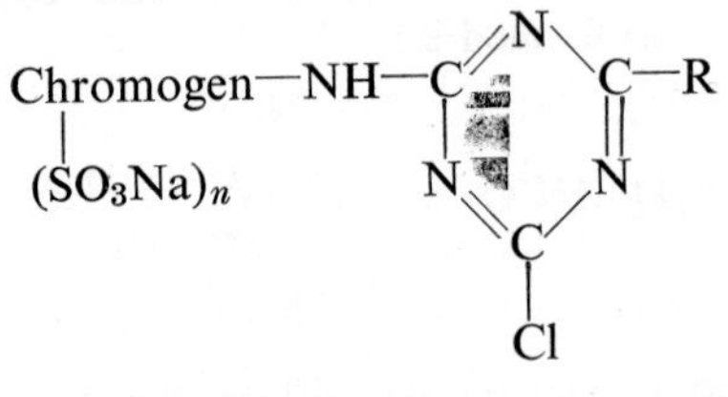

```
                                 N
                               // \
Chromogen—NH—C      C—R
                            |       ||
   |                        N       N
(SO3Na)n                  \\    /
                                 C
                                 |
                                Cl
```

where n is 2 or 3 and R is NH_2 or NHPh or $—NHC_6H_4SO_3Na$ or $—NHC_6H_4COONa$.

Such dyes ordinarily contain solubilising groups, *e.g.*, SO_3H, and these cause part of the dye instead of reacting with the fabric, to undergo hydrolysis, and then dye the fabric in the manner of a direct dyestuff, *i.e.*, not reactively and therefore not superbly fast.

A.C.N.A. have found that very satisfactory results are obtained by making reactive dyes which are derived from cyanuric chloride and which are able to lose a part of their solubilising groups during or after the chemical reaction with the fibre.

Their " main " claim is for dyestuffs having the general formula

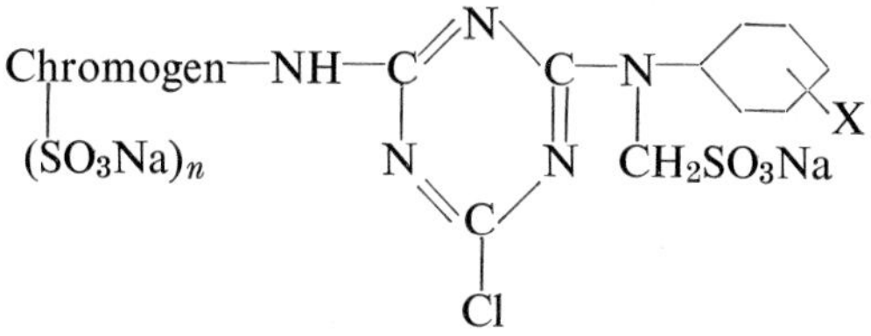

```
                              N
                           //   \
Chromogen—NH—C            C—N—C6H4—X
     |                    |       ||  |
  (SO3Na)n              N       N  CH2SO3Na
                           \\   /
                              C
                              |
                              Cl
```

where n is 2–4 and X is H, CH_3, OCH_3 or OCH_5.

These dyes have only one reactive chlorine atom. As might be expected they have to be dyed hot. The chromogen may be an azo dyestuff, an anthraquinone dye or a metallised azo dye, or even a phthalocyanine derivative. A typical member (anthraquinonoid) is:

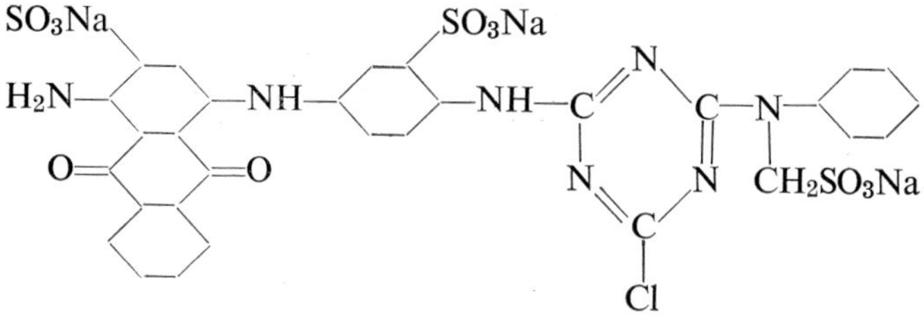

```
SO3Na                      SO3Na
                                       N
H2N—(anthraquinone)—NH—C6H3—NH—C     C—N—C6H5
O=               =O              |     ||  |
                                 N     N  CH2SO3Na
                                    C
                                    |
                                    Cl
```

To apply such a dyestuff, two parts of the dyestuff having the above formula mixed with 2 parts of sodium carbonate and 20 parts of urea, are dissolved in 80 parts of water. A cellulosic fabric is impregnated with the solution and expressed to 75 per cent of its own weight retention of liquor and dried down to 15 per cent liquor. At this stage the fabric has been impregnated, nearly dry, with about 1·5 per cent of its own weight of dye of the constitution shown above. The impregnated fabric is then heated (dry) for 5 min. at 140–160° C., then rinsed in cold water, then in hot water and given a light saponification for 15–20 min. in a 3 per cent solution of sodium carbonate containing a little of a surfactant. The dry heating fixes the dye, *i.e.*, causes it to react with the cellulose; then the light saponification removes some of the sulphonate solubilising groups that remain on the dye and thereby make it faster. The cellulosic fabric is thereby dyed in a brilliant blue shade with good fastness. The characteristic features of these dyes seem to be:

1. The inclusion in the structure of a triazine nucleus with only *one* reactive chlorine atom; in this respect they are similar to the Procion H type and they have to be applied hot. It requires two reactive chlorine atoms in the triazine nucleus to give the high degree of reactivity necessary for dyeing in the cold.

2. The dyes, either before or perhaps preferably after application to a cellulosic fibre, are saponified. This reduces the number of solubilising groups in the dye, makes it less soluble and therefore faster.

In a later specification 1,049,084 (1966) the same inventors describe some very similar dyes in which the chlorine atom on the triazine group has been replaced by a group associated with a chloride ion, of the formula:

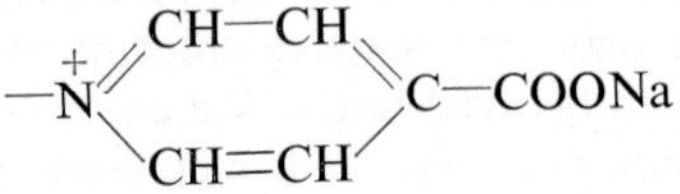

$$-\overset{+}{N}\begin{matrix} \diagup CH-CH \diagdown \\ \diagdown CH=CH \diagup \end{matrix} C-COONa$$

Such quaternised reactive dyestuffs as they are called can be obtained fairly easily from those of the first invention by reaction with a salt of isonicotinic acid. These quaternised reactive dyes (" quaternised " because they contain a quaternary nitrogen atom) have the general formula:

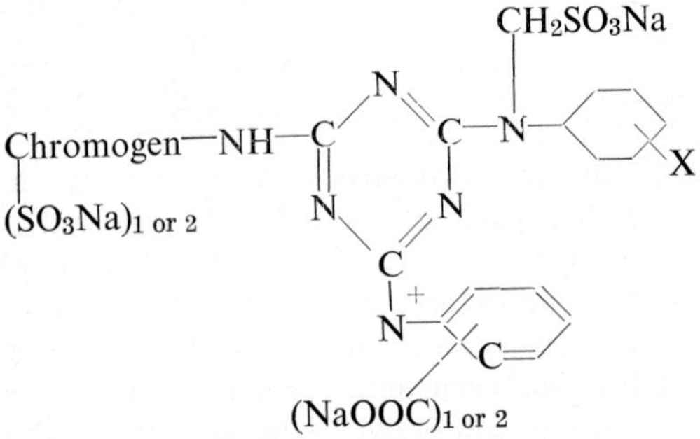

$$\underset{(SO_3Na)_{1\ or\ 2}}{\text{Chromogen}}-NH-\text{[triazine ring: C, N, C, N, C, N]}-\overset{CH_2SO_3Na}{N}-C_6H_4-X; \quad \text{triazine C}-\overset{+}{N}\text{(pyridinium)}-(NaOOC)_{1\ or\ 2}$$

Dyeing of cellulosic materials with these quaternised dyes can be carried out at 60° C. for, say, 2 hr. with defined additions of sodium sulphate (ASS) and sodium carbonate. Bright colours of very good fastness are claimed.

The Reacna dyes are supplied easily soluble so that they make the dyeing process easy. Then after having been dyed on the fibre they are made less soluble by the removal of their solubilising (sulphonate usually) groups. The British agents for A.C.N.A. are Allied Colloids of Bradford.

Primazine Reactive Dyes. The Primazines contain the acrylamide radicle:

$$-NH-CO-CH=CH_2$$

They are made by Badische Anilin und Soda-Fabrik (B.A.S.F.). The vinyl group, which is activated by the proximity of the amide group, will react additively with a fibre. But more interesting are the Basazol reactive dyes also made by B.A.S.F. These are somewhat different from the general run of reactive dyestuffs and will be described in outline at least. It may, however, be observed that B.A.S.F. have described the use of mixtures of Primazines and Basazols; the latter can be used in conjunction with several types of dye.

Basazol Dyes. Basazol dyes are not in themselves reactive, but they are intended for use with a fixing agent (Fixing Agent P) which contains at least two reactive groups which will react with both the dye and a cellulosic fibre in the presence of sodium bicarbonate, during steaming.

In what appears to be a relevant B.P. 923,162, Badische A.S.F. describe the use of a non-coloured compound having two or more $CH_2{=}CH{-}CO$ groups attached by way of nitrogen (compare the characteristic Primazine reactive group) during the dyeing process. The preferred compound seems to be hexahydro-1,3,5-triacryloyl-*s*-triazine (*cf.* M. A. Gaadsten and M. W. Pollock, *J. Amer. Chem. Soc.* **70**, 3079 (1948).)

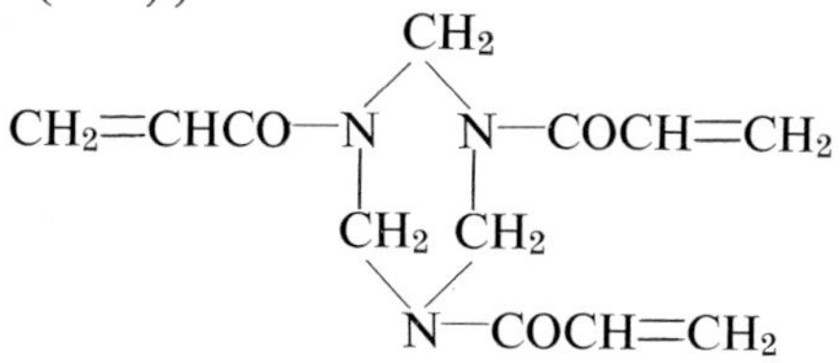

$$\begin{array}{c} CH_2 \\ \diagup \quad \diagdown \\ CH_2{=}CHCO{-}N \qquad N{-}COCH{=}CH_2 \\ | \qquad\qquad | \\ CH_2 \quad CH_2 \\ \diagdown \quad \diagup \\ N{-}COCH{=}CH_2 \end{array}$$

although probably the simplest is methylene *bis*-acryloylamide.

$$CH_2{=}CHCONHCH_2NHCOCH{=}CH_2$$

The general idea is that the Fixing Agent will combine with an ordinary dyestuff (not necessarily a reactive dyestuff) and also with the fibre bridging the two. The fibre may be acrylic, a polyester such as Terylene or Kodel, or may be nylon 6.

Instead of having as a general reaction

$$\text{Fibre} + \text{Reactive Dye} \longrightarrow \underset{\text{(covalent union)}}{\text{Dyed Fibre}}$$

we have

$$\text{Fibre} + \text{Fixing Agent} + \text{Non-reactive Dye} \longrightarrow \underset{\text{(covalent union)}}{\text{Dyed Fibre}}$$

An example of the process of dyeing with a Basazol and a Fixing Agent may make clear what is undoubtedly an interesting extension of the use of reactive dyes, thus:

> A 2 per cent solution (100 parts) of a copper phthalocyanine dye (Basazol) in water is mixed with 1 part of aqueous caustic soda solution of 38° Bé and 8 parts (all by vol.) of a 10 per cent

solution of hexahydro-1,3,5-triacryloyl-*s*-triazine in dimethyl formamide. This mixture is padded on to a rayon staple fabric. The fabric is dried and then steamed at 100° C. It is then rinsed (only very little dye is washed out) and soaped to give a turquoise dyeing of very good wet fastness.

The process is really very simple. The hard work is done by the dye manufacturer leaving an easy job for the dyer. This is the modern trend; the thought and design are put into a process further and further back. Similarly, the fibre manufacturer saves the dyer some hard work, and perhaps deprives him of some of his livelihood, when he spins colour pigmented yarn. Still another development which is of interest in this connection is that fibres can be made reactive, so that they will combine covalently with ordinary unreactive dyestuffs (p. 789).

Reactivity. A rough comparison of the various groups of reactive dyes suggests that reactivity diminishes (most reactive first, least reactive last) in the following order: Procion M (dichloro-*s*-triazinyl), Levafix E (dichloroquinoxaline), Remazol (vinyl sulphone), Procion H and Cibacron (monochlorotriazinyl), Drimarene and Reactone (chloropyrimidine), Primazine (acryloylamino). The Remazols show the best fastness towards acid agents. The Procions and Cibacrons are the fastest to alkaline conditions.

There are other ranges of reactive dyes besides those described above, but the principles involved are much the same in most of them, except perhaps the Basazols. There are, for example, the Drimalans made by Sandoz which rely for their reactivity on the presence of a ω-chloracetyl group: $—COCH_2Cl$.

The General Aniline Company in America make a range known as Genafix dyes.

Dyeing Variables. It must be borne in mind that there are many factors which influence reactive dyeing. They are:

1. Diffusion of the dyestuff, before it is made reactive, into the fibre.
2. The affinity or substantivity of the fibre for the dyestuff.
3. The intensity of reactivity of the dye—as we have just seen, some are more reactive than others.
4. The electrolyte content (usually salt or Glauber's salts) of the dye-bath.
5. The *p*H value of the dye-bath.
6. The temperature of the dye-bath.
7. The dyeing time.
8. The volume ratio of liquor : goods.

9. The concentration of the dye in the bath.
10. The influence (if any) of the material on the dyeing process.

Most of these variables apply to ordinary dyeing as well as to reactive dyeing. In reactive dyeing the degree of reactivity (3) is important in determining many of the other dyeing conditions. For example, if a dye is intensely reactive it may be a cold dyer, if it is less reactive, but still of course reactive, it may be a hot dyer. Such an assembly of variables is formidable, and one might wonder how the dyer ever does make good and reproducible dyeings. In fact, most of the variables, notably (4) electrolyte content, (5) *p*H value, (6) temperature, (7) time, (8) liquor ratio and (9) concentration of dye in the bath, can quite well be kept constant by the dyer, and indeed they do have to be kept so if reproducible dyeing is to be done. In fact, too, these " variables " are a boon to the good dyer who can fix his conditions as he wants; if there were no such variables, if for example all dyeings had to be done at one time and temperature, with one standard make-up of dye-bath and so on, life would be almost impossible for a dyer. The formidable list of variables he uses instead to his own advantage. Such variables as diffusion and affinity can be more troublesome. Dyeing is still an art as well as a science, and the best of dyers has his disappointments and unpleasant surprises. It is a matter for admiration for most of us that so much of the fibre, yarn and fabric that goes into a dye-house comes out nearly perfect. It is a sure sign that the boss dyer is a good man. A good head dyer sees the dyemakers' representatives regularly, and hears them out, for they are technical and experienced men with the vast technical resources of large chemical complexes behind them.

One last point about reactive dyeings. Most other dyeings are reversible; often the dye is held to the fibre by mechanical and physical forces, and can be removed therefrom. If a direct dye is adsorbed by cellulose it can be desorbed under suitably chosen conditions, and the dye stripped from the fibre. But reactive dyeing is irreversible, and it is for that reason that wet fastness is so good. Much of the success of modern dyeing is due to the manufacturers who make the dyes and to the engineers who make the dyeing machines.

Reactive Fibres. We have seen that nearly all of the reactive dyes consist of an ordinary dye, often an azo dye, to which has been attached a chemically reactive group which will react with a fibre. A modification of this procedure has been noticed in the Basazol dyes, in which an ordinary dye, a fibre and a bridging or fixing agent are brought together; the bridging agent combines covalently with both dye and fibre, and thereby unites these two covalently. A third possibility would appear to be to use ordinary dyes and a reactive fibre and still get covalent union between the two. Reactive fibres are not yet commercially available as reactive dyes are, but at the Chemical Institute of the National Academy of Sciences in Esthonia a reactive fibre has already been made, although only on an experimental scale. It is an acrylic fibre with quite satisfactory physical characteristics and with the entirely novel possession in its structure of reactive groups; these are aldehyde groups, the most reactive of all the common chemical groups, and they will react with any dyestuff which carries a free amino group, as a large proportion of azo and anthraquinoid dyes do. Possession by the fibre of the aldehyde group therefore gives it the ability to react and form covalent linkages with a large range of existing ordinary dyestuffs. Such dyeings ought to be as fast as those that are made on ordinary fibres with reactive dyes. The fundamental reaction is:

$$\underset{\text{Reactive fibre.}}{RCHO} + \underset{\text{Amino dye.}}{H_2ND} \longrightarrow \underset{\text{Covalently dyed fibre.}}{RCH{:}ND} + H_2O$$

The preparation of such a reactive fibre was first described by A. R. Kol'k, Z. A. Rogovin and A. A. Konkin in *Khim. Volokna*, 1963, **4**, 12, and this description together with later work was discussed by the author in *International Dyer*, **134**, 787–789 and 903, 905, 907, 909. This *Dyer* review offers in English much more information than is given here, and should be referred to by those interested in the subject. One cannot rule out the possibility that in the future all man-made fibres may have reactive groups built into them, and that reactive dyes, squeezed out by reactive fibres, will find use and that a steadily diminishing use only amongst the natural fibres.

The methods of obtaining covalent union between dye and fibre seem to be:

Reactive dye + Ordinary fibre ⟶ Covalently dyed fibre

Ordinary dye + Ordinary fibre + Reactive bridging compound ⟶ Covalently dyed fibre

Ordinary dye + Reactive fibre ⟶ Covalently dyed fibre.

Regarding the three possibilities from first principles (which are enduring) there is not much to choose between them. Looking at the matter from the standpoint of commercial and industrial expediency, which is more transient, there is probably much to be said for the reactive dye and ordinary fibre. But even at this level there is room for argument, and reactive fibres cannot be overlooked.

The fibre made by Kol'k *et al.* was from the product of co-polymerisation of acrylonitrile as the major constituent and methyl acrolein as the minor constituent (about 13 per cent of the fibre weight). The chemical reaction by which the co-polymer was made could be expressed:

$$\underset{\text{Acrylonitrile.}}{CH_2{:}CHCN} + \underset{\text{Methyl-acrolein.}}{CH_2{:}C(CH_3)CHO} + \underset{\text{Acrylonitrile.}}{CH_2{:}CHCN} \longrightarrow \underset{\text{Reactive polymer.}}{-CH_2CH(CN)-CH_2-C(CHO)(CH_3)-CH_2-CH(CN)-}$$

and it was allowed to proceed until the polymer had a molecular weight of about 35,000. The co-polymer is dissolved to a 15 per cent solution in dimethyl formamide and the solution is extruded through spinnerets into a coagulating bath of 60 dimethyl formamide/40 water. The resulting fibres are then stretched in near-boiling water to about seven times their length and are then relaxed; they have a tenacity of 3 gm./den. and an extensibility of 10 per cent, which values are reasonable for a useful fibre. Two possibilities now present themselves:

1. The fibre can be reacted with a dye intermediate such as H-acid and then with another intermediate such as diazotised benzidine, to complete the formation of the dye on the reactive fibre thus—

$$\underset{\text{Reactive fibre.}}{-CH_2-CH(CN)-CH_2-C(CH_3)(CHO)-CH_2-} + \underset{\text{H-Acid.}}{H_2N,\ OH\text{-naphthalene-}(SO_3H)_2} \longrightarrow$$

$$-CH_2-CH(CN)-CH_2-C(CH_3)(CH{=}N\text{-naphthalene}(OH)(SO_3H)_2)-CH_2-$$

Fibre covalently bound to dye intermediate.

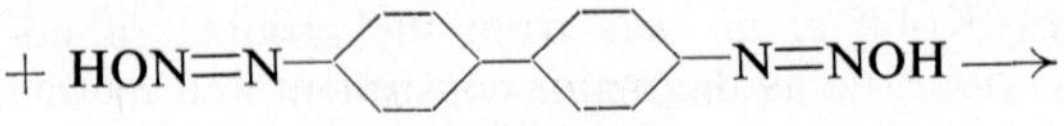

+ HON=N—⬡—⬡—N=NOH ⟶

Diazotised benzidine.

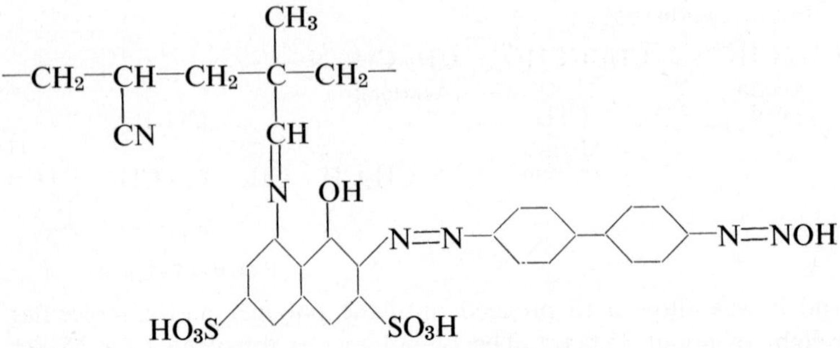

CH_3

—CH_2—CH—CH_2—C—CH_2—

CN CH

N OH

N=N—⬡—⬡—N=NOH

HO_3S SO_3H

Covalently dyed fibre.

At this stage the fibre is light brown in colour. It can, if desired, be reacted with another molecule of H-acid which attaches itself to the free end of the benzidine molecule and yields a reactively (covalently) dyed fibre with the constitution—

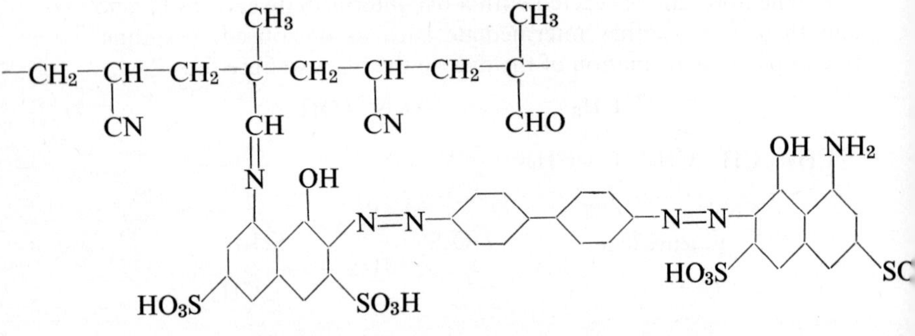

CH_3 CH_3

—CH_2—CH—CH_2—C—CH_2—CH—CH_2—C—

CN CH CN CHO

OH NH_2

N OH

N=N—⬡—⬡—N=N—

HO_3S SC

HO_3S SO_3H

This fibre is black; other colours can be obtained by a suitable choice of intermediates and amines. In this formula, the fibre is the top horizontal part, the dye is the rest.

2. The reactive fibre can be reacted with an already-made dyestuff. This can, for example, be the dye which the Esthonians refer to as Azo Black S but which in the U.K. is known as Disperse Black 1 (C.I. 11365). The dyeing is carried out at 100° C. for 3 hr. with 0·5 per cent acetic acid in the bath. The reaction may be expressed—

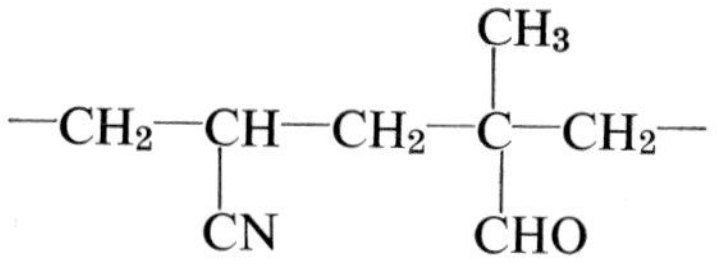

$$\begin{array}{c} \quad\quad\quad\quad\quad\quad CH_3 \\ -CH_2-\underset{CN}{CH}-CH_2-\overset{|}{\underset{CHO}{C}}-CH_2- \end{array}$$

Reactive fibre.

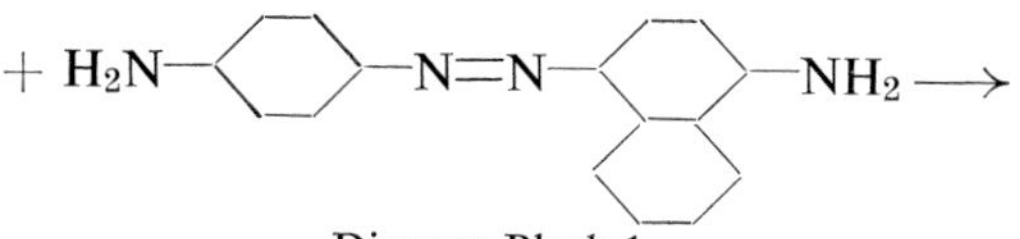

$$+ H_2N-\langle C_6H_4 \rangle-N{=}N-\langle C_{10}H_6 \rangle-NH_2 \longrightarrow$$

Disperse Black 1.

$$\begin{array}{c} \quad\quad\quad\quad\quad\quad CH_3 \\ -CH_2-\underset{CN}{CH}-CH_2-\overset{|}{\underset{\underset{\| }{CH}}{C}}-CH_2- \\ \quad\quad\quad\quad\quad\quad N-\langle C_6H_4 \rangle-N{=}N-\langle C_{10}H_6 \rangle-NH_2 \end{array}$$

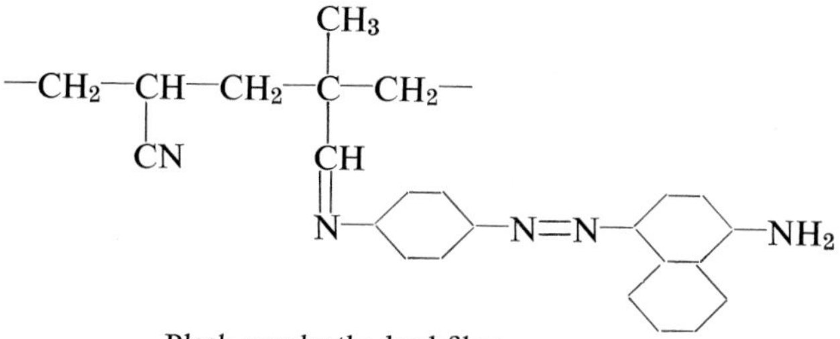

Black covalently dyed fibre.

Not all of the reactive aldehyde groups do react with the dye. In the example given above, about one-quarter of the fibre's aldehyde groups did react with the dye, but most often the proportion is considerably lower. Even although the aldehyde content of the fibre may not be more than 1–2 per cent and although only a small proportion of this reacts with the dye, there is still enough dye combined to give good coloration. If, for example, in the black dyeing described above there are 2 per cent aldehyde groups in the fibre and one-quarter of these react, then the quantity of dye picked up is:

$$\frac{2 \times 0{\cdot}25 \times \text{mol. wt. of dye}}{\text{mol. wt. of fibre repeat}} = \frac{2 \times 0{\cdot}25 \times 262}{53} = 2{\cdot}5\%$$

If a fibre picks up 2·5 per cent of its own weight of dye it should be dyed a good medium-heavy shade, but it is perhaps hardly enough for a good black. In this calculation the fibre repeat has been taken as

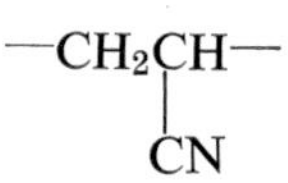

$$-CH_2\underset{\underset{CN}{|}}{CH}-$$

which constitutes about seven-eighths of the fibre; the correction if it were made for the modification (with methyl acrolein) of the fibre would no more than reduce the dye pick-up from 2·5 to 2·4 per cent.

Another possibility is that reactive fibres containing aldehyde groups can be cross-linked with diamines to improve their cohesiveness and strength thus:

$$\underset{\text{Reactive fibre.}}{\text{Fibre—CHO}} + \underset{\text{Hexamethylene diamine.}}{NH_2(CH_2)_6NH_2} + \underset{\text{Reactive fibre.}}{\text{OCH—Fibre}} \longrightarrow$$

$$\underset{\text{Cross-linked fibre.}}{\text{Fibre—CH}{=}N(CH_2)_6N{=}\text{CH—Fibre}}$$

Reactive fibres may be fibres of the future.

Aridye Pigment Padding. This method of pigment padding is one that is particularly suitable for those fibres such as the true synthetics which are very difficult to dye by conventional methods, either because they have very low moisture regains or because there are no chemically reactive groups in their molecules with which the dyestuff molecules can combine. Essentially, Aridye pigment padding consists of passing fabric through an aqueous dispersion of a pigment *and a resin*, mangling the fabric, drying it and then usually baking it. This polymerises and insolubilises the resin which holds the pigment to the fabric.

The Aridye padding colours are used for pigment padding. Typical formulae for dyeing pastel and medium shades are given in the following table:

Component.	Proportion for	
	Pastel shade.	Medium shade.
Aridye Padding Colour	0·5	2·5
Aridye Padding Clear (the resin)	2·5	5·0
Sodium alginate (2 per cent solution)	2·5	2·5
Diammonium phosphate	0·16	0·16
Water to	100	100

The sodium alginate is used to thicken the liquor, the diammonium phosphate to catalyse the polymerisation of the resin. The constituents are mixed, the ammonium phosphate being added last, predissolved in a little of the water. The fabric is padded in the liquor on a two or three bowl padder, and is then dried immediately and gently, usually on steam-heated cans. Finally the fabric is baked to cure the resin and hold the pigment firmly to the fabric; this gives maximum fastness. However, for many purposes, and particularly with light-weight fabrics, sufficient fastness to washing can be obtained without baking.

Aridye padding colours have been successfully applied to viscose, cuprammonium and acetate rayons, nylon, Terylene (Dacron), and Orlon. Because the pigments used are insoluble in water (they are not dissolved, but *dispersed* in water for the dyeing operation), and because the resin simultaneously applied is insolubilised in the baking process, the wet fastness of pigment padded dyeings is excellent. Light fastness is also excellent and the pigments are brilliant in shade. Aridye pigment colours are available in blue, yellow, red, scarlet, green, violet, orange, brown, navy, white and black, and they can be mixed with one another in any proportions to produce compound shades. Uniformity of shade is very good, even on fabrics made from irregular yarn, which if dyed by conventional methods would give very streaky and barry fabric. Furthermore, perfectly solid shades can be dyed on viscose/acetate rayon and viscose rayon/wool mixtures and on other unions.

Of particular interest is the fact that it is possible to add crease-resisting or dimensional stability to the dye-bath, and thus carry out the dyeing and finishing operations in one process. As an example of this, a fabric made from yarns of an 80 viscose rayon/20 wool blend can be simultaneously dyed a grey-green shade and made crease-resistant and dimensionally stable by padding with the following mixture:

Colours	Aridye Padding Green B	7	parts
	Aridye Padding Grey R	8	,,
	Aridye Padding Chartreuse	2	,,
Resin to hold pigment colours to fabric	Padding Clear 9914	33	,,
Thickening agent	2 per cent sodium alginate solution (" Manutex RS " is a suitable grade)	10	,,
Catalyst to polymerise the resin	Diammonium phosphate	0·64	,,
Resin to give crease-resistance and dimensional stability	Urea-formaldehyde resin	40	,,
Diluent	Water	to 400	,,

Aridye Pigment Colours for Printing. Aridye pigment colours have also been used very successfully and perhaps mainly for printing. If, after the application of the colour to the fabric, the fabric is dried and heat cured, the resin polymerises and binds the pigment firmly to the fibres, thus ensuring the maximum fastness to washing. Aridye colours can be used either for roller printing, which is the method always used for large production, or for screen printing, which is used where smaller quantities of any particular design of an intricate pattern, *e.g.*, for head scarves, are wanted.

A recent innovation is a range of metallic colours particularly suitable for screen printers; these give novel tinsel-like effects.

The importance of Aridye pigment colours is that they enable fabric to be dyed level shades, in good depth with excellent fastness, irrespective of the normal dyeing properties of the fabric. Fibres that are impossible to dye by ordinary means can be successfully dyed by Aridye pigment padding; the process is simply one of applying to the fabric an insoluble and bright pigment along with a resin in a low state of polymerisation; then after application the fabric is heated to cure and insolubilise the resin (it is polymerised by the heat) and the now-insoluble resin bonds the pigment firmly to the fabric. The only trouble that may perhaps be encountered is in cases where articles of apparel are hand-laundered; the abrasive action of this type of washing may remove some of the resin and with it some of the colour. The use of pigments for dyeing has been dealt with at some length because this is a method of coloration which is growing in importance. The Aridye colours which have been discussed to illustrate the method are not unique; other dyestuffs which qualify for mention in this context are the Helizarines, Oramines and the Acramins.

Dyeing Adjuncts

Anyone who carries out development work on dyeing soon runs into modifications which give better results. Most of them are limited in their application, but they are sometimes incorporated in the dyeing processes of the factory where the development was conducted. Some of the discoveries have much wider application; they are usually patented, and may be used. Needless to say, most of such improvements are found in use to introduce new deficiencies and troubles, and they do not survive for long. But there are some which survive and become useful. Three examples which may have possibilities for the future will indicate the nature of such developments.

Furfural. The Monsanto Company of St. Louis have described in B.P. 1,108,455 (1968) the use of furfural as a dyeing assistant. The fibre or fabric after having been immersed in a dye-bath, or if it is a carpet perhaps sprayed with a solution of dye, is exposed to the vapours of a boiling aqueous solution of furfural. The presence of the furfural vapour helps the dye to penetrate into the interior of the fibre, so that it becomes more homogeneously dyed and is not coloured just around its periphery. This, too, gives an improvement in fastness to wet processing and to rubbing; dye which is included in the interior of the fibre is very much faster than that which is concentrated on its periphery. It may also, because it increases the affinity of the fibre for the dye, permit the dyeing of polyester fibre to medium shades, without the use of either carriers (at atmospheric pressure), or of high temperatures in pressure dyeing equipment. Carriers are sometimes toxic, they smell, and they are not easy to remove completely from the dyed fibre. High-temperature pressure equipment is costly. The invention, the use of furfural, is applicable to nylon, polyester, polyacrylonitrile, spandex, rayons and even natural fibres; similarly it applies to a wide range of dyes, *e.g.*, basic dyes which can be used on some acrylic fibres and on silk, disperse dyes for acetate, polyamide, polyester and acrylic fibres, and premetallised dyes for acrylic and polyamide fibres.

In one example a modacrylic carpet was sprayed with an aqueous dye solution containing 0·25 gm. Calcozine Acrylic Red 3G (C.I. Basic Red 30) and 0·25 gm. Genacryl Blue 3G (C.I. Basic Yellow 28) per litre, adjusted to *p*H 9·5 with soda ash, and with sufficient of a natural gum to bring the viscosity of the dye solution to 175 centipoises. The carpet after having been sprayed with this solution was treated with the vapour from a boiling 9 per cent solution of furfural in water. In this way the carpet was dyed to a heavy shade, and penetration of the dye through the fibre was nearly perfect. This is a remarkable achievement, because spraying and steaming (without the furfural) would not have been expected to be very successful. An economic point of importance is that furfural

```
HC—CH
||   ||
HC   C·CHO
  \ /
   O
```

costs no more than 15*d.*/lb. (mid-1968) in bulk lots.

Triethanolamine. Work carried out in Russia, at the Chemical Technical Institute at Ivanov, has shown that the addition of from 2·5 to 5 gm. triethanolamine per litre of dye-bath enables the dyeing time of such polyesters as Terylene, Tetoron, Dacron and Lavsan to be shortened. If disperse dyes such as Disperse Fast Orange B and Fast Green G are used for dyeing polyester in a Longclose pressure machine at 120° C. the dyeing time is reduced from 90 to 40–50 min., or at 130° C. from 60 to 30 min. The triethanolamine $(HOCH_2CH_2)_3N$ has the effect of loosening the compact crystal structure of the polyester so that the micropores in the fibre are enlarged and so that the dye can penetrate more easily into the fibre. Dyeing through the interior of the fibre is more homogeneous, and fastness, particularly to rubbing, is improved. This modification of the dyeing process is very simple and one which anybody with the time to make an experiment can try. It is of interest that other workers at Ivanov, but at a different establishment, the Melanzhevyi Combine, have reported advantages from the inclusion of triethanolamine in the bath when dyeing cotton with Sulphur Blue colours. Irrespective of which fibre is being used, the triethanolamine has an opening action on the fibre and makes it easier for the dye solution to penetrate into it.

Water-soluble Polyamide

A process has been described by Precision Processes (Textiles) Ltd., in B.P. 1,101,185 (1968), whereby a special cationic water-soluble polyamide can be applied to many kinds of fibre, *e.g.*, polyester or nylon, and will confer on them the property of being dyeable at low temperatures, *e.g.*, 25° C. The water-soluble polyamide is suitably Kymene 709, the preparation of which was described by Hercules Powder Co. of Wilmington, Delaware in B.P. 865,727 (1961).

If such fibres as polyester or nylon are treated with a 20 per cent solution of Kymene 709, which is brought to *p*H 7 by the addition of soda ash, and are mangled to 80 per cent expression, dried at 80° C. and then entered in a dye-bath containing 2 parts Cuprofix Navy CGRL (C.I. Direct Blue 160) and 1,000 parts water at 25° C., dyeing takes place. It is allowed to continue for 15 min., and the fibre is rinsed and a satisfactory level shade of good wash fastness is obtained. This particular dyeing was done with knitted polyacrylonitrile fibre. Such devices open new avenues of thought to the dyer. Dyeing for so long was a traditional art with well-defined guide lines which it was asking for trouble to cross, that still very many excellent dyers are far too conservative. The influx of man-made fibres and generally the great advances of technical chemistry have really revolutionised the possibilities of dyeing, and even the most traditional of dyers might well take the time to have a look at the new possibilities.

Solvatochromism

This term is used to indicate the colour shift of a dye caused by the solvent. A dye dissolved in chloroform may give a solution of a different colour from the same dye dissolved in benzene. A spectrophotometer is used to measure the wavelength of light that is most absorbed by the solutions. In work on some cyanine dyes that was carried out in Russia, it was found that if the wavelength of maximum absorption of a solution of a dye was plotted against the expression $(n^2 - 1)/(2n^2 + 1)$, where n is the refractive index of the solvent, then the plots fell more or less on a straight line. In Fig. 265 these " straight " lines are shown for the two dyes 3,3-diethyl-

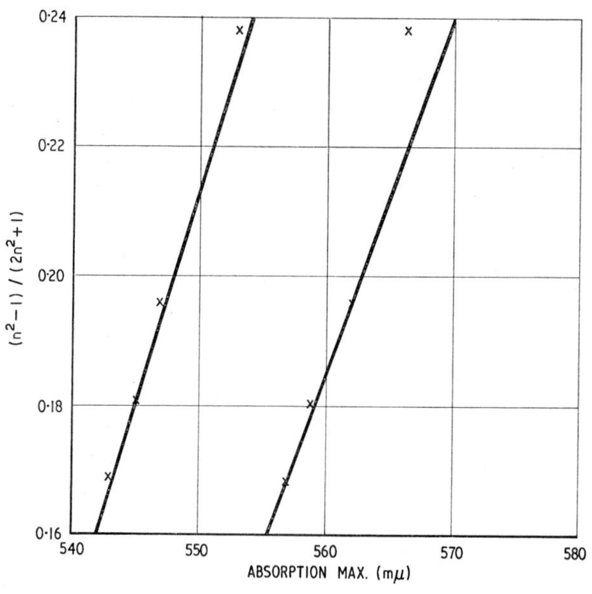

[*after E. B. Lifshits.*]

FIG. 265.—Solvatochromism. The colour of a dye (indicated horizontally) changes with the refractive index (of which a function is shown vertically) of its solvent.

thiacarbocyanine and 3,3-diethyl-9-methyl-thiacarbocyanine. The wavelengths of about 540 mμ (5,400 Å) correspond to the green part of the spectrum, and as the wavelength increases this corresponds to a yellowing of the shade. There is a progressive yellowing as the solvent is changed from methyl alcohol [$n = 1{\cdot}330$, $(n^2 - 1)/(2n^2 + 1) = 0{\cdot}169$] to ethyl alcohol [$n = 1{\cdot}361$, $(n^2 - 1)/(2n^2 \times 1) = 0{\cdot}181$] to butyl alcohol [$n = 1{\cdot}401$, $(n^2 - 1)/(2n^2 + 1) = 0{\cdot}195$] and benzyl alcohol [$n = 1{\cdot}539$, $(n^2 - 1)/(2n^2 + 1) = 0{\cdot}238$]. The values for each of these four solvents for each of the two dyes can be recognised in Fig. 265. The colour change is thought to be due to a reversible chemical interaction between dye and solvent. But not much dyeing is done from organic (instead of water) solutions of dyes. Such a process has long been recognised as having possibilities, but not much progress is made.

White Dyes or Optical Bleaches

It has long been customary to improve the whiteness of garments by blueing with ultramarine or some similar dyestuff. This causes more of the yellow and red components of the light which falls on the garments to be absorbed and so removes yellowness and gives a bluish white. The disadvantage is, of course, that some of the light falling on the garment is removed by the " blue ", so that less light is reflected by the " blued " garment than before. This disadvantage has been avoided in a group of materials marketed by some of the dye manufacturers under such names as Tinopal (Geigy), Uvitex (Ciba) and Calcofluor (American Cyanamid Co.). These substances are apparently colourless, but they have the property of increasing the wavelength or reducing the frequency of some of the ultra-violet light which falls on them as a component of daylight and of some kinds of artificial illumination, and thereby converting it into visible light usually in the violet or blue end of the spectrum. Thus invisible (U.V.) light is changed to blue visible light and as a consequence more visible light may be reflected from a fabric that has been dyed with one of the white dyes than was incident on the fabric. The white dyes are fluorescent, they fluoresce in light containing U.V. They are useful not only on whites but also on pastel shades where they will give a brilliance of pinks and blues hard to obtain otherwise. The use of Uvitex on Terylene has been mentioned (p. 412) and there are three different Calcofluor whites used to whiten or brighten the chemical fibres such as Acrilan, Arnel and Orlon 42.

These white dyes are useful, too, for applications to prints on a white ground which has been tinted by the dyestuffs running; the white dye restores the whiteness of the ground. Most of them are complex organic substances; some of them are derivatives of stilbene; they have a fairly good affinity for their appropriate fibres so that they will withstand washing.

Du Pont's Type 91 nylon 66 contains a fluorescent optical whitener which has been added to the polymer; the yarn is recommended for use, undyed, in foundation garments. Washfastness of the white is good; light fastness is not so good, it is adequate for intimate apparel but not for outerwear. Enka make a " Blanc de Blancs " nylon 6 which also incorporates an optical bleach; its fastness seems to be rather better and this may be due to the superiority of nylon 6 over nylon 66 in ability to withstand ultra-violet light or sunlight. Blanc de Blancs contains 0·45 per cent titanium oxide and is therefore less glossy than Type 91 which contains only 0·35 per cent: but the higher pigmentation has nothing to do with the fastness of the whitening agent.

Unions

Because artificial fibres are so often mixed with natural fibres, and indeed with each other, it is desirable to consider briefly the principles underlying the dyeing of these unions.

Wool and Viscose Staple. This is a popular mixture, especially for knitted wear. First, a good scour must be given to remove the oil from the wool, and it is essential that this scour should be really good if bright shades are to be obtained. If the residual oil exceeds 0·4 per cent on the weight of the wool some flattening of shade in pastel colours is inevitable, although medium and heavy shades will not be seriously affected. Another point to be borne in mind is that if any considerable quantity of oil is left on the wool, then the dyed material will " rub " or mark off. The reason for this is that most dyestuffs are readily oil-soluble, that the dyestuff dissolves in the oil, and when the fibre is rubbed, the dyed oil is removed and marks whatever it is rubbed with. Wool/viscose mixtures can be dyed either by a two-bath or a one-bath method.

Two-Bath Method. In the two-bath method the wool is first dyed with a level-dyeing acid colour or a chrome colour, the fabric is rinsed to remove acid; the viscose component is then dyed in a fresh bath containing a direct colour or colours which do not stain wool. Very many direct colours do stain wool and, as the wool has already been dyed to shade, it is essential to choose carefully a colour for the viscose which will not affect the wool. This restricts considerably the choice of colour available.

One-Bath Method. The one-bath process is most widely used. A direct colour is used for the cellulose and a neutral dyeing acid dyestuff for the wool. Most neutral dyeing acid dyestuffs only slightly stain the cellulose and, correspondingly, many direct colours do not stain wool heavily. Trials have to be made in the laboratory to find suitable proportions of colour to match a shade. Most shades can be matched in good all-round fastness with a yellow, a red, and a blue for the cellulose and the same for the wool. Thus six dyestuffs may be required in a bath. The following have been recommended:

Durazol Fast Yellow	6 RS	to dye the cotton or viscose.
Durazol Fast Red	6 BS	
Durazol Fast Blue	GS	
Coomassie Yellow	RS	to dye the wool or regenerated protein.
Coomassie Red	BGS	
Solway Sky Blue	BS	

For very bright shades other dyestuffs will be required.

Cotton and Viscose. Both dye with direct colours, but, as the two fibres have very different affinities, viscose having the greater affinity, two-tone effects often result. When these are required they can be obtained, but the great bulk of the union trade is for solid colours. (If the viscose is pigmented, there is an apparent lightening of its shade, and solid colours with cotton are more easily obtained.) The only solution to the difficulty is to choose colours carefully, taking the advice of the dyestuff manufacturers, and to make preliminary dyeings in the laboratory. The difference in lustre of fibres such as cotton and viscose is so great that it is often almost impossible to obtain a solid appearance if bright viscose has been used. If mercerised cotton is used, matters are easier to control, because mercerised cotton has a much higher affinity for direct colours than has natural cotton.

Cuprammonium and Cotton. Because of the even higher affinity of cuprammonium for direct dyestuffs—higher that is, than that of viscose—cuprammonium/cotton mixtures should always be made with mercerised cotton, as otherwise the cuprammonium takes up nearly all the dyestuff, and starves the cotton, so that it is almost impossible to secure solidity. Ladies' stockings were made from cuprammonium silk with mercerised cotton tops before nylons came.

Cellulose Acetate and Wool. Some crêpe dress fabrics are made from cellulose acetate and wool, and are extremely attractive. Dispersed acetate colours have to be used, and these stain the wool heavily. Neutral dyeing acid dyestuffs are used for the wool portion. A wetting-out agent should be added to the bath to reduce the tendency for particles of the dispersed colour to aggregate on the wool.

Nylon and Wool. A good solid black (logwood) can be obtained on nylon/wool mixtures by mordanting with 2 per cent bichromate and 3 per cent formic acid (85 per cent), and dyeing with 12 per cent of fully oxidised Hematine Crystals " ZA ". In practice, the scoured nylon/wool material is entered at 60° C. into a bath containing the bichromate and gradually raised to the boil; the formic acid is then added in portions and boiling continued for 1 hour. The material is then thoroughly rinsed to remove uncombined mordant. It is entered at 60° C. into the hæmatine bath, raised to the boil, boiled for $1\frac{1}{2}$ hours and rinsed. The black has a fastness to light of 6 and a fastness to milling of about 4. Severe milling reduces the depth of shade on the nylon more than on the wool, and if fastness to severe milling is important it is better to dye the nylon and wool separately before blending. Separate dyeing is also advisable if the blend contains more than 20 per cent nylon, as with high nylon contents the blacks appear a little greyish; the reason is that the nylon dyes to a thinner shade than the wool.

Orlon Type 42 and Wool. Blends of Orlon and wool have been widely used for knitted fabrics; the Orlon content is usually 40–50 per cent. Probably the main advantage over all-wool is that of dimensional stability, and this is achieved without undue sacrifice of handle because Orlon staple is itself warm and lofty to the touch. Blends have to be treated for dyeing purposes as two-fibre yarns or fabrics. Acid dyes applied by the cuprous ion technique are unsuitable as the wool takes nearly all the dyestuff, disperse dyes are not suitable because the wool is stained badly whilst the Orlon is dyed. The best method is to dye the Orlon with basic colours under conditions such that the wool is only lightly stained, *e.g.*, 95–100° C. in the presence of 2 per cent glacial acetic acid and 0·2 per cent Lissapol N, so that that dyestuff which does go on to the wool has an opportunity to migrate to the Orlon, this transfer being facilitated by both the acetic acid and the Lissapol N. One other precaution to be taken is to see that no great excess of dyestuff is used, otherwise the low saturation uptake (p. 513) of the Orlon will mean that plenty of dyestuff is available to dye the wool. If the four precautions:

1. presence of acetic acid,
2. presence of Lissapol N,
3. avoidance of large excess of dyestuff,
4. continuance of dyeing for at least 60–90 min. to give time for dye transfer

are taken, staining on the wool will be only slight. Next, the wool component is dyed with acid levelling colours which in the absence of cuprous ion hardly affect the Orlon. A practical advantage is that in a 50/50 blend the Orlon component being dyed first need only approximate to the desired shade, final matching being carried out when dyeing the wool component, this being much easier than shading the Orlon component. Because basic dyes have such good washing fastness on Orlon there is much to be said for the application of acid milling colours which are faster, instead of acid levelling colours, to the wool component; then the wool component will be as fast as the Orlon and as the fastness of the blend is no better than that of its least fast component, the blend will be improved in respect of its coloration.

Orlon Type 42 and Viscose Rayon (or cotton). Light colours are obtained by dyeing the Orlon with disperse colours and the cotton with direct colours, both in the same bath. Medium colours are best obtained by dyeing the Orlon with basic colours at 95–100° C., removing the stain from the cellulose component by a scour at 60° C. and then dyeing it with direct dyes, or with vats if required.

Orlon Type 42 and Nylon. The method is similar to that for Orlon and wool but transfer of the basic dye from the stained nylon to the Orlon takes place so much more easily that Lissapol N need not be used, and 0·75 per cent acetic acid is sufficient. The Orlon is dyed first with basic colours, then the nylon with acid colours.

Cellulose Acetate and Viscose (or Cotton). This type of union is particularly suitable for cross-dyeing effects. To obtain these, a dispersed colour is used for the acetate, and a direct colour, which does not stain the acetate, is used for the viscose. The selection of a non-staining dyestuff is particularly important if the cellulose acetate is to be reserved, *i.e.*, left white. When the viscose is dyed to medium or deep shades it is almost certain that the cellulose acetate will be stained, and to clear the stain the fabric is treated at 35° C. with a dilute acidified solution of potassium permanganate, followed by treatment with an acidified solution of sodium bisulphite to clear any brown manganese residues. Two-colour dyeing is carried out in one bath.

Cellulose acetate/viscose or cellulose acetate/cotton mixtures are sometimes used for furnishing materials, and in such cases vat dyestuffs have to be used. This is a very delicate operation, and satisfactory only with those vats (which are not the best of the vats) that will dye from *weak* alkaline solution, and even then the solution has to be buffered by the addition of β-naphthol and the dyeing time kept as short as possible. In cases where really excellent fastness is required on cellulose acetate there is much to be said for the use of spun-dyed yarns—*i.e.*, yarns spun from dope to which a colouring matter has been added. Then, when two-colour effects are required the problem is reduced to one of selecting a dye for the viscose which will not stain cellulose acetate.

Setting

Thermoplastic fabrics—and these include cellulose acetate, triacetate, nylon, Terylene, Dynel, Rhovyl, Vinyon and Saran—can be set to any shape by treatment in steam, and will then be dimensionally stable until heated to a temperature as high as or higher than that at which they were set. The process can be used to treat fabrics in the flat state, so that when afterwards dyed on the winch they will not take permanent rope-creases. Furthermore, it is of great use for setting hose, so that when subsequently worn and washed they will not go out of shape. The process is of more use for Tricel than it is for cellulose acetate, which is liable to be delustred by the treatment. The theory behind nylon heat-setting is that the long polyamide chains are held together by interchain bonds, some of which are stronger than others, and that when the temperature is raised some of these interchain bonds break and some parts of the molecular chains have a greater freedom and can relax. Then, still at the same temperature, new bonds form, which suit the relaxed molecular arrangement better. These bonds hold the fibre in position and the molecular structure is stable or set with respect to temperatures up to this temperature at which the nylon was set. The higher the setting temperature the more interchain bonds break, and the greater the relaxation, but care has to be taken not to heat the fibre so high as to damage it by incipient chemical decomposition. The setting is better, and it is easy to see why, if the nylon fibre or fabric is allowed to shrink a little during setting. The test for a well-set nylon fabric is that it should not shrink more than 1 per cent during half an hour in boiling water.

Nylon Fabric. Two processes are employed for setting nylon fabric, either woven or knitted. They are:

1. Steam setting in which the fabric after tentering is subjected to the action of wet steam at a temperature of about 120° C.
2. Dry setting in which the fabric is exposed to dry air heated to about 210° C.

If fabric should become discoloured in the setting process it can be bleached with 40 per cent peracetic acid; this is not corrosive to stainless steel. It has been claimed (British Pat. 684,837) that a better set can be obtained by treating the fabric in molten metal or mercury at a temperature over 160° C. and then in steam. The fabric, suitably held, is immersed in the hot metal.

Nylon Hose. In the U.K. nylon hose are made about half fully-fashioned, half seamless. In the U.S.A. 60 per cent are seamless and in both countries the proportion of seamless is increasing. When these are knitted they are in tube form, the tube narrowing a little towards the ankle due to a closer knit; the toe is cut off and seamed. The hose are steamed (wet steam) at 125° C. for 50 sec. on boards (Fig. 266) and then, when they are removed, they are put into string bags and dyed in a rotary reversing dyeing machine at 85° C. After dyeing for perhaps two hours, the hose are put back on to forms as

[*Samuel Pegg & Son*]

FIG. 266.—A drying and finishing machine suitable for nylon, pure silk or rayon hose.

shown in Fig. 266, which are hollow and steam-heated, to dry and finish the hose. If the pre-setting treatment was omitted the stockings would be shapeless and when dyed they would develop " crows' feet "—*i.e.*, over-dyed parts where they were bundled. The pre-setting removes stitch distortions and gives a shape to the stocking which persists through the dyeing and finishing, wear and washing, and also ensures uniform dyeing, because it prevents creasing of the hose. Nylon hose made in this way are not very inferior to fully-fashioned hose—sometimes a mock seam is stitched in to counterfeit the fully-fashioned appearance. Shapes vary from one maker to another; some prefer a thinner ankle than others. The machine actually shown in Fig. 266 is for drying and finishing. The pre-setting machine is not dissimilar in appearance; the main difference is that in pre-setting the shapes are solid and the stockings are treated in a steam chamber, whereas in drying the shapes are hollow and the stockings are dried by the heat from the steam inside the hollow shapes.

Machines are made, *e.g.*, the Pegg Paramount pre-boarder, Fig. 267, which carry out the pre-boarding operation automatically; not

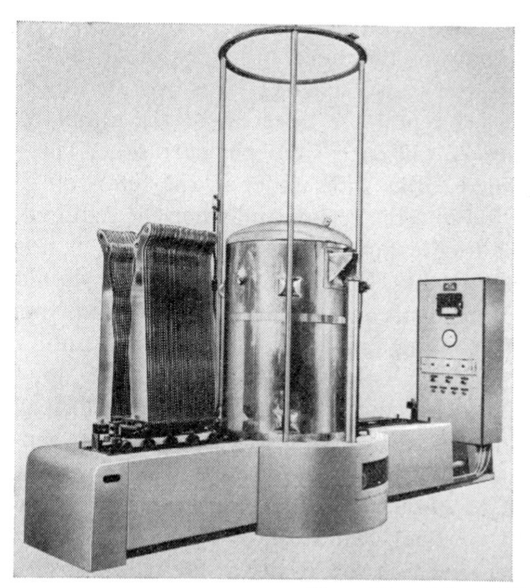

[*Samuel Pegg & Son, Ltd.*]

FIG. 267.—Pegg's Hi-Speed Pre/Post Boarding Machine for steam-setting stockings to shape.

only does this save labour, but it also ensures uniformity of treatment from batch to batch. The stockings are pulled manually on to metal frames and these are pushed into position. Then under automatic control the frames slide into a steam chamber, which closes, the steam is automatically turned on and after the determined time, say two minutes, the exhaust valve opens, the doors open, and a carriage holding the loaded forms is ejected. The stockings are removed by hand from the frames (or forms) and the operation is complete.

The machine shown in Fig. 266 has been superseded by newer

designs, but the way in which the stockings have to be put on and pulled off the shapes is the same to-day as is shown so clearly in Fig. 266. Is there a standard shape? There is no general agreement about the shape of the perfect leg; different finishers have their own ideas about perfection, and the shapes are fabricated to meet their views. If someone new to the business comes along and is too modest to define the perfect leg shape, then the manufacturers will give him a nice compromise between what their different customers do like. Amateur or connoisseur, one can hardly go wrong at this stage. One of the newer machines is Pegg's Hi-Speed Pre/Post Boarding Machine (Fig. 267). It looks different, but it does much the same work as the earlier machine shown in Fig. 266. It is designed to steam-set nylon or other synthetic hose or half-hose before (pre) or after (post) dyeing. It is operated by two men, one on either side, who pull the hose on to the aluminium forms, of which there are 72 (36 each side) per carriage. The retort opens, the stocking-filled forms slide under it, the retort closes automatically; steam is fed into the retort, and when the required temperature is reached the timer starts. For setting nylon 6 hose the steam temperature should be 110° C.; for nylon 66 it should be 130° C., and the time of exposure is 30 seconds. The steam supply stops, an exhaust opens, cooling fans start and continue until the drying is complete, the retort is unlocked and the top lifts up. The treated hose is wheeled away, and a new lot of forms which the operatives have filled whilst this processing was going on are slid in. Everything is automatic. The output is about 25–45 dozen pairs per hour, corresponding to about 6 cycles per hour, and if the machine is worked on two 8-hour shifts for a 5-day week, then the average production per week is 2,800 dozen pairs. Which is better, to steam-set before or after dyeing? The post-setting, *i.e.*, after dyeing, gives a better finish, but some of the heavier types of hose dye better if they are set before dyeing. Some people do both, steam-set the hose before dyeing and do it again after having dyed them. This probably pays—the hose sell on their appearance; pre-boarding should help level dyeing, and post-boarding should give a good sales appearance. In Fig. 267 the 72 forms are unmistakable, the round cylindrical vessel is the retort in which the steaming is done, and its cover lifts up to the top of the frame (about 13 ft. high) at the end of each cycle and then falls back into the operating position shown in Fig. 267. The rectangular box-shaped unit is the controller. The services needed are steam at 80–100 lb./in.^{2}, and compressed air at 80 lb./in.^{2}, as well as electric power for the motors. Steam is normally wet, *i.e.*, it has not been superheated and it is probably essential that it should be wet. But it must not be too wet; it should be fairly dry so as not to throw too much of a load on to the subsequent drying operation. There are two lots of 72 forms, 144 in all, one lot being treated whilst the next lot is being loaded with stockings, or with socks. With suitable formers, stretch tights can also be heat-set. The machine is used not only for all-nylon stockings but also for blends of nylon with cotton or wool. It is also used for " stretch " hose of the Helanca and other types. It takes men's and women's and kiddies' hose.

[*Samuel Pegg & Son, Ltd.*]

FIG. 268.—Pegg's Dye boarding Machine.

The Dyeboarding Machine. A logical development of the machine described above is one which simultaneously dyes and steam-sets the stockings. Such a machine is the Pegg Dyeboarder shown in Fig. 268. It is very similar to the boarding machine shown in Fig. 267, but has additionally on the extreme right a dye feed system. This consists of twin mechanically agitated concentrate tanks. One of these tanks supplies the machine, whilst the other is being cleaned and recharged without disturbing production. The large (lower) single-feed tank with a heater dilutes the dye concentrate ready for supply to the machine. The cycle of operations is normally (1) pre-board, (2) dye, (3) post-board, (4) dry. The makers say that uniformity of dyeing is good, and that when required colours can be changed within a matter of minutes. The introduction of this machine represents a big step forward in the dyeing and finishing of seam-free nylon hose.

The quality of nylon nowadays is so high that there is very seldom a broken thread in processing. Fully-fashioned hose are knitted on machines which knit 50 simultaneously, taking about 30 minutes to complete them; if an end does break, the machine is not stopped to repair it but that particular stocking is scrapped. Seamless stockings are run off very much more quickly, in a matter of 5 minutes. A recent development has been to give warmth to nylon hose by dipping their feet in a phenol solution to swell the fibre. Sometimes it is said that nylons do not wear like they used to do when introduced. But at that time they were 30 or 45 denier, were used carefully and occasionally; to-day they are often only 15 denier, are used regularly and without particular care.

Antron nylons. Trilobal nylon has been used for hose: the following example is illustrative of this and is useful as indicating design and finishing details of American practice in hose manufacture. Fully-fashioned ladies' hose were knitted on a Reiner hosiery machine. The yarn used for the leg was 15 denier, 2 fil denier $\frac{1}{4}$Z twist nylon 66 trilobal section containing 0·3 per cent TiO_2. Standard 40–13–$\frac{1}{2}$Z nylon with round filaments was used for the welt fabric and 30–10–$\frac{1}{2}$Z round filament for splicing.

The hosiery were seamed, pre-boarded 5 min. in 27 lb./in.^{2} steam, scoured in 2 per cent sodium lauryl sulphate + 2 per cent caustic soda. They were dyed gold, finished with a standard resin emulsion hosiery finish, and postboarded dry at 200° C. These stockings were extremely brilliant; similar 15–2–$\frac{1}{4}$Z round nylon gave hose that by comparison were extremely dull and had low colour clarity.

Embossing

A direct corollary of this property of taking a steam set is that fabrics made from thermoplastic fibres may be embossed by passing over heated cylinders engraved with the required design. In this way, for example, cellulose acetate fabrics can be embossed with a crêpe design to give a very good imitation of a crêpe-de-chine.

Crisp Finish on Nylon

A durable crisp finish is sometimes required on nylon fabrics. A urea-formaldehyde resin of the type BT 322 (B.I.P. Chemicals Ltd.) is suitable; it is similar to the BT 321 (p. 658) which is applied to viscose rayon fabrics except that it has a lower solids content.

A 3 per cent solution (1·35 per cent resin solids) containing a little accelerator is padded on to the nylon fabric, which is then dried and cured on the stenter for 2–3 min. at 140° C. Sometimes melamine-formaldehyde resins are used on nylon but the U.F. resins are cheaper and less inflammable.

Finishing

Rayon fabrics, after dyeing, are hydro-extracted to remove as much moisture as possible. Very delicate fabrics that might be damaged by "whizzing" in the hydro-extractor or centrifuge are

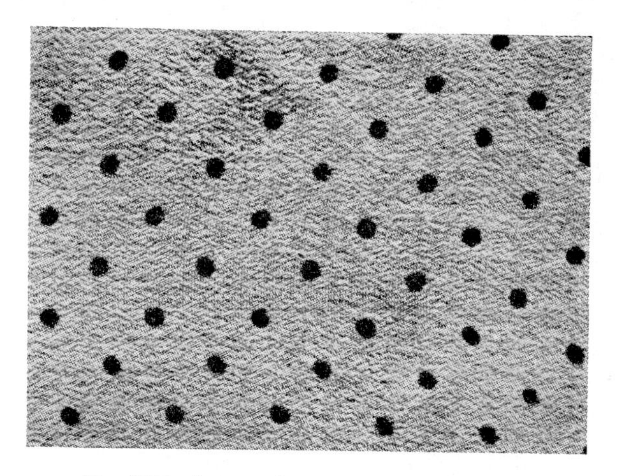

FIG. 269.—A spot print on a rayon crêpe fabric. Note the excellent "figure".

hydro-exhausted. The hydro-exhauster consists of a slotted pipe connected to a vacuum pump and, as the fabrics are passed slowly over the slot, most of the water is sucked out of them through the slot. As it is far cheaper to remove moisture mechanically than by heating, as much as possible is removed in the hydro-exhaustion.

Fabric is finally dried either: (1) by passing over steam-heated cylinders or (2) in a stenter. The former method is cheaper, but is inclined to give a hard or boardy handle which has subsequently to be "broken" by passing through a rubber mangle. The stentering method is more expensive, but gives a better handle, and is generally more suitable for rayon fabrics. The fabric is held out full width by being clipped at either side on an endless chain. It is carried through a cabinet in which hot air is blown through the fabric until it is dry; the fabric is then lapped or rolled at the end of the stenter. The air temperature in the stenter may be about 75° C. and the fabric may take several minutes to pass through the stenter.

Knitted fabric also may be finished in either of these two ways, but one additional method is available for tubular fabric. In this method it is pulled over a stainless steel cylinder, and at the top of this, after an interval of several feet, it passes through nip rollers. Part of the drying takes place in the passage up the tube, and hot air which is blown up the tube " balloons " the fabric out and escapes by finding its way through the fabric, so that drying is completed in the gap between the end of the tube and the nip. (Fig. 270.)

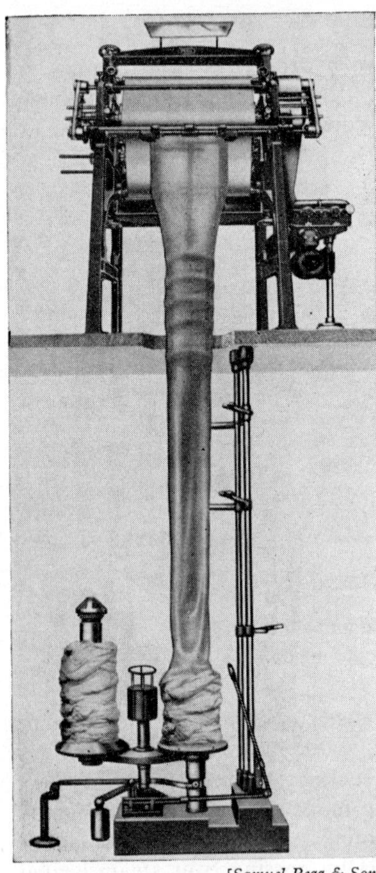

[*Samuel Pegg & Son*]

FIG. 270.—A drying and finishing machine for all classes of circular and warp knitted fabrics.
Continuous output is obtained by cartridge loading.

The machine shown in Fig. 270 has been available in principle for a very long time; for more than twenty-five years it has been very popular and has done a remarkably good job. Its basic virtue has been that the circular fabric is ballooned out by air injected into the fabric and that this air, in passing through and escaping from the tube of fabric, has dried it. To balloon circular fabric is a very desirable process indeed. The makers, anxious to be progressive, look on the machine nowadays as obsolescent and prefer to substitute for it the new continuous fabric drying and finishing machine for tubular knitted fabric, shown in Fig. 271. But some of their customers, successful dyers and finishers, prefer the old machine (Fig. 270) and insist on having it. It should be said at once that the new machine retains the valuable characteristic of ballooning the fabric. This can be seen in Fig. 272, where the fabric, probably having been dyed in rope form, is being untwisted and straightened automatically. The whole machine is shown in operation in Fig. 271; the continuity of the operations of dyeing and finishing is helpful economically, and the output of the combined machine com-

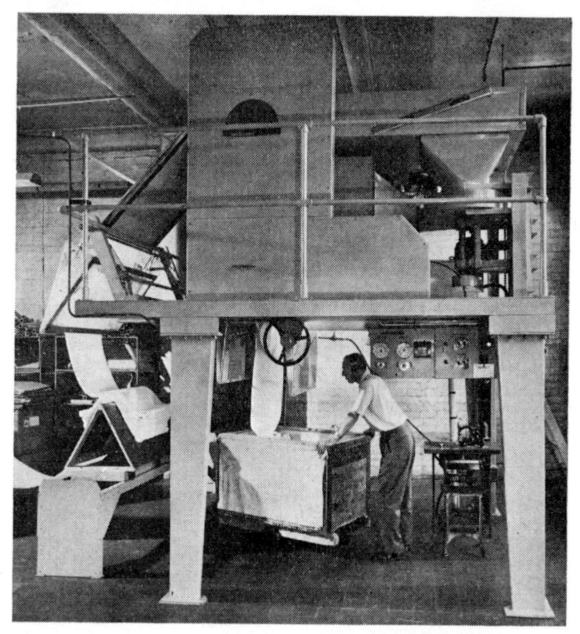

[*Samuel Pegg & Son, Ltd.*]

FIG. 271.—Continuous drying and finishing machine for tubular knitted fabric.

pares favourably with conventional calendering. No separate calendering is necessary. Pieces of fabric to feed it can be sewn end to end without stopping the machine.

Hanks of yarn can be dried on poles which are hung between travelling chains which carry them through a hot air dryer. Antron and trilobal Dacron should be given a tensionless finish to bring out their best.

Cutting Tubular Fabric

Such tubular fabric as polyester which has been dyed in a high-temperature pressure winch is very often cut open for finishing. The dyed fabric may be hydro-extracted and then guided through calender rolls and hand slit on the delivery side—a two-man

[*Samuel Pegg & Son, Ltd.*]

FIG. 272.—Detail of Fig. 271 showing the fabric being ballooned as it is dried.

operation. The guiding is necessary because the fabric may have become twisted during dyeing and hydro-extraction. There are some slitting machines available which will take only rolled fabric,

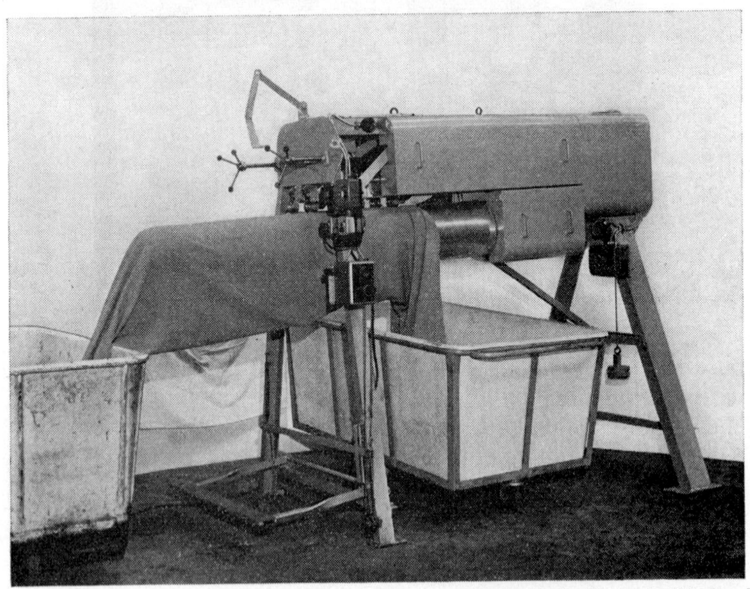

[*Engineering Workshops Ltd., Bury*]

FIG. 273.—The E.W.L. Slitting Machine.

this provision demanding an extra operation. But in the E.W.L. Slitting Machine, shown in Fig. 273 and made by Engineering Workshops Ltd. of Bury, provision is made to rotate the fabric to take out the twist and keep the wales straight. If reference is made to this illustration, it will be seen that the fabric is drawn from the truck on the left and is aligned over the rotating nose of the machine whilst a stretcher inside the fabric holds it taut. The tubular fabric is slit open by a wheel and then falls into the truck on the right. Fabric speed is about 35 yd./min. on fabrics that are easy to view, and lower on difficult patterns. One man can operate the machine. The machine shown is a prototype and is being fitted with a plaiting device.

FURTHER READING

R. S. Horsfall and L. G. Lawrie, " The Dyeing of Textile Fibres ", 2nd Edn. London (1946).

A. J. Hall, " Textile Finishing ", Iliffe Books Ltd, London (1966).

C. L. Bird, " The Theory and Practice of Wool Dyeing ", 2nd Edn. Bradford (1952). This gives an excellent description of dyeing *machinery*.

J. Boulton, *J. Soc. Dyers and Col.*, **50**, 381 (1934) and **54**, 268 (1938) (level dyeing on viscose).

G. S. J. White and T. Vickerstaff, " Colour " (an interesting discussion of colour vision), *J. Soc. Dyers and Col.*, **61**, 213 (1945).

Reports on the Dyeing Properties of Direct Dyes, *J. Soc. Dyers and Col.*, **62**, 280 (1946) and **64**, 145 (1948).

A. V. Ivitsky, " Pressure Dyeing in Germany ", B.I.O.S. Report 1806. H.M.S.O.

J. Boulton and T. H. Morton, " The Dyeing of Cellulosic Materials ", *J. Soc. Dyers and Col.*, **56**, 145 (1940).

E. Elod, " Theory of the Dyeing Process ", *Trans. Faraday Soc.*, **29**, 327 (1933).

M. R. Fox, " Vat Dyestuffs and Vat Dyeing ". London (1946).

American Viscose Corp., " Rayon Technology ", 190–251. New York (1948).

Morton Sundour Fabrics Ltd., *et al.*, British Pat. 620,584.

C. M. Whittaker and C. C. Wilcock, " Dyeing with Coal-tar Dyestuffs ". 5th Edn., London (1949).

Standfast Dyers and Printers Ltd., British Pat. 655,415.

A. J. Hall, " A Handbook of Textile Finishing ", 2nd Edn., London (1957).

J. Boulton, " The Application of Dyes to Viscose Rayon Yarn and Staple ", *J. Soc. Dyers and Col.*, **67**, 522 (1951).

H. A. Thomas, " Technique of Dyeing Rayons ", Manchester (Revised 1949).

N. B. Furvik, A. Bernsköld and N. Gralén, " Heat Setting of Nylon Fabrics ", *J. Text. Inst.*, **46**, T 662–T 667 (1955).

" How to Pressure Dye Tufted Carpets ", *Text. World* (Jan. 1956).

P. L. Meunier, " Practical Aspects of Dyeing in the Burlington Beam Machine ", *Amer. Dyest. Rep.* (21 May 1956).

J. A. Fowier, " The Dyeing of Textile Fibres above 100° C. ", *J. Soc. Dyers and Col.*, **71**, 443–450 (1955).

A. S. Fern, " The Dyeing of Terylene Polyester Fibre with Disperse Dyes above 100° C. ", *J. Soc. Dyers and Col.*, **71**, 502–513 (1955).

W. Bradley, " Recent Progress in the Chemistry of Dyes and Pigments ", *Roy. Inst. Chem.*, Monograph No. 5 (1958).

T. Vickerstaff, " Reactive Dyes for Textiles ", *J. Soc. Dyers and Col.*, **73**, 237–245 (1957).

R. C. Cheetham, " The Application of Reactive and Direct Dyes to Rayon Staple by Continuous Methods ", *J. Soc. Dyers and Col.*, **76**, 95–103 (1960).

D. F. Scott and T. Vickerstaff, " Reactive Dyes for Nylon ", *J. Soc. Dyers and Col.*, **76**, 104–113 (1960).

B. Kramrisch, " Recent Developments in the Dyeing of Acrylic Fibres ", *J. Soc. Dyers and Col.*, **77**, 237–243 (1961).

I. Ya. Kalontarov, " Diffusion of Dichlortriazin Dyestuffs into the Polyamide Fibre, Kapron ". *Dokladi Akad. Nauk Tadjik*, S.S.R., **7** (3), 19–23 (1964).

Reviewed by R. W. Moncrieff in *Text. Manufacturer*, 497, December (1964).
" Basic Dyestuffs Return to Favour ", *Skinner's Record*, **37**, 200–292 (1963).
Ciba Review No. 120, " Cibacron Dyes " (1957).
A. R. Kol'k, Z. A. Rogovin and A. A. Konkin, *Khim. Volokna*, **4**, 12 (1963) and *Eesti NSV Teaduste Akad. Toimetised, Fuusik. Mat. Ja Tehnikateaduste Seer.* **13** (3), 241–253 (1964). Reviewed by R. W. Moncrieff in *International Dyer*, **134**, 787–789 and 903, 905, 907 and 909 (1965).
E. B. Lifshits, " Solvatochromism of Cyanine Dyes ", *Dokl. Akad. Nauk. S.S.S.R.*, **179** (3), 596–599, reviewed by R. W. Moncrieff in *Text. Manufacturer*, 328 (August 1968).
I. D. Rattee, " Reactive Dyestuffs for Cellulose ", *Endeavour*, **20**, 154 (1961).
H. Luttringhaus, " Dyeing with Vinyl-Sulfone Reactive Dyes ", *Amer. Dyestuff Reporter*, **50**, 248 (1961).
H. Schumacher, " Die Reacton-Farbstoffe in Färberie und Druck ", *Melliand Textilberichte*, **41**, 1548–1554 (1960).
B. C. M. Dorset, " Factors in the Production of Fast Reactive Dyestuffs ", *Text. Manfr.*, 153–159 (April 1964).
E. A. Syson, " Development of Automation in the Dyeing Industries ", *Text. Manfr.* 207–211 (May 1967).
B. H. Mel'nikov, V. G. Radugin and P. V. Moryganov, " Rapid Dyeing of Polyester Fibre with Disperse Colours ", *Tekst. Prom.*, **28** (4), 63–65 (1968).
Dyeing of Wound Packages, *Internat. Textile Bulletin*, World Edition Dyeing, Printing, Finishing (February 1968).
G. Schetty, " The Irgalan Dyes—Neutral-dyeing Metal-complex Dyes ", *J. Soc. Dy. Col.*, **71**, 705–724 (1955).

CHAPTER 48

IDENTIFICATION AND ESTIMATION OF MAN-MADE FIBRES

It is necessary to be able to identify man-made fibres with certainty. It would be inviting disaster to attempt to dye and finish a yarn or fabric without definite knowledge of the material from which it is made, for dyestuffs that will dye one fibre may only stain another, and dyeing temperature and time conditions safe for one fibre might result in considerable damage being caused to another. The uses to which a yarn or fabric may suitably be put are also determined by the nature of the material from which it is made, and the worker will repeatedly find himself at a loss if he is not able, with assurance, to identify fibres.

A great variety of tests have been suggested, many of which require considerable equipment to apply satisfactorily, but the simple tests are really adequate to give a decision as to the identity of fibres, provided that reasonable care is exercised and that authentic specimens of fibres are available for comparison.

Authentic Samples

It is imperative to have a collection, which it is no difficult matter to build up, of samples of each kind of fibre, so that the behaviour on test of an unknown fibre may be compared with it. Verbal descriptions of the behaviour of a given fibre on test are not nearly so illuminating as observation of the tests actually carried out on fibres of known origin. This applies particularly to the examination of the fibres under the microscope. The reader will be well advised to make a collection, for reference purposes, of every different kind of fibre that he comes across.

Fabric

If a piece of woven fabric has to have its constituent fibres identified, it is necessary to dissect threads out of it, warp and weft ways. Both yarns, warp and weft, may be the same, but one must test them separately on the assumption that they may be different. Similarly, any coloured or other effect threads that may be woven into the fabric should also be dissected out and examined separately, for it is not an uncommon experience to find that the coloured threads are of a different fibre from the bulk of the warp. The same consideration regarding the separate examination of coloured ends applies to the examination of knitted fabrics.

Yarns

If a yarn is doubled or folded, it should be separated into its single components; in nearly every case they will be found to be similar, but infrequently they may be different. Core yarns are sometimes made in which a strong, continuous-filament, low-denier nylon yarn is covered with viscose or worsted. Occasionally, too, one comes across yarns which have, as one component, an elastic thread of rubber or a spandex fibre covered with nylon or cotton.

Sometimes, too, a yarn is made from a mixture of fibres, and this provides one of the most difficult cases for analysis. Blends of almost any pair of fibres are commonplace to-day. Sometimes, mechanical separation of the components in a single spun yarn can be effected, but the process is laborious, and usually undertaken only as a last resort.

Preliminary Examination

Having divided the fabric or yarn into its constituent parts, the first thing is to have a good look at them. At the very least, it is possible to say if the yarns are continuous filament or staple fibre, and if the latter is the case, an estimate of the staple length should be made, noting particularly if it is uniform or variable.

Continuity of filament indicates that the fibre is either real silk or one of the man-made fibres. It excludes the natural fibres, wool, goat-hairs, cotton, linen and the other bast fibres. Staple fibre—*i.e.*, discontinuity of filaments—offers no positive indication of the nature of the fibre, as there is clearly nothing to prevent any fibre which is customarily used in the continuous form from being cut and used as staple. A variation in length of staple, if considerable, usually indicates the presence of a natural fibre such as wool, goat-hair or cotton, whilst real uniformity of fibre length indicates that the fibre is almost certainly either schappe (staple fibre real silk) or a man-made fibre. Shortness of staple fibre, of about 1 in., means almost certainly that the fibre is cotton, although some viscose is made with a staple length of $1\frac{7}{16}$ ins. The staple length of wool and goat-hairs is very variable. The length of the strand—in which form they are usually woven—of linen and the bast fibres is high, although their ultimate fibres are of the same order of length as those of cotton.

The brightness of the fibres should be noted. This alone will give an indication of whether the fibre is natural or man-made, the latter being brighter; the matt effect, which the incorporation of a delustrant in the man-made fibre gives, is characteristic.

If a few inches of the yarn are available, hand-breaks will reveal if the fibre is very strong, indicating the probable presence of one of the highly oriented man-made fibres, or possibly real silk. At this stage, too, test the wet strength; if a tap is not conveniently situated, run the thread a few times between the lips, not without the application of saliva. The retention of very high strength indicates the presence of nylon, or other synthetic, whilst a very considerable reduction indicates the presence of reconstituted protein fibre.

All this observation of appearance and breaking of the yarn is carried out in less time than it takes to read and, with the application of a little common sense, will give some informative hints.

Burning Test

Once the yarn has been given a preliminary examination, a little of it should be burnt. Approach the flame (a match does quite well) to the fibre and carefully note its behaviour. The inferences that may be drawn are as follows:

Infusible. Asbestos.

Incombustible, but melts and shrinks from the flame. Glass.

Melts and chars but does not ignite: Teflon, Rhovyl, Saran.

Burns leaving skeleton of ash; (*a*) with smell of burning feathers: tin-weighted silk; (*b*) without smell of burning feathers: alginate, the ash being calcium oxide.

Burns without forming a bead, with burning-feathers smell. Real silk, wool, casein—*i.e.*, any protein fibre.

Burns and forms a bead—shrinks from flame. Nylon, Perlon or Terylene (hard, round bead which cannot be crushed between fingers) or Vinyon (irregular bead) or cellulose acetate (black bead easily crushed between the fingers and some smell of acetic acid can be noted during the burning of cellulose acetate) or an acrylic (bead like cellulose acetate but usually no smell of acetic acid). Verel burns only with difficulty.

Burns without forming bead and with papery smell. Viscose, cuprammonium, cotton, acetylated cotton, linen or one of the other bast fibres.

Staining Test

If the samples are white or light in colour, a staining test will give useful results. It is carried out with Shirlastain A, a product

Fibre.	Burning test.	Staining test. Shirlastain A.
Acrilan.	Like cellulose acetate but no smell of acetic acid.	Cold: Pink. Boil: Khaki.
Alginate.	Burns leaving ash skeleton without smell of burning feathers.	Salmon-pink.
Asbestos.	Infusible.	White.
Cellulose acetate.	Burns and forms bead. Shrinks from flame. Black bead, easily crushed. Smell of acetic acid.	Greenish-yellow.
Cellulose triacetate.	Burns and forms bead. Shrinks from flame. Black bead, easily crushed. Smell of acetic acid.	Off-white or slightly yellow.
Cotton.	Burns without forming bead and with papery smell.	Pale purple.
Creslan.	Burns slowly but brightly with black residue.	Cold: Faint yellow. Hot: Orange brown.
Cuprammonium.	Burns with papery smell. No bead.	Bright blue.
Darvan.	Burns rather fast, forms brownish bead.	Cold: Faint pink. Hot: Very faint yellow.
Dynel.	Difficult to burn. No bead.	Cold: Very pale dull pink. Boil: Pale yellow.
Glass.	Incombustible but melts and shrinks from flame.	White.
Nylon.	Burns and forms bead. Shrinks from flame. Hard round bead which cannot be crushed. Smoke usually white without soot.	Cream to yellow.
Orlon.	Like cellulose acetate but no smell of acetic acid.	Cold: Hint of pink, hardly detectable. Boil: Very pale dull yellow.
Real silk.	Burns, forming no bead, with burning-feathers smell.	Brown.
Regenerated proteins.	Burns without forming a bead, with burning-feathers smell.	Casein, Ardil, Soybean and zein. } Yellow-orange.
Rhovyl.	Shrinks from flame but will not ignite. Forms brownish-black residue.	Cold: No coloration. Hot: Very faint yellow.
Saran.	Burns and forms bead. Shrinks from flame. Difficult to burn.	White.

Microscopical exam.		Solubility test.
Longitudinal.	X section.	
Striations.	Round.	Soluble in ammonium thiocyanate (70%) 10 min. boil.
Featureless.	Serrated and very irregular.	Dissolves in very dilute alkali or in Calgon solution.
Marked fibrillation.	Clusters of fibrils.	Insoluble.
Featureless.	Three-lobed (occasionally 2 and 4 lobed).	Dissolves in conc. HCl when warmed to 35° C. Dissolves in H_2SO_4 (80%)—cold. Dissolves in meta-cresol. Soluble in 100% acetone and in 80%.
Featureless.	Three-lobed (occasionally 2 and 4 lobed).	Soluble in methylene chloride; insoluble in 80% and 100% acetone.
Flat and convoluted.	Collapsed tubular.	Dissolves in H_2SO_4 (80%)—cold.
Featureless.	Round.	Insoluble acetone, methylene chloride, amyl acetate.
Fine.	Round.	Dissolves in H_2SO_4 (80%)—cold.
Featureless.	Flattish, curly.	Insoluble acetone; soluble dimethyl formamide.
Featureless.	Irregular ribbon shape.	Soluble in acetone (100% but not in 80%).
Featureless.	Round.	Insoluble in all common solvents. Etched by HF.
Featureless.	Round.	Dissolves in H_2SO_4 (80%)—cold. Dissolves in meta-cresol. Dissolves in formic acid.
Striations.	Dog-bone or clover leaf.	Soluble in ammonium thiocyanate (70%) 10 min. boil.
Featureless.	Triangular with rounded corners.	Dissolves very rapidly in conc. HCl warmed to 35° C. Dissolves in H_2SO_4 (80%)—cold.
Faint striations.	Round with pittings.	Casein fibre very largely dissolves in caustic soda 5% at the boil.
Featureless.	Strongly indented.	Insoluble acetone; soluble in 50/50 acetone/carbon bisulphide.
Featureless.	Near round.	Insoluble in acetone. Soluble in dioxan.

Fibre.	Burning test.	Staining test. Shirlastain A.
Teflon.	Will not support combustion. Melts with decomposition.	No coloration.
Terylene and Dacron.	Like nylon, but smoke is black and sooty.	Cold: Very pale purple. Boil: Pale fawn.
Verel.	Difficult to ignite and flame often goes out. Forms bead.	Cold: Very faint pink. Hot: Maize.
Vinyon.	Burns and forms bead. Shrinks from flame. Irregular bead.	White.
Viscose.	Burns with papery smell. No bead.	Bright pink.
Wool.	Burns without forming bead with burning-feathers smell.	Yellow (bright).
Zefran.	Burns with bright flame.	Cold: Very faint pink. Hot: Nigger brown.

developed by British Cotton Industries Research Association and marketed by Shirley Developments Ltd. A sample of the fibre or yarn is degreased in some suitable solvent, such as ether or carbon tetrachloride, dried, wetted with water, and immersed in a dish of Shirlastain A for one minute at room temperature, then held under the cold-water tap to wash it thoroughly, squeezed between the fingers and examined. The colour of the stain gives a good indication of the nature of the fibre. If necessary, the test can be repeated in hot Shirlastain, as this further test sometimes gives more information. The colours given in the cold, unless otherwise stated, are:

White. Vinyon, Saran, Rhovyl.
Off-white. Cellulose triacetate, acetylated cotton.
Cream to yellow. Nylon (hot test: copper brown).
Yellow. Wool (bright yellow, copper brown if hot), kapok.
Yellow-orange. Regenerated protein fibres, casein, Ardil, soybean, zein. (*Note.*—Chlorinated wool gives a brownish-yellow in the cold but black if treated hot.)
Brown. Real silk and some of the bast fibres. Teflon, itself brown, is not stained.
Greenish-yellow. Cellulose acetate.

Microscopical exam.		Solubility test.
Longitudinal.	X section.	
Featureless.	Round.	Insoluble all solvents.
Featureless.	Round.	Soluble in meta-cresol 10 min. boil. Insoluble in formic acid and H_2SO_4 (70%).
Featureless.	Pea-nut shaped.	Soluble in acetone (warm); insoluble but plasticized by 80% acetone (warm).
Usually fine. False lumen may be seen.	Dumb-bell.	Soluble in 100% acetone. Soluble in methylene dichloride. Insoluble in 80% acetone.
Striated.	Highly serrated.	Dissolves in H_2SO_4 (80%)—cold.
Imbricated scales.	Round-elliptical.	Dissolves in caustic soda (5%) at the boil. Turns dark brown or black on treatment with lead plumbite.
Featureless.	Round.	Insoluble acetone, methylene chloride, amyl acetate.

Bluish-pink. Viscose.
Bright blue. Cuprammonium.
Salmon pink. Alginate.
Pink. Acrilan (khaki at the boil), Darvan (faint yellow hot), Verel (maize hot), Zefran (nigger hot).

So that the student may not be misled, colours given with some other fibres are given below.

Tin-weighted silk. Orange-red, bluish-red, Bordeaux, depending on the weight of tin present.
Raw silk. Very dark brown.
Degummed silk. Brownish-orange.
Tussore silk. Chestnut.
Cotton. Pale purple.
Mercerised cotton. Mauve.
Flax. Brownish-purple.
Ramie. Lavender.
Sisal and jute. Golden brown.

The highly oriented synthetics, Terylene, Orlon and Dynel, are not much more than off-white when stained. They develop colours in Shirlastain A at the boil (see pp. 820, 822). In 1961 two new stains were introduced to make it easier to distinguish between the thermoplastic fibres. These were:

Shirlastain D which, like Shirlastain A which has survived since it was developed in 1939, is supplied as a liquid ready for use. The primary function of Shirlastain D is to distinguish between cotton (bright blue) and viscose rayon (bright green). But if a mixture of

15 ml. Shirlastain A
5 ml. Shirlastain D
5 ml. dilute (10 per cent) sulphuric acid

is used, the fibre being stained at the boil for two minutes and then washed in warm water, it gives distinctive colours on some of the thermoplastic fibres.

Shirlastain E which is best used as a freshly made solution—2 grams of the powder (as it is supplied) per litre of 1 per cent sulphuric acid, *i.e.*

100 ml. dilute (10 per cent) sulphuric acid
900 ml. water
2 gm. Shirlastain E.

This stain also is applied to the fibre at the boil for 2 minutes, the sample then being washed in warm water. The colours that are given are as follows:

Fibre.	Shirlastain E (2 min. at boil).	Shirlastain A–D mixture (2 min. at boil).
Cellulose acetate (ordinary or secondary).	Orange-brown (persimmon).	Emerald green.
Cellulose triacetate.	Light brown.	Yellow, slightly greenish.
Nylon 66.	Brown.	Nearly black.
Perlon (nylon 6).	Very dark brown.	Very dark green.
Terylene.	Bright yellow.	Light grey.
Acrilan (Cl).	Olive green.	Dark purple.
Courtelle.	Yellow (brownish cast).	Light green.
Orlon 42.	Bright red.	Light blue.
Dynel.	Light brown.	Bright green.
Fibravyl (PVC).	Light brown.	Light grey.
Courlene X3.	Yellow.	Light grey.

A plate reproducing these colours was published in *J. Textile Institute* (November, 1961) and is useful for reference. But the best of all is a set of authentic known samples of the different fibres stained with the different stains and stored (in the dark to avoid fading) for reference.

Burning and staining may be enough for anyone familiar with the art to identify the fibre. With practice, fibre identification is easy; to the unpractised it is extremely difficult and one must carry out microscopic examination (less informative with man-made than with nature's fibres) and solubility tests. There is no substitute for practice, especially to-day with a dozen acrylics in common use and blends still commoner.

Microscopical Examination

The longitudinal examination of fibres enables one to detect wool and hair fibres and cotton with certainty. A few fibres are laid on a slide, so that they do not cross, and covered with a cover-glass. Care should be taken to see that they do not cross, otherwise they cannot all be in focus together. A magnification of 300 is very convenient for this kind of work. Wool and goat-hairs immediately reveal themselves by showing imbricated scales. If the wool has been over-chlorinated the scale structure may have been partly damaged. Cotton shows itself as flat, twisted fibres. Cotton and wool can usually be identified with certainty with the longitudinal microscopic appearance alone, although it is sometimes no easy matter to distinguish between wool and some of the goat-hairs, like mohair. In addition, badly over-chlorinated wool sometimes appears to be almost devoid of scales. So far as the man-made fibres are concerned, most of them offer the view of a smooth cylinder, sometimes striated, sometimes featureless, and little can be deduced from this. The only indication of value is the filament size, for very fine filaments of artificial fibres are usually those that have been made by a stretch-spinning or stretching process—*viz.*, cuprammonium, Terylene, nylon, Saran or other synthetic or Fortisan.

Cross-Sectional View. The cross-sectional view of the filaments is most informative. The method of cutting cross-sections is as follows: A bundle of fibres is threaded through a 0·75-mm. diameter hole in a metal plate 0·5 mm. thick in the form of a microscope slide. The bundle is firm, but not tightly packed in the hole, and can readily be threaded in with needle and cotton. The projecting tips of the fibres are removed from both sides of the plate by a *new* sharp razor-blade (do not waste time with old blades; a few cuts and the razor can be discarded). A smear of black mounting medium is applied to the second surface cut, and the excess removed by wiping with a clean, soft cloth. A drop of glycerine is applied, then the cover-glass. In many cases the mounting medium is not absolutely necessary. The cross-sections of the different fibres are as follows:

Round. Glass, nylon, cuprammonium, Terylene, Dacron, Creslan, Perlon, Acrilan, Teflon, Zefran, Courtelle, Kodel.
Round with pittings. Regenerated protein fibres, *e.g.*, casein.
Round-elliptical. Wool.
Highly serrated. Viscose.
Serrated and very irregular. Alginate.
Dumb-bell. Vinyon N, Orlon Type 42, Dynel, Verel.
Nearly round. Saran, polynosics.
Three-lobed (occasionally 2- and 4-lobed). Cellulose acetate, Orlon Type 81.
Triangular with rounded corners. Real silk, Antron, Dacron 62.
Collapsed tubular. Cotton.
Kidney-shaped. Vinylon, Darvan.

Solubility Tests

Acetone. The only fibres soluble in 100 per cent acetone are cellulose acetate, Dynel, Verel (warm) and Vinyon (and Pe Ce, which is unlikely to be met with). When making the test it is advisable to boil a little of the yarn under examination in acetone in a test-tube and then pour the acetone into another vessel full of water. If any cellulose acetate or Vinyon was present it is precipitated (white when the acetone solution is poured into the water). This device is necessary, because if the yarn contains only a small proportion of cellulose acetate it may not be easy to see that it has dissolved, but there is no mistaking the precipitate obtained on pouring the acetone solution into water. Rhovyl is swollen by acetone.

In 80 per cent acetone, 20 per cent water, Vinyon, Verel and Dynel are insoluble, although cellulose acetate is still soluble.

Methylene Dichloride. This solvent readily dissolves Vinyon and also cellulose *tri*acetate. It will not dissolve secondary cellulose acetate.

Meta-cresol. This dissolves nylon and cellulose acetate and Vinylon, but not Vinyon. It is therefore easy by solubility to distinguish between Vinyon and the others.

Sulphuric Acid 80 per cent, water 20 per cent (*N.B.* add acid to water, remember ATW alphabetical order). This reagent, in the cold, will not dissolve Vinyon, regenerated proteins such as casein, or wool, but dissolves most other fibres, including real silk, viscose, cotton, cuprammonium, cellulose acetate (yellow coloration) and nylon. It affords a means of distinguishing fibres of cellulosic origin from those of protein origin.

Formic Acid 88 per cent. This will dissolve nylon and (hot) Vinylon but will not dissolve Orlon, Acrilan or Creslan.

Concentrated Hydrochloric Acid warmed to 35° C. will dissolve cellulose acetate and real silk, the latter very rapidly, but no other fibres.

Caustic Soda 5 per cent, water 95 per cent at the boil dissolves wool, and very largely dissolves casein fibres, but no other fibres dissolve, although many swell or shrivel up. This reagent is useful to differentiate between wool and regenerated protein fibres, owing to the greater difficulty encountered in dissolving the latter. Casein, too, can readily be distinguished from wool by treatment with sodium plumbite solution, which is made by dissolving lead acetate in caustic soda. The wool turns dark brown or black, due to the formation of lead sulphide from the lead acetate and the sulphur in the wool. Casein, on the other hand, contains little or no sulphur, and does not give the brown or black coloration in sodium plumbite.

Flotation Test

This method is capable of yielding reliable results, although it is of more use with natural fibres than with synthetic. Fibres are pushed, with a glass rod, under the surface of the liquid chosen to be used as the flotation medium and viewed in a good light against a black background. If *ortho*-dichlorbenzene is chosen as the liquid, nylon floats on it, Vinyon dissolves, casein floats, and cellulose acetate keeps its own level—*i.e.*, neither sinks nor floats—and viscose sinks.

Density Gradient Tube. An amusing modification of the flotation test is provided by the density gradient tube. A glass tube (Fig. 274) about 2 in. in diameter and 15 in. high is set up vertically, the lower half filled with a liquid of high density, such as carbon tetrachloride (density 1·60), and the upper half carefully filled with a liquid of low density, such as xylol (density 0·87). The two liquids diffuse into each other, and after about two days there is a nearly linear density gradient in the middle of the tube. The density at various heights of the tube is determined by introducing

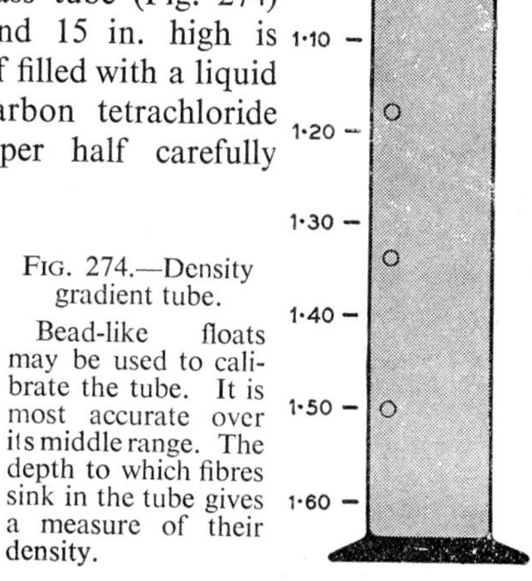

FIG. 274.—Density gradient tube.

Bead-like floats may be used to calibrate the tube. It is most accurate over its middle range. The depth to which fibres sink in the tube gives a measure of their density.

floats or fibres of known density, and the tube then calibrated for density in the same way as a thermometer is calibrated for temperature. When this has been done, fibres of unknown constitution are put in the tube, and the depth to which they sink gives a measure of their density. The specific gravities of some of the textile fibres are as follows:

Fibre	S.G.	Fibre	S.G.
Glass	2·56	Ardil	1·30
Asbestos	2·10–2·80	Cellulose acetate	1·30
Teflon	2·20	Casein {Lanital, Aralac}	1·29
Alginate	1·75	Courpleta	1·28
Saran	1·71	Real Silk (boiled off)	1·25
Viscose	1·53	Vicara	1·25
Fortisan	1·52	Zefran	1·19
Cuprammonium	1·52	Darvan	1·18
Cotton	1·50	Orlon	1·18
Linen	1·50	Acrilan	1·17
Hemp	1·48	Courtelle	1·17
Jute	1·48	Creslan	1·17
Terylene or Dacron	1·38	Nylon	1·14
Verel	1·37	Dralon	1·14
Vinyon	1·37	Pan	1·14
Real Silk (raw)	1·33	Crylor	1·12
Wool	1·32	Urylon	1·07
Camel hair	1·32	Polethylene Marlex or Ziegler type	0·96
Mohair	1·32	Polyethylene ("old" high pressure)	0·92
Dynel	1·31	Polypropylene	0·90
Vinyon N	1·31		
Soybean	1·31		
Zein	1·31		

From the results of these tests it should be possible to identify with certainty any man-made fibre. The salient features of the methods of identification are shown in the tables on pp. 820–823, which, whilst not containing so much information as is given in the text, contain sufficient for the identification of any common man-made fibre.

Detection of Titanium Dioxide

Titanium dioxide, if suspected to be present as a delustrant, can be detected by burning away the yarn and letting the ash dissolve in concentrated sulphuric acid. When cool, the solution should be diluted with cold water, and if on the addition of hydrogen peroxide a yellow colour is observed, then titanium is present. Alternatively, the yarn may be fused with potassium bisulphate on platinum foil and the residue dissolved in boiling dilute sulphuric acid. In this case, too, hydrogen peroxide is added, and a yellow (or orange) colour indicates the presence of titanium.

Estimation of Fibres

If a quantitative analysis is required, some thought must be given to the problem. Once the component fibres of a yarn have been identified, it is sometimes fairly easy to see how they can be evaluated quantitatively, but in other cases no simple method is available, and recourse has to be made to counting fibres under the microscope. If there are only two kinds of fibre present, and one is soluble in a solvent which does not attack the other, the matter is easy.

Solubility Methods

If, for example, cellulose acetate and viscose are found, the cellulose acetate can be dissolved by acetone out of a weighed hank (about 0·5 gram in weight), the residue (viscose) dried, conditioned and re-weighed. Weighings such as this are very conveniently and rapidly made on a Torsion Denier Balance (Fig. 1). Similarly, cellulose acetate can be separated from wool, cotton, silk, cuprammonium, nylon, casein and Saran by the same method. It can be separated from Vinyon by dissolving the cellulose acetate in 80 per cent acetone, in which Vinyon is insoluble. Similarly, Vinyon may be separated from all fibres except cellulose acetate, Verel and P.V.C. by dissolving in 100 per cent acetone (or in methylene dichloride from all except cellulose triacetate).

The cellulosic fibres (cotton, viscose and cuprammonium) may be separated from Vinyon, wool and the regenerated proteins by dissolving in cold 80 per cent sulphuric acid. Real silk and nylon also dissolve in this reagent.

Real silk can be separated from man-made fibres (except cellulose acetate) by dissolving in concentrated hydrochloric acid at 35° C.

Vinyon and nylon can be separated with *m*-cresol in which only the nylon is soluble, or with methylene dichloride as already described.

In making quantitative estimations by solubility methods, two points must be borne in mind, if accurate results are to be secured.

1. *All* weighings must be made *either* bone-dry *or* at standard conditions. It is by no means easy to weigh some fibres bone-dry because they pick up moisture so quickly from the atmosphere during transfer from oven to weighing bottle, and in addition it takes more than one hour in an oven to bring the fibres (wool, casein, *etc.*) to a state of real dryness. If a conditioned room, maintained at 65 per cent relative humidity and 70° F., is available it is best to do all weighings in that. It is true that it takes some time for fibres to gain their equilibrium moisture content, but much less time with small samples than with bulk lots. The analyst should never weigh the original sample as received, and then weigh the extracted sample after oven-drying but before conditioning. In practice, some compromise

has often to be reached; the conditioned room is the best way out if it is available; but, if not, care should be taken to weigh all samples under similar conditions.

2. When possible, check the result by another test in which the second fibre is dissolved instead of the first. If, for example, a mixture of wool and viscose is being analysed, make two experiments: (*a*) dissolve the wool in 5 per cent caustic soda at the boil and weigh the residual viscose; (*b*) dissolve the viscose in 80 per cent sulphuric acid in the cold and weigh the residual wool. If the two experiments give concordant results, confidence will be felt in the analysis. Wool and polypropylene afford another example; if the mixture is heated in 50 times its own weight of 1,1,2,2-tetrachlorethane and raised to the boil (146° C.), all the polypropylene fibre dissolves and leaves the wool practically unchanged. Another sample of the mixture can be heated in alkali; all the wool will dissolve, the polypropylene will be unaffected.

Separation of Proteins from Non-Proteins

Of late a method has been described for the analysis of mixtures of protein and non-protein fibres; it has been adopted as a Tentative Textile Standard by the Textile Institute.

The method consists of the use of a normal solution of sodium hypochlorite to which 5 grams per litre of caustic soda have been added. A specimen of about 0·4 gram of the fibre is stirred in about 50 ml. of the alkaline hypochlorite solution for 15 minutes at room temperature, and then filtered through a fritted glass crucible; the residue is washed, dried and weighed. The proteins, and only the proteins, dissolve and are removed in the filtrate. There are, however, two modifications that may be necessitated by the presence of certain fibres:

1. If Tussah silk is present, the stirring of the fibre in the alkaline hypochlorite solution must be continued for 30 instead of 15 minutes—the Tussah dissolves more slowly than the other proteins.

2. If Vicara is present, it must be pretreated in 4 per cent caustic soda in the cold for 30 minutes, filtered and washed and then treated as described above. The pretreatment is necessary because Vicara is much more chemically resistant and resists the action of hypochlorite much better than the other protein fibres. If this pretreatment were not given only about one-fifth of the Vicara present would be dissolved by the standard treatment; if it is given then all of the Vicara will be dissolved.

There is a slight loss of cellulosic fibre during the separation, and if it is cotton the weight of the dry residue should be multiplied by the factor 1·03, and if it is viscose rayon by the factor 1·01, to compensate for this loss. The new method is applicable to the following protein fibres : wool, chlorinated wool, raw and degummed cashmere, Ardil B, Ardil F, Fibrolane BX, Fibrolane BC, Lanital, soy-bean fibre (none made nowadays) and with the necessary modification to Vicara. It cannot be used in the presence of cellulose acetate or cellulose triacetate as these lose weight due to saponification. " Blends containing as little as 1 per cent of a non-protein fibre mixed with a protein fibre may be analysed with a satisfactory degree of precision " and reproducibility of results is good.

Counting Methods

In other cases no such simple solubility separations are possible. Sometimes a flotation test will give an indication; in this case one takes a representative group of, say, twenty-five fibres and observes how many sink and how many float. In such a case as wool and casein the simplest method is to cut a cross-section of the yarn and count, under the microscope, the number of round, uniform, pitted fibres (casein) and of rather irregularly round, unpitted fibres (wool). An estimation of the relative filament diameters must also be made to complete the quantitative analysis.

If the fibres are white, staining tests can be helpful. Take a number of fibres, immerse in Shirlastain A, then count the number of each colour, and either weigh them or determine their diameter with a microscope and average staple length by direct measurement. Wool and casein can be differentiated by sodium plumbite: take a sample of the yarn, stain it with sodium plumbite, then dissect it and separate into black and light fibres and weigh the two groups. More exact determination can be made by estimating the sulphur content chemically.

If alginate is one of the compounds, a sample of the yarn can be ashed and the heavy ash weighed, and from it the alginate content may be calculated; alternatively the alginate may be dissolved in dilute alkali.

Acrylics

The acrylic fibres are easy enough to identify as a group by the following tests :

1. They burn with the formation of a bead, like cellulose acetate, but usually there is no smell of acetic acid as there is with cellulose acetate.

2. They are insoluble in acetone and methylene chloride but soluble in dimethyl formamide (D.M.F.) at room temperature. To estimate an acrylic in a blend with cotton or wool, it is best to remove the acrylic with hot D.M.F.

3. Low specific gravity 1·12–1·19.

In practice the middle test (2) is the most useful, but Verel is exceptional in that it is soluble in acetone.

To distinguish one of the acrylics from another is not easy at present.

Verel. This is the easiest to identify, because of its solubility in acetone; the only other fibres that are soluble in acetone are Dynel, cellulose acetate (not triacetate) and Vinyon HH. Verel is insoluble in *cyclo*hexanone at room temperature, whereas Dynel and Vinyon HH are soluble. Cellulose acetate is soluble in glacial acetic acid at room temperature but Verel is not. A quantitative procedure can be used for separating complicated fibre blends by employing the solvents consecutively and filtering after each treatment—for example a mixture of Verel, Dynel, cellulose acetate and wool could be separated by first using acetic acid to dissolve the cellulose acetate, following with *cyclo*hexanone to dissolve the Dynel, and then using acetone to dissolve the Verel, the wool being left unaffected. Usually, in a technique such as that described above, about 0·5 gram of fibre is used and the specimen should be dissected into yarns if necessary, these untwisted and cut into short lengths of 3–5 mm. The sample is dried at 50° C. for 24 hours, weighed accurately and placed in a 125-ml. glass-stoppered flask containing 50 ml. of the selected solvent. The flask is shaken occasionally over a period of 2 hours to dissolve the soluble fibre. The fibre remaining undissolved is collected on a tared sintered-glass filter and washed with two 25-ml. portions of solvent. The residual fibre can be given a final wash with ethyl alcohol to speed the drying of the fibre, which is completely dried at 50° C. for 24 hours and weighed.

The only danger in detecting Verel is that it might be missed in the acrylic group because of its solubility in acetone. There is, however, a specific qualitative test which can be used to identify Verel or to demonstrate its absence unambiguously. It is:

Place 0·2 gram of scoured fibre in a test tube containing 5 ml. pyridine and heat on a steam bath for 3 minutes. If Verel is present the fibre turns a deep reddish brown and the solution becomes pale pink. Only one other known fibre, Saran, develops a red colour in pyridine, but Saran is soluble in pyridine whereas Verel is not. Clearly this affords a very good test for Saran, too.

Darvan. This fibre, unlike Verel, is typical of the acrylics in its insolubility in acetone and solubility in D.M.F. B. F. Goodrich Chemical Co., the makers, have suggested the following staining tests for identification of Darvan:

1. Identification stain G.D.C. (General Dyestuff Corp.) gives a pale mint green colour.
2. Calco Identification Stain No. 2 (American Cyanamid Co.) gives a light purplish grey colour.

Acrilan. Acrilan with its nearly round cross-section is easy to distinguish from Orlon which is lobed, but Acrilan and Courtelle have similar cross-sections which cannot be distinguished. These two can, however, be separated by the following dye test:

Treat the fibre for two minutes at the boil in a 40 : 1 bath containing 2 per cent Calcocid Alizarine Blue SAPX or other level dyeing acid dye and 4 per cent sulphuric acid based on the weight of the fibre (both dyestuff and acid). Rinse and dry. The Acrilan is dyed a dark blue, whereas Orlon or Courtelle will be undyed or only stained. This test is one that is used by Chemstrand in their own research laboratories.

Particular Separation Problems

These are only general indications, and the analyst will have to treat each problem on its merits. Most of the mixtures found in practice are easy to separate—*e.g.*, viscose and cellulose acetate, nylon and wool, cellulose acetate and cotton, Terylene and wool, and so on. Occasionally difficulties arise—*e.g.*, in the natural fibres—wool and goat-hairs are difficult to estimate, and wool and casein is a mixture which is quite common and is, as already indicated, by no means easy. Such artificial-fibre mixtures as cuprammonium and viscose are rarely met—if they were, one would try the microscope or staining count. An exact determination could probably be made by chemical estimation of the copper in the ash. Cotton and staple viscose may be met; if it is, then a mechanical separation of the fibres followed by weighing is probably the simplest method, although microscope and staining counts could also be employed. A differential solubility method may also be tried, for whereas viscose is soluble in cold 60 per cent sulphuric acid, cotton is not. In addition, a solution of sodium zincate dissolves viscose rayon but not cotton, and this separation gives reliable results and is regularly used by some analysts. This illustrates how the degradation to which the cellulose has been subjected in viscose manufacture has shortened the chain length of the molecules, and so increased the solubility. Nylon and real silk can be separated with boiling glacial acetic acid, which dissolves only the former; concentrated hydrochloric acid dissolves real silk readily, and nylon only slowly. The two tests together should enable a reliable analysis of real silk/nylon mixtures to be made.

Perlon and Nylon. Perlon (nylon 6) is soluble in boiling dimethyl formamide, whereas nylon (nylon 66) is not.

Acetate and Triacetate. These two fibres can be separated from other fibres by treatment with 90 methylene chloride/10 ethanol in which they are soluble. The only other fibre known to be affected by this mixture is Dynel which softens and swells. The loss in weight of a fibre mixture (provided Dynel is absent) represents the proportion of acetate and triacetate present. In order to separate the two acetates the fibres should be immersed in benzyl alcohol at 50° C. for 1 hour; the acetate completely dissolves, the triacetate does not dissolve at all.

Need for Determining Conditioned Weight. Analyses are most usefully reported as the weights of the fibres at standard conditions. A viscose/nylon mixture which was 70 viscose/30 nylon at standard conditions (65 per cent R.H. 70° F.) would be 68 viscose/32 nylon when bone-dry. The reasoning from which this follows is:

The dry weight of the viscose is $0{\cdot}88 \times 70 = 0{\cdot}616$ units

" " " " " nylon is $0{\cdot}97 \times 30 = 0{\cdot}291$ units

The percentage nylon is $\frac{0{\cdot}291}{0{\cdot}907} \times 100 = 32$ (bone-dry).

Inasmuch as all yarns are purchased on the basis of conditioned weight, the analysis is most useful if it is returned on the basis of conditioned weights.

FURTHER READING

N. Eyre, "Testing of Yarns and Fabrics", Textile Manufacturer Monograph, No. 4 (1947).

"The Identification of Textile Materials", 3rd. Edn. The Textile Institute, Manchester (1952).

P. Larose (density), *Can. J. Research*, B **16**, 61 (1938).

Carter and Consden (counting methods), *J. Text. Inst.*, **37**, T227 (1946).

Howlett, Morley and Urquhart (solubility methods), *J. Text. Inst.*, **33**, T75 (1942).

J. H. Skinkle, "Textile Testing", 2nd. Edn., 238–257. New York (1949).

J. M. Matthews, "Textile Fibres", 6th Edn. London (1954).

W. Garner, "Textile Laboratory Manual", 87 *et seq.* London (1949).

"British Rayon Manual", 53–75. Manchester (1947).

R. Preston and N. Warburton, "A Method for the Quantitative Determination of Ardil in Admixture with Wool", *J. Text Inst.*, **44**, T 298 (1953).

Tentative Textile Standard, No. 39, 1956, "Standard Method of Test for the Quantitative Chemical Analysis of Mixtures of Protein and Non-protein Fibres", *J. Text. Inst.*, **47**, P. 278–P. 279 (1956).

CHAPTER 49

ECONOMIC AND SOCIAL ASPECTS OF MAN-MADE FIBRES

All the significant industrial developments of man-made fibres, both regenerated and synthetic, have come within the last seventy years. The basic materials used by the *haute couture* of Paris at the beginning of the twentieth century were the same as had been used by the dressmakers in the courts of the Pharaohs. All the new fibres have come within living memory. Their coming has made a considerable difference to us not only industrially and economically but also socially.

Impact on Older Fibres

Generally the advent of the new fibres has not been to reduce the quantity of natural fibres that has been used, but until 1966 rather the reverse. Some of the first uses to which viscose and cellulose acetate rayons were put were in mixtures with cotton; brocades in which a rayon weft was brought to the surface over a cotton warp were very popular in the 'twenties, and they were rapidly followed by the suede-de-chine rayon cotton crêpes, which enjoyed enormous popularity. Cellulose acetate brought the possibility of cross-dyed shot effects which were also popular. The new fibres provided new products which incorporated a proportion of a natural fibre, and thereby provided new outlets for it. It was, however, only when the production of man-made fibres in staple form became considerable that the possibilities of using really large quantities of natural fibres with the artificial came along. The early mixture fabrics had been mixtures of yarns, *i.e.*, rayon yarns and cotton yarns had been woven into one fabric; rayon warp stripes were included in cotton shirtings and casements, cellulose acetate weft was run into a cotton warp, viscose crêpe weft was woven into cotton warps, and so on. But with the advent of staple rayon it became possible to mix the fibres and to make *yarns* which were mixtures of say cotton and rayon, or wool and rayon, and these new yarns opened up great new applications for the natural fibres. Indeed, many of the new fibres are used mainly in blends with natural fibres; such are viscose and cellulose acetate rayons, Terylene and Dacron, and Orlon. Almost all of the new fibres have proved to be very suitable for blending with natural fibres. Wool, for example, is expensive, and has sometimes been scarce, and the blending with it of the reconstituted protein fibres has stretched supplies in shortage periods, and has furthermore in some cases enabled fabrics and garments that look and behave like all-wool to enter a lower price range. Furthermore, most of the natural fibres have defects of one kind or another, and the man-made fibres when properly applied in blends help to correct these deficiencies. Examples are provided by the admixture of nylon and Terylene to wool, so increasing the strength and durability of the wool.

Another factor which is advantageous in blends of man-made and natural fibres, is that the price of man-made fibres (usually made from timber and chemicals) is much more stable than that of the natural fibres, which varies from year to year, according to the harvest; consequently the price of blends is more stable than that of the natural fibres alone.

One very important point is that the true synthetic fibres have properties very different from those possessed by any natural fibre; low moisture regain, very high strength, chemical resistance, and biological resistance are a few of these, and they have made it possible for fabrics to find quite new uses such as for protective clothing, outdoor furnishings, industrial filters, *etc.*; the synthetics should be not so much competitive with as complementary to the natural fibres.

There is, of course, another side to the picture: when the rayon jappes and satins and marocains came along and were made into lingerie and dresses, there must have been countless cases where a woman bought a rayon garment *instead of* a cotton or silk one; when to-day we buy a spun rayon sports shirt instead of one of cotton or wool; when a carpet is made from Verel or nylon or Acrilan or polypropylene instead of from wool; when a Dynel filter cloth lasts ten times as long as a cotton or wool filter cloth formerly did, there is no gainsaying the fact that less cotton or wool is being used in those instances. The production of high-tenacity filament rayon (viscose type filament) and nylon cord largely used for motor-tyre production has displaced about its own weight of cotton which was formerly used. At some stage, too, a man-made fibre may reach a stage when it is *better* than a natural fibre; it is probably true to say that, on the whole, thinking of industrial as well as apparel uses, nylon is a better fibre than natural silk; it is not better in every way, but it is better in more and perhaps in more important ways than it is poorer. When such a stage is reached the repercussions of the advent of the new fibre on the natural fibre may be very detrimental; nylon has practically ousted real silk from the stocking market.

Cotton. Cotton might have been expected to be the first natural fibre to suffer from the growth of staple fibre, but until 1965–66 it continued to increase in production. This increase was fostered by several factors; undoubtedly the most important of these is that cotton with its wonderfully intricate fibrillar structure is intrinsically a better fibre than rayon—it is stronger when wet, has a more pleasing appearance and wears better. Those multiple stores which substituted polynosic rayon for cotton in women's underwear such as knickers have had to go back to cotton to recover the wearing qualities that they lost when they gave up cotton. Cotton is a good fibre and, taking one thing with another, is better than any rayon yet made. However, the increased production of rayon has forced down the price of cotton; unfortunately devaluation of the pound has forced it up again, but for many years the competitive force of rayon brought down and kept down the price of cotton. Finally, there is the ever-growing character of the world population and the greater wealth of the population; these factors tend to support any good fibre. But for the 1966–67 season there was a drop of 10 per cent in the world production of cotton, and at the end of 1965 there was a world stock of 13,360 million lb. of cotton. These figures might suggest that cotton production has reached and passed its peak. But there have been other occasions in the past when the use of a natural fibre has seemed to be falling and then has recovered. So, it seems likely, will cotton recover, and its production in future years should top the 1965–66 record of 25,455 million lb. Jute and hemp are safer in the foreseeable future because of their low cost; the only setbacks to which they are liable are those caused by disruption of work in the political upheavals of some of the African countries. But these have brought in their train an increase in the price of jute which is now vulnerable to competition from polypropylene.

Most of the cut-back in cotton growing occurred in the United States as a result first of all of a voluntary acreage reduction programme and subsequently of legislation (Agricultural Act of 1965), which broadly meant that only those growers who agreed to a reduction in acreage would qualify for direct price-support payments; only those who grew less would get the guaranteed price for it. In the Soviet Union the production of cotton has increased from 3,274 million lb. in 1962–63 to 4,445 million lb. in 1966–67. It is worthy of note that whereas in America the yield of cotton per acre is about 500 lb. in a season, it is well over 700 lb. in the Soviet Union and exceeds 1,000 lb. in Israel. In India it is of the order of 110–120 lb. per acre. One other significant change in the picture of cotton production is that Tanzania has increased her cotton production from 86 to 174 million lb. per season between 1963 and 1967.

Production figures for the main cotton-producing countries are shown in the Table below.

Production of Cotton by Countries

Country.	Million lb. produced in:		
	1962–63.	1965–66.	1966–67.
United States	7,136	7,132	4,620
Soviet Union	3,274	4,206	4,445
China	2,055	2,772	2,772
India	2,366	2,142	2,270
Mexico	1,152	1,234	1,052
Egypt	1,008	1,146	1,019
Brazil	1,080	1,171	908
Others	4,976	5,652	5,629
World Total	23,047	25,455	22,715

In America there has been an increasing shift away from cotton growing in government agricultural policy. Perhaps this was dictated by the existence of a world stock of 14,500 million lb. cotton in 1966, of which 8,000 million lb. was in the U.S.A. The five main consuming countries are the U.S.A., China, Russia, India and Japan. The United Kingdom and the Western European countries are nowhere near the top.

As a result of the American cut-back in production, their stock in 1967 fell to 6,000 million lb. and the world stock to 12,500 million lb. The short-term results of the policy may therefore be considered rewarding. Long term, the decision to cut down cotton production must be looked on as mistaken. Cotton is a wonderful fibre, and it will be a long time before the best of our polynosic rayons can equal it. The basic objection to cotton growing is that it needs a lot of labour; people are not over-fond of work and have to be paid highly to do it. Cotton is therefore an expensive crop to pick.

Wool. Wool, which has always seemed a necessity and because of its quality so sure to maintain its success, has had a setback. The synthetic fibres are eating into its applications, particularly in carpets, furnishings and hand-knitting wools. Just as the rayons brought down the price of cotton, so now are the synthetic fibres bringing down the price of wool. All over the world there are stocks of unsold wool, and if the growers and merchants want to sell it they will have to make it cheaper. The world stock of wool in 1967 (clean basis) was 350 million lb. Production in 1966–67, also on a clean basis, was 3,400 million lb. In 1966–67 the world population of sheep was 940 million. Of this number,

Australia had 158 million lb.
U.S.S.R. had 130 ,, (this had fallen from 140 million lb. in 1962–63).
China had 59 ,,
New Zealand had 57 ,, (*The Times*, 12 June 1968 reported a New Zealand stock of wool of 700,000 bales, *i.e.*, 180 million lb. clean basis).

Silk. The world stock of real silk is about 30 million lb., and annual production about 75 million lb.

Flax. Annual production is still high at 1,600 million lb. (practically none in Ireland). She imports what she needs from Belgium or Russia.

Jute. For comparison the world's production of jute is about 8,000 million lb. per year.

Production of Natural Fibres. Efforts amongst the natural fibres to push up the price cannot be successful whilst there are enormous supplies of staple rayon, fibranne or Zellwolle ready to be pushed into any gap in the market. The provision of abundant supplies of fibre and the elimination of seasonal or artificial shortages is a contribution that man-made fibres have made to the textile market and thereby to us all. Reference should be made to the following figures of world production of some natural fibres in millions of lb.:

Fibre.	Yearly average 1920–29.	Yearly average 1930–39.	Yearly average 1946–50.	Yearly average 1951–55.	Yearly average 1956–60.	Yearly average 1961–65.
Cotton	10,800	13,000	12,900	17,800	21,500	23,900
Wool (clean)	From 1920 onwards about 2,000		2,200	2,500	2,800	3,200
Flax	From 1925 to 1934 averaged about 1,350		1,200	1,800	1,350	1,480
Silk	From 1920 to 1938 averaged about 140		34	50	71	71

The total production of those natural fibres given in the above table is about 28,700 million lb. per annum over the period 1961–65. In addition, there was an annual production of about 9,600 million lb. of sacking and cordage fibres, mainly jute and hemp, making a grand total of about 38,000 million lb. per annum. In 1965 comparable figures were 30,000 natural and 10,000 sacking and cordage totalling 40,000 million lb. and in 1967, 28,000 natural and 11,000 sacking and cordage, totalling 39,000 (rather less cotton). What we are beginning to see is the synthetic fibres (not the rayons) nibbling into the production and consumption of cotton. Despite its setback, cotton still accounts for three-fifths of all natural fibres and for two-fifths of *all* fibres. Sacking and cordage fibres account for one-quarter of all natural fibres, and they are of very great importance in the newly independent countries.

Production of Rayon

This started with Count Chardonnet in the 'eighties, and by 1900 man-made fibre production amounted to 2 million lb. per annum; as late as 1923 it was only 35 million lb. By 1935 it had risen to more than 1,000 million lb. and it consisted entirely of viscose, cellulose acetate and cuprammonium rayons. In 1967 it was about the 13,000 million lb. mark, so that of a total world production of 51,000 million lb. of fibres of all kinds, man-made fibres accounted for about 25 per cent.

The incidence of the staple form of man-made fibres undoubtedly accelerated the consumption and correspondingly the production of rayon. In 1921 staple fibre world production was only 0·2 million lb., but by 1929 it had reached 5 million lb. World production figures for continuous filament and for staple rayon (viscose, cuprammonium and acetate) show how the importance of staple fibre has grown. To-day more than half of the world's rayon production is in staple form.

Average yearly world production of rayon.

Year or period.	Filament yarn (million lb.).	Staple fibre (million lb.).	Total. (million lb.).
1935	935	138	1,073
1936	1,021	300	1,321
1937	1,197	625	1,822
1938	998	930	1,928
1939–48	1,184	1,046	2,230
1949–58	2,072	2,259	4,331
1959	2,421	3,142	5,563
1960	2,511	3,237	5,748
1961	2,560	3,400	5,960
1962	2,650	3,660	6,310
1963	2,713	4,024	6,737
1964	2,929	4,299	7,228
1965	3,017	4,313	7,330
1966	3,016	4,311	7,327
1967	2,919	4,227	7,146
1962–67 average	2,874	4,139	7,013

It will be seen that in 1967 the total rayon filament production was 2,919 and staple 4,227 million lb. The breakdown for 1967 shows what part of this was acetate and what part was high tenacity rayon:

	Continuous filament.	Staple fibre.	Total.
Viscose and cuprammonium .	1,278	4,125	5,403
Acetate	671	102	773
High tenacity (tyre) viscose .	970	—	970
	2,919	4,227	7,146

Cuprammonium figures are not available separately, but they are very small indeed compared with the viscose. Japan is probably the world's biggest cuprammonium producer with a 1966 output of 46 million lb. This is part of Japan's move to better fibres.

Cotton is the world's foremost fibre with 23,000 million lb., and is followed by jute with 8,000 million lb. and rayon with about 7,000 million lb. Some years ago rayon did achieve second place, but the production of jute in India and Pakistan has lately made great strides, and that fibre has pushed rayon out of second place. In 1967 rayon is no longer produced equally with jute and although cotton is a beautifully made fibre and one that has been used and admired since history began, it has felt the impact of the growth in use and esteem of rayon. It has lost some markets to rayon but has nevertheless increased in general use, with the general increase in world prosperity. In 1967 over three times as much cotton as rayon is still produced, but the man-made fibres are eating insatiably into the fibres market and, moreover, cotton has been able to maintain its production only by a considerable reduction in price. The recent reduction in cotton production can be attributed to the rapid advance of the synthetic fibres, rather than of the rayons. Indeed, the synthetics are beginning to take their toll of the rayons too.

The main producing countries of rayon (viscose, cellulose acetate and cuprammonium) are given in the Table below, production figures (in millions of lb.) being for the years 1962–66.

Production of Rayon (all kinds) by Country

Country.	Production of rayon in million lb. in:				
	1962.	1963.	1964.	1965.	1966.
U.S.A.	1,270	1,350	1,430	1,530	1,520
Japan	900	1,020	1,080	1,100	1,120
U.S.S.R.	540	590	670	730	800
West Germany . .	580	580	650	640	620
U.K.	440	490	550	540	500
Italy	420	440	470	410	400
East Germany . .	310	310	310	310	320
France	280	310	330	290	280

India	130	150	170	180	190
Poland	170	170	170	180	180
Czechoslovakia	150	150	160	150	160
Austria	120	130	140	150	140
Netherlands	100	110	130	130	120
Spain	130	140	140	120	110
Canada	90	100	110	110	100
Brazil	90	90	90	90	100
Belgium	80	80	80	80	90
China	50	60	70	80	80
Sweden	70	70	80	80	70
Finland	40	50	50	60	70
Rumania	10	10	10	40	60
Mexico	50	60	60	60	60
Yugoslavia	50	40	50	50	50
Norway	40	40	50	50	50
Switzerland	50	50	50	50	50
Argentina	30	20	40	40	30
Egypt	30	30	30	30	30
Others	90	100	100	100	100
Totals	6,310	6,740	7,270	7,380	7,400

(The agreement between the totals and those in the top Table on p. 841 is quite satisfactory.) The " others " include: Colombia, Iraq, Pakistan, Australia, South Korea, Greece and Turkey, Hungary, Portugal, Chile, Venezuela, Peru, Uruguay, Taiwan and Cuba, the biggest contributors being at the beginning, the smallest towards the end.

The order in which the countries arrange themselves is not in the main very different from what it was six years earlier (see Table on p. 687 of 4th Edition). The most rapid climber has been China; her production of rayon has been:

1957	0·4 million lb.
1958	5·8 ,,
1959	13·5 ,,
1960	19 ,,
1963	59 ,,
1966	84 ,,

Vickers-Zimmer have supplied some of the plants for China and have had a few of their men there incarcerated. Probably this is just a symptom of a rapidly developing industry in a country which is politically conscious. Rumania also has graduated from the " others ", and India has gained five places in the table—her production rose from 128 million lb. in 1955 to 194 million lb. in 1966. Most significant of all, the U.S.S.R. has overtaken West Germany and the United Kingdom, and looks set to overtake Japan by 1971.

These are the climbers: Russia, China, India and Rumania. In four years from 1962 to 1966 world rayon production increased by 1,090 million lb., and these four countries were responsible for 400 million lb. of it.

If ever there was a well-marked trend it is that the textile industry, man-made fibres and all, is moving to the East. It is interesting to compare the rayon production for the big world blocs. This is done in the following Table for the years 1963 and 1966.

Production of Rayon in million lb. in 1963 and 1966

Bloc.	1963.	1966.
Commonwealth (U.K., Canada, India, Australia) .	751	803
United States	1,349	1,519
Common Market (West Germany, Italy, France, Belgium, Luxembourg, Netherlands) . . .	1,529	1,504
Soviet Bloc (U.S.S.R., East Germany, Poland, Czechoslovakia, Rumania, Hungary, China) .	1,299	1,604
Japan	1,020	1,124

Such figures provide food for thought. So does the fact that Germany produced 935 million lb. in 1966. Evidence of industry, but in an Industry that will inevitably go to the East.

Split between Filament and Staple by Country

The total production of rayon by country is shown in the Table on pp. 841, 842. The split between continuous filament and staple fibre for each of the twenty-seven countries listed is not necessary for a fair overall impression of the industry to be obtained. The split between filament and rayon for seven countries for the last two years available, 1966 and 1967, is given below.

Country.	Production of rayon in million lb.:					
	Continuous filament.		Staple fibre.		Total.	
	1966.	1967.	1966.	1967.	1966.	1967.
U.S.A.	800	735	719	653	1,519	1,388
Japan	297	302	827	852	1,124	1,154
West Germany .	172	145	444	391	616	536
U.K.	200	200	296	317	496	517
Italy	192	199	206	201	398	400
France	123	109	155	131	278	240
Belgium	34	37	50	47	84	84

The agreement of totals with the figures shown in the larger Table on pp. 841, 842 is satisfactory.

[*Kurashiki Rayon Co., Ltd.*]

(*a*)

[*Kurashiki Rayon Co., Ltd.*]

(*b*)

[*Kurashiki Rayon Co., Ltd.*]

(*c*)

[*Kurashiki Rayon Co., Ltd.*]

(*d*)

FIG. 275.—Japan is the world's biggest producer of staple rayon, and is now the second largest producer of synthetic fibre and makes about one-fifth of the world's total production of synthetic fibres. In 1966 production of Vinylon alone reached 120 million lb., all Japanese.

The Japanese consume large quantities of man-made fabrics. (*a*) A student in a Vinylon/rayon blended uniform. (*b*) The little girl is dressed in fabric woven from spun Vinylon and spun rayon. (*c*) The late Dr. T. Tomonari, a leader of the Japanese man-made fibre industry and responsible for the development of Vinylon. (*d*) The factory girl who helps to make Vinylon is dressed in a Vinylon/cotton blend.

Strong Yarns

Where strength is the first consideration, viscose has the advantage over acetate and cuprammonium. About one-third of the world rayon production is high-tenacity viscose (strong yarn). The biggest single use is for tyre cords, which is all continuous filament. Carpet yarns, largely staple, also use a lot of high-tenacity viscose.

Production of Protein Man-made Fibres

Ardil and Vicara, the two most promising protein man-made fibres were both discontinued in 1957; probably sales of either never exceeded 5 million lb. per year, and such a modest production was unattractive to the manufacturers. It is a great pity because not only were they both good fibres, but they also represented a line of development—amongst the proteins—that would be beneficial to the man-made fibres industry as a whole. Fibrolane the U.K. casein fibre, and Merinova the Italian one, Lanital in Belgium and one or two others in Eastern Europe keep going. They are kept going for special purposes and seem to be stationary. They make little headway, nor apparently do the manufacturers push them. The protein man-made fibres have disappointed us all. There is no doubt that Ardil in particular approached more closely to wool than anything else has done. Wool is itself so very good that it is difficult for anything man-made to get within striking distance. Some time there will be built a synthetic " wool " from mixtures of synthetic amino-acids. That it has been so long in coming is a reproach to the top management of our big chemical fibre firms. Perhaps the polyamino-acids described in Chapter 42 will eventually close the gap.

Production of Synthetic Fibres Individually

Nylon is still far and away the biggest contributor; production figures for it are shown in the Table on p. 846.

The 1966 production of nylon represented 49 per cent of the world synthetic-fibre production. In 1964 the nylon output of 1,980 million lb. had constituted 53 per cent of world synthetic fibre output, and the 1959 output of 800 million lb. had represented 58 per cent of all synthetic fibres. Although nylon has grown and is continuing to grow fast, some of the others, particularly polyester and polyacrylonitrile fibres, are growing faster. By the end of 1969 (or perhaps 1970) nylon world production should be about 5,000 million lb., and this will only be 40 per cent of all synthetic fibres. If all goes well 1970 should see a world production of 12,000 million lb. of synthetic fibres.

Year.	Nylon, world production (million lb.).
1938	2
1940	4
1945	24
1950	100
1951	150
1952	160
1953	180
1954	260
1955	350
1956	440
1957	600
1958	580
1959	800
1960	950
1961	1,100
1962	1,340
1963	1,620
1964	1,980
1965	2,250
1966	2,660*

* Of this 2,390 filament, 270 staple and tow.

The figures for nylon include both nylon 6 and nylon 66. The biggest quantities are made in America with du Pont and Monsanto leading in nylon 66 and Allied Chemical in nylon 6. In the U.K. British Nylon Spinners, now I.C.I. Fibres Ltd., made most of the U.K. 1966 total of 183 million lb. of nylon. A breakdown by country for 1963 and 1966 production of nylon might run something like:

Country.	Nylon production (million lb.).	
	1963.	1966.
U.S.A.	670	1,070
Japan	180	320
West Germany	120	195
France	100	115
U.K.	135	185
U.S.S.R.	80	185
Italy	110	165
Canada	50	80

with Switzerland, Netherlands, East Germany and Belgium making smaller but considerable contributions.

The polyester fibres Terylene and Dacron and the similar fibres under other names may have accounted in 1966 for about 1,302 million lb. (351 filament, 951 staple and tow) made up mainly:

Country.	Fibre.	Polyester production (million lb.).	
		1963.	1966.
U.S.A.	Dacron	210	500
Japan	Tetoron	140	270
U.K.	Terylene	55	105
France	Tergal	40	60
West Germany	Trevira, Diolen	65	165
Canada	Terylene	20	30
Italy	Terital	20	40
Netherlands	Terlenka	15	35
U.S.S.R.	Lavsan	—	15

The acrylics in America have increased considerably. So has Courtelle at Grimsby in the U.K. It is certain that the American plants have not run to capacity. It is likely that 1963 and 1966 production of acrylics of all kinds ran somewhat as follows. Only one-half per cent of the acrylic fibre spun is in the form of continuous filament.

Country.	Acrylics production (million lb.).	
	1963.	1966.
U.S.A.	210	355
Japan	80	220
U.K.	35	90
U.S.S.R.	10	15
Others	110	330

Production of all Synthetic Fibre

The world production of synthetic (*i.e.*, non-cellulosic) fibres in 1963 was 3,700 million lb. and in 1966 it was 5,900 million lb., made up of:

Fibre.	Synthetic fibres, world production (million lb.).	
	1963.	1966.
Nylon	2,000	2,700
Polyester	700	1,300
Acrylics	600	1,000
Textile glass	200	500
Vinylon	80	120
Others	120	300

Of this, 2,500 million lb. was staple fibre. In 1969 world production of synthetic fibres may rise to 10,000 million lb.

The world production of synthetic fibres (that is, man-made, but not cellulosic in origin) has grown over the years as shown below:

Year.	Synthetic fibre production (million lb.).		
	Cont. filament.	Staple.	Total.
1955	405	182	587
1958	592	332	924
1961	1,095	735	1,830
1964	2,152	1,569	3,721
1967	3,238	2,973	6,211

For comparison the world production of rayon, including acetate, in 1967 was 2,900 filament and 4,200 staple, about 7,100 in all, so that there is still, or was in 1967, more rayon than synthetics made. It does not need much courage to predict that the synthetic production will outstrip that of rayon in 1968 or 1969 at the latest. It is interesting to note that in the dearer synthetics the proportion that is used as continuous filament is greater than it is in the rayons.

In the U.S.A. synthetics overtook rayon production in 1965; in less advanced countries, such as Japan, a larger proportion of their man-made fibres production is still rayon, but even in these countries it cannot be long before the synthetics will account for the lion's share of man-made fibre production.

Figures are published by most countries of their production of cellulosic rayon and of their total synthetic (non-cellulosic) fibre production; but as a rule, no breakdown is given of the synthetics, no information about how much nylon has been made, how much polyester and so on. The figures given in the bottom Table of p. 847 do no more than indicate the order of the quantities produced; the 1966 world production of synthetic fibres, including textile glass, was undoubtedly 5,900 million lb., but the breakdown into nylon, acrylics, *etc.*, has had to be estimated. Japan publishes her production figures fibre by fibre; not so the West, except for Italy.

The synthetic fibres, as distinct from the rayons, have made most progress in America. In the U.S.A. in 1967 the synthetics production was 2,330 million lb., about one-eighth higher than in the previous year. The increase in yarn and monofilament was only 4 per cent, whereas the synthetics staples (including tow) surprisingly increased by as much as 24 per cent. This was reflected, too, in the consumption of nylon; yarn and monofil actually fell by 2 per cent to 945 million lb., whereas nylon staple (including tow) rose by 2 per cent. Although nylon is the giant amongst synthetic fibres, it is being caught up by the polyester and acrylic fibres. Polypropylene, too, is increasing rapidly.

Total Fibres World Production. We have considered the world production of natural fibres, of rayons (protein man-mades are negligible) and of the true synthetics. Let us put these together and see what the picture is:

Fibres.	1951 production (million lb.).	1956 production (million lb.).	1964 production (million lb.).	1966 production (million lb.).	Estimated 1968 production (million lb.).
Natural . . .	22,300	25,200	29,700	27,700	29,000
Natural for sacking and cordage. .	6,800	7,200	9,900	10,700	11,000
Rayons . . .	4,040	5,250	7,300	7,400	8,000
True synthetics .	180	690	3,700	5,500	10,000
Textile glass . .	40	110	200	200	—
Total . . .	about 33,300	about 38,500	50,800	51,500	58,000

More simply:

Fibres.	Production in thousand million lb.				
	1951.	1956.	1964.	1966.	1968 (estimated)
Natural . . .	29	32	40	38	40
Man-made . .	4	6	11	13	18

It is a heartening picture. For the future, growth will continue. Over the period 1951–66, man-made fibres production has grown from 4 to 13 thousand million lb.; rayon has grown from 4 to 7·5 and synthetics from a hundred million or so to 5·5 thousand million lb. In the recent short period of 1962–66 rayons have increased only from 6·3 to 7·5, whereas the synthetics have increased from 2·4 to 5·5, all in thousand million lb.; in four years the synthetics have more than doubled. This rapid growth rate has sometimes brought a surplus of production. In Japan production of polyester had to be curtailed, and the rate of expansion slowed down from over a third in 1964 to 13 per cent in 1965. In this instance the result was that stocks dropped rapidly so that the control could soon be relaxed, and in the next year, in 1966, output of polyester rose by as much as 24 per cent. In the U.S.A. and in the U.K. there have been big price reductions as a result of the greatly increased production of synthetic fibres. The greater production has generally been taken up, but prices have had to be lowered to increase the breadth of the field in which the synthetic fibres could be used. Over a long period the price reductions have been very considerable: in the U.K. 3-denier staple nylon was 135*d.*/lb. in March 1954 and 109*d.*/lb. in March 1964, but only 82*d.*/lb. in July 1967. Terylene 3-denier staple cost 144*d.*/lb. prior to October 1955, but in October 1967 was down to

59*d.*/lb. Such falls are quite remarkable in the field of man-made fibres.

Turning to the production of natural fibres, we see that the discouragement of cotton growing in the U.S.A. has cut down production of this fibre very considerably since 1965–66. What the short-term reactions to this curtailment will be is impossible to say. But there are two factors which lend support to the view that in the long term cotton will regain what it has lost, and more; they are:

1. Cotton is a superlatively good fibre, and the naturally increasing demand of a growing and wealthier world population for good things makes it difficult to see how cotton can fail to recover. It will be sought as never before.
2. Russia continues to grow increasing quantities of cotton; in 1964–65 her crop was 3,950 million lb., in 1966–67 it was 4,450 million lb., about one-fifth of the world total production. Soviet cotton production and consumption will continue to increase.

Figures for the future production of fibres must necessarily reflect: (*a*) the steadying down of rayon production; (*b*) the continued acceleration of synthetic-fibre production; (*c*) the growth of cotton.

The first two lend themselves easily to prediction, the third is less commensurable. There are already clear enough indications that can no longer be overlooked that the growth of the synthetics has in the last few years taken its toll of cotton consumption. Perhaps the estimates in the next Table may prove not to be too far away from actuality. Compared with somewhat similar estimates that were made when the 4th Edition of this book was revised in 1962–63, the new figures show a much bigger rate of growth for synthetics and a rather lower rate of growth for natural fibres. This, of course, is an extension of what we have already experienced in the period 1963–67: synthetics have grown faster than the most hopeful of prophets could have expected, and there has been a falling off, which will probably prove to have been only very temporary, in cotton production. One feels intuitively that the estimates for the growth of man-made fibres in the next decade (those shown in the next Table), adventurous as they are, will prove to be underestimates. All one can say is that on the evidence available to-day they seem to be fair enough. If these estimates are to be beaten, as they quite probably will be, then it can be said that mankind must be very pleased indeed with his man-made fibres. But let us have sufficient humility to remember that cotton and wool and linen and silk are still the fibres for the connoisseur. Man-made fibres are for the masses, and they grow with them. Over the period 1951–67 their annual increment of growth has been very nearly 10 per cent. Already by 1967 one-quarter of all fibres were man-made fibres.

Estimated Fibre Growth in the 1967–76 Decade

Fibres.	Estimated world production of fibres (thousand million lb.).		
	1968.	1971.	1976.
Natural	40	42	44
Man-made	18	21	26

An extension suggests that man-made fibre production should exceed natural fibre production before 1990, and that by that time nearly all of it will be synthetic, very little rayon. The figures may be fanciful but the trends on which they are based are not. Blue sky for some, but not necessarily for the U.K. In U.K. the 1963 production was 720 million lb., about 7 per cent of world production. And in 1966 it was 890 million lb., which was fractionally less than 7 per cent. Rather does the future of all fibre production lie in the East.

U.K. Consumption of Fibres

World figures are more interesting than those of a single country, but perhaps for a few minutes we may be parochial and have a quick look at the directions in which the consumption of fibres in our own country is changing. Figures which are (production + imports − exports) are as follows:

Consumption (million lb.) of Fibres in the U.K.

Fibres.	Average 1951–60.	1963.	1966.
Wool (clean)	460	460	390
Cotton	760	500	470
Rayon	340	390	410
Synthetics	50	200	350
Flax	100	80	80
Real silk	0·7	0·4	0·3
Jute, sisal, etc.	470	490	430
Total	2,180	2,120	2,130

These figures include apparel, household and industrial uses. The population increases, but people's clothing tends to be scantier, their houses tend to be smaller with less fibre needed for carpets, curtains and upholstery. They are making do with less and less of the natural fibres, but are buying more and more of the synthetic fibres. Partly this is the result of advertising, partly that the synthetics are light in weight and strong, so that apparel can be briefer and lighter. The estimated population of the U.K. (England, Wales, Scotland and N. Ireland) in mid-1966 was 55 million, so that we use about 39 lb. of fibre apiece each year, and of this about 14 lb. is man made (rayon and synthetics).

Two things are necessary for the postulated enormous increases in the production of synthetic fibres; they are:

1. The fibres must be improved until they are better all round, and not just in a few specialised properties, than the natural fibres.
2. The cost of the intermediates such as adiponitrile (for nylon), terephthalic acid (for Terylene) and acrylonitrile (for Orlon, Dynel, Acrilan, *etc.*) must be greatly reduced.

In the author's view there is not the least doubt that both of these requirements will be fulfilled within the coming ten or twenty years, with the single reservation that it will probably be impossible to make a fibre as attractive and useful as is wool. Since 1954, when this was written, there have been big falls in the prices of some of the intermediates. To-day (1968) terephthalic ester has been costed at 14·5*d.*/lb., and if the costing shown on p. 404 is even approximately right, then polyester fibre can be produced for 3*s.*/lb. It is amazing how cheaply fibres can be made when once a good process gets going well. Viscose rayon was at one time spun for less than a shilling per lb. At the end of 1967 Terylene staple cost as little as 59*d.*/lb. The trend is undoubtedly downwards and will probably continue until the prices of synthetic fibres are lower than those of the natural. Nothing very remarkable is needed to bring about such a state of affairs; for a long time now synthetic rubber has been cheaper than natural, and yet for some decades the synthesis of rubber was almost impossible, and then for a long time, difficult. So it will be with the fibres; we may well see them all cheaper than cotton. A disservice is done to the fibres industry when a new fibre is introduced, with extravagant self-praise, at a high price, a price just under that of silk. Somebody has to pay for all the praise and publicity, and it will surely be the men and women who are cajoled into paying high prices for fabric and garments made from the fibre. In the end the fibre will find its own level.

The outlay of capital that is necessary to put a new fibre on production must be considered. It has been well said that " the route from test tube to wardrobe is long and hazardous ". First of all, research work is carried out with ordinary laboratory equipment: this may well go on for several years before anything of outstanding promise is found. Then the polymer has to be made on a pilot-plant scale and spun; the most suitable spinning conditions have to be found, and sometimes almost a whole new spinning technique has to be developed. Fabrics are made from the yarn that is spun and these are made into garments and worn. If all is well so far, a decision may be taken to build a large plant. This may seem early, but if the attempt is made to carry out sales development with pilot-plant quantities, the results can be fatally slow and may be misleading. Actual consumer acceptance on a fairly big scale is necessary as well as market surveys and textile industry opinion. In fact, therefore, a large plant costing from £10–15 million may be designed and built whilst the fibre is still undergoing change, perhaps even whilst its dyeing properties are still unsatisfactory. The preliminary research and development work, before any decision to build a factory is made, may cost £500,000 a year in a large concern and may well run to £2 million in all. Despite these apparently prohibitive costs, there are manufacturers who are bold and sufficiently far-seeing to make the outlay. Their old Terylene factory cost I.C.I. Ltd. £10 million. In the development of nylon du Pont spent $45 million on research, $21 million on sales investigation, *etc.*, and $196 million on plant—more than £50 million in all. There are in America several companies that are now investing, or have invested, more than $40 million in staple fibre plants, for fibres whose chemical constitution and properties may yet be subject to modification. All this is to the good; it is far better than hoarding hundreds of millions, and the primary function of money is to constitute a means of exchange. Provided that the work and materials devoted to such a project as fibre manufacture are devoted to the development of a useful product, such work and materials are used to good advantage and the sum of money spent, colossal though it may seem, is simply a reflection of the endeavour put into the project. It does, however, mean that the production of new synthetic fibres is in the hands of large companies and corporations, and that the synthetic fibre field is one in which it is becoming increasingly difficult for small organisations to engage with much hope of success. Fortunately this does not apply to the first stages of research, and the possibility is still open to anyone of making a new fibre-forming polymer in a laboratory vacuum distillation apparatus, or perhaps even in a test tube.

One other result of the high capital cost of plant for making synthetic fibres is that in order to justify it, the utmost possible production must be got out of the plant, and this can only be done by working it continuously, *i.e.*, by shift work with no week-end stoppages. Usually this is accomplished by having four teams of operatives and supervisors each working eight-hour shifts and so arranged that they have a two or three days' break every ten days. As there are 168 hours in a week, continuous working with four teams means an average working week for the individual of 42 hours. Quite apart from the high capital cost of the plant, the process of spinning itself, whether of synthetics or of rayon, usually calls for continuous working; if the plant were to be stopped for the night it would take all of the next day to get it started again. Spinning a fibre from a solution or from a melt is essentially a process which must be kept going.

To make and market a new fibre still, in 1968, costs a lot of money. Grilon S/A of Domat/Ems, who are not amongst the chemical giants, succeeded in making and marketing Grilene (p. 435), which at the time it was marketed was a novel polyester–polyether with unique dyeing properties. It is true that Grilon have since changed their fibre back to standard polyester, a rather regrettable move, but their initial achievement did show what can be done. If the problem confronting a manufacturer is not that of making a new fibre but simply of starting to make an existing fibre, then the Vickers-Zimmer plants (p. 65) can be bought so cheaply that they make the older figures of the chemical giants look ludicrous.

Nowadays, moderate sums can set up a synthetic-fibre factory, and that is a very good thing, because synthetic-fibre manufacture in Britain and America had tended to become the preserve of a small ring, or at least of a small number, of very large organisations. In Japan, where progress in the manufacture of synthetic fibres has been very rapid, there has been a free-for-all, with a large number of manufacturers competing against one another, to their own undoubted strengthening and prosperity. The Vickers-Zimmer do-it-yourself factories represent a welcome development in the fibres scene.

Soviet Union Production

Production of man-made fibres in the Soviet Union is mainly as follows:

Viscose. A large production of about 750 million lb. per year, both continuous filament and staple, both bright and dull lustres, in natural and in spun-dyed colours. Quality is much the same as viscose anywhere else, with tenacity approaching 2 gm./den. There is also some higher tenacity material made. It is made at six different factories, and the strength and quality of the viscose cord that is made from the H.T. rayon varies from one to another. At the Svetlogorskoe factory the 450-den. rayon had a tenacity of 4·5 gm./den. and an extension at break of 9 per cent. The cord made from it had a breaking load of 12·9 kg. and an extension at break of 17 per cent under standard conditions. Bone-dry its strength was 15·7 kg. and its extension at break 13 per cent. It is evidently good H.T. rayon, and its output is increasing by 5 per cent annually. Deniers of the standard (not high tenacity) viscose rayon run from 35 to 150 for weaving, and in tow form from 1,600 to 6,000. Soviet returns for the year 1966 are, for viscose production:

Continuous textile filament	102 million lb.
High-tenacity cord	222 million lb.
Staple fibre	405 million lb.

Experimentally, a cellulosic fibre with a high strength and low water imbibition, and, too, reduced swelling properties is being made, evidently a polynosic rayon. There is also some high tenacity viscose production for tyre cord, notably at Barnaul.

Acetate. Made in continuous filament in deniers of 90 to 150. In 1966 Soviet production of acetate was 45 million lb., of which 44 million lb. was continuous filament. In 1961 it had been only 7·5 million lb. In 1970 it is expected that the acetate production will approach 80–90 million lb.

Triacetate. Made only in very small quantities of the order of 1 million lb. in 1966, but growing.

Cuprammonium. Made in continuous filament, staple fibre and tow. Apparently the staple fibre is available not only in natural but also in spun-dyed colours. Production in 1967 about 40 million lb.

Regenerated Protein Fibres. Not made.

Synthetic Fibres. Main attention is directed to four fibres: Kapron, Anid, Lavsan and Nitron.

Kapron is nylon 6 made from caprolactam. Production in the U.S.S.R. started in 1950; in 1956 it was 25 million lb.; in 1961 about 45 million lb., with every indication that it would increase very fast because many new plants are going into production. The tyre market is one of Kapron's objectives. Production in 1964 about 90 million lb. and in 1966 about 185 million lb.

Anid is nylon 66; since 1956 it has been made on pilot plant scale and in 1966 large-scale manufacture of it started at Kursk. The development of Anid has apparently been retarded by a shortage of adipic acid; as the Russians make this acid from petroleum by way of benzene and *cyclo*hexanol it is rather odd that there should have been a shortage but whilst it lasted a fibre known as 669 was made from:

1 part hexamethylene diadipamide (6 and 6 carbon atoms)
2 parts caprolactam (6 carbon atoms)
1 part hexamethylene diazelaamide (6 and 9 carbon atoms)

in order to stretch the adipic acid supply. The azelaic acid has apparently been imported from Japan, where it is made (p. 397) from rice bran. It is almost incredible, but it has been reported that even in 1968 the Russians are still short of adipic acid.

Lavsan is polyethylene *tere*phthalate. Raw petroleum is converted at Novo Kinibyshev to *p*-xylol; at Stalingorsk it is converted to dimethyl *tere*phthalate, and the fibre Lavsan is spun at Kursk. Probably for some years most of the production of Lavsan will come from plant supplied by I.C.I.

Nitron is a polyacrylonitrile fibre with only a small amount of copolymerised additional monomer. Production in 1961 was of the order of 15–20 million lb. and rising rapidly. The complete acrylic plant which Courtaulds have contracted to supply to and start for Russia will increase the Russian acrylic fibre capacity considerably.

The Five Year Plan. The current plan runs from 1965–70. Until about 1956 the fibres industry in Soviet Russia lagged behind her other industries. At the end of 1957 when the first sputniks went up, Russia's synthetic fibres were at least ten years behind those in the U.K. or U.S.A. Even at the Soviet Trade Exhibition in London in 1961, what took the eye on the textiles stand were the fabulous carpets from Azerbaijan; there were no very promising synthetics. Round about 1956–58 decisions were made to push the lagging fibre industry forward; the main features were:

1. Production to concentrate on Kapron, Anid, Lavsan and Nitron roughly equivalent to Western Perlon, Nylon 66, Terylene and Courtelle, respectively.
2. An intensive and costly research programme to be undertaken to discover and develop new fibres to pilot plant stage.
3. In order to accelerate development, to buy from the West ready made and started plants for caprolactam, polyester, acrylonitrile and fibres from it, and possibly a complete plant for the Japanese modacrylic Kanekalon. All this was to save time; why

spend the time of research teams rediscovering what the West already knows, when it can be bought for £50 million or so?

4. To increase the production of man-made fibres from 350 million lb. in 1958 to 1,400 million lb. by 1965. Of this, the true synthetic fibre production is scheduled to rise from about 30 million lb. to about 400 million lb. The programme got off to a good start: 1960 production of all man-mades was 460 million lb. and 1961 was 550 million lb. At this stage Russian production of rayon was about 40 per cent of U.S.A. production; of true synthetic fibres Russian production was only 6 per cent of American and in 1963 still only 8 per cent. But new plants were springing up and going into production all over Russia. In 1963 Soviet production of all man-made fibres was 700 million lb. It is interesting to note that whilst the plan was to increase the production of man-made fibres from 350 million lb. in 1958 to 1,400 million lb. in 1965, what was actually achieved was 1,020 million lb. in 1966. The plan was to raise the production of synthetics from 30 million lb. in 1958 to 400 million lb. in 1965. What has actually been achieved is 220 million lb. in 1966.

Soviet 1966 Production of Man-made Fibres. In 1966 this was about 1,033 million lb., which was 1·83 times what it had been in 1961. Of this 1966 total the synthetic fibres accounted for 217 million lb., which was more than four times what they had been in 1961. This gain is really very great, and it is interesting to compare it with the rate of growth in other industrial countries.

Country.	Production (million lb.) of synthetic fibres in:		Ratio, 1966 : 1961.
	1961.	1966.	
U.S.S.R.	53	217	4·1
West Germany	147	479	3·3
Italy	99	316	3·2
Japan	345	1,036	3·0
U.S.A.	766	2,112	2·8
U.K.	149	402	2·7
France	115	246	2·1

It still remains that in 1966, in actual production of synthetic fibres, Russia was at the bottom of the league. However, her rate of growth was much the fastest. At what rate of compound interest does your money quadruple in five years? At about 32 per cent per annum. For every 3 lb. of synthetic fibres Russia made in one year she made 4 lb. in the following year. The 1966 Soviet production was made up as follows (in million lb.):

Rayons		
Viscose		
Textile continuous filament	98	
High-tenacity cord	223	
Staple fibre	406	
	727	
		727
Cuprammonium	44	44
Acetate		
Textile continuous filament	44	
Staple fibre	1	
	45	
		45
Total rayons		816
Synthetics		
Nylon (Kapron and a little Anid)	187	
Polyester (Lavsan)	16	
Acrylic	10	
P.V.C.	4	
Total synthetics	217	217
Total man-made fibres		1,033

These figures for Russian 1966 production of man-made fibres have been taken directly from work published in several papers in Russian journals and are believed to be accurate.

Fibres in Small Scale Production. Additional to the main manufacture of Kapron, Anid, Lavsan and Nitron, there are a number of fibres that have already reached small scale manufacture but which are marking time to allow maximum effort to be concentrated on the main four. They comprise:

Enant which is nylon 7 and is described on p. 377, also small quantities of Pelargon (nylon 9) and Rilsan (nylon 11) (p. 378).

Okson which is a polyester, mainly polyethylene *tere*phthalate, with a little plasticiser, probably a dihydric phenol. Better dyeability than Lavsan but poorer heat resistance.

Khlorin which is a chlorinated polyvinyl chloride fibre. Also, with the addition of 5 per cent cellulose nitrate known as Vinitron and having improved heat resistance. Another modification is Acetokhlorin which is Khlorin to which 15 per cent cellulose diacetate is added. Either addition is made to the spinning solution.

Soviden which is the equivalent of Saran (p. 467).

Saniv, a modacrylic similar to Verel (p. 490).

Kanekalon, the Japanese modacrylic made from 60 parts acrylonitrile, 40 parts polyvinyl chloride. The 15 tons per day acrylonitrile and 30 tons per day Kanekalon plant will cost Russia £10 million.

Sovinol, Vinol, fibres similar to the Japanese Vinylon (p. 484).

Yeranite, a polyvinyl acetate.

Ftorlon, possibly polymerised vinylidene fluoride. The polymer is wet spun from 12 per cent solution in acetone, has the chemical resistance of Teflon but greatly inferior heat resistance (softens at 125° C.). Specific gravity is 2·16 (Teflon 2·2) and strength is much higher than that of Teflon. The Russians were very proud of this fibre and have so far kept its composition secret, but not much has been heard of it since 1962.

Research and Development. The following fibres are in the stage of research or development:

Isotactic polypropylene, similar to Ulstron (p. 554).

Polyenanthamide (Enant, p. 377) has been graft polymerised on to carboxymethyl cellulose, in an attempt to get the dyeability of rayons and the strength of nylon in one and the same fibre.

Silicon (in the chain)—containing polyamides which have higher elastic recovery than similar polymers without the silicon. Here may be the way to avoid the flat-spot in nylon tyres that worries nylon manufacturers throughout the world.

Polyamides with sulphur replacing some of the oxygen in the main chain. Polyamides have also been made with phosphorus in the chain.

Acrylonitrile-protein fibres. This may be a way to get really wool-like fibres.

Metallic chain backbone polymers capable of withstanding heat.

Assessment. What does this information amount to? What is the overall picture? A fibre industry that until 1955 was very backward, probably ranked as unimportant compared with heavy industry. Then in the next few years a determination to advance it, to quadruple its output in seven years. Money is laid out to buy existing knowledge and ready made plant. Intense endeavour is put into research, not so much on the fibres that have been made in the West as on entirely new fibres. In speed of development the Russian picture is reminiscent of the fantastically rapid growth of the Japanese man-made fibre industry. By 1971 Russian production of man-made fibres should exceed that of the U.S.A.; before 1975 Russia should be the world leader in the field not only in quantity but also in variety and novelty. After that, leadership will probably pass to Asia. But the Russians think differently and they plan to produce 7,000 million lb. of man-made fibres in 1980. For comparison the U.S.A. production of man-made fibres in 1967 was about 3,600 million lb.

Raw Materials for Russian Man-made Fibres

The essential raw material for rayon is cellulose. The future may bring changes so that grass, leaves, and any plant material can be used, but the situation to-day is that wood-pulp is the universal form for fibres. At one time, fifty years ago, cotton was preferred for some fibres, but cotton to-day is dearer than the rayon that one could make from it. Wood-pulp is the essential for rayon. There are chemicals needed, too; mainly sulphuric acid, carbon disulphide and caustic soda for viscose rayon (pp. 155 *et seq.*), and acetic acid, acetic anhydride and acetone for cellulose acetate.

Timber. Of the industrial countries, the U.S.A. and the U.S.S.R. are richest in timber. In America it is envisaged that timber can be grown in the far north where it is too cold to grow wheat. In Russia about 32 per cent of the country or 2,700,000 square miles is forest containing timber estimated at 80,000 million cubic metres, nearly one-third of the world's timber resources. Each year about 400 million $m.^3$ is used, and this is only half of the natural annual increment due to growth. Efforts are being made in Russia to use more of the wood of deciduous or foliage trees for pulp so as to conserve the more valuable conifer wood for timber. The sight of all the timber lying around the docks at Helsinki might lead one to regard Finland as a big potential rayon manufacturer. Her production of rayon rose from 40 million to 70 million lb. over the period 1962–66; this is still only 1 per cent of world output, but it will probably increase.

Sulphuric Acid. Sulphuric acid is made in great quantities in all industrial countries. In 1965 Russia made 8·5 million tons, about 45 per cent of this from pyrites, 25 per cent from elemental sulphur and the remainder from hydrogen sulphide and other gases evolved in the working of non-ferrous metals.

Carbon Disulphide. World output of carbon disulphide in 1966 was estimated at more than 1·2 million tons, but the Russian share of this is not known; the main raw materials that are used for it are charcoal and elemental sulphur.

Caustic Soda. Caustic soda is made in enormous quantities throughout the world, and it is estimated that 30 per cent of all the world output is used in fibre manufacture. In 1965 the output of it in various European countries in millions of tons of 92 per cent material was: U.S.S.R. 1·30; West Germany 1·28; U.K. 0·91 (1964); Italy 0·79. Electrolysis of common salt is the preferred method of making caustic soda in all of these countries.

Viscose Chemicals. Thus, all the basic chemicals required in the viscose process are available in good supply in Russia, just as they are in the other advanced industrial countries. The output of rayon in the U.S.S.R. in 1966 was about 800 million lb., and it is certain to increase rapidly; at that figure it represented about 11 per cent of world production, and was lower only than the output in the U.S.A. and Japan; by the time this is printed the U.S.S.R. may well have overtaken Japan in rayon production. Russia has an enormous initial advantage in the possession of great cellulose resources in her forests. The highly stretched strong fibres, especially those that are to be made into tyre cord, demand a high-quality cellulose containing 95–98 per cent alpha cellulose; such high-grade material is available.

Acetate Rayon. The manufacture of acetate rayon took place first in those countries with an established chemical industry. The U.S.S.R. made only very little acetate for a long time, but as her chemical industry has grown so has the production of acetate increased, in fact by more than six times over the period 1958–65, a quite remarkable growth. Apparently, increasing quantities of acetic acid are being made from natural gas and oil; the acid can easily be dehydrated to the anhydride with the help of various catalysts. Acetone is also produced in Russia as elsewhere by the dehydrogenation of isopropyl alcohol:

$$(CH_3)_2CHOH \xrightarrow{-2} (CH_3)_2C{=}O$$

There is another method by which acetone is made, and that is from cumene (isopropyl benzene), and as this method yields phenol as a by-product, it is much favoured, there being (incredibly) a shortage of phenol in Russia. All the chemicals that are necessary for acetate rayon are available; in 1966 the U.S.S.R. made 40 million lb. of acetate continuous filament. This was only about one-twentieth of her total rayon output, but the proportion is increasing.

Synthetics. The basic materials required for synthetic fibres are caprolactam for nylon 6, acrylonitrile for the acrylics and mod-acrylics and terephthalic acid for polyesters, together with adipic acid if nylon 66 is to be made. There are, as has been indicated earlier, various ways of making these materials, but the preferred methods, the cheapest, all depend on a supply of natural oil or gas, in fact on a petrochemicals industry. The manufacturer of chemicals to-day can make almost anything from anything else provided that the required elements are present, but more and more he tends to make chemicals from oil. The development of oil and natural gas, of which Russia has vast resources, plays a determining part in the choice of the methods of synthetics. In the period 1958–66 the U.S.S.R. output of oil rose from 113 million to 265 million tons and of natural gas from 29·9 billion to 165 billion m.[3] The 1966 outputs were second only to those of the U.S.A. It is true that most of this oil and gas is used as fuel, but even so, some of it is used for chemicals; the use of natural gas in the Soviet chemical industry increased by 21 times over the period 1958–65. The growth is colossal; there is an abundance of raw materials for petrochemicals, and these are what is really essential for a thriving and fast-growing synthetic-fibres industry. What America's mineral wealth has meant to the Americans is clear enough to those who lack oil; in an earlier age what Britain's coal did for Britain was to make her the dominant world power. The Russians have more mineral wealth than any other country, and it will do a lot more for them than build up a synthetic-fibres industry.

Having written this, it is exasperating to have to say that Russia is still short of phenol and adipic acid, both of which can be made easily enough from the benzene which can be won from oil. There are still weak spots in the Soviet economy, and no doubt time will fill them in, but it is still surprising to find them still with us.

Kapron. Russia's nylon 6 is made from caprolactam, and although phenol has undoubtedly been used, and still is to some extent, to make the caprolactam, the method that they prefer is the oxidation of cyclohexane.

Polyester. The polyester fibre Lavsan is made from glycol and terephthalic acid, the latter coming from *p*-xylene, which in turn comes from oil. In Russia some time ago it was estimated that polyester could be made at a cost 30–35 per cent lower than poly-caprolactam. At the time this seemed surprising, because polyesters were then generally dearer than polyamides, but the dramatic fall in the price of our own Terylene staple (59*d.* at the end of 1967) shows how right they were. The U.S.S.R. is well equipped to make polyesters.

Acrylics. Four methods have been used in Russia for the manufacture of acrylonitrile:

1. From ethylene oxide and hydrocyanic acid

$$\underset{\diagdown\ O\ \diagup}{CH_2 - CH_2} + HCN \longrightarrow CH_2{:}CHCN + H_2O$$

2. From acetylene and hydrocyanic acid

$$HC{:}CH + HCN \longrightarrow CH_2{:}CHCN$$

3. From acetaldehyde and hydrocyanic acid

$$CH_3CHO + HCN \longrightarrow CH_2{:}CHCN + H_2O$$

4. From propylene and ammonia

$$CH_2{:}CHCH_3 + NH_3 \longrightarrow CH_2{:}CHCN + 3H_2$$

The first of these four methods was used initially in Russia, but it has gradually been displaced, and by the end of the present five-year period 70 per cent of Russia's acrylonitrile will be made from propylene and ammonia (4) and the remainder from acetylene and hydrocyanic acid (2). The cost of acrylonitrile made by method (4) is likely to be less than half of that made by method (1). It is all in line with the growth of Russia's petrochemicals industry; propylene is one of the big by-products of oil refining and one for which it is not always easy to find a remunerative use.

Modacrylics. There are several chemicals that are used in the production of acrylic and modacrylic fibres as co-monomers with acrylonitrile. Such are, for example, methyl acrylate, methyl methacrylate, vinyl pyridine, acrylamide and vinyl acetate. All these have been used, probably first, in the West. One that is used in small quantities in Russia's acrylic fibre Nitron is itaconic acid. They all increase the fibre cost because they are dearer, *e.g.*, itaconic acid is twenty times the price of acrylonitrile, but on the other hand they improve the performance of the fibre, usually in respect of its affinity for dyestuffs.

Timber and Oil. All round, Russia has all that is needed for man-made fibres and is appreciative of the possibility of producing polyester cheaply, and may well develop that fibre faster than nylon.

The polyacrylonitrile fibres are booming in Russia as they are everywhere else in the world. They are potentially very cheap and they may outvie the polyesters.

In brief, timber for rayon, oil for the synthetics.

Prices of Fibres

Fibres are valuable commodities; the value (sales price) of a year's production of all fibres, natural and man-made, is about £7,000 million. Prices of natural fibres have been as follows:

Fibre.	Price in pence per lb.									
	1934–38 average.	1949.	1950.	1951.	1953.	1955.	1959.	1965.	1967.	Dec.* 1967.
Cotton. American middling . .	6·3	24·9	35·8	46·5	31·8	31·8	21·5	23·2	22·6	26
Wool (clean). Crossbred 50 s . .	13	35·5	54	152	72	77	59	74	57	46
Raw silk. Jap 13/15	102	225	341	468	506	428	354	610	907	1,057
Flax. Belgian water-retted Grade "B".	8·7 (1938)	—	—	—	23·8	23·2	15·8	20·9	18·9	20·8

* After sterling devaluation.

The price of wool deserves special notice; it has fallen sharply. Throughout most of 1964 and 1965 the price of clean 50 s crossbred was around 73*d*./lb. Between August 1966 and August 1967 it fell from 72 to 59*d*./lb., and this fall has continued so that in the early months of 1968 it was only 48*d*./lb. The price of 64 s merino (clean), a finer wool, has, in the same period, remained much steadier; in 1964 it was about 110*d*./lb. and in early 1968 it was 108*d*./lb. This apparent stability is a little illusory, because by November 1967 the merino price had fallen to 87*d*./lb., and then when Britain devalued her currency it shot up to 101*d*./lb. The surplus of crossbred wool was such that the price, which had fallen to 42*d*./lb. at the time of devaluation (November 1967), only picked up to 46*d*./lb. There were large surpluses of wool in July 1967 at the end of the season; then the stock held by the New Zealand Commission as a result of purchasing to maintain prices was 161 million lb. clean wool, equal to one-third of a season's clip. In Australia, Argentina and Uruguay there were large stocks. In mid-1967 there was a world stock of 330 million lb. (clean basis) of wool. The stocks have gone up, the prices have gone down, the world clip (production) has not greatly changed. Has the time come that the synthetics are beginning to eat into the use of wool for apparel? It may have, and to the author it is a disquieting thought. There is so much about wool that is very good indeed; there is nothing like wool. Surely the wool industry will pick up. It has always been prone to ups and downs; a pre-war price of 13*d.*/lb. for clean 50 s crossbred and a 1951 price of 152*d.*/lb. for the same, illustrate the vagaries. It may, of course, be that just as a large supply of cheap rayon, Rayonne and Zellwolle, has brought down the price of cotton from 46*d.* in 1951 to 22*d.* in 1965, so the synthetic fibres as they have come down in price are forcing the wool price down; if a wool garment is the same price as a textured acrylic garment, then wool is preferred, but if it is twice the price the acrylic may be preferred. How far the sheep farmers will go along with this trend remains to be seen; probably they will shift from crossbreds to very fine merinos; it is still impracticable to make a babywool out of anything but merino wool.

But in 1967 the U.K. wool production fell by 2·1 per cent to 80·2 million lb., and the British sheep population fell by 3·5 per cent. No wonder that the British Wool Marketing Board are worried about the advance of man-made fibres (*Daily Telegraph*, 30 August 1968).

In 1960 the total value of the world production of fibres was not very different at £5,000 million from what it was in 1955. A very approximate breakdown for 1960 and for 1967 is:

Fibre.	£ thousand million.	
	1960.	1967.
Cotton	2	2·5
Wool	1	0·7
Rayon	1	1·0
Synthetics	0·5	1·8
Others (silk, flax, hemp and jute)	0·7	1·0

There has certainly been a big increase in the value of a year's output of fibres. It is the synthetics that have contributed most to this. There are two main tendencies to note in connection with the prices of artificial fibres, *viz.*:

1. Price is usually high to begin with, but as production grows, the fibre becomes considerably cheaper. In 1928 cellulose acetate 75 denier rayon in the U.K. cost more than two and a half times as much as it did in 1938. In America rayon prices more than halved between 1930 and 1939. Nylon was very much more expensive when it was first produced than it is to-day. A practice is already evident amongst manufacturers of new fibres of keeping the cost fairly low to begin with so as to give the fibre a good start; a loss may be made on the first

pilot plant production, but it is good policy to incur it, by selling fibre at a price that will be economic later when large manufacture starts. This latter tendency is of fairly recent introduction; it did not apply in the early days of man-made fibres when it was no light task to make anything at all that was saleable, and when this had been achieved the highest price that might reasonably be paid was asked.

2. Once a man-made fibre is on large-scale production its price is likely to be very steady, whereas natural fibres which depend on the quantity or quality of the cotton crop or the wool clip often fluctuate in price with disconcerting frequency. Most manufacturers prefer to work with a fibre whose price will be stable; too many of them have purchased wool at fantastic prices and later after the price has fallen, have been forced to sell their fabrics at a loss.

Variations in the prices of natural fibres are really astonishing; the figures in the Table on p. 864 are illustrative. Even if observations are restricted to the post-war period the variations are very considerable. Now let us consider, from the Table below, how the prices of rayon staple and rayon continuous filament yarn have varied over a similar period:

Fibre.	Price in pence per lb.									
	1939.	1949.	1950.	1951.	1952.	1954.	Jan. 1957.	April 1962.	Aug. 1968.	Nov. 1969.
Viscose. $1\frac{1}{2}$ denier $1\frac{7}{16}$ in. staple . .	10·2	17·8	18·6	24·9	26·8	24·0	24·0	22·8	23·0	23.0
Viscose. 150 denier 27 fils weft on pirn	32·0	50·5	62·0	68·0	68·0	68·0	74·0	75·5	75·0 (cones)	80·5 (cones)
										69·0 (cake)

From July 1953 until April 1959 the price of viscose staple was steady at 24*d.*; then it was reduced to 22¾*d.* In 1965 it was 22*d.*, but as a result of devaluation it was increased to 23*d.* in late 1967 or early 1968. The low price of staple makes the filament yarns look very dear, and so they are: they are held at artificially high prices by trade agreements. From January 1951 until January 1957 the price of 150 denier 27 fils viscose rayon on pirn was steady at 68*d.* In January 1957 most filament rayons were increased in price by about 10 per cent, and in April 1961 by another 2½ per cent. All the same, the man-made fibres are far and away steadier in price than the natural. There is very little continuous-filament rayon sold on pirn by the spinner nowadays, and it is for that reason that the latest prices quoted are for viscose rayon on cones.

So far as concerns the quantities of a fibre that will be used, price is one of the most important factors. Viscose staple at 23*d.* per lb. is a little cheaper than American middlings cotton at 26*d.*, and there can be no doubt at all that it is the low price of viscose staple which has brought about its use on the present large scale. In the U.S.A. viscose staple fibre was 25 cents per lb. in 1939; towards the end of 1950 its price reached 40 cents per lb., and this price was maintained until November 1952, when it was reduced to 37 cents per lb. and in May 1953 again reduced to 34 cents. In March 1962 it was about 27 cents with Zantrel, the polynosic variety, selling at 42 cents. American middlings cotton in America has followed very similar price trends. Undoubtedly the expansion of the rayon industry has forced down the price of cotton and when it has forced it down to the stage that it is no longer economic to grow and pick it then the cotton crop will decline. Rayon is at a tremendous advantage in that its raw material is timber. Large rayon producers such as the American Viscose Corporation, appreciating the need to have some control over their raw materials supply, have purchased large tracts of forest to safeguard their future supply of timber. Whereas in 1939 raw cotton was much cheaper than viscose staple, and cotton yarn was much cheaper than either spun rayon yarn or continuous filament yarn, the position since the war has been to the contrary, as the following figures exemplify:

Fibre.	Price on 30th Aug., 1939 (pence per lb.).	Price on 5th Feb., 1952 (pence per lb.).	Price on 7th Jan., 1957 (pence per lb.).	Price on 27th Oct., 1965 (pence per lb.).	Price in Oct., 1967 (before devaluation) (pence per lb.).	Price in Jan., 1968 (after devaluation) (pence per lb.).
Raw cotton. American middling	5½	44	27¾	23¼	22¼	26
Rayon staple. Fibro	10	27	24	22	22	23
Cotton yarn. 36 s American cop twist	9¾	72½	55	57¾	54½	67
Spun rayon yarn. 36 s Fibro on ring tube	17¾	51½	49¼	52	48	51
Rayon continuous filament yarn. 150 den. on cake	30½	54	60	62½	62½	66½

Comparison of the last five columns provides one more illustration not only of the influence which rayon has exerted to reduce the price of cotton but also of the relative steadiness of the prices of rayon.

Fibre prices have been disturbed by devaluation. Most simply, those that are imported from non-devaluing countries should go up, the others should be unchanged. The first changes were much along these lines, thus:

Fibre.	Pence per lb. in October 1967.	Pence per lb. in December 1967.
Cotton American middlings	22·2	26·0
Wool 64 s merino (clean basis)	89	103
Silk Jap 20/22	950	1,057
Flax Belgian retted	18·1	20·8
Sisal East African No. 1	6·9	7·5
Jute Export Lightnings	12·6	14·3
Viscose rayon $1\frac{1}{2}$ den. staple	22·0	22·0
Nylon 3 den. staple	82	82
Terylene 3 den. staple	59	59

But the effects of devaluation are greater than this Table would indicate. Already in January 1968 Terylene staple had gone up to 62*d.*, and Courtelle 2 den. staple, which had been 80*d.* since 1962, went up in January 1968 to 82*d.* Other increases have followed. Jute is becoming expensive, and will probably have to adjust itself to competition from polypropylene, which is potentially the cheapest of synthetic fibres and is already being used for baling twine.

Prices of synthetics are still much higher than those of other fibres. Prices ruling late in 1969 were as shown below. Since the 1967 devaluation of sterling the U.S. cent has been equal to the U.K. penny. Accordingly an American price of 86 cents per lb. (*e.g.*, staple fibre nylon) is equivalent to 86*d.* per lb.

Nylon (United Kingdom)	Staple fibre, 82*d.* per lb.
	210 den. filament yarn, 101*d.* per lb.
	60 ,, 108*d.* ,,
	30 ,, 116*d.* ,,
	15 ,, 207*d.* ,,
Nylon (U.S.A.)	Staple fibre, 86 cents per lb.
	70 den. filament yarn, 171 cents per lb.
	40 ,, 201 ,,
	15 den. monofil yarn, 274 cents per lb.
	,, multifil (3 fils) yarn, 391 cents per lb.
Terylene (United Kingdom)	Staple fibre, 67*d.* per lb.
	75 den. filament yarn, 145*d.* per lb.
	50 ,, 150*d.* ,,
Dacron (U.S.A.)	Staple fibre, 61 cents per lb.
	40 den. filament yarn, 190 cents per lb.
Orlon (U.S.A.)	Staple fibre, 86 cents per lb.
Acrilan Hi-Bulk (U.S.A.)	Staple fibre, 110 cents per lb.
Dynel (U.S.A.)	Staple fibre, 75 cents per lb.
Courtelle (U.K.)	Staple fibre, 3 den. $65\frac{1}{2}$*d.* per lb.
Ulstron (U.K.)	Staple fibre, 60*d.* per lb. (colours 6*d.* extra).

Very fine deniers of nylon are expensive, but their peculiar suitability for stockings enables them to command this price. Antron trilobal nylon filament is no dearer than regular round-section nylon in continuous filament, although staple is 10 cents dearer, *i.e.*, 96 instead of 86 cents per lb. in 6 denier. When the previous edition of this book was prepared Dacron was rather dearer than nylon. Since then the prices of both have come down, but those of the polyester faster than those of nylon, so that now polyester is cheaper than nylon, and in staple form is much cheaper. The intermediates for both nylon and Dacron are relatively expensive, and it is not easy to see how they can ever become really cheap fibres comparable in price with rayon. There can, however, be little doubt that their prices will be considerably reduced as manufacture increases. They have, in fact, been considerably reduced. It is interesting to compare the U.S.A. price for 70 den. weaving quality nylon yarn of $3·17 when it was introduced in 1938 with its price in 1968 of $1·20. There is a reduction of 60 per cent, but even so, what was a highly skilled technical synthesis in 1938 is the commonplace of 1968. People who want to spin nylon 66 usually make the chemicals themselves; nylon salt, hexamethylene diammonium adipate, can be bought from Badische, but perhaps nowhere else. Caprolactam, the basis of nylon 6, can be bought from many suppliers, but its price of around half a crown per lb. means that the fibre made from it is never going to compete on a price per lb. basis with rayon or with cotton. But on a use basis there may be competitiveness, because the strength and durability of nylon enable lighter fabric constructions to be used than are feasible with rayon, and usually with cotton.

The acrylates have rather different possibilities; it should not be beyond the ultimate resources of the chemical engineers to bring the price of acrylonitrile down to say 10 cents per lb.; if they can do this, then it should be possible to sell staple fibre acrylates at about 50 cents (say 4*s.*), and if this is done the ultimate production and consumption of acrylates may rival that of rayon. In 1961 the price of acrylonitrile did come down to 14½ cents per lb. In 1965 it was 17 cents; in 1968 it was back to 14½ cents and so it is in late 1969.

Early in 1962 the two modacrylic fibres, Verel and Dynel, were selling at 75 cents in 3 den. staple and at 65 cents in coarse 15–16 den. intended for the carpet trade. In 1965 the coarse denier was 62 cents, and so it was in 1968. Some fibres are made only abroad. Dynel, for example, is made only in America, and when the fibre comes into the U.K. there is an import duty of 7·8*d.*/lb., which is a burden when selling the fibre.

End-uses

What are the fibres to be used for? Relatively few of the people who make them have much idea about how they are used. World figures are difficult to come by, although clearly enough the type of application in a developing and backward country will be different from that in a rich and advanced country like America. Sometimes it seems that the Americans make life unnecessarily difficult and complex for themselves by making and possessing so many things which are too frivolous to be credible. Consequently their fibre end-uses may be, in a measure, theirs and only theirs. However that may be, they have classified their end-uses for man-made fibres and for all fibres under six headings. Some of their findings are shown below.

End-uses of Fibres

Kind of end-use.	Percentage of all fibres given up to it.		Percentage of man-made fibres given up to it.	
	In 1954.	In 1966.	In 1954.	In 1966.
Women's and children's wear	19	20	28	21
Men's and boys' wear .	21	20	11	11
Home furnishings . .	21	28	12	29
Other consumer uses .	10	11	10	12
Industrial uses . .	23	18	33	23
Exports	6	3	7	4

These end-uses do not account for such things as non-woven fabrics and military uses, but they account for a lot. The most significant inference to be drawn from this Table is that a large part of the increase in man-made fibres is going into furnishings. Less of it (relative to the volume of production) is going into apparel, and this may not be at all a bad thing.

Future Trends of Man-made Fibres

The technique of rayon manufacture is well established; rayon has demonstrated its versatility and general utility; people like it and it flourishes. In the main, its development will probably show no great changes in the relatively near future. Doubtless new raw materials will be increasingly used, *e.g.*, bamboo and esparto grass instead of timber; the use of cotton linters, already declining, will probably cease. Continuous spinning methods, already highly developed, will doubtless completely supersede the older processes. Stronger fibres will be forthcoming; the possibilities of fibres of the type of Fortisan which was made in the early 'thirties with a tenacity of 7 grams per denier have never been properly developed and improved fibres of this kind will come some time. Cellulosic fibres, which have more versatile dyeing properties, will probably be found useful. Taking a broad view, it might be suggested that viscose and acetate rayons as we know them to-day will continue for a long time pretty much as they are, and that the development of high-tenacity fibres will provide the main line of advance in the cellulosic field. The introduction of the polynosic rayons may be a milestone in rayon manufacture; viscose rayon, in them, may acquire the character and toughness of cotton. It is too early to say, there is too little experience, but if all goes well with them, manufacturers may turn a significant proportion of their viscose rayon capacity to the polynosic kind. But already it can be seen that the durability of the polynosic rayons is not so good as that of cotton.

Protein artificial fibres have had a bad setback and one cannot feel much confidence in the return of regenerated protein fibres. Some day synthetic protein fibres will come. That day may be imminent.

The synthetic field is wide open, despite the intensity of the work that is carried on; despite Carothers' wonderful achievement in synthesising nylon; despite the painstaking and successful work of his successors which has produced Terylene, Dynel and the other acrylates—despite all this, the surface of the field has been only scratched. True, those fibres that have been made have had special properties that have proved valuable and that have enabled fibres to find new uses in industry and at home, fibres that have led to sheerer stockings and fabrics than ever before, fibres that have been blended with natural fibres to improve the latter's strength and durability, but even so not one of these fibres is a really good all-round fibre. For special purposes they are sometimes ideal, but there is not one of them we can imagine being used so universally as wool or cotton or rayon.

All the synthetic fibres yet manufactured are strong and durable and chemically and biologically resistant, but none of them has the capacity for absorbing moisture that is so desirable for intimate wear, none of them has the natural waterproof character (non-wicking) of wool, none of them has the ease of coloration of cotton or wool, none of them has the enormous wet extensibility and excellent recovery therefrom of wool. So far, the chemists have synthesised those fibres that were easiest to synthesise, and not one of them is like a natural fibre; in the future they will turn their attention to imparting hydrophilic properties to those fibres they have already made by substitution of suitable chemical groups in the molecular chains, and still more profitably they will start to synthesise fibres from new intermediates that will give fibres with hydrophilic properties. Certainly a change in direction of synthetic fibre research away from hydrophobic and towards hydrophilic fibre synthesis is overdue. As will have been seen in Chapter 42, an offensive against these problems has already been mounted.

The World's Need of Man-made Fibres

Every second on this Earth three people die and four are born. The world's population (1967) is about 3,300 millions and is increasing at the rate of $1\frac{1}{2}$ per cent a year; at this rate it will be near the 5,000 million mark by the year A.D. 2000. The surface area of the world is about 128,000 million acres, of which 91,000 million are water, so that to-day there is 1 person to 12 acres of land on an average and as the "personal land area" decreases, problems of food supply will become more pressing, and food will always come before fibres. It takes 11 acres of good land to grow a ton of cotton and 20 acres to run the 200 sheep necessary to give a ton of greasy (or half a ton of clean) wool. Cotton and wool make big demands on the land.

Timber is an easier crop to grow than cotton and it will grow in cold regions that will not grow corn. One acre of land will have an annual increment of a half-ton of woody growth of pine and of this about half is cellulose, so that five acres of timber should provide a ton of viscose rayon a year. The saving over cotton is considerable and is all the more valuable in that much less labour is needed to harvest it—wood pulp is cheap in land and labour compared with cotton, and moreover can be grown in sub-Arctic regions. The increasing population of the world needs more food; it was Malthus who in 1798 forecast a world population (already exceeded) of 2,000 million by the year 2000 and who identified limitation of population with the available food supply. There is every reason (provided that a growing world population is still considered desirable) to use land that will grow corn to grow corn and not cotton, and to use land that will grow only timber to make viscose rayon.

But when the argument of greater needs of a growing population is pushed as it sometimes is to justify the making of *synthetic* fibres from oil, coal, natural gas, water, air, lime and salt, it seems to be a little precocious. The real justification for making these fibres is that they have very different properties from the natural fibres and the rayons, to which they are complementary. It may ultimately prove that the supplies of timber will be scarce, but it is inconceivable that supplies of cellulosic material of one kind or another will ever be scarce when all our vegetation is rich in it. It is difficult to foresee within the next decade or two the widespread displacement of rayon for apparel and in the home by the synthetic fibres. But Courtaulds' Teklan fibre has such good non-flam properties and is still so kindly to the hand that one must hope that it will be widely used for kiddies' frocks and nightwear. It will be the coming of such fibres with special properties and undeniable rights to be used that will gradually push rayon out. In the home they may find progress hard but industrially there is no shadow of doubt that the synthetic fibres have a great future. Price is at the bottom of this difference. Initial cost counts more with the housewife than with the industrialist.

The world's need of man-made fibres, indeed of any fibres, is made up of how much fibre each one of us uses. The first thought that we might all need about the same is not helpful. In North America they use 33 lb. each a year; in Africa only 4 lb. each. Doubtless climate plays a part, for in Russia they use 19 lb. but in Latin America only 9 lb. each. Throughout the world the average *per capita* consumption of apparel fibres was 11 lb. in 1960 having risen from 8¾ lb. in 1950. The 11 lb. that the average human used in a year was made up of:

1 lb. wool
7½ lb. cotton
2 lb. rayon
½ lb. nylon or other synthetic.

By 1967 the remarkable growth of the synthetic fabrics had made a significant difference to this *per capita* use of apparel fabrics. Indeed, for that year the figures were:

1 lb. wool
7 lb. cotton
2¼ lb. rayon
1¾ lb. synthetics (nylon, polyester, acrylics, etc.).

The apparel fibres used in a year by the average person now amount to 12 lb. Multiply these figures by 3·3 thousand million (the world population) and the results bear a strong, if not excessively close, resemblance to the world production of the fibres (p. 849).

The Social Repercussions of Man-made Fibres

Great good has come from the development and manufacture of the new fibres. The greatest good has come, appropriately enough, to the people who have made the fibres, and by this is meant not the few who by outstanding ability have amassed personal fortunes, but the hundreds of thousands of men and women who have given their labour and their thought to make the fibres. In the United Kingdom, the change that has taken place in the last forty or fifty years in the conditions of the textile workers has been revolutionary.

The author can remember, as a child, the women and girls clattering at some ungodly hour up the hill in Preston to t'mill, clog-irons ringing on the cobbled street, heavy black shawls enveloping heads and clutched round empty stomachs; long day after long day of incessant work for little more than bare subsistence was their lot. That mill, in its time reckoned a good one to work at, has woven no cloth this last thirty years. Times have changed, and in the textile industry for the first thirty-five years of this century the one development that overshadowed all others in the textile field was the growth of the viscose rayon industry, the advent of artificial fibres. In 1913 a few bobbins of artificial silk in a cotton-weaving factory constituted a novelty; at the best they would provide a narrow effect stripe in a shirting or a dress material. In 1928 artificial silk required huge factories with tens of thousands of employees to spin, wind and weave the rayon. These employees worked under vastly better conditions than their predecessors in the cotton factories. They had a shorter day and better wages with a high reward for outstanding skill; a weaver of rayon fabrics, provided she wove them well, might earn four times as much as a winder. Factory conditions were good; there were canteens, medical supervision and welfare facilities. Special 'buses transported the workers to and from the factories; their standard of living and degree of comfort far exceeded those of the Lancashire and Yorkshire weavers of the 'teens of the century. A real revolution in the life of many tens of thousands of workers had taken place, one due entirely to the growth of the artificial fibre industry and due in turn to the work of such people as Samuel Courtauld and Henry Dreyfus. Not only were the conditions for the tens of thousands improved, but great numbers of salaried jobs for technicians, chemists and physicists, accountants, clerks and so on emerged. Then came the nylon revolution, new ways of making fibres hitherto undreamt of emerged. These new techniques provided employment for more chemists, physicists, engineers and experts; fewer and fewer people worked on the older processes, on the loom and the spinning-frame. Engineers have designed automatic looms which require only one weaver to twenty-four looms; the length of yarn spun by an individual rayon or nylon spinner is astronomical compared with that spun on a spinning-wheel or even on modern worsted machinery. More and more men and women work farther back in the process, synthesising the chemicals, making the polymers, designing and building the spinning, winding, knitting and weaving machinery; all of this is more interesting work, work calling for a higher standard of training and for the greater use of intellectual capacity. The number of persons employed in the textile (excluding clothing) industries in the United Kingdom alone was 997,000 at the end of 1951. Tens of millions must be so employed throughout the world. The number of people employed in the U.K. textile industry has been subject to considerable changes. In the 'twenties it was about a million. In the 'thirties, affected badly by unemployment, it fell to about 900,000. When the Second World War came, the number had picked up to a million, but towards the end of this war it was only just over 600,000,

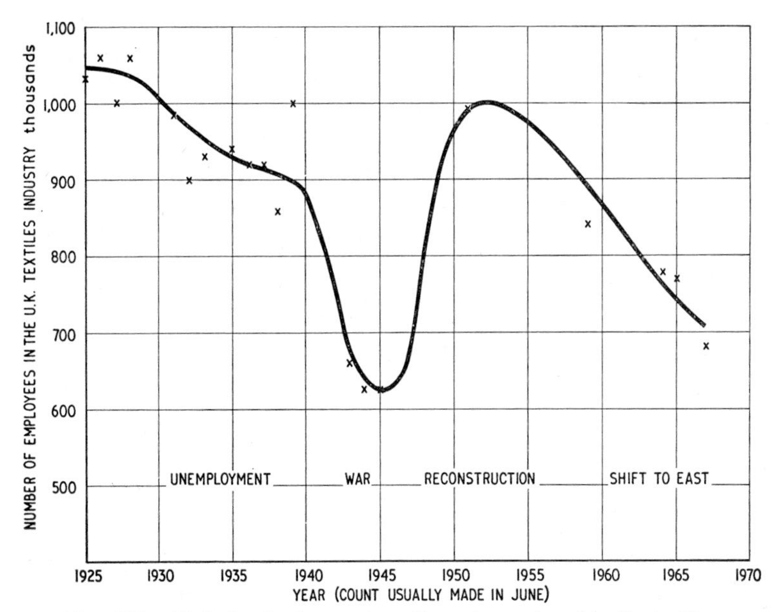

FIG. 276.—Variation in the number of people employed in the textiles (excluding clothing) industry in the the U.K.

and of these two-thirds were women. The industry revived, and the number employed rose again to about a million in 1951, but latterly has declined, and in 1967 was below 700,000. Some of the figures available are shown graphically in Fig. 276. There are mitigating circumstances: machines have been made which need less labour to look after them; people start work at a later age than formerly, and they give up work earlier; more people work in the design of the machines. But one thing stands out clearly and irrefutably, and that is that a much smaller proportion of our working population works on textiles than formerly. Is it a bad thing? Is it bad if our textiles industry move to the East? It is certainly a sad thing.

But if the workers in the textile industry itself have benefited the first and the most, the advantage that the rest of the world has gained from the new fibres has been not inconsiderable. It came a little later; even in 1928 rayon was still expensive, and most of the young women who wound it, warped it, and cut and sewed the fabric into dresses and underwear could not afford to buy it for their own use. Research went on in the rayon factories at an ever-increasing tempo; simultaneously with discoveries of new products there emerged discoveries of new processes. In many cases these enabled cost to be brought down; the chemicals used in the regeneration of rayon were recovered more and more completely. New and improved methods of preparing the raw materials were found, waste was almost eliminated, and all these improvements, all due to research, speedily brought down the price of rayon. Whereas in 1928 rayon garments had been beyond the purse of most of the factory girls, yet by 1938 they were their commonplace. Since the middle 'thirties girls and women from factory, shop, and office have worn clothes that in an earlier age would have been the envy of a queen. With the coming of beautiful materials, styles have changed; underwear has been streamlined and become briefer, to the advantage of health; dresses and frocks have developed character, a quality entirely missing from most of the day-to-day wear of working girls forty years ago. There has come with the new materials an incentive to smarten the personal appearance. The world at large has become a world of well-dressed women.

The Sale of Fibres Knowledge

But there are some other considerations that should not be disregarded. At the turn of the century there were thriving and very efficient technical schools throughout the industrial North; in Lancashire and Yorkshire particularly, men who worked in the mills during the day went to classes in the evening, and they were given a very clear understanding indeed of the textile processes amongst which they worked during the day. The technical schools did not hesitate to admit to their classes students from India, Russia, China, Japan, the Middle East and indeed the whole world. It was only a matter of a few decades before some of the countries from which these students had come were spinning and weaving cotton and wool for themselves; in the 'twenties the old kind of shipping order from China for 120 cases of 40 pieces each (say 200,000 yards of cotton cloth) just disappeared from the scene. The Lancashire manufacturers were soon thankful to receive orders for a few pieces of novelty fabrics, which often included a little artificial silk for figuring or decoration. Thankful, yes, but these were no proper substitutes to keep the looms going. In 1921, 42 of our textiles companies made a net profit of £13·6 million between them; in 1922 the same companies made only £7·1 million; the drop came very suddenly. By way of contrast, 83 of our brewers made a profit of only £2·1 million in 1921, but one of £7·8 million in 1922. The contrast highlights the sickness of our textiles industry in the 'twenties. But the man-made fibres were beginning then to be useful, and soon they grew in strength and they saved the situation. They were the best of reinforcements. The Industrial North had behaved with unstinted generosity in giving the technical knowledge that had been so slowly and hardly gained, and only a few decades later had lost her markets and much of her livelihood. Perhaps it was the right thing to do, and perhaps it would be done again. But as the writer sees it the decline of the textile industries, the prevalence of unemployment in the traditionally textile districts and the inevitable acceptance of a hard and uncultured and usually unpromising way of living resulted. Even now we hear of one mill after another closing. Tulketh, a fine spinning mill near Preston which the author has known from boyhood, closed late in 1967. All such events impoverish the Industrial North, and since the late 'twenties such events have been commonplace. The North now is impoverished, and it is saddening to return fifty or sixty years later to the once prosperous scenes of one's childhood.

If we take a very big step forward and look at the position of the synthetic fibres there seems to be reason for disquiet. If a country's industry can make one of the synthetic fibres it is broadly true to say that it can make them all; differences of processing and spinning from one fibre to another are of detail rather than of principle. It is the possession of a basic chemical industry which counts. Accordingly, when in the not very distant past a prominent American or British fibre producer made a deal with a Japanese manufacturer for the supply of manufacturing plant and the loan of technical personnel to start him up, and with the supply of technical information and licence rights, it was tantamount to the sale of knowledge that would enable the purchasing country to compete with the seller in the whole range of man-made fibres. Japan so acquired Exlan, which is roughly the equivalent of American Cyanamid's Creslan. Now Japan makes every fibre that is made in the West; the last gap was filled in 1965, when Teflon or PTFE went into large-scale production. Furthermore, the Japanese have Vinylon, a fibre of their own, and a very good one. Perhaps one must add that they probably do not yet have an equivalent of Nomex. And yet when the end of the war came in 1946 for Japan she had nothing, and if she had been left to her own devices she would still have had very little in the way of man-made fibres. But something was sold, help was freely given, and now the countries that gave it, the U.S.A. and the U.K., have to make laws to control fibre imports, including synthetic-fibre imports, from Japan.

An interesting development is that Japan is now becoming the seller, she is entering on the last stage. She has already sold fibre knowledge to Russia and has started to sell to China. Communist China ordered a plant capable of making 30 million lb. per annum of Vinylon, at a cost of about £10 million. The contract was secured by the Nichibo Co. in 1964, but its execution was held up pending approval from the Japanese Ministry of International Trade and Industry. This was forthcoming in 1965, and it was arranged that there should be a down payment by China of 25 per cent, with the balance spread over five years, with interest at 6 per cent per annum. What Lancashire did some fifty or sixty years ago with her cotton technology, Japan is evidently willing to do to-day with her man-made fibres technology. Events move much faster nowadays, and it will not be a few decades, but perhaps only a few years, before the Japanese manufacturers are despondently watching their erstwhile best customers making their own Vinylon and probably doing it very nicely. To sell technology to China is to ask for trouble if you want to stay in business.

It has to be added that although in Britain the Industrial North has suffered disastrously, there are some other areas of the country which have greatly benefited from changes in the textile industry. As cotton has fallen away, so rayon and later nylon, Terylene and Courtelle have grown, and grown fast, with correspondingly great gains to the people of Derby, Coventry, Pontypool and Grimsby. Courtaulds, who did most of the man-made fibres development in this country, have latterly sold a great deal of knowledge and plant to Russia, through their subsidiary Prinex, and I.C.I., too, have made similar deals.

At the beginning of 1967 Prinex signed a £9·4 million contract with the Soviet Government for the supply of plant and machinery for an acrylic fibre factory at Polotsk in Byelorussia, a plant which will probably make 100 million lb. of fibre a year. Previously, Prinex had completed five contracts for the Soviet Union, and all the factories are now in production. To sell knowledge to the developing countries where labour is cheap, and where new ideas and knowledge are welcomed as the new countries strive to become industrialists, is as sure a way as any of giving away your livelihood.

It all helps the inevitable translation of fibre production to the East. Meanwhile the proceeds of the sales help to support Britain's welfare activities, just as during the war the sale of Courtaulds' American factories helped to pay for the war.

There are many who believe that it is good to pass on our knowledge to poorer countries where the people have less technological experience and often lack, too, the ability to develop new ideas. Always there should be new processes and products coming along in the old-established countries where research and development are carried out on a large scale; always these gifted countries should keep a few jumps ahead and make their own livelihood by new methods. True, these in time will go to the copyists, but by then there should be a new lot and so on. This is an enlightened view, and one with which no one who has engaged in research would wish to quarrel. Research workers need an incentive, and there is no finer incentive than that of having to earn a living. But there will be hard times, intermittent periods when there is nothing much new to pay for our corn and meat.

On the other hand, consider the enormous research expenditure of I.C.I. The fibres they make are nylon, which came from du Pont in America, Terylene, which came from C.P.A. in England, and polypropylene, which came from Montecatini in Italy. There was, too, the ill-fated Ardil, a lovely fibre that I.C.I. took over from two professors, started to manufacture and then gave up. We should have liked to have seen new fibre after new fibre coming from I.C.I. With her wealth of technical staff, including a high proportion of scientists with first-class honours degrees, and with her finely appointed laboratories and development departments, at Harrogate in particular, success should surely have come. It may be that the period of waiting is nearly over; let us hope so and hope, too, that some synthetic protein fibre may soon be made. That way lies the break-through that this country so badly needs. Du Pont, the American chemical giant, have done well. They made nylon, Orlon, father of all subsequent acrylic fibres, and now they have Nomex. This is an impressive achievement which must encourage their chemists and physicists as well as their other employees.

Competition

Is it conceivable that if in this country there had been, say, a dozen independent fibre synthesis companies working competitively, small companies who would have had to compete to succeed, there would have been no new fibres over a period of years? It is not; the loss of competition is a great deprivation. In Japan there are twenty independent and competitive makers of synthetic fibres. As the author sees it, man-made fibre development there in the last ten years has far outstripped similar development in this country. Internal competition is essential, and it is all the more essential if in this country we are to continue our gifts of technical knowledge to the developing countries and our sale of it to other industrial countries who started later than we did.

If an industry is to become or even to remain prosperous it must have the benefit of competitive stimulus. It is, too, in the public interest to preserve competition. In July, 1965 the Board of Trade asked the Monopolies Commission to investigate whether monopoly conditions existed in the industry responsible for the supply of man-made cellulosic fibres. Their report to the Board was made on 12 March, 1968, and is summarised on p. 9 of the Annual Report (of the Monopolies Commission) by the Board of Trade for the year ended 31 December, 1968 (London, H.M.S.O., 1969; 2*s.* 3*d.*). There we can read:

" *The Conditions defined in the Monopolies Legislation*

" The Commission found that monopoly conditions did prevail because Courtaulds Ltd. supplied more than one-third of all the unprocessed man-made cellulosic fibres in the United Kingdom. The Commission further found that the following were things done by Courtaulds as a result of or for the purpose of preserving the monopoly conditions:

" (*i*) the practice of supplying cellulosic fibres to certain customers on terms or under conditions which were not obtainable by those customers' competitors;

" (*ii*) the regulation of imports of cellulosic fibres into the United Kingdom through arrangements with EFTA producers; and

" (*iii*) the policy of extensive participation in the textile industry.

" *The Public Interest*

" The Commission concluded that the monopoly conditions operate and may be expected to operate against the public interest."

Our local or home competition in the fibres industry has been largely eliminated by mergers and takeovers. The accountants point out in advance the savings that will accrue from a merger: one manager instead of two, one counting house instead of two and so on. Some results that they do not dwell on are: one expensive computer instead of none, a lot of men to nurse it instead of none, three or more accountants instead of two and so on. The accountants anticipate, with confidence, very considerable savings. It is perhaps never said explicitly, but there is an underlying assumption that all this will lead to lower costs. In fact there is no certainty that it will do so.

Mergers work to the detriment of most of the people concerned and should be discouraged. Employees soon have less choice of employers, customers have less choice of suppliers and shareholders get lower dividends. The biggest loss of all is the incentive to produce new products. If there are no effective competitors, there is no need to worry about what they are up to.

The Board of Trade has powers to control or stop mergers and takeovers. Only if competition is allowed to survive by such means as this will our man-made fibres industry flourish; encouragement should be given to proliferation. Thought should be directed not inwards but outwards. In wildlife, if a little territory is through misadventure left unoccupied some creature soon moves in and lives there. So will competition invade this country if home competition is discouraged. The subject in relation to protection in the textile industry was discussed in the House of Commons in December 1967 (report in *The Times* of Thursday, 14 December 1967): one Government speaker said, " The Government would not want to protect an industry which was not growing steadily more competitive, and here the position was not wholly satisfactory. Although the industry had increased its productivity in recent years, it looked as if there was still a considerable gap between it and foreign competitors. This seemed to be borne out by disappointing export performance. Britain has lost ground in export markets to countries where wages were as high or higher than in Britain."

Those with the interests of the country at heart cannot help feeling glad that the danger is recognised in government circles. It is true that that competition in which Britain has been failing is with foreign countries, and it is to that that the Government directed their indictment. But it is equally true that this failure was the result of lack of practice in competition at home.

A revival of the man-made fibres industry is still not impossible in this country, but it will come only from a competitive spirit. Given real competition, it could come in our laboratories and testing rooms within three years and in the saleroom within six. Without home competition we shall be left behind by those abroad who nurture fierce competition in their own countries. The Vickers-Zimmer do-it-yourself synthetic fibre-making is a new sort of competition to come into Britain, and a very welcome one indeed; one which has already become significant in our synthetic-fibre industry. If this activity is greatly increased it might well save the country's synthetic-fibre industry; it would not then be long before some of the newcomers to (say) polyamide making and spinning would be doing research on better fibres. We should be thankful that such pioneers can no longer easily be bought up.

One form of competition is that which comes from abroad; competition fills any vacant territory and quickly, too. The Americans, cognizant of the weak position of an uncompetitive British fibres industry, have moved in. Du Pont make their acrylic fibre Orlon at Londonderry and their snap-back Lycra at Maydown; Chemstrand make their acrylic Acrilan at Coleraine, also in Northern Ireland, and they also make some nylon 66 Blue C at Dundonald in Scotland. British Enkalon, an offshoot of the Netherlands Enka firm, are also in Antrim making nylon 6 and polyester. Retribution has overtaken Chemstrand, now taken over by Monsanto, but even so, Monsanto always had a share in Chemstrand. The chemical giants keep eating into the industry. Why do the Americans choose Northern Ireland? The most likely answer is that labour there may be more tractable and less troublesome. How different from the days when Courtaulds were building viscose spinning factories in the U.S.A. and British Celanese got American Celanese going.

Those indigenous and independent firms making fibres in Great Britain and Northern Ireland are: Courtaulds—who make viscose, acetate, triacetate, Courtelle, Celon (nylon 6), Teklan, Fibrolane, Evlan, tyre cord and alginate; I.C.I. Ltd.—who make nylon 66 (through British Nylon Spinners), Terylene and polypropylene; Greengate and Irwell Rubber Co.—who make some monofils; Plasticizers Ltd.—who make nylon 6 and monofils, as already indicated.

A variety of American giants have, as already mentioned, produced establishments in Northern Ireland, and there are various subsidiary companies who make snap-back yarns in England. Lansil Ltd. of Lancaster, one of our oldest makers of acetate yarn, were taken over by Monsanto of America in 1962.

But the great bulk of man-made-fibre production is in the hands of Courtaulds and I.C.I. From the standpoint of the vitality of our fibre industry it would be better if it were more widespread.

Much depends on top management. In this country there has been a trend to make managing directors and chairmen out of accountants and financiers. It may pay off for a while, but in the end technical management succeeds best; it does the work to-day that will be filling the firm's order books five or ten years hence. An accountant cannot do that, he may even try to prevent it being done in order to conserve liquid assets.

It is desirable that manufacturers put no obstacles in the way of outside criticism. We all know how difficult it is to assess fairly what we have made ourselves. The outsider can bring to bear a detachment that may be beyond us.

The critic's function is to look for and report on the good as much as on the bad. Especially does he look for what is new. He will give a fair deal because he knows that if his work is to be good it must always be fair. Any personal bias he may feel is vigilantly excluded. His report on what he sees and handles must always be on the basis of objective appraisal. But manufacturers become unsure of themselves without the constant vigilance which competition and criticism engender.

In the last twenty years the fibres scene has changed, and changed for the worse, in this country. In Japan, Russia, China and India too it has changed, and there greatly for the better. In other countries also, intermediately, but in few has it made so little progress as in the U.K. Devaluation did some good to our industry, and particularly our textile industry, but its effect is already fading, and will continue to fade, until we find that our wages are worth as much as they once were.

The Remedy

The remedy is clear; it is to foster the competitive spirit again. But how is it to be done? It is apparently easy to nationalise an industry but unthinkably difficult to denationalise it. Fibre making has not been nationalised, but it has been very highly concentrated. In the fibres industry the following measures would operate in the right direction and help to restore the industry to its earlier strength. Proliferation is essential:

1. Forbid the two giants from absorbing more firms. This would be Board of Trade action, and the Monopolies Commission has already made a start in this direction. Under the Monopolies and Mergers Act 1965 it is the Board of Trade which decides which mergers shall be investigated by the Monopolies Commission.
2. Release, so far as possible, such organisations as have already been captured.
3. Encourage those consumers of fibres who are unhappy with the present restriction of their sources of supply to make their own. The Vickers-Zimmer organisation makes this a practicable thing to do. And the Board of Trade presumably knows how to encourage and discourage where appropriate.
4. Discourage foreign fibre makers so that they do not come and set up plants in this country. Better to let our own people make what is needed. It is useless to allow the American giants to make mountains of nylon in Northern Ireland and then try

to put import curbs on some cotton shirtings or even shirts coming from Hongkong, East Africa or Portugal.

5. Ascertain what reasons are behind the closures of such old-established concerns as Tulketh and try to keep them going.

6. Organise combined teams of Board of Trade officials and practical small manufacturers (not high-grade employees or officials of the giants) to visit mills in the Industrial North and see where the Government can usefully lend a helping hand.

7. Introduce legislation to make all mergers and takeovers illegal without specific Board of Trade approval, which would only be granted in very exceptional, or better still in no, circumstances.

8. Gradually pursue a policy of lifting import controls. This would have to be done carefully and slowly, but the trend should be to lift rather than impose import duties. A competitive industry should be able to meet outside competition.

9. Forbid the sale of fibres knowledge and fibre spinning plants abroad.

10. Translate the careful vetting of admission which prevails at our uranium factories to the intrinsically just as valuable fibre research factories.

11. Greatly increase the influence of small practical manufacturers in the policy and running of our Research Associations.

If home competition with a multiplicity of small competing firms could be re-established this country might once again lead the world in the manufacture of man-made fibres. Until 1935 we did so lead the world, and given more technical top management and a bit of tough government we might well do so again. Inevitably in the end fibres will move to the East, but we might advantageously delay the process.

The accumulation of erstwhile customers which a large organisation can achieve is illustrated by the list of the major subsidiaries of Courtaulds. In the U.K. alone they have seven, wholly owned. Samuel Courtauld & Co. Ltd. and British Celanese Ltd. are well known, and the first may have ante-dated today's giant. But there are additionally the following:

Name.	Main Trading.
Furzebrook Knitting Co. Ltd	Wash knit fabrics
Gossard (Holdings) Ltd.	Foundation garments
Meridian Ltd.	Underwear, knitwear, hosier.
Aristoc Ltd.	Hosiery
Pinchin Johnson & Associates Ltd.	Paints and finishes

Furthermore, they have at least a three-quarters interest in the following three:

Bairnswear Ltd.	Knitwear, children's wear, rug wools and hand knitting yarns
Kayser Bondor Ltd.	Hosiery, lingerie
British Cellophane Ltd.	Packaging films and materials.

The adventure into the paint trade has always seemed a bit odd, but the activities of the other subsidiaries fall into line nicely with Courtaulds' original activities.

There are many who would, and probably do, applaud such diversification within the textile industry. It may be true that, in some measure, subsidiaries are permitted to buy their yarns where they like, even from suppliers not in the combine. As smooth a finish as possible will be given to the trading arrangements. But they do discourage the competitive spirit, and as such are detrimental to our national interests.

If a giant nylon spinner wants to go into the hosiery business and start making stockings out of his own fibres there is perhaps no tenable reason why he should not do so. But if instead he decides to buy up organisations that are already making stockings, perhaps partly from his own fibres, then such an action would probably be contrary to the national interest. It may not seem ethically wrong, but if it is something which destroys or seriously reduces competition, then it is something to be shunned and something at which people will increasingly look askance.

The Small Firms

In the U.K. the small independent fibre manufacturers have almost disappeared. They have been eaten up by the large companies, who found the competition that they had offered tiresome. In the light of the make-them-yourself fibres move, the position may change once again to one in which there are still more and still smaller manufacturers of fibres than ever before. And this will be to the good of fibre users, to the public who buy garments and furnishings and bed-linen made of the fibres. It will do no good to the fibre giants. In America this make-them-yourself move has already antagonised some of the largest fibre makers; in the U.K. the move has been tolerated so far. Efforts, supported by specious arguments about rationalising the industry, will doubtless be made to buy up or absorb those who are brave enough to manufacture on their own account. And they will need support of some kind.

In 1968 the Confederation of British Industry published a booklet " Britain's Small Firms; their Vital Role in the Economy ".

According to this, 97 per cent of all U.K. manufacturing establishments have less than 500 workers, and all together they employ 50 per cent of the manufacturing labour force and account for 45 per cent of all industrial sales. The sales value per person employed averaged £2,566 per annum in the small firms (under 500 employees) and £3,105 per annum in the larger firms; these last figures are derived ultimately from the Report on the Census of Production for 1958, Part 133. In Germany, France, Sweden and Norway the part played by small firms in their national economies is rather greater than in Britain. All that these figures show is that the small firms are very numerous and that they can hold their own if given a fair chance. Tax changes that have been introduced in recent years, particularly in 1968, discriminate against small firms, especially family businesses. What the C.B.I. says is that, " Private businesses should be encouraged to plough back as much profit as they consider necessary for their expansion ". That way, says the author, although not necessarily the C.B.I., lies the way to a great revival of British Industry. It must be accepted that when half of our working people are employed in only 3 per cent of our factories there is a serious shortage of the competitive stimulus. The competitive spirit should be strengthened. More small independent firms should be encouraged to spin synthetic fibres.

FURTHER READING

U.N. Statistical Yearbook (populations).

" Industrial Fibres ", 17th post-war issue, Commonwealth Economic Committee (1968); and 18th (1969).

Ind. Eng. Chem. Textile Fibers Symposium, **44**, 2101 (1952).

R. W. Moncrieff, " Appraisal of Man-made Fibres ", *Text. Manufacturer*, 188–91, May (1965).

R. W. Moncrieff, " You're Safer with Synthetics ", *Man-made Textiles*, June, July, August (1964).

" Russian Man-made Fibres: Growth and Innovation ", *The Times Review of Industry*, 51, February (1963).

R. W. Moncrieff, " Soviet Challenge in the Man-made Fibres Industry ", *City Press*, 8, February (1963).

R. W. Moncrieff, " China to Make Polyvinyl Alcohol Fibre ", *Text. Weekly*, 521–2 (19 March 1965).

Textile Organon (a monthly). Recent issues.

E. P. Ivanova and E. M. Mogilevskii, " Development of the Chemical Fibres Industry in the U.S.S.R. ", *Khim. Volokna*, 1967 (5), 4–9, and for English translation see *N.L.L. Translators' Bulletin*, **10** (4), 365–80 (1968).

E. P. Ivanova, " World Production of Chemical Fibres in 1966 ", *Khim. Volokna*, 1968 (1), 63–9.

COMMERCIAL MAN-MADE FIBRES

Name of fibre.	Type.	Manufacturer.
ACELBA	Acetylated viscose	Société de la Viscose Suisse, Switzerland.
ACELE	Cellulose acetate, now known as "acetate"	du Pont de Nemours & Co., U.S.A.
ACETA	Cellulose acetate filament	Farbenfabriken Bayer, Germany.
ACETAT RHODIA	Acetate staple	Deutsche Rhodiaceta A.G., Germany.
ACETATE	Generic name for cellulose acetate	
ACETATE, Type C	Crimpable cellulose acetate	du Pont de Nemours & Co., U.S.A.
ACETOKHLORIN	85 Khlorin/15 cellulose acetate co-spun	U.S.S.R.
ACRIBEL	Acrylic	Fabelta, Belgium.
ACRILAN	Modified polyacrylonitrile, staple	Monsanto Textiles, U.S.A. and N. Ireland.
ACRYBEL	Former name for Acribel	Fabelta, Belgium.
ADANA	Polyester	Turkey.
AGIL	Polyethylene	U.S.A.
AGILON	Stretch elastic nylon filament	Deering Milliken Research Corp., U.S.A.
AKULON	Nylon 6	A.K.U., Holland.
AKVAFLEX	Polyolefines	Norway.
ALASTIN	Viscose filament	Fabelta, Belgium.
ALASTRA	Viscose filament	Fabelta, Belgium.
ALBENE	Cellulose acetate	Deutsche Rhodiaceta A.G., Germany, and France.
ALBUNA	Cellulose acetate dull, filament	Snia Viscosa, Italy.
ALGINATE	Calcium alginate filament	Courtaulds Ltd., U.K.
ALON	High tenacity acetate staple	Toho Rayon Co., Japan.
ALPRONA	Polypropylene	Poland.
AMILAN	Nylon 6	Toyo Rayon Co., Japan.
AMILAR	Early name for Dacron	du Pont de Nemours & Co., U.S.A.
AMPLUM	Viscose bright	Algemene Kunstzijde N.V., Holland.
ANID	Nylon 66	U.S.S.R.
ANILANA	Acrylic	Poland.
ANTRON	Nylon with trilobal cross-section	du Pont de Nemours & Co., U.S.A.
ANTRON 24	Trilobal nylon 66 for upholstery	du Pont de Nemours & Co., U.S.A.
ARALAC	Casein (discontinued)	Atlantic Research Associates, U.S.A.
ARDIL B	Groundnut protein, almost white, staple (obsolete)	I.C.I. Ltd., U.K.
ARDIL F	Groundnut protein (fawn), staple (obsolete)	I.C.I. Ltd., U.K.
ARDIL K	Groundnut protein, carpet quality, staple (obsolete)	I.C.I. Ltd., U.K.
ARGENTEA	Viscose filament, bright	Snia Viscosa, Italy.

Name of fibre.	Type.	Manufacturer.
ARNEL	Cellulose triacetate, filament and staple	Celanese Corp. of America, U.S.A.
ARTILANA	Viscose (protein added)	Artilana Schlutius, Germany.
ASAHI	Viscose filament	Asahi Chemical Industry Co., Ltd., Japan.
ASTRALENE	Stretch (twist) Terylene	Dobsons (Silk Throwsters) Ltd., U.K.
ASTRALON	Stretch (twist) nylon	Dobsons (Silk Throwsters) Ltd., U.K.
ATLON	Cellulose acetate	Toho Rayon Co. Ltd., Japan.
AUSTRYLON	Nylon 66	Austria.
AVILA	Viscose straw, plain ribbon-like	Feldmühle Ltd., Switzerland.
AVILOC	Adhesive treated rayon yarn or cord	American Viscose.
AVISCO	General name for products of American Viscose Corp., U.S.A.	
AVISCO PE	Polyethylene	American Viscose Corp., U.S.A.
AVLIN	Viscose staple	American Viscose Corp., U.S.A.
AVRIL	High wet modulus viscose rayon (cross-linked)	American Viscose Corp., U.S.A.
AVRON	Medium tenacity viscose	American Viscose Corp., U.S.A.
AVRON XL	High tenacity viscose staple	American Viscose Corp., U.S.A.
AZLON	Generic name for reconstituted protein fibres	
AZOTON	Cyanoethylated cotton	
BAN-LON	Stretch elastic nylon, filament	J. T. Bancroft & Sons Inc., U.S.A.
BAYER-ACRYL	Old name for Dralon	
BAYER-ATLAS DRAHT	Nylon 6 for ropes	Farbenfabriken Bayer A.G., Germany.
BAYER-PERLON	Polycaprolactam filament and staple	Farbenfabriken Bayer A.G., Germany.
BELIMAT	Viscose filament, dull	Fabelta, Belgium.
BEMBERG	Cuprammonium filament	J. P. Bemberg A.G., Germany, American Bemberg Corp., U.S.A.
BEMSILKIE	Cuprammonium	Asahi Chemical Industry Co. Ltd., Japan.
BESLON	Polyacrylonitrile	Toho Rayon Co., Ltd., Japan.
BEXAN	Polyvinylidene chloride	B.X. Plastics, U.K.
B.H.S.	Modacrylic	Courtaulds Ltd., U.K.
BIJOHAI	Viscose filament	Teikoku Rayon Co. Ltd., Japan.
BLACKBIRD	Viscose filament	Nippon Rayon Co. Ltd., Japan.
BLANC DE BLANCS	Whiter (fluorescent) nylon 6	American Enka Corp., U.S.A.
BLUE C ELURA	Snap-back	Monsanto, U.S.A., and Polythane Fibres Ltd., U.K.
BLUE C NYLON	Nylon 66	Monsanto, U.S.A. and N. Ireland.
BOBINA	Viscose bristles	Kunstseidefabrik, Germany.
BOBINA-PERLON	Perlon bristles	Kunstseidefabrik, Germany.
BOBOL	Viscose staple	Snia Viscosa, Italy.
BODANA	Viscose filament (2½ den) bright and dull natural and spun-dyed	Feldmühle Ltd., Switzerland.
BODANELLA	Viscose filament (2 den) dull	Feldmühle Ltd., Switzerland.

Name of fibre.	Type.	Manufacturer.
BODANITA	Viscose filament (2 den) dull, high twist	Feldmühle Ltd., Switzerland.
BODANYL	Polycaprolactam filament	Feldmühle Ltd., Switzerland.
BOLTAFLEX	Polyvinylidene chloride round and flat monofils	Bolta Products, U.S.A.
BOLTATHENE	Polyethylene	Monsanto, U.S.A.
BREDA	General name for products of N.V. Hollandsche Kunstzijde Industries, Breda, Holland	
BREDANESE	Viscose, semi-dull	N.V. Hollandsche Kunstzijde Industries, Breda, Holland.
BRENKA	Viscose filament	British Enka Ltd., U.K.
BRENKONA	Viscose filament, bright	British Enka Ltd., U.K.
BRIGLO	Viscose filament, bright	American Enka Corp., U.S.A.
BRI-NYLON	Nylon 66	British Nylon Spinners Ltd., U.K., now I.C.I. Fibres Ltd.
BRITBEM	Cuprammonium filament (discontinued)	British Bemberg Ltd., U.K.
BRITENKA	Viscose filament, bright	British Enka Ltd., U.K.
BUBBLFIL	Hollow viscose	du Pont de Nemours & Co., U.S.A.
BURLANA	Acrylic	Bulgaria.
CADON	Multilobal nylon 66	Monsanto, U.S.A.
CALCIUM ALGINATE	Calcium alginate	Courtaulds Ltd., U.K.
CALGITEX	Calcium–sodium alginate gauze	Medical Alginates Ltd.
CALYX	Viscose matt	North British Rayon Co. Ltd. U.K.
CANTONA	Viscose, semi-dull	Algemene Kunstzijde N.V., Holland.
CANTRECE	Polyamide	Canada.
CAPROLAN	Polycaprolactam filament	National Aniline Division Allied Chem. & Dye Corp., U.S.A.
CAPRON	Polycaprolactam	Klin, U.S.S.R.
CARANA	Polyamide	Canada.
CARBOLAN	Carbonated polyacrylonitrile	Nippon Carbon Co., Yokohama.
CARBOXYMETHYL CELLULOSE	Experimental alkali-soluble fibre	U.S. Dept. of Agriculture, U.S.A.
CAROLAN	Cellulose acetate filament and staple	Mitsubishi Acetate Co. Ltd., Japan.
CASHMILAN	Polyacrylonitrile staple	Asahi Chemical Industry Co. Ltd., Japan.
CASLEN	Casein monofil	Rubberset Co., U.S.A.
CASOLANA	Casein staple	Vereen Melkwolfabrik, Holland.
CELACLOUD	Acetate filling fibre	Celanese Corp. of America, U.S.A.
CELAFIBRE	Cellulose acetate staple	British Celanese Ltd., U.K.
CELAFIL	Cellulose acetate, ruptured	British Celanese Ltd., U.K.
CELAIRE	Nylon–acetate blend	Celanese Corp. of America, U.S.A.
CELANESE	Cellulose acetate filament	British Celanese Ltd., U.K.
CELAPERM	Cellulose acetate, spun-dyed	Celanese Fibers Co., U.S.A.
CELATOW	Acetate tow	Celanese Fibers Co., U.S.A.
CELCOS	Cellulose acetate, partly saponified	Celanese Corp. of America, U.S.A.

Name of fibre.	Type.	Manufacturer.
CELCOSA	Viscose film	La Cellulose de Condé, France.
CELLESTRON	Cellulose acetate film	Dai Nippon Celluloid Co. Ltd., Japan.
CELON	Polycaprolactam	Courtaulds Ltd., U.K.
CELTA	Hollow filament viscose	Société de la Viscose Suisse, Switzerland.
CELTREL	Polyester	Colombia.
CELTRON	Polyester	Venezuela.
CHARDONNET	Regenerated cellulose from nitrocellulose, long obsolete	Société Anonyme pour la fabrication de la Soie de Chardonnet, France.
CHEMLINE, Type AO5	Nylon for industrial and marine ropes	Chemstrand, U.S.A.
CHEMLON	Polyamide	Czechoslovakia
CHESLON	Bulked acetate and nylon	U.K.
CHEVIOT	Viscose filament, bright	North British Rayon Co. Ltd., U.K.
CHEVISOL	Viscose filament, bright	North British Rayon Co. Ltd., U.K.
CHINLON	Nylon 6	China.
CHROMSPUN	Cellulose acetate spun-dyed, filament and staple	Tennessee Eastman Co., U.S.A.
CIFALON	Polyamide	Portugal.
CISALFA	Viscose proteinised	Cisa Viscosa, Italy.
CLORENE	Polyvinylidene chloride (monofil)	Société Rhovyl, France.
COLCESA	Viscose filament, bright spun-dyed	Glanzstoff-Courtaulds G.m.b.H., Germany.
COLCORD	High tenacity viscose tyre cord	Glanzstoff-Courtaulds G.m.b.H., Germany.
COLNOVA	Viscose filament pearl, spun-dyed	Glanzstoff-Courtaulds G.m.b.H., Germany.
COLOMAT	Viscose filament matt, spun-dyed	Glanzstoff-Courtaulds G.m.b.H., Germany.
COLORAY	Viscose staple, spun-dyed	Courtaulds (Alabama) Inc., U.S.A.
COLVA	Viscose staple	Glanzstoff-Courtaulds G.m.b.H., Germany.
COLVADUR	High tenacity viscose staple	Glanzstoff-Courtaulds G.m.b.H., Germany.
COLVALAN	Crimped viscose staple	Glanzstoff-Courtaulds G.m.b.H., Germany.
COMISO	Viscose tow	Beaunit, U.S.A.
CONYMA	Viscose	Nyma Rayon Works, Ltd., Holland.
COPET	Polyamide	Argentine.
CORDAMEX	High tenacity viscose	Cellulosa y Derivados, S.A., Mexico.
CORDURA	High tenacity viscose filament	du Pont de Nemours & Co., U.S.A.
CORNYL	Nylon 6 for tyre cord	A/B Svenskt Konstsilke, Sweden.
CORVAL	Cross-linked viscose	Courtaulds (Alabama) Inc., U.S.A.
COURLENE	Polyethylene filament	Courtaulds Ltd., U.K.
COURLENE X3	Polyethylene filament	Courtaulds Ltd., U.K.
COURNOVA	Polypropylene monofils	British Celanese, U.K.
COURPLETA	Cellulose triacetate, filament and staple	Courtaulds Ltd., U.K.
COURTELLE	Modified polyacrylonitrile	Courtaulds Ltd., U.K.

Name of fibre.	Type.	Manufacturer.
COURTOLON	Bulked nylon 66 yarn	Courtaulds Ltd., U.K.
COVA	Viscose straw	Société de la Viscose Suisse, Switzerland.
CREMONA	Vinylon	Kurashiki Rayon Co. Ltd., Japan.
CREPESYL	Viscose filament	North British Rayon Co. Ltd., U.K.
CRESLAN	Modified acrylic	American Cyanamid Co., U.S.A.
CRESLAN, Type 63	Readily dyeable modification of filament Creslan	American Cyanamid Co. Ltd.
CRILENKA	Acrylic	Spain.
CRIMPLENE	Textured polyester yarn	Licensees of I.C.I.
CRINOL	Viscose monofil	Société de la Viscose Suisse, Switzerland.
CRINOVYL	Polyvinyl chloride staple	Société Rhovyl, France.
CRISPELLA	Viscose, ribbon shaped filaments	Hollandsche Kunstzijde Industries, Holland.
CROLAN	Polyester	Mexico.
CRYLENE	Acrylic	Italy.
CRYLOR	Modified polyacrylonitrile filament and staple	Société Rhodiaceta, France.
CUMULOFT	Textured nylon 66	Chemstrand (now Monsanto), U.S.A.
CUPIONI	Cuprammonium slub yarn	J. P. Bemberg A.G., Germany.
CUPRACOLOR	Spun-dyed cuprammonium	American Bemberg, U.S.A.
CUPRAMA	Cuprammonium staple	Farbenfabriken Bayer A.G., Germany.
CUPRAMA TX	Cuprammonium, for carpets	Farbenfabriken Bayer A.G., Germany.
CUPRAMMONIUM	Regenerated cellulose made by cuprammonium process	
CUPRESA	Cuprammonium filament	Farbenfabriken Bayer A.G., Germany.
CYDSA	Polyamide	Mexico.
DACRON	Polyethylene *tere*phthalate filament and staple	du Pont de Nemours & Co., U.S.A.
DAIFUKI	Viscose filament	Toyo Rayon Co. Ltd., Japan.
DANUDUR	High tenacity viscose staple	Süddeutsche Chemiefaser A.G., Germany.
DANUFIL	Staple viscose, natural and spun-dyed	Süddeutsche Chemiefaser A.G., Germany.
DANUFLOR	Staple viscose for worsted system	Süddeutsche Chemiefaser A.G., Germany.
DANULON	Nylon 6	Hungary.
DARLAN	Polyvinylidene dicyanide staple	B. F. Goodrich Chem. Co., U.S.A. (*see* Darvan).
DARVAN	New name for Darlan	Celanese Fibers Co., U.S.A.
DARYL	Triacetate filament	Fabelta, Belgium.
DAWBARN	Polyolefine	Dawbarn Bros. Inc., U.S.A.
DAYAN	Nylon 6	Perlofil S.A., Spain.
DECORA	Viscose filament spun-dyed	Société de la Viscose Suisse, Switzerland.
DEDERON	Nylon 6	V.E.B. Kunstseidenwerk, East Germany and Roumania.
DELCRON	Polyester	Mexico.
DELFION	Nylon 6	Soc. Bombini, Italy.
DELUSTRA	Viscose filament, dull	Courtaulds Ltd., U.K.

Name of fibre.	Type.	Manufacturer.
DIAFIL	Viscose filament	Teikoku Rayon Co. Ltd., Japan.
DICEL	Cellulose acetate filament	British Celanese Ltd., U.K.
DICROLENE	Polyester	Argentine.
DIEN	Polybutadiene	Bataufsche Petroleum Maatschappij, Holland.
DIMAFIL	Polyamide Nylon 6	Plasticizers Ltd., Drighlington.
DINITRILE A	Early name for Darlan	
DIOLEN	Polyethylene terephthalate	Vereinigte Glanzstoff-Fabriken A.G., Germany.
DLP	Polyolefines	Dawbarn Bros. Inc., U.S.A.
DOLAN	Modified polyacrylonitrile staple	Süddeutsche Zellwolle, Germany.
DORLON	Polyurethane (Perlon U) monofils	Farbenfabriken Bayer A.G., Germany.
DRALON	Modified polyacrylonitrile staple	Farbenfabriken Bayer A.G., Germany.
DRAWINELLA	Cellulose acetate staple	A. Wacker G.m.b.H., Germany.
DRYLENE	Polyethylene monofil	Plasticizers Ltd., Drighlington.
DUCILO	Nylon 66	Argentine.
DULESCO	Viscose dull (obsolete)	Courtaulds Ltd., U.K.
DULKONA	Viscose filament matt	British Enka Ltd., U.K.
DUL-TONE	Viscose dull	Industrial Rayon Corp., U.S.A.
DURACOL	Indicative of Spun-dyed, *e.g.*, DURACOL-FIBRO	Courtaulds Ltd., U.K.
DURAFIL	High tenacity viscose staple	Courtaulds Ltd., U.K.
DURAFLOX	High tenacity viscose staple	Spinnfaser A..G, Germany.
DURASPON	Polyurethane	U.S.A.
DURETA	Cuprammonium filament	J. P. Bemberg A.G., Germany.
DYNEL	Copolymer of vinyl chloride/acrylonitrile staple	Carbide & Carbon Chemicals Corp., U.S.A.
EFYLON	Polyamide (made from waste fibre) (Nylon 6)	Hungary.
ELANA	Polyester	Poland
ENANT	Polyoenanthic lactam	U.S.S.R.
ENGLO	Viscose filament dull	American Enka Corp., U.S.A.
ENKA	General name for products of Algemene Kunstzijde N.V., Holland	
ENKALENE	Polyethylene *tere*phthalate	Algemene Kunstzijde N.V., Holland.
ENKALON	Nylon 6	American Enka Corp., U.S.A., and British Enkalon Ltd., N. Ireland.
ENKASA	Casein	Algemene Kunstzijde N.V., Holland.
ENKATRON	Multilobal nylon 6	American Enka Corp., U.S.A.
ENKA-5000	Viscose tyre yarn	American Enka Corp., U.S.A.
ENKONA	Viscose	Algemene Kunstzijde N.V., Holland.
ENVILON	Polyvinyl chloride filament	Toyo Chemical Co. Ltd., Japan.
ERILAN	Protein	Hungary.
ESSEVI	Viscose filament	Snia Viscosa, Italy.
ESTANE	Polyurethane spandex	B. F. Goodrich Chemical Co., U.S.A.
ESTERA	Cellulose acetate staple	Dai Nippon Celluloid Co. Ltd., Japan.

Name of fibre.	Type.	Manufacturer.
ESTRON	Cellulose acetate filament	Tennessee Eastman Co., U.S.A.
ETHOFIL	Experimental cyanoethylated cellulose	Dow Chemical Co., U.S.A.
ETHYLON	Polyethylene	Kureha Kasei Co. Ltd., Japan.
EUROACRYL	Acrylic	Italy.
EVLAN	Modified cellulose staple for carpet yarn.	Courtaulds Ltd., U.K.
EVLAN M	Similar to Evlan but with better abrasion resistance	Courtaulds Ltd., U.K.
EXLAN	Polyacrylonitrile staple	Japan Exlan Co. Ltd., Japan.
FABELCORD	High tenacity viscose filament	Fabelta, Belgium.
FEATHERAY	Hollow filament viscose	Hartford Rayon Corp., U.S.A.
FIBER E	Crimpable viscose	du Pont de Nemours & Co., U.S.A.
FIBER G	Early name for Cordura	du Pont de Nemours & Co., U.S.A.
FIBER V	Early name for Dacron	du Pont de Nemours & Co., U.S.A.
FIBERFRAX	Aluminium silicate refractory fibrous material	Carborundum Co., U.S.A.
FIBERGLAS	Glass filament and staple	Owens-Corning Fiberglas Corp. U.S.A.
FIBRANA	Viscose staple	Spain.
FIBRANNE	Viscose staple (generic French name)	
FIBRAVYL	Polyvinyl chloride staple	Société Rhovyl, France.
FIBREGLASS	Glass	Fibreglass Ltd., U.K.
FIBRELTA	Viscose staple	Fabelta, Belgium.
FIBRENKA	Viscose staple	Algemene Kunstzijde Unie N.V., Holland.
FIBRO	Viscose staple	Courtaulds Ltd., U.K.
FIBROCETA	Cellulose acetate staple	Courtaulds Ltd., U.K.
FIBROLANE A	Casein staple	Courtaulds Ltd., U.K.
FIBROLANE BX	Casein staple	Courtaulds Ltd., U.K.
FIBROLANE BC	Casein staple chromium combined	Courtaulds Ltd., U.K.
FIBROLANE C	Groundnut protein staple	Courtaulds Ltd., U.K.
FILMTEX	Polyolefines	Norway.
FIRESTONE	Polyethylene	Firestone Products, U.S.A.
FIRESTONE-NYLON	Nylon 66 tyre cord	Firestone Products, U.S.A.
FLATTOYO	Viscose ribbon straw (for wallpaper)	Toyo Rayon Co. Ltd., Japan.
FLIMBA	Viscose staple, spun-dyed	Société de la Viscose Suisse, Switzerland.
FLISCA	Viscose staple	Société de la Viscose Suisse, Switzerland.
FLOCK	Viscose very short staple (also a generic name for this material)	Société de la Viscose Suisse, Switzerland.
FLOTEROPE	Polypropylene	U.S.A.
FLOX	Viscose staple	Spinnfaser A.G., Germany.
FLUFLENE	Stretch Terylene filament	Wm. Frost & Sons Ltd., U.K.
FLUFLON	Stretch nylon filament	Wm. Frost & Sons Ltd., U.K.
FORCEL	Polyester	Peru.
FORLION	Nylon 6	Soc. Orsi Mangeli, Italy.

Name of fibre.	Type.	Manufacturer.
FORTINESE	Stretch spun cellulose acetate	Celanese Corp. of America, U.S.A.
FORTISAN	Cellulose (acetate stretched and saponified) filament	Celanese Fibers Corp., U.S.A.
FORTISAN 36	High tenacity cellulose, heavy denier	Celanese Corp. of America, U.S.A.
FORTREL	Polyester	Fiber Industries Inc. (Celanese), U.S.A.
FRANKELON	Polyamide	Italy.
FT(ARCT)	Stretch nylon (false twist)	Deering Milliken Research Corp., U.S.A.
FTORLON	Fluorocarbon (probably polyvinylidene fluoride)	U.S.S.R.
FURON	Polyamide	Poland.
GARFLON	Polyamide	India.
GERRIX	Glass	Gerresheimer Glasshüttenwerke, Germany.
GIZOLAN	Cellulosic	Hungary.
GLANZSTOFF	Viscose	Vereinigte Glanzstoff-Fabriken A.G., Germany.
GLANZSTOFF-KORDREYON	High tenacity viscose rayon for tyres	Vereinigte Glanzstoff-Fabriken A.G., Germany.
GLANZSTOFF-PERLON	Nylon 6 staple	Vereinigte Glanzstoff-Fabriken A.G., Germany.
GLASWOLLE	Glass	K. G. Schuller & Co., Germany.
GLOSPAN	Polyurethane snap-back	Globe Mnfg. Co., U.S.A., and Globe Elastic Thread Co. Ltd., U.K.
GOSSAMER	Polyester sewing thread	Coats, Patons, U.K.
GRILENE	Polyester-polyether	Grilon S.A., Switzerland.
GRILLON	Polycaprolactam	Nippon Rayon Co. Ltd., Japan.
GRILON	Polycaprolactam	Grilon S.A., Switzerland.
GRISINTEN	Polyester	East Germany.
GUSEI-ICHIGO	Polyvinyl alcohol	Japan.
HEATHERDINE	Stretch nylon 66 with ruptured filaments	G. H. Heath & Co., U.K.
HELANCA	Stretch elastic nylon filament and Terylene	Heberlein & Co., A.G., Switzerland.
HELCON	Polyamide	Italy.
HERCULAN	Polypropylene staple	Hercules Powder Co., U.S.A.
HI-BULK ACRILAN	Acrilan which crimps on near boiling	Chemstrand Corp. (now Monsanto), U.S.A.
HI-CRIMP VICARA	Crimped zein protein	Virginia Carolina Chem. Corp., U.S.A.
HIGH BULK ORLON	Orlon Type 42 which crimps on near boiling	du Pont de Nemours & Co., U.S.A.
HIGH BULK TRICEL	Bulked Tricel	Courtaulds Ltd., U.K.
HIGHTEL	Polynosic viscose staple	Teikoku Rayon Co. Ltd., Japan.
HILON	Polyamide	Italy.
HI-NARCO	Viscose for tyres	North American Rayon Corp., U.S.A.
HIPOLAN	Polynosic rayon	Japan.
HIRALON	Polyethylene	Japan.
HIRLON	Polyamide	Argentine

Name of fibre.	Type.	Manufacturer.
HISAFLEX	Polyamide	Argentine.
HISILON	Polyamide	Argentine
HISILON	Polyamide	Monte Video, Uruguay.
HOSTALEN	Polypropylene	Germany.
HSIEN-THIN	Polyethylene	Taiwan.
HYFIL	High modulus carbon	R.A.E.
IGG-VESTAN	Polyvinylidene chloride	Internationale Galalith G.m.b.H., Germany.
INTERSPAN	Polyurethane	Colombia.
IRC, Type 6	Polycaprolactam staple	Industrial Rayon Corp., U.S.A.
IRIDEX	Viscose spun-dyed filament	Kirklees Ltd., U.K.
IRIDYE	Viscose yarn-dyed	Kirklees Ltd., U.K.
ISOVYL	Polyvinyl chloride	Société Rhovyl, France.
IVOREA	Viscose filament dull	Snia Viscosa, Italy.
JAYANKA	Polyamide	India.
JEDMAT	Viscose filament matt	North British Rayon Co. Ltd., U.K.
JEDSOL	Viscose filament matt	North British Rayon Co., Ltd., U.K.
JEDSYL	Viscose filament bright	North British Rayon Co. Ltd., U.K.
KALIMER	Polyester	Italy.
KANEBIAN	Vinylon	Kanegafuchi Spinning Co. Ltd., Japan.
KANEKALON	Acrylonitrile/vinyl ester co-polymer (modacrylic staple)	Kanegafuchi Chem. Co. Ltd., and Kanekalon Co. Ltd., Japan.
KANELION	High tenacity rayon staple	Kanegafuchi Spinning Co. Ltd., Japan.
KANSU	Acrylic	China.
KAPRON	Another spelling of Capron	
KASILGA	Viscose filament	Kunstsilkefabrikken, Norway.
KELHEIM H	Hollow filament viscose	Süddeutsche Zellwolle A.G., Germany.
KHLORIN	Polyvinyl chloride	U.S.S.R.
KIKANSEI	Viscose filament	Toyo Spinning Co. Ltd., Japan.
KIRKSYL	Viscose filament	Kirklees Ltd., U.K.
KODEL	Polyester	Eastman Chemical Products, U.S.A.
KOLON	Polyamide	Korea.
KOSEI-SILK	Fibre spun from waste silk protein	Fuji Spinning Co. Ltd., Japan.
KREHALON	Polyvinylidene chloride	Kuahi Kasei, Japan.
KREHALON S	Polyvinyl chloride	Kureha Kasei, Japan.
K.R.P.	Viscose staple	Kokohu Rayon & Pulp Co. Ltd., Japan.
KRYLION	Acrylic	Italy.
KURALAY	Polyester	Japan.
KURALON	Polyvinyl alcohol (Vinylon) staple	Kurashiki Rayon Co. Ltd., Japan.
LAMÉ	Metallic	Standard Yarn Mills, U.S.A.
LAMITA	Viscose tow spun-dyed	Société de la Viscose Suisse, Switzerland.
LAMO	Viscose tow	Société de la Viscose Suisse, Switzerland.

Name of fibre.	Type.	Manufacturer.
LANABETA	Viscose proteinised	Snia Viscosa, Italy.
LANALPHA	Viscose proteinised	Snia Viscosa, Italy.
LANCOLA	Dope-dyed textured acetate yarn	Lansil Ltd., Lancaster.
LANESE	Cellulose acetate staple	Celanese Corp. of America, U.S.A.
LANITAL	Casein	Les Textiles Nouveaux, Belgium.
LANON	Polyester	V.E.B. Thuringisches Kunstfaserwerk, Germany.
LANSIL	Cellulose acetate filament	Lansil Ltd., U.K.
LANTUCK NR	Non-woven fabric	West Point Mfg. Co., U.S.A.
LANUSA	Viscose wool-like rayon (discontinued)	Badische Anilin & Soda Fabrik A.G., Germany.
LAVSAN	Polyester	U.S.S.R.
LEACRIL	Italian Acrilan	A.C.S.A., Italy.
LILIENFELD	Original high tenacity viscose	Name of a process.
LILION	Polycaprolactam	Snia Viscosa, Italy.
LIRELLE	High wet modulus rayon	Courtaulds North America Inc.
LONZONA	Acetate filament	Lonzona G.m.b.H., Germany.
LUMITE	Polyvinylidene chloride	Chicopee Mfg. Corp., U.S.A.
LUMIYARN	Metallised polyester filament	Toyo Chemical Co. Ltd., Japan.
LUREX	" Butyrate " metallic filament	Dobeckmun Co., U.S.A.
LUREX MF	" Mylar " metallic filament	Dobeckmun Co., U.S.A.
LUREX MM	" Mylar " discontinuous-metallic filament	Dobeckmun Co., U.S.A.
LUSTRAFIL	Viscose filament	Lustrafil Ltd., U.K.
LUXEL	Polyester	Argentine.
LYCRA	Snap-back	du Pont, U.S.A., and Maydown, Northern Ireland.
M-24	Early name for Verel	
MADAME BUTTERFLY	Viscose filament	Toyo Rayon Co. Ltd., Japan.
MAKROLAN	Acrylic	Yugoslavia.
MALON	Acrylic	Yugoslavia.
MANYRO	Polyvinyl alcohol	
MARACRAY	Polyester	Venezuela.
MARIMUSUME	Viscose staple	Nippon Rayon Co. Ltd., Japan.
MARLEX	Polyethylene	Phillips Petroleum Co., U.S.A.
MATAPOINT	Viscose straw	Société de la Viscose Suisse, Switzerland.
MATESA	Cuprammonium filament	American Bemberg Corp. U.S.A.
MELANA	Acrylic	Rumania.
MERAKLON	Polypropylene	Montecatini, Italy.
MERINOVA	Casein staple	Snia Viscosa, Italy.
MERON	Bulky nylon	U.S.S.R.
MERYL B8	Viscose	Comptoir des Textiles Artificiels, France.
METALIAN	Anti-static yarn	Teijin, Japan.
METLON	Butyrate metallic	Metlon Corp., U.S.A.
METLON-with-MYLAR	Metallic in Mylar	Metlon Corp., U.S.A.
MEUBALESE	Cellulose acetate slub filament	British Celanese Ltd., U.K.
MEWLON	Polyvinyl alcohol filament and staple (Vinylon)	Dai Nippon Spinning Co., Ltd., Japan.

Name of fibre.	Type.	Manufacturer.
MIHARAHYO	Viscose filament	Teikoku Rayon Co. Ltd., Japan.
MIKRON	Polyvinyl alcohol	
MINALON	Cellulose acetate filament and staple	Shin Nippon Chisso Hiryo Co. Ltd., Japan.
MIRALON	Bulking process for nylon	
MISRNYLON	Polyamide	Egypt.
MISROPHANE	Viscose film	Société M.I.S.R. pour la Rayonne, Egypt.
MOPLEN	Polypropylene	Montecatini, Italy.
MOVYL	Polyvinyl chloride	Montecatini, Italy.
MOYNEL	Polynosic viscose rayon staple same as S.C. 28	Courtaulds (Alabama) Inc., U.S.A.
MYLAR	Polyester film	du Pont de Nemours & Co., U.S.A.
N-53	Modified polyacrylonitrile	Kunstzijde-spinnerij, Nyma, Holland.
NAILON	Nylon 66	Rhodiatoce S.p.A., Italy.
NAILONSIX	Polyamide	Brazil.
NARCO	Viscose filament	North American Rayon Corp., U.S.A.
NEFA-PERLON	Polycaprolactam	Glanzstoff Fabriken A.G., Germany.
NEOCHROME	Courtelle spun-dyed	Courtaulds Ltd., U.K.
NIP	Polyvinyl chloride filament	Nichay Plastics Co., Japan.
NIPLON	Nylon 66	Dai Nippon Spinning Co. Ltd., Japan.
NIRLON	Nylon 6	Indian Rayon Corp., India.
NITLON	Acrylic	Japan.
NITRILON	Uncertain, possibly polyvinylidene dicyanide	U.S.S.R.
NITRON	Polyacrylonitrile	Leningrad, U.S.S.R.
NIVION	Polyamide	Italy.
NOMEX	Aromatic polyamide	du Pont, U.S.A.
NORTHYLENE	Polyethylene	Norddeutsche Seekabelwerke A.G., Germany.
NOVACETA	Cellulose acetate	S.A. Novaceta, Italy.
NRC NYLON	Nylon 6	Nippon Rayon Co., Japan.
NUFIL	Polypropylene film tape	I.C.I. Ltd.
NUPRON	Viscose	Industrial Rayon Corp., U.S.A.
NYCEL	Polyamide	Jalisco, Mexico.
NYCRON	Polyester	Brazil
NYLENKA	Polycaprolactam filament and staple	American Enka Corp., U.S.A.
NYLFIL	Polyamide	Mexico.
NYLFRANCE	Nylon 66	Rhodiaceta, France.
NYLOFT	Polyamide Nylon 6	I.R.C., U.S.A.
NYLON	Generic name for fibre-forming polyamides	du Pont de Nemours & Co., U.S.A.; British Nylon Spinners, U.K.; Chemstrand Corp., U.S.A., and others.
NYLON 4	Polypyrrolidone	General Aniline & Film Corp., U.S.A.
NYLON 6	Polycaprolactam (Perlon and many other names)	British Celanese Ltd., U.K.; Farbenfabriken Bayer A.G., Germany, and many others.
NYLON 6-T	Polyamide incorporating terephthalic acid	
NYLON 7	Enant	U.S.S.R.

Name of fibre.	Type.	Manufacturer.
NYLON 11	Polyundecanolactam, *e.g.*, Rilsan	
NYLON 66	Polyhexamethylene adipamide filament, and staple and tow	du Pont de Nemours & Co. U.S.A.; British Nylon Spinners, U.K., and others.
NYLON Type 91	Whiter (fluorescent) nylon 66	du Pont de Nemours & Co., U.S.A.
NYLON Type 250	Deep dye nylon 66	du Pont de Nemours & Co., U.S.A.
NYLON 1 B-610	Rubber-like nylon	du Pont de Nemours & Co., U.S.A.
NYLON 610	Polyhexamethylene sebacamide filament	du Pont de Nemours & Co., U.S.A.
NYLON Type 707	Heavy denier nylon 66 for ropes	du Pont de Nemours & Co., U.S.A.
NYLON Type 764	Nylon 66 tyre yarn	du Pont de Nemours & Co., U.S.A.
NYLON R	Nylon 11	
NYLSUISSE	Nylon 66	Société de la Viscose Suisse, Switzerland.
NYMACRYL	Acrylic	Kunstzijde-spinnerij, Nyma, Holland.
NYMACRYON	Acrylic	Kunstzijde-spinnerij, Nyma, Holland.
NYMCRYLON	Acrylic	Netherlands.
NYMKRON	Acrylic	Netherlands.
NYMPLEX	Polyethylene	Netherlands.
NYPEL	Polyamide	U.S.A.
OKSON	Polyester (modified)	U.S.S.R.
OLANE	Polypropylene filament and staple	Avi-Sun, U.S.A.
OLETENE	Polyethylene	France.
OPACETA	Cellulose acetate filament matt	Courtaulds Ltd., U.K.
OPELON	Snap-back	Toyo Rayon Co., Japan.
OPLEXMATT	Viscose filament matt	Harbens Ltd., U.K.
ORLON, Type 39	Modified acrylic heavy denier for woollen spinning	du Pont de Nemours & Co., U.S.A.
ORLON, Type 41	Modified acrylic (obsolete)	du Pont de Nemours & Co., U.S.A.
ORLON, Type 42	Modified acrylic staple and tow	du Pont de Nemours & Co., U.S.A.
ORLON, Type 81	Polyacrylonitrile filament	du Pont de Nemours & Co., U.S.A.
ORLON-CANTRECE	Filament polyacrylonitrile	du Pont de Nemours & Co., U.S.A.
ORLON SAYELLE	Bicomponent acrylic with built-in crimp	du Pont de Nemours & Co., U.S.A.
ORTALION	Polyamide	Bemberg, Italy.
PAN	Modified polyacrylonitrile filament	Farbenfabriken Bayer A.G., Casella Farbwerke Mainkur A.G., Germany.
PANAKRYL	Polyacrylonitrile	Hungary.
PARAMAFIL	Viscose staple	Nitto Spinning Co. Ltd., Japan.
PARAMOUNT	Viscose bright warp	North British Rayon Co., U.K.

Name of fibre.	Type.	Manufacturer.
PCU	Polyvinyl chloride	Badische Anilin & Soda Fabrik, Germany.
Pe Ce	Polyvinyl chloride	
Pelargon	Nylon 9	U.S.S.R.
Perfil	Polyethylene	Australia.
Perfilon	Polyamide, nylon 6	Chile.
Perlenka	Nylon 6	A.K.U., Holland.
Perlglo	Viscose filament, semi-dull	American Enka Corp., U.S.A.
Perlofil	Polycaprolactam	Spain.
Perlon	Polycaprolactam filament and staple	Farbwerke Hoechst A.G. Deutsche Rhodiaceta A.G., & Phrixwerke A.G., Germany.
Perlon 6	Another name for Perlon	
Perlon-Hoechst	Nylon 6 filament and staple	Farbwerke Hoechst A.G., Germany.
Perlon T	Nylon 66	I.G. Farbenindustrie, Germany.
Perlon U	Polyurethane	Germany.
Permalon	Polyvinylidene chloride	Pierce Plastics Inc., U.S.A.
Perro	Blend of 1 part viscose staple, 2 parts Peruvian cotton, same strength wet as dry	
Phrilan	Crimped viscose staple	Phrixwerke A.G., Germany.
Phrilon	Polycaprolactam	Phrixwerke A.G., Germany.
Phrix-Perlon	Nylon 6 staple	Phrixwerke A.G., Germany.
Phrix-Reyon	Viscose	Phrixwerke A.G., Germany.
Phrix JT	Viscose staple jute-type (30 den)	Phrixwerke A.G., Germany.
Phrix STW	Viscose staple	Phrixwerke A.G., Germany.
Piviacid	New name for Pe Ce	
Platilon	Polyethylene	Toyo Chemical Co. Ltd., Japan.
Plexon	Plastic coated core usually of viscose or glass (discontinued)	British Plexon Ltd., U.K.
Pluton	Organic heat-resistant fibre	Minnesota Mining & Mfg. Co., U.S.A.
Polan	Polyamide	Poland.
Politen-Omni	Polyethylene	Mexico.
Polufen	Fluorocarbon	U.S.S.R.
Polyamino-triazoles	Made from sebacic dihydrazide	Not yet marketed.
Polycrest	Polypropylene	Uniroyal Fibres.
Polycron	Polyester	Chile.
Polyethylene	Generic name for fibres made by polymerising ethylene, *e.g.*, Courlene, Wynene	
Polygol	Polyamide	Chile.
Polymer R	Nylon 11	Rehoboth Research Lab., Israel.
Polynak	Acrylic	Leningrad, U.S.S.R.
Polysplit	Polypropylene	Sweden
Polythene	Polyethylene	
Pontova	Viscose straw	Société de la Viscose Suisse, Switzerland.
Potassium Titanate	Refractory fibrous material	du Pont de Nemours & Co., U.S.A.
Powlon	Acrylic	Japan.
Prelana	Modified acrylic	Eastern Germany.
Premier	Viscose	Industrial Rayon Corp., U.S.A.

Name of fibre.	Type.	Manufacturer.
PREYLONN	Polyamide	Argentine.
PRIMEL	Polyester	Colombia.
PROLENE	Polypropylene filament and staple	Industrial Rayon Corp., U.S.A.
PROLON	Generic name for reconstituted protein fibres	
PROMILAN	Polyamide	Argentine
PROTEX	Generic name for reconstituted protein fibres	
PUSAN	Polyester	Korea.
PVC-BORSTEN HOECHST	P.V.C. bristle	Farbwerke Hoechst A.G., Germany.
PYLEN	Polyethylene	Toyo Rayon Co. Ltd., Japan.
QUARTZ	Vitreous silica refractory fibres	Björksten Research Labs., U.S.A.
RAILAN	Viscose proteinised	Snia Viscosa, Italy.
RAINBOW	Viscose staple	Teikoku Rayon Co. Ltd., Japan.
RATUJAL	Viscose straw	Société de la Viscose Suisse, Switzerland.
RAYCELON	A twinned yarn made from 5 parts viscose rayon and 3 parts Celon	Courtaulds Ltd.
RAYFLEX	High tenacity viscose filament	American Viscose Corp., U.S.A.
RAYOLANDA	Modified viscose	Courtaulds Ltd., U.K.
RAYONNE	Generic French name for filament viscose	
REDON	Modified polyacrylonitrile crimped staple	Phrix G.m.b.H., Germany.
REEVON	Polyethylene	Reeves Bros. Inc., U.S.A.
REFRASIL	Silica	British Refrasil Co. Ltd., U.K.
RELON	Nylon 6	Relonwerk, Roumania.
RETRACTYL	Polyvinyl chloride staple	Société Rhovyl, France.
REYMET	Metal foil-film laminate cut	Reynolds Metals Co., U.S.A.
RHIAKNOT	Acetate fancy yarn	Deutsche Rhodiaceta A.G., Germany.
RHODELIA	Acetate textured yarn	Deutsche Rhodiaceta A.G., Germany.
RHODIA	General name for products of Société Rhodiaceta, France	
RHODIA FASER	Cellulose acetate staple	Deutsche Rhodiaceta A.G., Germany.
RHODIAFIL	Cellulose acetate filament	Deutsche Rhodiaceta A.G., Germany.
RHODIALIN	Cellulose acetate	Deutsche Rhodiaceta A.G., Germany.
RHODIANIL	Nylon 66	Brazil.
RHONEL	Cellulose triacetate	France.
RHOVENYL	Fibravyl/nylon blend	Société Rhovyl, France.
RHOVYL	Polyvinyl chloride filament	Société Rhovyl, France.
RHOVYL MOSS	Helanca type Rhovyl	Société Rhovyl, France.
RHOVYLINE	Fibravyl blends with other fibres	Société Rhovyl, France.

Name of fibre.	Type.	Manufacturer.
RHOVYLON	Thermovyl/nylon blend	Société Rhovyl, France.
RILSAN	Nylon 11	Organico S.A., France; and Snia Viscosa, Italy.
ROFIL	Polyethylene	U.K.
ROLAN	Polyacrylonitrile	Rumania.
RONBEL	Graft polymer of acrylonitrile and viscose; wool-like staple	Kanegafuchi Spinning Co. Ltd., Japan.
ROTWYLA	Viscose rayon	Rottweiler Kunstseidenfabrik A.G., Germany.
ROTWYLA SUPERCORD	High tenacity viscose tyre cord	Rottweiler Kunstseidenfabrik A.G., Germany.
ROVANA	Ribbon shaped Saran monofil	Dow Chemical Co., U.S.A.
ROVICELLA	Viscose straw, bast like	Feldmühle Ltd., Switzerland.
ROYALENE	Polyolefines	U.S. Rubber Co., U.S.A.
SAABA	Stretch nylon (false twist)	Leesona Corp., U.S.A.
SANDERIT	Polycaprolactam monofil	Fr. Sander-Nachf., Germany.
SANFLON	Polyamide	India.
SANIV	Modacrylic 40 acrylonitrile/60 vinylidene chloride	U.S.S.R.
SARAN	Polyvinylidene chloride	Saran Yarns Co., U.S.A. (Polymer made by Dow Chemical Co., U.S.A.).
SARELON	Experimental groundnut protein	U.S. Dept. of Agriculture, U.S.A.
SARFA	Viscose filament twisted	Feldmühle Ltd., Switzerland.
SARILLE	Crimped viscose staple	Courtaulds Ltd.
SASTIGA	Viscose filament (4 den) bright and dull, natural and spun-dyed	Feldmühle Ltd., Switzerland.
SC 28	Polynosic rayon staple	Courtaulds Ltd., U.K.
SCALDURA	Polynosic viscose	Fabelta, Belgium.
SERACETA	Cellulose acetate filament	Courtaulds Ltd., U.K.
SHALON	Polystyrene	U.S.A.
SHALON	Polyethylene	U.S.A.
SHINKO	Viscose staple	Mitsubishi Rayon Co. Ltd., Japan.
SHIRO DIAFIL	Viscose filament, dull	Teikoku Rayon Co. Ltd., Japan.
SILENE	Cellulose acetate	Novaceta S.p.A., Italy.
SILKOOL	Soybean protein (discontinued)	Showa Sangyo Co., Japan.
SILON	Nylon 6	S.N.C., Czechoslovakia.
SIL-TEMP	Silica from glass by acid leaching	Haveg Industries Inc., U.S.A.
SKENANDOA	Viscose	Skenandoa Rayon Corp., U.S.A.
SKYBLOOM	High crimp rayon staple	American Enka Corp., U.S.A.
SLOVINA	Viscose proteinised	Bata Slovenska, Czechoslovakia.
SM 27	Polynosic rayon staple	Courtaulds (Alabama) Inc., U.S.A.
SNIAFIL	Viscose filament	Snia Viscosa, Italy.
SNIALON	Polyamide	Argentine.
SOLVRON	Water-soluble polyvinyl alcohol	
SOVIDEN	Saran type	U.S.S.R.
SOVINOL	Polyvinyl alcohol	U.S.S.R.

Name of fibre.	Type.	Manufacturer.
SOYBEAN	Soybean protein staple (discontinued)	Drackett Products Co., U.S.A.
SOYLAN	Soybean protein staple (discontinued)	Drackett Products Co., U.S.A.
SPANDELLE	Snap-back fibre	Firestone, U.S.A.
SPANDEX	Generic term for polyurethane snap-back fibres	
SPANZELLE	Snap-back fibre	Courtaulds, U.K.
SPONTEX	Viscose sponge	Société Novacel, France.
SPUNGO	Viscose sponge	Feldmühle Ltd., Switzerland.
SPUNIZED	Stretch nylon (stuffer box)	Spunize Co. of America, U.S.A.
SPUNSTRON	Polypropylene, staple fibre for ropes	I.C.I. Ltd.
STARNEL	Scandinavian name for Arnel	
STEELON	Polycaprolactam	Gorzowski Zaklady, Poland.
STILON	Same as Steelon	Italy.
STYROFLEX	Polystyrene	Norddeutsche Seekabelwerke A.G., Germany.
SUDALON	Polyamide	Venezuela.
SUIKO	Viscose staple	Toyo Rayon Co. Ltd., Japan.
SUPER AMILAN	Nylon 66 monofil	Toyo Rayon Co. Ltd., Japan.
SUPER CORDENKA	High tenacity viscose filament	A.K.U., Holland.
SUPER CORDURA	High tenacity viscose filament	du Pont de Nemours & Co., U.S.A.
SUPER NARCO	High tenacity viscose filament	North American Rayon Corp., U.S.A.
SUPER RAYFLEX	High tenacity viscose filament	American Viscose Corp., U.S.A.
SUPER TENAX	High tenacity viscose filament	Toyo Rayon Co. Ltd., Japan.
SUPER TOYOTENAX	High tenacity viscose filament	Toyo Rayon Co. Ltd., Japan.
SUPER VISTRON	High tenacity viscose filament	Nippon Rayon Co. Ltd., Japan.
SUPERBREDA	Viscose filament	Hollandsche Kunstzijde Industries, Holland.
SUPERENKA	Viscose filament	American Enka Corp., U.S.A.
SUPERLOFT	Stretch nylon (false twist)	Leesona Corp., U.S.A.
SUPRACOL	Spun-dyed viscose staple	Aktieselskapel Borregaard, Norway.
SUPRAL	Viscose staple, cotton type	Aktieselskapel Borregaard, Norway.
SUPRALAN	Viscose staple, wool type	Aktieselskapel Borregaard, Norway.
SUPRALON	Polyamide	Yugoslavia.
SYNFOAM	Stretch nylon	
SYNTHOFIL	Polyvinyl alcohol	Wacker-Chemie G.m.b.H., Germany.
SYNTHON	Generic name for synthetic fibres, never widely accepted	
T-1700	Polyester elastic fibre (similar to spandex)	Tennessee Eastman Co., U.S.A.
TACRYL	Polyacrylonitrile	Sweden.
TANIKALON	Polyethylene	Japan.

Name of fibre.	Type.	Manufacturer.
TAPILAN	Polyamide	Argentine.
TASLAN	Texturing process applied to nylon, viscose, etc.	Licensed by du Pont de Nemours & Co., U.S.A.
TECA	Cellulose acetate staple	Tennessee Eastman Co., U.S.A.
TEFLON	Polytetrafluoroethylene filament and staple	du Pont de Nemours & Co., U.S.A.
TEIJIN	Viscose filament	Teikoku Rayon Co. Ltd., Japan.
TEIJIN ACETATE	Cellulose acetate filament	Teikoku Rayon Co. Ltd., Japan.
TEIJIN CORD	High tenacity viscose filament	Teikoku Rayon Co. Ltd., Japan.
TEKLAN	Modacrylic	Courtaulds Ltd., U.K.
TEMPRA	High tenacity viscose filament	American Enka Corp., U.S.A.
TENASCO	High tenacity viscose filament	Courtaulds Ltd., U.K.
TENAX	High tenacity viscose filament	A.K.U., Holland.
TENDAN	Viscose filament	Nippon Rayon Co. Ltd., Japan.
TENKYO	Viscose filament	Teikoku Rayon Co. Ltd., Japan.
TERENE	Polyester	India.
TERGAL	Polyethylene *tere*phthalate	Société Rhodiaceta, France.
TERIBER	Polyester	Barcelona, Spain, also Farbwerke Hoechst, West Germany.
TERITAL	Polyethylene *tere*phthalate	Montecatini. Societa Generale per l'Industria Mineraria e Chimica, Italy.
TERLENKA	Polyethylene *tere*phthalate	Algemene Kunstzijde Unie N.V., Holland and British Enkalon Ltd.
TERON	Polyester (Fortrel)	Rumania.
TERSUISSE	Polyester	Switzerland.
TERYLENE	Polyethylene *tere*phthalate filament and staple	I.C.I. Ltd., U.K.
TESIL	Polyester staple and filament	Silon National Corp., Czechoslovakia.
TETORON	Polyester	Toyo Rayon Co. Ltd., Japan.
TEVIRON	Polyvinyl chloride	Teikoku Rayon Co. Ltd., Japan.
TEXNYL	Nylon 6 for apparel	A/B Svenskt Konstsilke, Sweden.
TEXTILION	Polyamide	Brazil.
TEXTRALIZED	Ban-Lon stretch nylon (hot knife)	J. Bancroft & Co., U.S.A.
THERMOVYL	Polyvinyl chloride staple	Société Rhovyl, France.
THIOLAN	Polyvinyl alcohol	Eastern Germany.
THIOZELL	Casein staple	Lodz, Poland.
THORNEL	Graphite high modulus	Union Carbide Corp., U.S.A.
TIOLAN	Casein	Spinstoff G.m.b.H., Germany.
TOHALON	High tenacity acetate staple	Toyo Rayon Co. Ltd., Japan.
TOLON	Polyvinyl chloride bristle	Toyo Chemical Co. Ltd., Japan.
TOPEL	Cross-linked viscose rayon	Courtaulds (Alabama) Inc., U.S.A.
TORAYLON	Acrylic	Japan.
TOVIS	Viscose crimped staple and tow	Toyo Rayon Co. Ltd., Japan.
TOYOBO	Viscose staple	Toyo Rayon Co. Ltd., Japan.
TOYOFLON	Polytetrafluoroethylene	Toyo Rayon Co., Ltd., Japan.
TOYOLAN	70 Viscose/30 Amilan blend	Toyo Rayon Co. Ltd., Japan.
TOYOTENAX	High tenacity viscose	Toyo Rayon Co. Ltd., Japan.

Name of fibre.	Type.	Manufacturer.
TRALBÉ	Polyester	France.
TRAVIS	European name for Darvan	Celanese-Farbwerke Hoechst, Germany.
TRELON	Polycaprolactam	V.E.B. Thuringisches, Germany.
TREVIRA	Polyethylene *tere*phthalate	Farbwerke Hoechst A.G., Germany.
TRI-A-FASER	Triacetate staple	Deutsche Rhodiaceta, Germany.
TRIALBENE	Cellulose triacetate	Rhodiaceta, France.
TRICEL	Cellulose triacetate filament and staple	British Celanese Ltd., U.K.
TRICELON	A twinned yarn made from Tricel and Celon	Courtaulds Ltd.
TRILAN	Cellulose triacetate	Canadian Celanese, Canada.
TRITOR	Monofil polypropylene	Plasticizers Ltd.
TROFIL	Polyethylene	West Germany.
TROPIC	Viscose straw	Société de la Viscose Suisse, Switzerland.
TRYLKO	Polyester sewing thread	English Sewing Ltd.
TUDENZA	Viscose fine-filament	Courtaulds Ltd., U.K.
TUFCEL	Polynosic viscose staple	Toyobo, Japan.
TURLON	Polyamide	Turkey.
TUSSON	Cuprammonium slub yarn	J. P. Bemberg A.G., Germany, and American Bemberg Corp., U.S.A.
TYCORA	Textured yarn of several kinds	Textured Yarn Co., U.S.A.
TYGAN	Polyvinylidene chloride	Fothergill & Harvey Ltd., U.K.
TYRENKA	Viscose high tenacity filament	British Enka Ltd., U.K.
TYREX	High tenacity viscose for tyres	Members of Tyrex Inc.
TYRON	Viscose high tenacity filament	Industrial Rayon Corp., U.S.A.
ULON	Polyamide	Taiwan.
ULSAN	Polyamide	Korea.
ULSTRON	Polypropylene filament	I.C.I. Ltd., U.K.
ULTREMA	Viscose filament	Société de la Viscose Suisse, Switzerland.
UNEL	Polyamide	Canada.
UNIROYAL	Polypropylene	Uniroyal Fibres.
URYLON	Polyurea	Toyo Koatsu, Japan.
USTEX	Modified cotton	United States Rubber Co., U.S.A.
VANDUARA	Gelatine spun 1894 (historical interest only)	
VANYLON	Polyamide	Colombia.
VELICREN	Acrylic	Italy.
VELON	Polyvinylidene chloride	Firestone Industrial Products, U.S.A.
VENECRON	Polyester	Venezuela.
VENUS	Viscose staple	Fuji Spinning Co. Ltd., Japan.
VEREL	Modified acrylic staple	Tennessee Eastman Co., U.S.A.
VERI DULL	Viscose dull	Skenandoa Rayon Corp., U.S.A.
VESTOLAN	Polyethylene	West Germany.
VIBREM	Slubbed viscose filament	Snia Viscosa, Italy.
VICARA	Maize protein staple and tow (obsolete)	Virginia Carolina Chemical Corp., U.S.A.

Name of fibre.	Type.	Manufacturer.
VIDILON	Polyamide	Bulgaria.
VINAL	Generic name for fibres based on polyvinyl alcohol	
VINAL 5F	Vinylon	Kurashiki Rayon Co. Ltd., Japan.
VINAL FO	High tenacity polyvinyl alcohol	Kurashiki Rayon Co., Japan, and Air Reduction Co., U.S.A.
VINALON	Polyvinyl alcohol	Korea.
VINCEL	Polynosic rayon	Courtaulds Ltd., U.K.
VINITRON	95 Khlorin/5 cellulose acetate co-spun	U.S.S.R.
VINOL	Polyvinyl alcohol	U.S.S.R.
VINYLIDENE	Japanese generic name for vinylidene chloride/vinyl chloride co-polymer fibres	
VINYLON	Polyvinyl alcohol filament and staple	Kurashiki Rayon Co. Ltd., Japan.
VINYON CF	Co-polymer of vinyl chloride/vinyl acetate filament	American Viscose Corp., U.S.A.
VINYON E	Elastic type of Vinyon (discontinued)	American Viscose Corp., U.S.A.
VINYON HH	Co-polymer of vinyl chloride/vinyl acetate staple	American Viscose Corp., U.S.A.
VINYON N	Co-polymer of vinyl chloride/acrylonitrile filament (discontinued)	Carbide & Carbon Chemicals Corp., U.S.A.
VISADA	Viscose	Breda, Holland.
VISCOR	High tenacity viscose filament	Société de la Viscose Suisse, Switzerland.
VISCOSE	Generic name for regenerated cellulose made by " viscose " process	
VISTRA XT	Viscose, wool-like staple	I.G. Farbenindustrie, Germany.
VISTRALEN	Viscose, wool-like	I.G. Farbenindustrie, Germany.
VISTRON	High tenacity viscose	Nippon Rayon Co. Ltd., Japan.
VITEL	Polyester resin from which Vycron is made	
VITROCELLE	Viscose, film	Ets. Dalle Frères et Lecomte, France.
VITRON	Glass	L.O.F. Glass Fibers Inc., U.S.A.
VONNEL	Polyacrylonitrile staple	Mitsubishi Vonnel Co., Japan.
VYCRON	Polyester	Beaunit Mills, U.S.A.
VYRENE	Snap-back	U.S. Rubber Co., Dunlop Rubber Co.
W-63	Another name for Lirelle	Courtaulds, North America Inc.
WIKILANA	Viscose, fish protein added	Mecheels-Hiltner, Germany.
WIPOLAN	Casein	Lodz, Poland.
WISTEL	Polyester	Italy.
WOLACRYL	Polyacrylonitrile	V.E.B. Eastern Germany.
WOLCRYLON-FASER	Modified polyacrylonitrile	V.E.B., Eastern Germany.
WOLPRYLA	Acrylic	East Germany.
WOOLIE AMILAN	Stretch elastic nylon 66	Toyo Rayon Co. Ltd., Japan.
WOOLON	Polyvinyl alcohol	
WYNENE 1	Polyethylene monofil	National Plastic Products Co., U.S.A.

Name of fibre.	Type.	Manufacturer.
WYNENE 6	Polystyrene bristle	National Plastic Products Co., U.S.A.
X-51	Modified polyacrylonitrile (discontinued)	American Cyanamid Co., U.S.A.
X-54	Original name for Creslan	American Cyanamid Co., U.S.A.
XTRA DULL	Viscose dull	North American Rayon Corp., U.S.A.
YAMBOLEN	Polyester	Bulgaria.
YERANITE	Polyvinyl acetate	U.S.S.R.
YULON	Polyamide	Yugoslavia.
Z 54	Polynosic rayon	France and Belgium.
ZANTREL	Polynosic rayon	U.S.A.
ZARYL	Polynosic staple viscose	Fabelta, Belgium.
ZEFRAN	Nitrile co-polymer staple and tow	Dow Chemical Co., U.S.A.
ZEHLA	Staple viscose	Spinnstoff-fabrik Zehlendorf A.G., Germany.
ZEHLA-PERLON	Nylon 6 filament and staple	Spinnstoff-fabrik Zehlendorf A.G., Germany.
ZEIN	Protein fibre (Vicara) (obsolete)	Virginia Carolina Chemical Co., U.S.A.
ZELLWOLLE	Generic German name for viscose staple	
ZETEK	Darvan	B. F. Goodrich, U.S.A.
ZYCON	Zein fibre for felt hats	Virginia Carolina Chemical Co., U.S.A.

A few names, which although not properly those of fibres, but of related products and which might be confused with fibre names, have been included.

BIBLIOGRAPHY

SUGGESTIONS for further reading have been made at the end of each chapter. Some books which the student will find of general interest and utility are as follows:

W. T. Astbury, "Fundamentals of Fibre Structure". (1933).
A very readable and short book from which the principles of X-ray analysis and molecular-chain structure can be derived.

H. E. Fierz-David, "Abriss der chemischen Technologie der Textilfasern". Basle (1948).
This gives a useful summary for students who can read German. Stronger on the dyeing and finishing side than on that of fibre manufacture.

W. Garner, "Textile Laboratory Manual". 2nd Edn. London (1952).
A useful companion for *practical* work.

P. H. Hermans, "Physics and Chemistry of Cellulose Fibres". London (1949).
Excellent, but very advanced; suitable only for specialists in cellulosic rayons.

R. Hill, "Fibres from Synthetic Polymers". London (1953).

R. G. Horsfall and L. G. Lawrie, "The Dyeing of Textile Fibres". 2nd Edn. London (1946).
A standard text-book which is widely appreciated in the industry. Deals with natural as well as man-made fibres.

H. F. Mark, S. M. Atlas and E. Cernia, "Man-Made Fibers: Science and Technology", 3 vols. Interscience, New York and London (1967).

J. M. Matthews, "Textile Fibres", 6th Edn. London (1954).
A classic. Much more space is devoted to natural than to man-made fibres. Every textile student should be familiar with Matthews.

R. Meredith, "The Mechanical Properties of Textile Fibres". Amsterdam (1956).

J. M. Preston, "Fibre Science". 2nd Edn. Textile Institute, Manchester (1953).
A collective effort written by a number of authors. Excellent discussion of fibre properties, orientation, X-rays, optical properties. Particularly suitable for the advanced student who can be selective.

G. S. Ranshaw, "The Story of Rayon". London (1949).
A semi-popular account of rayon manufacture, particularly good on the manufacture of viscose rayon.

A. X. Schmidt and C. A. Marlies, "Principles of High Polymer Theory and Practice". New York (1948).

J. V. Sherman and S. L. Sherman, "The New Fibres". New York (1946).
The student should make himself familiar with this book; it is particularly good in the wealth of patent references it provides. Suitable for the advanced reader who can be selective.

C. M. Whittaker and C. C. Wilcock, "Dyeing with Coal-tar Dyestuffs". 5th Edn. London (1949).

H. J. Woods, "Physics of Fibres—An Introductory Survey". London (1955).

R. L. Wormell, "New Fibres from Proteins". London (1954).

INDEX

(*Bold numerals denote main entries*)